Simulation Technologies in Networking and Communications

Selecting the Best Tool for the Test

Simulation Technologies in Networking and Communications

Selecting the Best Tool for the Test

Edited by Al-Sakib Khan Pathan
Muhammad Mostafa Monowar • Shafiullah Khan

CRC Press
Taylor & Francis Group
Boca Raton London New York

CRC Press is an imprint of the
Taylor & Francis Group, an **informa** business

CRC Press
Taylor & Francis Group
6000 Broken Sound Parkway NW, Suite 300
Boca Raton, FL 33487-2742

First issued in paperback 2017

© 2015 by Taylor & Francis Group, LLC
CRC Press is an imprint of Taylor & Francis Group, an Informa business

No claim to original U.S. Government works

ISBN-13: 978-1-4822-2549-5 (hbk)
ISBN-13: 978-1-138-03417-4 (pbk)

Visit the Taylor & Francis Web site at
http://www.taylorandfrancis.com

and the CRC Press Web site at
http://www.crcpress.com

Dedicated to
Every righteous, knowledgeable, and knowledge
seeker among humankind

The Editors

Contents

SECTION I

SECTION II

SECTION III

SECTION IV

SECTION V

Preface

Simulation is a widely used mechanism to validate the theoretical model of networking or communication systems. It is believed that claims made based on simulations are more or less reliable—at least in the sense that something *beyond theory* is provided. But, how reliable simulation technologies really are is a question asked when the practicality is evaluated with real-world implementation trials. It is a fact that all simulators for the same system do not have the same inherent working method. Different simulators developed for networking and communications technologies have different underlying mechanisms that may significantly affect the simulation scenarios. The same scenario could give different results if different simulators are used for evaluation. Hence, the question "which one is the best for which situation?" is raised. There is no clear verdict on this issue. Therefore, the selection of a particular simulator should be left for the researcher to decide, if that shows relatively better results for his system or model or proposed solution. From this perspective, simulation is considered as *something better than nothing* to validate a claim or to show something to establish a ground for the proposal. A practical scenario could be, again, very much different and a theoretical model may often be a more solid proof than simulations in the sense that theory would prove the core idea with a solid foundation, and then it would be left for practical testing and performance measurement.

The core objective of this book is to compile different perspectives about the simulation of various networking and communications technologies. Some contributors argue that theoretical modeling is preferable while some show with case studies how simulation could help evaluate different scenarios. The book is divided into five sections based on the topics the 22 chapters deal with, which were selected for inclusion in this book after a rigorous review process of a total of 37 proposals. Sections I to III contain five chapters each, Section IV has four chapters, and Section V three chapters.

To understand the content of the book, it should be noted that we have not provided any verdict on the best suitable tool for simulation but have provided analyses of different kinds of networks and systems from different perspectives. The book provides answers from different vantage points: what to simulate, where to simulate, whether to simulate or not, when to simulate, and how to simulate for various issues. Such a book cannot provide exhaustive information about simulation technologies for every communication- or networking-related field but we hope that the content would guide the readers in finding specific directions for some research topics.

Al-Sakib Khan Pathan, PhD
Department of Computer Science
International Islamic University Malaysia
Gombak, Malaysia

Muhammad Mostafa Monowar, PhD
Department of Information Technology
King AbdulAziz University
Jeddah, Saudi Arabia

Shafiullah Khan, PhD
IIT
Kohat University of Science and Technology
Kohat, Pakistan

MATLAB® is a registered trademark of The MathWorks, Inc. For product information, please contact:

The MathWorks, Inc.
3 Apple Hill Drive
Natick, MA 01760-2098 USA
Tel: 508-647-7000
Fax: 508-647-7001
E-mail: info@mathworks.com
Web: www.mathworks.com

Acknowledgments

We are very much grateful to the Almighty Allah to have allowed us the time to complete another work of this kind. The entire process has been lengthy, demanding nonstop working hours, interaction with several people in various ways, and firm determination. We are thankful to all the authors, reviewers, and critics who helped us shape the book in the best way possible. Apart from the 3 editors of this book, 57 authors from 16 different countries have contributed the chapters, which shows that there were responses from around the globe. We thank all of them for their valuable contributions.

Editors

Al-Sakib Khan Pathan earned his PhD in computer engineering in 2009 from Kyung Hee University, South Korea. He earned his BSc in computer science and information technology from Islamic University of Technology (IUT), Bangladesh, in 2003. He is currently an assistant professor in the computer science department at International Islamic University Malaysia (IIUM), Malaysia. Until June 2010, he was assistant professor at the computer science and engineering department in BRAC University, Bangladesh. Prior to holding this position, he worked as a researcher at the Networking Lab, Kyung Hee University, South Korea, until August 2009. His research interests include wireless sensor networks, network security, and e-services technologies. He is a recipient of several awards, including best paper awards, and has several publications in these areas. He has served as a chair, organizing committee member, and technical program committee member in numerous international conferences/workshops such as GLOBECOM, GreenCom, HPCS, ICA3PP, IWCMC, VTC, HPCC, and IDCS. He was awarded the IEEE Outstanding Leadership Award and Certificate of Appreciation for his role in the IEEE GreenCom'13 conference.

Dr. Pathan is currently an area editor of IJCNIS, editor of IJCSE, Inderscience, associate editor of IASTED/ACTA Press IJCA and CCS, guest editor of many special issues of top-ranked journals, and editor/author of 10 books. One of his books has been included twice in Intel Corporation's Recommended Reading List for Developers, second half of 2013 and first half of 2014; three other books are included in IEEE Communications Society's (IEEE ComSoc) Best Readings in Communications and Information Systems Security, 2013, and a fifth book is in the process of being translated from English to simplified Chinese. Also, two of his journal papers and one conference paper are included under different categories in IEEE Communications Society's (IEEE ComSoc) Best Readings Topics on Communications and Information Systems Security, 2013. He also serves as a referee of numerous renowned journals. He is a senior member of the Institute of Electrical and Electronics Engineers (IEEE), United States; IEEE ComSoc Bangladesh Chapter, and several other international professional organizations.

Muhammad Mostafa Monowar is currently working as an assistant professor in the Department of Information Technology at King Abdulaziz University, Kingdom of Saudi Arabia. He also served as an associate professor in the Department of Computer Science and Engineering at the University of Chittagong, Bangladesh, until February 2012. He earned his PhD in computer engineering in 2011 from Kyung Hee University, South Korea. He earned his BSc in computer science and information technology from the Islamic University of Technology (IUT), Bangladesh, in 2003. His research interests include wireless networks, especially ad hoc, sensor, and mesh networks; routing protocols; MAC mechanisms; IP and transport layer issues; cross-layer design; and Quality of Service (QoS) provisioning. He served as a guest editor of some special issues of IJCSE and is currently associate editor of IJIDS. He has also served as an editor of a book published by CRC Press, Taylor & Francis Group,

United States. He has served as a program vice chair of IEEE HPCC 2013, and also program committee member in several international conferences/workshops such as IEEE IDCS, DNC, ICCCS, and CNTA. He is also a referee of several renowned journals, such as *Annals of Telecommunications* and *Journal of Communications and Networks*.

 **Shafiullah Khan** is currently assistant professor at the Institute of Information Technology, Kohat University of Science and Technology (KUST), Islamic Republic of Pakistan. He earned his PhD in communication networks in 2011 from Middlesex University, United Kingdom. He earned his BIT in information technology from Gomal University, Islamic Republic of Pakistan, in 2005. His research interest includes wireless networks and wireless network security. He served as a guest editor of many special issues and is currently an editor-in-chief of IJCNIS. He has also served as an editor of several recently published books.

Contributors

Khandakar Ahmed
School of Electrical and Computer Engineering
RMIT University
Melbourne, Victoria, Australia

Shabbir Ahmed
Department of Computer Science
and Engineering
University of Dhaka
Dhaka, Bangladesh

Dabiah Alboaneen
Department of Computer, Communications
and Interactive Systems
School of Engineering and Built Environment
Glasgow Caledonian University
Glasgow, United Kingdom

Saiful Azad
Department of Computer Science
American International University-Bangladesh
Dhaka, Bangladesh

Krzysztof Bąkowski
Faculty of Electronics and
Telecommunications
Poznań University of Technology
Poznań, Poland

Ali Balador
Department of Computer Engineering
Polytechnic University of Valencia
Valencia, Spain

Ezio Biglieri
Department of Information and
Communication Technologies
Universitat Pompeu Fabra
Barcelona, Spain

and

Department of Electrical Engineering
University of California
Los Angeles, California

Jorge I. Blanco
Engineering Faculty
Minuto de Dios University
Bogota, Colombia

Fernando Boavida
Department of Engenharia Informática
Universidade de Coimbra
Coimbra, Portugal

Christos Bouras
Computer Technology Institute and Press
and
Department of Computer Engineering
and Informatics
University of Patras
Patras, Greece

Carlos T. Calafate
Department of Computer Engineering
Polytechnic University of Valencia
Valencia, Spain

Juan-Carlos Cano
Department of Computer Engineering
Polytechnic University of Valencia
Valencia, Spain

Filip Čertík
Faculty of Electrical Engineering
and Information Technology
Institute of Telecommunications
Slovak University of Technology
Bratislava, Slovakia

Savvas Charalambides
Department of Computer Engineering
and Informatics
University of Patras
Patras, Greece

Garland Chow
Sauder School of Business
The University of British Columbia
Vancouver, British Columbia, Canada

Mathieu Déziel
Communications Research Centre Canada
Ottawa, Ontario, Canada

Michalis Drakoulelis
Department of Computer Engineering
    and Informatics
University of Patras
Patras, Greece

Anuj Kumar Dwivedi
School of Studies in Computer Science and IT
Pandit Ravishankar Shukla University
Raipur, Chhattisgarh, India

Mark A. Gregory
School of Electrical and Computer Engineering
RMIT University
Melbourne, Victoria, Australia

Khandaker Tabin Hasan
Department of Computer Science
American International University-Bangladesh
Dhaka, Bangladesh

Georgios Kioumourtzis
Ministry of Public Order and Citizen Protection
Center for Security Studies
Athens, Greece

Jinwoo (Brian) Lee
Faculty of Science and Engineering
Queensland University of Technology
Brisbane, Queensland, Australia

Victor C.M. Leung
Department of Electrical and Computer
    Engineering
The University of British Columbia
Vancouver, British Columbia, Canada

Lev B. Levitin
Department of Electrical and Computer
    Engineering
Boston University
Boston, Massachusetts

Jun Li
Communications Research Centre Canada
Ottawa, Ontario, Canada

Lee Luan Ling
School of Electrical and Computer Engineering
State University of Campinas—UNICAMP
São Paulo, Brazil

David Luengo
Department of Circuits and Systems
    Engineering
Technical University of Madrid
Madrid, Spain

Pietro Manzoni
Department of Computer Engineering
Polytechnic University of Valencia
Valencia, Spain

Luca Martino
Department of Mathematics and Statistics
University of Helsinki
Helsinki, Finland

Farzana Mithun
ICT Team
United Nation World Food Programme
Dhaka, Bangladesh

Bartosz Musznicki
Chair of Communication and Computer
    Networks
Poznań University of Technology
Poznań, Poland

Dip Nandi
Department of Computer Science
American International University-Bangladesh
Dhaka, Bangladesh

Athanasios D. Panagopoulos
School of Electrical and Computer
    Engineering
National Technical University of Athens
Athens, Greece

Simon Perras
Communications Research Centre Canada
Ottawa, Ontario, Canada

Maciej Piechowiak
Institute of Mechanics and Applied Computer
Science
Kazimierz Wielki University
Bydgoszcz, Poland

Pedro Vale Pinheiro
Department of Engenharia Informática
Universidade de Coimbra
Coimbra, Portugal

Bernardi Pranggono
Department of Computer, Communications
and Interactive Systems
School of Engineering and Built
Environment
Glasgow Caledonian University
Glasgow, United Kingdom

Mashiour Rahman
Department of Computer Science
American International University-Bangladesh
Dhaka, Bangladesh

Julio Ramírez-Pacheco
Department of Basic Sciences
and Engineering
University of Caribe
Cancún, Mexico

Marcin Rodziewicz
Faculty of Electronics and
Telecommunications
Poznań University of Technology
Poznań, Poland

Rastislav Róka
Faculty of Electrical Engineering
and Information Technology
Institute of Telecommunications
Slovak University of Technology
Bratislava, Slovakia

Yelena Rykalova
Department of Electrical and Computer
Engineering
Boston University
Boston, Massachusetts

Kaveh Shafiee
Department of Electrical and Computer
Engineering
The University of British Columbia
Vancouver, British Columbia, Canada

Houbing Song
Department of Electrical and Computer
Engineering
and
West Virginia Center of Excellence
for Cyber-Physical Systems
West Virginia University
Montgomery, West Virginia

Kostas Stamos
Computer Technology Institute and Press
and
Department of Computer Engineering
and Informatics
University of Patras
Patras, Greece

Jeferson Wilian de Godoy Stênico
School of Electrical and Computer Engineering
State University of Campinas—UNICAMP
São Paulo, Brazil

Huaglory Tianfield
Department of Computer, Communications
and Interactive Systems
School of Engineering and Built Environment
Glasgow Caledonian University
Glasgow, United Kingdom

Homero Toral-Cruz
Department of Sciences and Engineering
University of Quintana Roo
Quintana Roo, Mexico

Deni Torres-Román
Department of Electrical Engineering
Center for Research and Advanced Studies
of the National Polytechnic Institute
Guadalajara, Mexico

Mylène Toulgoat
Communications Research Centre Canada
Ottawa, Ontario, Canada

Ken Umeno
Graduate School of Informatics
Kyoto University
Kyoto, Japan

Leopoldo Estrada Vargas
Department of Electrical Engineering
Center for Research and Advanced Studies
    of the National Polytechnic Institute
Guadalajara, Mexico

Om Prakash Vyas
Indian Institute of Information
    Technology-Allahabad
Allahabad, Uttar Pradesh, India

Cheng-An Yang
Department of Electrical Engineering
University of California
Los Angeles, California

Krzysztof Wesołowski
Faculty of Electronics and
    Telecommunications
Poznań University of Technology
Poznań, Poland

Kung Yao
Department of Electrical Engineering
University of California
Los Angeles, California

Piotr Zwierzykowski
Chair of Communication and Computer
    Networks
Poznań University of Technology
Poznań, Poland

Section I

1 Analysis and Simulation of Computer Networks with Unlimited Buffers

Lev B. Levitin and Yelena Rykalova

CONTENTS

ABSTRACT

We present theoretical and simulation results for the performance of a multiprocessor network modeled as a ring and as a toroidal square lattice of nodes with local processors that generate messages for output ports/buffers. The output buffers are assumed to have unlimited capacity, and the service time is deterministic. Two models are considered. One assumes that every processor generates messages with constant rate per time slot and per output port/buffer. The other model considers that the generation rate of a node depends on the intensity of the flow of arriving messages. Explicit expressions for the distribution of queue lengths, the average number of messages in the buffers, the average latency, and the critical network load depending on the distance between the source and the destination are obtained for the constant rate model. Simulation results show excellent agreement with theoretical predictions based on the assumption of independent queues. This proves the validity of Jackson's theorem for the deterministic time model.

1.1 INTRODUCTION

Modern massively parallel computers (MPCs) are characterized by a scalable architecture. These computers offer corresponding gains in performance as the number of processors is increased. Such computers often consist of self-contained processing nodes, with associated memory and other supporting devices. This design approach has many advantages. The repetition of identical components leads to scalability, modularity, greater reliability, and opportunities for fault tolerance. However, parallel computing in such systems requires extensive communications between otherwise independent nodes so that data and instructions are redistributed periodically to keep all processors busy performing useful tasks. Because memory is not shared between node processors, interprocessor communications are achieved by passing messages between nodes through a communications network. This network is implemented as a set of interconnected routers, each connected to its local processor. According to Ken Batcher (Kent University), "A supercomputer is a device for turning compute-bound problems into I/O-bound problems." Indeed, several of the most advanced supercomputers, such as Titan(Cray XK7), Sequoia, Mira, and Vulcan (IBM), have a 2D and 3D toroidal interprocessor network topology. These topologies are driven by the applied problems they were designed to solve, from weather forecasting to nuclear fusion, from cryptanalysis to biological macromolecules, and from quantum chromodynamics (QCD) to the nature of turbulence. Large-scale problems can be mapped to those topologies more efficiently. Such network configurations provide convenient modularization and low latencies for small messages. This implementation of a network reduces the path length between nodes and simplifies routing algorithms for static or dynamic routing. The need for prediction of networks behavior properties, such as relationships between network load, buffer's capacity, queue lengths, latency, and the point at which network saturates, becomes urgent. In the situation when the experimental approach (such as running on real machines or some combination of the software-hardware emulations) is difficult, expensive, and it is hard to achieve a controlled environment where parameters of the network can be separated and changed for analyzing their effect on the network performance, the analysis by use of theoretical models and controlled simulations is crucial.

Many papers (see, e.g., [1–20]) have been devoted to analyzing computer communication networks as networks of queues. The first most important result was obtained by Jackson. In [3], published in 1963, he proved that for an open network of single-server queues with exponential arrival/departure rates, *the equilibrium joint probability distribution of queue lengths is identical with what would be obtained by pretending that each individual service center is a separate queuing system independent of the others*. Subsequently, Gordon and Newell [4] and Buzen [5] have shown that the state distribution for the *M/M/m* queuing network has a product form for the first come, first served (FCFS) queuing discipline. (The *M/M/m* queuing network consists of nodes that have a Poisson flow of incoming messages, exponential distribution of service time, and m servers.) The Baskett, Chandy, Muntz, and Palacios (BCMP) theorem [6] extends this property of the state distribution for cases where the service rate is not necessarily exponential, but has a distribution with a rational Laplace transform, and the queuing discipline is one of the following four cases: FCFS, processor sharing (PS), infinite server (IS), or last come, first served (LCFS). For FCFS, the service time distribution must be a negative exponential. Subsequently [7–9], three other classes of networks with exponential service times have been shown to have product form distributions. Networks with this property have been analyzed further in [10–14].

Critical phenomena in computer networks have been known to exist in Internet networks and multiprocessor networks. It was observed by Leland et al. [21] and later by others [22–24] that the traffic in Internet systems displays self-similarity, long-range dependent characteristics, and power-law decay of the packet density fluctuations [23,25].

Network latency grows explosively as the network approaches its saturation point, which is similar to critical phenomena in physical systems. The early works [26–28] used simulations to discover those phenomena. Much current research is focused on the analysis of Internet network behavior near saturation, but critical phenomena in supercomputer networks have not been studied as well as those in the Internet network. Several papers analyze the phase transition in packet traffic in regular networks topology [29–32,33] and in parallel computer networks [34,35]. Other types of Internet topologies and the problem of traffic optimization were considered in [36–38]. In [39], a phase transition from low-efficiency traffic to high-efficiency flow at a critical value of a routing parameter was observed. A multiprocessor network in the form of a ring and a 2D toroidal (wrapped-around) square lattice has been analyzed in many papers, for example, in [31,36,40,41].

Analytical and numerical techniques have been used to predict and analyze the behavior of networks when they approach the critical *high-density* or *jammed* phase [42]. Performance evaluation of near-critical networks is a very difficult problem that can be solved analytically only for simplified models. Most results are based on simulations only (see [26,27]). Similar to models of complex physical systems and traffic flow [43,44], cellular automata (CA) models are widely used in network simulations to simplify modeling [25,29,45,46]. In some cases, researchers apply simulation techniques developed initially for physics problems, for example, fluid simulations [47–54], fractional Brownian motion [55,56], and percolation [57–59] (see [60–65] for more discussion) to reveal network behavior. However, analytical solutions for simplified models provide an important basis for greater insight into the behavior of complex systems and especially the qualitative changes in their performance, in particular, critical phenomena.

Communication networks constitute a special class of *large systems*, characterized by many components interacting with each other according to specific rules. There is a deep analogy between the saturation of networks and critical phenomena in fluids and magnetic systems. Many authors have noticed this analogy; however, progress in the description of network behavior using the tools of statistical physics has been modest because of substantial differences in paradigms in these two areas.

An important example is Jackson's theorem and its generalizations [1]. For a broad class of networks, the joint probability distribution of the states of all the nodes simplifies to the product of the marginal probability distributions of the state of each node. This situation corresponds to the *ideal gas* in statistical physics, where the interaction between particles can be considered negligible. However, the nodes in a network, unlike molecules of an ideal gas, interact strongly with each other. Therefore, it is surprising that the *independent queues* model provides an exact description for so many various networks, while the similar *mean-field theory* in statistical physics yields exact results only in rare special cases.

Another distinction is that, although phase transitions in physical systems occur only in the *thermodynamic limit* in which the number of particles approaches infinity, transitions in networks, such as from the steady state to the saturation state, occur even in systems with a finite number of components (nodes). It should be pointed out that, unlike physical systems, networks are governed by Markov processes that are usually not time reversible. Therefore, a network's steady state is, strictly speaking, not an equilibrium (no detailed balance is observed). Thus, networks experience dynamic transitions, rather than equilibrium phase transitions, which can explain the fact that the transitions take place even in finite systems. The dynamic transitions in networks seem to be similar to phase transitions in sand pile models, which have been the subject of intensive research [66–71].

In spite of these differences, critical phenomena in networks and physical systems display many similar characteristics, such as first-order and continuous (second-order) phase transitions, large fluctuations and the emergence of long-range order near the critical point, and the existence of domains of different phases. Our research is motivated, in particular, by the intention to analyze critical phenomena in network models that, on the one hand, reflect certain properties found in practical situations and, on the other, go beyond the applicability of Jackson's theorem.

We consider both theoretical modeling and the simulation of 1D and 2D networks. The operation of the network is presented as a sequence of discrete-time intervals (time slots). The features of the model are different from those considered in the literature. In particular, the service time is deterministic and, hence, does not have a rational Laplace transform. The combination of properties as described in Section 1.2 makes our model closer to a real supercomputer network, such as toroidal interprocessor networks used for parallel computation. We are particularly interested in the approach to the critical point as the message rate is increased and network saturates or fails to keep pace with demand. The difference between our work and previous studies is, in particular, the choice of features such as discrete time and deterministic service time.

We use the independent queues approximation in the theoretical analysis. All the model parameters (the distance between the source and the destination, network size, and network load) are fixed during the simulations. Our simulations show that the network size has no effect on the network behavior if the distance between the source and the destination is smaller than half the network size. In contrast, the distance between the source and the destination and the network load has a crucial effect.

Section 1.3 treats homogeneous networks with infinite buffers. We derive analytical expressions for the queue length distribution, the average number of messages in the buffers, and the latency (average delivery time) for both the ring and 2D toroidal topologies, assuming that the distribution of network states has a product form. It is shown that the network experiences a continuous transition to saturation, and the critical exponent for both average number of messages and the latency is equal to 1. The critical network load is found and shown to be inversely proportional to the distance between the source and the destination. Although there is no known proof that our network model has to obey Jackson's theorem, our simulation results demonstrate very good overall agreement with the theoretical predictions and are consistent with the assumption of independent queues. However, for loads close to the critical value, the network behavior seems to deviate slightly from that predicted by the model.

Section 1.4 considers networks with heterogeneous activity, where the probability of the message generation depends on the incoming flow to every node in the network. This type of network displays behavior completely different from that of homogeneous networks with infinite buffers, once the load approaches the critical value. For heterogeneous networks, the emergence of two phases in dynamic equilibrium with each other, similar to systems in statistical physics, and different values of the critical exponent show that we are dealing with a collective phenomenon, which cannot be accurately described within the framework of the mean-field theory. In this case, the independent queues hypothesis is not applicable, and Jackson's theorem no longer holds.

Section 1.5 outlines directions for future research.

The conclusions in Section 1.6 summarize our findings.

1.2   NETWORK MODEL

1.2.1   1D Case

Each node in our model for the ring topology consists of a local processor, a router, and buffers of infinite capacity. Each node has two output ports/buffers, each connected to one of its two neighbors. The following conventions are made:

1. At any time slot, a message intended for each of the output ports can be generated by the local processor at every node, independently of others, with probability λ.
2. All messages are sent to a destination at a distance exactly l hops from the source. (The distance between neighbors is equal to one hop.)
3. Any message generated at a node or arriving from a neighboring node is placed immediately in the output buffer in the direction of the shortest path to the destination.
4. At any time slot, if a buffer is not empty, exactly one message is transferred to the neighboring node, and it appears at that node at the next time slot. Thus, the service time is equal to one time slot.
5. If there is more than one message in the buffer, a message to be transferred is chosen at random with equal probabilities.
6. At the time slot when a message reaches its destination, it is immediately consumed and leaves the network.

1.2.2 2D CASE

In the 2D case with a torus topology, each node has four neighbors and, correspondingly, four output ports/buffers. In addition to the conventions listed in Section 1.2.1, the following rules are implemented:

1. At any time slot, a message is generated at every node for each of the four output buffers independently with probability λ. The destination for every generated message is chosen out of $2l - 1$ possible destinations for each buffer with equal probabilities. (The total number of destinations for each node is $4l$.)
2. If there is a choice between intermediate nodes on a shortest path to the destination, each of them is chosen with probability $1/2$.
3. Both the service-in-random-order (SIRO) priority discipline and one in which a newly generated message (new first order [NFO]) is sent first are implemented.

The proposed model differs from networks for which the product form of the limiting state probabilities has been proven.

1. Time is discrete, and arrivals occur with specific probabilities, depending on the distance l(non-Poisson and nonbinomial), as given in Equation 1.14.
2. We consider both SIRO and NFO queuing disciplines.
3. The service time is deterministic and does not belong to the class of service time distributions with rational Laplace transforms.

1.2.3 TIME DIAGRAM OF OPERATIONS

In all models, henceforth, we use the same discrete-time queuing conventions as in [72], with a slightly modified definition of the time slot. Namely, the state of a node is recorded in the middle of a time slot, after messages are received and generated. Departures occur at the end of the time slot. A time diagram of operations performed at each node during two consecutive time slots and message propagation from a node to a neighboring one is given in Figure 1.1.

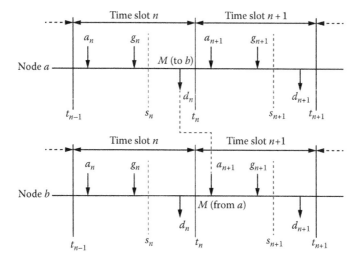

FIGURE 1.1 Conventions for arrivals, departures, and state of a discrete-time queue. Here, d_n and a_n are departures and arrivals within time slot n, respectively, s_n is the state of a buffer at the end of the time slot n, and M is a message transferred from node a to node b.

1.3 NETWORK SIMULATOR

1.3.1 THEORETICAL ANALYSIS: RING

We consider the evolution of the queue at each node buffer as a Markov chain where the next state depends on the present state of the buffer, while assuming the steady-state probability distribution for states of all other buffers. This approach is similar to mean-field theory in statistical physics. Our goal is to obtain analytical expressions for the distribution of the number of messages in a buffer at the steady state of the network and to determine the critical value of the load that results in saturation. Let us call the grade of a message the number k of hops the message has made toward its destination. For a given value of l, there exist messages of l different grades in the system: $k = 0, 1, \ldots,$ $l - 1$ because the messages of grade l disappear from the system. Note that a message of grade k that leaves a node appears at the next node as a message of grade $k + 1$. The state of a buffer can be described as a vector $(n_0, n_1, \ldots, n_{l-1})$ where n_k is the number of messages of grade k. Denote the limiting (steady-state probabilities) of a state $p(n_0, n_0, \ldots, n_{l-1})$ and

$$\sum_{k=0}^{l-1} n_k = n. \tag{1.1}$$

We first consider networks with SIRO queuing discipline. In this case, the probability that a message of grade k will be transferred is $n_k = n$. In the steady state, at any time slot, the expected number of messages generated at a node must be equal to the expected number of messages of grade 0 that leave the queue. Also, in general, the expected number of messages of grade k when they enter the next node must be equal to the expected number of the messages of grade k that leave the next node, and so on. Note that the messages that leave a node as grade k messages enter a neighboring (next) node at grade $k + 1$. Therefore,

$$\lambda = \sum_{n=1}^{\infty} \sum_{n_0+\cdots+n_{l-1}=n} \frac{n_0}{n} p\left(n_0, n_1, \ldots, n_{l-1}\right) = \sum_{n=1}^{\infty} \sum_{n_0+\cdots+n_{l-1}=n} \frac{n_1}{n} p\left(n_0, n_1, \ldots, n_{l-1}\right)$$

$$= \cdots = \sum_{n=1}^{\infty} \sum_{n_0+\cdots+n_{l-1}=n} \frac{n_{l-1}}{n} p\left(n_0, n_1, \ldots, n_{l-1}\right), \tag{1.2}$$

where the sum is taken over all combinations of values of $n_0, n_1, \ldots, n_{l-1} = n$ such that $n_0 + n_1 + \cdots + n_{l-1} = n$. If we sum the l expressions in Equation 1.2 and realize that the zero message state ($n = 0$) never sends any messages, we have

$$\sum_{n_i, n>0} \sum_{k=0}^{l-1} \frac{n_k}{n} p\left(n_0, n_1, \ldots, n_{l-1}\right) = 1 - p\left(0, 0, \ldots, 0\right) = l\lambda. \tag{1.3}$$

Equation 1.3 is the global flow equation for the network. It follows that the probability of the zero state is

$$p(0, 0, \ldots, 0) = 1 - l\lambda. \tag{1.4}$$

It can be shown (see [74]) that the limiting probabilities of all states with the same total number n of messages of all grades are equal:

$$p\left(n_0, n_1, \ldots, n_{l-1}\right) = \frac{n!(l-1)!}{(n+l-1)!} P(n), \tag{1.5}$$

where
n is given by Equation 1.1
$n!(l-1)!/(n+l-1)!$ is the number of different states with n messages
$P(n)$ is the total probability of all states with n messages

Henceforth, we will merge all states with n messages into one state n.
The transition probability matrix is

$$P = \begin{bmatrix} q & p & r & 0 & 0 & \cdots & \cdots & 0 & \cdots \\ q & p & r & 0 & 0 & \cdots & \cdots & 0 & \cdots \\ 0 & q & p & r & 0 & \cdots & \cdots & 0 & \cdots \\ \cdots & \cdots & \cdots & \cdots & \cdots & \cdots & \cdots & \cdots & \cdots \\ 0 & \cdots & \cdots & q & p & r & 0 & 0 & \cdots \\ \cdots & \cdots & \cdots & \cdots & \cdots & \cdots & \cdots & \cdots & \cdots \end{bmatrix}, \tag{1.6}$$

where
$$p = (1-\lambda)(l-1)\lambda + \left(1-(l-1)\lambda\right)\lambda$$
$$q = (1-\lambda)\left(1-(l-1)\lambda\right)$$
$$r = (l-1)\lambda^2$$

The state diagram with transition probabilities is shown in Figure 1.2. It is seen that the Markov chain is *almost time reversible*: starting with state 2, the flows from state i to state $i + 1$ and back are equal.

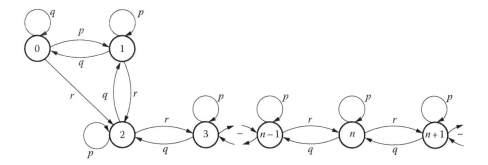

FIGURE 1.2 State diagram of the Markov chain for 1D ring network topology and distance l between the source and destination.

After simplification, the balance equations for the limiting probabilities are

$$P(0) = (1-\lambda)(1-(l-1)\lambda)\big[P(0)+P(1)\big],$$

$$P(1) = \Big[(1-\lambda)(l-1)\lambda + (1-(l-1)\lambda)\Big]\big[P(0)+P(1)\big] + (1-\lambda)(1-(l-1)\lambda)P(2),$$

$$P(2) = \lambda^2(l-1)P(0) + \lambda^2(l-1)P(1) + \lambda(1-(l-1)\lambda)P(2) + (1-\lambda)(1-(l-1)\lambda)P(3),$$

$$P(n) = \lambda^2(l-1)P(n-1) + \Big[\lambda(1-(l-1)\lambda) + (1-\lambda)(l-1)\lambda\Big]P(n)$$

$$+ (1-\lambda)(1-(l-1)\lambda)P(n+1), \quad (\text{for } n \geq 3),$$

$$\sum_{n=0}^{\infty} P(n) = 1, \tag{1.7}$$

where $P(0) = p(0, 0, \ldots, 0)$ is given by Equation 1.5. The solution of system (1.7) is

$$P(0) = 1 - l\lambda,$$

$$P(1) = (1-\lambda)\left[\frac{1}{(1-\lambda)(1-(l-1)\lambda)} - 1\right],$$

$$P(n) = (1-l\lambda)\frac{\lambda^{2(n-1)}(l-1)^{n-1}}{(1-\lambda)^n(1+\lambda-l\lambda)^n}, \quad (\text{for } n \geq 2). \tag{1.8}$$

Hence, the distribution of the length of the queue is geometric, starting with $n = 2$. Finally, the average number of messages in a queue is

$$\bar{n} = \sum_{n=1}^{\infty} nP(n) = \frac{\lambda^2(l-1)}{1-l\lambda} + l\lambda, \tag{1.9}$$

and the average latency is

$$\tau = \frac{\bar{n}}{\lambda} = \frac{\lambda(l-1)}{1-l\lambda} + l, \tag{1.10}$$

in accordance with Little's theorem [73]. It follows that the critical load is

$$\lambda_c = \frac{1}{l}, \tag{1.11}$$

and the critical exponent in Equations 1.9 and 1.10 is equal to 1.

It is shown in [74] that in case of non-SIRO queueing disciplines (first in, first out [FIFO], last in, first out [LIFO], NFO, etc.), the probabilities of states $p(n_0, ..., n_{l-1})$ are, in general, no more equal. However, expressions (1.8 through 1.11) remain true.

1.3.2 THEORETICAL ANALYSIS: TORUS

An important difference between the 1D and 2D cases is that in the Markov chain for the 1D case, there are transitions from a state with n messages only to states with $n - 1$, n, and $n + 1$ messages ($n \geq 1$). In contrast, in two dimensions, there exist transitions from a state with n messages to states with $n - 1$, n, $n + 1$, $n + 2$, and $n + 3$ messages ($n \geq 1$). It can be shown following the same line of arguments as in the 1D case that in the 2D case under the SIRO queueing discipline, the limiting probabilities of all states with the same total number n of messages of all grades are equal. This is not true for other queueing disciplines. However, the expected values of incoming and outgoing messages of each grade remain equal under steady-state conditions for all known queueing disciplines (including SIRO, FCFS, LCFC, NFO). Thus, for each buffer, the global flow equation is valid:

$$l\lambda = 1 - p(0,...,0) = 1 - P(0). \tag{1.12}$$

It follows that the balance equations for the *merged* probabilities $P(n)$ can be written in terms of total average flows in and out of the merged states with total number of messages n. Interestingly, the balance equations for $P(n)$ turn out to be the same for both SIRO and NFO queuing disciplines.

If we denote by $a_i (i = 0, 1, 2, 3)$ the probability that exactly I messages arrive to a particular buffer from neighboring nodes during one clock cycle, it follows from the description of the model that

$$a_0 = \left(1 - \frac{(l-1)\lambda}{2}\right)\left(1 - \frac{(l-1)\lambda}{4}\right)^2,$$

$$a_1 = \frac{(l-1)\lambda}{2}\left(1 - \frac{(l-1)\lambda}{2}\right)\left(\left(1 - \frac{(l-1)\lambda}{2}\right) + \left(1 - \frac{(l-1)\lambda}{4}\right)\right),$$

$$a_2 = \frac{(l-1)^2\lambda^2}{16}\left(\left(1 - \frac{(l-1)\lambda}{2}\right) + \left(4 - \frac{(l-1)\lambda}{4}\right)\right), \tag{1.13}$$

$$a_3 = \frac{(l-1)^3\lambda^3}{32}.$$

Because a message intended for a buffer can be also generated independently with probability λ by the local processor, the probability of the total number of arrivals being equal to i is

$$w_i = \lambda a_{(i-1)} + (1 - \lambda)a_i, \tag{1.14}$$

where
$i = 0, 1, 2, 3, 4$
$a_1 = a_4 = 0$

After some simplifications, the balance equations for each buffer can be written as

$$P(0) = 1 - l\lambda = w_0 \left[P(1) + P(0) \right],$$

$$P(1) = w_0 P(2) + w_1 P(1) + w_1 P(0),$$

$$P(2) = w_0 P(3) + w_1 P(2) + w_2 P(1) + w_2 P(0),$$

$$P(3) = w_0 P(4) + w_1 P(3) + w_2 P(2) + w_3 P(1) + w_3 P(0),$$

$$P(4) = w_0 P(5) + w_1 P(4) + w_2 P(3) + w_3 P(2) + w_4 P(1) + w_4 P(0),$$

$$P(n) = w_0 P(n+1) + w_1 P(n) + w_2 P(n-1) + w_3 P(n-2) + w_4 P(n-3), \quad (\text{for } n \geq 5),$$

$$\sum_{n=0}^{\infty} P(n) = 1. \tag{1.15}$$

The flow balance equations can be written in the following form for states $n \geq 4$:

$$(1 - \lambda)a_0 P(n) = \left(\lambda a_1 + a_2 + a_3 \right) P(n-1) + \left(\lambda a_2 + a_3 \right) P(n-2) + \lambda a_3 P(n-3). \tag{1.16}$$

Hence, the characteristic equation is cubic:

$$(1 - \lambda)a_0 x^3 - \left(\lambda a_1 + a_2 + a_3 \right) x^2 - \left(\lambda a_2 + a_3 \right) x - \lambda a_3 = 0. \tag{1.17}$$

The general solution of Equation 1.15 has the form

$$P(n) = A x_1^n + B x_2^n + B^* x_3^n, \quad (n \geq 4), \tag{1.18}$$

where

A, B, and B* are functions of λ and l

A is real, while B and B* are complex conjugate

x_1 is the real root of Equation 1.18

x_2, x_3 are two complex conjugate roots of the characteristic equation

The coefficients A, B, and B* can be calculated from the values of $P(0)$, $P(1)$, $P(2)$, and $P(3)$ that are calculated from Equations 1.12 and 1.15. In particular,

$$P(1) = (1 - l\lambda) \left[\frac{32}{(4 - (l-1)\lambda)^2 (2 - (l-1))(-1 + \lambda)} - 1 \right]. \tag{1.19}$$

Explicit expressions for larger n are too long to be written here. However, the generating function $\varphi(z)$ can be readily obtained:

$$\varphi(z) = \sum_{n=0}^{\infty} z^n P(n) = P(0) \frac{(z-1)\left(w_0 + z w_1 + z^2 w_2 + z^3 w_3 + z^4 w_4 \right)}{z - \left(w_0 + z w_1 + z^2 w_2 + z^3 w_3 + z^4 w_4 \right)}. \tag{1.20}$$

By the use of the generating function technique, we obtain explicit expressions for the average number $\bar{n}$ of messages in the buffer and the average latency τ:

$$\bar{n} = \frac{\lambda^2 (l-1)(11 + 5l)}{16(1 - l\lambda)} + l\lambda, \tag{1.21}$$

$$\tau = \frac{\lambda(l-1)(11+5l)}{16(1-l\lambda)} + l. \tag{1.22}$$

It follows that the critical load is

$$\lambda_{crit} = \frac{1}{l}, \tag{1.23}$$

and the critical exponent is equal to 1.

1.3.3 SIMULATIONS

1.3.3.1 Convergence to the Steady-State Regime

It is important to estimate the time it takes for the system to reach the stationary regime. We expect that the time depends on the load λ and increases as λ approaches the critical point.

The total contribution of the initial probability distribution to the distribution of the queue at time T is bounded from above by γ^T, where T is the number of time slots that have elapsed from the beginning of the simulation and is the largest eigenvalue smaller than 1 of the transition probability matrix. Calculations for the transition probability matrix P given by Equation 1.6 yield an expression for in terms of the transition probabilities p, q, and r:

$$\gamma = p + 2\sqrt{qr}. \tag{1.24}$$

The time T sufficient to make the effect of initial conditions not larger than ε is found from the equation

$$\gamma^T = \varepsilon. \tag{1.25}$$

Plots of $T = T(\lambda)$ for ε equal to 10^{-2}, 10^{-3}, and 10^{-4} are shown in Figure 1.3 for $l = 5$ ($\lambda_c = 0.2$). It is seen that, in the interval $0.19980 \le \lambda \le 0.19999$, $T(\lambda)$ increases from 6×10^4 up to 2.2×10^7 for $\varepsilon = 10^{-4}$.

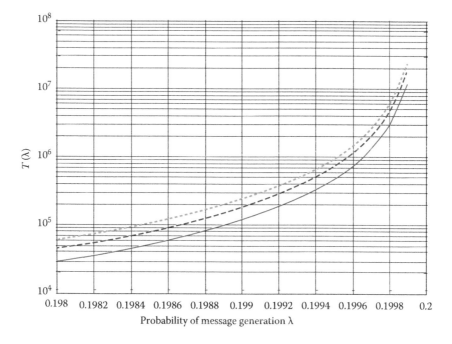

FIGURE 1.3 Simulation time $T(\lambda)$ required for approaching the steady-state regime closer than $E = 10^{-2}$ (magenta solid line), 10^{-3} (blue dashed), 10^{-4} (green dotted).

Time T is related to the characteristic relaxation time of the fluctuations in the system. The average lifetime of fluctuations T_r can be obtained from Equation 1.24 by setting $\varepsilon = 1/e$. For λ close to λ_c, this yields

$$T_r = \frac{4(l-1)\lambda_c^2}{l^2(\lambda_c + \lambda)^2}. \tag{1.26}$$

1.3.3.2 Ring Topology

Simulations have been done for rings of length 256 for several values of l starting with $l = 2$. The data were collected after achieving steady state with intervals of 500 time slots over a total time up to 10^6 time slots. The empirical values of $\bar{n}$ were calculated as $\bar{n} = \sum_{i=1}^{N} n_i$, where n_i is the number of messages in a particular buffer at a particular sampling time $= 10^6 \times 256/500 = 512{,}000$. The inverse value $1/\bar{n}$ of the average number of messages as a function of the load λ is shown in Figure 1.4.

The empirical values for the latency were collected independently, and they are in excellent agreement with the values of the queue length, as required by Little's theorem. Table 1.1 contains the values of the latency τ. The overall agreement of the queue length $\bar{n}$ with the theoretical values given by Equation 1.9 is very good (see Figure 1.4a), but, for load values exceeding $0.5\lambda_c$, the network behavior deviates slightly from the independent queues model. The critical exponent for the queue length is slightly less than 1. As seen from Figure 1.5, the average number of messages follows a power law:

$$\bar{n} = a(\lambda_c - \lambda)^{-\beta}, \tag{1.27}$$

where
  λ_c is the critical load
  β is the critical exponent

The data from simulations with $l = 5$ show that for SIRO queuing, $\lambda_c = 0.2$ and $\beta = 0.9701 \pm 0.066$ and for FCFS queuing, $\lambda_c = 0.2$ and $\beta = 0.956 \pm 0.014$. Thus, the difference between the results for SIRO and FCFS disciplines is statistically insignificant.

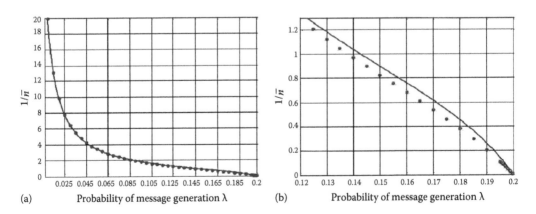

FIGURE 1.4 The inverse length of the queue for $l = 5$. Red line, theoretical values given by Equation 1.9; blue dots, simulation results for SIRO service policy; green crosses, simulation results for FCFS service policy. (a) The plot with the total range of λ. (b) The plot for large values of λ to show details.

TABLE 1.1

Simulation Data for the Average Latency τ_{SIRO} and τ_{FCFS} and for Their Standard Deviations σ_{SIRO} and σ_{FCFS}, Respectively

λ	τ_{SIRO}	σ_{SIRO}	τ_{FCFS}	σ_{FCFS}	λ	τ_{SIRO}	σ_{SIRO}	τ_{FCFS}	σ_{FCFS}
0.005	5.018	0.029	5.019	0.032	0.1900	25.765	0.805	25.265	0.554
0.035	5.182	0.040	5.175	0.036	0.1950	47.871	1.719	47.0621	1.615
0.065	5.432	0.054	5.427	0.043	0.1960	58.655	2.391	57.672	2.545
0.100	5.933	0.070	5.923	0.056	0.1970	77.608	3.616	75.518	3.570
0.120	6.448	0.090	6.428	0.068	0.1980	112.206	8.162	111.671	8.138
0.130	6.822	0.109	6.805	0.084	0.1990	216.485	24.839	216.684	25.092
0.140	7.345	0.116	7.292	0.087	0.1993	311.015	24.915	306.089	25.100
0.150	8.060	0.146	7.976	0.115	0.1994	359.760	29.510	356.844	32.258
0.160	9.133	0.177	9.044	0.138	0.1995	446.355	45.639	422.342	47.523
0.170	11.010	0.235	10.809	0.162	0.1996	520.444	72.958	520.911	64.398
0.180	14.649	0.362	14.426	0.255	0.1997	689.338	129.887	723.710	127.117

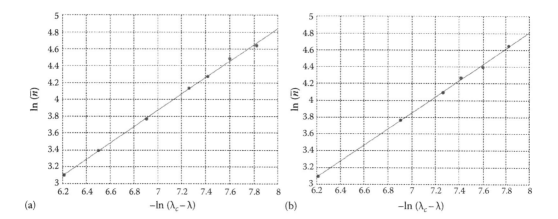

FIGURE 1.5 The dependence of τ and λ for $l = 5$ in logarithmic scale. Blue dots, data from simulation; red line, the best linear approximation. (a) SIRO service policy; (b) FCFS service policy.

The fluctuations of the number of messages in the network during the simulation for different message generation rates are shown in Figure 1.6. Note that the nature of the fluctuations in time changes considerably when the load approaches its critical value. The appearance of *slow* fluctuation waves serves as an indicator of the proximity of the critical point. Figure 1.7 shows the same phenomena from a different perspective. It is seen that correlations between queue lengths in time increase with λ and become very long near the critical point. On the other hand, no long-range correlations in space are observed.

1.3.3.3 Torus Topology

The simulations were done for a 16 × 16 toroidal square lattice for distances $l = 2$ and $l = 5$. Sufficient time was allowed for the network to come to the steady-state regime before samples were taken.

Figure 1.8 shows the inverse values of the average number of messages, $\bar{n}$, as a function of the load λ for distance $l = 5$. The standard deviations of the mean values were calculated, but, because of large sampling sizes, they are too small to be shown in the plot. Figure 1.8b shows details of the values of $\bar{n}$ for large λ. It is seen that, even for the probability of message generation close to the saturation point, mean-field theory gives very accurate predictions of the network behavior.

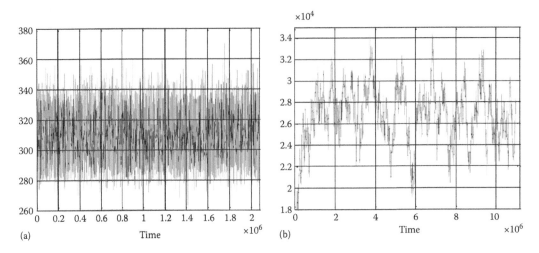

(a)   Time   ×10⁶   (b)   Time   ×10⁶

FIGURE 1.6 The number of messages in the network during the simulation. (a) $\lambda = 0.15$ and (b) $\lambda = 0.1996$.

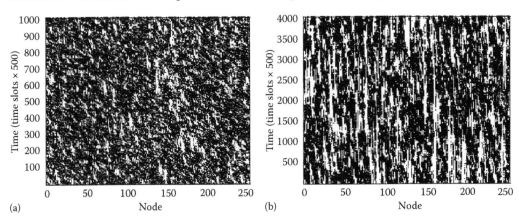

(a)   Node   (b)   Node

FIGURE 1.7 The instantaneous number of messages n in the network during simulation taken every 500 time slots. (a) $\lambda = 0.15$, white nodes, $n \leq 1.2$; black nodes, $n > 1.2$. (b) $\lambda = 0.1996$; white nodes, $n \leq 104$; black nodes, $n > 104$.

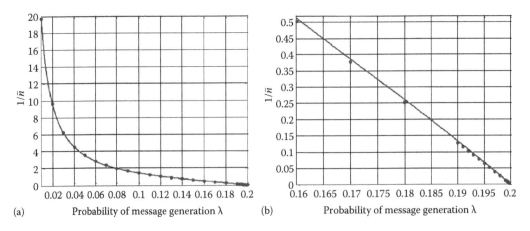

(a)   Probability of message generation λ (b)   Probability of message generation λ

FIGURE 1.8 The inverse length of the queue for $l = 5$. Red line, theoretical values given by Equation 1.21; blue dots, numerical simulation results for a torus of 256 nodes. (a) The plot with the total range of λ. (b) The plot for large values of λ to show details.

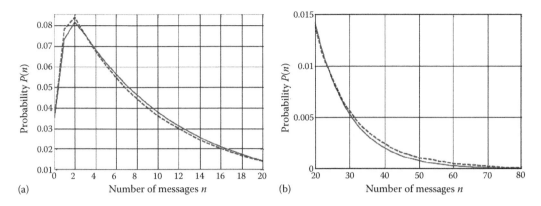

FIGURE 1.9 Probability distribution of the queue length; $l = 5$, $\lambda = 0.193$, torus size, 256 nodes. Red line, predicted theoretical values; blue dashed line, numerical simulation results. (a) The queue length from 0 to 20. (b) The queue length from 20 to 80.

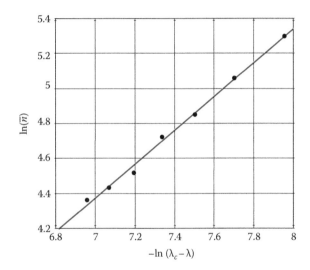

FIGURE 1.10 The dependence between $\bar{n}$ and λ in logarithmic scale for $l = 5$.

The probability distribution of the number of messages n in a buffer is given in Figure 1.9. The deviations from the theoretical values found from Equation 1.15 are small. However, they result in a small, but systematic, discrepancy between the theoretical and simulation values of the queue length $\bar{n}$, which leads to an exponent in the power law in Equation 1.27 different from 1. The data from the simulation with $l = 5$ show that $\lambda_c = 0.2$ and $\beta = 0.960 \pm 0.02$ (see Figure 1.10).

Figure 1.11 demonstrates the time behavior of the network. Figure 1.11a shows the data for $\lambda = 0.15$, which is still far from the critical point $\lambda_c = 0.2$. Fluctuation patterns for a larger value, $\lambda = 0.199$, are shown in Figure 1.11b. It is seen that the fluctuation amplitude reaches about 10% of the average. As λ approaches λ_c, fluctuations become much larger in amplitude (approximately proportional to the number of messages in the network), which is in agreement with the exponential-type distributions shown in Figure 1.9 and consistent with Equation 1.18. As for the ring topology, the characteristic *wavelength* of fluctuations sharply increases near the critical point (see Figure 1.11). The absence of long-range correlations between nodes can be observed in Figure 1.12. Indeed, no *islands* of nodes with small or large number of messages can be seen.

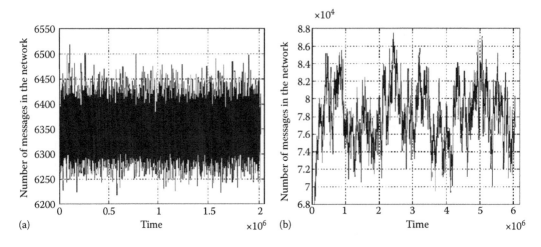

FIGURE 1.11   Number of messages in the network during simulation; $l = 5$. (a) $\lambda = 0.15$ and (b) $\lambda = 0.199$.

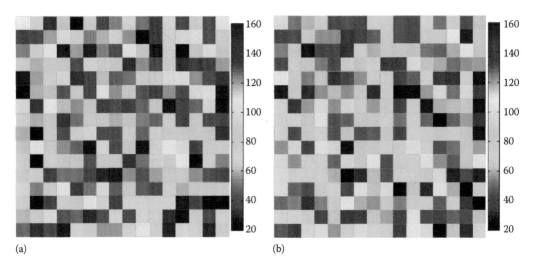

FIGURE 1.12   Snapshot of the number of messages in each router, $\lambda = 0.491$, $l = 2$. (a) Time $t = 4.95 \times 10^4$. (b) Time $t = 5 \times 10^4$.

1.4   NETWORKS WITH HETEROGENEOUS ACTIVITY

As we have shown, the behavior of networks with infinite buffers and equal probabilities of message generation for all nodes (homogeneous activity) is consistent with Jackson's theorem. Hence, we expect a continuous transition to saturation with the critical exponent equal to 1, as given by the mean-field theory. Small deviations from the mean-field theory observed in the simulations do not change the picture qualitatively. However, in many practical situations, networks can display much more complex behavior. To reveal patterns of such behavior, we need more complex models that cannot be adequately described within the framework of the mean-field approximation.

Apparently, to go beyond Jackson's theorem, at least one of two fundamental assumptions should be relaxed. In particular, the two following classes of networks can be considered:

1. Networks with heterogeneous (rather than homogeneous) activity, where there is a positive feedback between the incoming flow of messages and the message generation rate.
2. Networks with finite (rather than infinite) buffers.
   This section is devoted to the first type of networks. The general features of the model are explained in the following.

As for the case of the homogeneous network, each node in the model consists of a local processor, a router, and buffers of infinite capacity, but now, there are two different message generation rates. Let us assume that each processor in the ring network can generate messages either with probability λ_1, or $\lambda_2(\lambda_1 < \lambda_2)$, depending on the intensity of the flow of messages arriving at the corresponding node. If the average number of incoming messages per time slot over the fixed number M of the previous time slots is smaller than a fixed value $l\mu$, where $\lambda_1 < \mu < \lambda_2$, the node generates messages with rate λ_1 during next M time slot; otherwise, it generates messages with rate λ_2. The separation $\Delta\lambda = \lambda_2 - \lambda_1$ is kept constant, and the threshold value of μ is chosen so that the average fraction of nodes with rates λ_1 and λ_2 is approximately equal to 50%. In other words, μ serves as a reference point for assigning the generation rate for every router in the system for the next fixed period M.

Care was taken to ensure that the network achieves a steady state, in which the values of the queue lengths in each buffer and average number r of incoming messages per time slot have been measured and calculated after every M time slots within the interval $T = (t_1, t_2)$. In our simulations, $M = 500$, $t_1 = 10^6$, and $t_2 = 2.5 \times 10^6$ time slots.

1.4.1 RING TOPOLOGY

We consider a network of 500 nodes connected in a ring. The values of $\Delta\lambda$ used in simulations were chosen to be small compared to the critical value $\lambda_c = 1/l$ for the homogeneous case.

The number of nodes with rates λ_1 and λ_2 is c_1 and c_2, respectively. The numbers of messages in each queue were measured at every time checkpoint of M clock cycles over the total time interval $T = t_2 - t_1 = 1.5 \times 10^6$ time slots to express the average queue length $\bar{n}$ as a function of the overall average load in the network λ_{av}:

$$\lambda_{av} = c_1\lambda_1 + c_2\lambda_2 \tag{1.28}$$

The theoretical analysis of a network with heterogeneous activity is very difficult. A rough lower bound for $\bar{n}(\lambda_{av})$ can be obtained by assuming that the system still performs as if it satisfies the independent queues model for an effective load λ_{eff}, defined by the condition that

$$\bar{n}_{th}(\lambda_{eff}) = c_1\bar{n}_{th}(\lambda_1) + c_2\bar{n}_{th}(\lambda_2), \tag{1.29}$$

where $\bar{n}_{th}(\lambda_{eff})$ is given by Equation 1.9, and setting $\lambda_{av} = \lambda_{eff}$, instead of Equation 1.28. Obviously, this lower bound leads to the critical value λ_c given by Equation 1.11. Another approximation that is close to that defined by Equation 1.29 for smaller loads, but becomes an upper bound for $\bar{n}(\lambda_{av})$, when λ_{av} approaches the critical value, is obtained by considering that each of the phases of λ_1 and λ_2 obeys the independent queue model. This approximation is

$$n^*(\lambda_{av}) = c_1\bar{n}_{th}(\lambda_1) + c_2\bar{n}_{th}(\lambda_2), \tag{1.30}$$

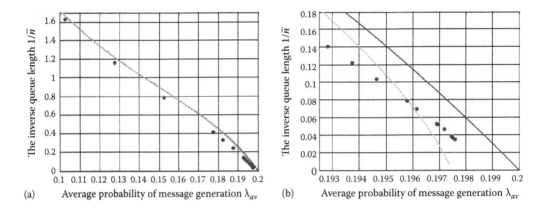

(a) Average probability of message generation λ_{av} (b) Average probability of message generation λ_{av}

FIGURE 1.13 The average number of messages in a single buffer $\bar{n}(\lambda_{av})$ taken every 500 time slots during the time interval $T = 1.5 \times 10^6$ from $t_1 = 10^6$ to $t_2 = 2.5 \times 10^6$ (blue dots). Red solid line, lower bound defined by Equation 1.29. Green dashed line, approximation given by Equation 1.30. (a) The entire range. (b) The range $\lambda_{av} \geq 0.19$ is presented to show the details in the region close to the critical point.

where λ_{av} is defined by Equation 1.28. The critical value resulted from Equation 1.30 is

$$\lambda_c^* = \lambda_c - c_1 \Delta\lambda. \tag{1.31}$$

In fact, the real behavior of the system is far from that described by the independent queues model. Figure 1.13a shows the function $\bar{n}(\lambda_{av})$, and Figure 1.13b shows the most interesting range closer to the critical point. It turns out that $\bar{n}(\lambda_{av})$ closely follows a power law of the form

$$\bar{n}(\lambda_{av}) = a(\tilde{\lambda}_c - \lambda_{av})^{-\beta}, \tag{1.32}$$

where $\tilde{\lambda}_c$ and b are new values of the critical load and critical exponent, respectively, as shown in Figures 1.13 and 1.14. Simulation data for the parameters $l = 5$ and $\Delta\lambda = 0.005$ yield $\tilde{\lambda}_c = 0.19875$ and $\beta = 0.8462 \pm 0.0028$.

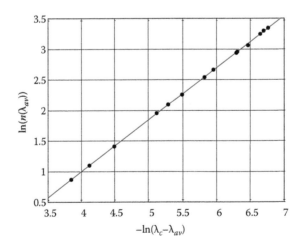

FIGURE 1.14 The dependence of $\bar{n}(\lambda_{av})$ on λ_{av}. It is seen that $\bar{n}(\lambda_{av})$ follows a power-law behavior for $\lambda \geq 0.177$ (see Equation 1.32), where $\lambda_c' = 0.19875$.

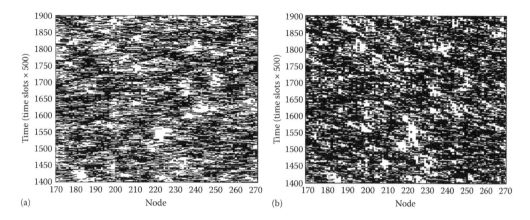

(a) Node (b) Node

FIGURE 1.15 Generation probabilities and number of messages in each node (only part of the ring network from router 170 to 270 is shown). Data were recorded every 500 time slots from $t_1 = 5 \times 10^4$ to $t_2 = 95 \times 10^4$ with $l = 5$. (a) Generation probabilities $\lambda_1 = 0.194$ (black nodes) and $\lambda_2 = 0.199$ (white nodes). (b) Number of messages n at each node. Black nodes, $n \leq 15$; white nodes, $n > 15$.

An interesting phenomenon found in the simulations is the emergence of a domain structure for λ near λ_c, which demonstrates the long-range dependence between queues in nodes located at distances substantially larger than l. The nodes with rate λ_1 and λ_2 form a two-phase system with well-defined domains of different generation rates. This picture changes in time and corresponds to domains of different lengths of the queues (see Figure 1.15a and b).

Our simulations also show that the appearance of two different phases becomes more pronounced for larger values of $\Delta\lambda$: $\Delta\lambda = 0.01$, $\Delta\lambda = 0.02$ (Figure 1.16b), and $\Delta\lambda = 0.055$ were explored. As we note in Figure 1.16c and d, the queue lengths within one phase are distributed randomly, without correlation, as in homogeneous networks.

It was also observed that with an increase of the generation rate and λ_2 approaching the critical value (close to the critical value for the homogeneous network), the fluctuations increase rapidly, the network becomes unstable, and, as a result, it goes to a low message generation rate or high message generation rate state completely. Small changes in the value of μ result in dramatic changes in the network behavior. This instability makes the data recorded in those simulations unreliable. Even when the difference between λ_1 and λ_2 is equal to $\Delta\lambda = 0.005$, we noticed that fluctuations close to the critical point become very large (see Figure 1.17a), and it is difficult to obtain reliable data (see also Figure 1.17b). In particular, the proportion of generation rates λ_1 and λ_2 changes dramatically during the simulation.

1.4.2 Torus Topology

In the 2D case, we consider a 16×16 square lattice with toroidal boundary conditions. We used the same parameters as for the ring topology: $l = 5$, $\Delta\lambda = 0.005$, $M = 500$, $t_1 = 10^6$, $t_2 = 2.5 \times 10^6$.

Using Equations 1.29 and 1.30, we note that, for the 2D torus, the model of independent queues gives a good approximation to the network behavior for small network loads. However, the behavior deviates substantially from the independent queues model for load values closer to the saturation point. The critical value seems to be smaller, as shown in Figure 1.18, similar to the 1D case (see Figure 1.13). Note that, for λ close to λ_c, it was not possible to keep the proportion between nodes with generation rate λ_1 and λ_2 equal to 50%. For example, for $\lambda_{av} = 0.197816$ in Figure 1.18b, the fraction of the routers with message generating rate λ_1 is equal to 76.3%. Therefore, the average number of the messages in a queue is less than expected. Still, data from simulations give

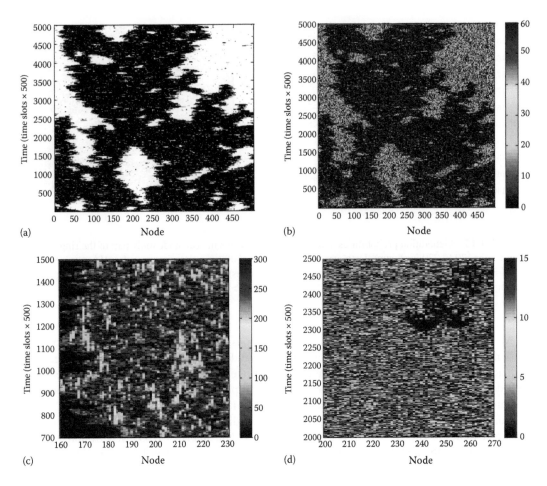

FIGURE 1.16 Data recorded every 500 time slots in the network with $\lambda_1 = 0.178$, $\lambda_2 = 0.198$, $\mu = 0.1885$. (a) Generation probabilities: the black nodes represent the routers with message generation rate λ_1; the white nodes represent the routers with message generation rate λ_2. (b) The number of messages in the network during the simulation: $\bar{n} = 14.05$, $n_{min} = 0$, and $n_{max} = 356$. (c) The number of messages in the network: a domain of nodes with generation rate λ_2. (d) The number of messages in the network: a domain of nodes with generation rate λ_1.

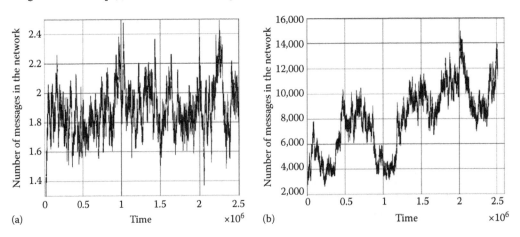

FIGURE 1.17 The average number of messages in the network during the simulation. (a) $\lambda_1 = 0.1945$, $\lambda_2 = 0.1995$, and $\mu = 0.198$. (b) $\lambda_1 = 0.178$, $\lambda_2 = 0.198$, and $\mu = 0.1885$.

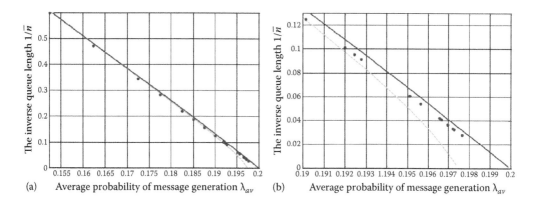

(a)   Average probability of message generation λ_{av} (b)   Average probability of message generation λ_{av}

FIGURE 1.18   The average number of messages in the network $\bar{n}(\lambda_{av})$. Blue dots, the data were collected every 500 time slots between $t_1 = 1 \times 10^6$ and $t_2 = 2.5 \times 10^6$. Red solid line, lower bound defined by Equation 1.21. Green dashed line, approximation given by Equation 1.30.

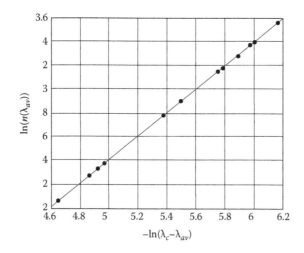

FIGURE 1.19   The dependence of $\bar{n}(\lambda_{av})$ on λ_{av}.

us the value of $\lambda_c = 0.19977$, less than for the homogeneous network, and $\beta = 0.9859 \pm 0.0018$ (Figure 1.19). The pattern of large fluctuations for the network loads close to the critical point is shown in Figure 1.20.

The most dramatic effect of the heterogeneous activity is the formation of domains of two different phases of low and high activity, respectively, and, correspondingly, domains of small and large queue lengths, as shown in Figure 1.21. The characteristic size of the domains is much larger than the interaction distance (distance l between the source and the destination), which indicates the emergence of long-range order when the network load is close to critical.

Unlike the 1D case, a 2D network allows us to study the topology of the domains of different loads and queue lengths. Using an analogy with physical two-phase systems (e.g., gas–liquid in equilibrium), one could expect the domains of different phases to be compact, so that their boundary is small compared to their area (in terms of the number of nodes). However, the simulation data (Figure 1.21) cannot be viewed as supporting this hypothesis, perhaps because of the relatively small size of the system.

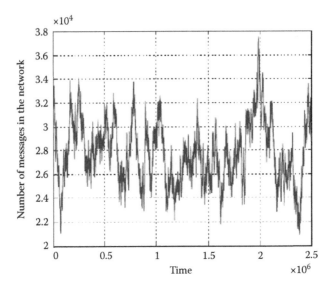

FIGURE 1.20 The average number of messages in the network during the simulation taken every 100 clock time slots: $\lambda_1 = 0.1949$, $\lambda_2 = 0.1999$, and $\mu = 0.1978$.

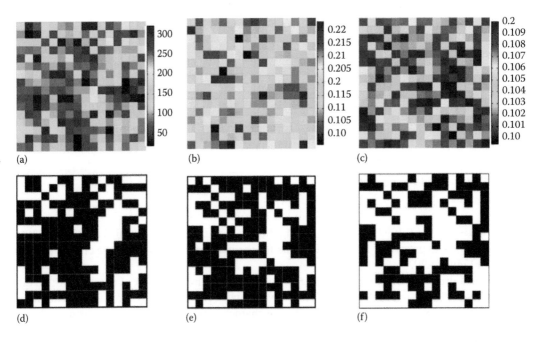

FIGURE 1.21 The heterogeneous network: $\lambda_1 = 0.1949$, $\lambda_2 = 0.1999$, and $\mu = 0.1978$. Snapshot of the probability, number of messages, and incoming flow to every router at time $t = 25.995 \times 10^5$. (a) Number of messages n in the network. (b) The actual generation rate in the routers. (c) The average incoming flow rate f. (d) Instantaneous number of messages n in routers. Black nodes, $n \leq 108$; white nodes, $n > 108$. (e) Generation probabilities $\lambda_1 = 0.1949$ (black nodes) and $\lambda_2 = 0.1999$ (white nodes) in every router. (f) Incoming flow rate f. Black nodes, $f \leq 0.19698$; white nodes, $f > 0.19698$.

1.5 FUTURE RESEARCH DIRECTIONS

The behavior of networks with infinite buffers considered in this chapter seems to follow closely Jackson's theorem. However, it should be borne in mind that a theoretical proof of Jackson's theorem is absent for these models. Moreover, there is a small, but systematic discrepancy between the theoretical and simulation results, which remains unexplained up to now. Indeed, one possibility is the presence of a very weak dependence between neighboring queues, while the other is that the discrepancy is simply an artifact that is due to the simulation procedure. It is important, therefore, to clarify the situation and to establish the limits of applicability of Jackson's theorem to the networks with discrete time and deterministic service time.

Another interesting problem is the effect of various queueing disciplines on the effect of various queueing disciplines on the network performance. Our results show that the choice of queueing discipline makes no difference in the theoretical model as far as the expected value of the delivery time and the distribution of messages are concerned. However, this result has been verified by simulations only for SIRO and NFO disciplines.

The dependence of the network performance characteristics on the network performance characteristics on the message length is another problem of practical importance that needs further investigation.

Our theoretical analysis is focused on the steady-state characteristics of the network performance. It is of great interest, both theoretically and practically, to get a better understanding of the network behavior in time, especially, in the region near the critical load value, where the appearance of large and *long* fluctuations has been observed in simulations as predictors of the critical transition.

Another interesting topic for future research would be to include all nodes in the range l from a source in a list of the possible destination for a message and find out how it affects critical network load and the critical exponent.

Both homogeneous and heterogeneous networks can be viewed as a specific class of complex systems. On one hand, the presence of a large number of individual agents that act randomly and, at large extent, independently of each other creates a situation very similar to that observed in system belonging to the realm of statistical physics. On the other hand, however, those agents interact rather strongly with each other. Moreover, as demonstrated in the heterogeneous network model, the system behavior can be controlled and modified by the imposition of certain external rules. The combination of these factors creates a complex interplay of *chaos* and *order* that is a great challenge for researchers in the future.

1.6 SUMMARY

The specific features of our models are discrete time and deterministic service time. The choice of the deterministic service time implies that the service time does not have a rational Laplace transform. We used the mean-field (independent queues) approximation in the theoretical analysis. We derived closed-form expressions for the average queue length, latency, and the critical load for both the ring and 2D toroidal topologies. Although there is no known proof that such a network has to obey Jackson's theorem, our simulation data show that this approximation gives an overall good prediction for the network behavior in our model of homogeneous networks with infinite buffers:

1. The average number of messages in a buffer and the latency (measured independently) is close to those predicted by the theory for both 1D and 2D networks over the entire range of the load parameter λ and various values of distance l. This fact is remarkable, because our system has properties that are different from those for which the product form of the state probability distribution has been proved (Jackson's theorem and its generalizations). The saturation phenomenon in the 1D and 2D interconnection networks was shown to be

a continuous (second-order) phase transition with a critical exponent slightly less than 1, which means that mean-field theory provides a correct description of the types of network we considered.

2. A slight systematic discrepancy is observed between the theoretical and empirical distributions of the queue length for λ close to the critical point. This fact suggests a small correlation between the queues, which may indicate that the model of independent queues is not exact for deterministic service time in discrete-time networks.

3. In contrast with physical systems for which critical phenomena are observed only in the thermodynamic limit, the saturation in networks occurs for finite networks and, moreover, the characteristics of the network performance seemingly do not depend on the size of the system provided that the distance l is smaller than the maximum distance between nodes.

4. In both 1D and 2D networks, the critical load is inversely proportional to the distance between the source and the destination. This behavior is not surprising, because the utilization of every link in the network (the fraction of time when the link is busy) is $\rho = 1 - P(0) = l\lambda$. The critical point corresponds to $\rho = 1$, which, of course, cannot be exceeded.

5. The choice of the queuing discipline based on the message grades or arrival time has no effect on the network behavior within the framework of the independent queues model.

6. The amplitude of the fluctuations of the number of messages in the network increases approximately proportional to the average (over time) of the number of messages. This fact demonstrates that the distribution of number of messages at different times (or in different parts of the network under the same conditions) is quite broad, as in a distribution of exponential type (rather than in a binomial or Poisson distribution), which is in agreement with our theoretical model. This breadth of the distribution should be taken into account in practical situations: because of large fluctuations, the network (or some parts of it) can become overloaded, even if the average number of messages is still in the *safe* range. Another characteristic feature of the fluctuations is that their characteristic *wavelength* in time increases when the load parameter λ approaches the critical value. Such large and *slow* fluctuations are typical for a physical system near the critical point as well. They may serve as a predictor of the phase transition coming on.

7. No long-range order is observed in the network even when the load is close to the critical value. This fact is in accord with all other results that demonstrate the validity of the independent queues theory for this type of network.

Networks with heterogeneous activity show patterns of behavior that are different from those of homogeneous networks:

1. Jackson's theorem is not applicable to this type of network. The model of independent queues gives a very rough approximation of the network performance, which is acceptable only for values of the load far from the critical region.

2. The transition to the saturation state is a typical continuous phase transition. The apparent nontrivial value for the critical exponent suggests that this phenomenon cannot be adequately described by the mean-field theory approximation.

3. The network behavior near the critical point is very sensitive to the threshold value μ. Small changes of μ can lead to the transition of the entire network to one of the values of the message generation rate (λ_1 or λ_2). In this case, the network behaves eventually as a network with homogeneous activity.

4. For the steady-state regime, when nodes with both λ_1 and λ_2 values of generation rate are present, the nodes form large homogeneous domains of two separate phases. The characteristic size of the domains in both 1D and 2D networks is much larger than the interaction radius l, which indicates the emergence of a long-range order when the load is close to critical.

5. The domains of different phases have different values of the queue lengths. The configuration of the regions of different average values of the number of messages in buffers follows very closely the shapes of the phase domains.

6. The size and the shape of the phase domains change slowly in time, which creates patterns of fluctuations of large amplitudes, but of rather *low frequency*, combined with *high-frequency* fluctuations of smaller amplitudes.

7. An interesting feature of this type of network is the existence of a steady-state nonsaturated regime even in the case when one of the load values, λ_2, is larger than the critical value for a similar homogeneous network. The explanation of this fact is that saturation does not occur in the nodes of the higher activity λ_2 because after a finite time interval, they always become nodes with lower activity λ_1.

REFERENCES

1. L. Kleinrock, *Queueing Systems*, Volume I: Theory, Wiley, New York, 1975.
2. L. Kleinrock, *Queueing Systems*, Volume II: Computer Applications, Wiley, New York, 1976.
3. J.R. Jackson, Job shop like queueing systems, *Management Science*, 10 (1), 131–142, 1963.
4. W.J. Gordon and G.F. Newell, Closed queueing systems with exponential servers, *Operations Research*, 15, 254–265, 1967.
5. J.P. Buzen, Computational algorithms for closed queueing networks with exponential servers, *Communications of the ACM*, 16 (9), 527–531, 1973.
6. F. Baskett, K.M. Chandy, R.R. Muntz, and F.G. Palacios, Open, closed, and mixed networks of queues with different classes of customers, *Journal of the ACM*, 22 (2), 248–260, April 1975.
7. J.P. Spirn, Queueing networks with random selection for service, *IEEE Transactions on Software Engineering*, SE-5, 287–289, 1979.
8. A.S. Noetzel, A generalized queueing discipline for product form network solution, *Journal of the ACM*, 26, 779–793, October 1979.
9. K.M. Chandy and J. Martin, A characterization of product form queueing networks, *Journal of the ACM*, 30, 286–299, 1983.
10. M. Reiser, Mean value analysis of queueing networks: A new look at an old problem, in *Proceedings of the Fourth International Symposium on Modelling and Performance Evaluation of Computer Systems*, vol. I, pp. 63–77, Vienna, Austria, February 1979.
11. M. Reiser and S.S. Lavenberg, Mean value analysis of closed multichain queueing networks, *Journal of the ACM*, 27 (2), 313–322, April 1980.
12. M. Reiser, Mean value analysis and convolution method for queueing dependent servers in closed queueing networks, *Performance Evaluation*, I (1), 7–18, January 1981.
13. C.H. Sauer and K.M. Chandy, *Computer Systems Performance Modeling*, Prentice Hall, Englewood Cliffs, NJ, 1981.
14. I.F. Akyildiz and A. Sieber, Approximate analysis of load dependent general queueing networks, *IEEE Transactions on Software Engineering*, 14, 1537–1545, 1988.
15. N. Nikitin and J. Cortadella, A performance analytical model for Network-on-Chip with constant service time routers, in *Proceedings of the 2009 International Conference on Computer-Aided Design (ICCAD'09)*, San Jose, CA, pp. 571–578. ACM, New York, 2009. doi:10.1145/1687399.1687506, http://doi.acm.org/10.1145/1687399.1687506.
16. E. Fischer, A. Fehske, and G.P. Fettweis, A flexible analytic model for the design space exploration of many-core Network-on-Chips based on queueing theory, in *SIMUL 2012, The Fourth International Conference on Advances in System Simulation*, Lisbon, Portugal, pp. 119–124, 2012.
17. Z. Youhui, X. Dong, S. Gan, and W. Zheng, A performance model for Network-on-Chip wormhole routers, *Journal of Computers*, 7 (1), 76–84, 2012.
18. A.E. Kiasari, Z. Lu, and A. Jantsch, An analytical latency model for Networks-on-Chip", *IEEE Transactions on Very Large Scale Integrated (VLSI) Systems*, 21 (1), 113–123, 2013.
19. Q. Tie, L. Feng, F. Xia, W. Guowei, and Y. Zhou, A packet buffer evaluation method exploiting queueing theory for wireless sensor networks, *Computer Science and Information Systems/Com SIS*, 8 (4), 1027–1049, 2011.

20. T.M. Nagendrappa and F. Takawira, Queuing analysis of distributed and receiver-oriented sensor networks: Distributed DS-CDMA-based medium access control protocol for wireless sensor networks, *IET Wireless Sensor Systems*, 3.1, 69–79, 2013.

21. W.E. Leland, M.S. Taqqu, W. Willinger, and D.V. Wilson, On the self-similar nature of Ethernet traffic, *ACM SIGCOMM Computer Communication Review*, 23 (4), 183–193, 1993.

22. J.H.B. Deane, C. Smythe, D.J. Jefferies, and G.G. Xh, Self-similarity in a deterministic model of data transfer, *Journal of Electronics*, 80 (5), 677–691, 1996.

23. M. Takayasu, H. Takayasu, and T. Sato, Critical behaviors and 1/f noise in information traffic, *Physica A*, 233, 824–834, 1996.

24. W. Willinger, M.S. Taqqu, R. Sherman, and D.V. Wilson, Self-similarity through high-variability: Statistical analysis of Ethernet LAN traffic at the source level, *Networking, IEEE/ACM Transactions on*, 5 (1), 71–86, 1997.

25. J. Yuan, Y. Ren, and X. Shan, Self-organized criticality in a computer network model, *Physical Review E*, 61, 1067–1071, 2000.

26. T. Ohira and R. Sawatari, Phase transition in a computer network traffic model, *Physical Review E*, 58 (1), 193–195, 1998.

27. A.Y. Tretyakov, H. Takayasu, and M. Takayasu, Phase transition in a computer network model, *Physica A: Statistical Mechanics and Its Applications*, 253, (1), 315–322, 1998.

28. H. Fukś and A.T. Lawniczak, Performance of data networks with random links, *Mathematics and Computers in Simulation*, 51, 101–117, 1999.

29. J. Yuan and K. Mills, Exploring collective dynamics in communication networks, *Journal of Research of the National Institute of Standards and Technology*, 107, 179–191, 2002.

30. D.K. Arrowsmith, R.J. Mondragón, J.M. Pitts, and M. Woolf, Phase transitions in packet traffic on regular networks: A comparison of source types and topologies, Report 08, the Mittag-Leffler Institute, Royal Swedish Academy of Sciences, Stockholm, Sweden, 2004.

31. G. Mukherjee and S.S. Manna, Phase transition in a directed traffic flow network, *Physical Review E*, 71, 066108, 2005.

32. D. De Martino, L. Dall'Asta, G. Bianconi, and M. Marsili, Congestion phenomena on complex networks, *Physical Review E*, 79 (1), 015101, 2009.

33. A.T. Lawniczak and X. Tang, Network traffic behaviour near phase transition point, *The European Physical Journal B—Condensed Matter*, 50 (1–2), 231–236, 2006. http://arxiv.org/pdf/nlin.AO/0510070.

34. S. Valverde and R.V. Solé, Self-organized critical traffic in parallel computer networks, *Physica A: Statistical Mechanics and Its Applications*, 312, 636–648, 2002.

35. T. Yokota, K. Ootsu, F. Furukawa, and T. Baba, Phase transition phenomena in interconnection networks of massively parallel computers, *Journal of the Physical Society of Japan*, 75, 2006.

36. S.H. Yook, H. Jeong, and A.-L. Barabasi, Modeling the Internet's large-scale topology, *Proceedings of the National Academy of Sciences*, 99, 13382–13386, 2002.

37. R.V. Solé and S. Valverde, Information transfer and phase transitions in a model of internet traffic, *Physica A: Statistical Mechanics and Its Applications*, 289, 595–605, 2001.

38. R.D. Smith, The dynamics of internet traffic: Self-similarity, self-organization, and complex phenomena, *Advances in Complex Systems*, 14 (06), 905–949, 2011.

39. S. Valverde and R.V. Solé, Internet's critical path horizon, *The European Physical Journal B, Condensed Matter and Complex Systems*, 38, 245–252, 2004.

40. M. Woolf, D.K. Arrowsmith, R.J. Mondragón, and J.M. Pitts, Optimisation and phase transition in a chaotic model of data traffic, *Physical Review E*, 66, 046106, 2002.

41. A.K. Chandra, Jamming of directed traffic on a square lattice, *Journal of Statistical Mechanics: Theory and Experiment*, 8, 5–17, 2006.

42. M. Takayasu, H. Takayasu, and K. Fukuda, Dynamic phase transition observed in the Internet traffic flow, *Physica A: Statistical Mechanics and Its Applications*, 277, 248–255, 2000.

43. O. Biham, A.A. Middleton, and D. Levine, Self-organization and a dynamical transition in traffic-flow models, *Physical Review A*, 46, R6124–R6127, 1992.

44. Y. Honda and T. Horiguchi, Self-organization in four-direction traffic-flow model, *Journal of the Physical Society of Japan*, 69, 3744–3751, 2000.

45. W. Wu, J. Yuan, X. Shan, and Y. Ren, Exploring collective behaviors with short-range correlation between routers, in *Proceedings of International Conference on Communication Technology of the 16th World Computer Congress*, Beijing, China, pp. 70–75, 2000.

46. Z. Ren, Z. Deng, and Z. Sun, Cellular automaton modeling of computer network, *Computer Physics Communications*, 144, 243–251, 2002.

47. B. Liu, Y. Guo, J. Kurose, D. Towsley, and W. Gong, Fluid simulation of large scale networks: Issues and tradeoffs, in *Proceedings of the International Conference on Parallel and Distributed Processing Techniques and Applications*, Las Vegas, NV. CSREA Press, Las Vegas, NV, pp. 2136–2142, 1999.

48. P. Jelenković and P. Momčilović, Finite buffer queue with generalized processor sharing and heavy-tailed input processes, *Computer Networks*, 40, 433–443, 2002.

49. C. Kiddle, R. Simmonds, C. Williamson, and B. Unger, Hybrid packet/fluid flow network simulation, in *PADS'03: Proceedings of the 17th Workshop on Parallel and Distributed Simulation, IEEE Computer Society*, Washington, DC, p. 143, 2003.

50. R.C. Hampshire, M. Harchol-Balter, and W.A. Massey, Fluid and diffusion limits for transient sojourn times of processor sharing queues with time varying rates, *Queueing Systems: Theory and Applications*, 53, 19–30, 2006.

51. V. Arunachalam, V. Gupta, and S. Dharmaraja, A fluid queue modulated by two independent birth–death processes, *Computers & Mathematics with Applications*, 60 (8), 2433–2444, 2010. doi:10.1016/j.camwa.2010.08.039.

52. D. Shah and D. Wischik, Fluid models of congestion collapse in overloaded switched networks, *Queueing Systems*, 69 (2), 121–143, 2011.

53. M.A. Yazici and N. Akar, Analysis of continuous feedback Markov fluid queues and its applications to modeling Optical Burst Switching, in *Proceedings of the 2013 25th International Teletraffic Congress (ITC)*, Shanghai, China, pp. 1–8, 2013. doi:10.1109/ITC.2013.6662952.

54. M.S. Telek and M.S. Vécsei, Analysis of fluid queues in saturation with additive decomposition, *Modern Probabilistic Methods for Analysis of Telecommunication Networks. Communications in Computer and Information Science*, 356, 167, 2013. doi:10.1007/978-3-642-35980-4_19.

55. I. Norros, On the use of fractional Brownian motions in the theory of connectionless networks, *IEEE Journal on Selected Areas in Communications*, 13 (6), 953–962, 1995.

56. T. Mikosch, S. Resnick, H. Rootzén, and A. Stegeman, Is network traffic approximated by stable Lévy motion or fractional Brownian motion? *The Annals of Applied Probability*, 12 (1), 23–68, 2002.

57. Y. Kiuchi, M. Tanaka, and T. Mishima, Application of percolation model for network analysis, *Proceedings of the ITC-CSCC2002, International Technical Conference on Circuits/System, Computers and Communications*, 2, pp. 1101–1104, 2002.

58. M. Franceschetti, O. Dousse, N.C. David, and P. Thiran, Closing the gap in the capacity of wireless networks via percolation theory, *Information Theory, IEEE Transactions on*, 53(3), 1009–1018, 2007.

59. H. Hooyberghs, B. Van Schaeybroeck, A.A. Moreira, J.S. Andrade Jr., H.J. Herrmann, and J.O. Indekeu, Biased percolation on scale-free networks, *Physical Review E*, 81 (1), 011102, 2010.

60. I. Norros, A mean-field approach to some Internet-like random networks, in *Teletraffic Congress, 2009. ITC 21 2009. 21st International*, Paris, France, pp. 1–8. IEEE, Piscataway, NJ, 2009.

61. K.H. Hui, D. Guo, R. Berry, and M. Haenggi, Performance analysis of MAC protocols in wireless line networks using statistical mechanics, in *Communication, Control, and Computing, 2009. Allerton 2009. 47th Annual Allerton Conference on*, Allerton, IL, pp. 1315–1322. IEEE, Piscataway, NJ, 2009.

62. P. Bogdan, M. Kas, R. Marculescu, and O. Mutlu, QuaLe: A quantum-leap inspired model for non-stationary analysis of NoC traffic in chip multi-processors," in *Proceedings of the 2010 Fourth ACM/IEEE International Symposium on Networks-on-Chip (NOCS'10)*, Grenoble, France, pp. 241–248. IEEE Computer Society, Washington, DC. doi:10.1109/NOCS.2010.34 http://dx.doi.org/10.1109/NOCS.2010.34.

63. S. Sarkar, K. Mukherjee, A. Srivastav, and A. Ray, Critical phenomena and finite-size scaling in communication networks? in *American Control Conference (ACC)*, Portland, OR, pp. 271–276. IEEE, Piscataway, NJ, 2010.

64. C.H. Yeung and D. Saad, Networking—A statistical physics perspective, *Journal of Physics A: Mathematical and Theoretical*, 46 (10), 103001, 2013.

65. S. Sarkar, K. Mukherjee, A. Ray, A. Srivastav, and T.A. Wettergren, Statistical mechanics-inspired modeling of heterogeneous packet transmission in communication networks, *Systems, Man, and Cybernetics, Part B: Cybernetics, IEEE Transactions on*, 42 (4), 1083–1094, 2012.

66. A. Vespignani, R. Dickman, M.A. Muñoz, and S. Zapperi, Absorbing-state phase transitions in fixed-energy sandpiles, *Physical Review E*, 62 (4), 4564–4582, 2000.

67. J.J. Ramasco, M.A. Muñoz, and C.A. da Silva Santos, Numerical study of the Langevin theory for fixed-energy sandpiles, *Physical Review E*, 69, 045105, 2004.

68. L. Dall'Asta, Dynamical phenomena on complex networks, *Physical Review Letters*, 96, 058003, 2006. doi:10.1103/PhysRevLett.96.058003.

69. M. Casartelli, L. Dall'Asta, A. Vezzani, and P. Vivo, Dynamical invariants in the deterministic fixed-energy sandpile, *European Physical Journal B*, 52 (1), 91–105, 2006.

70. A. Fey, L. Levine, and D.B. Wilson, Driving sandpiles to criticality and beyond, *Physical Review Letters*, 104, 145703, 2010. arXiv:0912.3206v1.
71. S.-C. Park, Absence of the link between self-organized criticality and deterministic fixed energy sandpiles, arXiv preprint arXiv:1001.3359, 2010.
72. M.E. Woodward, *Communication and Computer Networks: Modelling with Discrete-Time Queues*, IEEE Computer Society Press, Los Alamitos, CA, 1994.
73. J.D.C. Little, A proof for the queuing formula: $L = \lambda W$, *Operations Research*, 9, 383–387, 1961.
74. Y. Rykalova, Queues, latency and critical phenomena in interconnection networks, PhD Thesis, Boston University, Boston, MA, 2008.

2 Computer Networks with Finite Buffers

Beyond Jackson's Theorem

Lev B. Levitin and Yelena Rykalova

CONTENTS

ABSTRACT

We present theoretical models and simulation results for performance of a multiprocessor network modeled as a ring and as a 2D wraparound square lattice of nodes with local processors that generate messages with constant rate per time slot. The buffers can hold a limited number of messages. Explicit theoretical results based on first-order (independent queues) and second-order approximation of the queue distributions are obtained for small buffer sizes (1 and 2). For larger buffers, the problem appears analytically intractable and has been studied by simulations. The average queue lengths and average latency are obtained. The results show that the model of independent queues, which is valid for networks with infinite buffers, is still applicable for small generation rates but breaks down for larger loads, which violates Jackson's theorem.

2.1 INTRODUCTION

Over the last several decades, the efforts of many researchers were focused on analyzing computer communication networks as networks of queues (see [1–3] for more references). Most of the early important results [4–17] have been obtained for networks with infinite buffers. However, this assumption in many cases does not adequately depict real networks. In practice, buffers are finite and sometimes quite small. Therefore, it is very important to study such networks, in spite of analytical difficulties this may present.

There is a rise of interest in the performance of the networks with small buffers in recent years. Analysis shows that increase in buffer size does not improve the performance in wormhole routing significantly [6,18–20]. Moreover, latest research has proved that the use of smaller buffers does not decrease the link utilization [21] for Internet routers. In interconnection networks, buffer depth of five packets has been shown to offer optimal performance for optical packet-switched clockwork routing [22]. The use of smaller buffer sizes gives some advantage in speed and provides possibilities to use SRAM or optical packet switching (OPS).

Networks with infinite buffers can be viewed as the limiting case of finite buffer networks. Thus, comparison with the results obtained for infinite buffer networks allows us to elucidate the specific effects of buffer size restriction on the network behavior. We start, therefore, with invoking certain results related to the infinite buffers.

In [1–3], we presented a model of a multiprocessor network with unlimited buffers, in the form of a ring and a 2D toroidal (wraparound) square lattice. The operation of the network was represented as a sequence of discrete time intervals (time slots). The specific features of the model were different from those of models [4–17] known to obey Jackson's theorem [7]. However, our analysis proved that Jackson's theorem remains valid for networks with discrete time, deterministic service time, and various queueing disciplines. We obtain theoretical and numerical (simulation) results for the performance of the model. We showed that for the ring topology, the average number of messages in queue $\bar{n}$ and latency τ is given by closed form expressions (2.1) and (2.2), for ring and torus topology, respectively:

$$\bar{n} = \frac{\lambda^2(l-1)}{1-l\lambda} + l\lambda; \quad \tau = \frac{\lambda(l-1)}{1-l\lambda} + l. \tag{2.1}$$

$$\bar{n} = \frac{\lambda^2(l-1)(11+5l)}{16(1-l\lambda)} + l\lambda; \quad \tau = \frac{\lambda(l-1)(11+5l)}{16(1-l\lambda)} + l. \tag{2.2}$$

It can be seen from these equations that critical network load λ_c is the same for both topologies and equal to

$$\lambda_c = \frac{1}{l}. \tag{2.3}$$

The results show remarkably close agreement between the theory and simulation that demonstrates the validity of the independent queues hypothesis for networks discussed in [1,3].

In this chapter, we consider several models of networks with different size of buffers, starting with size 1 (only one message can be kept in a buffer). In general, theoretical analysis of such networks is a very challenging problem. Though the independent queues assumption is not expected to give an accurate description of the performance of networks with finite buffers, it still makes sense to analyze the networks in terms of the first-order probability distributions (as if the queues were independent). In order to reflect better the correlations between nodes, the second-order probabilities (i.e., joint probabilities of two neighboring nodes) have been obtained for certain models.

Comparison with simulation results shows that both the first-order probabilities and, especially, the second-order distributions yield a reasonably good description of the system behavior until the load reaches the critical region. The simulation results show that when the load approaches the critical value, the system behavior displays typical patterns of long-range dependences between nodes, on the scale much exceeding the *interaction radius* (the distance between the source and destination). The network performance near the critical point is beyond the framework of Jackson's theorem; rather, it is similar to the critical phenomena in systems described by statistical physics. Both second-order (continuous) and first-order phase transitions have been observed in networks with finite buffers. Fluctuations with very large amplitudes, slowly changing in time, and instabilities in critical region are characteristic features of the network behavior.

All models considered in the succeeding text have the following common properties:

1. All nodes in the network are both the routers and hosts: every node can generate and receive messages as well as store and send further passing messages.
2. Time is discrete: all nodes simultaneously send, receive, and generate messages within every time slot.
3. Each node generates at most one message per output port within every time slot with a certain probability. This probability may depend on the state of the node (the number of messages in the local output port).
4. Destination of a message is selected with equal probabilities among the nodes at the exact distance l from the source.
5. Service time is deterministic (therefore, it does not have a rational Laplace transform). In the absence of queues, a message received or generated during a time slot will appear at the next node in the next time slot.
6. If possible, a message (if there is one) is sent from the output port to the neighboring node at every time slot. This depends on the number of messages in the output port (for the ring topology) or in the router (for the torus topology) in the next (neighboring) node.

In the case of limited buffers, the next state of a buffer depends not only on its previous state (as in the case of infinite buffers) but also on the previous state of its neighbor. It can be expected that limitations of buffer size create strong dependences between states of nodes, which become more and more pronounced with the increase of the network load (message generation rate). From the beginning, it is obvious that, as in the case of networks with heterogeneous activity, the independent queues model results may be applicable to networks with finite buffers only in the range of low loads, when the probability of queue lengths achieving the buffer limit remains very small. The independent queues model that yields the first-order probability distribution of the queue lengths in an individual buffer still remains an important reference tool for comparison with simulation results. However, in order to be able to account, at least partially, for the correlation between nodes, we have undertaken also an analysis of the second-order distributions, that is, joint probability distributions of the queue lengths in two neighboring nodes. As shown in the following, this analysis, though it results in quite complex and cumbersome expressions, provides a better agreement with simulation results up to the critical transition region.

Section 2.2 deals with ring topology networks. The buffer capacity $m = 1$ and $m = 2$ is chosen for the theoretical model in order to obtain explicit analytical expressions for the steady-state probabilities (first-order probability distribution) of the states and the average queue length in the *mean field* theory similar to that in statistical physics. Explicit expressions for the second-order distributions have been also obtained. The general case of buffer limit m has been considered; however, since it does not seem possible to derive closed analytical solutions even in the first-order approximation, experimental results based on simulation have been obtained for the ring length of 500 routers and the distance between source and destination $l = 5$ hops, for the buffer sizes $m = 5$, 10, 20, and 50.

Section 2.3 is devoted to toroidal network. Since consistent theoretical analysis seems to be intractable in this case, we provide results of simulation studies for 16×16 toroidal square lattice and $l = 5$ hops. The total router buffer limit was set to $m = 200$. The comparison with the performance of a network with similar parameters but with unlimited buffers has been drawn.

Section 2.4 proposes a viewpoint in regard to systems considered in the chapter as a basis for the future research.

The overall results are reviewed and discussed in Section 2.5.

2.2 RING TOPOLOGY

In the case of buffer size $m = 1$, we assume that a node does not accept messages and does not generate messages when the number of messages in a buffer $n = 1$. If at the beginning of a time slot a node was in state 0 and then received an incoming message that was not consumed, it does not generate any new message.

2.2.1 BUFFER LIMIT M = 1

2.2.1.1 First-Order Probability Distribution

Under the assumption of independent queues, the queue state transitions form a simple two state Markov chain (Figure 2.1).

Denote by p_0 and p_1 the probabilities of having zero or one message in a buffer, respectively. From the balance equation $p_0 p_{01} = p_1 p_{10} = (1 - p_0) p_{10}$, we obtain

$$p_0 = \frac{1 - \lambda(l+1)}{1-\lambda}, \quad \overline{n}^{(1)} = p_1 = \frac{\lambda l}{1-\lambda}, \quad \lambda_c^{(1)} = \frac{1}{l+1}. \tag{2.4}$$

Here, λ is the *nominal* probability that a message is generated by a node within one time interval, $\overline{n}$ is the average number of messages in a buffer, and τ is the average latency. The *actual* generation rate λ_{act} in all our models with finite buffers is different from the *nominal* value λ. Indeed, in the model with $m = 1$, a new message can be generated in state 0 only if no passing massage has arrived, and in state 1, only if the message kept in the output buffer has been sent to the next node.

A calculation shows that in the first-order approximation,

$$\lambda_{act}^{(1)} = p_0 \lambda \left(1 + \frac{p_1}{l} \right) = \frac{\lambda(1 - \lambda(l+1))}{(1-\lambda)^2}. \tag{2.5}$$

The actual load λ_{act} changes nonmonotonically with increase of λ: it approximately equals λ when λ is small $\left(\lambda \gg \lambda_{act}^{(1)} \right)$, then reaches the maximum, and decreases for larger λ, turning into zero at $\lambda = \lambda_c^{(1)}: \lambda_{act}^{(1)} \left(\lambda_c^{(1)} \right) = 0$. Note that Little's theorem [23] that connects the average queue length and the latency includes the actual rate λ_{act}, and not λ. Thus, in the first-order approximation,

$$\tau^{(1)} = \frac{\overline{n}^{(1)}}{\lambda_{act}^{(1)}} = \frac{(1-\lambda)l}{1-\lambda(l+1)}. \tag{2.6}$$

FIGURE 2.1 First-order state diagram for finite buffers (limit is 1 message).

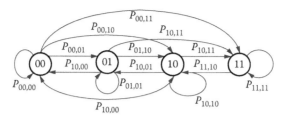

FIGURE 2.2 Second-order state diagram for finite buffers (limit is 1 message).

2.2.1.2 Second-Order Probability Distribution

The transition probabilities can be explicitly expressed in terms of the model parameters λ and l [24]. For two neighboring nodes and their combined states, the Markov chain has four states (Figure 2.2). The balance equations for this Markov chain are

$$p_{00}\left(p_{00,01} + p_{00,10} + p_{00,11}\right) = p_{01}p_{01,00} + p_{10}p_{10,00},$$

$$p_{00}\left(p_{00,10} + p_{00,11}\right) + p_{01}\left(p_{00,10} + p_{01,11}\right) = p_{01}\left(p_{10,01} + p_{10,00}\right), \tag{2.7}$$

$$p_{00}p_{00,11} + p_{01}p_{00,11} + p_{10}p_{10,11} = p_{11}p_{11,10}.$$

After solving the system of balance equations for this Markov chain, we obtain

$$p_0 = 1 - \frac{\lambda l^2(1-\lambda l)}{\lambda + l(1-3\lambda) - \lambda^2(l+1)(l-1)^2},$$

$$\overline{n}^{(2)} = p_1 = \frac{\lambda l^2(1-\lambda l)}{l - \lambda(3l-1) - \lambda^2(l+1)(l^2-1)}. \tag{2.8}$$

$$\lambda_{crit}^{(2)} = \frac{l^2 + 3l - 1 + \sqrt{(l^2+3l-1)^2 - 4l(l^2+3l-1)}}{2(l^2+l-1)},$$

$$\lambda_{act}^{(2)} = p_0\lambda\left(1 + \frac{p_1}{2l}\right), \tag{2.9}$$

$$\tau^{(2)} = \frac{\overline{n}^{(2)}}{\lambda_{act}^{(2)}} = \frac{2p_1 l}{\lambda(1-p_1)(2l+p_1)}.$$

$\lambda_{act}^{(2)}$ changes with increase of λ in the same manner as $\lambda_{act}^{(1)}$ but reaches zero at a smaller value $\lambda_c^{(1)}$. For comparison, the corresponding theoretical values for a system with infinite buffers are given by (2.1).

2.2.1.3 Simulations

The simulation was done for ring length 500 routers and distance between source and destination $l = 5$ hops. The values of $\overline{n}$ from simulation together with theoretical approximations $\overline{n}^{(1)}$ and $\overline{n}^{(2)}$ are shown in Figure 2.3. One can see that second-order theoretical analysis, compared to the first-order approximation, provides much better prediction of the network behavior for small network loads (up to $\lambda \approx 0.7\lambda_c$). It also can be seen that the second-order analysis results in better approximation of

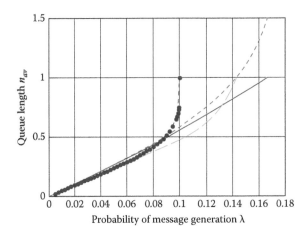

FIGURE 2.3 The queue length $\bar{n}$ as a function of the network load λ, $l = 5$, $m = 1$. Red solid line, the first-order approximation (2.4); green dashed line, the second-order approximation (2.6); blue line with bullets, simulation data; orange dotted line, theoretical values for infinite buffers given by (2.1).

the real critical point: second order $\lambda_c^{(2)} = 0.1435$ versus first order $\lambda_c^{(1)} = 0.1667$. Paradoxically, the first-order approximation gives an upper bound on $\bar{n}^{(1)}$ for small network loads when the probability of message generation $\lambda < 0.8\lambda_c$, while the second-order approximation provides a lower bound on $\bar{n}$ for all λ.

The real behavior of λ_{act} differs dramatically from both first-order and second-order predictions once λ approaches the critical region ($\lambda > 0.85\lambda_c$) (see Figure 2.4). Evidently, this fact is due to the emergence of a long-range order, which is not reflected in both approximate theories.

The empirical value of the latency τ, its analytical approximation $\tau^{(1)}$ and $\tau^{(2)}$, and the latency τ_{inf} for the network with infinite buffers are plotted in Figure 2.5.

A remarkable result is that the actual latency τ is considerably smaller than $\tau^{(1)}$ for $\lambda < 0.6\lambda_c$ and close to, but slightly smaller than, $\tau^{(2)}$ for $\lambda > 0.6\lambda_c$. The reason for this paradoxical behavior is that in the first-order approximation (independent queues), the probability of having a pair of neighboring nodes both in state 1, namely, $p_{11} = p_1^2$, is substantially larger than in the reality: a pair (11) is unstable at small λ and most probably will become (10) pair of states at the next time interval, since a node in state 1 does not accept messages. Thus, states 0 and 1 negatively correlated for small λ and

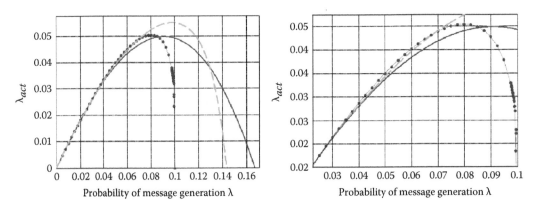

FIGURE 2.4 Actual average probability generation rate as a function of nominal generation rate λ, $l = 5$, $m = 1$. Red solid line, first-order approximation $\lambda_{act}^{(1)}$; green dashed line, second-order approximation $\lambda_{act}^{(2)}$; blue line with bullets, simulation data λ_{act}.

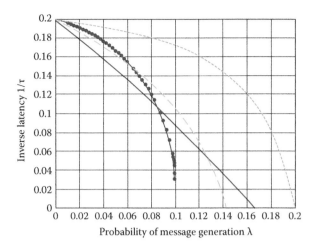

FIGURE 2.5 The inverse average latency τ as a function of the network load λ, $l = 5$, $m = 1$. Red solid line, the first-order approximation $\tau^{(1)}$ (2.4); green dashed line, the second-order approximation $\tau^{(2)}$ (2.7); blue line with bullets, experimental data; orange dotted line, theoretical values for infinite buffers given by (2.1).

positively correlated for large λ ($\lambda > 0.85\lambda_c$). We are tempted to compare this phenomenon with the antiferromagnetic–ferromagnetic transition in solid-state physics.

The network behavior near the critical point displays a pattern of the second-order phase transition. The average latency τ follows a power law (Figure 2.6):

$$\tau(\lambda) = a(\lambda_c - \lambda)^{-\beta}, \tag{2.10}$$

where
λ_c is the critical network load
β is the critical exponent

The data from our simulation with distance between the source and destination $l = 5$ hops yield $\lambda_c = 0.09981$, $a = 4.891 \pm 0.056$, and critical exponent $\beta = 0.2070 \pm 0.0094$. The value of the critical

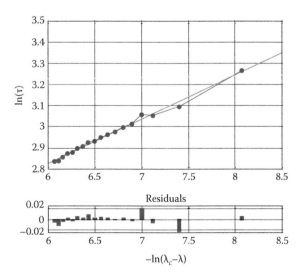

FIGURE 2.6 The dependence between τ and λ in logarithmic scale, $l = 5$, $m = 1$. Norm of residuals $= 0.037776$.

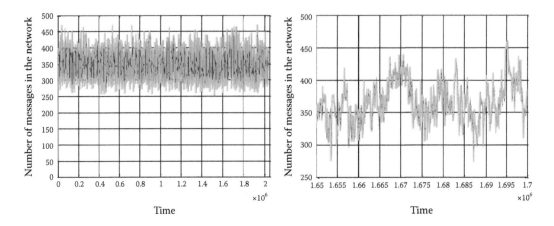

FIGURE 2.7 Number of messages in the network during simulation, $\lambda = 0.099$, $l = 5$, $m = 1$. Blue (black) line, the number of messages in network averaged over every 100 time slots; green (gray) line, snapshot of the number of messages every 100 time slots.

exponent is very far from the *trivial* value $\beta = 1$ given by the mean field theory (cf. (1) and (3)). As pointed out earlier, the fluctuations of the total number of messages in the network (which is equal to the number of saturated nodes in the case $m = 1$) have very large *low-frequency* component in the vicinity of the critical point (Figure 2.7).

2.2.2 Buffer Limit m = 2, Model 1

In the case of the network with buffer capacity $m = 2$, we consider two models:

1. A node does not accept messages and does not generate messages when $n = 1$ or $n = 2$. This model is analyzed in the present section.
2. A node does not accept messages when $n = 1$ or $n = 2$ and does not generate messages when $n = 2$. This model is analyzed in Section 2.2.3.

Unlike the model discussed in previous section, in the following discussion we assume that all decisions concerning the generation of a message or sending a message to the next node are based on the previous states of the nodes. Therefore, if in the previous time slot, the node was in state 1 there will be no attempt to generate a message, regardless whether the message was sent to the next router or not after we recorded the state of the node. On the other hand, if the state of node's neighbor in the previous time slot was 1 or 2, no message will be sent out to the neighbor.

2.2.2.1 First-Order Probability Distribution

Under the assumptions described earlier, only a router with no messages can accept incoming message or generate a message with probability λ. State diagram with transition probabilities for this model is shown in Figure 2.8.

From the system of the flow balance equation

$$p_0(p_{01} + p_{02}) = p_1 p_{10}, \quad p_0 p_{02} = p_2 p_{21}, \quad p_0 + p_1 + p_2 = 1, \tag{2.11}$$

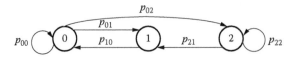

FIGURE 2.8 The Markov chain for buffer limit $m = 2$, model 1.

we have

$$p_0 = 1 - \lambda l, \quad p_1 = \lambda l - \lambda^2(l-1), \quad p_2 = \lambda^2(l-1),$$

$$\bar{n}^{(1)} = \lambda l + \lambda^2(l-1),$$

$$\lambda_{act}^{(1)} = p_0\lambda = \lambda - \lambda^2 l, \tag{2.12}$$

$$\tau^{(1)} = \frac{\bar{n}^{(1)}}{\lambda_{act}^{(1)}} = \frac{l + \lambda(l-1)}{1 - \lambda l}, \quad \lambda_c^{(1)} = \frac{1}{l}.$$

By comparing the expression for $\lambda_{act}^{(1)}$ and $\tau^{(1)}$ for this model and for the model of network with buffer limit $m = 1$, one can see that the first-order approximation gives a smaller value of $\lambda_{act}^{(1)}$, and, therefore, the larger latency $\tau^{(1)}$ for $m = 2$ when $\lambda < 0.4\lambda_c^{(1)}$, but the opposite is true for larger values of λ. The reason for that is that $\lambda_c^{(1)}(m = 2) > \lambda_c^{(1)}(m = 1)$. Curiously, the critical load $\lambda_c^{(1)}$ is the same as in the model with infinite buffers.

2.2.2.2 Second-Order Probability Distribution

The second-order approximation becomes more challenging for this model, as compared to the second-order approximation described in Section 2.2.1.2. Now the state transition diagram has nine states (Figure 2.9), which are combined states of two neighboring nodes. Note that the states are not symmetric, since the messages are always sent from left to right.

The global balance equations for such a Markov chain are highly nonlinear, since the transition probabilities depend on the second-order probabilities of the states. After tedious calculations, we obtain the second-order probabilities in terms of marginal probabilities p_0, p_1, and p_2 as

$$p_{00} = (1 - \lambda l)p_0,$$

$$p_{01} = \frac{\lambda + l^3(1-\lambda)^2\lambda - \lambda^2 - l\lambda(3-2\lambda) - l^2(1-3\lambda+\lambda^3)}{l(1-\lambda)(l^2\lambda - l - \lambda - (l-1)^2\lambda^2)} p_1,$$

$$p_{02} = \lambda l p_0 - \left[\frac{(1-\lambda l)}{(1-\lambda)(1-\lambda l+\lambda)(l-\lambda l+\lambda)} - \frac{1}{l(1-l\lambda+\lambda)} \right] p_1,$$

$$p_{10} = \lambda(1 - l\lambda + \lambda)p_0,$$

$$p_{11} = \left[\frac{l-1}{l} + \frac{l-1}{1-\lambda} + \frac{\lambda}{l} - \frac{1+\lambda^2}{l-\lambda l+\lambda} + \frac{2}{l(1-\lambda l+\lambda)} \right] p_1,$$

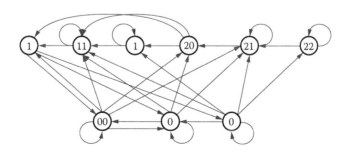

FIGURE 2.9 The second-order Markov chain for buffer limit $m = 2$, model 1. Nonzero transition probabilities are shown by arrows.

$$p_{12} = \left[l + \left(\lambda l p_0 + \frac{(l-1)\lambda(l-1+\lambda)}{l(l(1-\lambda)+\lambda)(1-\lambda)(1-(l-1)\lambda} p_1 \right) \right.$$

$$\left. \times \left(1 + \lambda - \lambda l + \frac{(l-1)(1-\lambda)\left(p^2 - (l-1)p_0\lambda^2\right)}{p_2} \right) \frac{1}{(l-1)p_0\lambda} \right] p_2,$$

$$p_{20} = \lambda^2(l-1)p_0,$$

$$p_{21} = \frac{(l-1)\lambda^2\left(l+\lambda+(l-3)\lambda l+(l-1)\lambda^2\right)}{l(1-\lambda)\left(l+\lambda-l^2\lambda+(l-1)^2\lambda^2\right)} p_1,$$

$$p_{22} = \lambda^2(l-1)\left[\lambda l p_0 - \left(\frac{(1-\lambda l)}{(1-\lambda)(1-\lambda l+\lambda)(l-\lambda l+\lambda)} - \frac{1}{l(1-\lambda l+\lambda)} \right) p_1 \right].$$

However, the explicit expressions for p_0, p_1, and p_2, as functions of parameters λ and l, are too monstrous to present here (see [20]). The equation for λ to be solved to find $\lambda_c^{(2)}$ is of degree 12 (!). Solving it numerically for $l = 5$, we obtain the smallest real nonnegative root: $\lambda_c^{(2)} = 0.1393$. The expression for the actual load has the same form $\lambda_{act}^{(2)} = p_0\lambda$ as in the first-order approximation but expression for p_0 is different [20].

Correspondingly, the average queue length $\bar{n}^{(2)}$ and average latency $\tau^{(2)}$ are

$$\bar{n}^{(2)} = p_1 + 2p_2,$$

$$\tau^{(2)} = \frac{\bar{n}^{(2)}}{\lambda_{act}^{(2)}} = \frac{p_1 + 2p_2}{p_0\lambda}. \tag{2.13}$$

2.2.2.3 Simulations

The approximations $\bar{n}^{(1)}$ and $\bar{n}^{(2)}$ for the average queue length, as well as experimental values, are shown in Figure 2.10. As in the previous model, the second-order approximation gives an excellent prediction for the network behavior up to the moment when the network becomes saturated. The network goes to the saturation state not only much earlier than the theoretical approximations predict, but it does so abruptly, sharply departing from the second-order approximation curve. Unlike the

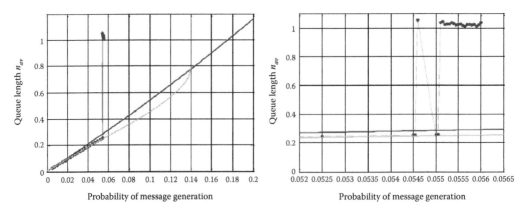

FIGURE 2.10 The queue length n as a function of the network load λ, $m = 2$, model 1. Red solid line, the first-order approximation (2.12); green dashed line, the second-order approximation (2.13); blue line with bullets, simulation data.

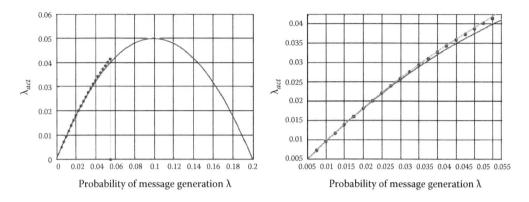

Probability of message generation λ   Probability of message generation λ

FIGURE 2.11   Actual average probability generation rate as function of nominal generation rate λ, $l = 5$, $m = 2$, model 1. Red solid line, first-order approximation $\lambda_{act}^{(1)}$; green dashed line, second-order approximation $\lambda_{act}^{(2)}$; blue line with bullets, simulation data λ_{act}.

previous model, there is no continuous transition to the saturation. The picture we have here closely resembles a liquid–gas first-order phase transition. Close to the critical point, the network behavior becomes very unstable, which makes it difficult to determine the critical point precisely. Roughly, the critical load is $\lambda_c = 0.0546$.

It is remarkable that the actual network load during the simulation follows very closely the second-order approximation of the load $\lambda_{act}^{(2)}$ until the point when network goes to the saturation state (Figure 2.11). However, the picture for the latency shown in Figure 2.12 is a little bit different. The curve received from the simulation data follows the second-order approximation of the latency $\tau^{(2)}$ until the value of $\lambda = 0.0225$, and, after that, it deviates slightly from the second-order approximation. The reason for that is that λ_{act} and $\bar{n}$ deviate from the theoretical values in the opposite directions (see Figure 2.10), and these deviations add up in the value of τ.

Typical patterns of fluctuations are shown in Figures 2.13 through 2.15 for two values of λ: far from the critical point (see Figures 2.13 and 2.15a) and close to λ_c (Figures 2.14 and 2.15b).

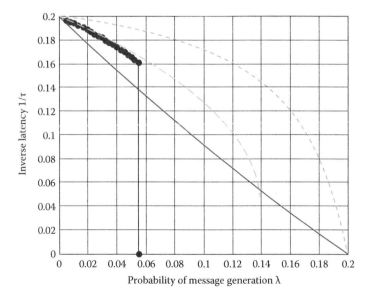

Probability of message generation λ

FIGURE 2.12   The inverse average latency τ as a function of the network load λ, $l = 5$, $m = 2$, model 1. Red solid line, the first-order approximation (2.9); green dashed line, the second-order approximation (2.14); blue line with bullets, simulation data; orange dotted line, theoretical values for infinite buffers given by (2.1).

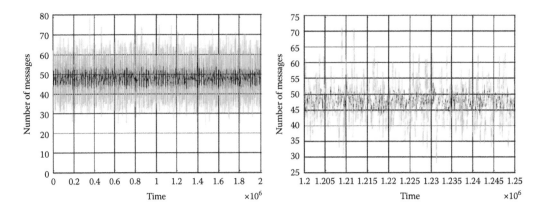

FIGURE 2.13   Number of messages in the network during simulation, $\lambda = 0.02$, $l = 5$, $m = 2$, model 1. Blue (black) line, the number of messages in network averaged over 100 time slots; green (gray) line, snapshot of the number of messages every 100 time slots.

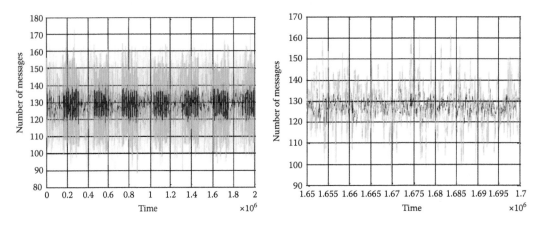

FIGURE 2.14   Number of messages in the network during simulation, $\lambda = 0.05455$, $l = 5$, $m = 2$, model 1. Blue line, the number of messages in network averaged over 100 time slots; green line, snapshot of the number of messages every 100 time slots.

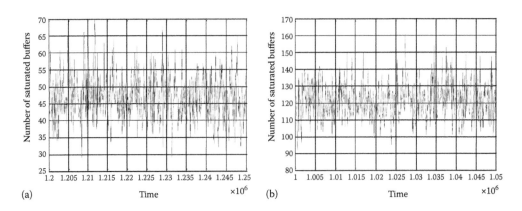

FIGURE 2.15   Snapshot of the number of saturated buffers in the network taken at every 100 time slots, $l = 5$, $m = 2$, model 1. (a) $\lambda = 0.02$; (b) $\lambda = 0.05455$.

2.2.3 Buffer Limit M = 2, Model 2

In this section, we consider the second model for the buffer capacity $m = 2$. Here, a node can generate a message when there is one message in the buffer (state 1), but it does not accept messages when $n = 1$ or $n = 2$.

2.2.3.1 First-Order Probability Distribution

The Markov chain for this model is shown in Figure 2.16.

In the mean field approximation, we assume that the states of neighboring nodes are independent, but the transition probabilities depend on the neighbor states. After simplification, the system of the nonlinear equations for flow balance can be written in the following form:

$$p_1 \left[p_0 + \lambda(1 - p_0) \right] = p_0(1 - p_0) \left[1 - \lambda \frac{l-1}{l} \right],$$

$$p_1(1 - \lambda) = (1 - p_0) \frac{l-1}{l} + \frac{\lambda}{l} \left[1 + p_0(l - 1) \right], \tag{2.14}$$

$$p_0 + p_1 + p_2 = 1.$$

The solution of this system is

$$p_0 = \frac{(1 - \lambda)^2 - \sqrt{(1 - \lambda)^4 - 4(1 - \lambda)\lambda(l - 1 + \lambda)}}{2(1 - \lambda)},$$

$$\tag{2.15}$$

$$p_1 = \frac{p_0(1 - p_0)}{l\lambda} - p_0, \quad p_2 = 1 - \frac{p_0(1 - p_0)}{l\lambda}.$$

The critical value $\lambda_c^{(1)}$ of the network load is obtained from the condition that the system of equations (2.15) has no real roots. At $\lambda = \lambda_c^{(1)}$, the discriminant in the expression for vanishes. Hence,

$$\lambda_c^{(1)} = \frac{1}{3} \left(8 + 18l + 6\sqrt{3l(8 - 13l + 16l^2)} \right)^{\frac{1}{3}} - \frac{1}{3} - \frac{4}{3}(3l - 1) \left(8 + 18l + 6\sqrt{3l(8 - 13l + 16l^2)} \right)^{-\frac{1}{3}}. \tag{2.16}$$

For example, for $l = 5$, the $\lambda_c^{(1)} = 0.05248$. The value of $\lambda_c^{(1)}$ given by (2.16) is much smaller than the critical value (2.3) in the case of infinite buffers. Moreover, in the case of infinite buffers, the probability p_0 of having zero messages in the queue vanishes when the load approaches λ_c. In the case of the finite buffer ($m = 2$), the mean field theory predicts $p_0 = (1 - \lambda_c)/2 \approx 1/2$ at saturation point (Figure 2.17).

The actual load is

$$\lambda_{act}^{(1)} = (p_0 + p_1)\lambda; \tag{2.17}$$

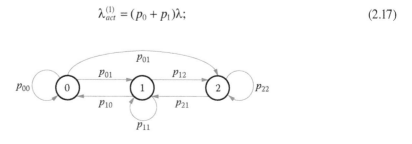

FIGURE 2.16 The Markov chain for buffer limit $m = 2$, model 2.

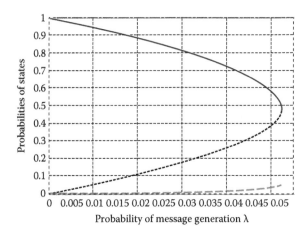

FIGURE 2.17   Mean field prediction for probabilities of states, $m = 2$, model 2: p_0—dark magenta solid line; p_1—dark blue dotted line; p_2—dark olive dashed line.

the average queue length

$$\bar{n}^{(1)} = p_1 + 2p_2; \tag{2.18}$$

and for the average latency,

$$\tau^{(1)} = \bar{n}^{(1)} \big/ \lambda_{act}^{(1)} = \left(p_1 + 2p_2\right)\big/\left(\left(p_0 + p_1\right)\lambda\right). \tag{2.19}$$

Their numerical values for the given distance $l = 5$ between the source and destination are plotted in Figures 2.18 through 2.22.

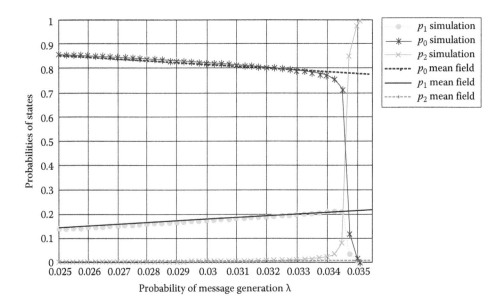

FIGURE 2.18   The probabilities of states, $l = 5$, $m = 2$, model 2. Mean field prediction for probabilities of states and numerical data.

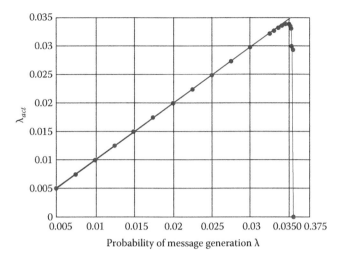

FIGURE 2.19   The inverse average latency τ as a function of the network load λ, $m = 2$, model 2. Red line, the first-order approximation $\tau^{(1)}$ (2.17); blue line with bullets, simulation data.

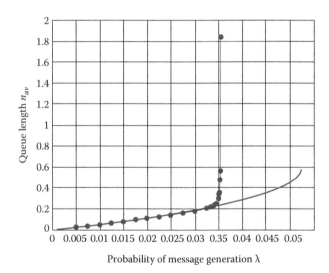

FIGURE 2.20   The queue length $\bar{n}$ as a function of the network load λ, $m = 2$, model 2. Red line, the first-order approximation $\bar{n}^{(1)}$ (2.18); blue line with bullets, simulation data.

2.2.3.2   Simulations

The values of actual network load and assigned network load are very close until $\lambda \approx 0.033$ (see Figure 2.19). The same behaviors can be seen in the average queue length and the average latency. The simulation data for probabilities of states with comparisons to the first-order approximation is shown in Figure 2.18. It can be seen that numerical data follows the theoretical prediction very closely until network load reaches value $\lambda \approx 0.0345$. Up to this point, the data from experiments show even a slightly better network performance (smaller queue length and latency) when the first-order approximation predicts. However, numerical experiments show a steep decline of the probabilities of state 0 and state 1 (Figure 2.18), as well as a sharp increase in the average queue length (Figure 2.20) just before the network reaches saturation and long before we can expect the critical transition according to the mean field theory. The network reaches saturation much earlier than the

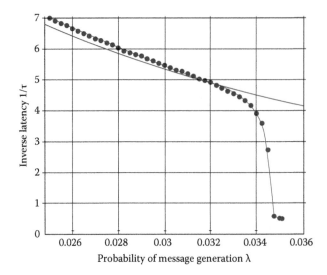

FIGURE 2.21　The inverse average latency τ as a function of the network load λ, $m = 2$, model 2. Red line, the first-order approximation $\tau^{(1)}$ (2.19); blue line with bullets, simulation data.

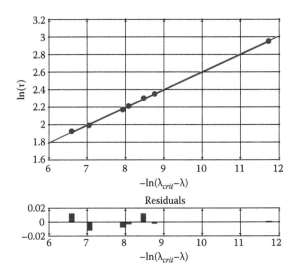

FIGURE 2.22　The dependence between τ and λ in logarithmic scale, $m = 2$, model 2.

theoretical model predicts, at $\lambda_c \approx 0.0354 < \lambda_c^{(1)} = 0.05248$. Thus, the mean field theory provides a very good agreement with simulation data up to larger value of λ $\left(\text{about } \lambda \approx 2/3\lambda_c^{(1)}\right)$, when correlations between the node states become prevailing and long-range order emerges (Figure 2.23).

The transition of the network to the saturation phase shows characteristics of the second-order phase transition, despite the fact that this transition is more abrupt then those described in Section 2.2.1. Large fluctuations in the number of messages in the network (Figures 2.24 and 2.25) and in the number of saturated buffers (Figure 2.26) were observed using this model. This is dramatically different from model 1 described in the previous section, which has a first-order phase transition (Figures 2.13 through 2.15).

By comparing model 2 for the buffer limit $m = 2$ described here with model from Section 2.2.2.1 ($m = 1$), one can see that increasing the buffer capacity does not necessarily lead to defer the saturation

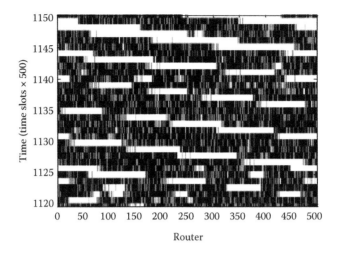

FIGURE 2.23 Number of messages n at each node taken at every 500 time slots, $\lambda = 0.0345$, $l = 5$, $m = 2$, model 2. Black nodes, $n = 0$; gray nodes, $n = 1$; white nodes, $n = 2$.

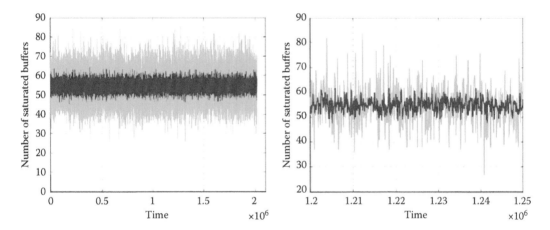

FIGURE 2.24 Number of messages in the network taken at every 100 time slots, $\lambda = 0.03535$, $l = 5$ hops, $m = 2$, model 2. Blue (black) line, number of messages in the network averaged over 100 time slots; green (gray) line, number of messages in the network taken every 100 time slots.

point. On the contrary, the former model reaches the saturation state under network load almost three times smaller than the latter. The explanation of this fact is that in model 2, a node can generate messages in both states 0 and 1, rather than in state 0 only, which results in a larger effective load. The shape of the latency curve (Figure 2.21), the fluctuation patterns, and the formation of domains of different phases (Figure 2.23) speak in favor of a continuous (second-order) phase transition. The empirical data on latency, as seen in Figure 2.22, agree very well with a power law (2.7). Calculations for l = 5 yield $\lambda_c = 0.03536$, $\alpha = 1.775 \pm 0.028$, and a critical exponent $\beta = 0.2026 \pm 0.0095$.

2.2.4 GENERAL CASE: BUFFER LIMIT M

2.2.4.1 Theoretical Analysis

Let us look at the case of the finite buffers with some arbitrary capacity m. A node does not accept messages when a buffer already holds $n = m - 1$ or $n = m$ messages, and a message is not generated when $n = m$. State diagram with transitional probabilities is shown in Figure 2.27.

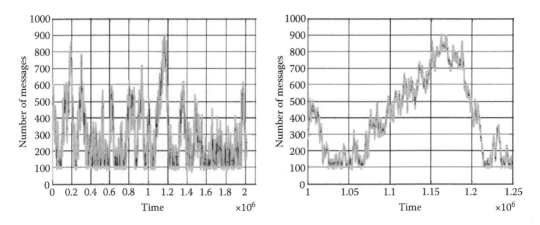

FIGURE 2.25 Number of messages in the network taken at every 100 time slots, $\lambda = 0.03535$, $l = 5$ hops, $m = 2$, model 2. Blue (black) line, number of messages in the network averaged over 100 time slots; green (gray) line, number of messages in the network taken every 100 time slots.

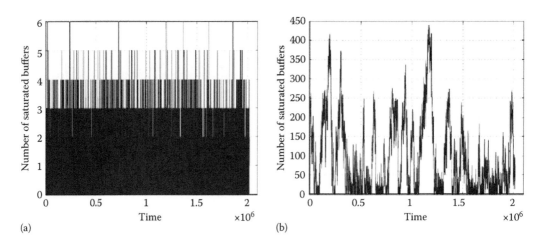

FIGURE 2.26 Snapshot of the number of saturated buffers in the network taken at every 100 time slots, $l = 5$, $m = 2$, model 2. (a) $\lambda = 0.02$; (b) $\lambda = 0.03535$.

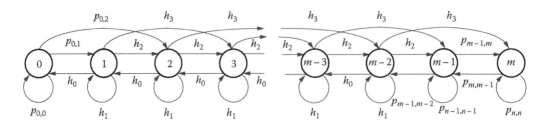

FIGURE 2.27 State diagram of the Markov chain for buffer capacity m.

The transition probabilities h_0, h_1, h_2, and h_3 are standard and have the following form:

$$p_{n,n-1} = h_0 = (1 - p_{m-q} - p_m)(1-\lambda)\left[p_0 + \frac{1-p_0}{l}\right],$$

$$p_{n,n} = h_1 = (1-\lambda)(1-p_0)\frac{l-1}{l}(1-p_{m-1}-p_m) + \lambda\left[p_0 + \frac{1-p_0}{l}\right](1-p_{m-1}-p_m)$$

$$+ (1-\lambda)\left[p_0 + \frac{1-p_0}{l}\right](p_{m-1}+p_m),$$

$$p_{n,n+1} = h_2 = \lambda\left[p_0 + \frac{1-p_0}{l}\right](p_{m-1}+p_m) + (1-\lambda)(1-p_0)\frac{l-1}{l}(p_{m-1}+p_m)$$

$$+ \lambda(1-p_0)\frac{l-1}{l}(1-p_{m-1}-p_m),$$

$$p_{n,n+2} = h_3 = \lambda(1-p_0)\frac{l-1}{l}(p_{m-1}+p_m).$$

(2.20)

The *special* flow balance equations are

$$p_1 p_{1,0} = p_0(p_{1,0} + p_{0,2}),$$

$$p_2 p_{2,1} = p_0 p_{0,2} + p_1(p_{1,2} + p_{1,3}),$$

$$p_3 p_{3,2} = p_1 p_{1,3} + p_2(p_{2,3} + p_{2,4}),$$

$$p_4 p_{4,3} = p_2 p_{2,4} + p_3(p_{3,4} + p_{3,5}),$$

$$p_{m-2} p_{m-2,m-3} = p_{m-4} p_{m-4,m-2} + p_{m-3}(p_{m-3,m-2} + p_{m-3,m-1}),$$

$$p_{m-1} p_{m-1,m-2} = p_{m-3} p_{m-3,m-1} + p_{m-2}(p_{m-2,m-1} + p_{m-2,m}),$$

$$p_m p_{m,m-1} = p_{m-2} p_{m-2,m} + p_{m-1} p_{m-1,m}.$$

(2.21)

The standard flow balance equation is

$$p_n p_{n,n-1} = p_{n-2} p_{n-2,n} + p_{n-1}(p_{n-1,n} + p_{n-1,n+1}) \quad \text{or}$$

$$p_n h_0 = p_{n-2} h_3 + p_{n-1}(h_2 + h_3), \quad \text{for } 3 \le n \le m-2.$$

(2.22)

The generation function can be written as

$$\varphi(z) = \sum_{n=0}^{m} z^n p_n = \sum_{n=1}^{m-2} z^{n-1} p_n p_{n,n-1} + z^{m-2} p_{m-1} p_{m-1,m-2} + z^{m-1} p_m p_{m,m-1}$$

$$+ \sum_{n=1}^{m-2} z^n p_n p_{n,n} + z^0 p_0 p_{0,0} + z^{m-1} p_{m-1} p_{m-1,m-1} + z^{m-1} p_m p_{m,m-1}$$

$$+ \sum_{n=1}^{m-2} z^{n+1} p_n p_{n,n+1} + z p_0 p_{0,1} + z^m p_{m-1} p_{m-1,m} + \sum_{n=1}^{m-2} z^{n+2} p_n p_{n,n+2} + z^2 p_0 p_{0,2}. \quad (2.23)$$

It follows from Equation 2.23 that

$$\varphi(z)\left(1-\frac{1}{z}h_0 - h_1 - zh_2 - z^2h_3\right) = \left(-z^{-1}p_0h_0 - z^{n-2}p_{n-1}h_0 - z^{n-1}p_nh_0\right) + \left(z^{n-2}p_{n-1}p_{n-1,n-2} + z^{n-1}p_np_{n,n-1}\right)$$

$$+\left(-z^0p_0h_1 - z^{n-1}p_{n-1}h_1 - z^np_np_{n,n}\right) + \left(-z^1p_0h_2 - z^np_{n-1}h_2 - z^{n+1}p_nh_2\right)$$

$$+\left(-z^2p_0h_3 - z^{n+1}p_{n-1}h_3 - z^{n+1}p_nh_3\right) + \left(zp_0p_{0,1} + z^np_{n-1}p_{n-1,n}\right) + z^2p_0p_{0,1}. \tag{2.24}$$

We can see from (2.20) that the standard transition probabilities h_0, h_1, h_2, and h_3 depend on probabilities of three states: p_0, p_{m-1}, and p_m. These probabilities are related by the global flow equation, which, by taking into account normalization condition, can be written as

$$\lambda l(1 - p_m) = (1 - p_0)(1 - p_{m-1} - p_m). \tag{2.25}$$

Therefore,

$$p_{m-1} = (1 - p_m)\left[1 - \frac{\lambda l}{1 - p_0}\right]. \tag{2.26}$$

The characteristic equation that follows from the expression (2.22) is

$$h_0x^2 - (h_2 + h_3)x - h_3 = 0. \tag{2.27}$$

Hence, the general solution for the limiting probabilities has a form $p_n = Ax_1^n + Bx_2^n$, for $3 \leq n \leq m - 2$, where x_1 and x_2 are roots of the characteristic equation (2.27):

$$x_{1,2} = \frac{h_2 + h_3 \pm \sqrt{(h_2 + h_3)^2 + 4h_0h_3}}{2h_0}. \tag{2.28}$$

The constant coefficients A and B and the values of x_1 and x should be determined from the relation (2.26) together with the system (2.21) and (2.22). However, the highly nonlinear character of Equations 2.21 and 2.22, as well as expressions for x_1 and x_2, in terms of p_0, p_{m-1}, and p_m prevents us from being able to find a closed form expressions for the steady-state probabilities even in this first-order approximation. Furthermore, even the use of the generation function (2.24) does not allow us to obtain explicit expressions for the average number of messages in a buffer. Therefore, we have to rely on simulation results and compare them with the theoretical values obtained for a network with infinite buffers.

2.2.4.2 Numerical Results

Simulations were done for buffer's limits $m = 5, 10, 20, 30, 40$, and 50 for the ring length 500 and the distance between source and destination $l = 5$. Data were recorded for the last 2×10^6 time slots. The data from simulation for buffer sizes $m = 20$ and 40 are presented in Tables 2.1 and 2.2.

The first series of simulations were done for the buffer limit $m = 5$. The data shows similarity in the behavior of this network and that described in the Section 2.2.3. One can see a continuous (second-order) transition to the saturated state (Figures 2.28 and 2.29), which is similar to but a little

TABLE 2.1

Finite Buffers: $m = 20$

λ	λ_{act}	$\bar{n}$	τ
0.01000	0.00999	0.050103	5.013002
0.02000	0.02000	0.101741	5.086307
0.03000	0.03000	0.154701	5.156302
0.04000	0.03999	0.208220	5.206649
0.05000	0.049998	0.264580	5.292598
0.06000	0.06001	0.322852	5.380111
0.07000	0.07000	0.382943	5.470539
0.08000	0.08001	0.448509	5.605532
0.09000	0.09001	0.517412	5.748228
0.10000	0.09999	0.592450	5.925255
0.11000	0.10999	0.677597	6.160750
0.12000	0.11999	0.775041	6.459440
0.13000	0.13001	0.888363	6.832912
0.14000	0.14000	1.026691	7.333464
0.15000	0.14999	1.208007	8.053991
0.16000	0.16001	1.465103	9.156271
0.17000	0.16999	1.864902	10.971104
0.17100	0.17101	1.919113	11.222099
0.17200	0.17200	1.977993	11.500161
0.17300	0.17299	2.041441	11.801295
0.17400	0.17400	2.107953	12.114593
0.17500	0.17500	2.178780	12.450009
0.17600	0.17601	2.252166	12.795500
0.17700	0.17702	2.340541	13.221923
0.17800	0.17801	2.428392	13.641863
0.17900	0.17898	2.527251	14.120444
0.17905	0.17903	2.535231	14.160831
0.17910	0.00000	20.000000	∞

Notes: Experimental data for the average length of buffer queues $\bar{n}$, real generation rate λ_{act}, and latency τ.

bit sharper than in the previous case. The behavior of the actual average generation rate λ_{act} as a function of nominal probability of message generation λ is shown in Figure 2.30. The second-order transition characterization is also supported by the characteristic fluctuations in the number of messages in the network during simulations and fluctuations in the number of saturated buffers (Figures 2.31 through 2.33). They both increase in amplitude and in *wavelength* as network load increases (Figure 2.32). To obtain more accurate estimate of the critical point, we have conducted several series of the experiments; however, large fluctuations near saturation point preclude more accurate estimate of the saturation point (Figures 2.34 through 2.38).

In contrast with behavior of networks with buffer limit $m = 5$, simulations with greater buffer limit show the first-order phase transition. An example for buffer limit $m = 10$ is presented in Figures 2.34 through 2.39. Simulation values for the average queue length n and latency τ are shown in Figures 2.34 and 2.35, respectively. One can see that initially (for small network loads), simulation data very closely follow theoretical prediction for the infinite buffers, then depart from it slightly, and, for larger loads, the network abruptly goes into the saturation state. Close-up in Figure 2.34 shows huge fluctuations in the average queue length for network load close to the critical value.

TABLE 2.2
Finite Buffers: $m = 40$

λ	λ_{act}	$\bar{n}$	τ
0.1000	0.00999	0.592906	5.929827
0.1100	0.11004	0.677795	6.160368
0.1200	0.11999	0.774550	6.455232
0.1300	0.12998	0.888499	6.835501
0.1400	0.14002	1.027425	7.337882
0.1500	0.15001	1.209170	8.060757
0.1600	0.16001	1.461264	9.132107
0.1700	0.17002	1.866281	10.977012
0.1810	0.18098	2.756137	15.229019
0.1820	0.18201	2.891215	15.885043
0.1830	0.18300	3.037871	16.600857
0.1840	0.18398	3.205781	17.425087
0.1850	0.18500	3.396810	18.360967
0.1860	0.18600	3.609457	19.405421
0.1870	0.18702	3.855659	20.616502
0.1880	0.18802	4.150999	22.077474
0.1890	0.18899	4.479163	23.701032
0.1891	0.18911	4.524785	23.926502
0.1892	0.18919	4.557253	24.088254
0.1893	0.18928	4.597342	24.289176
0.1894	0.18941	4.646119	24.529986
0.1895	0.18950	4.678089	24.686467
0.1896	0.18962	4.728524	24.936463
0.1897	0.18972	4.769971	25.142246
0.1898	0.18981	4.803398	25.306741
0.1899	0.00000	40.000000	∞

Notes: Experimental data for the average length of buffer
queues $\bar{n}$, real generation rate λ_{act}, and latency τ.

The dynamics of the number of saturated buffers during the simulation for subcritical values of λ is shown in Figure 2.39a and b. Initially, the number of saturated buffers rapidly grows. However, later on, the number of saturated buffers decreases and then stabilizes. A different behavior is shown in Figure 2.39c for the network load $\lambda = 0.1592$ that exceeds the network saturation limit. In this case, the number of saturated buffers rapidly increases and soon reaches the maximum possible value. The final queue length when saturation phenomena in network were observed appears to be different in different cases (Figures 2.40 and 2.41). By the time the simulation ends, all buffers in the network are expected to reach their capacity limit (i.e., they all hold m messages). However, since the saturation time fluctuates broadly from one realization to another, sometimes, we may have recorded data before the saturation is complete. Thus, computed averages may be slightly biased low. In addition, instability in the network causes fluctuations and leads to different outcomes for the different experiments near the critical network load. We expect that longer simulations will reduce both inaccuracies and improve the estimate of the critical point. Figures 2.40 and 2.41 show the dependence between average queue length and latency on network load λ. Figure 2.42 demonstrates the dependence of the critical value λ_c on the buffer limiting capacity.

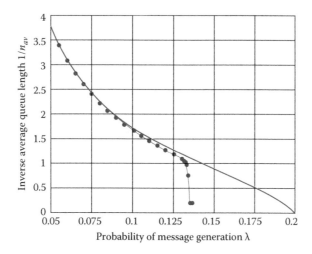

FIGURE 2.28 The inverse queue length $1/\bar{n}$ as a function of the network load λ, $m = 5$. Red solid line, the first-order approximation $\bar{n}^{(1)}$ for infinite buffers (2.1); blue line with bullets, simulation data.

The critical value increases steeply in the beginning but grows slowly for larger buffer sizes asymptotically approaching the critical load λ_c^{inf}. The numerical values obtained are in a good agreement with an empirical formula

$$\lambda_c(m) = \lambda_c^{(\text{inf})} - \frac{c}{m}, \tag{2.29}$$

where c is a constant. However, at the present time, we have no theoretical model to justify (2.29) (Figure 2.43).

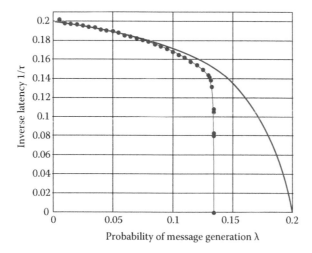

FIGURE 2.29 The inverse average latency τ as function of probability of message generation λ, $l = 5$, $m = 5$. Red solid line, the first-order approximation $\bar{n}^{(1)}$ for infinite buffers (2.1); blue line with bullets, simulation data.

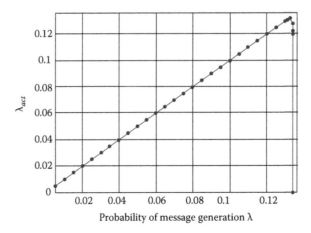

FIGURE 2.30 Actual average generation rate λ_{act} as a function of nominal probability of message generation λ, $l = 5$ hops, $m = 5$.

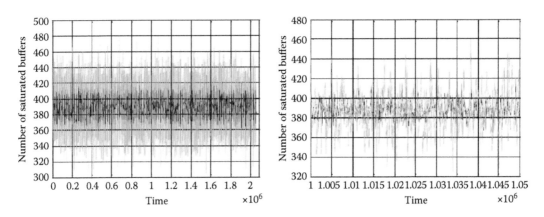

FIGURE 2.31 Number of messages in the network taken at every 100 time slots, $\lambda = 0.12$, $l = 5$ hops, $m = 5$. Blue (black) line, number of messages in the network averaged over 100 time slots; green (gray) line, number of messages in the network taken every 100 time slots.

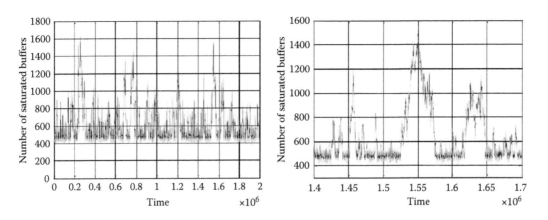

FIGURE 2.32 Number of messages in the network taken at every 100 time slots, $\lambda = 0.13405$, $l = 5$ hops, $m = 5$. Blue (black) line, number of messages in the network averaged over 100 time slots; green (gray) line, number of messages in the network taken every 100 time slots.

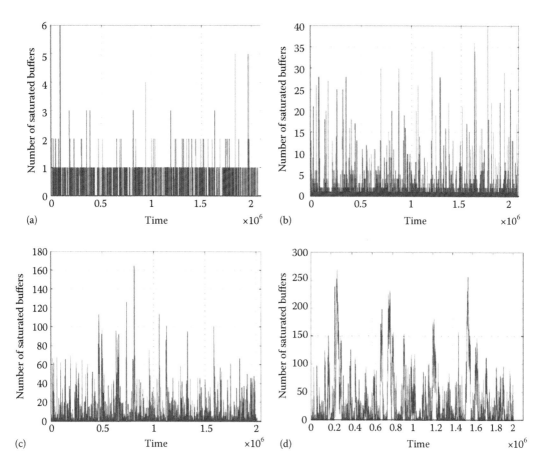

FIGURE 2.33 Snapshot of number of saturated buffers in the network taken every 100 time slots, $l = 5$ hops, $m = 5$. (a) $\lambda = 0.012$, (b) $\lambda = 0.013$, (c) $\lambda = 0.0133$, (d) $\lambda = 0.13405$.

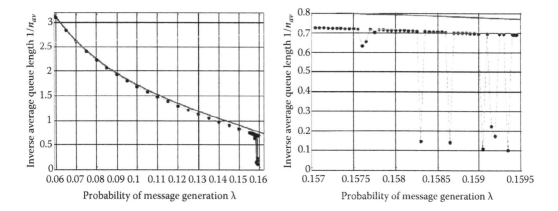

FIGURE 2.34 The inverse queue length n as a function of the network load λ, $m = 10$. Red (solid) line, the first-order approximation $\bar{n}^{(1)}$ for infinite buffers (2.1); blue (gray) line, simulation data.

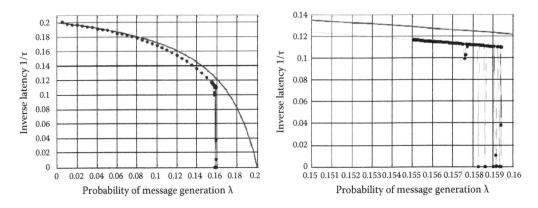

FIGURE 2.35   The inverse average latency τ as function of probability of message generation λ, $l = 5$, $m = 5$. Red line, the first-order approximation $\bar{n}^{(1)}$ for infinite buffers (2.1); blue line, simulation data.

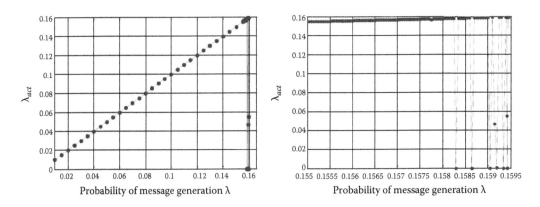

FIGURE 2.36   Actual average generation rate λ_{act} as a function of nominal probability of message generation λ, $l = 5$ hops, $m = 10$.

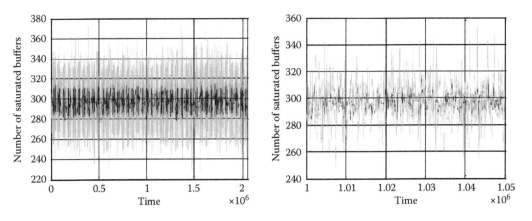

FIGURE 2.37   Number of messages in the network taken at every 100 time slots, $\lambda = 0.1$, $l = 5$ hops, $m = 10$. Blue (black) line, number of messages in the network averaged over 100 time slots; green (gray) line, number of messages in the network taken every 100 time slots.

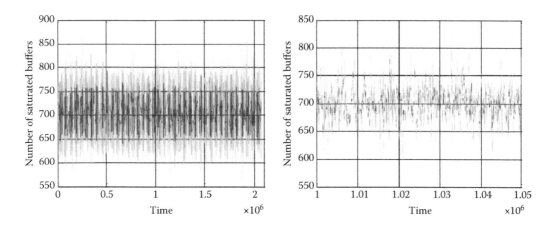

FIGURE 2.38 Number of messages in the network taken at every 100 time slots, $\lambda = 0.158$, $l = 5$ hops, $m = 10$. Blue (black) line, number of messages in the network averaged over 100 time slots; green (gray) line, number of messages in the network taken every 100 time slots.

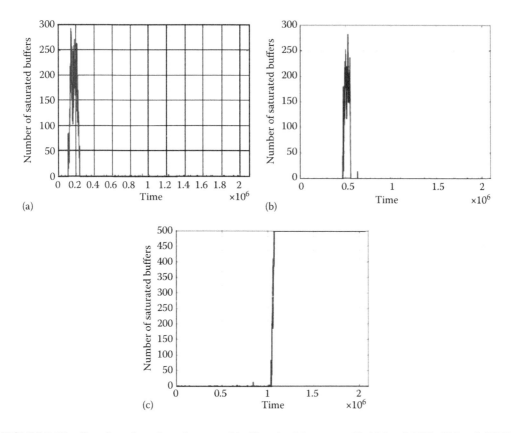

FIGURE 2.39 Snapshot of number of saturated buffers, $l = 5$ hops, $m = 10$. (a) $\lambda = 0.1576$, (b) $\lambda = 0.15765$, (c) $\lambda = 0.1592$.

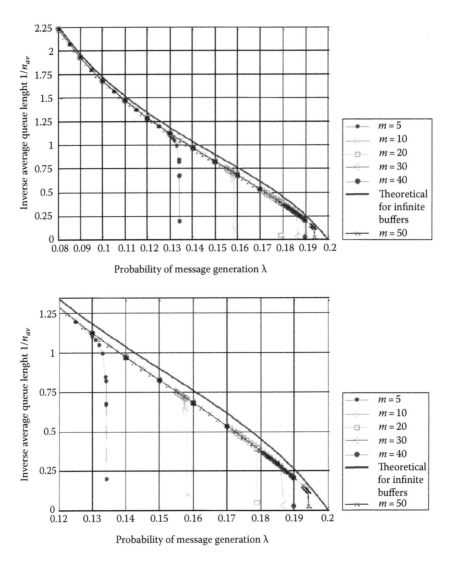

Probability of message generation λ

Probability of message generation λ

FIGURE 2.40 The inverse queue length n as a function of the network load λ for the different buffer capacity and distance between source and destination l = 5 hops. Solid red line, the theoretical prediction for buffers with unlimited capacity.

2.3 TORUS TOPOLOGY

We considered a 16 × 16 toroidal square lattice network. We used the assumptions that messages are always routed along the shortest path to the message destination, but we do not use static routing (there is no fixed path assigned to the message). When a router receives a passing message and there is a choice between two possible buffers to keep the message, one of the buffers will be chosen at random. Therefore, in the case of 2D torus network topology and finite buffers, we elected to make decision, whether to send a message to a neighboring node, based not on the queue in any single buffer at the next node along the message path, but rather on the total number of messages in the neighboring router. The limit on total number of messages held in the router has been set equal to be m = 8, 20, 40, 80, 120, 160, 200, and 400, but no limitations have been imposed on individual output buffers (four in each router). We assumed that router stops accepting incoming messages once it accumulates (m − 4) messages. Locally generated messages are generated individually and

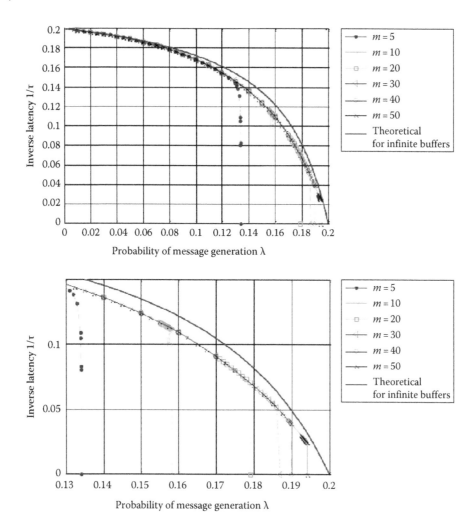

FIGURE 2.41 The inverse latency $1/\tau$ as a function of the network load λ for the different buffer capacity and distance between source and destination $l = 5$ hops. Solid red line, the theoretical prediction for buffers with unlimited capacity.

independently for all four directions, and once router accumulates m messages, it would stop generating messages. Critical values λ_c as depending on the router capacity m are shown in Figure 2.44. One can see that the obtained values follow the empirical formula (2.29) as in the case of the ring topology (Figure 2.45).

Detailed results for the router limit $m = 200$ are presented in the following. Numerical data for the average queue length $\bar{n}$ and latency τ are shown in the Figures 2.46 and 2.47. Surprisingly, until the network load comes close to the saturation value, the network behaves as if there was no buffer limit. Up to the saturation state, λ_{act} does not differ significantly from the nominal λ and we can see an excellent agreement between theoretical prediction (red line) and simulation results up to critical load. However, the critical load observed in experiments ($\lambda_c \approx 0.17725$) is substantially smaller than that predicted by the independent queues theory $\left(\lambda_c^{(\text{inf})} = 0.2\right)$. It also differs from that received in simulations for the ring topology for the same buffer limit $m = 50$ ($\lambda_c \approx 0.1942$ in case of the ring topology).

Simulation shows a sharp transition from the steady-state regime to the saturated state. Therefore, in the case of torus topology and finite buffers, the network behaves as undergoing a first-order

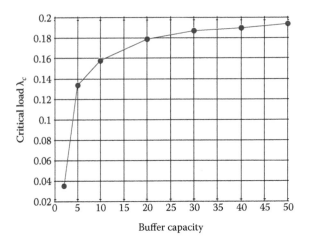

FIGURE 2.42   Dependence between the critical network load λ_c and the buffer capacity for $l = 5$ hops.

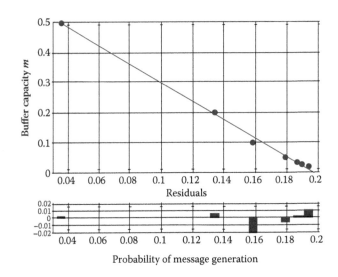

FIGURE 2.43   The dependence between buffer capacity m and network load λ. Red solid line, empirical data (2.29); blue line with bullets, data from simulation.

phase transition. Close to the critical point, the network behavior is very unstable and shows the same pattern as in the case of the ring topology with the first-order phase transition: in one experiment, the network may go into the saturation; in another one (even with a larger network load), it does not reach the saturation (Figures 2.48 through 2.50).

In cases when the network does reach the saturation, it also exhibits very peculiar behavior. Initially, it is similar to the case where saturation is not reached: the snapshot of the network with load $\lambda = 0.17506$ (Figure 2.51) is very similar to the one in Figure 2.52a and b. Then the state of the network changes dramatically when a cluster consisting of a few nodes holding more than 45 messages emerges (Figure 2.52c). This cluster triggers a *chain reaction*: the cluster begins growing rapidly and soon the whole network becomes saturated (Figure 2.52d). This example illustrates how unstable this network is, even though this instability may not be apparent initially in the beginning of the simulation. The same phenomenon is shown also in Figures 2.49 and 2.50 for the total number of messages in the network and for the number of

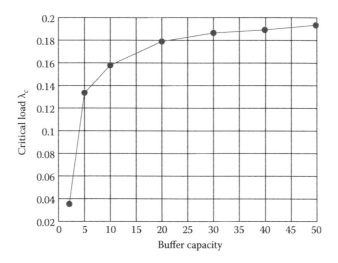

FIGURE 2.44   Dependence between the critical network load λ_c and the buffer capacity for $l = 5$ hops.

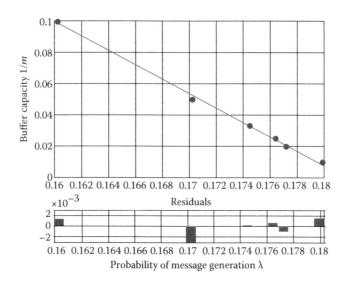

FIGURE 2.45   The dependence between buffer capacity m and critical network load λ_c. Red solid line, empirical formula (2.29); blue bullets, simulation data.

saturated buffers, respectively. One can see that there is no significant difference in the patterns of fluctuation in a network with load $\lambda = 0.17506$ (Figure 2.48) and in a network with the critical load $\lambda_c = 0.17507$ just before the saturation phase (Figure 2.49). But in the latter case, the number of messages in the network and the number of saturated buffers suddenly grow explosively (rapidly increasing from 0 to 1024 for the saturated buffers), and finally, the entire network transitions to the saturation state.

2.4   FUTURE RESEARCH DIRECTIONS

It would be important to put the results obtained for networks with finite buffer in the proper perspective, interpreting them from the standpoint of statistical physics. Our results show that the type of the phase transition in networks depends on the value of the parameter m, which is the buffer

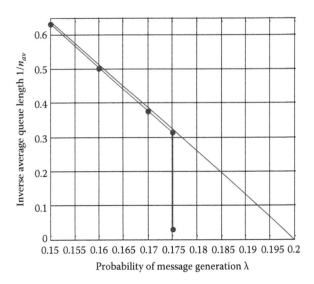

FIGURE 2.46   The inverse queue length $\bar{n}$ as a function of the network load λ, $l = 5$, $m = 200$. Red solid line, the first-order approximation $\bar{n}$ for infinite buffers (2.2); blue line with bullets, simulation data.

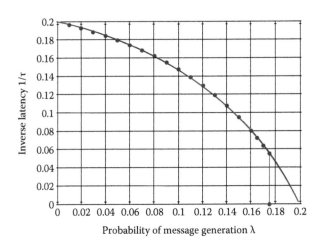

FIGURE 2.47   The inverse latency $1/\tau$ as a function of the network load λ for distance between source and destination $l = 5$ hops, $m = 200$. Red solid line, theoretical prediction for unlimited buffers; numerical simulation results for 16×16 torus are shown by blue line with dots.

capacity. The transition is one of the second order for $m = 1$ and $m = 2$ but becomes a first-order transition for $m \geq 10$. The classical liquid–gas phase transition as described by the van der Waals model of *real* gases (Figure 2.53) changes its type (from first order to the second order) with the increase of temperature. We would like to draw an analogy between critical phenomena in physical systems and those in interconnection networks.

Simulations data demonstrate the emergence of long-range dependences between nodes, far exceeding the *interaction radius* (equal to l) when the network reaches the critical region. The most interesting observation is that the network shows both first-order and second-order phase transitions. The type of the phase transition depends not only on the buffer limit but also on the assumptions for the message generation. In particular, with the buffer limit $m = 2$, we observed a second-order phase transition in model 1 but model 2 has a first-order transition. Similarly to the networks with heterogeneous activity, we observed large and *slow* fluctuations increasing in amplitude as the network

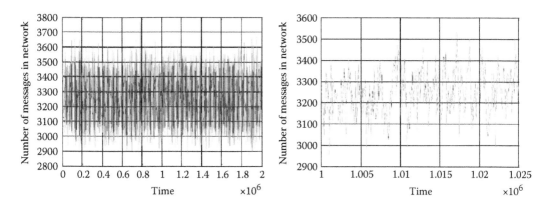

FIGURE 2.48 Number of messages in the network taken at every 100 time slots, $\lambda = 0.17506$, $l = 5$ hops, $m = 200$. Blue (black) line, number of messages in the network averaged over 100 time slots; green (gray) line, number of messages in the network taken every 100 time slots.

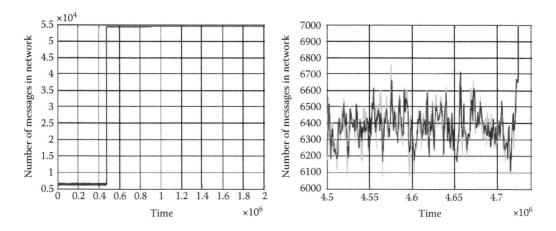

FIGURE 2.49 Number of messages in the network taken at every 100 time slots, $\lambda = 0.17507$, $l = 5$ hops, $m = 200$. Blue (black) line, number of messages in the network averaged over 100 time slots; green (gray) line, number of messages in the network taken every 100 time slots.

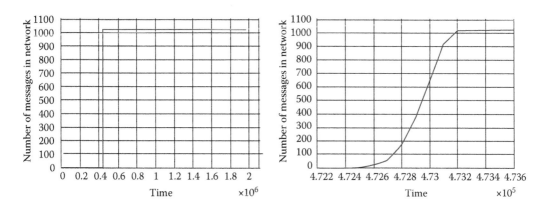

FIGURE 2.50 Snapshot of number of saturated buffers taken at every 100 time slots, $\lambda = 0.17507$, $l = 5$ hops, $m = 200$.

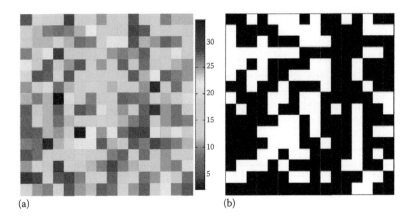

FIGURE 2.51   Snapshot of number of the messages in routers in $t = T - 500$, $l = 5$, $m = 200$, $\lambda = 0.17506$. (a) Actual number of messages in the router. (b) Black nodes, $m \leq 13$; white nodes, $m > 13$.

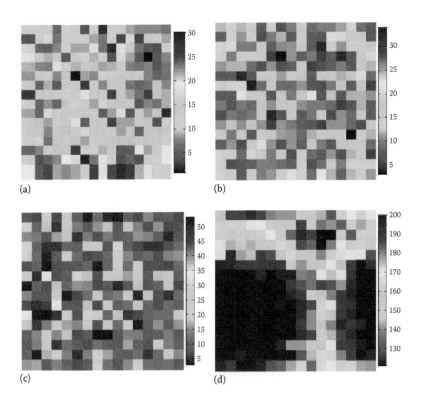

FIGURE 2.52   Snapshot of number of the messages in routers in $t = T - 500$, $l = 5$, $m = 200$, $\lambda = 0.17507$. (a) $t = 5000$. (b) $t = 55,000$. (c) $t = 472,500$. (d) $t = 473,000$.

approached the critical point. This phenomenon is due to the formation and *dissolution* of domains of different phases. Near the critical point, the domains become larger, and the processes of their formation and dissolution take longer time.

The ultimate goal of the theoretical research of networks should be to analyze them within the paradigm of *large systems*, alike physical systems consisting of very large number of interacting particles. Such a consistent description of communication networks would allow us to employ the powerful concepts, methods, and apparatus of statistical mechanics. However, the current stage

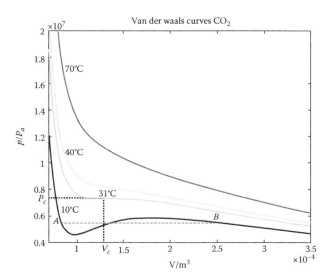

FIGURE 2.53 The van der Waals curves for four different temperatures (10°C, 31°C, 40°C, and 70°C) for the carbon dioxide molecule (CO_2). (Adapted from Citizendium, http://en.citizendium.org/wiki/Van_der_Waals_equation.) The dashed line from A to B is drawn in accordance with Maxwell's equal area rule. The liquid–gas transition is of the first order for temperatures lower than $T_c = 31$°C.

of theoretical analysis of networks seems to be rather far from this ideal situation. Networks are mostly open systems, and it is not clear how they can be characterized in terms of energy, temperature, and other physical concepts. This great challenge should be addressed by future research.

2.5 SUMMARY OF RESULTS AND DISCUSSION

In this chapter, we analyze the networks with finite buffers. A few theoretical models are proposed. For small buffer size $m = 1$ and $m = 2$, both first-order probability distributions (independent queue approximation) and second-order distributions (joint probabilities of pairs of neighboring nodes) have been obtained. The results show that theoretical predictions given by the second-order approximation provide a better description of the network behavior compared to the first-order (mean field) approximation. But theoretical analysis even for small values of m has proved to be very challenging, especially for the second-order approximation. The solution for the general case of the arbitrary buffer capacity m seems to be impossible to obtain in a closed form because of the nonlinearity of the balance equations, but we outline the procedure to obtain numerical solutions for given values of parameters λ and l. Simulations data demonstrate that

1. As in the case of networks with heterogeneous activity [3], Jackson's theorem is not applicable to the networks with finite buffers. The independent queues theory (first-order probability distributions) gives only a rough approximation of the network performance once the load approaches critical region. However, first-order distributions provide an adequate description for smaller values of message generation rate.
2. The second-order approximation that takes into account the joint probability distribution of the queue lengths in two neighboring nodes gives a considerably better prediction of the network behavior, but it also breaks down when λ is close to the critical point. Both first-order and second-order approximations yield critical values much larger (for small buffer sizes) than the empirical data.
3. Both first-order and second-order (continuous) phase transitions to the saturation state have been observed. The second-order phase transitions are characterized by non-trivial values of critical exponent. There is an impression that the type of the phase

transition depends critically on the details of the model. Also, systems with larger buffers tend to have first-order transition.

4. An interesting feature of the first-order phase transitions is that the network behavior follows the second-order approximation very closely until the load reaches the critical value. When it happened, the network goes to the saturation state abruptly. It is very difficult to predict when it can happen even from the simulation data: there are no significant increases in the average latency or queue length right before the critical point. Only the observation of the fluctuations of average number of the messages in the network during the simulation can serve as an indication that the network is close to saturation. These fluctuations increase both in amplitude and in *wavelength* just before the critical point.

5. It is remarkable that a network with finite buffers performs *better* (the latency is smaller) than according to the independent queues approximation for loads not too close to the critical value, as if there is a negative correlation between unsaturated and saturated nodes. This effect is due to a combination of two factors: the instability of the joint state of two neighboring saturated nodes and the decrease in the actual load value λ_{act}, since saturated nodes do not generate messages.

6. The value of λ_c increases with the increase of the buffer capacity approaching the critical value for a network with infinite buffer size.

7. Networks with different buffer sizes show almost no difference in its performance up to their critical loads. However, their latency and average queue length are clearly different from those for a network with infinite buffers for a broad range of load values for from the critical regions.

REFERENCES

1. Y. Rykalova, L. B. Levitin, and R. Brower, Performance model of a multiprocessor interconnection network, *Proceedings of the 10th Communications and Networking Simulation Symposium, CNS'07*, Norfolk, VA, 2007, pp. 100–105.
2. Y. Rykalova, L. B. Levitin, and R. Brower, Analysis and simulation of a model of multiprocessor networks, in *Proceedings of Advances in Computer Science and Technology, ACST 2007*, Phuket, Thailand, April 2007, pp. 200–205.
3. Y. Rykalova, L. B. Levitin, and R. Brower, Critical phenomena in discrete-time interconnection networks, *Physica A: Statistical Mechanics and Its Applications*, 389 (22), 5259–5278, November 15, 2010.
4. L. Kleinrock. *Queueing Systems*, Volume I: Theory. New York: Wiley, 1975.
5. L. Kleinrock. *Queueing Systems*, Volume II: Com. Applications. New York: Wiley, 1976.
6. J. Duato, S. Yalamanchili, and L. M. Ni. *Interconnection Networks: An Engineering Approach*. San Francisco, CA: Morgan Kaufmann, 2003.
7. P. Jackson, Job shop like queueing systems, *Management Science*, 10 (1), 131–142, 1963.
8. W. J. Gordon and G. F. Newell, Closed queueing systems with exponential servers, *Operations Research*, 15, 254–265, 1967.
9. J. P. Buzen, Computational algorithms for closed queueing networks with exponential servers, *Communications of the ACM*, 16 (9), 527–531, September 1973.
10. F. Baskett, K. M. Chandy, R. R. Muntz, and F. G. Palacios, Open, closed, and mixed networks of queues with different classes of customers, *Journal of the ACM*, 22 (2), 248–260, April 1975.
11. J. P. Spirn, Queueing networks with random selection for service, *IEEE Transactions on Software Engineering*, SE-5, 287–289, 1979.
12. A. S. Noetzel, A generalized queueing discipline for product form network solution, *Journal of the ACM*, 26 (4), 779–793, October 1979.
13. K. M. Chandy and J. Martin, A characterization of product form queueing networks, *Journal of the ACM*, 30 (2), 286–299, 1983.
14. M. Reiser, Mean value analysis of queueing networks: A new look at an old problem, in *Proceedings of the Fourth International Symposium on Modeling and Performance Evaluation of Computer Systems*, Vienna, Austria, vol. I, February 1979.

15. M. Reiser and S. S. Lavenberg, Mean value analysis of closed multichain queueing networks, *Journal of the ACM*, 27 (2), 313–322, April 1980.
16. M. Reiser, Mean value analysis and convolution method for queueing dependent servers in closed queueing networks, *Performance Evaluation*, I (1), 7–18, January 1981.
17. C. H. Sauer and K. M. Chandy. *Computer Systems Performance Modeling*. Englewood Cliffs, NJ: Prentice-Hall, 1981.
18. A. K. Kodi, A. Sarathy, and A. Louri. Adaptive channel buffers in on-chip interconnection networks—A power and performance analysis, *IEEE Transactions on Computers*, 57 (9), 1169–1181, September 2008. doi:http://dx.doi.org/10.1109/TC.2008.77.
19. M. Enachescu, Y. Ganjali, A. Goel, N. Mckeown, and T. Roughgarden, Routers with very small buffers, in *Proceedings of the 25th IEEE International Conference on Computer Communications*, Barcelona, Spain, 2006, pp. 1–11.
20. Y. Gu, D. Towsley, C. Hollot, and H. Zhang, Congestion control for small buffer high speed networks, in *Proceedings of the 26th IEEE International Conference on Computer Communications*, Anchorage, AK, May 6–12, 2007, pp. 1037–1045.
21. G. Appenzeller, I. Keslassy, and N. McKeown, Sizing router buffers, in *Proceedings of the 2004 Conference on Applications, Technologies, Architectures, and Protocols for Computer Communications, SIGCOMM'04*, Portland, OR, August 30–September 03, 2004. New York: ACM.
22. E. Bravi and D. Cotter, Optical packet-switched interconnect based on wavelength-division-multiplexed clockwork routing, *Journal of Optical Networking*, 6 (7), 840–853, 2007.
23. Little, J. D. C., A proof for the queuing formula: $L = \lambda W$, *Operations Research*, 9 (3), 383–320, 1961.
24. Y. Rykalova, Queues, latency and critical phenomena in interconnection networks, PhD thesis, Boston University, Boston, MA, 2008.

3 Superefficient Monte Carlo Simulations

Cheng-An Yang, Kung Yao, Ken Umeno, and Ezio Biglieri

CONTENTS

ABSTRACT

Ulam and von Neumann first formulated the Monte Carlo (MC) simulation methodology as one using random sequences to evaluate high-dimensional integrals. Since then, MC simulation methods are widely used to solve complex engineering and scientific problems. Unlike other deterministic methods, MC methods use statistical sampling to produce approximate solutions. A fundamental question of implementing MC simulation is how to generate random samples. It turned out that the generation of truly random sequences in a controlled manner is a nontrivial problem. Fortunately, in many applications, it suffices to use pseudorandom (PR) sequences. A PR sequence can be generated deterministically by some transformations, and it appears to be random from the statistical point of view. As the processed sample size N grows, the uncertainty of the MC solution is reduced. It is well known that the variance of the approximation error decreases as $1/N$. However, for computationally intensive simulations, MC methods may take an extremely long number of samples to obtain a solution with acceptable tolerance.

In this chapter, we describe the novel superefficient (SE) MC simulation method, originated by Umeno, which produces a solution whose variance of the approximation error decreases as fast as $1/N^2$. In order to achieve this SE result, one needs to consider the generation of the PR sequences using a particular class of chaotic dynamical systems. The greatest distinction between conventional and chaotic MC simulation is that the chaotic sequence has correlation between samples. For conventional MC simulation, good PR number generators produce near independent and identically distributed (iid) samples. Correlation between samples is generally considered to be a bad thing, because it may decrease the convergence rate of the simulation. However, if we select the chaotic mapping carefully, the correlation between samples may actually improve the convergence rate for certain integrands. In this chapter, we show how an appropriate choice of the correlation affects the variance of the approximation error. Let N denote the number of samples and σ_N^2 denote the variance of the approximation error of the chaotic MC simulation. It turns out that the convergence rate σ_N^2 has two contributors, one decaying as $1/N$ and the other as $1/N^2$. Eventually, the convergence rate will be dominated by $1/N$, which suggests that the chaotic MC simulation has the same performance as the standard MC simulation. However, if the dynamical system introduces the right amount of negative correlation such that the $1/N$ part becomes zero, the convergence rate becomes $1/N^2$, which is a huge improvement over the conventional MC simulation. Based on the Lebesgue spectrum representation of the integrand, a necessary and sufficient condition (NSC) for SE is given. Various examples based on the use of the Chebyshev polynomials are given. When the integrand does not meet the NSC, we describe an approximation SE (ASE) MC simulation method that is applicable to a wider class of problems than the original SE method and yields a convergence rate as fast as $1/N^\alpha$ for $1 \leq \alpha \leq 2$. The SE and ASE methods can also be generalized to two- and higher-dimensional MC simulation problems.

3.1 INTRODUCTION

Ulam and von Neumann first formulated the Monte Carlo (MC) simulation methodology as one using random sequences to evaluate high-dimensional integrals [1]. Since then, MC simulations have been used in many applications to evaluate the performance of various systems that are not analytically tractable.

The simplest, yet most important, form of MC simulation is used to approximate the integral

$$I = \int_\Omega A(x)dx, \tag{3.1}$$

where the integrand $A(x)$ is defined on the domain $\Omega = [a,b]$ for some real number $a < b$.

To do this, we first choose a probability density function (pdf) $\rho(x) \neq 0$ in Ω and define the function

$$B(x) := \frac{A(x)}{\rho(x)}, \tag{3.2}$$

The integral (3.1) is approximated by calculating the N-sample average

$$\frac{1}{N} \sum_{i=1}^{N} B(X_i) \approx \mathbf{E}[B(X_j)] = I, \quad j = 1, 2, \ldots, N. \tag{3.3}$$

where
   N is the sample size
   X_i's are independent iid random samples whose common pdf is $\rho(x)$
   $\mathbf{E}[.]$ denotes the expectation operator with respect to $\rho(x)$

By the strong law of large numbers, the summation (3.3) converges almost surely to I if the random samples are independent. Furthermore, the variance of the approximation decreases at rate $1/N$. That is,

$$\mathrm{Var}\left[\frac{1}{N} \sum_{i=1}^{N} B(X_i)\right] = \frac{1}{N} \mathrm{Var}[B(X_j)], \quad j = 1, 2, \ldots, N. \tag{3.4}$$

Note that (3.4) holds regardless of the dimension of the domain Ω of the integrand $A(x)$, which make MC simulation suitable for performing multidimensional integrations.

Umeno's superefficient MC (SE MC) algorithm [2] is a variation of standard MC based on chaotic sequences and exhibits a superior rate of convergence. Umeno and Yao's approximate SE (ASE) MC [3] removes some restriction in the original method to make the concept of superefficiency applicable to more general situations. In the following sections, we review the pseudorandom (PR) number generation used in conventional MC simulation and describe the concept of chaotic sequences and chaotic MC simulation. The correlation between samples of the chaotic sequence gives rise to the SE convergence rate, which makes chaotic MC simulation SE. We illustrate how to generate chaotic sequences from the practical point of view and how to apply SE simulation methods to a wide class of integrands using the notion of ASE MC. In the last section, we provide some concluding remarks and point to the directions for future research.

3.2 PSEUDORANDOM NUMBER AND CHAOTIC SEQUENCE

A fundamental question of implementing MC simulation is how to generate random samples. It turned out that the generation of truly random sequences in a controlled manner is a nontrivial problem. Fortunately, in many applications, it suffices to use PR sequences [4]. A PR sequence is generated deterministically by some transformations, and it appears to be random from the statistical point of view [5]. For example, the sequence of linear congruential PR numbers (LCPRN) $x_0, x_1, \dots$ is produced by the recursion

$$x_{n+1} = \left(ax_n + c\right) \bmod m, \qquad (3.5)$$

where $0 < a < m$, $0 \le c < m$, and $0 \le x_0 < m$ is the seed of the sequence [1]. When the parameters a, c, m, and x_0 are properly selected, the linear congruential recursion can produce a sequence of period m.

LCG is one of the oldest and a popular algorithm for generating PR sequences due to its simplicity and well-understood properties. Although an LCPRN sequence passes many randomness tests, LCG has some serious defects. Most notably, it exhibits correlation between successive samples. The Mersenne twister algorithm [6] is a better choice for generating high-quality PR numbers for reducing this correlation. For example, MATLAB® uses the Mersenne twister algorithm as the default uniform random number generator starting from its version 7.4 in 2007 [7].

We can think of the process of generating the PR sequence as applying a deterministic transformation on some state variable repeatedly. More precisely, let Ω denote the collection of all possible states of PR generator and $T: \Omega \mapsto \Omega$ denote the transformation. We select a seed or initial state $x_0 \in \Omega$ and generate the sequence $(x_1, x_2, \dots)$ by

$$x_{i+1} = T(x_i), \quad i = 0, 1, \dots. \qquad (3.6)$$

The output of the PR generator can be written as $y_i = g(x_i)$ for some suitable output function g.

Another way of generating PR sequences is through *dynamical systems*. Formally, a measure-preserving dynamical system is the quadruple (Ω, A, μ, T), where Ω is the state space, A is the σ-algebra on Ω, μ is a probability measure on A, and T is a mapping from Ω to itself such that

$$\mu(T^{-1}(E)) = \mu(E) \qquad (3.7)$$

for all measurable $E \in A$. A mapping T that satisfies (3.7) is called a measure-preserving transformation.

The initial state x_0 of a dynamical system at time 0 is a point in the domain Ω, and the evolution of the state is governed by a mapping T such that $x_{i+1} = T(x_i)$ for $i = 0, 1, \dots$. The sequence $(x_1, x_2, \dots)$

with seed $x_0 \in \Omega$ is called the *orbit* of the dynamical system under T. The *time average* of the integrand $B(x)$ with respect to the orbit is defined as

$$\langle B(x_i) \rangle_N := \frac{1}{N} \sum_{i=1}^{N} B(x_i). \tag{3.8}$$

A natural question to ask is whether (3.8) will converge or not as $N \to \infty$. More importantly, will it converge to the ensemble average I? Birkhoff's theorem [8] says that the time average of an integrable function $B(x)$ will converge to an integrable function $\bar{B}(x)$ almost surely, and $\mathbf{E}[B(X)] = \mathbf{E}[\bar{B}(X)]$, where the expectation is taken with respect to the measure μ. In general, $\bar{B}(x)$ is a function of the initial seed x_0. If a measure-preserving dynamical system has the property that every integrable function $B(x)$ has a constant time average, then it is called an *ergodic dynamical system*. By Birkhoff's theorem, this constant must agree with the ensemble average I. That is,

$$\langle B(x_i) \rangle_N \to \mathbf{E}[B(X)] \quad \text{pointwise as } N \to \infty. \tag{3.9}$$

In this chapter, we will focus on a special type of ergodic system, which has *chaotic* behavior in the sense of Auslander–Yorke [9] that (1) it has a dense orbit in the space Ω and (2) the orbits are unstable, meaning that orbits arising from different x_0, even if arbitrarily close to each other, grow apart exponentially.

For example, the *doubling map*

$$T_d(x) = \begin{cases} 2x & \text{if } 0 \leq x < 0.5, \\ 2x - 1 & \text{if } 0.5 \leq x < 1, \end{cases} \tag{3.10}$$

defined on $\Omega = [0,1)$ is known to be chaotic. The invariant pdf $\rho(x)$ is the uniform distribution on Ω, that is, $\rho(x) = 1$ on $0 \leq x < 1$ and 0 elsewhere. The doubling map is related to many other chaotic dynamical systems, like the Chebyshev dynamical system. The Chebyshev dynamical system of order p is defined on the domain $\Omega = [-1,1]$ with the mapping

$$T_p(y) = \cos(p \arccos(y)), \tag{3.11}$$

where p is a positive integer. The mapping $T_p(y)$ is in fact the pth-order Chebyshev polynomial of the first kind. The Chebyshev dynamical system of order 2 is related to the doubling map via the relation

$$y_i = \cos(2\pi x_i), \tag{3.12}$$

where $x_i = T_d(x_{i-1})$.

3.3   CHAOTIC MC SIMULATION

A chaotic MC simulation is a MC simulation with a PR sequence replaced by a chaotic sequence [2]. Furthermore, specifically, let T be a chaotic mapping, and $\rho(x)$ its invariant pdf. We first draw a seed x_0 from the invariant pdf $\rho(x)$ and use the chaotic mapping T to generate the sequence $x_1, x_2, \ldots, x_N$ by $x_{i+1} = T(x_i)$ for $i = 0, 1, \ldots, N - 1$, where N is the number of samples. The *time average*

$$\langle B(x_i) \rangle_N = \frac{1}{N} \sum_{i=1}^{N} B(x_i) \to I \tag{3.13}$$

will converge to the integral I defined in (3.1) as N approaches infinity [8].

3.3.1 STATISTICAL AND DYNAMICAL CORRELATION

The greatest distinction between conventional and chaotic MC simulation is that the chaotic sequence has correlation between samples. For conventional MC simulation, good PR number generators produce near iid samples. Correlation between samples is generally considered to be a bad thing, because it may decrease the convergence rate of the simulation. However, if we select the chaotic mapping carefully, the correlation between samples may actually improve the convergence rate for certain integrands. In the following, we show how the correlation can affect the variance of the approximation error.

For measure-preserving dynamical systems, any measurable function $B(x)$ on Ω forms a stationary random process $\{B(x_k)\}_{k \in \mathbb{N}}$, where $B(x_k) = B(T^k(x_0))$. For simplicity, denote $B(x_k)$ by B_k and $\langle B(x_i) \rangle_N$ by $\langle B \rangle_N$. Define the autocorrelation function

$$R(k) = \mathbf{E}(B_{k+i} - I)(B_i - I), \tag{3.14}$$

where the expectation is taken with respect to the invariant pdf $\rho(x)$, and $i = 1, 2, \ldots$ is arbitrary because $B(x_k)$ is stationary.

The variance of the approximation error $\langle B \rangle_N - I$ is given by

$$\sigma_N^2 := \mathbf{E}\left[(\langle B \rangle_N - I)^2\right] = \frac{1}{N}\text{Var}[B] + \frac{2}{N^2}\sum_{k=1}^{N}(N-k)R(k). \tag{3.15}$$

The first term on the right-hand side in (3.15) is called the *statistical correlation*, which depends on the integrand $B(x)$ and the pdf $\rho(x)$. The second term is called the *dynamical correlation*, which depends on the integrand as well as on the chaotic sequence [2].

Clearly, for iid random samples $x_1, x_2, \ldots$, we have $R(k) = 0$, and hence (3.15) reduces to the conventional case (3.4), where the convergence rate is $1/N$. If there are positive correlations between samples, the variance of the approximation error σ_N^2 will increase. On the other hand, negative correlations between samples might decrease σ_N^2.

It is therefore natural to ask what is the best achievable convergence rate of chaotic MC simulation. This leads to the notion of superefficiency of the chaotic MC simulation detailed in the next section.

3.3.2 SUPEREFFICIENT CHAOTIC MC SIMULATION

Rewrite the variance of the approximation error (3.15) as

$$\sigma_N^2 = \frac{1}{N}\underbrace{\left(\text{Var}[B] + 2\sum_{k=1}^{N}R(k)\right)}_{\eta} - \frac{2}{N^2}\sum_{k=1}^{N}kR(k). \tag{3.16}$$

This shows that the convergence rate of σ_N^2 has two contributors, one decaying as $1/N$ and the other as $1/N^2$. Eventually, the convergence rate will be dominated by $1/N$, which suggests that the chaotic MC simulation has the same performance as the standard MC simulation.

However, if the dynamical system introduces the right amount of negative correlation such that $\eta = 0$, the convergence rate becomes $1/N^2$, which is a huge improvement over the conventional MC simulation.

To obtain $\eta = 0$, and hence convergence rate $1/N^2$, one should suitably combine the sequence correlation with the integrand [2]. We say the chaotic MC simulation is SE if the variance of the approximation error decays as $1/N^2$ for $N \to \infty$. Umeno [2] also showed that the condition $\eta = 0$ for superefficiency is necessary as well as sufficient.

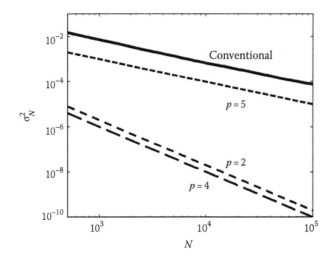

FIGURE 3.1 Variance of the approximation error σ_N^2 versus the sample size N. We compare the convergence rate of the chaotic MC simulation using Chebyshev mapping of order $p = 2, 4$, and 5 and the conventional MC simulation.

Example 3.1: Consider the integrand defined on p. 1447 of [12]:

$$A(x) = \frac{-8x^4 + 8x^2 + x - 1}{\pi\sqrt{1-x^2}}. \tag{3.17}$$

It satisfies the SE condition under Chebyshev dynamical systems (3.11) of order $p = 2$ and $p = 4$. Figure 3.1 shows the results of applying the chaotic MC simulation to find the integral of $A(x)$ using Chebyshev chaotic mappings and compares their convergence rates with the conventional MC simulation using uniform PR samples.

The numerical results verify that the chaotic MC simulation is SE when $p = 2$ and $p = 4$. On the other hand, both conventional and chaotic MC simulation with $p = 5$ have convergence rate $1/N$.

Using the properties of the Chebyshev polynomials, it can be shown that the variances σ_N^2 under $p = 2$ and $p = 4$ are $2/N^2$ and $1/N^2$, respectively, which indicate superefficiency. On the other hand, the case $p = 5$ yields a nonzero η in (3.16). In the next section, we present a very powerful characterization of superefficiency.

3.3.3 Condition for Superefficiency

The superefficiency condition $\eta = 0$ arising from (3.16) does not explicitly suggest any way to achieve it. Umeno [2] first gave a characterization of superefficiency in terms of the coefficients of the generalized Fourier series of the modified integrand $B(x) = A(x)/\rho(x)$ for Chebyshev and piecewise linear dynamical systems, but the results in [2] did not made clear whether those conclusions are also applicable to other dynamical systems. Yao [3] established the connection between superefficiency and the Lebesgue spectrum of ergodic theory [8], which puts the SE condition derived in Umeno's work in a general framework and generalizes Umeno's result to a wide range of dynamical systems, namely, those with a Lebesgue spectrum. This observation helps us explain the superefficiency systematically and hopefully leads to practical algorithms as detailed in Section 3.4.

In this section, we briefly introduce the concept of Lebesgue spectrum and the important characterization of superefficiency in terms of Lebesgue spectrum.

Definition 3.1

Let Λ be an index sets and $\mathbb{N}_0 = \{0,1,2,\ldots\}$. A dynamical system with mapping T is said to have countable one-sided Lebesgue spectrum if there exists an orthogonal basis containing the constant function 1 and the collection of functions $\{f_{\lambda,j}(x)|\lambda \in \wedge, j \in \mathbb{N}_0\}$ such that

$$f_{\lambda,j}(T(x)) = f_{\lambda,j+1}(x) \tag{3.18}$$

for all λ and j, where the index λ labels the classes and j labels the functions within each class. ∎

The Koopman operator induced by the transformation T is defined as $U_T f(x) := f(T(x))$. It is an isometry and it becomes unitary when T is invertible [10]. We may rewrite (3.18) as $U_T f_{\lambda,j} = f_{\lambda,j+1}$, which means U_T has invariant subspaces $W_\lambda = \mathrm{span}(f_{\lambda,0}, f_{\lambda,1}, \ldots)$ generated by $f_{\lambda,0}$'s. Therefore, the *least element* of the invariant subspace W_λ is the generating vector $f_{\lambda,0}$.

Note that the dynamical system has one-sided Lebesgue spectrum if and only if it is *exact* (see, e.g., [8,11]). If a dynamical system has Lebesgue spectrum, then it is strongly mixing [8] and hence chaotic in the sense of Auslander–Yorke [9]. All the dynamical systems we consider in this chapter have Lebesgue spectrum.

Since $\{1\} \cup \{f_{\lambda,j}(x)|\lambda \in \Lambda, j \in \mathbb{N}_0\}$ forms a complete orthogonal basis on the square integrable functions $L^2(\Omega)$, the generalized Fourier series expansion of an integrand $B(x)$ can be written as

$$B(x) = b_{0,0} + \sum_{\lambda \in \Lambda} \sum_{j \in \mathbb{N}_0} b_{\lambda,j} f_{\lambda,j}(x) \tag{3.19}$$

where $b_{0,0}$ is the coefficient corresponding the constant function 1, which is just the integral I of $B(x)$.

Theorem 3.1

Consider a dynamical system that has Lebesgue spectrum $\{f_{\lambda,j}(x)|\lambda \in \Lambda, j \in \mathbb{N}_0\}$ indexed by the sets Λ. The associated chaotic MC simulation is SE if and only if

$$d_\lambda := \sum_{j=0}^{\infty} b_{\lambda,j} = 0 \quad \text{for all } \lambda \in \Lambda, \tag{3.20}$$

where

$$B(x) = b_{0,0} \sum_{\lambda \in \Lambda} \sum_{j \in \mathbb{N}_0} b_{\lambda,j} f_{\lambda,j}(x) \tag{3.21}$$

is the generalized Fourier series of $B(x) = A(x)/\rho(x)$.

Proof

The autocorrelation function (3.14) can be written as

$$R(n) = \mathbf{E}\left[\sum_{\lambda \in \Lambda} \sum_{j=0}^{\infty} b_{\lambda,j} f_{\lambda,j}(T^n x) \sum_{v \in \Lambda} \sum_{i=0}^{\infty} b_{\lambda,i} f_{\lambda,i}(x)\right] = \sum_{v,\lambda \in \Lambda} \sum_{i,j=0}^{\infty} b_{\lambda,j} b_{\lambda,t} \mathbf{E}\left[f_{\lambda,j+n}(x) f_{\lambda,i}(x)\right]$$

$$= \sum_{\lambda \in \Lambda} \sum_{j=0}^{\infty} b_{\lambda,j} b_{\lambda,j+n}. \tag{3.22}$$

From (3.16), η can be expressed as

$$\eta = \sum_{\lambda \in \Lambda} \left(\sum_{j=0}^{\infty} b_{\lambda,j}^2 + 2\sum_{k=1}^{N} \sum_{j=0}^{\infty} b_{\lambda,j} b_{\lambda,j+k} \right) = \sum_{\lambda \in \Lambda} \left(\sum_{j=0}^{\infty} b_{\lambda,j}^2 + 2\sum_{j=0}^{\infty} \sum_{i=j+1}^{j+N} b_{\lambda,j} b_{\lambda,i} \right). \qquad (3.23)$$

As N goes to infinity,

$$\eta = \sum_{\lambda \in \Lambda} \left(\sum_{j=0}^{\infty} b_{\lambda,j}^2 + 2\sum_{i>j}^{\infty} b_{\lambda,j} b_{\lambda,i} \right) = \sum_{\lambda \in \Lambda} \left(\sum_{j=0}^{\infty} b_{\lambda,j} \right)^2. \qquad (3.24)$$

Therefore, $\eta = 0$ if and only if $\displaystyle\sum_{j=0}^{\infty} b_{\lambda,j} = 0$ for each $\lambda \in \Lambda$. ∎

Thus, the explicit condition for superefficiency is that the sum of coefficients in each class λ be zero. We say that an integrand $A(x)$ is *SE* (under the dynamical system with mapping T and invariant pdf ρ) if (3.20) holds for $B(x) = A(x)/\rho(x)$.

Example 3.2

Consider a variant of the integrand (3.17):

$$A(x) = \frac{-8x^4 + 8x^2 + (1+\epsilon)x - 1}{\pi\sqrt{1-x^2}} = B_\epsilon(x)\rho(x). \qquad (3.25)$$

Under the Chebyshev dynamical system (3.11), $B_\epsilon(x)$ can be expanded as

$$B_\epsilon(x) = (1+\epsilon)T_1(x) - T_4(x). \qquad (3.26)$$

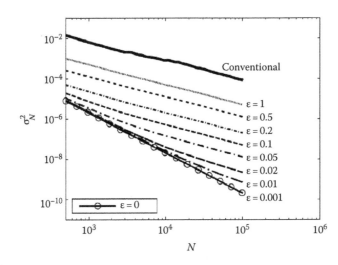

FIGURE 3.2 The variance of the approximation error σ_N^2 versus the number of samples N. The slope of the conventional MC simulation curve is -1, indicating its $1/N$ behavior. On the other hand, the slope of the SE MC simulation is -2 because σ_N^2 decays like $1/N^2$. Between these two extremes are the mismatched SE MC simulations with different size of ε. For $\varepsilon = 0.001$, the curve is almost identical to the SE curve. As ε becomes larger, the slope of the mismatched SE MC simulations gradually increases as N becomes larger.

If $p = 2$, the coefficients of the generalized Fourier series are $b_{0,1} = 1 + \epsilon$, $b_{1,2} = -1$ and zero otherwise. The sum of coefficients are $d_1 = \epsilon$ and $d_\lambda = 0$ for $\lambda \neq 1$. Therefore, $A(x)$ is SE if and only if $\epsilon = 0$. When $\epsilon \neq 0$, we have a *mismatched* SE MC simulation, which appears to be SE for small N but gradually loses superefficiency as N increases [12] (see Figure 3.2).

From the previous example, it is now clear why the integrand $A(x)$ in (3.17) under the chaotic mappings T_2 and T_4 (but not T_5) are SE. More specifically, the modified integrand under the Chebyshev dynamical system can be written as $B(x) = T_1(x) - T_4(x)$. When $p = 2$ and $p = 4$, the sum of coefficients in the class $\lambda = 1$ is $d_1 = 1 - 1 = 0$, and all other d_λ's are zero. Hence, the chaotic MC simulations under both chaotic mappings T_2 and T_4 are SE. On the other hand, when $p = 5$, the chaotic MC simulation has the same convergence rate as the conventional MC simulation, because $d_1 = 1$ does not satisfy the superefficiency condition.

3.3.4 MULTIDIMENSIONAL DYNAMICAL SYSTEMS

Note that the characterization of superefficiency (3.20) holds regardless of the dimension of the domain Ω as long as the system has Lebesgue spectrum. Nevertheless, high-dimensional dynamical systems arise naturally through the product of multiple 1D dynamical systems. In this section, we show that the Lebesgue spectrum of the product dynamical system has a special structure, and we derive the corresponding NSC for superefficiency. For simplicity, we consider two 1D dynamical systems $(\Omega_1, \mathcal{A}_1, \mu_1, T_1)$ and $(\Omega_2, \mathcal{A}_2, \mu_2, T_2)$. The product dynamical system (Ω, A, μ, T) is defined as the product probability space $(\Omega_1 \times \Omega_2, \mathcal{A}_1 \otimes \mathcal{A}_2, \mu_1 \otimes \mu_2)$ with the mapping $T(x, y) = T_1(x) T_2(y)$. It is not difficult to show that the product space is also a measure-preserving dynamical system [11]. Suppose both the dynamical systems $(\Omega_1, \mathcal{A}_1, \mu_1, T_1)$ and $(\Omega_2, \mathcal{A}_2, \mu_2, T_2)$ have Lebesgue spectrum with basis function $\mathcal{B}_1 = \{1\} \cup \{f^{(1)}_{\lambda_1, j_1} | \lambda_1 \in \Lambda_1, j_1 \in \mathbb{N}_0\}$ and $\mathcal{B}_2 = \{1\} \cup \{f^{(2)}_{\lambda_2, j_2} | \lambda_2 \in \Lambda_2, j_2 \in \mathbb{N}_0\}$, respectively. The complete orthogonal basis on $(\Omega_1 \times \Omega_2, \mathcal{A}_1 \otimes \mathcal{A}_2, \mu_1 \otimes \mu_2)$ is $\mathcal{B} = \mathcal{B}_1 \times \mathcal{B}_2$, which can be written explicitly as

$$\{1\} \cup \{f^{(1)}_{\lambda_1, j_1}(x)\} \cup \{f^{(2)}_{\lambda_2, j_2}(y)\} \cup \{f^{(2)}_{\lambda_2, j_2}(y)\} \quad \text{for} \quad \lambda_1 \in \Lambda_1, \lambda_2 \in \Lambda_2, j_1, j_2 \in \mathbb{N}_0. \tag{3.27}$$

Because of the constant function 1, the expression for B in (3.27) becomes very messy. It gets even more cumbersome for higher-dimensional spaces. For notational convenience, we define the redundant functions

$$f_{0,j}(x) := 1 \quad \text{for all } j = 0, 1, 2, \ldots. \tag{3.28}$$

This way, the constant function 1 can be indexed by $(0, j)$ for any nonnegative j. To make sense of this definition, we require that $b_{0,j} = 0$ for all $j > 0$ and $b_{0,0} = I$ is the integral of $B(x)$.

Clearly, the action of U_T on the basis function is

$$U_T f^{(1)}_{\lambda_1, j_1}(x) f^{(2)}_{\lambda_2, j_2}(y) = f^{(1)}_{\lambda_1, j_1}(T_1(x)) f^{(2)}_{\lambda_2, j_2}(T_2(y)) = f^{(1)}_{\lambda_1, j_1 + 1}(x) f^{(2)}_{\lambda_2, j_2 + 1}(y) \tag{3.29}$$

for all $\lambda_1 \in \Lambda_1$, $\lambda_2 \in \Lambda_2$ and $j_1, j_2 \in \mathbb{N}_0$. That is, the index of the basis function changes from $(\lambda_1, \lambda_2, j_1, j_2)$ to $(\lambda_1, \lambda_2, j_1 + 1, j_2 + 1)$ after applying U_T. From (3.30), we found that the least element in each invariant subspace associated with U_T has the form

$$f^{(1)}_{\lambda_1, j_1}(x) f^{(2)}_{\lambda_2, j_2}(y), \quad j_1, j_2 \in \mathbb{N}_0, \quad \text{and} \quad j_1 j_2 = 0. \tag{3.30}$$

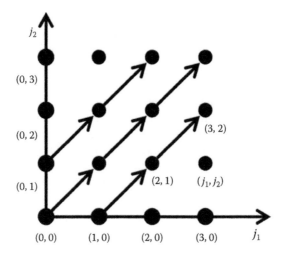

FIGURE 3.3   An illustration of the Lebesgue spectrum for 2D systems. Given λ_1 and λ_2, each circle represents a basis function indexed by $(\lambda_1, \lambda_2, j_1, j_2)$. Applying T on the basis function increases both j_1 and j_2 by 1. The least elements are on the boundary. They generate the invariant subspace.

That is, at least one of the index j_1 and j_2 of the least element must be zero so that no other function can precede it under U_T (see Figure 3.3).

Therefore, we define the index set $\Lambda = \Lambda_1 \times \Lambda_2$ and $J = \{j_1, j_2 \in \mathbb{N}_0 : j_1, j_2 \in \mathbb{N}_0 j_1, j_2 = 0\}$ to index the least elements in each of the invariant subspaces associated with U_T. The generalized Fourier expansion of an integrable function $B(x, y)$ is given by

$$B(x, y) = \sum_{\lambda \in \Lambda} \sum_{j \in J} \sum_{k=0}^{\infty} b_{\lambda, j+k\mathbf{1}} f_{\lambda, j+k\mathbf{1}}(x, y), \tag{3.31}$$

where $\mathbf{1}$ denotes the vector with all unity components. Similarly, for d-dimensional space, we define

$$\Lambda = \prod_{i=1}^{d} \Lambda_i \tag{3.32}$$

and

$$J = \{(j_1, j_2, \ldots, j_d) \in \mathbb{N}_0^d : j_1 j_2 \ldots j_d = 0\}. \tag{3.33}$$

The NSC for superefficiency is given by

$$d_{\lambda, j} = \sum_{k=0}^{\infty} b_{\lambda, j+k\mathbf{1}} = 0 \tag{3.34}$$

for each $\lambda \in \Lambda$ and $\mathbf{j} \in J$.

Example 3.3

2D Walsh system. The Walsh system is a complete orthonormal set associated with the doubling map (3.10). The Rademacher system on the unit interval equipped with Lebesgue measure is a set of orthonormal functions

$$\{x \mapsto r_n(x) = \text{sgn}(\sin 2^{n+1}\pi x) \big| x \in [0,1), n \in \mathbb{N}\}, \tag{3.35}$$

where sgn is the signum function. Note that the Rademacher system is not complete. The complete orthonormal basis is the given by the Walsh–Paley system

$$\{W_n(x) : [0,1) \mapsto \{-1,1\} [n \in \mathbb{N}_0\}, \tag{3.36}$$

where $W_0(x) = 1$ and $W_n(x) = r_{v_1}(x)r_{v_2}(x)\cdots r_{v_m}(x)$, where $n = 2^{v_1} + 2^{v_2} + \cdots + 2^{v_m}$ is the binary representation of n and $v_1 < v_2 < \cdots v_m$. To show that the Walsh–Paley system satisfies (3.18), observe that $r_n(T_d(x)) = r_{n+1}(x)$ and

$$W_n(T(x)) = r_{v_1}(T(x))r_{v_2}(T(x))\cdots r_{v_m}(T(x))$$
$$= r_{v_1+1}(x)r_{v_2+1}(x)\cdots r_{v_m+1}(x) \tag{3.37}$$
$$= W_{2n}(x).$$

Therefore, the complete orthonormal basis is given by

$$f_{\lambda,j}(x) := W_{\lambda 2^j}(x). \tag{3.38}$$

Because of the simple structure of the Walsh function, there is a fast way of evaluating it. Let $x = 0 \cdot b_1 b_2 \ldots b_p$ be the first p-bit binary representation of a number $x \in [0,1)$. Then

$$W_n(x) = -1^{\oplus_{i=1}^{p}(b_i \wedge d_{p-i+1})}, \tag{3.39}$$

where $\wedge$ is the logic AND and $\oplus$ is the exclusive OR.

Consider the integrand

$$B(x,y) = W_1(x)W_2(y) + 0.5W_6(x)W_1(y) - (1-\epsilon)(W_2(x)W_4(y) + 0.5W_{12}(x)W_2(y)) \tag{3.40}$$

on the product space $[0,1)^2$. By (3.38), we can verify that

$$d_{\lambda_1,j_1} = b_{\lambda_1,j_1}b_{\lambda_1,j_1+1}+ = 1 - (1-\epsilon) = \epsilon, \qquad \lambda_1 = (1,1), \quad J_1 = (0,1),$$
$$d_{\lambda_2,j_2} = b_{\lambda_2,j_2}b_{\lambda_2,j_2+1}+ = 0.5 - 0.5(1-\epsilon) = \epsilon, \quad \lambda_2 = (3,1), \quad J_2 = (1,0), \tag{3.41}$$

which satisfies SE condition (3.34) if $\epsilon = 0$ (see Figure 3.4).

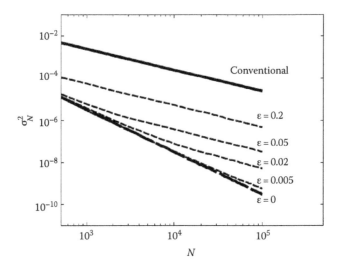

FIGURE 3.4   The variance of the approximation error σ_N^2 versus the number of samples N. The slope of the conventional MC simulation curve is -1, indicating $\alpha = 1$ behavior. On the other hand, the slope of the SE MC simulation is -2. The ε indicates the degree of mismatch of the superefficiency. As ε becomes larger, the slope of the mismatched SE MC simulations gradually increases as N becomes larger.

3.4   ASE MC SIMULATION

Applying chaotic MC simulation on SE integrands yields superior convergence rate of $1/N^2$ in contrast to the conventional convergence rate $1/N$. However, most integrands are not SE. This implies that chaotic MC simulation has no advantage over conventional MC simulation in general.

While most integrands do not satisfy the SE condition, Yao proposed the ASE algorithm [3] that modifies the integrand so that it is approximately SE, and by applying chaotic MC simulation on the modified integrand, we get a much faster convergence rate of $1/N^\alpha$ for convergence exponent α between 1 and 2 (a concept equivalent to ASE was proposed by Umeno in 2002 [13]).

A crucial step here consists of adding to $B(x)$ a function that has zero mean. This will not change the integral of $B(x)$ [3]. Therefore, if we know the sum of coefficients d_λ in (3.20) for each class λ, then the new integrand

$$B'(x) = B(x) - \sum_{\lambda \in \Lambda} d_\lambda f_{\lambda,0}(x) \tag{3.42}$$

will be SE without changing the integral of $B(x)$ (recall that the basis functions $f_{\lambda,j}(x)$'s have zero mean for all λ and j).

We call the function $d_\lambda f_{\lambda,0}(x)$ the *compensator* associated with class λ. By subtracting compensators from $B(x)$, we introduce negative dynamical correlation and make the chaotic MC simulation nearly SE.

In practice, since we do not know the sum of coefficients, it is not possible to construct infinitely many compensators to achieve perfect superefficiency. The idea of ASE algorithm is to approximate the sum of coefficients d_λ by its K_λ-term partial sum

$$\hat{d}_\lambda \approx b_{\lambda,0} + b_{\lambda,1} + \cdots + b_{\lambda,K_\lambda} \tag{3.43}$$

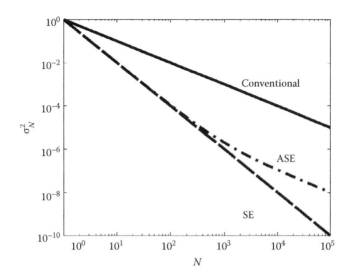

FIGURE 3.5 Illustration of the convergence rate of σ_N^2 versus the sample size N. The convergence exponent α is the negative slope of the curve. Conventional MC simulation has $\alpha = 1$. SE MC simulation has $\alpha = 2$. ASE algorithm has $\alpha \approx 2$ when N is small and will gradually decrease to $\alpha \approx 1$ when N becomes large. Note that even though decay exponent α of the ASE algorithm ultimately goes to 1, the error variance σ_N^2 is significantly smaller than the conventional case.

using conventional MC or chaotic MC simulations, where L_λ is some (hopefully not too large) positive integer. Then we form the modified integrand

$$\tilde{B}(x) = B(x) - \sum_{\lambda \in \Lambda_\omega} \hat{d}_\lambda f_{\lambda,0}(x), \tag{3.44}$$

where the index set $\Lambda\omega$ contains ω classes.

If the sum of coefficients $\tilde{d}_\lambda = d_\lambda - \hat{d}_\lambda$ of $\tilde{B}$ is close to zero, and $\tilde{\eta} = \sum_{\lambda \in \Lambda} \tilde{d}_\lambda = \epsilon > 0$ is small, then from (3.16), the variance of the approximation error can be written as

$$\sigma_N^2 = \frac{\epsilon^2}{N} + \frac{\zeta}{N^2} \tag{3.45}$$

for some ζ. The effective convergence rate can be expressed as $1/N^\alpha$, where $\alpha \in [1,2]$ is referred to as the convergence exponent, and it is defined as the negative slope of the N-σ_N^2 curve (see Figure 3.5).

$$\sigma_N^2 = \frac{\epsilon^2}{N} + \frac{\zeta}{N}$$

$$\alpha := -\frac{d\sigma_N^2}{dN} \frac{N}{\sigma_N^2}. \tag{3.46}$$

The conventional MC simulation has $\alpha = 1$ and the SE MC simulation has $\alpha = 2$. For ASE algorithm, α will decrease as the sample size N increases. Indeed, when N increases, α will gradually decrease to 1, because the term ϵ^2/N will eventually dominate the convergence rate. The more accurate the estimates $\hat{d}_\lambda$'s are, the slower α decreases to 1 (Figure 3.5).

3.4.1 Fixed-Accuracy ASE Algorithm

Yao first proposed the following two-stage ASE algorithm [3] (see Algorithm 3.1):

1. Approximate the sum of coefficients $\hat{d}_\lambda$'s in (3.43) using n-sample conventional or chaotic MC simulation for each $\lambda \in \Lambda_\omega$.
2. Subtract the compensators from the integrand $B(x)$ to form $\tilde{B}(x)$ as defined in (3.44) and apply chaotic MC simulation on $\tilde{B}(x)$.

Note that we need to spend n samples to estimate d_λ for each $\lambda \in \Lambda_\omega$ in stage 1. The quality of the estimates will affect how well the chaotic MC simulation performs in the second stage. To illustrate this point, we apply ASE using different sizes of n and compare their performance in the following example.

Algorithm 3.1 Approximate Superefficient Algorithm

	For $\lambda \in \Lambda_\omega$, do	Stage I: Estimate d_λ for each λ.
1.	$\hat{d}_\lambda \leftarrow \left\langle F_\lambda^{(K)}(x_i) B(x_i) \right\rangle_n$	
2.	$\tilde{B}(x) := B(x) - \sum_{\lambda \in \Lambda_\omega} \hat{d}_\lambda\, f_{\lambda,0}(x)$	Define the ASE integrand.
3.	$x \sim \rho$	Draw a seed from the invariant pdf.
4.	$S \leftarrow 0, \quad D_\lambda \leftarrow 0$ for each $\lambda \in \Lambda_\omega$.	Initialize accumulators.
	While #$iteration < N$ do	Stage II: Chaotic MC simulation.
5.	$x \leftarrow T(x)$	Compute the next state.
6.	$S \leftarrow S + \tilde{B}(x)$	
7.	**Return** S/N	

Example 3.4

Consider the Chebyshev dynamical system of order $p = 2$ and the integrand [12]:

$$A(x) = (1 - x^2)\exp(-x^2) = B(x)\rho(x). \tag{3.47}$$

Unlike the previous examples, where the integrand $B(x)$ could be expressed as finite sum of basis functions, the integrand (3.47) has infinitely many terms in its generalized Fourier series expansion. We choose $\Lambda_\omega = \{1,3,5,7,9\}$ and $K_\lambda = 5$.

We perform chaotic MC simulation for $N = 10^6$ samples using conventional ASE and progressive ASE (PASE) MC algorithms (to be defined shortly) (see Figure 3.6). As a benchmark, we compute the sum of coefficients using accurate numerical integration for the SE case. For ASE MC simulations, we use different number of samples n to estimate d_λ's to demonstrate the effect of inaccurate estimates and convergence rate. For PASE MC simulation, we estimate d_λ's at the same time as the chaotic MC simulation runs.

To better visualize the decay exponent α, we use least square method to find the slope of the curves in Figure 3.7 (recall that α is the negative slope of the curve in the log–log plot). From (3.16), if the integrand is nearly SE, then the decay exponent α will be around 2. For conventional MC simulation, $\alpha = 1$.

Note that for ASE simulations, we need to spend n random samples in the first stage for each class in Λ_L. The effective number of samples for ASE simulations should take those extra samples into consideration.

On the other hand, the PASE algorithm (see the next section) does not have this overhead, and its convergence rate is improving as N increases because the estimates of d_λ are getting more and more accurate.

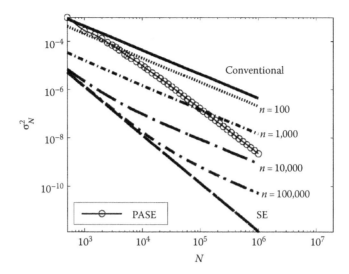

FIGURE 3.6 Variance of approximation error σ_N^2 versus the number of samples N. Throughout the entire simulation, both the conventional and SE MC simulations constantly have $1/N$ and $1/N^2$ behavior, respectively. The ASE MC simulations have $1/N^2$ behavior at first but gradually degrade to $1/N$. ASE simulations with larger values of n have better accuracy than the estimates $\hat{d}_\lambda$'s and lose superefficiency later. The PASE simulation has $1/N$ behavior at first but gradually improves to $1/N^2$, because the estimates $\hat{d}_\lambda$'s get more accurate as N increases.

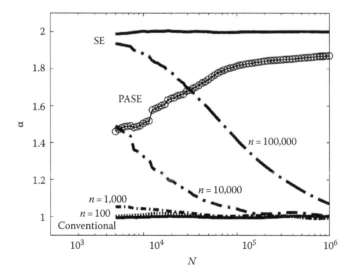

FIGURE 3.7 To see the decay exponent α more clearly, we use the least squares method to find the slope of the curves in Figure 3.6. The decay exponent for the SE MC simulation is 2 for the entire simulation. On the other hand, the exponent for the conventional MC simulation is 1. For the ASE MC simulation with sample size $n = 100,000$, it is superefficient at $N = 10^4$ but the decay exponent gradually decreases to 1.1 in the end of the simulation. With ASE MC, simulations with lower sample size n have smaller decay exponents, all of which decrease to 1 very fast. On the other hand, the PASE simulation has decay exponent around 1.4 in the beginning and gradually increases to nearly 1.9 in the end, indicating that the quality of the estimates $\hat{d}_\lambda$'s is getting better.

3.4.2   Progressive ASE Algorithm

ASE simulation is approximately SE for moderate sizes of N. However, from (3.45), it is clear that ASE simulation will eventually lose the $1/N^2$ convergence rate as long as $\epsilon \neq 0$.

In [12], Biglieri suggested computing $\hat{d}_\lambda$'s iteratively to improve the accuracy of the estimation. As opposed to the original ASE algorithm, which has fixed accuracy for the entire simulation, we proposed a PASE algorithm that keeps improving the accuracy of $\hat{d}_\lambda$'s as the chaotic MC simulation goes on. The idea is to use the samples $B(x_i)$ generated in the main chaotic MC simulation to estimate $\hat{d}_\lambda$'s continuously. Therefore, we get progressively better estimates of $\hat{d}_\lambda$'s and improve the decay rate (see Algorithm 3.2).

Algorithm 3.2 Progressive Approximate Superefficient Algorithm

1.	$x \sim \rho$	Draw a seed from the invariant pdf.
2.	$S \leftarrow 0, \quad D_\lambda \leftarrow 0$ for each $\lambda \in \Lambda_\omega$	Initialize accumulators.
	While #*iteration* $< N$ **do**	
3.	$\quad x \leftarrow T(x)$	Compute the next state.
4.	$\quad S \leftarrow S + B(x)$	
5.	$\quad D_\lambda \leftarrow D_\lambda + F_\lambda^{(L)}(x)B(x)$ for each $\lambda \in \Lambda_\omega$	
6.	**Return** $\left(S - \displaystyle\sum_{\lambda \in \Lambda_\omega} D_\lambda \left\langle f_{\lambda,0} \right\rangle_N \right) / N$	

3.5   IMPLEMENTATION ISSUES

3.5.1   Fast Generation of Compensators

One of the most important steps of the ASE algorithm is the estimation of the coefficients

$$d_\lambda = \sum_{j=0}^{\infty} \left\langle B(x), f_{\lambda,j}(x) \right\rangle. \tag{3.48}$$

Define the K partial sum of the basis functions as

$$F_\lambda^{(K)}(x) := \sum_{j=0}^{k-1} f_{\lambda,j}(x). \tag{3.49}$$

The coefficient d_λ can be approximated by

$$\hat{d}_\lambda^{(K)} := \left\langle B(x_k), F_\lambda^{(K)}(x_k) \right\rangle_n. \tag{3.50}$$

In the PASE algorithm, we need to evaluate (3.50) repeatedly. This can be challenging especially when K is large. In this section, we present a fast way of computing (3.50). Recall that a dynamical system has the Lebesgue spectrum if the basis function $f_{\lambda,j}(x)$ satisfies the condition

$$f_{\lambda,j}(T(x_n)) = f_{\lambda,j+1}(x_n). \tag{3.51}$$

By using the aforementioned property, the partial sum $F_\lambda^{(K)}(x_n)$ and $F_\lambda^{(K)}(x_{n+1})$ can be written as

$$F_\lambda^{(K)}(x_n) = f_{\lambda,0}(x_n) + f_{\lambda,1}(x_n) + \cdots + f_{\lambda,K-2}(x_n) + f_{\lambda,K-1}(x_n),$$
$$F_\lambda^{(K)}(x_{n+1}) = f_{\lambda,1}(x_n) + f_{\lambda,2}(x_n) + \cdots + f_{\lambda,K-1}(x_n) + f_{\lambda,K}(x_n). \tag{3.52}$$

Therefore, $F_\lambda^{(K)}(x_n)$ can be written as

$$F_\lambda^{(K)}(x_{n+1}) = F_\lambda^{(K)}(x_n) - f_{\lambda,0}(x_n) + f_{\lambda,K}(x_n). \tag{3.53}$$

This is to say that once we have computed $F_\lambda^{(K)}(x_0)$, $F_\lambda^{(K)}(x_n)$ can be computed easily regardless of K for all n. The procedure is summarized in Algorithm 3.3.

Algorithm 3.3 Fast Compensator Generation

	$x \leftarrow x_0$	Generate an initial state.
1.	$F \leftarrow F_\lambda^{(K)}(x)$	Compute the F at the initial state.
	Repeat	
2.	$F \leftarrow F - f_{\lambda,0}(x) + f_{\lambda,k}(x)$	Compute F recursively.
3.	$x \leftarrow T(x)$	Generate the next state.

3.5.2 GENERATING CHAOTIC SEQUENCE

As we have seen from the previous section, generating compensators can be implemented very efficiently: we only need to evaluate two terms $f_{\lambda,0}(x_n)$ and $f_{\lambda,K}(x_n)$ for each invariant class λ as in (3.53). Thus, the real question is how to evaluate these terms. In this section, we consider the doubling map and present a fast algorithm based on the one-sided Bernoulli shift. Note that this method is applicable to any dynamical systems that have topological conjugacy relation [14] with the doubling map, such as the Chebyshev dynamical system with $p = 2$.

Implementing a doubling map is particularly simple using a digital computer. Recall that the doubling map is defined as

$$T_2(x) = (2x) \bmod 1. \tag{3.54}$$

Consider a real number $x \in (0, 1)$, which has the binary representation

$$x = 0 \cdot b_1 b_2 \ldots, \tag{3.55}$$

where b_k's are either 0 or 1.

Applying T_2 on x is equivalent to performing a left shift to (3.55), that is,

$$T_2(x) = 0 \cdot b_2 b_3 \ldots. \tag{3.56}$$

If we want to implement such an operation using a digital computer, the first problem we will encounter is that the computer can only store a finite number of bits, say L bits of x. If we apply the mapping T_2 on x for L times, all the L bits will be flushed to the left and result in zero output.

More specifically, let

$$\underline{x} := 0 \cdot b_1 b_2 \ldots b_L \tag{3.57}$$

be the L-bit representation of x in(3.55). We use the notation $T_2^n(x)$ to mean applying T_2 on x for n times. We have the following:

$$
\begin{array}{llllllll}
\underline{x} & = & 0. & b_1 & b_2 & \ldots & b_{L-1} & b_L \\
T_2(\underline{x}) & = & 0. & b_2 & b_3 & \ldots & b_L & 0 \\
& \vdots & & & & & & \\
T_2^{L-1}(\underline{x}) & = & 0. & b_L & 0 & \ldots & 0 & 0 \\
T_2^{L}(\underline{x}) & = & 0. & 0 & 0 & \ldots & 0 & 0
\end{array}
\tag{3.58}
$$

Note that the numerical error between x and its finite binary representation $\underline{x}$ is bounded by $0 \leq x - \underline{x} \leq 2^{-L}$. Let $x_0 \in (0,1)$ define $x_k = T_2^k(x_0)$. The error between x_k and $T_2^k(\underline{x_0})$ is bounded by

$$0 \leq x_k - T_2^k(\underline{x_0}) \leq 2^{-L+k} \wedge 1, \tag{3.59}$$

where $x \wedge y$ is $min\{x, y\}$. This means that the naive implementation of T_2 leads to a numerical disaster: after applying T_2 for L times, the relative error $\dfrac{x_L - T_2^L(x_0)}{x_L}$ is always 100%.

The problem stems from the fact that digital computers can only store finite number of bits. On the other hand, storing an irrational number requires infinitely many bits.

A trick to get around this problem is to *generate* the lost bit after applying T_2: suppose we store the finite binary version $\underline{x_0}$ of an irrational number $x_0 \in (0,1)$. Each time we apply T_2 on the previous value, we add the term $b2^{-L}$ to compensate for the lost bit b. Starting from $\underline{x_0}$, we have

$$\underline{x_1} := T_2(x_0) = T_2(\underline{x_0}) + b_{L+1} 2^{-L}. \tag{3.60}$$

By adding the term $b_{L+1} 2^{-L}$, we restore the truncated version $\underline{x_1}$ of x_1. Similarly, we can restore the truncated version $\underline{x_2}$ of x_2 by

$$\underline{x_2} = T_2(\underline{x_1}) + b_{L+2} 2^{-L}. \tag{3.61}$$

The general step is

$$\underline{x_n} = T_2(\underline{x_{n-1}}) + b_{L+n} 2^{-L}. \tag{3.62}$$

This is what we called the *randomized doubling* procedure, and we denoted the operation in (3.62) by

$$T_2^*(\underline{x_{n-1}}) = T_2^{n*}(\underline{x_0}) = \underline{x_n}. \tag{3.63}$$

The error of the randomized doubling procedure is

$$0 \le x_n - T_2^{n^*}(\underline{x_0}) \le 2^{-L}, \tag{3.64}$$

which is always bounded by 2^{-L} regardless of n, in contrast to the exponentially large error in (3.59) of the naive implementation.

The remaining problem is how to generate the *lost bits* b_{L+k} for $k = 1, 2, ...$ If we select $x_0 \in (0,1)$ at random, then b_k's form a Bernoulli process, that is, b_k's are independent to each other and have the same chance to be 0 or 1. Therefore, the problem becomes how to generate an infinitely stream of independent bits.

From practical point of view, we do not need an infinitely long sequence of bits. Suppose that we will apply T_2 at most M times on x_0, it suffices to generate M independent bits. For example, the linear feedback shift register (LFSR) of length m is a simple algorithm to generate high-quality, period 2^m-1 PR bits (Figure 3.8).

Besides doubling x, we also need to double the number of the form λx as in (3.51) when $\lambda \ne 1$. Let

$$\lambda = e_0 + e_1 2 + e_2 2^2 + \cdots + e_n 2^n \tag{3.65}$$

be the binary representation of $\lambda \in \mathbb{N}$, where $e_i \in \{0,1\}$. λx can be written as

$$\lambda x = e_0 x + e_1 T_2(x) + e_2 T_2^2(x) + \cdots + e_n T_2^n(x) \bmod 1, \tag{3.66}$$

and $T_2(\lambda x)$ can be written as

$$T_2(\lambda x) = e_0 T_2(x) + e_1 T_2^2(x) + e_2 T_2^3(x) + \cdots + e_n T_2^{n+1}(x) \bmod 1. \tag{3.67}$$

A straightforward way of implementing (3.66) and (3.67) on a digital computer is to replace $T_2^n(x)$ by $T_2^{n^*}(\underline{x})$:

$$T_2(\lambda x) \approx e_0 T_2^*(\underline{x}) + e_1 T_2^{2^*}(\underline{x}) + e_2 T_2^{3^*}(\underline{x}) + \cdots + e_n T_2^{n+1^*}(\underline{x}) \bmod 1. \tag{3.68}$$

Clearly, if we implement (3.68) directly, it would be very inefficient because we need to perform numerous randomized doubling procedures. A better approach is to exploit the fact that the least significant bit (LSB) generated by $T_2^{k^*}(\underline{x})$ for $k = 1, 2, ..., n - 1$ have already been generated before. For example, suppose we want to compute $10x$ from doubling $5x$. Let x be given by (3.55). Since $5x = x + 4x$, we can imagine there are two shift registers S1 and S2 that hold the values of x and $4x$,

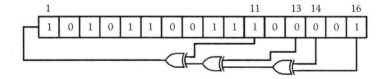

FIGURE 3.8 The example of a 16-bit LFSR. By connecting the 11th, 13th, 14th, and 16th bits to the XOR gates, it is able to produce a binary sequence of length $2^{16} - 1$, which can be used to generate the *loss bits*.

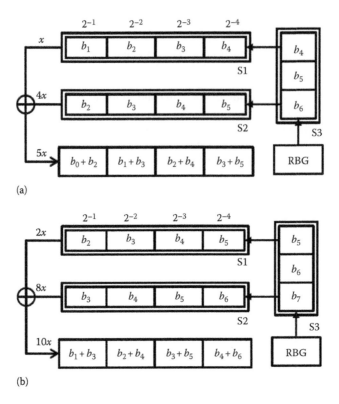

(a)

(b)

FIGURE 3.9   The illustration of doubling the number 5x. (a) Before doubling. (b) After doubling.

respectively (the first and the second row of Figure 3.9a). In addition, we introduce a shift register S3 that stores the previously generated bits (the rightmost column of Figure 3.9a).The cache in the bottom of Figure 3.9a stores the sum of x and $4x$. In the next cycle, S1 and S2 shift in their corresponding LSB from S3, and S3 shifts in the bit b_7 produced by the random bit generator (RBG) (see Figure 3.9b). Effectively, we do not need the shift registers S1 and S2 to store the values of x and $4x$. The entire operation can be implemented as shown in Figure 3.10. In general, we only need a shift register S to hold the current value of $10x$ and an extra shift register S3 to store the previously generated random bits.

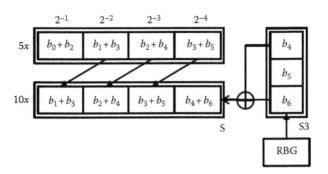

FIGURE 3.10   The actual doubling process for 5x.

The procedure of doubling the number λx is summarized in Algorithm 3.4. Note that the numerical error of (3.68) is given by

$$0 \leq T_2(\lambda x) - e_0 T_2^*(\underline{x}) + e_1 T_2^{2*}(\underline{x}) + e_2 T_2^{3*}(\underline{x}) + \cdots + e_n T_2^{n+1*}(\underline{x}) \bmod 1 \leq \log_2(\lambda) 2^{-L}. \qquad (3.69)$$

For a given λ, the error can be made arbitrarily small by increasing the number of bits L, and there is no error propagation.

Algorithm 3.4 Randomized Doubling

	$\lambda := e_0 + e_1 2 + \ldots + e_n 2^n$	Input: the multiplier.
1.	$\mathbf{b} \leftarrow (b_0, b_1, \ldots, b_n)$	Initialize buffer.
	Repeat	
2.	$s \leftarrow \sum_{i=0}^{n} e_i b_i$	Calculate the LSB.
3.	$x \leftarrow (x \ll 1) + 2^{-L} s$	Left shift x and restore the loss bit by adding the LSB.
4.	$\mathbf{b} \leftarrow (\mathbf{b} \ll 1) + (0, 0, \ldots, X)$	Left shift the buffer and generate a new bit X.

3.5.3 PARALLEL IMPLEMENTATION

It is possible to implement the parallel algorithm of the PASE in various ways. We argue that PASE is particularly suitable for heterogeneous computing based on the assumption that the most computationally intensive task of the simulation is the evaluation of the integrand $B(x)$.

Define $\mathbf{B} = (B(x_1), B(x_2), \ldots, B(x_M))^T$ and $\mathbf{F}_\lambda = (F_\lambda^{(K)}(x_1), F_\lambda^{(K)}(x_2), \ldots, F_\lambda^{(K)}(x_M))$ for each $\lambda \in \Lambda_0$, where $F_\lambda^{(K)}(x)$ is the K partial sum of $f_\lambda(x)$ as defined in (3.49). Recall that the estimate $\hat{d}_\lambda$ can be computed as the inner product

$$\hat{d}_\lambda = \left\langle B(x) F_\lambda^{(K)}(x) \right\rangle_M = \mathbf{B}^T \mathbf{F}_\lambda. \qquad (3.70)$$

By linearity of the time average operator, the integral can be written as

$$\left\langle \tilde{B}(x) \right\rangle = \left\langle B(x) \right\rangle - \sum_\lambda \hat{d}_\lambda \left\langle f_{\lambda,0} \right\rangle. \qquad (3.71)$$

Using the matrix notation, (3.71) can be expressed as

$$\left\langle \tilde{B}(x) \right\rangle = \frac{1}{M} \left(\mathbf{B}^T \mathbf{1} - \mathbf{d}^T \mathbf{f} \right), \qquad (3.72)$$

where $\mathbf{1} = (1, 1, \ldots, 1)^T$ is an M-vector of unit entries, $\mathbf{d}^T = \mathbf{B}^T \left(\mathbf{F}_{\lambda_1}, \ldots, \mathbf{F}_{\lambda_\omega} \right)$ and $\mathbf{f} = \left(\left\langle f_{\lambda_1,0} \right\rangle, \ldots, \left\langle f_{\lambda_\omega,0} \right\rangle \right)^T$.

Based on the assumption that the most computationally intensive task is the evaluation of B, it is reasonable to assign the task to powerful CPUs, which can execute complex instructions. On the other hand, the task of constructing F and f consists of lots of similar and lightweight computations, which can be computed efficiently by special-purpose processors such as graphics processing units (GPUs) (see Algorithm 3.5).

Algorithm 3.5 Parallel PASE

    0.   Generate x_0

        Repeat

    1.   Generate $x_1,\ldots,x_M$ from x_0

        In CPU

    2. $\mathbf{B} \leftarrow (B(x_1),\ldots,B(x_M))^T$

    3. $S \leftarrow S + \mathbf{B}^T\mathbf{1}$

    4.   Send $\mathbf{B}$ to GPU

        In GPU

    5. $\mathbf{F} \leftarrow (F_{\lambda_j}^{(K)}(x_i))_{ij}$

    6. $\mathbf{f} \leftarrow \mathbf{f} + \left(\displaystyle\sum_{m=1}^{M} f_{\lambda_i,0}(x_m) \right)_i$

    7. $\mathbf{d} \leftarrow \mathbf{d} + \mathbf{F}^T\mathbf{B}$

    8. $x_0 \leftarrow x_M$

        Until N samples have been generated

    9. $I \leftarrow \dfrac{1}{N}\left(S + \dfrac{M}{N}\mathbf{d}^T\mathbf{f} \right)$

3.6 CONCLUSIONS AND FUTURE WORKS

While conventional MC simulation yields the convergence rate $1/N$, SE MC has superior convergence rate $1/N^2$ for integrands of the SE type.

Since most integrands are not SE, we introduce the concept of ASE. The ASE and PASE algorithms are general and at least as fast as conventional MC simulation, while sometimes yielding near SE convergence rate. Furthermore, the introduction of the Lebesgue spectrum concept from ergodic theory allows us to systematically study the SE MC simulation. The discussions earlier are applicable also to multidimensional integrands. It is of great interest to find more applications to exploit the concept of SE and ASE.

REFERENCES

1. R. Y. Rubinstein, *Simulation and the Monte Carlo Method*, Wiley, New York, 1981.
2. K. Umeno, Chaotic Monte Carlo computation: A dynamical effect of random-number generations, *Japanese Journal of Applied Physics* 39(Part 1, No. 3A), 1442–1456, 2000.
3. K. Yao, An approximate superefficient MC method is better than classical MC method, Manuscript, 2009.
4. J. Hammersley and D. Handscomb, *Monte Carlo Methods*, Methuen & Co., London, U.K., 1964.
5. D. E. Knuth, *Seminumerical Algorithms*, vol. 2 of *The Art of Computer Programming*, 2nd edn., Addison-Wesley, Reading, MA, 1981.
6. M. Matsumoto and T. Nishimura, Mersenne twister: A 623-dimensionally equidistributed uniform pseudo-random number generator, *ACM Transactions on Modeling and Computer Simulation* 8(1), 3–30, 1998.
7. C. Moler, Random numbers, in *Numerical Computing with MATLAB*, SIAM, Philadelphia, PA, 2004, p. 265.
8. V. I. Arnold, *Ergodic Problems of Classical Mechanics*, vol. 50, Benjamin, New York, 1968.
9. J. Auslander and J. A. Yorke, Interval maps, factors of maps, and chaos, *Tohoku Mathematical Journal* 32(2), 177–188, 1980.

10. A. Lasota and M. Mackey, *Chaos, Fractals, and Noise: Stochastic Aspects of Dynamics*, Springer-Verlag, New York, 1994.
11. K. Petersen, *Ergodic Theory*, Cambridge University Press, Cambridge, 1989.
12. E. Biglieri, Some notes on "supereffiﬁcient" Monte Carlo methods, Manuscript, 2009.
13. K. Umeno, Device and method for numerical integration and program, Japanese Patent No. 3711270 (submitted on March 30, 2002, and granted on August 19, 2005).
14. K. Alligood, T. Sauer, and J. Yorke, *Chaos: An Introduction to Dynamical Systems*, New York, 1997.

4 High-Performance Tool for the Test of Long-Memory and Self-Similarity

Julio Ramírez-Pacheco, Deni Torres-Román, Homero Toral-Cruz, and Leopoldo Estrada Vargas

CONTENTS

ABSTRACT

A novel high-performance software tool for the test of *long-memory* and *self-similarity analysis* is presented. Its performance and robustness are compared with a well-known software tool (*Selfis*) using synthetic *self-similar* time series. Based on a thorough study and, in some cases, algorithm implementation study, it is shown that the tool presents better accuracy and robustness than *Selfis* in the *fGn* and *f-ARIMA*(0, *d*, 0) case. Its wide range of additional methodologies, ease of use, and intuitive functionality make it the most complete freely accessible software tool in the research community dealing with *self-similar* traces. Its immediate applicability can be found in the off-line analysis of measured traces in the computer networking area, which will help in the analysis and design of networks and network algorithmics and in the wide range of fields of science studying *long-memory* phenomena.

4.1   INTRODUCTION

Long-memory processes are stochastic processes used for modeling different types of random phenomena (e.g., phenomena in astronomy, economics, physics, chemistry, hydrology, and computer networking) [1–7]. In the computer networking area, it can be used to model aggregate traffic [8,9],VBR video traffic [10], WWW traffic [11,12], network delay [13], network jitter [14,15], etc. As *long-memory processes* become popular and are needed for performance evaluation studies, network design, dimensioning, and analysis, there is a need for techniques to estimate the index of *memory* of a process. The requirements for the estimation are minimum bias, robustness, and rapid estimation of *the long-memory* parameter in long time series. Several estimators have been proposed: time-domain, frequency-domain, and joint time-frequency algorithms (wavelet techniques). Another problem in the analysis of long-memory processes is the lack of *software* tools for the estimation and generation of *long-memory* sequences. The requirements for the tools are accuracy, robustness, and ease of use. In this area, the most common tool is *Selfis*, a *self-similarity* analysis tool [16].

Unfortunately, *Selfis* presents high bias in the estimation on several algorithms and is not robust when analyzing long time series. Motivated by this, we present a novel software tool for the test of long-memory called *SelQoS* (*self*-similarity and *QoS* analysis tool). *SelQoS* presents minimum bias estimates of the long-memory parameter and is robust under long time series. The techniques implemented in *Selfis* are also implemented in *SelQoS*. The modified Allan variance also called MAVAR [17,18] technique and other methodologies are additionally implemented in *SelQoS*. In order to show which one is the best tool for the test of long-memory and self-similarity, in terms of accuracy and robustness, we present a complete comparison study between *SelQoS* and *Selfis* using synthetic long-memory time series of type fractional Gaussian noise (*fGn*) and *f-ARIMA*(0, *d*, 0) (*fdn*).

The rest of the chapter is organized as follows. First, the theory of *self-similar* and *long-memory* stochastic processes, particular stochastic processes with long-memory, and their relationship are presented. Second, a description of current algorithms for the estimation of the long-memory parameter is provided. The implementations and sources of inaccuracies in the algorithms are also identified. Third, *SelQoS* and *Selfis* tools are described. Fourth, the accuracy and robustness of *Selfis* and *SelQoS* using several synthetic *long-memory* time series is compared. Finally, conclusions and future research directions are presented.

4.2   LONG-MEMORY AND SELF-SIMILARITY

Long-memory and *self-similarity* are two important properties of stochastic processes. The presence or absence of these properties is characterized by the invariance of distributional characteristics under scaling and the slow asymptotic decay of the autocorrelation function respectively. For the purposes of this chapter, a stochastic process, $\Psi = \{\psi(t), t \in T\}$, is a sequence of random variables $\psi(t_i), t_i \in T$. In computer networking, a stochastic process may model the sequence of delays $\mathcal{D}_1, \mathcal{D}_2, ..., \mathcal{D}_m$ of packets traveling along a route I; the aggregate traffic $X_1, X_2, ..., X_m$ at times $t_1, t_2, ..., t_m$, where, $t_1 < t_2, ..., < t_m$; etc. A stochastic process Ψ is called stationary if for arbitrary m, ξ, and $t_1, t_2, ..., t_m, m = 1, 2, ..., n; n = 1, 2, ...$ such that $t_i + \xi \in T$, the following holds:

$$\left\{\psi(t_1 + \xi), \psi(t_2 + \xi), ..., \psi(t_m + \xi)\right\}_{=}^{d} \left\{\psi(t_1), \psi(t_2), ..., \psi(t_m)\right\}, \tag{4.1}$$

where $\underline{d}$ denotes equality of all joint distribution functions. This chapter concentrates in finite-variance stationary stochastic processes. Stationarity can also be defined in the increment process of a random process Ψ. In this context, a stochastic process is said to have stationary increments if for arbitrary m, ξ, and $t_1, t_2, \ldots, t_m$, the following holds:

$$\left\{ \left(\psi \left(t_i + \xi \right) - \psi \left(t_{i-1} + \xi \right) \right) \right\}_{i=2}^{m} \overset{d}{=} \left\{ \left(\psi \left(t_i \right) - \psi \left(t_{i-1} \right) \right) \right\}_{i=2}^{m}. \tag{4.2}$$

A stochastic process Ψ may also have independent increments, that is, a stochastic process Ψ has independent increments if for any $m \geq 1$ and arbitrary $t_1, t_2, \ldots, t_m$; $t_1 < t_2 < \cdots < t_m$, the variables

$$\left\{ \left(\psi \left(t_m \right) - \psi \left(t_{m-1} \right) \right), \ldots, \left(\psi \left(t_2 \right) - \psi \left(t_1 \right) \right) \right\} \tag{4.3}$$

are independent.

4.2.1 LONG-MEMORY PROCESSES

Let $\Psi = \{\psi(t), t \in T\}$ denote a finite-variance stationary stochastic process; it is said that Ψ has *long-memory* (or long-range dependence) if its autocorrelation function $\{\rho(k), k \geq 0\}$ is regularly varying (at infinity) with parameter $\beta \in (0, 1)$, that is,

$$\rho(k) \sim L_1(k) |k|^{-\beta}, \tag{4.4}$$

where $L_1(.)$ is a slowly varying function $\left(\lim_{x \to \infty} \left(L(tx)/L(x) \right) = 1 \right)$. Defining a random variable via its autocorrelation function is equivalent to defining it in the frequency domain via the spectral density function. A finite-variance stochastic process has long-memory if its spectral density $f(\lambda)$ varies regularly (at $\lambda = 0$) with parameter $\alpha \in (0, 1)$, that is,

$$f(\lambda) \sim L_2 \left(\frac{1}{\lambda} \right) |\lambda|^{-\alpha}. \tag{4.5}$$

The parameters α and β are related according to $\alpha = 1 - \beta$, and frequently the exponents in (4.4) and (4.5) are given in terms of the *Hurst* exponent H, that is, $\beta = 2 - 2H$. A consequence of $\rho(k) \sim k^{-\beta}L_1(k)$ is a nonsummability of correlations [1,4,19], that is, a long-memory stochastic process satisfies $\sum_{k-\infty}^{\infty} \rho(k) = \infty$ or equivalently its spectral density possesses a pole at the origin. In contrast, a process is said to be short-range dependent (or with short-memory) if

$$\rho(k) \sim ck^{-\beta}, \tag{4.6}$$

where $\beta \in [1, 2)$ and $k \to \infty$.

The preceding relations give rise to summable autocorrelations, that is,

$$\sum_{k=-\infty}^{\infty} \rho(k) < \infty. \tag{4.7}$$

4.2.2 SELF-SIMILAR PROCESSES

Self-similar processes are stochastic processes for which its distributional properties do not change with time and space scaling [1,20]. Let $\Psi = \{\psi(t), t \in T\}$ denote a stochastic process; it is said that Ψ is *self-similar* iff there exists an $H > 0$ such that for any $m \geq 1$, $s > 0$, and arbitrary $t_1, t_2, \ldots, t_m$, the following holds:

$$\left\{\psi(t_1), \psi(t_2), \ldots, \psi(t_m)\right\} \overset{d}{=} \left\{s^{-H}\psi(st_1), s^{-H}\psi(st_2), \ldots, s^{-H}\psi(st_m)\right\}. \tag{4.8}$$

The parameter H is called the *Hurst-index* or the *self-similarity* parameter. An important type of self-similar process is the class of *Hsssi* (*self-similar* with parameter H and with stationary increments) processes. A process Ψ with stationary increments is said to be *self-similar* iff there exists an $H > 0$ such that for any m, $s > 0$ and for arbitrary $t_1, t_2, \ldots, t_m$, the following holds:

$$\left\{\left(\psi(t_i + \xi) - \psi(t_{i-1} + \xi)\right)\right\}_{i=2}^{m} \overset{d}{=} \left\{\left(\psi(st_i + \xi) - \psi(st_{i-1} + \xi)\right)s^{-H}\right\}_{i=2}^{m}. \tag{4.9}$$

A process satisfying Equation 4.9 is called an *Hsssi* process. *Hsssi* processes are very important since if $\Psi(t)$ denotes an *Hsssi* process, then $Z(j) = \Delta\Psi(j) = \Psi(j + 1) - \Psi(j)$ is a stationary time series or sequence. An important example of an *Hsssi* process is fractional Brownian motion. For the purposes of the chapter, the interest is in discrete second-order self-similar processes or sequences. Before defining discrete self-similarity, the block averaging operator is introduced.

Let $X = \{x_1, x_2, \ldots, x_n\}$ be a finite-length time series; the block averaging operator Γ_m over X is defined as

$$\Gamma_m\left(x_1, x_1, \ldots, x_n\right) = \left(\frac{1}{m}\sum_{i=1}^{m} x_{i+(k-1)m}, \quad k \geq 1\right), \tag{4.10}$$

where $m \geq 1$. As can be noted from (4.10), the block averaging operator $\Gamma_m(.)$ maps a sequence to another sequence $\left\{X_i^{(m)}\right\} = \Gamma_m\left(\{x_i\}\right)$. From this, let $X = \{x_i, i \in \mathbb{N}\}$ be a weakly stationary random series; it is said that X is discrete *self-similar* iff there exists an $H > 0$ such that for any $m \geq 1$,

$$m^{1-H}\Gamma_m\{x_i\} \overset{d}{=} \{x_i\}. \tag{4.11}$$

From this, it is easily seen that $m\Gamma_m\left(\{x_i\}\right) \overset{d}{=} m^H\{x_i\}$ resembles the continuous time definition $\psi(s.) \overset{d}{=} s^H\psi(.)$. Equation 4.11 is strict for the modeling of packet network traffic, then a weak condition of (4.11) is used, that is, second-order discrete *self-similarity*. A weakly stationary process X is called second-order *self-similar* if the second-order structure of $m^{1-H}\Gamma_m(\{x_i\})$ and $\{x_i\}$ are the same for any m, that is, if

$$\rho\left(\left(\Gamma_m\left(\{x_i\}\right)\right)_k, \left(\Gamma_m\left(\{x_i\}\right)\right)_{k+h}\right) = \rho\left(X_k X_{k+h}\right) \tag{4.12}$$

or equivalently

$$Var\, m^{1-H}\Gamma m\left(\{x_i\}\right) = Var(X), \tag{4.13}$$

where

$h \in \mathbb{R}$

$\rho(.)$ denotes the autocorrelation function

A more appropriate form of *self-similarity*, suitable for modeling packet network traffic, is the class of asymptotic second-order *self-similarity*. A process is called asymptotic second-order *self-similar* if for any $h \in Z$ and as $m \to \infty$,

$$\rho^m(h) = \frac{1}{2}\Big[(h+1)^{2H} - 2h^{2H} + (h-1)^{2H}\Big], \tag{4.14}$$

where $h \geq 1$ and $\rho^m(h)$ is the autocorrelation function of the block-averaged process $X_i^{(m)}$. Letting $k \to \infty$ in (4.14) results in $\rho^m(k) = H(2H-1)k^{2H-2}$, which is a long-memory process. Thus, a *long-memory process* is a subset of asymptotic second-order self-similar processes, that is, *lrd $\subset$ ass*.

4.2.3 SELF-SIMILAR STOCHASTIC PROCESSES

Two common examples of *self-similar* stochastic processes are fractional Gaussian noise [21] and fractional *ARIMA(p, d, q)* [22] stochastic processes. Fractional Gaussian noise is a stochastic process whose autocorrelation function $\gamma(h) = E(X_i X_{i+h})$ satisfies $\gamma(h) = 2^{-1}\{(h+1)^{2H} - 2h^{2H} + (h-1)^{2H}\}$, $h \geq 0$, that is, it is an exact second-order self-similar stochastic process. *f-ARIMA(p, d, q)* is defined according to

$$\Phi(B)X_i = \Theta(B)\Delta^{-d} \in_i, \tag{4.15}$$

where
$\Phi(B)$ and $\Theta(B)$ are the autoregressive and moving average coefficients, respectively
Δ^{-d} is the differencing operator, defined as $\Delta_{\in_i} = \in_i - \in_{i-1}$

The spectral density of *f-ARIMA(p, d, q)* processes is obtained as

$$f(\lambda) = \frac{\sigma^2}{2\pi}\left(2\sin\left(\frac{\lambda}{2}\right)\right)^{-2d} \frac{|\Theta(e^{j\lambda})|^2}{|\Phi(e^{j\lambda})|^2}. \tag{4.16}$$

The behavior of the spectral density (4.16) as the frequency tends to zero ($\lambda \to 0$) can be expressed as

$$f(\lambda) \sim f_{ARIMA}(0)|\lambda|^{-2d}. \tag{4.17}$$

From this, it is easily seen that (4.16) and (4.17) are *long-range* dependent or asymptotically self-similar processes with Hurst-index given by $H = d + (1/2)$.

4.2.4 LONG-MEMORY AND SELF-SIMILARITY RELATIONSHIP

The properties of *self-similarity* and long-range dependency are closely related concepts [4,9,19]. An asymptotic *self-similar* process is defined according to Equation 4.14; now let $k \to \infty$, the n $\rho^m(k) = H(2H-1)k^{2H-2}$, thus, Equation 4.14 implies that in the limit (as $k \to \infty$), an asymptotic self-similar process is *long-range* dependent. Similarly, a long-range-dependent process X_t can be constructed by the increment process of a *self-similar* process, that is, $X_t = (Y_t - Y_{t-1}, t = 1,2,...)$, for example, fractional Gaussian noise is obtained from the increment of a fractional Brownian motion process.

4.3 ESTIMATION OF THE LONG-MEMORY PARAMETER

This section briefly reviews the most common techniques for inferring the *long-memory* parameter in a discrete time series. Most of the techniques are currently implemented in *Selfis* and *SelQoS*.

4.3.1   R/S Statistic: Definition and Algorithms

The rescaled adjusted range, range over standard deviation, or simply R/S statistic is one of the old-est methods for estimating the *Hurst-index*, H, in *long-memory* and *self-similar* time series [23–27]. It was initially developed by E. Hurst while studying nonstandard behavior in the Nile River. The range in $(\tau_i, \tau_i + n)$ is defined for a process Y(t) according to the following relation:

$$R(\tau_i, n) := \left[\max_{0 \le u \le n} \left(W\left(\tau_i, n\right) \right) - \min_{0 \le u \le n} \left(W\left(\tau_i, n\right) \right) \right],$$ (4.18)

where
$W(\tau_i, n) = Y(\tau_i + u) - Y(\tau_i) - u E(\tau_i, n).$

$E(\tau_i, n)$ denotes the sample mean with index τ_i in the interval $(\tau_i, \tau_i + n)$

The range in (4.18) is adjusted by rescaling it with the sample standard deviation in the interval $(\tau_i, \tau_i + n)$. The result is the rescaled adjusted range statistic, R/S statistic, defined by

$$\frac{R}{S}(\tau_i, n) := \frac{R(\tau_i, n)}{S(\tau_i, n)},$$ (4.19)

where $S^2(\tau_i, n)$ is the sample variance in the interval $(\tau_i, \tau_i + n)$. Hurst found that for long-memory records, (4.19) behaves like a power law, that is,

$$E\left\{ \frac{R}{S}(\tau_i, n) \right\} \sim n^H, \quad H > 0.5.$$ (4.20)

In contrast, short-range dependent processes follow the following relation:

$$E\left\{ \frac{R}{S}(\tau_i, n) \right\} \sim n^{0.5}.$$ (4.21)

The different behavior of Equations 4.20 and 4.21 is known as the *Hurst effect*. From (4.20), it is easily seen that a log–log plot of the mean values of the R/S statistic values for varying n versus n should result in a straight line with slope H, equal to the *Hurst*-index.

4.3.2   Aggregate Variance Method

Consider the aggregated series of a length N time series, obtained using Equation 4.10. The sample variance of the block-averaged process,

$$Var\left(\Gamma_m\left(\{x_i\}\right)\right) = \frac{1}{N/m} \sum_{k=1}^{N/m} \left(X_k^{(m)} - \bar{X}\right)^2, \quad k = 1, 2, \dots, \frac{N}{m},$$ (4.22)

is mostly referred to the aggregated variance of time series $\{x_i, i = 1, 2, \dots, N\}$. In *long-memory* time series, the aggregated variance behaves asymptotically as

$$Var\left(\Gamma_m\left(\{x_i\}\right)\right) \sim cm^{-\beta},$$ (4.23)

where
   c is a constant
   $\beta = 2 - 2H$

From this result, a log–log plot of $Var\left(\Gamma_m\left(\{x_i\}\right)\right)$ versus m, for different values of m, and such that $m_{i+1}/m_i = C \in \mathbb{R}+$ results in a straight line with slope $\beta = 2H - 2$, from which H can be inferred.

4.3.3 ABSOLUTE MOMENT METHOD

The absolute moment method is very similar to the aggregated variance technique. The absolute moment for an aggregated series is defined as

$$AM^{(m)} = AM\left(\Gamma_m\left(\{x_i\}\right)\right) = \frac{1}{N/m}\sum_{k=1}^{N/m} \mid X_k^{(m)} - \bar{X} \mid. \tag{4.24}$$

Asymptotically, (4.24) behaves as $AM^{(m)} \sim m^{-\beta/2}$; thus, a log–log plot of $AM^{(m)}$ versus m results in a straight line with slope $-(\beta/2) = H - 1$ from which H can be inferred.

4.3.4 VARIANCE OF RESIDUALS METHOD

Let $\{X_i, i \in \mathbb{Z}+\}$ be a time series; define a new series by $\Upsilon_m = \{\Upsilon_1^m, \Upsilon_2^m, ..., \Upsilon_n^m, ...\}$, where each Υ_i^m is defined as $\Upsilon_i^m = \{X_{(i-1)m+1}, X_{(i-1)m+2}, ..., X_{im}\}$ and represents a partition of size m of the original series $\{X_i\}$. Now, to every Υ_i^m, compute the partial sum series $F(\Upsilon_i) = \{F_i^m(1), F_i^m(2), ..., F_i^m(m)\}$, where each $F_i^m(j) = \sum_{j=1}^{m} X_{(i-1)m+j}$ and $F^m = \{F(\Upsilon_2^m), ..., F(\Upsilon_i^m), ...\}$. Once the computation of the partial sum series F^m is obtained, a least squares line is adjusted to each $F(\Upsilon_i^m)$ obtaining a series $z^m = \{z_1^m, z_2^m, ..., z_i^m, ...\}$, called the least squares series, where each $z_i^m = a_i + b_i t$. The parameter b_i in z_i represents the slope of the series $F(\Upsilon_i^m)$. The variance of residual series $V_{res}^m = \{V_{res}^m(1), ..., V_{res}^m(i), ...\}$ of size m is then obtained with the help of F^m and z^m, where each $V_{res}^m(i)$ is obtained as

$$V_{res}^m(i) = \frac{1}{m}\sum_{j=1}^{m}\left(F_i^m(j) - z_i\right)^2. \tag{4.25}$$

In a *self-similar* stochastic process, the median of the variance of residual series behaves as $Med\left(V_{res}^m\right) \sim m^{2H}$ for large m. Thus, a log–log plot of the median of V_{res}^m versus m for varying m should result in a straight line with slope $2H$. The method is referred to as the variance of residuals methods.

4.3.5 PERIODOGRAM

This method is based on the periodogram that is defined in the following way:

$$I(\upsilon) = \frac{1}{2\pi N}\left|\sum_{j=1}^{N} X(j)e^{ij\upsilon}\right|^2, \tag{4.26}$$

where
  υ is the frequency
  X is a time series of length N

If the variance of X is finite, then $I(\upsilon)$ is an estimator of the spectral density of X, and a series with *LRD* will have a spectral density proportional to $|\upsilon|^{1-2H}$ close to the origin. Therefore, a *log–log* plot of the *periodogram* versus *frequency* should show a straight line with a slope of $1 - 2H$. In practice, the lowest 10% of the frequencies is used to compute the Hurst estimation since it is needed to work with frequencies around zero.

4.3.6 WHITTLE

The Whittle method [23,37–30] is a nongraphical maximum likelihood estimator strongly related to the periodogram through the following equation:

$$Q(\eta) := \int_{-\pi}^{\pi} \frac{I(\upsilon)}{f(\upsilon;\eta)} \, dv + \int_{-\pi}^{\pi} \log f(\upsilon;\eta) dv, \tag{4.27}$$

where
     η is a vector of unknown parameters
     $I(\upsilon)$ is the periodogram

$f(\upsilon; \eta)$ is the spectral density at frequency υ of the studied function; the value of vector η that minimizes the function Q is considered the *Whittle Estimator*; Equation 4.27 can be mathematically manipulated to obtain a discrete approximation with the same minimization properties:

$$Q^*(\eta) = \sum_{j=1}^{[(N-1)/2]} \frac{I(\upsilon)}{f^*(\upsilon_j;\eta)} \, dv, \tag{4.28}$$

where N is the series length. A semiparametric version, the one that is implemented in *SelQoS*, known as *Local Whittle* is also available [23,28,30]; while the *Whittle MLE* specifies the functional form of the spectral density at all frequencies, the *Local Whittle* assumes only the functional form where υ is near zero, namely,

$$f(\upsilon) \sim G(H) |\upsilon|^{1-2H} \quad as \ \upsilon \rightarrow 0, \tag{4.29}$$

and from (4.29), the task is reduced to minimize the function

$$R(H) = \log\left(\frac{1}{M} \sum_{j=1}^{M} \frac{I(\upsilon_j)}{\upsilon_j^{1-2H}} \right) - (2H-1) \frac{1}{M} \sum_{j=1}^{M} \log \upsilon_j. \tag{4.30}$$

Its computation involves the introduction of the parameter M, which is an integer less than $N/2$, and satisfying $\frac{1}{M} + \frac{M}{N} \rightarrow 0$ as $N \rightarrow \infty$.

4.3.7 ABRY–VEITCH

Let $d_x(i, j)$ denote the wavelet coefficients of a particular finite-length sequence $X = (x_1, x_2, \ldots, x_N)$; it is known that for *long-memory* stochastic processes, the variance at level i of the coefficients is given by

$$Var(d_x(i,.)) = \frac{\sigma^2}{2} V_\psi(H)(2^j)^{2H+1}, \tag{4.31}$$

where $V\psi(H)$ depends on the particular wavelet and the *Hurst-index* [31] and is defined by

$$V_\psi \left(H \right) = - \int_{-\infty}^{\infty} \gamma_\psi \left(\tau \right) |\tau|^{2H} d\tau, \tag{4.32}$$

taking that the logarithm at (4.31) should result in $\log(Var(d_x(i,.))) = (2H + 1)j + K$, where K is a constant. Abry and Veitch have suggested a *Hurst-index* estimator based on this particular behavior using Daubechies wavelets [32–34].

First, a time average μ_i of $d_x(i.j)$ is computed at a given scale, where μ_i is defined as $\mu_i = (n_i)^{-1} \sum_{j=1}^{ni} d_x^2(i, j)$, where n_i is the wavelet coefficient number at scale i and n the time series points. The estimated *Hurst-index* is then obtained from the slope of a linear regression method for

$$\log_2 \left(\mu_i \right) = \log_2 \left(\frac{1}{n_i} \sum_{j=1}^{n_i} d_x^2 \left(i, j \right) \right), \tag{4.33}$$

where $i = 1, 2, \ldots, [\log_2(n)]$.

4.3.8 SOURCES OF INACCURACIES

Time-domain and frequency-domain algorithms' accuracies depend greatly on some parameters such as cutoff selection in the linear regression and resolution. Resolution has a small effect on accuracy. Cutoff selection is more accurate sensitive than resolution and must be taken into account. A good cutoff section is important since it may give rise to biased estimations. The cutoffs are affected by the short-range dependence of the time series and the size of the blocks. Additionally, there may be different implementations of the same method, and different implementations give varying accuracy. The chapter studies only a different implementation of the *R/S* statistic.

4.4 HIGH-PERFORMANCE TOOL FOR SELF-SIMILARITY

SelQoS is a powerful C++-based software tool for the analysis and generation of *self-similar* time series. The motivation behind the design of *SelQoS* was to obtain accuracy in the estimation and robustness in the analysis of long time series, a feature that is absent in current *self-similarity* analysis software tools. For the estimation of the long-memory or self-similarity parameter, a two-step approach is followed, namely, opening the file and estimating the parameter. A similar approach is followed when other functionality of *SelQoS* is required. Current *Hurst-index* estimation algorithms implemented in *SelQoS* include, *R/S* statistic, absolute moment, aggregated variance, variance of residuals, and modified Allan variance in the time-domain case. Also, the local Whittle method, periodogram, and Abry–Veitch method are currently implemented. *SelQoS* also generates *self-similar* traffic based on the Paxson's *FFT* algorithm. Synthetic *self-similar* traffic can be used for testing the accuracy of algorithms for *Hurst-index* estimation as well as for checking the performance of methodologies under self-similar behavior. Figure 4.1 shows the interface of *SelQoS*. As can be observed, the software tool consists of three main parts: the menu, the toolbar, and the edit part. *SelQoS* complete functionality is found in the menu, while the most common methodologies are found in the toolbar. The edit part permits the manipulation of the time series such as cutting pieces of the series or aggregating parts of another. *SelQoS* performs basic statistical methodologies such as mean and variance, and more advanced ones such as autocorrelation function, power spectral density function, cross correlation function, histogram estimation, and empirical correlation function estimation, on an opened time series $\{X_i\}$.

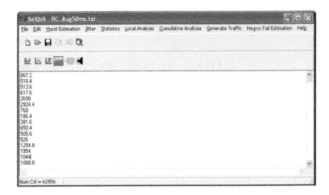

FIGURE 4.1　*SelQoS* GUI.

Additional methodologies implemented include *Fourier* and *Hilbert* transformation. More importantly, the results of these methodologies can be sent on a file for further analysis on a different software tool. An important feature of *SelQoS* is its ability to perform local and cumulative analysis plots on a specific functionality $H(.)$, where $\mathcal{H}(.) : \{X_k\} \to \mathbb{R}$ could be any basic statistical methodology or any *Hurst*-index estimation algorithm. Local analysis plot is performed on a time series $\{X_i\}$ by first partitioning the series into blocks of size m, for obtaining $\Upsilon^m = \{\Upsilon_1^m, \Upsilon_2^m, ...\}$, now to every Υ_1^m, compute $\mathcal{H}(\Upsilon_1^m)$.

From this, a new series, say $\mathcal{H}^m \left|= \{\mathcal{H}(\Upsilon_1^m), \mathcal{H}(\Upsilon_2^m), ..., \mathcal{H}(\Upsilon_i^m), ...\}\right.$ is obtained. Finally, plot the series $\mathcal{H}^m$ versus i to obtain the local analysis plot. The cumulative analysis plot is obtained by partitioning $\{X_i\}$ into blocks of size m to obtain $\Upsilon^m = \{\Upsilon_1^m, \Upsilon_2^m, ..., \Upsilon_i^m, ...\}$; the next step is to compute $\mathcal{H}(.)$ for $\Upsilon_1^m, \Upsilon_1^m \cup \Upsilon_2^m, ..., \cup_{j=1}^{\infty} \Upsilon_j^m$, $j = 1, 2, ...$, to obtain $\mathcal{H}_1^m, \mathcal{H}_1^{2m}, ..., j = 1, 2, ...$. Finally, plot $\mathcal{H}_1^{jm}$ versus jm for $j = 1, 2, ...$ to obtain the cumulative analysis plot. Time series' *Hurst*-index estimation in *SelQoS* is performed into two steps. First step involves the opening of a one-column vector of numbers (a text file). The nonconformance of this requirement would cause *SelQoS* to show an error message. Finally, once the opening of a correct file is selected, the second step involves either estimating the Hurst-index, computing a statistic, computing local or cumulative analysis plot, etc. In the final step, usually a new window is generated, where the plot of a methodology is shown. Figure 4.2 shows the new window generated when the estimation of the Hurst-index via the periodogram is performed. It is expected that *SelQoS* will be useful in the areas of science dealing with long-memory and self-similarity phenomena, which include economics, finance, hydrology, telecommunications, physics, and chemistry.

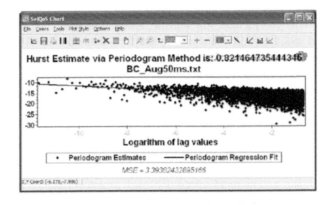

FIGURE 4.2　Estimation of the Hurst-index via periodogram method.

4.4.1 SELFIS

Selfis (*Self*-similarity analys*is*) is a java-based *self-similarity* analysis software tool developed by Thomas Karagiannis at University of California at Riverside. Current algorithms implemented in *Selfis* include *R/S* statistic, variance type, absolute moment, variance of residuals, local *Whittle*, periodogram, and *Abry–Veitch* method. Additionally, *Selfis* performs basic statistics computations, ACF, and power spectral density estimation, and some cleansing algorithms such as bucket shuffling. *Selfis* functionality is the same as in *SelQoS*. First, for *Hurst-index* estimation, the opening of a one-column file is selected; after this, the time series is plotted and shown in a special part of *Selfis*. Once the file is opened, a *Hurst-index* estimation methodology could be applied for obtaining $\hat{H}$. *Selfis* presents problems when applied to long time series; it was observed that when the series is longer than 200K points, *Selfis* experienced long delays and even never finishes the estimations. The interested reader can find additional information on *Selfis* in [16,35].

4.4.2 SIMILARITIES

SelQoS and *Selfis* share some similarities such as estimating the *Hurst-index* using the same set of estimation methods and following a two-step approach for *Hurst-index* estimation. Even though the estimation methods are the same, the implementation changes. The *R/S* statistic is implemented in *Selfis* using a dynamic block approach, while *SelQoS* implements the method using a static block approach. The dynamic block approach partitions the original length N time series in blocks of size m and computes the *R/S Statistic* for each block, that is, obtains $R/S(\tau_i, m)$, $\tau_i = im$, and $i = 0, 1, 2, \ldots,$ N/m. This is done for several block size values m. A log–log plot of *R/S Statistic* values versus m results in an estimate of the *Hurst-index* H. The static block approach first partitions the original length N time series into blocks of size K and then computes the *R/S* statistic for increasing subsets $M \leq N$ of the original series. The subsets of the original series must start at points $\tau_i = iN/K$ for $i = 0, \ldots, K - 1$ and such that $\tau_i + M \leq N$. Note that computation of $(R/S)(\tau_i, N)$ for different N values is performed, and the K blocks are identical. This means that when $M < N/K$, N/K estimates of the *R/S* statistic are obtained, and when $M = N$, only one *R/S Statistic* estimate is obtained. As before, a log–log plot of the *R/S Statistic* values versus K should result in a straight line with slope H. The other time estimation methods vary only in the way the cutoffs of the linear regression are taken and in the resolutions of aggregated series. The cut-offs in *SelQoS* are selected according to $10^{0.3}$ for the lower and $10^{\log_2(N)}$ for the upper cut-off. Cut-off selection affects the final estimation value.

4.5 COMPARISON

In this section, a complete comparison study of *SelQoS* and *Selfis* is presented. In order to perform the comparison study, a set of synthetic *fGn* and *f-ARIMA*(0, d, 0) self-similar time series were generated. For the *fGn* case, series were generated with $H = 0.6$, $H = 0.7$, $H = 0.8$ and $H = 0.9$. For each H, 50 series were generated. The generation was done using modified's Paxson *FFT* algorithm [36]. For the *f-ARIMA* case, time series were generated with $d = 0.1$, $d = 0.2$, $d = 0.3$, and $d = 0.4$. As in the *fGn* case, 50 series were generated for each d value. The generation of *f-ARIMA* and *fGn* series was done using S+ version 6.0. Each generated time series is 65,536 points in length. The selection of this length makes possible the study of the convergent behavior in a particular estimation method. Once the generation of the time series is accomplished, the estimation of the *Hurst-index* is computed. For each method, 50 *H*-estimations are obtained.

4.5.1 FRACTIONAL GAUSSIAN NOISE RESULTS

Figure 4.3 shows the perspective plot for the estimation of the *Hurst-index* using finite-variance *fGn* time series. The plot illustrates the estimation for *SelQoS* and *Selfis*. The axes in the plot should be interpreted as follows: *Number of estimations* axis represents the 50 time series used in

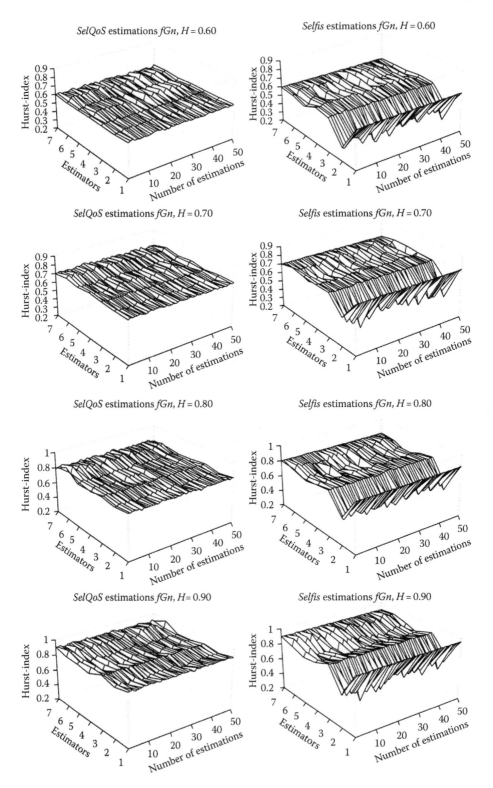

FIGURE 4.3 *Hurst-index* estimation for 50 *fGn* time series with $H \in \{0.60, 070, 0.80, 0.90\}$. The estimators used are as follows: 1, Abry–Veitch; 2, absolute moment; 3, periodogram; 4, *R/S* statistic; 5, variance method; 6, variance of residuals; and 7, Whittle.

the estimation methodology. *Estimators* axis represents the particular *Hurst* estimation method, and for this axis, 1 represents the *Abry–Veitch* method; 2, *absolute moment*; 3, *periodogram*; 4, *R/S* statistic; 5, *variance*-type; 6, *variance of residuals*; and 7, the local *Whittle* method. Finally, *Hurst-index* axis represents the particular *Hurst-index* value. As can be noted from Figure 4.3, *Selfis* presents a rougher surface in the perspective plot, which indicates bias in the estimations. Note that in this perspective plot, a fine sheet will be an indicator of high accuracy. The methods that made the surface rough in *Selfis* are the *absolute moment* and the *Abry–Veitch* method in a high degree and *variance*-type, *variance* of residuals, and *R/S* statistic in a lower degree. *SelQoS* on the other hand is less rough than *Selfis*. The methods that made the surface rough in *SelQoS* are *R/S* statistic, *variance*-type, *variance of residuals*, and *absolute moment* methods. Overall, the surface in *Selfis* is rougher than that of *SelQoS* and indicates that estimations are more biased. A better way of comparing the accuracy of the estimation methods is by means of statistics. The chapter makes use of three such statistics, namely, the $BIAS = H_0 - \overline{X}$, where H_0 is the nominal H value; the standard deviation $= \sqrt{\sigma^2}$, where $\sigma^2 = (N-1)^{-1} \sum_{i=1}^{N} (X_i - \overline{X})^2$; and $\sqrt{MSE} = N^{-1} \sum_{i=1}^{N} (x_i - H_0)^2$. The statistics are obtained for each estimation method using the set of 50 estimations. Table 4.1 presents the results for the *fGn* case. Based on Table 4.1, it is observed that for *fGn* the better algorithms in *SelQoS* are the *periodogram*, local *Whittle*, and *Abry–Veitch* method. In contrast, high accuracy in *Selfis* is obtained only on the *periodogram* and local *Whittle* method. The *Abry–Veitch* implementation in *Selfis* seems to be wrong since the $BIAS > 0.048$. Time-domain algorithms behave differently in both tools. The absolute moment algorithm behaves better in *SelQoS* for the *fGn* case, the bias increases for increasing *Hurst-index*, and the variation from mean is low. In the *variance*-type algorithm, for *Hurst-index* values less than 0.8, the estimations are not biased in *SelQoS*, while the estimations in *Selfis* for this algorithm are biased for every value of H and with high variation from the mean and nominal. In this methodology, *SelQoS* behaves better and with lower bias than *Selfis*. Unfortunately, as it was mentioned earlier, this estimator presents high bias when $H > 0.80$ in *SelQoS*. The variance of residuals is the most biased estimation method in *SelQoS*, and the bias is greater than in *Selfis*. In this method, *Selfis* presents low bias but high variability from the mean and nominal values. The reason behind the better accuracy of *SelQoS* for the *absolute* moment and *variance*-type could be in the selection of the cutoffs in the final linear regression. As the *R/S* statistic is implemented in a different fashion in *SelQoS* and *Selfis*, an additional comparison is presented for this method. The additional comparison is accomplished via the use of the convergent analysis. The convergent analysis measures the convergent behavior of a specific algorithm to specific values of the *Hurst-index*. The convergent analysis is the same as the cumulative analysis described earlier, and it must be applied to long time series. Figure 4.4 presents the convergent behavior (cumulative analysis) of the *R/S* statistic for both tools using *fGn* time series. As can be noted, the behavior of the *R/S* statistic in *Selfis* is irregular, and the algorithm does not seem to converge.

This is particularly true when the *R/S* statistic implementation in *Selfis* is applied to any *fGn* time series with any *Hurst-index* value. In *SelQoS*, in contrast, the algorithm seems to converge and the behavior is irregular at the first 20,000 points and then regular for the points that follow. This result suggests that the static block approach for implementing the *R/S* statistic is better than the dynamic block approach for *fGn*, and that for obtaining accurate estimations using the *R/S* statistic, a minimum of 30,000 points is necessary. This result is important since many *R/S* statistic algorithms seem to be implemented using the dynamic block approach; this is especially true in the algorithms implemented in Economics and Finance.

4.5.2 Fractional *ARIMA*(0, d, 0) Results

The attention is now turned to the estimations when using *f-ARIMA* time series. Figure 4.5 shows the perspective plot for the *f-ARIMA* case. Note that the surface in *Selfis* is rougher than its surface in the *fGn* case. It is noted the surface in *Selfis* and *SelQoS* is rougher, the lower the *Hurst-index* is.

TABLE 4.1

SelQoS and *Selfis BIAS*, σ, and $\sqrt{MSE}$ for 50 Synthetic Finite-Variance Fractional *Gaussian* Noise Time Series, Using Time-Domain, Frequency-Domain, and Wavelet Methodologies

	Fractional Gaussian Noise (*SelQoS* vs. *Selfis*)							
	Nominal *H*							
	SelQoS				Selfis			
	0.6	0.7	0.8	0.9	0.6	0.7	0.8	0.9
Absolute moment method								
BIAS	0.001	0.005	0.019	0.036	0.215	0.233	0.252	0.277
σ	0.013	0.015	0.018	0.018	0.045	0.052	0.038	0.054
$\sqrt{MSE}$	0.013	0.015	0.026	0.040	0.220	0.238	0.255	0.282
Periodogram								
BIAS	0.006	−0.004	−0.002	−0.003	0.001	−0.003	−0.001	−0.002
σ	0.007	0.006	0.008	0.008	0.007	0.006	0.008	0.008
$\sqrt{MSE}$	0.007	0.008	0.009	0.015	0.007	0.007	0.008	0.008
R/S statistic								
BIAS	−0.016	−0.004	0.010	0.040	−0.022	−0.004	0.025	0.068
σ	0.025	0.023	0.025	0.028	0.014	0.011	0.016	0.016
$\sqrt{MSE}$	0.030	0.023	0.029	0.049	0.026	0.012	0.030	0.070
Aggregated variance								
BIAS	0.001	0.004	0.020	0.036	0.025	0.042	0.062	0.089
σ	0.013	0.014	0.017	0.016	0.046	0.053	0.039	0.053
$\sqrt{MSE}$	0.013	0.015	0.026	0.040	0.052	0.067	0.074	0.104
Variance of residuals								
BIAS	−0.020	−0.024	−0.021	−0.106	−0.005	−0.001	0.005	0.005
σ	0.020	0.020	0.029	0.034	0.028	0.027	0.028	0.035
$\sqrt{MSE}$	0.029	0.032	0.036	0.035	0.028	0.027	0.028	0.035
Whittle								
BIAS	0.001	−0.002	0.001	0.002	0.001	0.000	0.000	0.000
σ	0.009	0.011	0.012	0.009	0.002	0.002	0.002	0.003
$\sqrt{MSE}$	0.009	0.011	0.013	0.009	0.002	0.002	0.002	0.003
Abry–Veitch								
BIAS	−0.000	−0.005	−0.002	−0.006	−0.048	−0.053	−0.056	−0.058
σ	0.007	0.006	0.008	0.007	0.002	0.002	0.002	0.003
$\sqrt{MSE}$	0.007	0.008	0.009	0.010	0.048	0.054	0.056	0.058

From Figure 4.5, it is observed that the time-domain algorithms are particularly affected with this behavior. The *periodogram*, *Whittle*, and the *Abry–Veitch* method are the most accurate in *SelQoS*, while in *Selfis*, the *periodogram* and the *variance* of residuals are the most accurate of its algorithms. *The variance*-type and the absolute moment behave better in *SelQoS* than in *Selfis*. Table 4.2 shows the statistics for the 50 *f-ARIMA* time series using the *Hurst* estimation methods. In this table, it is noted that most technques for *H*-estimation implemented in selfis are highly biased when *H* > 0.80. The variance of residuals seems to be the only technique which presents minimum biases in Selfis.

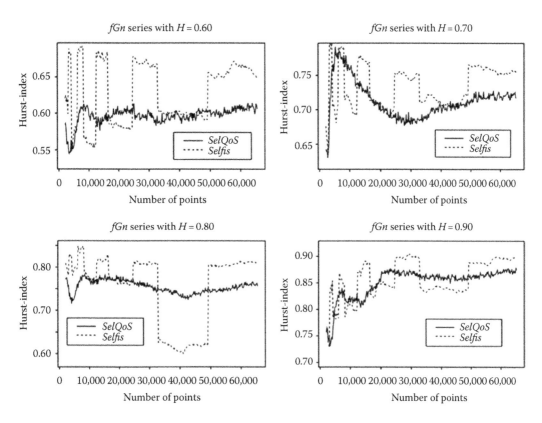

FIGURE 4.4 Convergent analysis for *fGn* time series using *R/S* statistic.

Note that although the bias in the *variance* of residuals in *Selfis* is low, the variability from the mean and nominal values is high, then, this algorithm presents in some estimations low bias and in others high bias. The absolute moment is highly biased in *Selfis*, and the variation from the mean and nominal is also high. In *SelQoS*, this algorithm behaves well when the *Hurst-index* is less than 0.8, and when $H = 0.9$, the bias is $BIAS = 0.036$. Similar results are obtained in *SelQoS* in the *variance*-type case. In contrast, the estimations in *Selfis* are still highly biased. Overall, in the *f-ARIMA* case, the estimations in *SelQoS* are more accurate than those of *Selfis*, accuracy defined in terms of bias, and deviation from the mean and nominal values. Only the variance of residuals method performs better in *Selfis*; nevertheless, this estimation deviates highly from the mean and nominal *H*. The attention is now turned to the convergent analysis of the *R/S* statistic when *f-ARIMA* time series are used. Figure 4.6 shows the convergent analysis of the *R/S* statistic of *SelQoS* and *Selfis*. As can be noted, the dynamic blocks *R/S* statistic implemented in *Selfis* continues to be chaotic and does not seem to converge to the *H* value. The erratic behavior is higher, the greater the *H* value is. *SelQoS*, on the other hand, presents convergent behavior when $H \leq 0.80$ and seems to have a nonconvergent behavior when $H > 0.80$. From this, it is noted that for both the *fGn* and the *f-ARIMA* case, the estimations in *SelQoS* are better than those of *Selfis*, and since the implementation of the *R/S* statistic is different in both tools, a convergent analysis suggests that the implementation in *SelQoS* is better than in *Selfis*. It is only the variance of residuals method that behaves better in *Selfis* for *fGn* and *f-ARIMA* case; however, the deviation from the mean and nominal values is high.

4.5.3 PERFORMANCE COMPARISON

This section presents a performance comparison of *SelQoS* and *Selfis* based on the response time of both tools to a *self-similar* time series of 65,536 points in length. The response time is applied

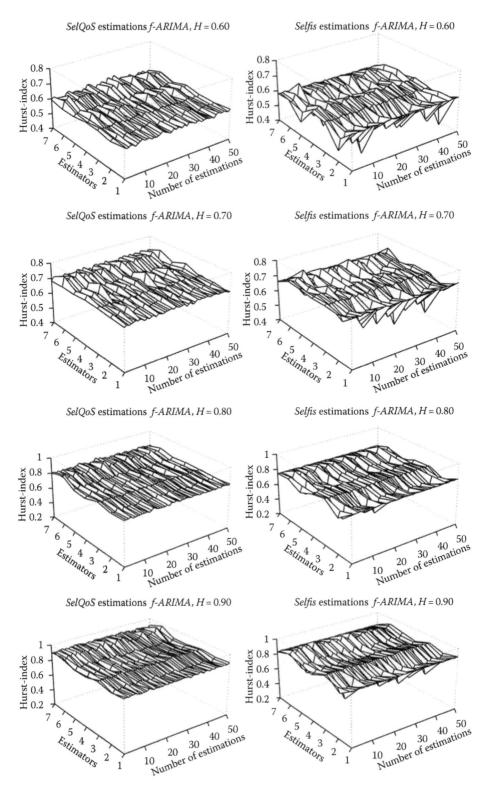

FIGURE 4.5 *Hurst-index* estimation for 50 *f-ARIMA* time series with $H \in \{0.60, 070, 0.80, 0.90\}$. The estimators used are as follows: 1, Abry–Veitch; 2, absolute moment; 3, periodogram; 4, *R/S* statistic; 5, variance method; 6, variance of residuals; and 7, Whittle.

TABLE 4.2

SelQoS and *Selfis BIAS, σ,* and $\sqrt{MSE}$ for 50 Synthetic Finite-Variance *f-ARIMA* (0, *d,* 0) Time Series, Using Time-Domain, Frequency-Domain, and Wavelet Methodologies

	f-ARIMA(0, d, 0) (SelQoS vs. Selfis)							
	Nominal *H*							
	SelQoS				*Selfis*			
	0.6	**0.7**	**0.8**	**0.9**	**0.6**	**0.7**	**0.8**	**0.9**
Absolute moment method								
BIAS	0.001	0.005	0.019	0.036	0.062	0.060	0.086	0.108
σ	0.013	0.015	0.018	0.018	0.050	0.048	0.052	0.056
$\sqrt{MSE}$	0.013	0.015	0.026	0.040	0.079	0.077	0.101	0.122
Periodogram								
BIAS	0.006	−0.004	−0.002	−0.003	0.001	0.002	0.001	0.000
σ	0.007	0.006	0.008	0.008	0.006	0.006	0.009	0.007
$\sqrt{MSE}$	0.007	0.008	0.009	0.015	0.006	0.006	0.009	0.007
R/S statistic								
BIAS	−0.016	−0.004	0.010	0.040	−0.016	0.002	0.027	0.064
σ	0.025	0.023	0.025	0.028	0.011	0.013	0.014	0.014
$\sqrt{MSE}$	0.030	0.023	0.029	0.049	0.019	0.013	0.031	0.066
Aggregated variance								
BIAS	0.001	0.004	0.020	0.036	0.047	0.044	0.068	0.084
σ	0.013	0.014	0.017	0.016	0.050	0.048	0.052	0.056
$\sqrt{MSE}$	0.013	0.015	0.026	0.040	0.068	0.065	0.086	0.101
Variance of residuals								
BIAS	−0.020	−0.024	−0.021	−0.106	0.005	0.005	0.009	0.004
σ	0.020	0.020	0.029	0.034	0.027	0.028	0.028	0.029
$\sqrt{MSE}$	0.029	0.032	0.036	0.035	0.028	0.028	0.029	0.029
Whittle								
BIAS	0.001	−0.002	0.001	0.002	0.019	0.038	0.054	0.068
σ	0.009	0.011	0.012	0.009	0.002	0.002	0.002	0.002
$\sqrt{MSE}$	0.009	0.011	0.013	0.009	0.020	0.038	0.054	0.068
Abry–Veitch								
BIAS	−0.000	−0.005	−0.002	−0.006	−0.030	−0.017	−0.006	0.002
σ	0.007	0.006	0.008	0.007	0.002	0.002	0.002	0.003
$\sqrt{MSE}$	0.007	0.008	0.009	0.010	0.030	0.018	0.007	0.004

to every common methodology. The performance comparison was done using a Pentium 4, 2.4 GHz machine with 256 MB in RAM. Table 4.3 shows the response time in milliseconds of several methodologies. It is noted from the table that most methodologies in *SelQoS* are faster than those of *Selfis*; the longer the time series, the longer the response time in a software tool. *SelQoS* performs well when the series is longer than 300,000 points; in contrast, *Selfis* does not even open the file. From this, it is noted that *SelQoS* is more robust than *Selfis* when analyzing long time series.

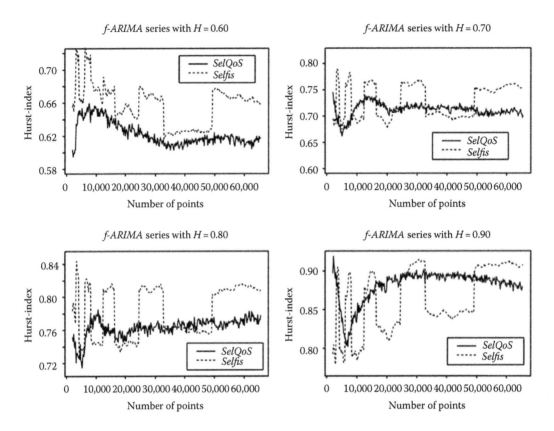

FIGURE 4.6 Convergent analysis for *f-ARIMA* time series using *R/S* statistic.

TABLE 4.3

***SelQoS* and *Selfis* Performance Comparisons Based on Response Times for the *Hurst* Estimation Methods and PSD**

	SelQoS vs. *Selfis*			
	SelQoS	***Selfis***	***SelQoS***	***Selfis***
Delay (ms)	*R/S* statistic		Absolute moment	
	2765	4750	797	3187
Delay (ms)	Variance-type		Variance of residuals	
	813	3141	1938	3625
Delay (ms)	Periodogram		Local Whittle	
	1062	3797	1953	2062
Delay (ms)	Abry–Veitch		Power spectral density	
	1016	2281	1062	3938

4.6   FUTURE RESEARCH DIRECTIONS

The field of self-similar stochastic processes and its applications is rich in research, and currently no tool or technique is regarded as the ultimate one for accurate yet robust estimations. Novel methodologies that may consist of combinations of standard and modern techniques are proposed. In the near future, it is expected to incorporate these techniques within *SelQoS* and therefore adapt their analyses to the ones of the literature. Specifically, information theory techniques are into consideration because of their ability to analyze a variety of signal characteristics and behavior. These novel techniques would permit *SelQoS* to be an up-to-date and robust software tool for self-similarity and long-memory analysis. Moreover, a support for stable distribution is into consideration. With this, *SelQoS* pursues to be the tool of choice for network traffic analysis.

4.7   CONCLUSIONS

In this chapter, a novel tool for the analysis/estimation of self-similarity is presented. *SelQoS* estimates the self-similarity parameter using a two-step approach and moreover provides an intuitive computation of several time, frequency, and time-scale algorithms. From simple to more advanced statistics, *SelQoS* permits to perform a complete analysis framework for self-similar and LRD time series. A comparison study with the most common self-similarity analysis tools was provided. It was shown that *SelQoS* not only provides accurate estimations but also outperforms Selfis when studying the self-similarity degree in synthesized *fGn* and *f-ARIMA* time series. In addition, *SelQoS* performed better in long time series, and its response in time was shorter. Based on a complete comparison study, on an algorithmic implementation comparison and by using a wide range of synthesized *fGn* and *f-ARIMA*, we can conclude that *SelQoS* behaves better than *Selfis* within these conditions. *SelQoS* is thus the most complete freely accessible stand-alone software tool for self-similarity and long-memory analysis, which will be helpful for the online analysis of computer network time series and which can also be applied in the wide range of fields of science dealing with long-memory and self-similarity behavior. *SelQoS* can be obtained under request to email: dtorres@gdl.cinvestav.mx. *SelQoS* is registered as SELQOS: A Tool for Analysis for Time Series, Self-similarity, Long-range Dependence, and Frequency Behavior, Office for Intellectual Property, Register Number 03-2008-010810221100-01, Mex, January 2008.

REFERENCES

1. Beran, J., *Statistics for Long-Memory Processes*, Chapman & Hall, (New York, 1994).
2. Rangarajan, G. and Ding, M., *Processes with Long-Range Correlations*, Springer, (Berlin, Germany, 2003).
3. Park, K. and Willinger, W., *Self-Similar Network Traffic and Performance Evaluation*, Wiley-Interscience (New York, 2000).
4. Beran, J., Statistical methods for data with long-range dependence, *Statistical Science*, 7(4), (1992), 404–427.
5. Mishra, R. K., Sehgal, S., and Bhanumurthy, N. R., A search for long-range dependence and chaotic structure in Indian stock market, *Review of Financial Economics*, 20(2), (2011), 96–104.
6. Serletis, A. and Rosenberg, A., The Hurst exponent in energy futures prices, *Physica A*, 380, (2007), 325–332.
7. Montanari, A., Rozzo, R., and Taqqu, M. S., Fractional differenced ARIMA models applied to hydrological time series: Identification, estimation and simulation, *Water Resources Research*, 33, (1997), 1035–1044.

8. Leland, W., Willinger, W., Taqqu, M. S., and Wilson, D., On the self-similar nature of Ethernet traffic, *IEEE/ACM Transactions on Networking*, 2(1), (1994), 1–15.
9. Leland, W., Willinger, W., Taqqu, M. S., and Wilson, D., On the self-similar nature of Ethernet traffic, *ACM SIGCOMM Computer Communication Review*, 25(1), (1995), 202–213.
10. Beran, J., Sherman, R., Taqqu, M. S., and Willinger, W., Long-range dependence in variable-bit-rate video traffic, *IEEE Transactions on Communications*, 43(234), (1995), 1566–1579.
11. Paxson, V. and Floyd, S., Wide area traffic: The failure of Poisson modelling, *IEEE/ACM Transactions on Networking*, 3(3), (1995), 226–244.
12. Crovella, M. and Bestavroz, A., Self-similarity in World Wide Web traffic: Evidence and possible causes, *ACM SIGCOMM Computer Communication Review*, 5(6), (1997), 835–846.
13. Borella, M., Uludag, S., Brewster, G., and Sidhu, I., Self-similarity of Internet packet delay, in *IEEE International Conference on Communications*, Montreal, Quebec, Canada, (1997), pp. 8–12.
14. Toral-Cruz, H., Pathan, A. S. K., and Ramírez Pacheco, J. C., Accurate modeling of VoIP traffic QoS parameters in current and future networks with multifractal and Markov models, *Mathematical and Computer Modelling*, 57(11–12), (2013), 2832–2845.
15. Estrada Vargas, L., Torres Roman, D., and Toral-Cruz, H., A study of wavelet analysis and data extraction from second-order self-similar time series, *Mathematical Problems in Engineering*, 2013(2013), 1–14.
16. Karagiannis, T. and Faloutsos, M., SELFIS: A tool for self-similarity and long-range dependence analysis, in *First Workshop on Fractal and Self-similarity in Data Mining: Issues and Approaches*, Edmonton, Alberta, Canada, (2002).
17. Bregni, S. and Primerano, L., The Modified Allan variance as time-domain analysis tool for estimating the Hurst parameter of long-range dependent traffic, in *Proceedings of IEEE GLOBECOM 2004*, Dallas, TX, (2004).
18. Bregni, S. and Jmoda, L., Accurate estimation of the Hurst parameter of long-range dependent traffic using Modified Allan and Hadamard variances, *IEEE Transactions on Communications*, 55(11), (2007), 2224–2224.
19. Beran, J. and Ghosh, S., Slowly decaying correlations, testing normality, nuisance parameters, *Journal of the American Statistical Association*, 86(415), (1991), 785–791.
20. Hassani, H., Leonenko, N., and Patterson, K., The sample autocorrelation function and the detection of long-memory processes, *Physica A: Statistical Mechanics and Its Applications*, 391(24), (2012), 6367–6379.
21. Mandelbrot, B. and Wallis, J., Fractional Brownian motions, fractional noises and applications, *SIAM Review*, 10(4), (1968), 422–437.
22. Hosking, J., Fractional differencing, *Biometrika*, 69(1), (1981), 165–176.
23. Adler, R., Feldman, R., and Taqqu, M. S., *A Practical Guide to Heavy Tails: Statistical Techniques and Applications*, Birkhauser, (Boston, MA, 1998).
24. Stolojescu, C. and Isar, A., A comparison of some Hurst parameter estimators, in *Optimization of Electrical and Electronic Equipment (OPTIM)*, Brasov, Romania, (2012), pp. 1152–1157.
25. Li, M. and Zao, W., On band limitedness and lag-limitedness of fractional Gaussian noise, *Physica A: Statistical Mechanics and Its Applications*, 392(9), (2013), 1955–1961.
26. Bryce, R. M. and Sprague, K. B., Revisiting detrended fluctuation analysis, *Scientific Reports*, 2, (2012), 315.
27. Taqqu, M. S., Teverovsky, V., and Willinger, W., Estimators for long-range dependence: An empirical study, *Fractals*, 3(4), (1995), 785–798.
28. Taqqu, M. S. and Teverovsky, V., Robustness of Whittle-type estimators for time series with long-range dependence, *Stochastic Models*, 13(4), (1997), 723–757.
29. Rangayyan, R. M., Oloumi, F., Wu, Y., and Cai, S., Fractal analysis of knee-joint vibroarthrographic signals via power spectral analysis, *Biomedical Signal Processing and Control*, 8(1), (2013), 23–29.
30. Taqqu, M. S. and Teverovsky, V., Semi-parametric graphical estimation techniques for long-memory data, in *Athens Conference on Applied Probability and Time Series Analysis*, (1996), pp. 420–432, Springer, New York.
31. Flandrin, P., Wavelet analysis and synthesis of fractional Brownian motion, *IEEE Transactions on Information Theory*, 38(2), (1992), 910–917.
32. Veitch, D. and Abry, P., A wavelet based joint estimator of the parameters of long-range dependence, *IEEE Transactions on Information Theory*, 45(3), (1999), 878–897.
33. Abry, P. and Veitch, D., Wavelet analysis of long-range dependent traffic, *IEEE Transactions on Information Theory*, 44(3), (1998), 2–15.

34. Abry, P., Helgason, H., and Pipiras, V., Wavelet-based analysis of non-Gaussian long-range dependent processes and estimation of the Hurst parameter, *Lithuanian Mathematical Journal*, 51(3), (2011), 287–302.
35. Karagiannis, T., Faloutsos, M., and Molle, M., A user-friendly self-similarity analysis tool, *ACM SIGCOMM Computer Communication Review*, 33(3), (2003), 81–93.
36. Rolls, D., Limit theorems and estimation for structural and aggregate teletraffic models, PhD thesis, Queen's University, Kingston, Ontario, Canada, (2003).

5 Cluster-Oriented Emulation Tool for Performance Evaluation of Very Large-Scale Networking Scenarios

Pedro Vale Pinheiro and Fernando Boavida

CONTENTS

ABSTRACT

Networked systems are becoming larger and more complex. Nevertheless, simulation tools have not accompanied this trend in what concerns their ability to deal with large-scale communication systems. This clearly affects their applicability and usefulness and poses a problem to networking researchers, which are faced with the lack of tools to study and assess their proposals in large-scale scenarios and realistic conditions. In order to address this problem, this chapter presents the mobSim network emulation tool, which was developed for networking systems emulation of very large-scale scenarios with tens of thousands of networks, routers, and end systems. Although its underlying operating principles are general, the current version of mobSim was engineered for the study of network mobility, as this is an area with several open research issues. After briefly identifying the characteristics and main limitations of existing simulation tools, mobSim is presented in detail, namely, its functionality, architecture, features, implementation, and use. mobSim's use is illustrated in a sample network mobility scenario, in which three different network mobility solutions are compared in terms of their ability to deal with varying traffic load, route optimization, and level of nesting. In the final part of the chapter, future research directions are identified.

5.1　INTRODUCTION

Inability to cope with very large-scale scenarios often limits simulations to scenarios that hardly reproduce realistic and/or meaningful situations, given the size and complexity of current networks. In fact, it can be said that none of the currently available simulation tools (e.g., Network Simulator version 2 [ns-2] [1], Network Simulator version 3 [ns-3] [2,3], OMNeT++ [4], or OPNET [5]) are fit for large-scale network simulations involving tens of thousands of networks and hosts. Often, the documentation of some of these tools alerts to the fact that scenarios with high number of nodes combined with some system load may lead to erroneous results. Nevertheless, networks and communication systems are becoming dramatically larger and more complex when compared to networks of merely a decade ago. We are thus faced with several simultaneous problems that pose significant obstacles to the study of communication systems. On one side, small-to-medium-scale test beds or implementation trials may not lead to significant results. On the other side, theoretical models for large-scale systems may not be feasible or will only be so if several simplifying assumptions are made, thus compromising their validity and applicability. Last but not least, as mentioned before, simulation tools are also unable to cope with large or very large-scale scenarios. Dealing with this type of scenarios leads to problems that are typical of the *big data*, such as generation, capture, storage, transfer, and analysis of very large data sets.

Due to the need for studying very large-scale networks and communication systems, the authors decided to develop an emulation tool capable of dealing with them. As a case study, the tool was tailored for studying network mobility, as several challenges remain in this area of networking research. Naturally, having in mind the necessarily large processing requirements and the fidelity of the simulation results, the emulator had to satisfy several key requirements, namely,

1. Be constructed and optimized for cluster operation, taking advantage of parallel processing capabilities
2. Allow for *fine-grain* emulation detail, from protocol header fields to the correct implementation of mobility mechanisms and from individual nodes to routers, networks, and global scenarios

3. Be highly flexible in terms of scenarios definition, including fixed and mobile networks, fixed and mobile nodes, topology, and dynamic behavior
4. Enable the use of the exact same parameters and conditions for the different solutions under evaluation

The developed emulation tool—named mobSim—is presented in this chapter. Being specifically designed for running in clusters, mobSim's architecture reflects this design option, its main components being a master node and several slave nodes, whose number is only limited by the number of available cluster nodes. Each slave runs IP-level code that could run in real routers and end nodes and can execute the code of a few hundred virtual devices. In practice, everything happens as if real devices were communicating. Real packets are actually constructed and sent between devices, routing tables exist, and mobility mechanisms follow the respective request for comments (RFCs).

With the goal of adequately presenting the mobSim emulation tool in mind, this chapter is organized as follows. Section 5.2 identifies the main problems at hand as well as previous related work. Section 5.3 provides an overview of the main network simulation tools and their characteristics in what concerns network mobility simulation. Section 5.4 presents mobSim, addressing its functionality, architecture, and features in detail. Section 5.5 describes the use of the emulator based on simple examples. Section 5.6 explores the full potential of the tool in large-scale scenarios with moderate-to-high traffic loads, presenting several emulation examples and their results. Section 5.7 identifies future research directions, and Section 5.8 presents the conclusions.

5.2 NETWORK MOBILITY: AN OVERVIEW

Network mobility deals with the problem of moving entire networks, as opposed to moving individual hosts. Currently, an increasing number of users and their respective applications explore mobility. In addition to host mobility, network mobility scenarios will become quite common, especially with the emerging Internet of Things, where all sorts of equipment are mounted on moving entities and are connected to the Internet.

Several solutions have been proposed for dealing with network mobility. Nevertheless, this is still an area of intensive research that, quite naturally, resorts to simulation tools for the study of various mobility scenarios. As the presented cluster-oriented emulation tool was developed with the network mobility case study in mind, the following subsections briefly present the main network mobility proposals for contextualization purposes, from [6]. A comprehensive survey on mobility can be found in [7].

5.2.1 NEMO BASIC SUPPORT PROTOCOL

The simplest network mobility solution is the NEMO Basic Support protocol, specified in RFC 3963 [8]. Its main objective is to enable totally transparent network mobility without requiring any modification to mobile nodes and/or legacy (LG) nodes. For this reason, in the context of this chapter, we will also refer to NEMO as the LG paradigm for network mobility. NEMO is an acceptable, nonoptimized network mobility solution, with minimum impact on networks and nodes at the cost of reduced performance and efficiency.

The underlying idea of NEMO is to readily allow network mobility, that is, without requiring any changes neither to mobile network nodes (MNNs) nor to correspondent nodes (CNs). Whenever a packet destined to the mobile network (mobile network prefix, MNP) arrives at the mobile network's home network, a home agent (HA) encapsulates the packet and sends it to the care-of address (CoA) of the mobile router (MR). The encapsulation creates a tunnel, known as the mobile router home agent (MRHA) tunnel. On receiving the packet, the MR deencapsulates it and hands it to the MNN. Packets from the MNN to the CN follow the reverse path.

Although it is extremely simple and fully compatible with LG devices, the NEMO solution has several drawbacks, leading to triangular routing, bottleneck at the home network, and amplified suboptimality in nested mobility scenarios, as discussed in [9–11]. All of these problems derive from the lack of route optimization procedures in NEMO.

5.2.2 NETWORK-BASED NETWORK MOBILITY SOLUTIONS

With the objective of overcoming NEMO's drawbacks—of which the main one is lack of route optimization—several network mobility solutions have been developed, having in common the fact that they place additional functionality inside the network, such as in MRs, other routers, or mobility agents on the Internet, in order to achieve some form of route optimization. These network-based (NB) solutions share a common feature with the NEMO solution: they try to minimize the impact of mobility on end systems.

Some examples of NB network mobility solutions are optimized route cache (ORC) [12,13], path control header (PCH) [14,15], global HA to HA [16], and, last but not least, Mobile IPv6 route optimization for network mobility (MIRON) [17–19].

Although the aforementioned solutions differ in minor details, they essentially use a similar approach, described in the following. Typically, in the case of local fixed nodes (LFNs) and local mobile nodes (LMNs), traffic to/from these nodes is route optimized by a network element, be it the MR, or a transit router, or even a specific mobility agent, which, in this context, we will simply refer to as an MR.

In these approaches, MRs have to process and maintain information on every route optimization concerning the packet flows that traverse them and have to perform the return routability (RR) mechanism whenever the mobile network changes its point of attachment. Given that the number of nodes and flows can be high, this may represent considerable load to the MR.

5.2.3 OPTIMIZED MOBILITY FOR ENHANCED NETWORKING

A different approach—optimized mobility for enhanced networking (OMEN)—was proposed by the authors of this chapter in [20,21], in which the mobility tasks are performed not by routers or other infrastructural elements but by the end systems themselves. For this reason, in the context of this chapter, we will also refer to OMEN as client-based (CB) network mobility paradigm.

OMEN mobile nodes are aware of their mobility and they take up an active role on the execution of route optimization procedures. Such approach frees MRs from the burden of optimizing the routes of a potentially high number of packet flows, allowing them to efficiently perform simple routing tasks.

In order to allow this, whenever an MR acquires a CoA, it announces it to its inner network, using the standard Neighbor Discovery (ND) protocol (RFC 4861) [22], so that every mobile node can use that CoA as if it were its own. The announcement is made in an optional field of the ND packet.

After the MR is assigned an IP address (the *IP assignment* packet) by a router (FR) of the foreign network, it sends a binding update (BU) packet to its HA and waits for the binding acknowledgement (BA). When this message is received, the MR proceeds to announce its CoA to the MNNs on its inside network, through an ND packet.

From this moment on, mobile nodes have all the needed information to perform route optimization on their own, without the intervention of their MR, using the standard MIPv6 protocol.

During the RR procedure, communication between MNNs and corresponding nodes (CNs) is assured by two essential fields of the MIPv6 protocol header, namely, the home address (HoA) option and the Type 2 routing header (T2RH), defined in RFC6275 [23].

In order to execute the RR procedure, the MNN sends two packets to its CN: a host test init (HoTI) packet sent through the MRHA tunnel, that is, using RFC 3963, and a care-of test init (CoTI) packet sent directly to the CN. Both packets contain a token that identifies the MNN. When the CN gets these two packets, it answers each of them, sending back a home test (HoT) message through the MRHA tunnel and a care-of test (CoT) message directly to the MNN. These packets contain the reply to the token send in the HoTI and CoTI messages, respectively. Once the RR procedure is complete, the MNN can initiate a BU/BA procedure, thus completing the route optimization.

After route optimization, the MNN and CN can communicate directly using the CoA address as the address of the MNN, complemented by the T2RH or HoA fields in order to identify the next hop or the source address, respectively.

OMEN does not suffer from several problems that exist in NEMO and in NB network mobility solutions. In NB solutions, route optimization is performed for all routes, whereas in OMEN, route optimization can be performed on a per-flow and per-mobile-node basis. Moreover, in OMEN, the decision to optimize a given route is taken by the mobile nodes and is taken only for flows that justify it, namely, medium and long duration flows.

Another important aspect is the issue of the load put on MRs by NB solutions. For small networks, this load may be acceptable. Nevertheless, in generalized network and node mobility scenarios such as the ones foreseen for the future Internet, this additional load on MRs may significantly affect their performance. By placing the decision and the task of route optimization in end systems, MRs can be kept as light as possible, being limited to performing normal switching and routing functions.

5.3 EXISTING NETWORK MOBILITY SIMULATION TOOLS

As mentioned in the previous section, host and network mobility are areas under intense research, for which it is very frequent to resort to some kind of simulation tool. This is due to two main factors: on one side, it is foreseen that a substantial part of the future Internet will be mobile and, on the other side, it is not viable to set up lab or real environments for the study of generalized mobility proposals in large-scale scenarios.

Nevertheless, studying existing or new proposals for network mobility through simulation is highly influenced and constrained by the available simulation tools that, as a rule, do not allow for large-scale simulations. On the other hand, in small- and medium-scale scenarios, the differences between various proposals may become irrelevant. In such scenarios, a bad proposal will not perform very badly and a good proposal will not significantly stand out.

In this section, a brief characterization of existing network simulation tools is presented, so that their potential for studying network mobility scenarios can be assessed.

5.3.1 NETWORK SIMULATOR VERSION 2, NS-2

The Network Simulator, version 2, simply known as ns-2 [1], is a widely recognized and extensively used simulation tool. ns-2 is a discrete event simulator, which uses C++ and TCL. The simulator is highly portable, with extensive documentation and some native support for node mobility [24]. Nevertheless, it has little support for network mobility, which has led researchers to use the MobiWan module as a basis to develop NEMO Basic Support protocol functionality [25].

ns-2 has some interesting capabilities in what concerns large-scale scenarios [26]. Nevertheless, this is vertical scalability, that is, the size of the scenarios is limited by the system hardware, more specifically by its memory. On the other hand, ns-2 documentation alerts to the fact that a high number of nodes combined with some system load (generation of packets at high rate or to/from a large number of nodes) may lead to erroneous results.

5.3.2 NETWORK SIMULATOR VERSION 3, NS-3

ns-3 [2,3] is an evolution of ns-2. Developed in C++ with optional support for a Python interface, it is a promising simulation tool, with extensive documentation. Its objective is to allow highly precise and reliable simulations in academic environment and, at the same time, to solve some limitations of ns-2, with emphasis on memory management and debugging. In order to improve fidelity and achieve as high performance as possible, ns-3 allows the integration of the tool with real system elements (e.g., kernel modules, interfaces, programs). In addition, it is possible to connect several ns-3 instances running in different machines. Nevertheless, it has limited support for host and network mobility, mainly at layer 2. There is some work on IP mobility simulation with ns-3 [27] but not as far as network mobility is concerned.

5.3.3 OMNeT++

Another well-known simulator is OMNeT++ [4], consisting of a framework and a set of libraries developed in C++. The modular architecture of the simulator makes it extensible. It is well documented and there are many external contributions, which demonstrate the intense activity around this tool. In what concerns mobility, it natively provides basic support for mobile IP. Additionally, there is an externally developed module for mobility, known as extensible MIPv6 (xMIPv6) [28,29], which, nevertheless, does not support network mobility. Scalability concerns were at the basis of the development of a parallel processing project for this simulator in 2003 [30]. However, there is very little documentation on this and it is not clear that it is officially supported.

5.3.4 OPNET MODELER

The OPNET Modeler [5] is an object-oriented discrete event simulator with support for Mobile IPv6. Although support for RFC 3963 is not native, it was developed by some extension projects [31,32]. OPNET's scalability is determined by the used hardware, having the ability to explore multicore processors or multiprocessor machines in order to speed up the simulations run time. On the other hand, OPNET does not have horizontal scalability, that is, the possibility to simultaneously run on different machines, being limited by the hardware of the single platform on which it runs.

5.3.5 SUMMARY

Table 5.1 presents a brief comparison of the previously mentioned simulation tools in what concerns mobility.

Clearly, for all of the referred simulators, there is a strong concern with fidelity. Support for host mobility is quite common, but in what concerns network mobility, there is still much room for improvement. There is also some concern with scalability. Nevertheless, none of the mentioned simulators are fit for very large-scale network mobility simulations involving tens of thousands of networks and hosts.

TABLE 5.1
Comparison of Network Mobility Simulation Tools

	ns-2	ns-3	OMNeT++	OPNET
Native support for host mobility (MIPv6)	✓	✗	✓	✓
Native support for network mobility(NEMO)	✗	✗	✗	✓
Scalability	Vertical	Horizontal	Vertical	Vertical

This and the need to study very large-scale networks and communication systems including network mobility were, thus, the main reasons for developing the mobSim emulation tool. It should be highlighted that although mobSim was constructed with network mobility in mind, the underlying principles are applicable for networking in general.

The following section describes mobSim in detail, addressing its functionality, architecture, features, and implementation. This will be complemented in Section 5.5 with information on how to use the tool.

5.4 MOBSIM

This section is organized into four subsections. Section 5.4.1 presents mobSim's functionality. This is followed by a description of the emulator's architecture in Section 5.4.2. Section 5.4.3 describes mobSim's features, while Section 5.4.4 details the tool's implementation.

5.4.1 MOBSIM'S FUNCTIONALITY

The initial mobSim motivation and case study was the comparison of the three network mobility solutions presented in Section 5.2. As these solutions partly use standardized protocols, it was essential to follow the respective RFC specifications as closely as possible. In the case of mobSim, the tool's implementation goes down to the level of protocol header fields, making it very close to a protocol implementation. Table 5.2 presents the list of standard functionality implemented by

TABLE 5.2
mobSim Standard Functionality

RFC	Implemented Standard Functionality
IPv6 basic support (RFC 2460)	Hop limit
	Next header implementations
	ICMPv6 (RFC4443)
	ICMP echo, reply, unreachable, time exceeded
	IPv6 encapsulation (RFC2473)
	Mobility header (RFC3775)
	Type 2 routing header (routing type equals 2 or MIPv6 final hop HoA)
	Home address option (specific for T2RH)
MIPv6 (RFC 4775)	Binding update
	Binding acknowledgement (binding accept, reject)
	Return routability procedure
	Home test init
	Care-of test init
	Home test
	Care-of test
	Nonce utilization (RFC3775 Section 5.2.2)
	Binding refresh
Neighbor Discovery (RFC 4861)	Router advertisement
	Router solicitation
Scalability	Bidirectional tunnel (MRHA tunnel)
	Binding update
	Binding acknowledgment
	Home agent implementation
	Mobile router implementation

mobSim. This functionality was used to construct implementations of the mentioned three network mobility solutions, namely, the NEMO basic support specification (RFC 3963) [8], a mix of the most important features of the NB proposals in [12,18,19], and the OMEN solution [6].

5.4.2 MOBSIM'S ARCHITECTURE

Figure 5.1 presents the overall architecture of the mobSim emulation tool. Being specifically designed for running in clusters, mobSim architecture reflects this design option, its main components being a master node and several slave nodes, whose number is only limited by the number of available cluster nodes. In turn, the master node is composed of three modules, namely, the controller, the data analyzer, and the results database. Master and slaves communicate via a high-speed network.

The controller is the core of the emulator. It is responsible for setting up the emulation scenarios according to the user commands, constructing the code that will run in each slave, sending the code to each slave, initializing the emulations, getting the results from the slaves, storing these results in the database, and shutting down the slaves at the end of the emulation. Due to the high number and importance of tasks performed by the controller, it is a dedicated module that does not perform emulation itself. The slave nodes perform emulation tasks.

mobSim creates an overlay network, where packets are sent to or received from virtual hosts, which operates over a real network and real systems. This is depicted in Figure 5.2.

Figure 5.3 illustrates communication between virtual equipment, mapping it to real hosts and network. When a virtual node sends a packet to another virtual node, it follows the path labeled with the number 1. Links between cluster nodes (labeled with the number 2) are physical links. Links labeled with the number 3 are virtual links, between virtual equipment.

If routing determines that the communicating devices reside in the same slave, the *localhost* interface is used to send the packets. If they reside in different slaves, the network is used. The controller determines which devices run in which slaves during simulations setup. At the end of simulations, the slaves send all the simulation data to the controller, which stores it in the results database.

The data analyzer module processes the simulation results in order to extract relevant information, such as packet round-trip times (RTTs), handoff times, tunnel setup times, and RR times.

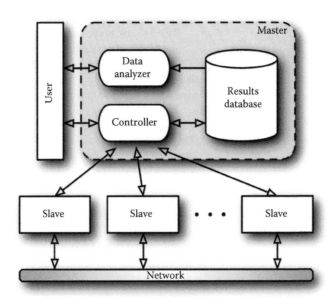

FIGURE 5.1 mobSim architecture.

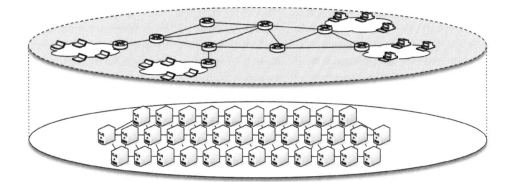

FIGURE 5.2 mobSim overlay.

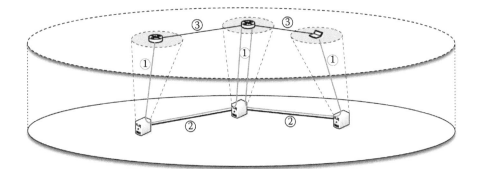

FIGURE 5.3 Example of communication between virtual equipment.

The analyzer also provides access to various logs that can be used for finer grain analysis (e.g., routes taken by packets, tunneling actions) and for debugging purposes. Access to the raw simulation data is also possible.

5.4.3 MOBSIM'S FEATURES

mobSim is extremely flexible and allows for any network topology; any number of networks, routers, and nodes; and any number of levels of nested mobility. Most parameters are configurable at node and/or network level. It is possible to define different values for different nodes and/or networks. The following text briefly presents mobSim's features in what concerns delays configuration, routing, traffic generation, dynamic behavior, supported types of equipment, and emulation control.

Configurable delays include the time for data link layer handoff, the time needed for acquiring a new IP address (through DHCP), the BU procedure time, the BA procedure time, line speed, HoTI time, CoTI time, HoT time, CoT time, encapsulation, and decapsulation times.

Routing tables for each virtual device are created automatically by the controller, based on the network topology defined by the user. The user can also define traffic characteristics, which nodes generate which traffic, as well as the dynamic behavior of the scenarios, that is, movement of nodes and networks. Parameters, topology, and dynamic behavior are kept exactly the same for the various mobility solutions under study.

mobSim supports two categories of devices: routing equipment and terminal equipment. The former includes top routers, routers, MRs, and HAs. The latter comprises LFNs, LMNs, visiting mobile nodes (VMNs), and CNs. CNs can be fixed or mobile (LFN, LMN, or VMM). By default, all devices use MIPv6, although MIPv4 can also be configured.

Because the emulator was developed for use in large clusters, in which resource usage must be optimized and CPU time is very expensive, special care was taken in order to minimize the time taken by emulations. For this reason, emulations setup is done off-line and initialization of the various slaves is done automatically by the master, which also controls emulation execution and shutdown. During emulations, there is no user intervention.

5.4.4 MOBSIM'S IMPLEMENTATION

mobSim was written in Perl [33] and its core has less than 2000 lines of code. The controller module comprises a set of scripts that are responsible for building the code for each scenario, send that code to the operational nodes (i.e., the slaves), and control the emulation process, namely, start it, stop it, and get the emulation results. The following subsections present the main mobSim scripts.

5.4.4.1 mobSim_master Script

This script is part of the controller module and creates the code that will be executed in each of the operational nodes. The script starts by reading the base scenario information, that is, all the needed IP addresses and all the default delay values. Subsequently, it obtains the list of all of the virtual devices that must be created. The central part of the script is presented in Figure 5.4.

Each *.conf file represents a virtual equipment. Unix imposes some limits to the rm and ls commands. These limits caused some problems during the initial phase of mobSim construction, as the program must be able to deal with tens of thousands of virtual devices, each one represented by its own text file. With such numbers of devices, the mentioned commands led to memory problems and the subsequent abnormal program termination. We also tried to solve the problem using the internal Perl command <$f/*.conf> for listing the files in a given directory, also without success. The solution was the use of the find command, which makes a very efficient system memory management and did not lead to any system error.

The mobSim_master script executes several consistency checks before creating a virtual device. For example, the script does not allow creating a device if its default gateway is not defined. The same happens if the IP address is not specified.

This script builds the code for each virtual device. After determining the needed functionality for a given virtual device, the script gets the respective, additional code by using the extracode option, as illustrated in Figure 5.5.

The obtained code is added to the device's code as illustrated in Figure 5.6.

The $code variable will then contain the correct code, whether or not the device has additional code. This code will then be sent to the respective operational cluster nodes for execution, as illustrated in Figure 5.7.

```
open FIND, "find $f -name \"*.conf\" | sort |";
while ($file=<FIND>) {
    [...]
}
close FIND;
```

FIGURE 5.4 Central part of the mobSim_master script.

```
if ($fconfs{"extracode"}) {
    open EXTRACODE, $f."/".$fconfs{"extracode"};
    @extracode=<EXTRACODE>;
    close EXTRACODE;
}
```

FIGURE 5.5 Getting additional code for virtual devices.

```
$code=""; $extracode=0;
open INSERTCODE, $conf{"bin _ dir"}."/".$fconfs{"prog"};
while (<INSERTCODE>) {
   if (/^#INSERT#CODE#HERE#/) {
        [...]
   }
   if ($extracode) {
        if (/^#END#INSERT#CODE#HERE#/) {
             [...]
        }
   }
   $code.=$ _ ;
}
close (INSERTCODE);
```

FIGURE 5.6 Insertion of additional code to virtual devices.

```
$prog{$aux1}=$aux."-".$aux1.".pl";
my $sock = create _ sock($auxipaddr{$fconfs{"ipaddr1"}},65000);
die "Couldn't contact destination port\n" unless ($sock);
print $sock "$prog{$aux1}\n";
print $sock $code;
close $sock;
```

FIGURE 5.7 Sending of device code to operational nodes.

5.4.4.2 create_ipaddresseslst Script

This is one of the most important scripts of the controller module, as it performs critical tasks such as route creation, mapping of real IP addresses to virtual IP addresses, and creation of configuration files for each of the virtual devices.

The script starts by obtaining a list of the available cluster nodes. This is a dynamic list, as it depends on the used cluster and its usage level at execution time. Using this information, the script establishes an association between real ID addresses and virtual IP addresses. This dynamic approach makes the emulator adaptable to various clusters and load levels, without the need for modifications or additional configuration.

In order to build the emulated scenario, the create_ipaddresseslst script reads the configuration file that contains information on the intended scenario (this file will be explained later in the text). With this information, this script now has all the needed information concerning all the networks and virtual equipment that should be instantiated.

Depending on the type of virtual equipment, the script adds several pieces of specific information, for example, one or two IP addresses (for hosts or routers) and default gateway. The code in Figure 5.8 illustrates this step.

As the emulator needs to know in advance every possible virtual IP address, this script additionally analyzes all node and router movements beforehand (i.e., during setup time) and reserves virtual IP addresses for these situations. For example, if an MR is scheduled to move to networks A, B, and C, then this script reserves topologically correct IP addresses in these networks, to be used as CoAs of the MR. The code that implements this address reservation is presented in Figure 5.9.

Now that the script has all the needed information, it can build the files for each specific virtual device. These files are of type *.conf. As already mentioned, the script also creates a file that establishes the association between real IP addresses and virtual IP addresses. The code that implements this is presented in Figure 5.10.

```
open AX, "$f/$s/$file";
while (<AX>) {
  [...]
  # Specific sections
  if ($type eq "toprouter") {
        [...]
  }
  if ($type =~ m"^(mobile)?router$") {
        [...]
  }
  if ($type eq "mobilerouter") {
        [...]
  }
  if ($type eq "node" or $type=~/^multiplenode:/) {
        [...]
  }
  if ($toprouter{$myinfo{"gateway"}}) {
  [...]
  }
}
close AX;
```

FIGURE 5.8   Virtual devices configuration.

```
open AX, "$scriptfile";
while (<AX>) {
  next unless (/^move;[^;]+;(.*)\n$/);
  $mrmove{$1}=1;
}
close AX;
```

FIGURE 5.9   Reserving virtual IP address for mobility purposes.

```
foreach $k (keys %final) {
  [...]
}
open IPADDRLST, ">$f/$paradigm/ipaddresses.lst";
print IPADDRLST "# All basic IPs needed\n";
print IPADDRLST "LOG:1 $myipaddris:1051\n";
foreach $k (keys %ip) {
  [...]
}
```

FIGURE 5.10   Association between real IP addresses and virtual IP addresses.

5.4.4.3   mobsim_2ports_udp Script

This is the basic script for all virtual devices that emulate equipment with routing capability. The script starts by opening two user datagram protocol (UDP) sockets with the ports specified by the create_ipaddresseslst script. This is done by the code presented in Figure 5.11.

As it is not possible to anticipate in which of the ports a packet will be first received, the script must listen to both ports in a nonblocking fashion, as illustrated by the code in Figure 5.12.

The function process _ packet($buf) is central to routing devices, as it is responsible for deciding what to do with each received packet: route the packet, generate an internet control message protocol (ICMP) error, or ignore the packet, depending on the various fields contained therein. This function is automatically modified by mobSim for the particular cases of MRs and HAs. In the case of normal, fixed routers, the function executes some basic packet processing, as illustrated in Figure 5.13.

```
$sock=create _ sock($conf{"port1"});
$check _ socks=new IO::Select($sock);
$save _ sock{$conf{'port1'}}=$sock;

$sock1=create _ sock($conf{"port2"});
$check _ socks->add($sock1);
$save _ sock{$conf{'port2'}}=$sock1;
```

FIGURE 5.11 Opening two UDP sockets.

```
while (@ready=$check _ socks->can _ read) {
    foreach $k (@ready) {
        [...]
    }
}
```

FIGURE 5.12 Getting packets from the UDP sockets.

```
sub process _ packet {
  my $packet=$ _ [0];

  if
($packet=~/h\[ipv6;[^;]+;[^;]+;[^;]+;[^;]+;[^;]+;[^;]+;([^;\]]+)/) {
        if ($conf{"ipaddr1"} eq $1 or $conf{"ipaddr2"} eq $1) {
            myprint("<TEXT>\n");
        } else {
            route _ packet($packet);
        }
    } else {
        myprint("Unknown packet: $packet. Droping it!\n");
        return;
    }
}
```

FIGURE 5.13 Basic packet processing.

The first if checks the packet header, making sure that it complies with the standard format. If this is not the case, the packet is discarded and the following information is added to the log file: "Unknown packet: $packet. Droping it!"

If the packet's destination address is the router address, the virtual device will not know what to do with it (routers are not valid packet destinations) and the function will generate the following error message, represented in Figure 5.13 as "<TEXT>": "This packet's destination was myself... I don't know what to do with it! $packet." The $packet variable contains the discarded packet and its value is also added to the log file.

Lastly, the route _ packet($packet) function is responsible for forwarding the packet to its next hop.

In the case there is the need to generate an ICMP unreachable error message, the function that does this is presented in Figure 5.14.

The function checks that the conditions for sending the error message are met, according to the corresponding RFC. ICMP unreachable messages are not sent in response to other ICMP unreachable messages. Moreover, the ICMP error message is only sent if the IP address of the respective packet sender can be determined. In Figure 5.14, the error messages returned by the function are represented as "<TEXT>" in order to simplify the figure, as these text messages are quite long.

It should also be noted that packets are transmitted/received using the UDP protocol, as this protocol adds a negligible overhead to the IP protocol. Nevertheless, logging information is sent via transmission control protocol (TCP) for reliability purposes. For this, a separate process is used, in order not to interfere with the UDP packet flows.

```
sub icmpunreach {
  my $packet=$_[0];

  if
($packet=~/h\[ipv6;[^;]+;[^;]+;[^;]+;[^;]+;[^;]+;[^;]+;[^;\]]+;1;/)
  {
        myprint("<TEXT>\n");
        return;
  } elsif ($packet =~
/h\[ipv6;[^;]+;[^;]+;[^;]+;[^;]+;[^;]+;([^;]+);[^;\]]+.*\]d\[([^\]]+)\]/) {
        my $msg=encode_base64($packet,"");
        sendpacket($1,
        "h[ipv6;0;0;0;0;64;$conf{'ipaddr1'};$1;1;0;0;0]d[$msg]");
        return;
  } else {
        myprint("<TEXT>\n");
  }
}
```

FIGURE 5.14   Generation of an ICMP unreachable error message.

5.4.4.4   mobsim_1port_udp Script

The objective of this script is to generate the code for virtual end systems and, thus, the script opens one UDP socket only, with the port specified by the create_ipaddresseslst script. Conceptually, the script is quite similar to the one described in the previous subsection.

End hosts have support for MIPv6 (RFC 6275), including the RR procedure. Similarly to the previous case, this script has the possibility of generating ICMP error messages.

5.5   MOBSIM USE

As with any other simulator or emulator tool, the use of the mobSim emulator requires the user to study, prepare, and define the various aspects of the simulations/emulations. Firstly, it is important that the user defines the scope and objectives of the emulation. Then, the tool must be configured so that it can run in an unsupervised fashion. Recall that due to the fact that mobSim was designed for taking advantage of the computational power of large clusters, no user interaction exists during execution time.

The following steps are required for preparation of mobSim emulations:

1. Definition of the overall emulation scenario
2. Configuration of virtual devices
3. Configuration of the emulation scenario
4. Emulations setup

These steps are described in the following subsections.

5.5.1   DEFINITION OF THE OVERALL EMULATION SCENARIO

This is the first step of any emulation although, in fact, it is not performed with the help of the mobSim tool. In this step, the user must define with as much detail as possible the objectives of the emulation, the intended target of the study, the general networking scenario, the various parameters to use in the elements that make up the defined scenario, the various tests that are going to be performed, and the type of data that should be collected for subsequent, off-line analysis.

As a way of example, suppose the objective of a given emulation is to assess various mobility solutions in an environment of nested mobility, that is, an environment in which mobile networks can connect to other mobile networks. Figure 5.15 presents one possible scenario for the study, with several networks, some fixed routers, and two MRs.

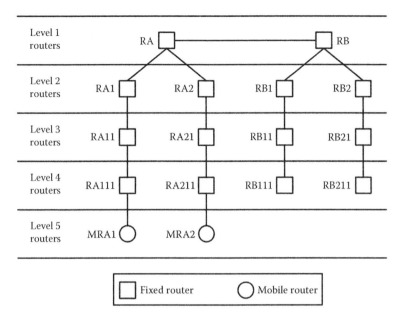

FIGURE 5.15   Sample scenario for the study of nested network mobility.

In this case, two top networks were created, A and B, represented in the figure by their routers, RA and RB. Each top-level router is connected to two level 2 routers. Each of these is connected to only one level 3 router and, again, each of these is connected to one level 4 router. RA111 and RA211 are, additionally, HAs for the MRA1 and MRA2 MRs, respectively. So, this simple small-scale scenario comprises fourteen fixed routers, two MRs, and several end systems not represented in the figure for simplification reasons.

This is, nevertheless, the initial scenario. During the emulation, MRs (and their respective mobile nodes) may move to other networks, and it may be the case that MRA1 moves inside the network of MRA2, while the latter is outside its home network, thus leading to a nested mobility scenario.

After defining the overall scenario, it is time to start using the mobSim tool, in order to configure the various pieces of equipment, to build the scenario, and to define the dynamic behavior to be emulated. These aspects are presented in the following subsections.

5.5.2   Configuration of Virtual Devices

Configuration of virtual devices is achieved through .conf files, which contain information on the various device-specific parameters. These allow the definition of the IP address, gateway address, log server, debug level, additional code file, among other relevant information. Normally, these files are automatically constructed by the controller module.

As a way of example, Figure 5.16 presents the configuration of the RA111 router, which also acts as HA to the MRA1 MR.

Configuration files can contain one or more of several directives. Specifically, the available directives are the following:

- prog—Type of code that will be used as base code for each virtual device; currently, the following options are mobsim_2ports_udp.pl, mobsim_1port_udp.pl, mobsim_2ports_tcp.pl, and mobsim_1port_tcp.pl.
- logaddr—Address of the log/debug server.
- debug—Debug level; the higher the value, the more debug information will be provided; during the normal mobSim execution, this should have the value of 0 (zero).

```
prog=mobsim_2ports_udp.pl
logaddr=LOG:1
debug=0
logfile=/tmp/vapirun/logs/log_ra111
extracode=HA.extracode

ipaddr1=ra111:ra111
ipaddr2=ra11:ra111
gateway=ra11:ra11
routing=mra1::ra111:mra1
```

FIGURE 5.16   Configuration of HA RA111.

- logfile—The specified file will be used for storing log information, which can also be used for debugging purposes.
- extracode—Specifies a file that contains code that should be added to the base code specified in the "prog" directive.
- ipaddr1—Virtual IP address of the network interface of end system devices or of interface 1 of router devices.
- ipaddr2—Virtual IP address of interface 2 of router devices.
- gateway—Address of the router that should be used for routing packet whose destination is not in the routing table of the virtual device.
- routing—Routing table of the virtual device; the format of each table entry is the following: <network address>::<next hop>. For instance, in the case of the scenario in Figure 5.15, mra1::ra111:mra1 means that packets destined to the mra1::/64 network should be sent to the ra111:mra1 device.
- delay_network—Packet routing/forwarding delay.
- delay_handoff—Handoff delay.
- delay_dhcp—DHCP address assignment delay.
- delay_bu—Delay associated with the BU procedure.
- delay_ba—Delay associated with the BA procedure.
- delay_rr—Delay associated with the RR procedure.
- delay_hoti—Delay associated with the HoT init procedure.
- delay_coti—Delay associated with the CoTI procedure.
- delay_hot—Delay associated with the HoT procedure.
- delay_cot—Delay associated with the CoT procedure.
- delay_mrha—Delay associated with the MRHA tunnel.
- delay_panareauth—Delay associated with the PANA re-authentication procedure.
- delay_omen_nd—Delay associated with the neighbor discovery procedure.

mobSim is constructed in such a way that new directives can be easily added. Moreover, the *extra-code* directive is a very powerful and flexible means of adding features to virtual devices in an entirely automated way.

The convention for representing virtual IP addresses is net _ prefix:host _ id, where net _ prefix stands for the network prefix and host _ id stands for the host ID inside the network. For instance, in Figure 5.16, we can see that the inner address of the RA111 router is ipaddr1 = ra111:ra111, meaning that the inner (lower, in the figure) interface is on the RA111 network and has a host ID of RA111. On the other hand, the address of the outer (upper) interface is ipaddr2 = ra11:ra111, meaning that this interface is on network RA11 and has a host ID of RA111. Moreover, RA111's gateway is gateway = ra11:ra11, that is, host RA11 or network RA11. Finally, in Figure 5.16, we can also see the line routing = mra1::ra111:mra1 that states that in order to reach the MRA1 network, router RA111 should send the packets to host MRA1 in the RA111 network.

5.5.3 CONFIGURATION OF THE EMULATION SCENARIO

The initial, static configuration of scenarios (i.e., global topology, number and types of equipment) is made through the use of .scn scripts. The configuration of a scenario comprises the definition of all the existing networks, their interconnection, and the various devices that they contain. Figure 5.17 provides the configuration script for the example presented in Figure 5.15.

The various virtual devices that make up the scenario are created based on the scenario configuration script. With the exception of lines starting by "#", which are comment lines ignored by mobSim, each line in the script identifies a virtual device and the respective configuration. The format of each line (i.e., of the device configuration directive) is the following:

<operation>|<value>[;<operation>|<value>[…]]

The <operation> field is used for identifying the type of equipment and also for identifying the various parameters associated with the particular device, such as IP address or gateway address. When used for identifying the type of equipment, it can have the following values:

- toprouter—Top or core router, normally used for interconnecting large networks; devices of this type know the routes for all other top networks.
- router—Virtual fixed device with routing capability.
- mobilerouter—Virtual mobile device with routing capability.
- node—Terminal equipment (portable device, sensor, etc.).
- multiplenode—Stands for multiple virtual devices with exactly the same characteristics; this is useful for creating nodes under the same MR, for example.
- LOGS—Central system for collecting emulation data, log data, and debugging data.

In the sample configuration presented in Figure 5.17, the RA top router is defined by the line toprouter|ra. The resulting configuration, automatically generated by the controller module, is presented in Figure 5.18.

```
# Network A
toprouter|ra
router|ra1;gateway|ra
router|ra11;gateway|ra1
router|ra111;gateway|ra11;extracode|HA.extracode
mobilerouter|mra1;gateway|ra111;extracode|mr.extracode
node|mnna1;gateway|mra1;prog|mobsim _ 1port _ udp.pl;extracode|mipv6.
extracode

router|ra2;gateway|ra
router|ra21;gateway|ra2
router|ra211;gateway|ra21;extracode|HA.extracode
mobilerouter|mra2;gateway|ra211;extracode|mr.extracode

# Network B
toprouter|rb
router|rb1;gateway|rb
router|rb11;gateway|rb1
router|rb111;gateway|rb11;extracode|HA.extracode

router|rb2;gateway|rb
router|rb21;gateway|rb2
router|rb211;gateway|rb21;extracode|HA.extracode
node|mnnb1;gateway|rb211;prog|mobsim _ 1port _ udp.pl;extracode|mipv6.
extracode
```

LOGS|00logs;prog=mobsim _ 1port _ tcp.pl;ipaddr1=LOG:1;logfile=/tmp/vapirun/logs/
log _ log;extracode=logs.extracode;debug=6;logaddr=

FIGURE 5.17 Configuration script for the example presented in Figure 5.15.

```
prog=mobsim _ 2ports _ udp.pl
logaddr=LOG:1
debug=0
logfile=/tmp/vapirun/logs/log _ ra

ipaddr1=ra:ra
ipaddr2=TOP:1

routing=rb::TOP:2;rb1::TOP:2;rb11::TOP:2;rb111::TOP:2;rb2::TOP:2;rb21::TOP:2;rb211::TOP
:2;ra1::ra:ra1;ra2::ra:ra2;ra1::ra:ra1;ra11::ra:ra1;ra111::ra:ra1;mra1::ra:ra1;ra2::ra:
ra2;ra21::ra:ra2;ra211::ra:ra2;mra2::ra:ra2
```

FIGURE 5.18　Configuration file for the RA top router.

The routing table is automatically generated and contains all the networks that exist in the scenario. The route for each network uses the router closest to RA as the next hop. For example, in order to route packets to the network of the RB111 router, RA knows that the next hop is the TOP:2 IP address, which is an address of the RB top router.

Router RA1 is created through the command router|ra1;gateway|ra. This command creates the RA1 virtual device and indicates that its gateway is the RA router. This command instructs the controller module to generate the RA1 router configuration file presented in Figure 5.19. The routing table is automatically constructed and contains routes for the networks of routers RA11, RA111, and MRA1.

In Figure 5.17, it is also possible to see the commands for the creation of two MNNs, MNNA1 and MNNB1. For this, the "node" command is used. For example, in the case of MNNA1, the script establishes that its gateway is the MRA1 router. Also, being an end host, the base code script should be mobsim _ 1port _ udp.pl. The additional option mipv6.extracode states the device has MIPv6 support and, consequently, is capable of performing route optimization. With these options, the generated configuration file is presented in Figure 5.20.

As can be seen in the example, different devices of the same type can have different characteristics, depending on the used options. This is also true for delay parameters, which can be defined per device. As already mentioned, delay parameters are the following: delay _ network, delay _ handoff, delay _ dhcp, delay _ bu, delay _ ba, delay _ rr, delay _ hoti,

```
prog=mobsim _ 2ports _ udp.pl
logaddr=LOG:1
debug=0
logfile=/tmp/vapirun/logs/log _ ra1

ipaddr1=ra1:ra1
ipaddr2=ra:ra1
gateway=ra:ra

routing=ra11::ra1:ra11;ra111::ra1:ra11;mra1::ra1:ra11
```

FIGURE 5.19　Configuration file for router RA1.

```
prog=mobsim _ 1port _ udp.pl
extracode=mipv6.extracode
logaddr=LOG:1
debug=0
logfile=/tmp/vapirun/logs/log _ mnna1

ipaddr1=mra1:mnna1
gateway=mra1:mra1
```

FIGURE 5.20　Configuration file for the MNNA1 mobile node.

```
LOG:1        2001:690:2180:120:a00:27ff:fe18:d7c4:1051
ra2:ra21     2001:690:2180:120:a00:27ff:fed4:ecfd:20100
ra1:ra1      2001:690:2180:120:a00:27ff:fed4:ecfd:20101
ra211:mra2   2001:690:2180:120:a00:27ff:fed4:ecfd:20102
rb21:rb211   2001:690:2180:120:a00:27ff:febe:9eb8:20100
```

FIGURE 5.21 Example of mapping between virtual IP addresses and real IP addresses.

delay _ coti, delay _ hot, delay _ cot, delay _ mrha, delay _ panareauth, and delay _ omen _ nd.

As mentioned before, real packets are sent between virtual devices, using the underlying network. Thus, it is necessary to map virtual IP addresses to real IP addresses. This mapping is done with the help of a table constructed by the controller module. Figure 5.21 presents such a table. For instance, in the figure, we can see that the virtual IP address ra1:ra1 is mapped to the real IP address 2001:690:2180:120:a00:27ff:fed4:ecfd, with port 20101.

5.5.4 EMULATIONS SETUP

As opposed to the scenario configuration scripts, which describe the static aspects of emulations/ simulations (e.g., topologies, networks, devices), emulations setup scripts define the intended dynamic behavior, including traffic generation characteristics, sources and destinations of the various packet flows, movement of nodes, and movement of MRs (and their respective networks).

Emulation setup is made through a set of scripts that define the various dynamic aspects of the overall scenario under study and, at the same time, guarantee that the exact same conditions (traffic types and intensity, flows, movement, etc.) apply to the different network mobility solutions being assessed.

Figure 5.22 presents the scenario that will be used in this subsection for illustrating the emulations setup. In this example, it is intended that MR MRA2 moves to a foreign network (the RB111 network) and, subsequently, the MRA1 router moves into MRA2's mobile network, thus creating a level-1 nested mobility scenario. The completion of these movements requires that first MRA2 establishes the necessary MRHA tunnel with its HA, followed by the establishment of the MRHA tunnel between MRA1 and its HA. After this, it is intended to check that the MNNA1 mobile node, at the MRA1 network, can communicate with the MNNB1 CN, at the RB211 network. For this, the "ping" program will be used.

Figure 5.23 presents the emulation setup script for the discussed sample scenario.

The first line names the scenario. In this case, the scenario is named "example." Subsequently, the network mobility solutions to be studied are identified. In this specific example, these are NEMO, NB, and OMEN. The following lines describe the dynamic behavior to be emulated.

The move;ra211:mra2;rb111:mra2 command indicates that the MRA2 router, with virtual IP address ra211:mra2, should move to the network of the RB111 router. At that point, it should use the rb111:mra2 virtual IP address. The emulation should only continue after this movement is completed. This is achieved through the use of the waitmove;40 command, which indicates that the emulation should resume only after the MRA2 router reports that the BU procedure is concluded.

Subsequently, the MRA1 router moves into MRA2's mobile network, where it will use the mra2:mra1 virtual IP address.

Lastly, "ping" is executed, as instructed by the ping;1;100000;mra1:mnna1;rb211:m nnb1 command. The command establishes that a ping packet is sent from the device with IP address mra1:mnna1 to the device with IP address rb211:mnnb1. The 100000 value indicates that there should be an interval of 100 ms between the first ping packet and the second. As only one packet is sent, this option is ignored. The emulation then waits for the reception of the ping response and then another ping cycle is executed.

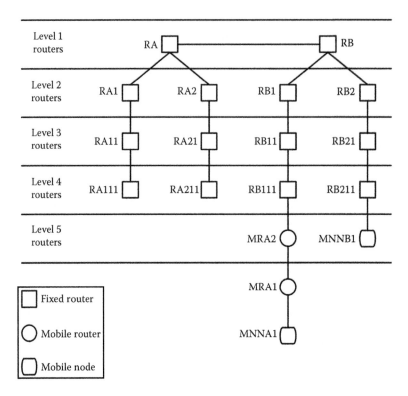

FIGURE 5.22 Sample scenario for illustrating emulations setup.

```
scenario;example
paradigm;nemo
paradigm;netbased
paradigm;omen

# MRA2 network move
move;ra211:mra2;rb111:mra2
waitmove;40

# MRA1 network move
move;ra111:mra1;mra2:mra1
waitmove;40

# Ping test
ping;1;100000;mra1:mnna1;rb211:mnnb1
waitping;40
ping;1;100000;mra1:mnna1;rb211:mnnb1
waitping;40
```

FIGURE 5.23 Emulation setup script for the discussed sample scenario.

Whenever an emulation run completes, it produces a set of reports based on the emulation setup script, such as the ones presented in Figure 5.24.

In a first phase, mobSim executes some configuration consistency checks and distributes the various virtual systems by the available cluster nodes. Subsequently, mobSim runs the emulations according to the emulation setup script. When the emulation finishes, it stops all virtual systems, including the controller. The emulation results are stored in the central log file.

Figure 5.25 shows the result of the example under consideration in this subsection, for the case of NEMO.

```
Cleaning:
2001:690:2180:120:a00:27ff:fed4:ecfd
2001:690:2180:120:a00:27ff:febe:9eb8
2001:690:2180:120:a00:27ff:fe18:d7c4
--------------------------------------------------------
Starting simulation for script ../confs/scripts/example.scr!
Scenario: example
Paradigms: nemo; netbased; omen;
Servers:
2001:690:2180:120:a00:27ff:fed4:ecfd;
2001:690:2180:120:a00:27ff:febe:9eb8;
2001:690:2180:120:a00:27ff:fe18:d7c4;
--------------------------------------------------------
Generating IPs: Done!
Nr of hosts per server:
   2001:690:2180:120:a00:27ff:fed4:ecfd:  8
   2001:690:2180:120:a00:27ff:febe:9eb8:  10
Total number of hosts: 18
Finding nodes: 3 Done!
Finding ports: 34 Done!
--------------------------------------------------------
starting controler daemons: DONE

========================================================================
= Starting nemo paradigm
========================================================================
! Starting simulator: 18 hosts with 34 ports: [0][34]Done!
! Actions:
nemo]     + move;ra211:mra2;rb111:mra2
nemo]     + waitmove;40:     Waiting for 1 moves: [0][1]Done!
nemo]     + move;ra111:mra1;mra2:mra1
nemo]     + waitmove;40:     Waiting for 2 moves: [1][2]Done!
nemo]     + ping;1;100000;mra1:mnna1;rb211:mnnb1
nemo]     + waitping;40:     Waiting for 1 pings: [1]Done!
nemo]     + ping;1;100000;mra1:mnna1;rb211:mnnb1
nemo]     + waitping;40:     Waiting for 2 pings: [2]Done!
- Shutting down the nemo paradigm. Waiting for 18 hosts: [0][18]Done!

========================================================================
= Starting netbased paradigm
========================================================================
! Starting simulator: 18 hosts with 34 ports: [0][34]Done!
! Actions:
netbased]     + move;ra211:mra2;rb111:mra2
netbased]     + waitmove;40:     Waiting for 1 moves: [0][1]Done!
netbased]     + move;ra111:mra1;mra2:mra1
netbased]     + waitmove;40:     Waiting for 2 moves: [1][1][2]Done!
netbased]     + ping;1;100000;mra1:mnna1;rb211:mnnb1
netbased]     + waitping;40:     Waiting for 1 pings: [1]Done!
netbased]     + ping;1;100000;mra1:mnna1;rb211:mnnb1
netbased]     + waitping;40:     Waiting for 2 pings: [2]Done!
- Shutting down the netbased paradigm. Waiting for 18 hosts: [0][18]Done!
```

FIGURE 5.24 Emulation reports. *(Continued)*

```
========================================================================
= Starting omen paradigm
========================================================================
! Starting simulator: 18 hosts with 34 ports: [0][34]Done!
! Actions:
omen]      + move;ra211:mra2;rb111:mra2
omen]      + waitmove;40:      Waiting for 1 moves: [0][1]Done!
omen]      + move;ra111:mra1;mra2:mra1
omen]      + waitmove;40:      Waiting for 2 moves: [1][2]Done!
omen]      + ping;1;100000;mra1:mnna1;rb211:mnnb1
omen]      + waitping;40:      Waiting for 1 pings: [1]Done!
omen]      + ping;1;100000;mra1:mnna1;rb211:mnnb1
omen]      + waitping;40:      Waiting for 2 pings: [2]Done!
- Shutting down the omen paradigm. Waiting for 18 hosts: [0][18]Done!
------------------------------------------------------------
Stopping the overall system (including myself!):
```

FIGURE 5.24 (*Continued*)   Emulation reports.

```
1344098198.22418 [mra2] changing IP from ra211:mra2
    to rb111:mra2 gw rb111:rb111
1344098199.15255 [mra2] MRHA active from ra211:ra211 to ra211:mra2
1344098200.25432 [mra1] changing IP from ra111:mra1
    to mra2:mra1 gw mra2:mra2
1344098201.27376 [mra1] MRHA active from ra111:ra111 to ra111:mra1
1344098205.63706 [mnna1] RES 1-PING _ mra1:mnna1 _ rb211:mnnb1: 0.20
1344098205.95269 [mnna1] RES 2-PING _ mra1:mnna1 _ rb211:mnnb1: 0.19
```

FIGURE 5.25   Results of the "ping" example for the NEMO case.

It is possible to see that MRA2 initiated its movement from its home network (RA211) to the foreign network (RB111) at time 1344098198.22418. At time 1344098199.15255, the MRHA tunnel was established and active. It is also possible to see that the two pings (1-PING and 2-PING) were executed and took 0.20 and 0.19 s, respectively.

Finally, mobSim enables us to check the path taken by the ping packets. This can be done by activating mobSim's debugging functionality, something that normally is only done for testing purposes, which allows the user to follow the path taken by the packets as they progress from source to destination.

It should be noted that in the scenario presented in Figure 5.22, there are two MRHA tunnels, one from MRA2 to its HA (i.e., RA211) and another one from MRA1 to its own HA (i.e., RA111). Moreover, the latter tunnel is inside the former tunnel, which means that the traffic should first go to RA211 and then to RA111.

Figure 5.26 presents the path taken by the ping packets from MNNA1 to MNNB1. In order to simplify the figure, for clarity reasons, packet headers and packet data were removed from the debugging output. Each line presents information provided by the device identified in the square brackets. Each device reports to where it has sent the packet. In the case of end nodes, such as MNNA1, they report the packet's final destination, whereas routing devices provide information on the next hop.

Thus, in Figure 5.26, it is possible to see that MNNA1 is sending a packet to MNNB1. This packet is received by the MRA1 router, as this is its default gateway. Subsequently, the packet is routed to MRA2's ingress interface (i.e., mra2:mra2). The packet must now be routed via the MRHA tunnel to MRA2's HA, that is, to RA211. This is actually the case, as we can see that the packet follows the following path: RB111, RB11, RB1, RB, RA, RA2, RA21, and RA211. Then, because the packet comes from the MRA1 router, it must go to MRA1's HA, that is, to the RA111 router, through the respective MRHA tunnel. So, the following path is used: RA21, RA2, RA, RA1, RA11, and RA111. Finally, when arriving at RA111, the packet is deencapsulated and routed normally to its destination (i.e., MNNB1) using the following path: RA11, RA1, RA, RB, RB2, RB21, and RB211.

```
[mnna1] Sending to rb211:mnnb1 (final destination)
[mra1] Sending to mra2:mra2
[mra2] Sending to rb111:rb111
[rb111] Sending to rb11:rb11
[rb11] Sending to rb1:rb1
[rb1] Sending to rb:rb
[rb] Sending to TOP:1
[ra] Sending to ra:ra2
[ra2] Sending to ra2:ra21
[ra21] Sending to ra21:ra211
[ra211] Sending to ra21:ra21
[ra21] Sending to ra2:ra2
[ra2] Sending to ra:ra
[ra] Sending to ra:ra1
[ra1] Sending to ra1:ra11
[ra11] Sending to ra11:ra111
[ra111] Sending to ra11:ra11
[ra11] Sending to ra1:ra1
[ra1] Sending to ra:ra
[ra] Sending to TOP:2
[rb] Sending to rb:rb2
[rb2] Sending to rb2:rb21
[rb21] Sending to rb21:rb211
[rb211] Sending to rb211:mnnb1
```

FIGURE 5.26 Path taken by packets from MNNA1 to MNNB1.

5.5.5 mobSim Sum Up

This section addressed several phases of mobSim's use for network mobility emulation, namely, definition of overall emulation scenarios, configuration of virtual devices, configuration of emulation scenarios, and emulations setup. Through simple examples, it was possible to see how mobSim allows for different levels of detail and high flexibility. Currently, the emulator is oriented toward the study of network mobility solutions, but its principles and functionality can easily be extended to any other area of networking research.

5.6 EMULATING LARGE-SCALE NETWORK MOBILITY SCENARIOS WITH MOBSIM

Despite the fact that IP mobility—either for individual hosts or for entire networks—has been the subject of intense research for roughly a decade and that mobility solutions have been in existence for some time, the truth is that the behavior of these solutions under realistic conditions is largely unknown. In this respect, and specifically for the more demanding case of network mobility, several questions come to mind. Are the existing network mobility paradigms adequate for generalized IP mobility? How will network mobility solutions behave in realistic scenarios and under considerable load? Will these solutions be scalable? This section addresses these questions by studying and comparing the three existing network mobility paradigms under consideration in this chapter—LG (i.e., NEMO), NB (i.e., ORC [12], MIRON [18,19]), and CB (i.e., OMEN). The study used the mobSim emulation tool in a scenario with real wireless links (established between some of the cluster nodes for this purpose), actual mobility, and varying load and identified the limitations of each paradigm, pointing to the very good potential of CB (more specifically, OMEN) network mobility solutions.

mobSim was already used and put to the test in two studies that compared the mentioned network mobility paradigms under somehow exaggerated and unrealistic situations: an extremely large-scale scenario with tens of thousands of mobile networks and up to sixteen levels of nested mobility [21] and a scenario with extremely high load [6]. Emulations of scenarios involving up to 22,800 routers,

11,250 networks, and more than 27,000 end nodes elicited a perfectly stable and coherent behavior of the tool. The work presented in the current section clearly distinguishes itself from the previous work, complementing and completing it by addressing realistic medium-to-large-scale scenarios with moderate-to-high traffic loads. It thus provides an insight not only on how mobSim can be used for emulating large scenarios but also on how each of the network mobility paradigms will behave if and when implemented and deployed on the Internet.

5.6.1 EMULATED SCENARIOS

Using mobSim, the base topology depicted in Figure 5.27 was created, with the objective of comparing the performance of the three network mobility paradigms under a variety of conditions. In this scenario, all routers were configured with the routes to every other router, so that communication between all networks was possible. Several types of networks exist in this scenario. The mobile test network is composed of an HA (the RA router), an MR (MRA), and a set of mobile nodes that, in total, can generate up to five hundred simultaneous packet flows (fl1 to fl500). Communication between RA and MRA is done through a wireless link. There are four other mobile networks (the pairs RB/MRB through RE/MRE) that do not contain any nodes and are only used for creating nested networks (i.e., mobile networks within other mobile networks).

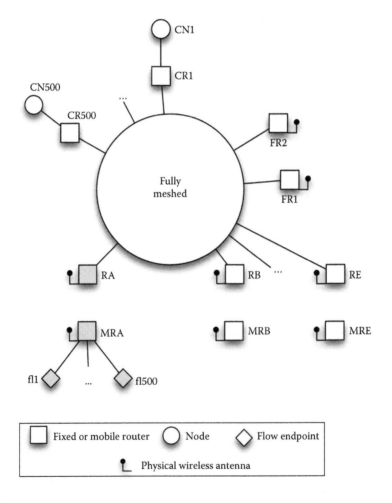

FIGURE 5.27 Emulation scenario base topology.

Network mobility scenarios without nesting are constructed by moving the MRA network to the FR1 network. Nested mobility scenarios are constructed by moving routers RB, RC, RD, and/or RE to the FR2 network (e.g., RB moves to FR2, then RC moves to RB, and then MRA moves to RC), in order to create up to 4 levels of nesting. Communication between any MR and its hosting network is always performed using a wireless link.

The corresponding networks are made up of a corresponding router (CR1 through CR500) and a CN (CN1 through CR500). In order to avoid bottlenecks in the fixed part of the network, a CN can only be the endpoint of a single flow.

It should be noted that we intentionally isolated the test scenarios from mobility-extraneous factors, such as background traffic, use of different traffic models, or the imposition of bottlenecks in the fixed part of the network. In this way, the presented results exclusively derive from the architectural options of each of the network mobility paradigms under study.

The tests that were carried out cover all the combinations of the following parameter values, for each of the three network mobility paradigms under study:

- Average packet inter-arrival times—50, 100, 250, and 500 ms.
- Number of packet flows—100, 200, 300, 400, and 500 flows.
- Ratio of packet flows for which route optimization (RO) was performed—2:10 (meaning 2 flows in 10 were optimized), 5:10, and 8:10.
- Nesting level—No nesting and the mobile network was in a foreign network, 1 level of nesting (i.e., the mobile network was inside a nonnested mobile network), 2 levels of nesting, 3 levels of nesting, and 4 levels of nesting.

Thus, 300 different tests were made, for each network mobility paradigm. As each test was performed 3 times for each of the 3 mobility paradigms, a total of 2700 tests were performed. The choice of the better result set was done using the Statgraphics tool (http://www.statgraphics.com/product/centurion.aspx), which provides Tukey's honestly significant difference (Tukey HSD) test, scatter diagrams, and analysis of variance (ANOVA).

In addition to the aforementioned, several delay parameters were used for all emulations. The chosen values are an approximation of actual values measured in a lab implementation and were the following: DHCP delay, 300 ms; RR delay, 200 ms; HoTI, CoTI, HoT, or CoT messages processing delay, 100 ms; MAC-layer handoff delay, 500 ms; MRHA tunnel setup delay, 10 ms; and BU or BA messages processing delay, 10 ms.

5.6.2 Emulation Results

We studied the responsiveness of the various paradigms to traffic load variation, route optimization, and level of nesting.

5.6.2.1 Traffic Load Analysis

Load variation was achieved in two different ways: by changing the packets' mean inter-arrival time and by changing the number of flows.

Figure 5.28 shows the average RTT of each of the mobility paradigms—LG, NB, and CB—as a function of the packets' mean inter-arrival time. Note that the average RTT for a given mean packet inter-arrival time is calculated using all emulations that were performed with this particular inter-arrival time, irrespectively of the values of the other parameters (number of packet flows, route optimization ratio, and level of nesting). The same is applicable to the remaining cases, for which, when we analyze a particular parameter value, we average the results of all emulations performed with that parameter value.

It should also be noted that for graphical intelligibility reasons, the columns pertaining to the LG paradigm were cut at 1 s in all figures. The actual average RTT values can be found in the numerical part of the figure.

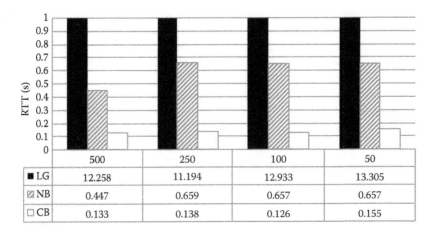

FIGURE 5.28   Average RTT as a function of the packet's mean inter-arrival time.

The first thing to note is that the worst performing paradigm is the LG paradigm, followed by the NB paradigm. The CB paradigm clearly outperforms the other two.

In the case of the LG paradigm, there is no route optimization and, thus, this leads to very poor RTT performance. Note that the extremely high RTT values result from the fact that several scenarios comprise various levels of nesting. This will be further addressed later, in the subsection on level of nesting analysis.

By introducing route optimization, the NB paradigm significantly reduces the average RTT. As the load increases (i.e., as packet inter-arrival times decrease), NB's performance is slightly affected at first, but it quickly stabilizes due to the route optimization factor.

Nevertheless, it should be noted that the CB's average RTT is substantially better than the one for NB in all situations. The reason for this is that in the NB case, all flows are optimized and almost the entire signaling load is put on MRs, which thus become critical bottleneck points.

On the other hand, in the case of the CB paradigm, MRs simply perform routing and, thus, are quite unaffected by these load levels.

A direct view of the influence of the load level expressed as a function of the number of existing packet flows can be seen in Figure 5.29. The differences between the various paradigms are, again, quite obvious. Significantly, in this figure, the performance degradation of the NB paradigm is clearly visible as the load increases, confirming the thesis that there is a bottleneck.

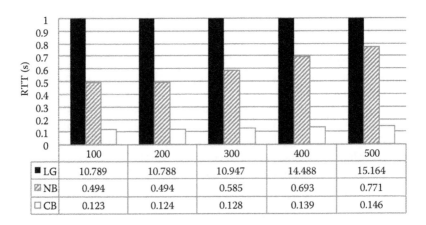

FIGURE 5.29   Average RTT as a function of the number of packet flows.

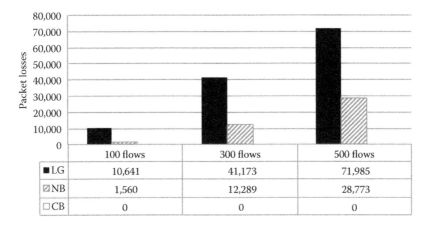

	100 flows	300 flows	500 flows
■ LG	10,641	41,173	71,985
▨ NB	1,560	12,289	28,773
□ CB	0	0	0

FIGURE 5.30 Total packet losses as a function of the number of packet flows.

As there is no bottleneck in the fixed part of the network (all flows go to different corresponding networks and use separate paths), it is apparent that it lies in the MR. Moreover, using exactly the same topology, the CB paradigm performs much better and does not show significant performance degradation, thus confirming that relieving MRs from mobility management and putting it in client systems pays off.

As a final check of the influence of load on the performance of each of the three paradigms, Figure 5.30 shows the total packet losses for the 100-flow, 300-flow, and 500-flow cases. A clear increase of the number of lost packets with the load increase is visible for the LG and NB paradigms. Interestingly, at these load levels, the CB paradigm does not yet exhibit any packet losses.

We did perform a specific CB emulation test in order to determine the number of flows at which losses would occur and we arrived at the conclusion that they start at 900 flows.

5.6.2.2 Route Optimization Analysis

For the route optimization analysis, three different ratios of route optimization were used, in the case of the CB paradigm: 2:10 (i.e., route optimization was performed for 2 out of 10 flows), 5:10; and 8:10. Note that for the LG and NB paradigms, the route optimization ratio should have no influence, as there is no route optimization in LG, and all flows are optimized in NB.

The obtained emulation results concerning the average RTT as a function of the route optimization ratio are presented in Figure 5.31.

The results confirm that for the LG and NB cases, there is no significant variation. Moreover, they also confirm that performing route optimization pays off, as the NB's RTT average values are significantly lower than the ones for LG.

In turn, the average RTT values for the CB network mobility paradigm are significantly lower than the NB ones. In fact, as the route optimization ratio increases, CB's average RTT decreases, as more flows see the respective routes optimized. It should be noted, however, that the decrease is not very sharp because, in fact, RTT values are already at minimal levels in the case of the CB paradigm.

5.6.2.3 Level of Nesting Analysis

A final analysis addressed the behavior of the various network mobility paradigms under different levels of nesting.

In the emulated scenario, the reference mobile network, that is, the one with the MRA MR (see Figure 5.27), could move to a fixed network or to mobile networks inside other mobile networks, as explained in Section 5.6.1. When MRA moves to a fixed network, it is said that there is 0 nesting. If it moves to a mobile network that is itself attached to a fixed network, then the level of nesting is 1. The maximum level of nesting considered in the study was 4.

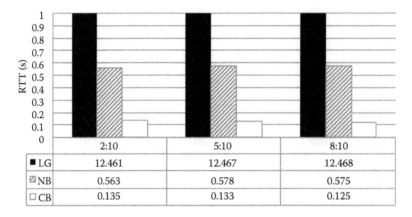

FIGURE 5.31   Average RTT as a function of the route optimization ratio.

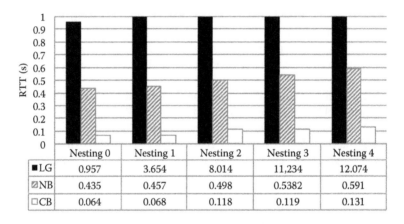

FIGURE 5.32   Average RTT as a function of the level of nesting.

The average RTT values as a function of the level of nesting are presented in Figure 5.32.

The first thing to notice is that there is a sharp increase in the average RTT with the level of nesting for the case of the LG paradigm. This, in fact, was expected and is a confirmation of the consistency of the emulation study, as in this paradigm, there is no route optimization. As the level of nesting increases, so increases the number of tunnels inside tunnels and, consequently, the number of networks that have to be traversed by the packets in order to reach the destination. The result is a dramatic increase in the RTT with the level of nesting. Another thing to notice is that in the case of the NB and CB paradigms, the level of nesting has a small, although quite perceptible, impact on the average RTT. Both paradigms use route optimization and, thus, after route optimization is done, the RTT remains at low values. There is, nevertheless, a slight increase, which is higher in the case of the NB paradigm.

The explanation for this higher increase in NB's average RTT lies in the amount of route optimization signaling and in the number of traversed wireless networks. In the NB case, all flows are route optimized, which requires considerable signaling traffic. Naturally, for higher levels of nesting, this traffic has to traverse a higher number of wireless networks, thus taking longer. In the case of the CB paradigm, this increase is smaller, as not all routes need to be optimized.

As a final remark, it is again clear that the CB network mobility paradigm leads to much better results than the NB paradigm, thus confirming that, regardless the perspective, the CB network mobility paradigm has significant advantages over the NB network mobility paradigm and, consequently, over the LG paradigm.

5.7 FUTURE RESEARCH DIRECTIONS

Three main areas for further research can be identified. These concern scalability, tool extension, and network mobility.

It is crucial that current and future networks and protocols are designed to scale better. For this, an important requirement is the ability to test and study environments and solutions before they are deployed. Naturally, this requires simulation or emulation tools, such as mobSim. Nevertheless, although mobSim was specifically engineered for very large-scale network mobility scenarios, it was not yet possible to test all of its potential due to constraints on the availability of the used cluster. In the near future, this potential will be explored in larger scenarios, with heavier loads, using a larger cluster, with the objective of assessing the behavior of the tool.

The extension of the tool for dealing with other networking areas and problems will also be the subject of upcoming work. mobSim can easily be extended in order to make it a general network emulation tool. Support for features such as dynamic routing (especially OSPF and BGP routing protocols), QoS routing solutions, multicast, and a variety of traffic generation options can greatly enhance the tool's applicability.

Finally, in the area of network mobility, for which the tool was originally developed, mobSim has very high potential in further understanding the characteristics of different network mobility paradigms and solutions. The development of extensions for the emulation of MultiPath TCP (MPTCP) [34], Locator/Identifier Separation Protocol (LISP) [35], Site Multihoming by IPv6 Intermediation (Shim6) [36], as well as mobility solutions identified in RFC 6301 [7] will greatly enhance the potential application area of mobSim.

5.8 CONCLUSION

In a world where networking is both ubiquitous and indispensable, the study of very large-scale networks and communication systems is extremely important. Nevertheless, existing network simulation tools cannot cope with scenarios of such dimension. Thus, it is indispensable to study new approaches and develop new tools that allow the research community to study those scenarios. mobSim is one such tool.

This chapter presented the mobSim network emulation tool. Although it was tailored for studying network mobility, the underlying approach and solutions can easily be extended to other areas. mobSim's main distinguishing feature is the ability to cope with very large-scale scenarios, of unprecedented size.

In view of the necessarily large processing requirements and the fidelity, mobSim had to satisfy several key requirements, namely, be constructed and be optimized for cluster operation, allow for *fine-grain* emulation detail, be highly flexible in terms of scenarios definition, and enable the use of the exact same parameters and conditions for the different solutions under evaluation.

In order to present mobSim's features and use, we concentrated on the emulation of network mobility. For this, we started by identifying the main problems at hand and the previous work. Subsequently, we overviewed the main network simulation tools and their characteristics in what concerns network mobility simulation. Then, mobSim's functionality, architecture, and features were presented in detail, as well as a description of mobSim's use based on simple examples. Finally, the tool was used to study network mobility in large-scale scenarios with moderate-to-high traffic loads, showing that it can easily deal with such scenarios, exhibiting a stable and coherent behavior.

mobSim opens up several interesting possibilities, either in terms of advanced training or research. The emulation and study of very large-scale systems is one of them. The extension of the tool for general purpose networking emulation is another one. Last but not least, further study of mobility solutions for the future Internet is also of interest.

REFERENCES

1. The Network Simulator—ns-2, http://www.isi.edu/nsnam/ns/, last accessed on January 10, 2014.
2. ns-3, http://www.nsnam.org/, last accessed on January 10, 2014.
3. ns-3, http://www.nsnam.org/wiki/index.php/Main_Page, last accessed on January 10, 2014.
4. OMNet++, http://omnetpp.org/, last accessed on January 10, 2014.
5. OPNET, http://www.opnet.com/, last accessed on January 10, 2014.
6. P. Pinheiro and F. Boavida, Some results on network mobility stress testing. In: *Proceedings of BCFIC 2012—Second Baltic Conference on Future Internet Communications*, Vilnius, Lithuania, April 25–27, 2012.
7. Z. Zhu, R. Wakikawa, and L. Zhang, A survey of mobility support in the internet, RFC 6301, Internet Engineering Task Force, July 2011.
8. V. Devarapalli et al., Network Mobility (NEMO) Basic Support Protocol, RFC3963, Internet Engineering Task Force, January 2005.
9. C.-W. Ng et al., Network mobility route optimization problem statement, draft-ietf-nemo-ro-problem-statement-03, Internet Engineering Task Force, September 2006.
10. C.-W. Ng et al., Network mobility route optimization solution space analysis, draft-ietf-nemo-ro-space-analysis-03, Internet Engineering Task Force, September 2006.
11. C. J. Bernardos et al., NEMO: Network mobility in IPv6, Upgrade vol. IV, issue no. 2, April 2005.
12. R. Wakikawa et al., Optimized route cache protocol (ORC), draft-wakikawa-nemo-orc-01, work in progress, Internet Engineering Task Force, November 2004.
13. R. Wakikawa et al., ORC: Optimized route cache management protocol for network mobility. In: *10th International Conference on Telecommunications*, Tahiti, Papeete, French Polynesia, vol. 2, pp. 1194–1200, February 2003.
14. J. Na et al., Route optimization scheme based on path control header, draft-na-nemo-path-control-header-00, work in progress, Internet Engineering Task Force, April 2004.
15. J. Na, Seoul National University, Supporting Route Optimization in Network MObility (NEMO), September 2004.
16. P. Thubert et al., Global HA to HA protocol, draft-thubert-nemo-global-haha-01.txt, work in progress, Internet Engineering Task Force, October 2005.
17. C. Bernardos et al., MIRON: MIPv6 route optimization for NEMO. In: *Fourth Workshop on Applications and Services in Wireless Network*, Boston, MA, August 2004.
18. C. Bernardos et al., Mobile IPv6 Route Optimisation for Network Mobility (MIRON), draft-bernardos-nemo-miron-00, work in progress, Internet Engineering Task Force, July 2005.
19. C. Bernardos, Route optimisation for mobile networks in IPv6 heterogeneous environments, PhD thesis, Universidad Carlos III de Madrid, Madrid, Spain, September 2006.
20. P. V. Pinheiro and F. Boavida, OMEN—A new paradigm for optimal network mobility. In: *Proceedings of WWIC 2008—Sixth International Conference on Wired/Wireless Internet Communications*, Tampere, Finland, May 28–30, 2008.
21. P. V. Pinheiro, S. Jain, and F. Boavida, A comparative study of network mobility paradigms. In: *Ninth International Conference on Wired/Wireless Internet Communications, WWIC 2011*, Barcelona, Spain, June 2011.
22. T. Narten et al., Neighbor Discovery for IP version 6 (IPv6), RFC 4861, Internet Engineering Task Force, September 2007.
23. C. Perkins, D. Johnson, and J. Arkko, Mobility Support in IPv6, RFC 6275, Internet Engineering Task Force, July 2011.
24. MobiWan: NS-2 extensions to study mobility in Wide-Area IPv6 Networks, http://www.inrialpes.fr/planete/mobiwan/, last accessed on January 10, 2014.
25. A. Z. M. Shahriar, Mohammed Atiquzzaman: Network Mobility in satellite networks, http://citeseerx.ist.psu.edu/viewdoc/download?doi = 10.1.1.125.149&rep = rep1&type = pdf, last accessed on January 10, 2014.
26. Tmix: Internet traffic generation, example, http://www.isi.edu/nsnam/ns/doc/node575.html, last accessed on January 10, 2014.
27. F. Mauchle, S. Frei, and A. Rinkel, Simulating mobile IPv6 with ns-3. In: *Proceedings of SIMUTools 2010—Third International ICST Conference on Simulation Tools and Techniques*, Torremolinos, Malaga, Spain, March 15–19, 2010. doi:10.4108/ICST.SIMUTOOLS2010.8682.
28. An Accurate and Extensible Mobile IPv6 (xMIPv6) Simulation Model for OMNeT++, http://www.kn.e-technik.tu-dortmund.de/content/view/232/lang,de/, last accessed on January 10, 2014.

29. F. Z. Yousaf, C. Bauer, and C. Wietfeld, An accurate and extensible mobile IPv6 (xMIPV6) simulation model for OMNeT++. In: *Proceedings of the First International Conference on Simulation Tools and Techniques for Communications, Networks and Systems & Workshops.* Marseille, France: ICST (Institute for Computer Sciences, Social Informatics and Telecommunications Engineering), pp. 1–8, 2008.

30. Parallel Simulation with OMNeT++, CTIE, Monash University, Melbourne, Victoria, Australia, http://ctieware.eng.monash.edu.au/twiki/bin/view/Simulation/ParallelSimulation, last accessed on January 10, 2014.

31. Academic Research and Teaching with OPNET Software, http://www.ctr.kcl.ac.uk/opnet/opnet.html, last accessed on January 10, 2014.

32. OPNET Contributed Models, https://enterprise1.opnet.com/tsts/4dcgi/Models_SearchSubmit?QueryModels_what = FindAll&QueryRecordsPerPage = 500, last accessed on January 10, 2014.

33. The Perl Programming Language, http://www.perl.org/, last accessed on January 10, 2014.

34. A. Ford, C. Raiciu, M. Handley, and O. Bonaventure, TCP extensions for multipath operation with multiple addresses, RFC 6824, Internet Engineering Task Force, January 2013.

35. D. Farinacci, V. Fuller, D. Meyer, and D. Lewis, The Locator/ID Separation Protocol (LISP), RFC 6830, Internet Engineering Task Force, January 2013.

36. E. Nordmark and M. Bagnulo, Shim6: Level 3 multihoming shim protocol for IPv6, RFC 5533, Internet Engineering Task Force, June 2009.

Section II

6 Fractal Traffic Modeling Applied in Network Simulation

Jeferson Wilian de Godoy Stênico and Lee Luan Ling

CONTENTS

ABSTRACT

For network engineering, simulation strategies have become indispensable for almost all network issues, especially in network design and performance analysis. Network traffic modeling and analysis are two essential topics for today's multimedia and high-speed networks. Better understanding of traffic characteristics and behavior allows maximum beneficial use of network capacities and therefore achieving better network performance. In this chapter, we focus our attention to fractal-theory-based traffic modeling and simulations procedures for this end. Initially, we introduce some main features of monofractal processes, emphasizing the characterization of the Hurst parameter. Then we extend our discussion to multifractal theory and processes, especially two functional parameters called the Hölder exponents and multifractal spectrum. In sequel, we compare the difference between monofractal and multifractal processes and how each one of these two process types influences network performance. For traffic characterization, we list a set of statistical parameters extracted from different real network traffic data and evaluate their relevance for traffic modeling and network simulations, in the sense of how to create a more realistic scenario for simulation of a network. Finally, we conclude this chapter by suggesting some possible future research themes.

6.1 INTRODUCTION

Different network services and applications, such as multimedia applications [1], video conference [2], and telemedicine [3], among others, demand different software and hardware requirements, distinctive interactions among users, and varying quality of service (QoS) guarantees of information transport [4–6]. In order to achieve accurate measurements and allocation of modern network resources in communication, it has become imperative to learn precisely involved network traffic types and to create and use good traffic models for network design and resource allocation. However, nowadays, accurate traffic modeling is by no means a trivial task and has become even more difficult due to constant changes in traffic characteristics.

The traffic characterization, referred here, is a study that aims to identify and analyze some major similarities and differences among different traffic flows in network simulations and analysis. More precisely, it consists of determination of the values of predefined measurable parameters extracted from a given network traffic flow by means of traffic engineering and/or statistical investigation. Through this characterization procedure, it is possible to learn the so-called traffic behavior that is essential not only for network device/equipment development but also network planning, sizing, simulation, and performance analysis.

6.1.1 Fractal Modeling of Network Traffic

The word *fractal*, having it origin in Latin as *fractus*, means irregular or broken and was originally used by Mandelbrot [7] to describe objects that were too irregular to fit the traditional Euclidean geometry. In a deterministic sense, a geometric shape is fractal or self-similar if it preserves the same appearance when observed on various spatial or temporal scales. A well-known classical example of deterministic fractals is the von Koch curve. Figure 6.1 illustrates the construction process of the von Koch curve invented by Sweden mathematician Helge von Koch* in 1906.

Until quite recently, fractals were a subject receiving little attention from the research community. This situation has changed rapidly when many natural objects and phenomena such as clouds surfaces, topography surface, and turbulence in fluids [7] can be better described and represented via fractals. Thereafter, the fractal geometry has become a very useful tool for the scientific community.

* School of Mathematics and Statistics, University of St Andrews, Scotland, JOC/EFR MAY 2000 Copyright Information, http://www-history.mcs.st-andrews.ac.uk/Biographies/Koch.html.

FIGURE 6.1 The von Koch curve is a fractal obtained from the limit of an infinite number of subdivisions. (Extracted from http://mathworld.wolfram.com/KochSnowflake.html; http://www-groups.dcs.st-and.ac.uk/~history/ Mathematicians/Koch.html; http://fractalfoundation.org/resources/fractivities/koch-curve/.)

The research and investigation on communication network traffic based on the fractal theory have been intensified by researchers after the publication of the work of Leland et al. [8]. This research work established an evident relationship between the fractal theory and modern network traffic processes. Experimentally, the authors detected some fractal properties such as self-similarity and long-range dependence (LRD) in traffic traces collected at the Ethernet network of Bellcore Morristown Research and Engineering Center.

Some later studies also claimed the presence of fractal characteristics in traffic flows generated by transmission of variable rate video [9], high-speed networks [10], World Wide Web [11], and others. Moreover, it was noted that the fractal properties, especially the LRD, strongly influence the network performance [12] but cannot be appropriately described by most traditional Markovian stochastic models. LRD can be revealed through observed self-similar characteristics in traffic processes that, by their turns, are the consequence of the heavy tail distribution of the size or duration of the sessions or connections after traffic aggregation [13,14].

Different mathematical models have been proposed to represent network traffic with self-similar properties. Among them, the model of fractional Brownian approach is widely accepted, mainly because of their relative simplicity. However, experimental investigations showed that while on time scales of the order of 100 ms and more the traffic behavior was well represented by the self-similar models, on smaller time scales, the self-similar models fail to effectively match real traffic characteristics. This finding has led the search for more comprehensive traffic models capable of providing a more faithful description of the network traffic.

The investigations into WAN TCP/IP traffic reported in [15,16] found that statistical properties of this traffic observed in small time scales could be adequately described by using the multifractal analysis. These properties are consequences of the action of prevailing transport used by the networks and the end-to-end existing congestion control mechanisms of the Internet, which determine the behaviors of the information flows between different layers in the hierarchy of TCP/IP protocols [15].

The multifractal analysis here considered mainly aims for the study of traffic characteristics different from those observed in monofractally behaved traffic flows more precisely, while studies of self-similarity investigating into the low-frequency contents of a signal, the multifractal analysis focus statistically on the high-frequency contents. One of the motivations for studying high-frequency components is that they are responsible for large and instantaneous impacts on networks that are much more critical and deserve further understanding than long-range correlations [17]. The surges of these high-frequency components are consequences of some network applications that demand fast responses, for example, real-time traffic control mechanisms.

Some more elaborated traffic models have been proposed; most of them are based on the multifractal analysis. Among the existing multifractal models, the most representative ones are the following:

- The multifractal wavelet model (MWM) is a model based on wavelet transform using the Haar wavelet [16].
- The adaptive wavelet-based multifractal model (AWMM) is one that simplifies the synthesizing process of the MWM model, by incorporating some additional parameter information that describes the corresponding scaling function and moment factor of the multifractal process [18].
- The multifractal model based on Newton binomial (MMNB) [10]. The construction of this model relies on a conservative multiplicative binomial cascade with its multipliers determined by a Newton binomial equation. The major feature of this model is its robust capability of capturing major multifractal properties represented by the corresponding scaling function and moment factor.
- The variable variance Gaussian multiplier model (VVGM) is a modeling version that takes into account the Gaussian nature of the multiplier distributions used to generate binomial multiplicative cascade [19].
- The variable scale parameter Cauchy multiplier (VSCM) is a multiplicative cascade model structurally similar to VVGM, however, having its multipliers holding Cauchy distribution, instead of Gaussian formats [20].
- The multifractional Brownian motion (mBm) [21] is a traffic model that generalizes the definition of fractional Brownian motion (fBm) by making the scaling exponent H no longer constant but time dependent [22].

Based on the fact that the fractal theory is capable of describing major statistical properties of network traffic processes, the main objectives of this chapter are to compile a robust reading material to understand modern network processes and to introduce some important traffic parameters that characterize faithfully network traffic flows and can be easily applied to network simulations and network design.

The chapter will be structured as follows. In Section 6.2, we focus on the presentation of the main features of a monofractal process in a simple way, emphasizing the characterization of the Hurst parameter. In Section 6.3, we introduce some definitions of multifractal processes and present some mathematical and statistical parameters and function that are traditionally used to characterize and analyze multifractal traffic processes. In Section 6.4, we discuss in detail traffic characterization by introducing many parameters, both mono- and multifractal ones. In addition, using some real network traffic trace, we explicitly present the numerical values of these traffic characterization parameters and discuss and compare their relevance in network simulation. In Section 6.5, we provide future research directions. Finally, in Section 6.6, we present our conclusions.

6.2 MONOFRACTAL PROCESSES

The concept of self-similarity was introduced by Kolmogorov [23] as early as in 1941 with intention to reveal a class of processes that are scalable in time or space without altering their statistical properties. However, this concept was effectively adopted and used only in the 1960s by Mandelbrot and van Ness [24] in their research work. Mandelbrot [7] connected the self-similarity concept with *fractals* that were some observed phenomena and statistical properties preserved on different temporal or spatial scales.

From the viewpoint of network analysis and design underlying on classical Markovian queuing theory, many studies have shown that self-similar traffic can degrade significantly network performance. This is due to the fact that traffic bursts tend to be longer and frequent and provoke slow decay of buffer occupation at network switching nodes [14,25,26]. Some immediate consequences and observed facts are the following:

- The rate of packet loss due to buffer overflow grows rapidly with the degree of self-similarity and this loss rate cannot be reduced significantly simply by increasing buffer sizes in switches.
- High incident burst rates of self-similar traffic reduce the efficiency of statistical multiplexing due to increase in effective use of link transmission capacity.
- Information transfer delays tend to increase significantly due to augmented buffer occupancy rates in function of the degree of self-similarity of traffic.

6.2.1 SELF-SIMILAR STOCHASTIC PROCESSES

Self-similar stochastic processes are particularly interesting in traffic modeling; they allow parsimonious process characterization by means of a single parameter, namely, the Hurst parameter or self-similar parameter. Figure 6.2a shows the achievement of the increments of a self-similar stochastic process under different scales of aggregation (10, 100, 1,000, and 10,000 ms).

Although the process curves are not exactly equal, they are visually similar. More precisely, the statistical properties are preserved under different time scale aggregation; in other words, the alternation of the periods of bursts and the smooth segment is preserved on all scales of observation. To contrast, such a feature is not present in a stochastic process that is not self-similar, for instance, a Poisson process shown in Figure 6.2b. In this case, the aggregation results in a white Gaussian noise and definitely is unable to preserve the same statistical properties of the process on different time scales.

Mathematically, a self-similar stochastic process can be defined as follows.

Definition 6.1

(Self-similar stochastic process) A random process $Y(t), t \in \mathbb{R}$, is said to be self-similar with self-similarity parameter denoted by $H \in (0,1)$ if, for all $a > 0, t \geq 0$, the process $Y(t)$ and $a^{-H}Y(at)$ are identically distributed, that is,

$$Y(t) \overset{d}{=} a^{-H}Y(at) \tag{6.1}$$

where $\overset{d}{=}$ represents equality in distribution.

The self-similarity parameter H is also known as the Hurst parameter, in honor of British hydrologist Harold Edwin Hurst [27]. According to Definition 6.1, the process $Y(t)$ must be nonstationary, that is, the mean and variance are time dependent. In the context of traffic modeling, the process $Y(t)$ represents the cumulative or total traffic to time instant t, therefore sometimes also called the accumulation process. A cumulative can be associated with another one, known as the increment process, which indicates the cumulative traffic volume at the $[t_0; t_0 + t]$ interval. Thus, the self-similarity concept can also be defined in terms of the corresponding increment process [24]. ∎

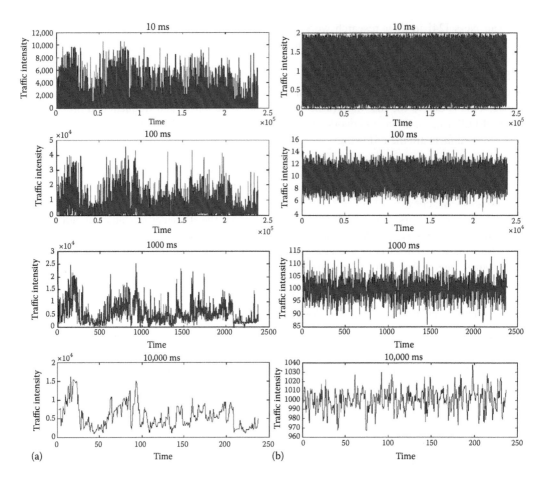

FIGURE 6.2 Comparison between increments of self-similar stochastic process and a common stochastic process from the point of view of preserving the statistical properties at different scales aggregate (10, 100, 1,000, and 10,000 ms). (a) Self-similar stochastic process. (b) Common stochastic process. (Graph generated in MATLAB®, http://www.mathworks.com/products/matlab/.)

Definition 6.2

The increment process $X(t), t \in \mathbb{R}$, associated with a given process $Y(t)$, is said self-similar with self-similarity parameter $H \in (0, 1)$ if, for all $a > 0$ and $t_0 \geq 0$, we have

$$X(t) = Y(t_0 + t) - Y(t_0) \overset{d}{=} a^H \left[Y(t_0 + at) - Y(t_0) \right]$$ (6.2)

∎

6.2.2 SECOND-ORDER SELF-SIMILAR PROCESS

Strict self-similarity may be a too restricted condition for process modeling in many practical applications. For network traffic modeling, for instance, the second-order statistical information of traffic processes has been more relevant and useful. Therefore, investigations on a link between self-similarity and the second-order statistical information have resulted in a definition of less restricted self-similar processes, called second-order self-similar processes [25].

Definition 6.3

A stochastic process $X(t), t \in \mathbb{Z}$, is said self-similar of second order with parameter $H \in [1/2, 1)$, if it is possible to express their autocorrelation as

$$R_X(t) = \frac{1}{2}\left[(t+1)^{2H} - 2t^{2H} + (t-1)^{2H}\right]$$ (6.3)

The main characteristic of a second-order self-similar process is that the aggregated versions of this process present nondegeneration behavior of their corresponding autocorrelation. Figure 6.3 illustrates the behavior of the autocorrelation function $R_X(t)$ of processes for different values of H.

Note: From Equation 6.3, notice that the H parameter expresses the decaying speed of the autocorrelation function of the process. The asymptotic behavior of $R_X(t)$ can be approximately written as [28]

$$R_X(t) \cong H(2H-1)t^{2H-2}, t \to \infty$$ (6.4)

A close look at Equation 6.4 and varying the H parameter, we get the following five cases:

1. For $1/2 < H < 1$, the autocorrelation function $R_X(t)$ of the process $X(t)$ behaves approximately as $ct^{-\beta}$, with $0 < \beta < 1$, constant $c > 0$, and $\beta = 2 - 2H$. If we integrate the discrete version of this autocorrelation function, we get

$$\sum_{t=-\infty}^{\infty} R_X(t) = \infty$$ (6.5)

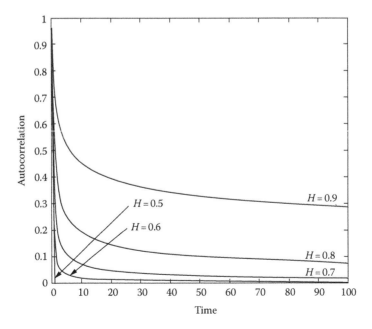

FIGURE 6.3 Behavior of the autocorrelation function of second-order self-similar process in function of the self-similarity parameter H. (Graph generated in MATLAB®.)

In other words, the autocorrelation function is nonsummable due to its slow decaying rate or its hyperbolically decay behavior. When a stochastic process presents a hyperbolical decaying behavior, it is said that the process holds the property of LRD.

2. For $H = 1/2$, $R_X(t) = 0$ for $t \neq 0$. In other words, the process has nonzero correlation only at the origin ($t = 0$ different) and is short-range dependence (SRD).

3. For $0 < H < 1/2$, the autocorrelation function is presented as follows:

$$\sum_{t=-\infty}^{\infty} R_X(t) = 0 \tag{6.6}$$

4. For $H = 1$, we have $R_X(t) = 0$, for all $t \geq 1$. This is a case that is not interesting for practical application.

5. For $H > 1$, we have prohibited the situation because, by definition, the process $X(t)$ should be stationary. ∎

6.2.3 Methods for Estimating the Hurst Parameter

To estimate the Hurst parameter is not straightforward due to the influence of several factors; they are the adopted estimation approaches, sample size, time scale, and data structure, among others [29]. In other words, the Hurst parameter cannot be directly calculated, but only estimated indirectly through a procedure involving related function tools. There are several estimation methods for this end, but they are basically classified into two categories, methods operating in the time domain and frequency domain.

The estimation methods that operate within the time domain consist of investigating the existence of power law relationship between a certain statistical property of the time series and aggregation block size (m). The time series is said to be holding the LRD property if the power law relation is linear on a log–log scale. More precisely, the slope of the corresponding straight line defines the estimated value of the Hurst parameter. Some well-known estimation methods of the Hurst parameter include the R/S method—(*rescale adjust range*) [30], absolute moments [31], variance of residual [32].

Through the frequency domain (sometimes called the wavelet domain), the estimation methods examine either energy or spectrum of the time series follows power law behavior. The most mentioned methods in the literature are periodogram [33], Whittle [34], and wavelet Abry–Veitch [35].

6.2.4 Evaluating the Estimators

The authors in [36] made a comparative study on some major Hurst parameter estimation approaches and tentatively established some relationship between the Hurst parameter and LRD. Some major conclusions are the following:

- There is no single estimator that can provide a definitive answer. For example, Whittle is the most accurate when LRD exists, but can be misled in showing LRD by periodic non-LRD data.
- LRD may exist, even if the estimators have different estimates value, provided that the estimates show the $1/2 < H < 1$.
- LRD is unlikely to exist, if there are several estimators that cannot produce sufficient estimations of the Hurst exponent (e.g., low confidence intervals).
- Periodicity can obscure the analysis of a signal giving partial evidence of LRD.
- A visual inspection of the signal can be very useful providing a qualitative analysis and revealing many of its features, like periodicity.

The authors in [29] recommended using the R/S or periodogram method to detect the presence of self-similar characteristics in the time series as well as to obtain approximate estimates of the Hurst parameter. However, we recommend the following operational procedure for robust analysis and accurate estimation of the Hurst parameter experimentally:

- Make a visual inspection of the given traffic trace; that is, plot the traffic trace graphically on various time scales and identify time scales that self-similar characteristics being highly revealed and an approximate period of occurrence.
- Choose appropriate time scales and apply the R/S and periodogram methods to confirm the existence of self-similarity and LRD property for the given traffic trace.
- Apply a more elaborated estimation method, for instance, the Whittle methods, to obtain more accurate estimates.

Next, we show the result of evaluating the proposed Hurst parameter analysis and estimation through a real wireless network traffic trace collected during the ACM SIGCOMM08 conference [37]. We use a java-based software tool called "SELFIS" to extract necessary information for our self-similarity and LRD analysis [38,39]. Notice that this computational tool ("SELFIS") is available for free downloading in [40].

Figure 6.4 illustrates the original and its seven different time-scaled aggregated versions of the "ACM SIGCOMM08" traffic trace. A quick visual inspection on these plots hints the presence of self-similarity as well with a period behavior in the analyzed traffic trace. Then, using "SELFIS" to process the 10 ms scale aggregated traffic samples, we obtained the estimates of the Hurst parameter for the following estimation methods: aggregate variance, R/S, periodogram, absolute moments, variance of residual, Abry–Veitch estimator, and Whittle estimator. Figure 6.5 shows a part of the traffic trace through the tool's user graphic interface, while Figure 6.6 shows the results of the application of the involved estimation methods both graphically and numerically.

Notice that the analysis of the results provided by the R/S and periodogram methods confirms the presence of self-similarity and LRD in the traffic trace. We adopt the Hurst parameter estimate ($H = 0.92$) obtained by the Whittle method as the most reliable one. Some additional comments about the performed experiment are the following:

- The estimated Hurst parameter value ($H = 0.92$) is very close to 1; this result indicates the presence of a high level of self-similarity in the traffic trace.
- The estimated value obtained by the *variance residuals* method violates the rule of stationary traffic; therefore, the method cannot be used as an appropriate alternative for the confirmation of the presence of self-similarity in the traffic.
- The estimation results provided by the *absolute moment* method show a low degree of self-similarity, which is far away from the true value ($H = 0.92$) by the Whittle method. This result suggests that the *absolute moment* method may not be reliable for Hurst parameter estimation.

The experimental investigation results demonstrate that self-similar traffic holds scale-invariant behavior with structural similarities across a range of time scales, which may be characterized by the Hurst parameter. However, this scaling invariant property or the Hurst parameter provides only global behavior of a traffic process. A more complete or detailed description of today's network traffic requires also the knowledge of its dynamic nature, dynamic behavior on both large and small time scales simultaneously. In this sense, more sophisticated analysis tools are needed in order to capture not only traffic's long-term dependence characteristics but also its instantaneous or dynamic properties in small time scales. The traffic analysis tool that we recommend in this chapter for this end is based on the multifractal theory or multifractal analysis that explores a new and more sophisticated class of stochastic processes called multifractal processes.

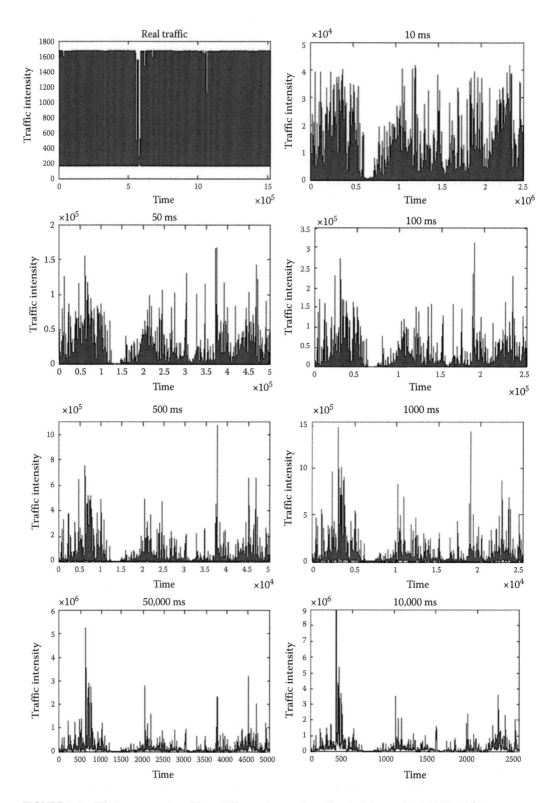

FIGURE 6.4   Wireless network traffic at different time scales. (Graph Generated in MATLAB®.)

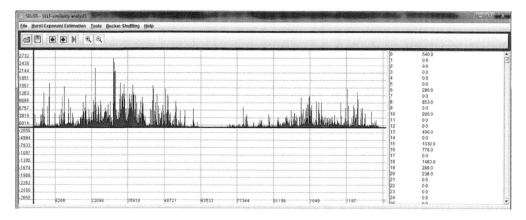

FIGURE 6.5 Tool interface "SELFIS."

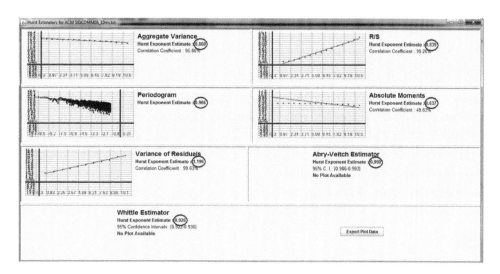

FIGURE 6.6 Estimated values for the Hurst parameter using different methods.

6.3 MULTIFRACTAL PROCESSES

In contrast to the self-similar or monofractal behavior in modern network traffic, recent studies have claimed that measured TCP/IP and WAN ATM traffic flows exhibit a more complex scaling behavior, which is consistent with multifractals [16]. Multifractal-based traffic modeling has more general and broad covering than the monofractal-based one (e.g., self-similar and long-range dependent), providing a more accurate and detailed description of network traffic series in different time scales [15]. In the context of network traffic, multifractal-based approaches can be viewed as a natural extension and refinement of self-similar fractal processes. Indeed, while the self-similarity is characterized by a single scale law of behavior that remains globally constant in time (characterized by the Hurst parameter), multifractals allow multiple laws of behavior that is time scale dependent and thus is highly flexible in describing the irregular phenomena of time processes. These irregular phenomena are typically caused by network-specific control mechanisms operating on smaller time scales and thus possibly impact on traffic dynamics in network connections.

6.3.1 MULTIFRACTAL ANALYSIS

The multifractal analysis allows the description of the process in local and global behavior in terms of measures of parameters, distributions, and functions through statistical or geometric methods. To locally characterize singular structures present in a signal, it is necessary to define and quantify its local regularity. The Lipschitz exponent, also known as Hölder exponent [41], provides uniform regularity measures on time intervals as well as also at isolated points. The local regularity measures are obtained via a geometric characterization or, more precisely, given by the statistical distribution of Hölder exponents, known as the multifractal spectrum [7].

From the point of view of the fractal theory, the degree of self-similarity (the Hurst parameter) now changes in time, that is, $H = H(t)$ is a function that depends on time in an unpredictable manner. In this sense, we have the following definition of multifractal processes.

Definition 6.4

A stochastic process $X(t)$ is called multifractal if it has stationary increments and satisfies

$$E\left(\left|X(t)\right|^q\right) = c(q)t^{\tau(q)+1} = c(q)t^{\tau_0(q)} \tag{6.7}$$

for some positive values $q \in Q$, $[0,1] \subseteq Q$, $\tau(q)$ (scaling function), and $c(q)$ (moment factor) are functions on domain Q and are independent of t. The function $\tau(q)$, also known as the partition function, is concave with $\tau(0) = -1$ [7].

In particular, if the process $\{X(t), t \geq 0\}$ is monofractal with a constant index H, then the two following expressions are valid: $X(t)t^H X(1)$ and $E\left(\left|X(t)\right|^q\right) = t^{Hq}E(|X(1)|^q)$. Thus, using Equation 6.7 for monofractal processes $\{X(t), t \geq 0\}$, we have

$$\tau(q) = H_q - 1 \tag{6.8}$$

and

$$c(q) = E\left(|X(1)|^q\right) \tag{6.9}$$

In other words, a monofractal process is completely characterized by the H index and its scaling function $\tau(q)$ becomes linear as a function of q.

A thorough description of the peculiar structure of multifractals demands that the degree of the local regularity of the multifractal process be precisely quantified. The pointwise Hölder exponent is one of these measure functions the most widely used for this purpose, defined as follows. ■

Definition 6.5

(Pointwise Hölder exponent) Let α be a real number and C be a constant, both strictly positive, and $x_0 \in \mathbb{R}$. The function f: $\mathbb{R} \to \mathbb{R}$ is $C^\alpha(x_0)$ if we can find a polynomial P_n of degree $n < \alpha$ such that

$$\left|f(x) - P_n(x - x_0)\right| < C \, | x - x_0 |^\alpha \tag{6.10}$$

The pointwise Hölder exponent h of the function f at x_0 is defined as

$$h = Sup\left\{\alpha >: f \in C^\alpha(x_0)\right\} \tag{6.11}$$

Notice that such a polynomial P_n can be found even the expansion Taylor series of f around x_0 does not exist. ∎

6.3.1.1 Holder Exponent and Network Traffic

According to Definition 6.5, the Hölder exponent of a time process at a particular time instant t_0 is related to the regularity level of the signal at that time point. In the context of network traffic, the exponent estimates the degree of local variation of traffic processes. More precisely, here, we show how traffic traces vary in terms of number of bytes or packages, on a range $[t_0; t_0 + \Delta t]$ of size Δt at instant t_0. The Hölder exponent α can be interpreted as a real number that locally controls the multiscale behavior of a process. The network traffic is said having local multiscale behavior with Hölder exponent $\alpha(t_0)$ at time t_0 if the traffic process rate behaves according to $(\Delta t)\alpha^{(t_0)}$ when $\Delta t \rightarrow 0$. In terms of traffic behavior, when $\alpha t_0()$ has a smaller value near to zero, the traffic burst intensity becomes larger around t_0. On the other hand, when $\alpha(t_0)$ approaches to one, low intensity of traffic rate variation is generally observed.

For illustration purposes, Figures 6.7a and b depicts the real traffic trace "ACM SIGCOMM08" on the time scale of 10 ms and their Pointwise Hölder exponents, respectively. Note that traffic burst intensity is inversely related to the value of the Hölder exponent, that is, the lower (higher) value the Hölder exponent holds, the more (less) intense will be the traffic bursts.

6.3.1.2 Multifractal Spectrum

In developing a robust multifractal model, a key criterion used is the multifractal spectrum; that is, the distribution of Hölder exponent values (α) of the multifractal process [41]. Hölder exponents is used to express the degree of local singularity (or the level of the smoothness) of the process [42]. In an alternative interpretation, the multifractal spectrum determines the fractal dimension of the set of time instants that compose the multifractal process [16].

The multifractal spectrum is a 1D curve, usually with a concave profile, representing the total amount of signal points for each singularity level (exponent value α). Different methods of analysis result in different spectra. The multifractal analysis based on geometric description leads to the Hausdorff spectrum [43], while a statistical-based description grants the large deviations spectrum [43]. The Hausdorff spectrum denoted by $f_h(\alpha)$ or simply f_h is defined as the dimension of a set of points having the same Hölder exponent value; the estimation of its value is difficult to be done in practice. The large deviations spectrum denoted by $f_g(\alpha)$, or simply f_g, provides a statistical description of how fast the probability of observing a Hölder exponent value different from the expected one tends to zero when the time resolution approaches to infinity. More specifically, f_g is associated with the rate function defined by the large deviations principle (LDP) [43]. In the literature, there is also a third type of spectrum known as the Legendre spectrum that provides a simpler way to obtain the large deviations spectrum under certain conditions [44].

In Figure 6.8, three multifractal spectral curves of the "ACM SIGCOMM08" traffic trace are plotted, representing the aforementioned Hausdorff, large deviations, and Legendre spectrum. Notice that the concave profile of these spectral curves confirms the presence of multifractal features of the analyzed traffic trace.

6.3.1.3 Multiscale Diagram

Local statistics behavior of a stochastic process can be obtained through wavelet-based analysis, instead of seeking the time-based local regularity exponents. This wavelet-based approach has become another important tool in multifractal analysis, summarizing important process information in the so-called wavelet partition function.

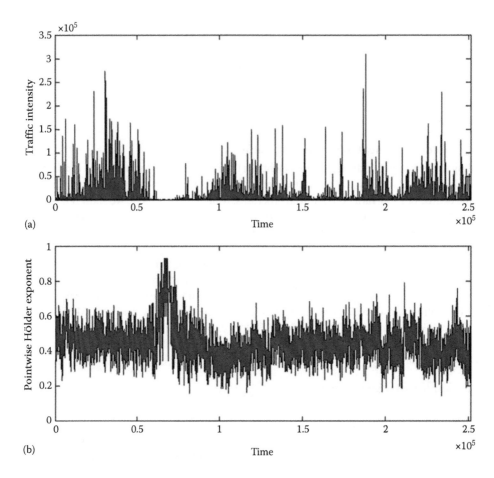

FIGURE 6.7   (a) "ACM SIGCOMM08" traffic samples; (b) the corresponding pointwise Hölder exponent.

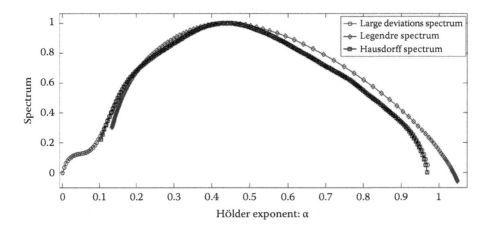

FIGURE 6.8   Multifractal spectrum curves of the ACM SIGCOMM08 traffic trace.

Definition 6.6

The wavelet partition function is defined as [45]

$$S_j(q) = E \mid W_{j,k} \mid^q \tag{6.11a}$$

where $W_{j,k}$ are Haar wavelet coefficients.

The asymptotic tendency of the partition function can be observed through the approximation operation given by Equation 6.12 letting $j \to \infty$:

$$\log_2 S_j(q) \sim q \cdot cte + j\alpha_q \tag{6.12}$$

Thus, the slope of $\log_2 S_j(q)$ as a function j provides an estimate for α_q. For monofractal processes (self-similar processes), one expects that α_q varies linearly with respect to q, while this relation becomes nonlinear for strict multifractal processes.

Multiscale diagrams (MDs), as the proper name suggests, provide some clues about the presence of scaling properties [46]. For traffic multifractal analysis, MD show is used to analyze a process presenting whether monofractal or multifractal behavior on interval $[j_1, j_2]$.

For this experimental investigation, two parameters were defined [47]: $\sigma_q = \alpha_q - q/2$ and $h_q = \sigma/q$. The MD expresses the exponent parameter σ_q as a function of q, while the linear MD (LMD) provides the normalized version of σ_q in terms of q. If a process is monofractal, its corresponding MD results in a straight line (σ_q being linear of q) and h_q a constant. For multifractals, the MD of σ_q is no more linear of q, neither constant h_q [42].

For illustration, both MD and LMD diagrams for the wireless network "ACM SIGCOMM08" traffic trace are plotted in Figure 6.9. The nonlinear MD curve and the varying LMD value suggest that the analyzed traffic trace be highly probably multifractal. ∎

6.3.1.4 Autocorrelation Function

The autocorrelation function can indicate the presence or absence of LRD in processes. Moreover, autocorrelation function reflects the second-order statistic characteristics of time series. Figure 6.10 plots the autocorrelation function curves of the ACM SIGCOMM08 traffic trace on various time scales.

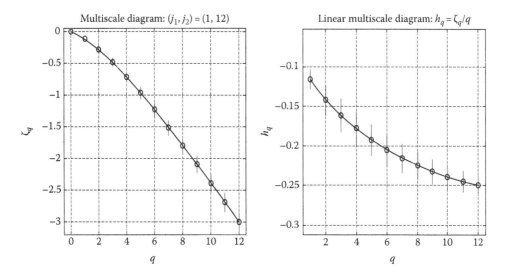

FIGURE 6.9 The MD and LMD for ACM SIGCOMM08 traffic trace.

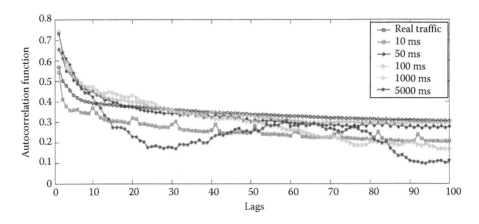

FIGURE 6.10   Autocorrelation function for ACM SIGCOMM08 traffic traces on different time scales.

Clearly, the autocorrelation curves plotted in Figure 6.10 demonstrate the presence of LRD in this traffic time series due to the fact that the autocorrelation curves do not vanish even for large lag and the curve decay rate of this statistical dependence is considerably slower than an exponential decaying rate.

6.4   TRAFFIC CHARACTERIZATION

6.4.1   SINGLE-SOURCE TRAFFIC

Single-source traffic may be interpreted as data flow that occurs exclusively between two network users. More precisely, it consists of data flowing from a unique source to a determined destination. This type of transmission strategy is also called *unicast* or point-to-point communication.

Single-source traffic can be found in all network types, either connected or without connection. In connectionless networks such as IP networks, single-source traffic is identified via its IP address and TCP port numbers [48]. In general, traffic generated by a single source presents a more complex and varying behavior than aggregated traffic originated from a multiplexing procedure. The behavior of single-source traffic depends highly on the type of involved network applications and services. It is more likely that a sequence of data bursts is found in single-source traffic, and as a consequence, the traffic flow holds fractal properties. Network applications and services that frequently generate single-source burst traffic are video and compressed audio streams, terminal emulation, video conferencing, games via networks, automation processes, and telemedicine, among others.

Another situation that may contribute to the high complexity of single-source traffic is the simultaneous use of multiple applications by a unique user, thus generating the so-called multimedia traffic (*multimedia* is the term used to describe the use of various types of media representation, registration, and use of information) [49].

6.4.2   AGGREGATE TRAFFIC

Traffic aggregation can be carried out in two distinct modes, vertical or superposition aggregation and horizontal aggregation. Vertical aggregation is a multiplexing procedure performed by network elements such as switches and routers, so that various traffic sources is able to simultaneously share the common transmission or processing resources or capacity [48]. Horizontal aggregation, by its turn, consists of a rescheduling of traffic over time [50].

6.4.2.1 Vertical Aggregation

Vertical traffic aggregation consists of temporal multiplexing of various traffic flows, taking place at network devices (switches and router) where temporal multiplexing of various traffic flows is performed. The very first vertical traffic aggregation occurs at ingress nodes of networks localized in a user–network interface, executing multiplexing of traffic flows from different sources or users.

Consider a *buffer* system, connected to N traffic sources, operating at a service rate of C bytes/s as illustrated in Figure 6.11. Let $A_i(t)$ denote the total traffic load from source I during time interval $[0, t]$, $i = 1, ..., N$. We assume that the N input traffic flows are independent and identically distributed with stationary increments and $E\left[A_i(t)^2\right] < \infty$ [50,51]. The total number of bytes transmitted through the output channel on time interval $[0, t]$ is denoted by $X_0(t)$ and b represents the buffer capacity in bytes.

The vertically aggregate traffic process denoted by $X(t)$ is defined as

$$X(t) = \sum_{i=1}^{N} A_i(t) \tag{6.13}$$

Two exclusive situations happen with this buffer system, they are

- $X(t) - X_0(t) > b$ representing a situation where there is loss of information at the buffer system on time interval $[0, t]$
- $X(t) - X_0(t) < b$ representing a situation where no loss of information occurs on time interval $[0, t]$

The total input traffic process defined by Equation 6.13 in accordance with the assumptions already assumed can be represented by the following standard procedure:

$$\frac{1}{\sqrt{N}}\left(X(t) - Nrt\right) \tag{6.14}$$

where the individual traffic streams average is given by $r = E[A_i(1)]$ that approaches a Gaussian process $Y(t)$ with mean zero and the same autocovariance function towards convergence of their finite dimensional distributions. If the process $A_i(t)$ has independent increments, the limit process $X(t)$ is a process with independent stationary increments. Additionally, if $X(t)$ has continuous paths, then $X(t)$ is a Brownian motion and the cumulative process of the approximate Gaussian traffic model can be expressed as [51]

$$X(t) = Nrt + \sqrt{N}Y(t) \tag{6.15}$$

Note that $Y(t)$ is characterized by Gaussian Brownian motion.

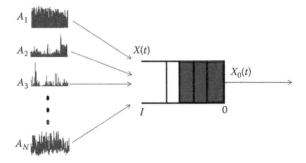

FIGURE 6.11 Statistical multiplexing of a buffer system.

6.4.2.2 Horizontal Aggregation

Let $A(t)$ and α be a cumulative traffic process and a constant, respectively. The process $A^{(\alpha)}(t)$ is called the horizontal aggregate version of $A(t)$ on time scale α and defined as

$$A^{(\alpha)}(t) = A(\alpha t) \tag{6.16}$$

In other words, $A^{(\alpha)}(t)$ is a rescaled version of $A(t)$ at the α scale rate. Let $X^{(\alpha)}(t)$ be a normalized process of $A^{(\alpha)}(t)$ defined as

$$X^{(\alpha)}(t) = \frac{A^{(\alpha)}(t) - r\alpha t}{\sqrt{Var\left(A^{(\alpha)}(1)\right)}} \tag{6.17}$$

where $r = E[A(1)]$ and if $A(t)$ has independent and stationary increments and finite variance, by the CLT, the normalized process $X^{(\alpha)}(t)$ (from $A^{(\alpha)}(t)$), represented by Equation 6.17, converges to a Brownian motion [50,51].

In 1951, Donsker defined the *invariance principle* [52], also known as functional limit theorem. Given a sequence of random variables, $Y_1, Y_2, \ldots$, identically distributed with mean μ and variance σ^2, let the following partial sums $R_0 = 0$ and $R_n = Y_1 + Y_2 + \cdots + Y_n$. The stochastic $S_n(t)$ defined on the interval [0, 1] is represented as

$$S_n(t) = \begin{cases} \dfrac{R_i - \mu i}{\sigma \sqrt{n}}, & \text{if } t = \dfrac{1}{n} \ (i = 0, \ldots, n) \\ \text{any linear interpolation} \end{cases} \tag{6.18}$$

If $\sigma < 1$, the process $S_n(t)$ converges in distribution to Brownian motion over the interval [0, 1] when $n \to \infty$. In other words, we can interpret the Brownian motion as a limit process in which time scale was increased. For example, if $A(t)$ is a Poisson process with parameter λ for large t, we approximate $A(t) \approx \lambda t + (\lambda)^{\frac{1}{2}} W(t)$, that is, $A(t)$ is proportional to λ [50,51].

Among the various generalizations of the *invariance principle* we mention, the one that on a limit scale time leads to a process $X(t)$ is not degenerative. This process is a process called self-similar where for a number H (defined in Section 6.2) such that for any $\alpha > 0$, the process $X(\alpha t)$ has the same finite dimensional distributions as $\alpha^H X(t)$ [50,51].

In the case of fBm process, $A(t)$ is Gaussian and self-similar with parameter H. The limit scale time $X(t)$ is also Gaussian with stationary increments and variance given by $Var(X(t)) = t^{2H}$ [9,29].

6.4.3 PARAMETERS OF TRAFFIC CHARACTERIZATION

In this section, we list some statistical parameters derived from traffic flows. These parameters are widely cited in the literature and used as traffic parameters in network simulations and traffic analysis.

Consider a random process X, $\{X = X(t), t \in \mathbb{R}, t > 0\}$, that represents the intensity of the traffic arrival rate to a buffer at time instant t.

There are two different representations widely used: a time series and a continue time random processes. Based on the first representation, a traffic process can be seen as a sequence of discrete entities (packets, cells, or frames), which can be characterized as a point process [53] in which discrete events occur sequentially at instants of arrivals $(T_1, T_2, \ldots, T_n, \ldots)$ observed since $T_0 = 0$. In this case, a traffic process can described as a nonnegative random sequence $\{Y_n, n \geq 0\}$ with $Y_n = T_n - T_{n-1}$ representing the interarrival time between two consecutive observed events [53]. Based on the second type of representation, a traffic process becomes a counting process $\{N(t), 0 \leq t < \infty\}$, which is a nonnegative-valued continuous-time stochastic process where $N(t) = \max\{n: T_n \leq t\}$ representing the total number of arrivals of entities in the time interval $[t_1, t_2]$.

Next, we present the main traffic parameters.

6.4.3.1 Average Rate

The average traffic rate, denoted by $\bar{X}$, represents the average throughput of traffic entities observed on a given time interval $[t_1, t_2]$ defined as [54]

$$\bar{X} = \frac{1}{(t_2 - t_1)} \int_{t_1}^{t_2} X(t) \, dt \tag{6.19}$$

6.4.3.2 Peak Rate

The peak rate, denoted by P, is defined as the maximum throughput of traffic entities in a given time interval $[t_1, t_2]$ [53,54].

6.4.3.3 Peak-to-Mean Ratio

Peak-to-mean ratio (PMR), here denoted by β, is defined as the ratio of peak rate to the average rate observed on time interval $[t_1, t_2]$ [53,54]:

$$\beta = \frac{P}{\bar{X}} \tag{6.20}$$

Notice that β takes into account only the first-order statistical properties of traffic.

6.4.3.4 Coefficient of Variability

Considering the random sequence $\{Y_n\}$ introduced before, the coefficient of variation (CV) is defined as the ratio between the standard deviation and the mean value of traffic interarrival times [53]:

$$CV = \frac{\sigma\{Y_n\}}{E\{Y_n\}} \tag{6.21}$$

6.4.3.5 Index of Dispersion of Counts

Index of dispersion of counts (IDC) is the ratio of the variance to the average of traffic counting process $N(t)$ observed on time interval $[t_1, t_2]$ [55]:

$$IDC_{t_1, t_2} = \frac{Var\{N_{t_1, t_2}\}}{E\{N_{t_1, t_2}\}} \tag{6.22}$$

IDC is a second-order statistical parameter because it takes into account a time interval delimited by two time instances.

6.4.3.6 Maximum Burst Size

The maximum burst size, denoted by L, is the maximum number of traffic entities transmitted continuously at the peak rate. As a result, BT is the maximum period during which the source can transmit in peak rate [54]. The relationship between the two parameters is given by the following equation:

$$BT = (L - 1) \left(\frac{1}{\bar{X}} - \frac{1}{P} \right) \tag{6.23}$$

where
   P is the peak rate
   $\bar{X}$ is the average rate, observed on time interval $[t_1, t_2]$

6.4.3.7 Time Burst Mean

The time burst mean, denoted by τ, is the average time in which the source transmits at peak rate [53–55].

6.4.3.8 Number of Traffic Sources

The number of sources in a given time interval $[t_1,t_2]$, denoted by N, is the number of sources that generate a given amount of data traffic on the network. If traffic is vertically aggregate, we have N traffic sources. For a single source, we have $N = 1$. The number of sources can be determined using their source address contained in the preamble/header of packets or frames (addresses: IP, MAC, IPX, Decnet, among others).

6.4.3.9 Self-Similarity Parameters

The intensity of self-similar phenomenon is given by the Hurst parameter value (H) as described in Section 6.2 and estimated using the "SELFIS" tool [40].

6.4.3.10 Hölder Exponent

As mentioned in Section 6.3, the multifractal analysis [10,18] deals with the description of the singular structure of signals. This local information is given by the Hölder exponent (α) at each time instance. The global structural information of a process is through the characterization of the geometric or statistical distribution of Hölder exponents, called multifractal spectrum. To estimate Hölder exponents in this work, we use the fractal traffic analysis tool called "FRACLAB" developed by INRIA [56].

6.4.4 REAL TRAFFIC ANALYSIS

This section is devoted to presenting experimental investigation carried out on real network traffic traces collected from different sources and analyzing their main characteristics mainly based on the traffic parameters listed in the previous section. These traffic traces were collected at Petrobras [57], Bellcore, Digital Equipment Corporation (DEC), and Lawrence Berkeley Laboratory (LBL) [58]. Each traffic trace contains two pieces of information, the starting time instance and size of each frame.

The traffic traces were collected in various locations of the Petrobras network between the year 2000 and 2003, via a data analyzer (Acterna™ Model DA350) operating with a 32 ms time resolution [57]. We adopted the following notation to distinguish these locations:

- "S" denoting aggregate traffic traces captured at application servers
- "I" denoting v traffic traces captured at Internet access routers
- "R" denoting aggregate traffic traces captured at enterprise IP traffic routers
- "CLI" denoting traffic traces captured at a point of an end-user network, possibly with traffic data containing several applications such as e-mail, Internet, and database, among others
- "FTP" denoting single-source data traffic files
- "MTX" denoting single-source audio/video traffic files.

Figure 6.12 depicts the network scenario where these real traffic trace files were collected.

The traffic traces collected at DEC and the LBL are WAN TCP/IP traffic traces captured at Internet access points. In this work, we investigated the following traffic traces, namely, dec-pkt-1, dec-pkt-2, dec-pkt-3, dec-pkt-3, lbl-pkt-4, and lbl-pkt-5. Notice that these traffic files have been widely studied [10,15,18,19] and are available at a dedicated website [58].

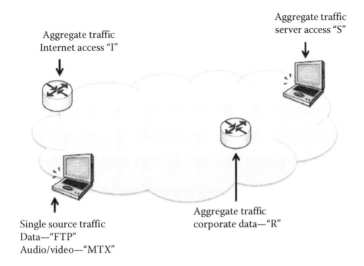

FIGURE 6.12 Network scenario of real traffic trace.

6.4.4.1 Traffic Parameters

Table 6.1 shows the numerical values of the set of traffic characterization parameters introduced in Section 6.4.3 for the following traffic traces: 3_7_I_1, 4_7_R_1, 10_7_S_1, 13_7_MTX_1, 13_7_FPT_1, 13_7_CLI_1, dec_pkt_1, dec_pkt_2, dec_pkt_3, lbl_pkt_3, lbl_pkt_4, and lbl_pkt_5.

Analyzing the numerical values of the traffic characterization parameters listed in Table 6.1, we have the following conclusions and comments for each characterization parameter:

- Average rate ($\bar{X}$) is mainly affected by the total amount of information generated by applications. Aggregate traffic tends to have a high average rate as a result of large number of involved sources. However, some traffic sources like *downloads* and *stream video* in general also hold high average rates. In other words, this parameter depends highly on the nature of involved applications or services.
- Peak rate (P) is much more affected by the nature of applications of average rate ($\bar{X}$). Abrupt changes in service typically results in high peak rates.

TABLE 6.1
Traffic Characterization Parameters

Traffic	$\bar{X}$	P	β	CV	IDC	L	τ
3_7_I_1	427,044.07	1,541,850.22	3.61	3.82	95.27	20.20	0.005
4_7_R_1	896,736.76	1,266,033.93	1.41	25.85	5.54	88.79	0.004
10_7_S_1	319,596.01	19,466,666.66	60.91	2.06	45.95	4.79	0.052
13_7_MTX_1	776,596.34	1,524,906.81	1.96	11.37	158.03	45.83	0.012
13_7_FTP_1	756,463.69	1,256,025.75	1.66	5.99	5.92	44.65	0.001
13_7_CLI_1	339,293.06	1,643,510.67	4.84	3.57	619.80	20.06	0.042
dec_pkt_1	930,776.66	1,301,159.21	1.39	709.16	688.06	92.16	0.003
dec_pkt_2	880,951.73	1,266,033.93	1.44	25.04	9.35	51.99	0.004
dec_pkt_3	792,163.82	1,457,133.18	1.84	12.72	87.53	46.75	0.011
lbl_pkt_4	875,586.46	1,524,906.81	1.74	21.69	17.76	51.67	0.003
lbl_tkp_5	841,236.04	1,370,896.74	1.63	25.22	250.36	49.65	0.004

- PMR (β) has important significance since it represents the magnitude relationship between the peak rate and average rate. We have observed that heavy traffic (jamming) usually holds a high average and a low PMR.
- Coefficient of variability (*CV*), defined as the ratio of the standard deviation to the average rate, is a statistical parameter expressing the degree of variation of a traffic process.
- Maximum burst size parameter (*L*) and time burst mean (τ) are two measures that enhance the burst behavior of traffic. As mentioned before, the first parameter is a function of peak rate, while the latter is closely related to average rate.
- *IDC* captures the variability of traffic on different time scales. Obtaining a reliable estimate of the *IDC* is not a trivial task; there is a need to adjust appropriately the measurement time interval. A good recommended rule is that the block size does not exceed 10% of the sample size. In general, the aggregate traffic holds high *IDC* for single-source traffic [53–55].

6.4.4.2 Self-Similarity and Multifractal Analysis

6.4.4.2.1 Hurst Parameter

To evaluate the self-similarity of traffic, we adopted the Whittle method implemented in the "SELFIS" tool to estimate the Hurst parameter [38–40]. Table 6.2 shows the estimates of the Hurst parameter for each investigated traffic flow. According to Table 6.2, in general, aggregated traffic traces present a Hurst parameter value (*H*) larger than 1/2, which implies self-similar characteristics or LRD. Self-similar characteristics also occurred in single-source traffic, but much weaker than the aggregate ones. However, for single-source traffic such as FTP and *streams* of audio and video, the H parameters had their numerical values that were less than 1/2.

6.4.4.2.2 Multifractal Spectrum

Multifractal spectrum analysis is a useful traffic analysis tool capable of quantifying and characterizing singular behavior of traffic data. In other words, it distinguishes between mono- and multifractal processes. For multifractal cases, multifractal spectrum analysis quantifies statistically the degree and complexity of singularities. In other words, when data holds a common singularity measure, the process is a monofractal process or a linear fractal. However, if the signal holds varying singularity values, it is regarded as a nonlinear or multifractal process.

Two classical examples of physics that illustrate multifractal behavior generation are the Navier–Stokes equations of fluid dynamics [59] and the binomial multiplicative processes arising

TABLE 6.2

Hurst Parameter Estimates (Whittle Method)

Traffic	H Value
3_7_L_1	0.552
4_7_R_1	0.613
10_7_S_1	0.586
13_7_MTX_1	0.497
13_7_FTP_1	0.067
3_9_CLI_1	0.903
dec_pkt_1	0.696
dec_pkt_2	0.774
lbl_pkt_3	0.782
lbl_pkt_4	0.922
lbl_tkp_5	0.782

from turbulence [23]. For network traffic analysis, in 1997, Taqqu, Willinger, and Teverovsky published their research paper entitled *Is the Network Traffic Self-Similar or Multifractal?* [60]. Since then, a lot of research work has been done to tentatively conclude or answer this question [10,15,18,19]. Taqqu et al. [60] claimed that in general, a self-similar model is suitable for LAN and WAN traffic data, except when the data aggregation level is not significant. Véhel and Sikdar [61] experimental showed the existence of multifractal characteristics in TCP traffic data. Feldmann et al. [15] demonstrated multifractal behavior in network traffic through the scale phenomenon.

The existence of multifractal characteristics of a process can be confirmed through the shape of the multifractal spectrum, as already presented in Section 6.3. More precisely, spectral plots have concave parabolic formats with the following features:

- $f(\alpha) \leq \alpha(t)$, for all $f(\alpha)$ and $\alpha(t)$ where $\alpha(t)$ is the Hölder exponent value at time t
- $f(\alpha) \leq D_0$, for all $\alpha(t)$

where $D_0 = f(\alpha_0)$ being the maximum value of $f(\alpha)$.

In this work of multifractal analysis, the spectra of real traffic data files were generated by using a computational tool called "FRACLAB," developed by INRIA [56]. The generation of spectral curves is based on the application of the weak fractal formalism and Legendre spectrum approach. Figures 6.13 through 6.20 show the estimated Legendre spectra generated by FRACLAB of the following traffic traces: 3_7_I_1, 4_7_R_1, 10_7_S_1, dec_pkt_1, dec_pkt_2, lbl_pkt_3, lbl_pkt_4, and lbl_pkt_5.

From the obtained shapes and nonlinear behavior of these Legendre spectral curves, clearly all these traffic traces hold multiscaling or multifractal properties.

For illustrative purposes, Table 6.3 lists four principal parameters of a multifractal spectrum (the minimum, maximum, and mean vales of the Holder exponent and spectral bandwidth) as well as the Hurst parameter values (estimated by R/S method [30], periodogram method [33], and Whittle method [34]) of the investigated traffic trace data.

Table 6.3 shows clearly that all investigated traffic traces hold monofractal characteristics with distinct levels of self-similarity according to their corresponding H values. These traffic traces also exhibit multifractal behavior according to their multifractal spectral shapes.

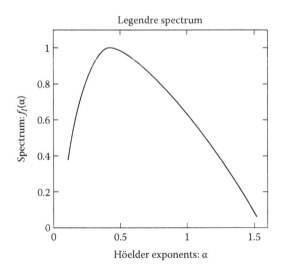

FIGURE 6.13 Legendre spectrum of 3_7_I_1.

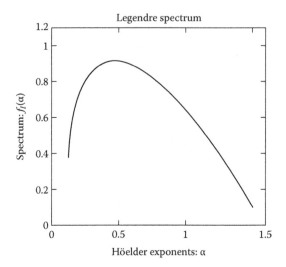

FIGURE 6.14   Legendre spectrum of 4_7_R_1.

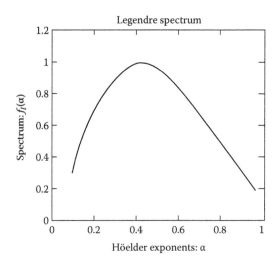

FIGURE 6.15   Legendre spectrum of 10_7_S_1.

In terms of multifractal spectrum analysis [7], it is interesting to note that the triple $(\alpha_{min}, \alpha_{max}$, and $\omega)$, which denote the minimum and maximum exponent values and spectral bandwidth, provides significant information to describe the multifractal spectrum of given traffic trace data.

Although, real traffic data seems to fit well to multifractal modeling model in many scales of resolution, according to Mannersalo and Norros [62], one should be very careful when applying such a model. Véhel and Sikdar [61] suggested that the detected multifractal behavior and the observed self-similar characteristics of WAN traffic can coexist simultaneously, that is, at microscopic levels, the process has multiplicative nature, while at macroscopic level, the process becomes additive.

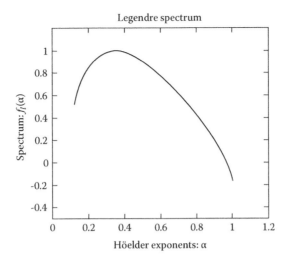

FIGURE 6.16 Legendre spectrum of dec_pkt_1.

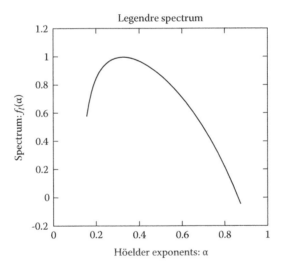

FIGURE 6.17 Legendre spectrum of dec_pkt_2.

Feldmann et al. [15] present some evidence of a complicated mixture of additive and multiplicative components within the TCP sessions. Additionally, she noticed that there is scaling transition zone between the multifractal and self-similar behaviors typically having its value on the order of packet traveling time in the network (round-trip delay).

According to the performed experimental fractal analysis, we conclude that the accurate characterization to determine whether traffic is monofractal (Hurst parameter) or multifractal (Hölder exponents) is not a trivial task. Traffic classification and characterization in fact depend on involved network applications and services as well as impacts caused in terms of network performance [10,15,16,18,19].

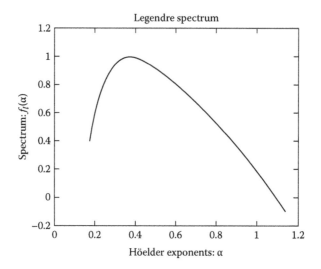

FIGURE 6.18 Legendre spectrum of lbl_pkt_3.

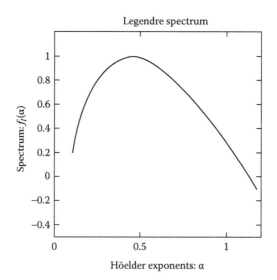

FIGURE 6.19 Legendre spectrum of lbl_pkt_4.

6.5 FUTURE RESEARCH DIRECTION

In this section, we list some future research themes:

- Seeking for more elaborated traffic models. Modern network traffic is very complex, possibly holding distinct properties such as LRD, self-similarity, and multifractal nature. A highly elaborated traffic model should be capable of generating simulated traffic data capable of exhibiting these characteristic with fidelity.
- Traffic shaping and control. Investigation on traffic shaping and control that permits to possibly alter the original Hurst parameter and Hölder exponent values will be an important research issue especially for fractal nature traffic.

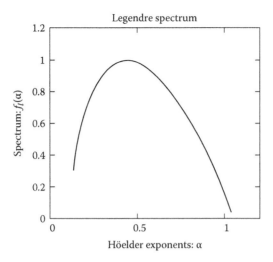

FIGURE 6.20 Legendre spectrum of lbl_pkt_5.

TABLE 6.3
Fractal Characterization (Hurst Parameter and Hölder Exponent)

Traffic	Hurst Parameter			Hölder Exponent			
	R/S	Periodogram	Whittle	α_{min}	α_{max}	α_{mean}	ω
3_7_L_1	0.561	0.639	0.552	0.1095	1.5287	0.2810	1.4192
4_7_R_1	0.726	1.092	0.613	0.1299	1.4064	0.3028	1.2765
10_7_S_1	0.669	0.868	0.586	0.0961	0.9737	0.2305	0.8776
dec_pkt_1	0.791	0.840	0.696	0.1251	1.0041	0.5055	0.8790
dec_pkt_2	0.781	0.750	0.774	0.1574	0.8751	0.3653	0.7177
lbl_pkt_3	0.639	0.797	0.782	0.1728	1.1380	0.3469	0.9652
lbl_pkt_4	0.721	0.926	0.922	0.1054	1.1774	0.3377	1,0720
lbl_tkp_5	0.602	0.831	0.787	0.1331	0.9726	0.3183	0.8395

- Traffic analysis, characterization, and prediction as a function of network dynamic behavior.
- Explore new traffic characteristics and parameter in order to get even more faithful traffic models beyond multifractality.
- Investigating on new service strategies at nodes/router making traffic data streams at the output of buffering systems much smoother or low *peakiness*.

6.6 CONCLUSION

Current communication networks are found with massive growth and evolution, operating under various types of protocols simultaneously. The increasing complexity and heterogeneity of networks make them difficult to be characterized and theoretically analyzed and consequently fully and realistically simulated. However, simulation is still the most promising tool for understanding the dynamics issues of network as well as network traffic although simulation models can cover most crucial aspects and characterize faithfully networks or network traffic behavior [9–11].

Today, some of the biggest challenging issues in network performance analysis are scalability, latency, transmission bandwidth, and connection failure. Frequently, these problems are directly related to user behavior and network profile or network behavior. In order to better reproduce real characteristics of a network, we divide the traffic properties into two main categories [15]. The first category relates traffic characteristics with changes in the behavior of users or sessions, for example, size of transferred files from web sessions, FTP, e-mail, and interarrival time of sessions. The second category relates traffic characteristics with variation of the profile or network behavior, for example, delay, loss packet, and topology.

Based on the given classification, we presented a brief discussion of some relevant network aspects and traffic characteristics found in current communication networks that should be taken into account to create realistic scenarios for network simulation.

Using the fractals theory, initially, we introduced and discussed some concepts about mono-fractal processes and self-similarity. Then, we presented various estimation methods for the Hurst parameter being used to measure the degree of self-similarity from traffic data. Next, we formally defined multifractal processes and showed how to characterize communication network traffic data by means of multifractal modeling processes. Explicitly, we presented some computational algorithms that were elaborated to verify the existence of some specific properties of multifractal processes such as specific particularities like the Hölder exponent and the multifractal spectrum.

Both monofractal and multifractal theories developed over network traffic analysis were extensively applied and used experimentally for real traffic data characterization. We ended this fractal traffic investigation with the following conclusion: (1) fractal characteristics of real network traffic are time scale dependent and (2) it is crucial using an appropriate fractal model for traffic characterization, network design, and network performance analysis. As mentioned before, a multifractal model should be associated with traffic data observed at very small time intervals where real-time variations of traffic data impact greatly network dynamic behavior.

A monofractal model, by its turn, is based only on the Hurst parameter that is a global characterization parameter for burst levels and LRD properties observed in traffic trace data. A multifractal model, on the other hand, relies on a set of Hölder exponents or multifractal spectrum analysis.

Therefore, before adopting a computational tool or statistical model for the communication network simulation, performing accurate characterization of network traffic is essential. Each traffic characterization parameter possibly provides some clue about how network performance can be affected by the involved traffic data. Moreover, good traffic characterization greatly helps to create more realistic scenarios for robust and efficient network simulations.

REFERENCES

1. Hwang, J. N. *Multimedia Networking: From Theory to Practice*. Cambridge Press, New York, 2009.
2. Salah, K., Calym, P., and Buhari, M.I. Assessing readiness of IP networks to support desktop video-conferencing using OPNET. *Journal of Network and Computer Applications*, 31 (4), 921–943, 2008. doi:10.1016/j.jnca.2007.01.001.
3. Navarro, J. G. C. and Polo, M. T. S. Implementing telemedicine through eListening in hospital-in-the-home units. *International Journal of Information Management*, 30 (6), 552–558, December 2010. doi:10.1016/j.ijinfomgt.2010.04.005.
4. Ivascu, G. I., Pierre, S., and Quintero, A. QoS routing with traffic distribution in mobile ad hoc networks. *Computer Communications*, 32 (2), 305–316, February 2009. doi:10.1016/j.comcom.2008.10.012.
5. Huang, J., Huang, X., and Ma, Y. An effective approximation scheme for multiconstrained quality-of-service routing. In: *IEEE Global Telecommunications Conference (Globecom 2010)*, Miami, FL, pp. 1–6, December 2010. doi:10.1109/GLOCOM.2010.5683581.
6. Frikha, A. and Lahoud, S. Performance evaluation of pre-computation algorithms for inter-domain QoS routing. In: *18th International Conference on Telecommunications—ICT*, Ayia Napa, Cyprus, pp. 327–332, May 2011. doi:10.1109/CTS.2011.5898944.
7. Mandelbrot, B. B. *The Fractal Geometry of Nature*, 1st edn. W. H. Freeman and Company, San Francisco, CA, 1982.
8. Leland, W., Taqqu, M., Willinger, W., and Wilson, D. On the self-similar nature of Ethernet traffic (extended version). *IEEE/ACM Transactions on Networking*, 2 (1), 1–15, February 1994. doi:10.1109/90.282603.

9. Beran, J. Long-range dependence. *Wiley Interdisciplinary Reviews: Computational Statistics*, 2 (1), 26–35, January/February 2010. doi:10.1002/wics.52.

10. Stenico, J. W. G. and Lee, L. L. A new binomial conservative multiplicative cascade approach for network traffic modeling. In: *27th IEEE International Conference on Advanced Information Networking and Applications—IEEE AINA 2013*, Barcelona, Spain, pp. 794–801, March 2013. doi:10.1109/AINA.2013.18.

11. Parka, C., Hernández, F. C., Le, L., Marron, J. S., Park, J., Pipiras, V., Smith, F. D., Smith, R. L., Trovero, M., and Zhu, Z. Long-range dependence analysis of internet traffic. *Journal of Applied Statistics*, 38 (7), 1407–1433, September 2011. doi:10.1080/02664763.2010.505949.

12. Norros, I. A Storage model with self-similar input. *Queueing Systems*, 16 (3–4), 387–396, 1994. Kluwer Academic Publishers, Boston, MA. doi:10.1007/BF01158964.

13. Crovella, M. E. and Bestavros, A. Self-similarity in World Wide Web traffic—Evidence and possible causes. *IEEE/ACM Transactions on Networking*, 5 (6), 835–846, December 1997. doi:10.1109/90.650143.

14. Park, K., Kim, G., and Crovella, M. E. On the relation between file sizes, transport protocols and self-similar network traffic. In: *International Conference on Network Protocols*, Columbus, OH, pp. 171–180, October/November 1996. doi:10.1109/ICNP.1996.564935.

15. Feldmann, A., Gilbert, A., and Willinger, W. Data networks as cascades: Investigating the multifractal nature of Internet WAN traffic. In: *Proceedings of the ACM Sigcomm'98 Conference on Applications, Technologies, Architectures and Protocols for Computer Communication*, Vancouver, British Columbia, Canada, pp. 42–55, 1998. doi:10.1145/285237.285256.

16. Riedi, R. H., Crouse, M. S., Ribeiro, V. J., and Baraniuk, R. G. A multifractal wavelet model with application to network traffic. *IEEE Transactions on Information Theory*, 45 (3), 992–1082, April 1999. doi:10.1109/18.761337.

17. Duffield, N. G. and O'Connel, N. Large deviations and overflow probabilities for the general single server queue, with applications. *Mathematical Proceedings of the Cambridge Philosophical Society*, 118 (2), 363–374, September 1995. doi:10.1017/S0305004100073709.

18. Vieira, F. H. T. and Lee, L. L. Adaptive wavelet based multifractal model applied to the effective bandwidth estimation of network traffic flows. *IET Communications*, 3 (6), 906–919, June 2009. doi:10.1049/iet-com.2008.0078.

19. Krishna, P. M., Gadre, V. M., and Desai, U. B. *Multifractal Based Network Traffic Modeling*. Kluwer Academic Publishers, Boston, MA, Springer 2003 (December 31, 2003).

20. Xu, Z., Wang, L., and Wang, K. A new multifractal model based on multiplicative cascade. *Information Technology Journal*, 10, 452–456, 2011. doi:10.3923/itj.2011.452.456.

21. Peltier, R. and Véhel, J. L. Multifractional Brownian motion: Definition and preliminary results. Technical Report 2695, INRIA, Rocquencourt, France, 1995. Available in: http://hal.inria.fr/docs/00/07/40/45/PDF/RR-2645.pdf.

22. Vieira, F. H. T., Bianchi, G. R., and Lee, L.L. A network traffic prediction approach based on multifractal modeling. *Journal High Speed Network*, 17 (2), 83–96, 2010. doi:10.3233/JHS-2010-0334.

23. Kolmogorov, A. N. The local structure of turbulence in incompressible viscous fluid for very large Reynolds' numbers. *Doklady Akademiia Nauk SSSR*, 30, 301–305, 1941. Available in: http://www.astro.puc.cl/~rparra/tools/PAPERS/kolmogorov_1951.pdf.

24. Mandelbrot, B. B. and Van Ness, J. W. Fractional Brownian motion, fractional noises and applications. *Society of Industrial and Applied Mathematics Review*, 10 (4), 422–437, 1968. doi:10.1137/1010093.

25. Park, K. and Willinger, W. *Self-Similar Network Traffic and Performance Evaluation*. Wiley-Interscience, New York, Published Online: January 2002. doi:10.1002/047120644X.fmatter_indsub.

26. Paxson, V. and Flyd, S. Wide area traffic: The failure of Poisson modeling. *IEEE/ACM Transactions on Networking*, 3 (1), 226–244, 1995. doi:10.1109/90.392383.

27. Hurst, H. E., Black, R. P., and Simaika, Y. M. *Long-Term Storage: An Experimental Study*. Constable, London, U.K., 1965.

28. Beran, J. *Statistics for Long-Memory Process*. Chapman & Hall, New York, 1994.

29. Sheluhin, O. I., Smolskiy, S. M., and Osin, A. V. *Self-Similar Processes in Telecommunications*. John Wiley & Sons, Ltd, West Sussex, England, 2007.

30. Mandelbrot, B. B. and Wallis, J. R. Computer experiments with fractional Gaussian noises: Part 1, averages and variances. *Water Resources Research*, 5 (1), 228–267, February 1969. doi:10.1029/WR005i001p00228.

31. Taqqu, M. S., Teverovsky, V., and Willinger, W. Estimators for long-range dependence: An empirical study. *Fractals Complex Geometry, Patterns and Scaling In Nature and Society*, 3 (4), 785–788, 1995. doi:10.1142/S0218348X95000692.

32. Taqqu, M. S. and Teverovsky, V. On estimating the intensity of long-range dependence in finite and infinite variance time series. In: R. J. Adler, R. E. Feldman, and M. S. Taqqu (eds.). *A Practical Guide to Heavy Tails: Statistical Techniques and Applications*, pp. 177–217. Birkhäuser, Boston, MA, 1998.

33. Geweke, J. and Porter-Hudak, S. The estimation and application of long memory time series models. *Journal of Time Series Analysis*, 4 (4), 221–238, July 1983. doi:10.1111/j.1467-9892.1983.tb00371.x.

34. Whittle, P. Gaussian estimation in stationary time series. *Bulletin of the International Statistical Institute*, 39, 105–129, 1962. doi:10.1007/BF02590998.

35. Abry, P. and Veitch, D. Wavelet analysis of long-range-dependent traffic. *IEEE Transaction Information Theory*, 44 (1), 2–15. 1998. doi:10.1109/18.650984.

36. Karagiannis, T., Faloutsos, M., and Riedi, R. H. Long-range dependence: Now you see it, now you don't! In: *IEEE Global Telecommunications Conference—Globecom'02*, Taipei, Taiwan, Vol. 3, pp. 2165–2169, November 2002. doi:10.1109/GLOCOM.2002.1189015.

37. Schulman, A., Levin, D., and Spring, N. CRAWDAD data set UMD/Sigcomm2008 (v. 2009-03-02), Downloaded from: http://crawdad.cs.dartmouth.edu/umd/sigcomm2008 (last accessed January 2014).

38. Karagiannis, T., Faloutsos, M., and Molle, M. A user-friendly self-similarity analysis tool. *Special Section on Tools and Technologies for Networking Research and Education, ACM SIGCOMM Computer Communication Review*, 33 (3), 81–93, 2003 (to appear). Available in: http://alumni.cs.ucr.edu/~tkarag/papers/ccr03.pdf.

39. Karagiannis, T. and Faloutsos, M. SELFIS: A tool for self-similarity and long-range dependence analysis. In: *First Workshop on Fractals and Self-Similarity in Data Mining: Issues and Approaches (in KDD)*, Edmonton, Alberta, Canada, July 23, 2002. Available in: http://alumni.cs.ucr.edu/~tkarag/papers/kdd02.pdf.

40. SELFIS—Downloaded from http://alumni.cs.ucr.edu/~tkarag/Selfis/Selfis.html (last accessed February 2014).

41. Seuret, S. and Lévy-Véhel, J. The local Holder function of a continuous function, *Applied and Computational Harmonic Analysis*, 13 (3), 263–276, November 2002. doi:10.1016/S1063-5203(02)00508-0.

42. Zhang, Z. L., Ribeiro, V., Moon, S., and Diot, C. Small-time scaling behaviors of internet backbone traffic: An empirical study. In: *INFOCOM 2003. 22nd Annual Joint Conference of the IEEE Computer and Communications Societies*, San Francisco, CA, Vol. 3, pp. 1826–1836, March/April 2003. doi:10.1109/INFCOM.2003.1209205.

43. Falconer, K. *Fractal Geometry: Mathematical Foundations and Applications*, 2nd edn. Wiley, Chichester, U.K., 2003.

44. Xiong, G., Zhang, S., and Shu, L. The Legendre multifractal spectrum distribution based on WTMM. In: *International Workshop on Chaos-Fractals Theories and Applications—IWCFTA*, Kunming, China, pp. 481–485, October 2010. doi:10.1109/IWCFTA.2010.67.

45. Ribeiro, V. J., Riedi, R. H., Crouse, M. S., and Baraniuk, R. G. Multiscale queueing analysis of long-range dependent traffic. In: *Proceedings. 19th Annual Join Conference of the IEEE Computer and Communications Societies. IEEE INFOCOM 2000*, Tel Aviv, Israel, Vol. 2, pp. 1026–1035, March 2000. doi:10.1109/INFCOM.2000.832278.

46. Abry, P., Flandrin, P., Taqqu, M. S., and Veitch, D. Wavelets for the analysis, estimation and synthesis of scaling data, pp. 39–88, 2002; http://www.cubinlab.ee.unimeld.edu.au/darry/MS_code.html (last accessed February 2014).

47. Pavlov, A. N. and Anishchenko, V. S. Multifractal analysis of complex signals. *Physics Uspekhi*, 50 (8), 819–834, 2007. doi:10.1070/PU2007v050n08ABEH006116.

48. Tanenbaum, A. S. *Computer Networks*, 4th edn. Prentice Hall PTR, Upper Saddle River, NJ, 2003.

49. Xu, Y., Chang, Y., and Liu, Z. Calculation and analysis of compensation buffer size in multimedia system. *IEEE Communications Letters*, 5 (8), 355–357, August 2001. doi:10.1109/4234.940990.

50. Kilpi, J. and Norros, I. Testing the Gaussian approximation of aggregate traffic. In: *Proceedings of the Second ACM Sigcomm Workshop on Internet Measurement*, Marseille, France, pp. 49–61, 2002. doi:10.1145/637201.637207.

51. Norros, I. On the use of fractional Brownian motion in the theory of connectionless networks, *IEEE Journal on Selected Areas in Communications*, 13 (6), 953–962, 1995. doi:10.1109/49.400651.

52. Donsker, M. D. An invariance principle for certain probability limit theorems. *Memoirs American Mathematical Society*, 6, 1–12, 1951.

53. Garavello, M. and Piccoli, B. *Traffic Flow on Network*, 1st edn. American Institute of Mathematical Sciences, Springfield, MO, June 2006.

54. ATM Forum, Traffic Management Specification, version 4.0, AF-TM-0056.000, April 1996.

55. Mi, N., Casale, G., Cherkasova, L., and Smirno, E. Burstiness in multi-tier applications symptoms, causes, and new models. In: *Middleware 2008*, Leuven, Belgium, Lecture Notes in Computer Science, Vol. 5346, pp. 265–286, 2008. doi:10.1007/978-3-540-89856-6_14.

56. FRACLAB, INRIA—Institut National de Recherche en Informatiqueet en Automatique—http://fraclab. saclay.inria.fr/ (last accessed February 2014).

57. Perlingeiro, F. R. and Lee, L. L. A new bandwidth estimation approach for fractal processes. *IEEE Latin America Transactions*, 3 (5), 60–70, December 2005. doi:10.1109/TLA.2005.1642440.

58. Paxson, V. and Floyd, S. Wide area traffic: The failure of Poisson modeling. *IEEE/ACM Transactions on Networking*, 3(3), 226–244, June 1995. doi: 10.1109/90.392383.

59. Xu, H. and He, Y. Some iterative finite element methods for steady Navier–Stokes equations with different viscosities. *Journal of Computational Physics*, 232 (1), 136–152, January 2013. doi:10.1016/j. jcp.2012.07.020.

60. Taqqu, M., Teverovsky, V., and Willinger, W. Is network traffic self-similar or multifractal? *Fractals Complex Geometry, Patterns and Scaling in Nature and Society*, 5 (1), 63–73, March 1997. doi:10.1142/ S0218348X97000073.

61. Véhel, J. L. and Sikdar, B. A multiplicative multifractal model for TCP traffic. In: *Proceedings. Sixth IEEE Symposium on Computer and Communications*, Hammamet, Tunisia, pp. 714–719, 2001. doi:10.1109/ ISCC.2001.935454.

62. Mannersalo, P. and Norros, I. Multifractal analysis: A potential tool for teletraffic characterization? COST 257, TD(97)32 (European Cooperation in the field of Scientific and Technical Research), 2000. Available in: http://www3.informatik.uni-wuerzburg.de/cost/TDs/257td9732.pdf.

7 Tutorial on Random Number Generators in Discrete Event Simulators

Shabbir Ahmed and Farzana Mithun

CONTENTS

ABSTRACT

Computer simulations have been effectively used to predict the behaviors of many natural systems in almost all disciplines. The ability to simulate the complex real-world stochastic systems largely depends on the use of random numbers. Although we use the notion *random* to indicate the underlying uniform distribution, the stochastic nature of most physical systems (e.g., traffic intensity in a transport system) often exhibits the properties of various other statistical distributions (e.g., *Poisson*, *exponential*). Improper use of stochastic models leads to unrealistic simulation outcomes, which deviate far from truth. Therefore, in order to use the right randomness to appropriate stochastic models, it is of utmost importance to understand about random numbers and to learn how simulators generate these numbers. This chapter focuses on the random number generation process from statistical distributions. Although simulators usually come with some random number generator functions, the methods discussed in this chapter can be used to generate random numbers from any user-defined statistical distribution, which is often more desirable to model a natural phenomenon than the bundled random number generators of existing simulators. Utilizing Python programming language, this chapter teaches the basics of various random number generation techniques. In addition, this chapter shows how to simulate real-world scenarios by leveraging SimPy (a discrete event simulator framework for Python). Using these simple simulation examples, we will show how inappropriate random variables can affect the simulation outcomes.

7.1   INTRODUCTION

The events of this world are not random. We can refer to Einstein's famous quote, "*God does not play dice with the Universe!*" [1]. Another notable quote by the Dutch philosopher Baruch Spinoza [2], "*Nothing in Nature is random ... a thing appears random only through the incompleteness of our knowledge.*" The feeling of *randomness* only occurs due to human ignorance of what follows in the sequence of events. This is *subjective randomness* and implies that randomness only persists in human minds, at least not in this objective world [3,4]. Knuth published a substantial detailed section about *randomness* and *random numbers* in [5]. These are usually known as *pseudorandom* numbers because of the fact that they are theoretically deterministic. Fortunately, the *pseudorandom numbers* are often good enough to serve most practical purposes (again due of the incompleteness of human knowledge). A great deal of literature [4–9] and many others can be found on the generation of random variables. Readers are urged to go through those.

Any algorithm to simulate a real-world behavior requires a random number generator. Apart from the simulation domain, random number generators have their places in *statistical sampling* to *cryptography* to even in *recreation* [5]. Although, a computer cannot produce a truly random number using a deterministic algorithm, it can generate *pseudorandom* numbers that are sufficient to model many real systems. It is worth mentioning that there exists no random number generator that is good for every application [5]. Therefore, in order to have a plausible simulation result, it is extremely important to have a thorough understanding of various random number generators and their application domains, since improper usage of random numbers in simulations can lead to unrealistic results. In this chapter, we briefly discuss some common classes of random number generators and their effects on simulation outcomes (especially when improper random numbers have been chosen).

Also, learning the innards of random numbers is vital when one has to code his or her own simulator. Of course, there are many commercial (e.g., *Qualnet*, *ModelSim*) as well as free simulators (e.g., *NS3*, *Omnet++*) available for simulating real-world scenarios. However, the learning curves to utilize these simulators are usually steep. For this reason, sometimes, it is faster to write a simulator from scratch rather than using an existing simulator. For example, SimPy, a Python-based simulation framework, gives us the luxury to write our own simulator with consummate ease. Python's interpreted programming environment and simple syntax make it a perfectly suitable tool for teaching random numbers as well as simulation methodologies to the students. Leveraging the power of Python, this chapter primarily discusses the process of random number generation—the single most important aspect of any simulator. Next, with the help of SimPy, we have shown how to simulate simple real-world scenarios and demonstrated how a poor choice of random sequence can affect the simulation outcome.

7.1.1 CHAPTER OUTLINE

The organization of this chapter is as follows: Section 7.2 starts with the discussion of uniform random number generation techniques (viz, *linear congruential generators [LCGs]*, *multiple generators*, and *Mersenne twister* algorithms), which lays the foundation for generating nonuniform random numbers from any given probabilistic distribution. Next, in Section 7.3, we delve into nonuniform random number generators (*inversion* and *acceptance–rejection method* in particular). Section 7.4 presents a brief note about the methodologies of *randomness* testing. Section 7.5 shows how to simulate real-world scenarios leveraging the power of SimPy and demonstrates the disparities of simulation outcomes due to variations in random number generators. Finally, Section 7.6 concludes the chapter.

7.2 UNIFORM RANDOM NUMBERS

The generation of uniform random numbers on an interval (i.e., $U(0, 1)$) is the basis for generating random numbers from various probabilistic distributions. Therefore, in this section, we start our discussion with the generation techniques of uniform pseudorandom numbers.

In fact, pseudorandom number generators use mathematical formulae or precalculated tables to generate a list of numbers that appear to be random. In other words, the random sequence of numbers generated by a pseudorandom number generator can be reproduced at a later date if the starting point (i.e., *seed)* is known. von Neumann [10] said it best: "Anyone who considers arithmetical methods of producing random digits is, of course, in a state of sin."

The basic idea in generating pseudorandom numbers is to produce a sequence of numbers $x_1, x_2, x_3,...$ using a recurrence of the form

$$x_i = f(x_{i-1}, x_{i-2},...,x_{i-n}).$$

where n initial numbers (known as seed blocks) are needed to initiate the recurrence. The crucial part is to find a function f that generates numbers that are *as random as possible* [6]. If we divide the *pseudorandom* numbers by their upper bound, we obtain the $U(0, 1)$ variates. However, we should note that due to the finite precision of computers, we cannot generate a true continuous random variable.

An undesired characteristic of a pseudorandom number generator is its *periodicity*. That means the sequence will repeat itself at some later point. However, modern pseudorandom number generators exhibit large enough periods to serve most practical purposes [11]. For example, in 1998, Matsumoto and Nishimura [12] invented the first random number generator that has a period of $2^{19,927} - 1$, which is believed to exceed the number of electron spin changes since the creation of the universe [13].

In the following subsection, we will discuss *LCG* (which is one of the most common pseudorandom numbers generators), *multiple recursive generators* (a generalized version LCG), and *Mersenne twister* (one of the most widely used random number generator). For more comprehensive discussions about these and similar other techniques, readers are referred to [14,15].

7.2.1 Linear Congruential Generator

The following algorithm shows the *LCG* [16–18], one of the oldest and best-known techniques to generate $U(0, 1)$ variates. Basically, this method produces a sequence of random numbers calculated with a linear equation (Equation 7.1). In other words, the next number produced is a linear function of previous number.

$$X_{n+1} = \left(AX_n + C\right) \bmod M \tag{7.1}$$

where

X is the sequence of pseudorandom variates
$M, 0 < M$, is the modulus
$A, 0 < A < M$, is the multiplier
$C, 0 \le C \le M$, is the increment
$X_0, 0 < X_0 < M$, is the seed value

Python implementation of the LCG is shown in Listing 1.

Listing 1: LCG

```
def pnrg():
    A = 25173
    C = 13849
    M = 32768

    pnrg.seed = (A*pnrg.seed + C) % M
    return (pnrg.seed/float(M))
```

Output:

```
>>> pnrg.seed = 1
>>> pnrg()
0.19085693359375
>>> pnrg()
0.864227294921875
```

The LCG seems to produce *random* numbers. But is it a good approximation of $U(0, 1)$? Let us delve into this matter.

The period of an LCG is at most M, because this is the modulus. That means a 16-bit random number generated in this way can have a period of maximum 2^{16} (i.e., 65,536), which is definitely not adequate for a *Monte Carlo* simulation (discussed later in this section). In fact, the period of LCG varies widely varies with the choice of A and seed values and is much less than M in most cases. However, according to the *Hull–Dobell* theorem [19], an LCG exhibits a full period for all seed values if and only if the following conditions are met (considering a nonzero C) [16]:

1. C and M are relatively prime.
2. $A - 1$ is divisible by all prime factors of M.
3. $A - 1$ is a multiple of 4 if M is a multiple of 4.

Interested readers are referred to [5] for a proof.

7.2.2 MULTIPLE RECURSIVE GENERATORS

Multiple recursive generators can be considered as the generalization of linear congruential methods [13]. They are based on the following recurrences:

$$x_n = \left(a_1 x_{n-1} + \cdots + a_k x_{n-k} + c\right) \bmod m.$$

where k is a fixed integer. Here, the nth term of the sequence depends on previous k terms.

A Fibonacci-lagged generator is a special case of multiple recursive generators when

$$x_n = \left(x_{n-37} + x_{n-100}\right) \bmod 2^{30}.$$

This generator has been invented by Knuth [5] and is also known as *Knuth-TAOCP-2002* or simply *Knuth-TAOCP.**

7.2.3 MERSENNE TWISTER

The *Mersenne twister* [12,20], invented by Matsumoto and Nishimura, is one of the widely used pseudorandom number generators. The nomenclature comes from the fact that its period length is chosen to be a Mersenne[†] prime. It was the first pseudorandom number generator that produces high-quality pseudorandom numbers efficiently. In fact, it is one of the first pseudorandom number generation algorithms whose period is so large that it is believed that it exceeds the number of electron spin changes since the birth of the universe [13].

The Mersenne twister algorithm is based on matrix LCGs over finite binary field F_2. A variable x is represented by a vector (e.g., of 32 bits). The following linear recurrence is used for the $(n + i)$th term:

$$x_{i+n} = x_{i+m} \oplus \left(x_i^{upp} \mid x_{i+1}^{low}\right) A.$$

where
  $n > m$ are integer constants
  x_i^{upp} means the upper $\omega - r$ (r) bits of x_i
  A is a $\omega \times \omega$ matrix of the finite field F_2

---

* TAOCP stands for *The Art of Computer Programming*, Knuth's famous book.
† A *Mersenne prime* takes the form $M_n = 2^n - 1$. The most commonly used Mersenne twister algorithm uses the Mersenne prime $2^{19,927} - 1$.

Therefore, $\left(x_i^{upp} | x_{i+1}^{low}\right)$ appends the upper $\omega - r$ bits of x_i with lower r bits of x_{i+j}. After a right multiplication with A, the result is added with x_{i+m} bit using *XOR* operation. Once the initial seed $(x_0, x_2,..., x_{n-1})$ is provided, the algorithm produces random integers in $0,1,...,2\omega-1$ [13].

Listing 2 shows a Python implementation of the Mersenne twister algorithm and is provided by Yasar Arabac [21].

Listing 2: Implementation of Mersenne twister algorithm

```
# Create a length 624 list to store the state of the generator
MT = [0 for i in xrange(624)]
index = 0

# To get last 32 bits
bitmask_1 = (2 ** 32) - 1

# To get 32. bit
bitmask_2 = 2 ** 31

# To get last 31 bits
bitmask_3 = (2 ** 31) - 1

def initialize_generator(seed):
    "Initialize the generator from a seed"
    global MT
    global bitmask_1
    MT[0] = seed
    for i in xrange(1,624):
        MT[i] = ((1812433253 * MT[i-1]) ^ ((MT[i-1] >> 30) + i)) &
bitmask_1


def extract_number():
    """
    Extract a tempered pseudorandom number based on the index-th value,
    calling generate_numbers() every 624 numbers
    """
    global index
    global MT
    if index == 0:
        generate_numbers()
    y = MT[index]
    y ^= y >> 11
    y ^= (y << 7) & 2636928640
    y ^= (y << 15) & 4022730752
    y ^= y >> 18

    index = (index + 1) % 624
    return y

def generate_numbers():
    "Generate an array of 624 untempered numbers"
    global MT
    for i in xrange(624):
        y = (MT[i] & bitmask_2) + (MT[(i + 1 ) % 624] & bitmask_3)
        MT[i] = MT[(i + 397) % 624] ^ (y >> 1)
        if y % 2 != 0:
            MT[i] ^= 2567483615

if __name__ == "__main__":
    from datetime import datetime
    now = datetime.now()
    initialize_generator(now.microsecond)
    for i in xrange(100):
        "Print 100 random numbers as an example"
        print extract_number()
```

```
import random
random.seed()                    #initialize the random number generator

n = 30000                        # 30000 random points
count = 0

for i in range(n):
    x = random.random()
    y = random.random()

    if x**2+y**2 <= 1:
        count = count + 1

print "pi = ", 4*float(count)/n
```

7.2.4 EXAMPLE: UNIFORM RANDOM NUMBERS IN MONTE CARLO SIMULATIONS

In this subsection, we outline the *Monte Carlo method* in order to show one of the usages of uniform random numbers in simulation. In particular, we will estimate the value of π using the *Monte Carlo method* [22].

Consider Figure 7.1, which shows a circle inscribed in a unit square. If we randomly place some points (e.g., throw darts) on the unit square, then the following equation holds:

$$\frac{\text{Area of the square}}{\text{Area of the circle}} \approx \frac{\text{Total no. of random points}}{\text{No. of random points inside the circle}}. \tag{7.2}$$

The ratio of the area of the unit square and the circle is

$$\frac{\text{Area of the square}}{\text{Area of the circle}} = \frac{1}{\pi \times 0.5^2} = \frac{4}{\pi} \tag{7.3}$$

$$\Rightarrow \pi = 4 \times \frac{\text{Area of the circle}}{\text{Area of the square}}.$$

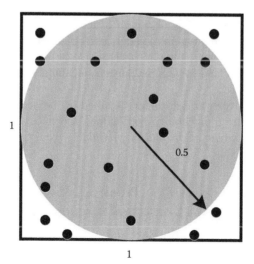

FIGURE 7.1 Illustration of the Monte Carlo method: randomly placed points over the entire.

From Equations 7.2 and 7.3, the value of π can be estimated as the following equation:

$$\pi \approx 4 \times \frac{\text{No. of random points inside the circle}}{\text{Total no. of random points}}. \tag{7.4}$$

In this way, we can estimate the value of π using uniform random variables. A Python listing for the Monte Carlo simulation is presented in Listing 3.

Listing 3: Monte Carlo simulation

```
import random
random.seed()                   #initialize the random number generator

n = 30000                       # 30000 random points
count = 0

for i in range(n):
    x = random.random()
    y = random.random()

    if x**2+y**2 <= 1:
        count = count + 1

print "pi = ", 4*float(count)/n
```

After placing 30,000 random points (i.e., when, $n = 30,000$), the estimate for π is within 0.07% of the actual value. This occurs at an approximate probability of 20%.

In the next section, we discuss some classes of random number generators from probability distributions.

7.3   RANDOM NUMBERS FROM NONUNIFORM PROBABILITY DISTRIBUTIONS

As mentioned earlier, in simulating a real-world scenario, we often require random numbers from a particular probability distribution rather than from the uniform distribution. There are various techniques to accomplish that task. Discussing all of the techniques is beyond the scope of this chapter. Therefore, we will discuss two most widely used techniques to generate random numbers from probability distributions. In particular, we will discuss the *inverse transformation* method, the *acceptance–rejection* method, and some miscellaneous techniques that uncover the properties of those probability distributions.

Before we discuss the inverse transformation method or the acceptance–rejection method to generate random variables from probability distributions, it is necessary to outline the key differences among the terms *probability distributions*, *probability mass function (pmf)*, and *cumulative distribution function (cdf)*.

7.3.1   Note on Probability Distribution, Probability Mass Function, and Cumulative Distribution Function

7.3.1.1   What Is a Probability Distribution?

A discrete random variable can take only a countable numbers of distinct values (e.g., 0, 1, 2, 3, …). The probability distribution of a discrete random variable is simply a list of probabilities associated with each of its possible values [23,24]. For example, if X denotes the numbers one gets in rolling a

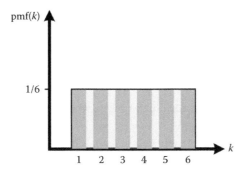

FIGURE 7.2 pmf of the rolling dice example.

dice, then the values that X can have are 1, 2, 3, 4, 5, 6 each with probability of 1/6. The distribution of X is

$$\text{Distribution of } X = \left\{ \left(1, \frac{1}{6}\right), \left(2, \frac{1}{6}\right), \left(3, \frac{1}{6}\right), \left(4, \frac{1}{6}\right), \left(5, \frac{1}{6}\right), \left(6, \frac{1}{6}\right) \right\}.$$

From a probability distribution, we can define the pmf of a discrete random variable X as

$$p_x(k) = P(X = k).$$

for any value of k that X can accept.

From our dice rolling example, the pmf (see Figure 7.2) [17,18] is

$$p_x(k) = \begin{cases} \dfrac{1}{6}, k = 1 \\[2mm] \dfrac{1}{6}, k = 2 \\[2mm] \dfrac{1}{6}, k = 3 \\[2mm] \dfrac{1}{6}, k = 4 \\[2mm] \dfrac{1}{6}, k = 5 \\[2mm] \dfrac{1}{6}, k = 6 \end{cases}.$$

7.3.1.2 Continuous Distributions

In contrast to a discrete random variable that can take a finite number of discrete values, a continuous random variable takes an infinite number of possible values. As a consequence, the probability of observing a single value in a continuous distribution is 0. An example might be helpful to explain this [7]. Suppose, we place a point randomly at the interval (0, 1). Let D be the landing spot of the point. The probability of the point landing in (0.6, 0.7) is the same as for (0.2, 0.3), (0.123, 0.223), and so on. This is because of the fact that all the subintervals are of equal length and the landing of the point on any subinterval is equally probable. Therefore,

$$P(u \leq D \leq v) = v - u,$$

for any value of $0 \leq u \leq v \leq 1$.

Now, we can see that for any individual point c (i.e., when $u = v$), the probability becomes 0:

$$P(D = c) = v - u = 0.$$

This might look like counterintuitive. One way to explain this is to consider that there are infinitely many points, and if they all had nonzero probabilities, then the sum of the probabilities would become infinite instead of 1. So, they must have zero probabilities.

7.3.1.3 Notion of Probability Mass Function in Continuous Distribution

Unlike discrete distributions, the notion of zero individual probabilities for continuous distributions creates a problem in defining the term *distribution* of the variables. This problem can be handled by the *cdf*, defined as

$$F_X(t) = P(X \leq t), \quad -\infty < t < \infty,$$

which is basically the probability that the random variable X takes on a value less than or equal to x. The probability that X lies in an interval $(a, b]$ is

$$P(a < X \leq b) = F_X(b) - F_X(a).$$

In our aforementioned *random point placement* example, the cdf becomes

$$F_D(t) = \begin{cases} 0, & \text{if } t \leq 0 \\ t, & \text{if } 0 < t < 1. \\ 1, & \text{if } t \geq 1 \end{cases}$$

Figure 7.3 shows the graphs of F_D.

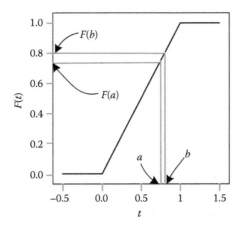

FIGURE 7.3   cdf of the random point placement example.

Note that for a discrete random variable, the cdf can be calculated by summing its pmf. In continuous world, we use integration instead of summation. Analogous to the discrete world, the cdf of a probability can be calculated by integrating its probability function. For continuous distributions, we do not term the probability function as *pmf*, rather *probability density function (pdf)*, which is the derivative of the cdf.

7.3.2 INVERSE TRANSFORMATION METHOD

7.3.2.1 For Continuous Distribution

Suppose we want to simulate a random variable X from an arbitrary distribution whose cdf is F_X. This can be done with $F_X^{-1}(U)$, the inverse* of $F_X(U)$ (considering U has a $U = (0, 1)$ distribution). A simple proof [25] is given in the following:
Consider the random variable $Y = F_X^{-1}(U)$:

$$F_Y(y) = P(Y \le y)$$

$$= P\left(F_X^{-1}(U) \le y\right)$$

$$= P\left(U \le F_X(y)\right)$$

$$= F_X(y) \qquad \text{[see the following intuitive proof].}$$

So, Y has the same distribution as X.
Intuitive proof of the same is depicted here.
Figure 7.4 shows the cdf (F_X) of a distribution. Let us select a fixed value x_0.
Now, $X_1 \le x_0$ if and only if $U_1 \le F(x_0)$.
Therefore,

$$P(X_1 \le x_0) = P(U_1 \le F(x_0))$$

$$= F(x_0) \quad \left[\text{by the definition of CDFs}\right].$$

To summarize, the procedure for generating a random variable form a distribution, F_X:

1. Find a formula for the inverse function F_X^{-1}.
2. Generate a uniform random variable $U = (0, 1)$.
3. Return the random number $x = F_X^{-1}(u)$.

In the next subsection, we present an example to produce random variables from exponential distribution.

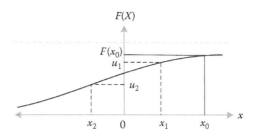

FIGURE 7.4   An intuitive proof that $P(U_1 \le F(x_0)) = F(x_0)$.

---

* Here, *inverse* is in the sense that square and square rooting operations are inverse of each other.

7.3.2.2　Example: The Exponential Family

Suppose we want to generate exponential random variables with parameter λ. The cdf of exponential distribution is

$$F_x(t) = 1 - e^{-\lambda t}.$$

Writing $u = F_X(t)$ and solving for t, we have

$$t = F_X^{-1}(U) = \frac{1}{\lambda} \ln(1 - u).$$

A Python code to generate random variables form exponential distribution using the inverse transformation method is presented in Listing 4.

Listing 4: Inverse transform method for a continuous distribution (e.g., exponential)

```python
import math
import random
lmb = 0.5

def my_exp_rnd(lmb):
    rnd = -1.0/lmb*math.log(1 - random.random())
    return rnd
```

Output:

```
>>> my _ exp _ rnd(lmb)
0794645778600887
>>> my _ exp _ rnd(lmb)
3.2432570141485084
>>>
```

Note, if a random variable Y has a $U(0, 1)$ distribution, then so has $1 - Y$. Therefore, in the aforementioned equation, one might consider to write $U(0, 1)$ instead of $1 - U(0, 1)$. However, it should be noted that $U(0, 1)$ may return 0 but never returns 1. So, if we use $U(0, 1)$, then there is a probability that we get a $\ln(0)$, which is undefined, and our program will raise error.

7.3.2.3　For Discrete Distributions

Suppose the random variable X is discrete with cdf, $F_X = P(X \le x)$, for all real numbers x. The pmf $p(x_i) = P(X = x_i)$, where $x_1, x_2, \ldots$ are the possible values that X can take on (Figure 7.5).

If X is a discrete random variable whose pmf is nonzero at points $\cdots \le x_{i-1} \le x_i \le x_{i+1} \le \cdots$, then the inverse function is given by

$$F_X^{-1}(p) = x_i \quad \text{where,} \, F_X(x_i - 1) \le p \le F_X(x_i).$$

Therefore, the procedure to generate random variables from a discrete distribution using inverse transform method can be summarized as follows:

1. Generate $U \sim U(0, 1)$, that is, random number from uniform distribution.
2. Find the smallest integer I, so that $U \le F(x_I)$.
3. Return $X = x_I$.

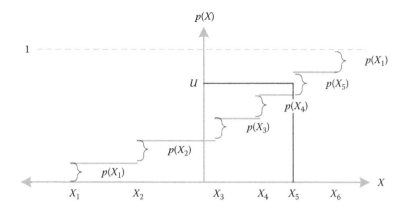

FIGURE 7.5   Inverse transform applied to a discrete distribution.

7.3.2.4   Example of Inverse Transform Methods on Discrete Distribution

Suppose $\{X_1 = 1,\ X_2 = 2,\ X_3 = 3\}$ be random variables with associated probabilities $\{P_1 = 0.2,\ P_2 = 0.3,\ P_3 = 0.5\}$. We need to generate a random variable from the aforementioned distribution.

The cdf of the distribution:

$$F(x) = \begin{cases} 0, & x < 1 \\ 0.2, & 1 \le x < 2 \\ 0.5, & 2 \le x < 3 \\ 1, & 3 \le x \end{cases}.$$

Therefore, the procedure to generate the desired random variable would be as follows:

1. Generate $U \sim U(0, 1)$.
2. If $U < 0.2$, then $X = 1$.
3. If $U < 0.5$, then $X = 2$.
4. If $U < 1$, then $X = 3$.

Figure 7.6 depicts a case that returns 3, when $0 < U \le 1$. A Python code for the aforementioned example is given in Listing 5.

Listing 5: Inverse transform method for a discrete distribution

```
import random
random.seed()

def myDist(u):
        if(u < 0.2):
                return 1
        elif(u < 0.5):
                return 2
        elif(u < 1):
                return 3
```

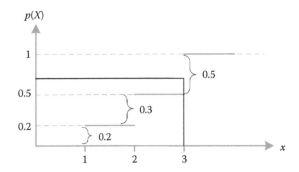

FIGURE 7.6 Example scenario: returns 3 when $0 < U \leq 1$.

Output:

```
>>> a = random.random()
>>> a
0.3224956892686738
>>> myDist(a)
2
>>> a = random.random()
>>> a
0.6923335755477685
>>> myDist(a)
3
>>>
```

Although the inverse transform method is simple, it is not necessarily practical, because the computation of F_X^{-1} may not be easy. As a result, the inverse transformation method is used in practice only when there exists an explicit formula for F_X^{-1} in closed form. In the next section, we will describe another method to compute random variables (viz, the *accept rejection method*), which do not have this limitation.

7.3.3 ACCEPTANCE–REJECTION METHOD

7.3.3.1 Continuous Distribution

As mentioned earlier, the inverse transform method can be utilized practically only when a closed form of the inverse of cdf of the arbitrary distribution can be found. In the previous section, we have worked with the exponential distribution whose inverse cdf could be found easily. Therefore, we were able to use the inverse transform method on the exponential distributions. However, that is not the case for most of the distributions. In this section, we will learn about the *acceptance–rejection method* [26], which can be used to generate random variables from known and computable distribution whose cdf cannot be inverted easily. The acceptance–rejection method is similar for both continuous and discrete distribution; hence, we only discuss the method for continuous distribution.

Suppose we want to generate random variables from a distribution function $f(x)$. The basic idea of the acceptance–rejection method is to find a distribution function $g(x)$ that bounds $f(x)$ over a finite interval and for which one can compute random variables easily. That is, there is some c such that for all x,

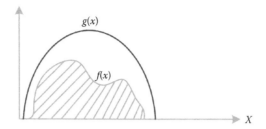

FIGURE 7.7   Illustration of the acceptance–rejection method. If a point $u,g(x)$ lies in the shaded region, it is accepted; rejected otherwise.

$$f(x) \le cg(x).$$

Basically, one generates a point in the 2D plane under the function $g(x)$ (see Figure 7.7). If the point lies in between the shaded region (i.e., under the function $f(x)$), it is accepted as the desired random number. Discarded otherwise.

Astute readers may find the similarities of the acceptance–rejection method with Monte Carlo integration, in a way that the number of rejected points depends on the ratio of the area of $g(x)$ to that of $f(x)$. Therefore, a good guess (i.e., as close as possible with $f(x)$) of the function $g(x)$ is essential in order to minimize the number rejection points.

Algorithm of the acceptance–rejection method:

1. Generate a random variable y from the density function $g(x)$.
2. Generate a random variable u from $U(0, 1)$.
3. If $u \le f(y)/(c \times g(y))$, return y, else go back to step 1.

7.3.3.2   Example of Acceptance–Rejection Method

Consider the U-shaped density function $f(x) = 12(x-0.5)^2$ on $(0, 1)$, for which we need to calculate the random variable [7]. For simplicity, we take our $g(x)$ to be $U(0, 1)$ and take c to be the maximum value of $f(x)$, which is 3 (see Figure 7.8).

Listing 6 shows the Python code that implements the algorithm of the acceptance–rejection method to generate random variables from the aforementioned distribution.

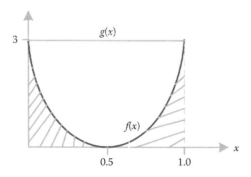

FIGURE 7.8 $f(x)$ is a U-shaped distribution on $(0, 1)$. We need to calculate random variables from it. $g(x)$ is a $U(0, 1)$ distribution. If the random variable falls under the curve of $f(x)$ (i.e., in the shaded region), it will be accepted, otherwise rejected.

Listing 6: Implementation of acceptance–rejection method

```
import random
random.seed()

def myrandom():
    While 1:
        y = random.random()
        u = random.random()
        if(u <= 4*(y - 0.5)**2):
            return y
```

Output:
```
>>> myrandom()
0.45907346850
```

7.3.4 Miscellaneous Methods

There are many other methods to generate random variables from a given distribution. Some techniques exploit the salient properties of the distributions to generate random variables from them. In this section, we discuss some of those for both continuous and discrete cases.

7.3.5 Example from Continuous Distribution

7.3.5.1 Example: The Erlang Distribution

The *Erlang distribution* is a continuous probability distribution that has relationship to the *exponential* and *gamma* distribution. Originally, it was developed by Erlang in order to examine the number of telephone calls made concurrently at the switching stations [27]. The density function of Erlang distribution is given in the following:

$$f(x:k,\lambda) = \frac{1}{(k-1)!}\lambda^k t^{k-1} e^{-\lambda x}, \quad \text{where } x > 0.$$

It has two parameters: the shape k and the rate λ. It is also the distribution of the sum of k independent exponential random variables with mean μ. In other words, if $Z_i \sim \exp(\lambda)$ and independent, then the random variable X appears to be

$$X = Z_1 + Z_2 + Z_3 + \cdots + Z_k.$$

The Listing 7 shows a Python code snippet to generate random variables from Erlang distribution by leveraging this property.

Listing 7: Random number generator from Erlang distribution

```
import random
random.seed()

def my_erlang_rv(lmb, k):
    rv = 0
    for i in range(k):
        rv = rv + exponential_rv(lmb)
    return rv
```

Output:

```
>>> my_erlang_rv(3, 20)
```

In the aforementioned code snippet, the exponential_rv() is the function that generates exponentially distributed random variables (see Listing 4 for a Python code of an exponentially distributed random number generation technique).

7.3.6 EXAMPLE FROM DISCRETE DISTRIBUTION

7.3.6.1 Poisson Distribution

Poisson distribution is used to express the probabilities concerning the number of events (e.g., number of phone calls per hour) per unit time if the events occur independently at a rate λ. Unlike *normal* distribution, Poisson distribution is skewed to the left of the median. Assume that interevent times are exponentially distributed random variables with rate λ, then the number of events $N(t)$ up to time t has Poisson distribution with parameter λt. The pmf is of Poisson distribution is

$$P\big(N(t) = k\big) = \frac{e^{-\lambda t}(\lambda t)^k}{k!}.$$

where $k = 0, 1, 2, 3,\ldots$

The basic idea to generate random variables from Poisson distribution is to calculate the cumulative probability $S(n)$ of observing n or less events for all $k \geq 0$ and generate uniform random number p_u. Then the first n for which $p_u \geq S(n)$ is the desired random variable from the Poisson distribution.

This method is often referred to as the *cumulant method* and works for most probability distributions. However, this method is practically useful when the calculation of $S(n)$ is easy. Using the Poisson distribution, the sum $S(n)$ can be calculated as

$$S(n) = \sum_{k=0}^{n} \frac{e^{-\lambda t}(\lambda t)^n}{k!}.$$

However, this method of calculating $S(n)$ is painfully slow because of the *factorial* and *exponent* parts of the expression. Fortunately, there exists an optimized version that leverages the following property of the distribution:

$$P(n+1) = P(n)\frac{\lambda}{k} \quad \text{where } P(0) = e^{-\lambda}.$$

Enforcing an upper line on k is required for the unlikely event when $p_u = 1$ and program might get stuck into an infinite loop. Listing 8 shows a Python implementation of the aforementioned scheme.

Listing 8: Random number generator for the Poisson distribution

```
import random
random.seed()

def poisson_rv(alpha):
    count = 0
    sum = 0
    while 1:
     count = count + 1
     sum = sum + exponential_rv(alpha)
     if (sum > 1):
         return (count - 1)
```

where exponential_rv(α) means generating an exponentially distributed random variable with a rate α.

Listing 9: An optimized version of the Poisson random variable generator

```
import math
import random

def poisson_rv(lmb):
    max_k = 1000        #upper limit
    u = random.random()
    p = math.exp(-lmb)
    sum = p
    if (sum >= u):
        return 0
    for k in range(max_k):
        p = p * lmb / float(k)
        sum = sum + p
        if (sum >= p):
            break
    return k
```

Note that further optimization (Listing 9) on speed is possible if we precalculate all $S(n)$ and store them in a lookup table.

7.4 NOTE ON RANDOMNESS TEST

In the previous section, we have discussed the techniques behind various random number generators. Now, it is time to talk about the quality of those generators. While discussing about pseudorandom number generators for uniform distribution, we have mentioned the desirable and undesirable properties of those. We have seen that choosing the parameters of a pseudorandom number generator plays a crucial role in the quality of the random variables. Over the past, many empirical statistical tests have been developed to seek for short-term or long-term correlations between the number sequences. The goal of all these tests is to generate pseudorandom numbers that behave like *true* random numbers. In particular, these test whether the output sequence $x_1, x_2, ...,x_n, ...$ can be considered as *independent and identically distributed (iid)* uniform variates [13].

This section briefly mentions some of these tests (e.g., *frequency, gap, runs, autocorrelation tests*). Readers are referred to [28–30] for a rigorous discussion on this subject. There are several tools available for randomness checking. *Dieharder* [29] is one of the widely used industry standard tools, which incorporates various statistical tests to check the randomness of a sequence. If a random number generator does not pass these tests, we should not use it.

The tests for randomness can be categorized into two classes, *empirical* and *theoretical*.

7.4.1 EMPIRICAL TESTS

The basic idea of the empirical tests is to apply some statistical tests on some sample random numbers. Some examples of the empirical tests include [31] the following:

- *Birthday spacings*: If we choose random points on a large interval, the spacings between the points should be asymptotically exponentially distributed [29].
- *Overlapping permutations* [5]: Analyze the sequences of five consecutive random numbers. The 120 possible ordering (i.e., permutation, 5! = 120) should occur with the same probability.
- *Frequency test* [5]: The frequency of the numbers in a random sequence should be approximately equal. For example, if we generate a random sequence with zeros and ones, the number of zeros and ones should be approximately equal to each other. It compares the

distribution of the set of generated random numbers against a uniform distribution. Usually, the Kolmogorov–Smirnov fit (K-S) test (or the chi-square test) is used for comparison.

- *Runs* [5]: This tests the runs up and down or the runs above or below the mean by comparing (usually by using the *chi-square* method) the actual values to expected values.
- *Autocorrelation test*: It tests the correlation between numbers.
- *Gap test*: The idea is to count the number of digits present between the repetitions of a particular digit. This value is then compared with the expected number of gaps using K-S fit test.

7.4.2 THEORETICAL TESTS

This class of tests does not generate any sample random variables, but rather use the analytical properties of the random number generator. Theoretical tests are better in a sense that these tests give us more insight about the random number generator. However, unlike the empirical tests, these cannot be used to test every random number generators, mainly due to the fact that it might be hard to find such tests for a particular generator under observation. Some examples of theoretical tests are as follows:

- *Spectral and lattice tests*: These tests measure the spacing of hyperplanes (the smaller the better). George Marsaglia (1968), in his paper *Random numbers fall mainly in planes* [32], established the lattice structure of LCGs. This means that if we produce n-tuples from the coordinates obtained from consecutive use of the generator, those will lie on a small number of equally spaced hyperplanes in n-dimensional space.

Are these tests enough? Is there any *ultimate* test? The short answer is *no*. Therefore, one should run his or her simulation with different pseudorandom number generators. If the results agree within the error limits, the results are more likely acceptable.

7.5 IMPORTANCE OF CHOOSING THE APPROPRIATE RANDOM NUMBERS IN SIMULATIONS

As discussed earlier, most physical phenomena in the real world are not random. Therefore, in simulations, we often do not pick random numbers from the uniform distribution, rather we choose random numbers from a probability distribution (i.e., a biased form of uniform distribution). In this section, we discuss how simulation results might be affected by choosing random numbers from inappropriate probability distributions. Besides depicting the importance of choosing appropriate random numbers, this section also shows how to write one's own simulator using SimPy, a discrete event simulator framework written in Python.

First, we will introduce SimPy, and then with the help of this simulator framework, we will show how to implement simple simulations. Next, by utilizing SimPy, we will simulate some real-world scenario and will compare the result variations due to the random numbers from different distributions.

7.5.1 SIMPY BASICS

SimPy [33] is a process-oriented discrete event simulator framework written in Python programming language. By virtue of a process-oriented simulator, SimPy produces more modular code than event-oriented simulators (e.g., Omnet++ [34]). Besides, due to its small foot print, the learning curve is not as stiff as that of more sophisticated simulators (e.g., NS3 [35]). For these reasons, it is an excellent tool for researchers and the students studying simulation basics.

Unlike most process-oriented simulators that use *threads* to implement each process, SimPy makes novel use of Python's *generators* [36] (can be considered as *extremely lightweight threads*).

Programmers can leverage the *generators* to make a function prematurely exited and then later reentered at *the point of last exit*. The execution of the function is resumed by each new call to the function immediately after the last exit point.

The next subsection discusses the major components of SimPy. A brief albeit comprehensive tutorial can be found in [8,33].

7.5.1.1 Major Components of SimPy

The three major SimPy terms are briefly discussed in the following [8]:

- *Process*: The behavior (which evolves in time) of active components (e.g., customers, vehicles) is modeled by *processes* (also known as *process functions*). For example, a customer who is waiting to be served can be modeled by a process.

  All processes live in an *environment*. The processes interact with others via *events*. Processes are implemented with Python generators. During the lifetime of the processes, they can create events and *yield* (a Python keyword that indicates the exit/entry point of a function) them, wait for them to be triggered [37]. A process remains in a *suspended* state when it yields an event. Multiple processes can wait for the same event. SimPy resumes them according to the same order in which they yielded that event [37].
- *Events*: SimPy is basically an asynchronous event dispatcher. A scenario can be modeled as a collection of events that need to be scheduled at some particular simulation time. The events are sorted at the event dispatcher according to their priority, simulation time, and event id. It keeps track of the events and current simulation time. An event maintains a list of callback functions (i.e., the process functions). When an event is triggered, the callback functions are executed by the event loop.

  An important event type is the *Timeout* [37]. This event puts a process into sleep for a specified amount of simulated time.
- *Resource*: It simulates something (i.e., possibly shared resources) to be queued for. As an example, consider a banking system where clients are served by an ATM. The ATM is a resource for which the clients are waiting for.

Following are some of the major SimPy operations, which we need to know in order to write simulation codes with SimPy:

- *env.process()*: After the creation of the process, this function marks the process as runnable. *env* is an object of *Environment* class (more on this is discussed later in this section).
- *env.run()*: This function is used to start the simulation.
- *yield env.event()*: Puts the current running process into suspended state for a certain amount of time. *yield* is a Python operator whose first operand is a function to be called. In this case, the *event()* function of SimPy library is called to turn the process into suspended state.
- *yield resource.request()*: The *request()* function of the SimPy library (i.e., an object of *Resource* class) keeps the process waiting in a queue for a given resource. In case there are no other jobs (i.e., *processes*) waiting for that resource, the process starts using the resource immediately.
- *resource.release()*: Indicates that the process does not need the given resource. As a result, the next process waiting for that resource is going to be served.

There are many other useful functions in the SimPy library. A detailed description of the aforementioned functions and more others can be found from [37]. Also note that the examples presented here requires SimPy version 3, the newer SimPy framework.

Next, we will present some examples that demonstrate the basic simulation capabilities of SimPy. Let us start with a simple random walk model in SimPy.

7.5.2 EXAMPLE: A SIMPLE RANDOM WALK

Listing 10 simulates a simple random walk model that illustrates the simplicity of the SimPy simulation framework. Although the example does not cover most of the features of SimPy, it provides a gentle introduction to the SimPy beginners. Please note that for the sake of simplicity, the object-oriented paradigm (OOP) was not followed in these examples. However, one should adhere to the OOP in order to model more complex scenarios.

Suppose a node (i.e., any object) can move either right or left. At each move, it covers random units of distance (in this case, the distance is 0–3 units). Also, suppose each move occurs at a fix interval (here the interval is 5 s). We want to simulate the node's movement if the node is allowed to move for a fixed amount of time (say 21 s). Listing 10 shows how to simulate this random walk model in SimPy.

Listing 10: Random walk simulation in SimPy

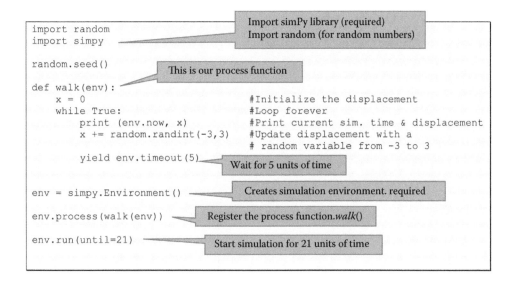

Output from the program:

(0, 0)
(5, 1)
(10, −2)
(15, −4)
(20, −6)

Figure 7.9 shows the flow chart of the basic random walk simulation. We need a process function that will run asynchronously. Here, *walk()* is our desired process function. The state diagram of the process function *walk()* is shown in Figure 7.9. This function defines the actual random movement when it is called. The first thing to do after defining a process function is to create an instance of the Environment class. The instance is passed to the process function. It creates a process generator that needs to be started and added to the environment via *Environment.process()* [37]. At this stage, the execution of the process is scheduled. And finally, the simulation is started by invoking the *run()* of the Environment class and passing an end time for the simulation. A sample output and its graphical representation is shown in Figure 7.10.

Figure 7.11 shows a plot of the random walk simulation. *X*-axis represents the number of steps taken, whereas the *Y*-axis shows the cumulative displacement. A LCG (see Listing 1) with different

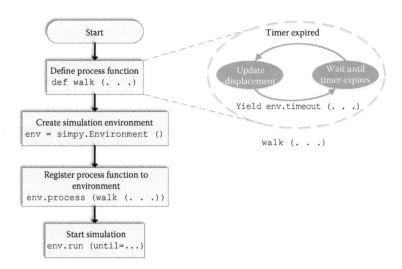

FIGURE 7.9 Flow and state diagram of the random walk simulation code in SimPy.

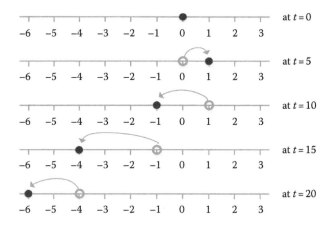

FIGURE 7.10 Random walk example.

parameters is utilized to produce all the graphs. As can be seen from the figure, the displacement is biased towards +ve end and tends to be linear when the random number generator is used with bad parameters (in this case, $A = 1{,}277$, $C = 0$, and $M = 131{,}072$). On the other hand, the same random number generator algorithm can produce better random sequences when a good choice of parameters (e.g., $A = 16{,}807$, $C = 0$, and $M = 2{,}147{,}483{,}647$) is provided. This example illustrates how a bad sequence of random variables produces unrealistic results.

7.5.3 EXAMPLE: CAR FUELING SIMULATION

This example has been taken from the official SimPy tutorial [38]. Suppose we want to simulate a refueling station (Figure 7.12). A gas station has a limited number of refueling machines

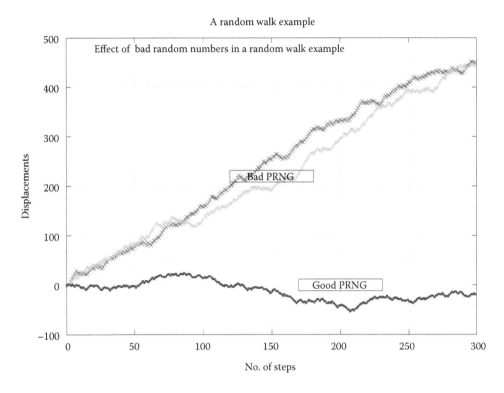

FIGURE 7.11   Plot of a random walk example, showing the effect of choosing bad random number generators.

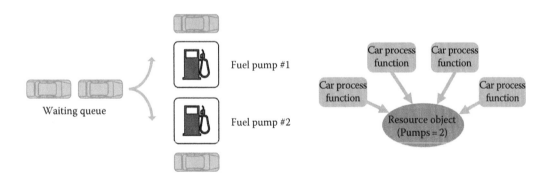

FIGURE 7.12   Car fueling example.

and the refueling process requires some random amount of time (due to the fact that each car requests for a different amount of fuel). A car (modeled as a *process function*) arrives at the refueling station at a random time. If at least one of the refueling machines (modeled as an object of *Resource()* class) is available, the fueling process begins and continues until the request of the car is fulfilled. In case all the refueling machines are busy, the car waits in the queue (see Figure 7.13).

The Listing 11 shows how to model the aforementioned scenario in SimPy.

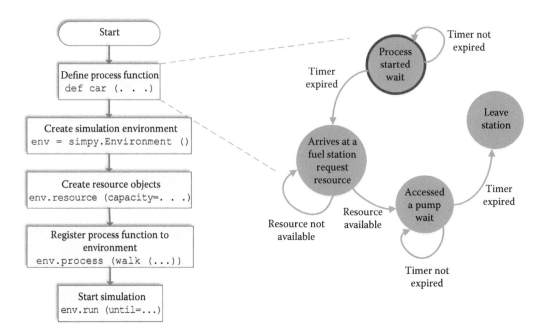

FIGURE 7.13   Flow and state diagram of the car fueling example.

Listing 11: Car fueling simulation in SimPy

```
import simpy
import random
random.seed()
                            Process function

                                                    Arrival time for this car is a random
                                                    integer ranging from 1 to 100
def car(env, name, fs):
    arrival_time = random.randint(1,100)

                                            Wait until arrival_time
    yield env.timeout(arrival_time)         models the arrival time of cars

    print('%s arrived fuel station at %d' % (name, env.now))
    with fs.request() as req:               Request a resource, e.g., fuel pump
        yield req
        print('%s accessed a fuel pump at %s' % (name, env.now))
        service_time = 10                   Refueling time
        yield env.timeout(service_time)
        print('%s left the fuel station at %s' % (name, env.now))

                            Create two resource objects
env = simpy.Environment()

fs = simpy.Resource(env, capacity = 2)

                                            Create processes (e.g., car)
for i in range(10):
    env.process(car(env, 'Car %d:' % i, fs))

env.run(until=50)          Start simulation for 50 units for time
```

Output from the program:

```
Car 1: arrived fuel station at 9
Car 1: accessed a fuel pump at 9
Car 7: arrived fuel station at 17
Car 7: accessed a fuel pump at 17
```

```
Car 2: arrived fuel station at 18
Car 1: left the fuel station at 19
Car 2: accessed a fuel pump at 19
Car 7: left the fuel station at 31
Car 2: left the fuel station at 32
Car 8: arrived fuel station at 44
Car 8: accessed a fuel pump at 44
Car 4: arrived fuel station at 46
Car 4: accessed a fuel pump at 46
Car 9: arrived fuel station at 47
```

A sample output and its corresponding resource utilization diagram (partial) is shown in Figure 7.14. In the example given earlier, we have introduced the *Resource* class of SimPy, which allows us to model shared resources (fuel pump in this case). The parameter *capacity* determines the number of resources (e.g., 2 in this example). The *request()* method of the *Resource* class generates an event that keeps us waiting until the resource becomes available. Once our process captures the resource, it remains unavailable to other processes unit current incumbent releases it (see Figure 7.13). The *with* statement (as in our example earlier) causes an automatic release of the resource (i.e., the resource is released automatically when the *with* block ends).

Two key points to note in this simulation: (1) the arrival time of the cars is assumed to be a random integer (ranging from 1 to 100) and (2) the fueling time (i.e., service time) is a considered to be a constant. This model is clearly not realistic because the fueling time is not constant for every car; it too should be modeled as a random variable. Therefore, a simple solution to this problem is to generate a random variable using the *random()* function of Python and scale it with a constant (approximated from real-world experience) and use it as fueling time (as shown in the following):

```
service _ time = C*random.random()
```

Is our model realistic? Not yet. Note that the built-in functions *random()* and the *randint()* generate random variables form uniform distribution, which is not a good approximation of arrival time of the cars in a gas station. So, which probabilistic distribution to choose to model cars' arrival in a gas station? Unfortunately, there is no definite answer for that. The arrival pattern of cars varies hourly throughout the day. Moreover, the pattern of week days varies considerably from that of weekends.

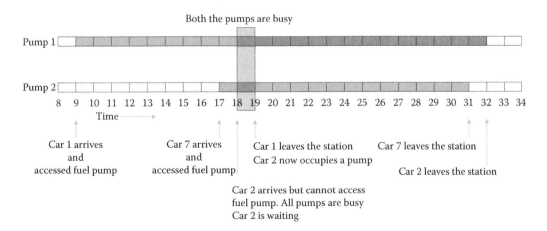

FIGURE 7.14 Occupancy diagram of the fuel pumps.

Empirical studies show that these types of patterns can be modeled more accurately through random variables from exponential distributions. Therefore, our Python code to generate arrival time of the cars should be

```
arrival _ time = C*random.expovariate(LAMBDA)
```

where *expovariate()* is the Python's built-in exponentially distributed random number generator function.

7.5.4   EXAMPLE: UNIFORM DISTRIBUTION IN MODELING A DoS ATTACK

In the arena of network security, a *denial of service (DoS)* attack is an attempt to disrupt the normal operation by exhausting the resource of a service (e.g., website or web service, DNS service). The general technique of DoS involves an attacker to send an overwhelming number of service requests (usually with crafted request packets) to the victim (e.g., a web service), so that the service becomes busy processing those. As a consequence, a legitimate request may be denied of the required service (due to lack of resources). The attacker usually uses impersonated source IP addresses (randomly generated) while crafting the service request packets.

An interesting empirical finding [39,40] in computer security reveals that DoS attacks usually follow the pattern of a uniform distribution. In particular, the distribution of randomly generated fake IP addresses often resembles a uniform distribution. Therefore, if we want to model a DoS attack, we should select a random number generator from a uniform distribution. Also, a web server can sample the IP addresses from requesting clients and take necessary measures against possible DoS attack if the incoming IP address pattern shows resemblance to a uniform distribution. The following code snippet (Listing 12) demonstrates how to generate random IPs from a uniform distribution:

Listing 12: A DoS attack simulation with SimPy

```
from random import randint, random
import simpy

victim_ip = '1.2.3.4'
seed()
                                    Generates random IP addresses
def random_IP():
    ip='.'.join([str(randint(1,256)),str(randint(1,256)),\
        str(randint(1,256)), str(randint(1,256))])
    return ip
                                    This function sends crafted request pkts to server
                                    (not implemented)
def send_fake_pkt(src, dst):
    """This function sends a crafted request packet to a destination"""
                                    Process function
def DOS_Attack(env):
    source_ip = random_IP();
    send_fake_pkt(source_ip, victim_ip)        Watits for random amount of time
    yield env.timeout(random.random())         Simulates interval between requests
    print('Attacking from %s at %f' % (source_ip, env.now))

env = simpy.Environment()
                                    Create 10000 attack processes
for i in range(10000):
    env.process(DOS_Attack(env))

env.run(until=10)
```

Output from the program:

```
Attacking from 7.217.88.179 at 0.981676
Attacking from 138.34.237.73 at 0.981943
Attacking from 7.63.73.239 at 0.984446
Attacking from 23.68.18.22 at 0.984725
Attacking from 124.227.147.78 at 0.987119
Attacking from 171.128.176.32 at 0.987243
Attacking from 16.95.88.187 at 0.988132
Attacking from 97.247.149.213 at 0.988897
...
```

There are two important things to note in this code: (1) each attack is implanted as a *process* function, and (2) the random IP address is produced using Python's built-in function *randint()*, which generates uniform random integers within a given range. A logical explanation of the phenomenon, that the fake random IP addresses exhibit the properties of a uniform distribution, comes from the fact that attacker simply uses the default random number generator (which is a uniform variate for most programming languages) to generate these fake IP addresses.

Simulating a DoS attack can provide us an insight about how good or resilient our web services are. The code given earlier only implements the attack part. However, we can use SimPy's *Resource()* class to model web services and get statistics about the outage of the web services (see Figure 7.15).

A modified version of the aforementioned code, which implements web services as SimPy *Resource()* objects, is shown in Listing 13. Note the similarities with the *car fueling* example in Listing 11.

When someone sends a request to a web server, it forks a child process to handle the request. During the processing of the request, if another request packet comes to the web server, it forks another child process, and so on. However, there is a limit (primarily due to limited memory and scarcity of other network resources) on how many child processes a web server can create concurrently. Therefore, SimPy's *Resource* class is just what we need to model web resources. Addition of the *Resource* objects is the only difference between this example and the previous one.

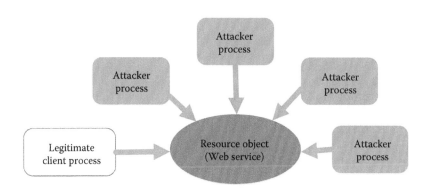

FIGURE 7.15 Web services are modeled using SimPy's resource objects. DoS attacker processes and legitimate client processes contend each other for web resources.

Listing 13: Simulating the resilience of web services under DoS attack

```
import simpy
from random import randint, random

victim_ip = '1.2.3.4'    #This is our web server under attack

def DOS_Attack(env, ws):                          Attack process
    arrival_time = random()
    yield env.timeout(arrival_time)        Simulates the resource occupied time by
    source_ip = random_IP()                each attack process with a random variable
    send_fake_pkt(source_ip, victim_ip)

    #access a web service                   Hold a resource during attack
    with ws.request() as req:
        yield req                           Simulates the web service unavailable
        print('Attack at %s' % env.now)     period as a uniform random variable
        attack_duration = random()
        yield env.timeout(attack_duration)
                                            Request process for
def client(env, ws):                        a legitimate client
    arrival_time = random()
    yield env.timeout(arrival_time)         Clients' arrival time is modeled
    source_ip = self_IP()                   as uniform random variables

    with ws.request() as req:
        yield req
        print('Legitimate request at %s' % env.now)
        service_duration = random()         The duration of a client
        yield env.timeout(service_duration) is modeled as uniform

env = simpy.Environment()                   Web service is modeled as a resource
ws = simpy.Resource(env, capacity=100)

for i in range(10000):
    env.process(DOS_Attack(env, ws))        Create 10000 attack processes

for i in range(100):
    env.process(client(env,ws))            Create 100 legitimate

env.run(until=10)
```

Output from the program:
...
```
Attack at 3.78888629546
Attack at 3.80271710895
Attack at 3.80280588383
Attack at 3.80397476446
Attack at 3.80817892241
Attack at 3.80896510518
Attack at 3.81855583842
Attack at 3.82909467536
Attack at 3.8327927507
Attack at 3.83376423844
Attack at 3.83424268633
Attack at 3.83708639202
Attack at 3.84077935915
Legitimate request at 3.84261894304
Attack at 3.85186507668
Attack at 3.8565976999
Attack at 3.85743772585
```
...

The implementations of the functions *send_fake_pkt()* and *random_IP()* are the same as those in Listing 12, hence omitted in Listing 13. Also the implementation of the function *self_IP()*, which returns the IP address of the client, is not shown because it is not relevant to the primary objective of this simulation. Note that the service time of the clients (i.e., *service_duration*) and the attacker (i.e., *attack_duration*) is modeled as uniform random variables ($U(0, 1)$), which may not be realistic. The service time for a burst of packets (usually originating from same attacker) should be similar because of fact that the request technique and network resources of the packets are the same. Therefore, an empirical constant might be a better candidate for modeling the *attack_duration* than a uniform random variable.

Figure 7.16 shows the simulation results under various conditions. There is a noticeable difference between the simulation outcomes of the two different scenarios. In the scenario #1, the arrival time of an attack packet, the arrival time of a client's legitimate request, the attack duration (i.e., resource occupied time per attack), and the service time for a client's request, all are modeled with uniform random variables. Whereas in scenario #2, while every other parameters remain the same as those of scenario #1, the arrival time of a client's request and its service time have been approximated with exponentially distributed random variables (with $\lambda = 0.5$). The number of resource is set to 100 (same as the number of clients). The plot has been produced by varying the number of attackers from 0 to 10,000 and measuring the number of valid requests that are served. The simulation has been run for 10 time units. A global counter variable is included in the aforementioned code (not shown in Listing 13) in order to record the service availability statistics (i.e., how many legitimate clients were able to access within 10 time units).

The plot suggests that the web service can withstand approximately 2000 attacks per second (considering *seconds* as the time unit). However, after that threshold point (when all web resources become exhausted), its service degrades. The interesting point to note that the service degrades more rapidly when the arrival and service time of clients' request packets are modeled as exponentially distributed random variables (as in scenario #2). In fact, according to several empirical studies [41], there are reasons to believe that the result from scenario #2 is more realistic and more accurate than that from scenario #1. Hence, we should be extremely careful about the choice of random variables in computer simulations.

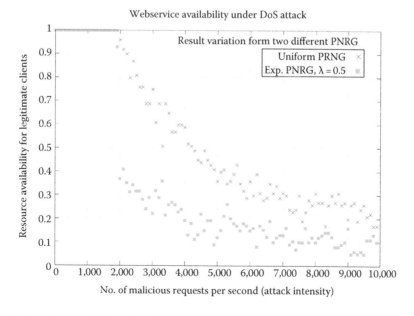

FIGURE 7.16 Service availability of a web server during a DoS attack, showing a considerable variation in results due to the effect of different random number generators.

7.6 CONCLUSION

A random number generator is an immanent part of computer simulations. A working knowledge of various random number generation techniques is essential because improper usage of random numbers often led to unrealistic simulation outcomes. This chapter is an attempt to provide an overview of random number generation techniques and highlight the importance of choosing the right random variables in simulation and modeling.

Starting with uniform random number generators, we have focused on various nonuniform random number generation techniques. In particular, we have talked about the *inversion method*, *acceptance–rejection method*, and miscellaneous techniques that exploit the properties of a given statistical distribution. For the sake of completeness, a brief note about the *randomness* checking is also provided. Apart from the discussion of random numbers, this chapter also presents a step-by-step tutorial to use SimPy (a discrete event simulator framework) in simulating real-world scenarios. Moreover, leveraging the power of SimPy, we have simulated some real-world scenarios and shown the effect on the outcomes due to the selection of improper random variables.

At this point, we hope that the reader has developed a better understanding of various random number generators and recognized the importance of choosing the appropriate random number generators in their simulations. As final advice on using random variables in simulations, one should always run his or her simulations with different types of random number generators. If the outcomes are not within error bounds, one should delve into his or her models before jumping into conclusions.

REFERENCES

1. W. Hermanns, *Einstein and the Poet: In Search of the Cosmic Man*, Branden Press, Brookline, MA, 1983.
2. B. Spinoza, Ethics Part I, [Online]. Available: http://en.wikipedia.org/wiki/Ethics [accessed February 1, 2014].
3. D. Biebighauser, *Testing Random Number Generators*, University of Minnesota—Twin Cities, Minneapolis, MN, 2000.
4. D. J. Bennett, *Randomness*, Harvard University Press, Cambridge, MA, 1999.
5. D. E. Knuth, *The Art of Computer Programming: Seminumerical Algorithms*, Vol. 2, 3rd edn., Addison-Wesly, Reading, MA, 1997.
6. H. G. Katzgraber, *Random Numbers in Scientific Computing: An Introduction*, Lecture given at the International Summer School of Modern Computational Science, Oldenburg, Germany, 2010, [Online]. Available: http://arxiv.org/abs/1005.4117v1 [accessed February 1, 2014].
7. N. Matloff, From algorithms to Z-scores: Probabilistc and statistical modeling in computer science, 2010, http://heather.cs.ucdavis.edu/matloff/public_html/probstatbook.html [accessed July 15, 2014].
8. N. Matloff, Introduction to discrete event simulator, [Online]. Available: http://heather.cs.ucdavis.edu/~matloff/156/PLN/DESimIntro.pdf [accessed February 1, 2014].
9. Y. A. Rozanov, *Probability Theory: A Concise Course*, Dover Publicaions, New York, 1977.
10. J. Von Neumann, Various techinques used in connection with random digits, *Applied Mathematics Series*, 12 (1), 36–38, 1951.
11. Random.Org, [Online]. Available: http://www.random.org/randomness [accessed February 1, 2014].
12. M. Matsumoto and T. Nishimura, Mersenne Twister: A 623-dimensionally equidistrubuted uniform random number generator, *ACM Transaction on Modeling and Computer Simulations*, 8 (1), 3–30, 1998.
13. C. Dutang and D. Wuertz, A note on random number generation, [Online]. Available: http://cran.r-project.org/web/packages/randtoolbox/vignettes/fullpres.pdf [accessed February 1, 2014].
14. W. H. Press, S. A. Teukolsky, W. T. Vetterling, and B. P. Flannery, *Numerical Recipes in C*, Cambridge University Press, Cambridge, U.K., 1995.
15. J. B. T. Morgan, *Elements of Simulation*, Cambridge University Press, London, U.K., 1984.
16. Linear congruential generator, Wikipedia, [Online]. Available: http://en.wikipedia.org/wiki/Linear_congruential_generator [accessed 1, February 2014].
17. D. Lehmer, Mathematical methods in large-scale computing units, in *Second Symposium on Large-Scale DIgital Calculating Machinery*, Cambridge, MA, 1951.
18. S. K. Park and K. W. Miller, Random number generators: Good ones are hard to find, *Communications of the ACM*, 31 (10), 1192–1201, 1988.

19. F. Severance, *System Modeling and Simulation*, John Wiley & Sons, Ltd., Chichester, U.K., 2001.
20. Mersenne twister, Wikipedia.org, [Online]. Available: http://en.wikipedia.org/wiki/Mersenne_twister [accessed 1, 2 2014].
21. Mersenne twister, [Online]. Available: http://code.activestate.com/recipes/578056-mersenne-twister/ [accessed February 1, 2014].
22. Monte Carlo, [Online]. Available: http://en.wikipedia.org/wiki/Monte_carlo_method [accessed February 1, 2014].
23. V. J. Easton and J. H. McColl, Statistics glossary v1.1, [Online]. Available: http://www.stats.gla.ac.uk/steps/glossary/ [accessed February 1, 2014].
24. D. H. Morris and S. J. Mark, *Probability and Statistics*, 4th edn., Pearson, Boston, MA, 2011.
25. N. Matloff, Random number generation, [Online]. Available: http://heather.cs.ucdavis.edu/~matloff/156/PLN/RandNumGen.pdf [accessed February 1, 2014].
26. Inverse transform method, [Online]. Available: http://en.wikipedia.org/wiki/Inverse_transform_method [accessed February 1, 2014].
27. Erlang distribution, [Online]. Available: http://en.wikipedia.org/wiki/Eralng_distribution [accessed February 1, 2014].
28. J. E. Gentle, *Random Number Generation and Monte Carlo Methods (Statistics and Computing)*, 2nd edn., Springer, New York, 2004.
29. R. G. Brown, DieHarder: A Gnu public licensed random number tester, 2008, [Online]. Available: http://www.phy.duke.edu/~rgb/General/dieharder.php [accessed February 1, 2014].
30. P. L'Ecuyer and R. Simard, TestU01: A C library for empirical testing of random number generators, *ACM Transaction on Mathematical Modeling*, 33 (4), 22, 2007.
31. Diehard tests, Wikipedia, [Online]. Available: http://en.wikipedia.org/wiki/Dieharder [accessed February 1, 2014].
32. G. Marsaglia, Random numbers fall mainly in planes, *Proceedings of National Academy of Science*, 61 (1), 25–28, 1968.
33. SimPy, [Online]. Available: http://simpy.readthedocs.org/ [accessed February 1, 2014].
34. OMNeT++ network simulation framework, [Online]. Available: http://www.omnetpp.org [accessed February 1, 2014].
35. NS3, [Online]. Available: http://www.nsnam.org [accessed February 1, 2014].
36. Python, [Online]. Available: http://www.python.org [accessed February 1, 2014].
37. Basic concepts of SimPy, [Online]. Available: http://simpy.readthedocs.org/en/latest/simpy_intro/basic_concepts.html [accessed February 1, 2014].
38. S. Scherfke, SimPy: An introduction and example with electric vehicles, 2012, [Online]. Available: archive.euroscipy.org/file/9023/raw/SimPy.pdf [accessed February 1, 2014].
39. J. J. Marchette, Statistical methods for network and computer security, 2003, [Online]. Available: http://orion.math.iastate.edu/IA/2003/foils/marchette.pdf [accessed February 1, 2014].
40. Y. Wang, *Statistical Techniques for Network Security*, IGI Global, Hershey, PA, 2008.
41. Exponential distribution, Wikipedia, [Online]. Available: http://en.wikipedia.org/wiki/Exponential_distribution [accessed February 1, 2014].

8 Accurate and Reliable Simulation Framework for Mobile Computing Systems Using Stochastic Differential Equations

Athanasios D. Panagopoulos

CONTENTS

ABSTRACT

Mobile computing systems have offered spectacular achievements and opportunities for applications and services in information and communication technologies. These attainments are based mainly on progress in the following fundamental areas: communication and information theory, signal processing, and computer and related technologies. The statistical approach to the consideration of most problems related to information transmission has become dominant since any real signal, propagation media, interference, and even information itself all have an intrinsically random nature. In this chapter, a stochastic radio framework that uses multidimensional stochastic differential equations (SDEs) and captures the properties of wireless channels is presented. Furthermore, this radio framework is applied to simulate mobile computing systems (terrestrial and satellite) in terms of predicting outage and interference statistics and other telecommunication parameters. More particularly, some complex wireless networks such as multihop mobile computing systems and coexistence of broadband satellite communication networks and terrestrial systems are investigated. Finally, a section for further research directions and some conclusions is given.

8.1 INTRODUCTION

Mobile computing systems have offered spectacular achievements and opportunities for applications and services in information and communication technologies [1]. These attainments are based mainly on progress in the following fundamental areas: communication and information theory, signal processing, and computer and related technologies. The statistical approach to the consideration of most problems related to information transmission has become dominant since any real signal, propagation media, interference, and even information itself all have an intrinsically random nature. The radio channel/interference statistical behavior is changing over time and is really hard to be predicted as a result of the movement of users, which are simultaneously transmitters and receivers, and the dynamic change of the propagation environment [2–4].

In mobile computing systems, the reliable communication is constrained by the radio propagation effects, the interference, and the noise that are inherently present in all wireless communication systems [2]. An accurate description of such impairments is very important for the design of high reliable and resilient modern radio communication systems. The simulation approaches that are employed in order to realistically represent the channel and the interference models are very crucial in the design process of a mobile computing system and are an important step for their testing and performance evaluation. Another very important technical challenge in modern wireless networks is the integration of the heterogeneous radio access technologies (RATs).

Accurate radio channel models that will describe unifyingly the space–time variation of the miscellaneous radio links are necessary. The specific communication requirements that aim to improve the performance of radio systems are often conflicted and hard to be formulated in a way convenient for optimization. On one side, the models that are used for the radio system performance and evaluation must accurately reflect the main features of the radio channel and the radio interference under investigation. On the other hand, simple description is required to be applicable for the massive numerical simulations that are necessary to test the performance of complex radio communication systems. In realistic radio communication systems, nonstationary and non-Gaussian radio channel and interference are present. Consequently, simple simulated channel models fail to represent accurately the time-varying phenomena of the radio channel. All of these, coupled with increased complexity of the radio systems (multiantenna systems—MIMO, multiuser, complex cooperative systems, large-scale radio networks, etc.) and reduced design time, require simplicity and accuracy for the complete radio communication system.

In this chapter, a stochastic radio framework that uses multidimensional SDEs [5–12] and captures the properties of wireless channels is presented. Furthermore, this radio framework is applied to simulate mobile computing systems (terrestrial and satellite) in terms of predicting outage and interference statistics and other telecommunication parameters. More particularly, some complex wireless networks such as multihop mobile computing systems and coexistence of broadband satellite communication networks and terrestrial systems are investigated [13–15]. Finally, a section for further research and some conclusions has been added.

8.2 MOBILE COMPUTING SYSTEMS

8.2.1 INTRODUCTION AND GENERAL DESCRIPTION

Mobile computing systems can be defined as the radio communication systems in which the interaction between the nodes is implemented even when the nodes are mobile. More generally, it is the inclusion of the mobility capability in all the next-generation networks including millimeter wave (MMW) radio systems and satellite communication networks. A mobile computing system should consider and involve mobile communication, mobile hardware, and mobile software.

Nowadays, there is a plethora of independent RATs that can be considered as mobile computing systems and each supporting distinct coverage, mobility, data rates, and quality-of-service (QoS). The next-generation mobile communication networks have been envisaged as a convergence platform, where heterogeneous RATs will leverage on a converged all-IP core network to create an adaptive self-resilient network, such that services may be provisioned optimally through the most efficient access network.

The implementation of ITU's vision for optimal connection, anywhere, and anytime as published in Recommendation ITU-R M.1645 states that the future mobile computing systems should create a coalition of different RATs, each connected to a common IP-based backhaul network (either wireless at MMW frequencies or fiber optics scenario). One of the most important ideas in all the vision on future mobile communication systems is that they constitute a multiaccess network environment, where multiple heterogeneous access networks will be available and all the mobility constraints will be satisfied. This assumption brings up the issue of selecting the most appropriate access network to cover specific applications' requirements and suggests the investigation of the relevant network selection approaches.

The most important parameters of mobile computing systems are the bandwidth and the coverage radius, the power consumptions, the security policies for the access in the mobile computing system, intersystem and intrasystem interference, and finally the connectivity of the mobile terminals.

Wireless cellular consumer communication systems have already been spread all around world. Global System for Mobile Communications and Universal Mobile Telecommunication System are the major RATs that have been adopted by mobile operators in order to provide mobile telephony, multimedia, and Internet services to mobile users. Classically, mobile speech as well as video telephony services is supported widely by mobile networks.

Another important metric for mobile computing systems is the quality of experience (QoE) that describes the user experience of the mobile services and is drawn subjectively by perceptual quality algorithms [16]. The whole concept of the coexistence of mobile computing systems is that services are ubiquitously delivered over multiple wireless access technologies in a heterogeneous wireless network environment. A significant issue is the ranking of the alternative access networks and the selection of the most efficient and suitable one in order to meet the QoS requirements of a specific service, as this is defined by the user. With this way, the user receives enhanced QoE.

In order to accurately and reliably evaluate the performance of mobile computing systems, we propose the following simulation framework using SDEs.

8.2.2   QoS Enhancement in Mobile Computing Systems and Technical Challenges

While the QoS in mobile computing systems may be enhanced through innovative computing methods and new radio communication technologies, future trend should focus also on the mobile terminal cooperation and on the efficiency of resource reservation and allocation. Certain cooperation communication protocols and new cooperative techniques should involve the dynamic cooperation of mobile terminal nodes so that resources can be exchanged and common interests can be served (throughput optimization and power consumption minimization). The concept of mobile terminal cooperation introduces a new form of diversity that results in an increased reliability of the communication, leading both to the extension of the coverage and to the lifetime maximization.

The employment of cooperation among mobile terminal belonging to heterogeneous mobile computing systems has many technical challenges. Cooperative mobile computing systems have different needs compared to the individual mobile computing systems and generally to the cellular wireless data networks.

In principle, the mobile computing systems need to optimize their overall performance of power, task distribution, and resource usage. Network services should aim to achieve such an optimization. In practical implementations of the new cooperative techniques, these characteristics are often in conflict with each other, and then trade-off solutions, tailored to the specific scenario, are needed. The most important factors and technical challenges are briefly described: scalability, resource constraints (bandwidth spectrum, power), and mobile hardware and software constraints. One of the most important technical challenges involves the security and the privacy issues of the mobile computing system users. Finally, cooperation incentives have been proposed in order to enhance the cooperation. Some mobile terminals may refuse to cooperate in order to conserve their limited resources resulting in traffic disruption or overall QoS degradation in a mobile computing system. Mobile terminals exhibiting such behavior are termed selfish. Incentive mechanisms intend to provide a framework that forces players to cooperate for the best interest of all the participants. In other words, they provide a motive so that each individual mobile terminal prefers to work along with others, sometimes sacrificing their own resources and sometimes benefiting from the resources of others. The incentive mechanisms are usually distinguished into credit exchange systems and reputation-based systems.

8.3 CHANNEL MODELING AND MOBILE COMPUTING SYSTEMS EVALUATION

Wireless propagation channel modeling plays an important for the evolution and the evaluation of the mobile computing systems [4]. In recent years, the wireless systems have been accelerating at an extraordinary pace and have become merged into a global integrated network providing telecommunication, multimedia, navigation, broadcasting, and services to the users. The reliable communication is constrained by the radio propagation effects of the wireless links. The accurate design requires a comprehensive knowledge of the various propagation media and phenomena that differ on the frequency and the type of application. The choice of the relevant channel models is crucial in the design process and constitutes an important step in performance evaluation and testing of mobile computing systems.

Nowadays, frequencies ranging from 100 MHz to 100 GHz (from VHF to W band) are used for mobile radio communication systems. A number of propagation mechanisms such as reflections, diffractions, and scattering are the most important ones. These effects are usually caused by the local environmental features either at the transmitter's side or at the receiver's side. Additionally, large objects such as buildings and mountains may also intervene in the wireless link characteristics causing significant time spreading. Moreover, as we use higher frequencies especially for frequencies above 10 GHz, the tropospheric phenomena (rainfall rate, fog, turbulence, clouds) cause significant fading to the propagated signal. In modern mobile computing systems, the earlier-prescribed frequency bands are used for area coverage, including outdoor-to-outdoor, outdoor-to-indoor, and indoor-to-indoor links. They can be used for point-to-point links, point-to-multipoint links, and also for backhaul links. Fading can be categorized into two main categories: Large-scale fading is due to motion in a large area and can be characterized by the distance between transmitter and receiver and the atmospheric effects for higher frequencies. Small-scale fading is due to small changes in position (as small as half wavelength) or due to changes in the environment (surrounding objects, people crossing the line of sight between transmitter and receiver, opening or closing of doors, vegetation, wind turbulence for higher frequencies, etc.). In most of the cases, the received signal may be seen as the product of two random variables, the large-scale fading component and the small-scale fading component.

Very recently, computing systems, in the upcoming 5G network [17], operate at MMW frequencies providing high data rate multimedia services. At these frequencies, both atmospheric and multipath effects will be present. These propagation mechanisms in most of the references in the literature have been assumed statistically independent. Nevertheless, in a novel interesting analysis, addressing the possible correlation between these two fading effects was reported, which was further supported by the assumptions in [18–20]. This novel assumption has been further reinforced by some recent relevant findings originated from a different scientific field, namely, remote sensing and hydrology, where it was shown that there is a correlation between multipath amplitude variations with local rainfall records and land surface model predictions for soil moisture fluctuations.

The propagation channel plays the most significant role for the accurate simulation of the most important telecommunication parameters and the evaluation of the resource allocation algorithm. These are called *figures of merit* and are quantities that characterize the performance of mobile computing systems for specific applications and are characterized in different OSI layers.

In the physical layer, the most important metrics for the performance of mobile computing systems are as follows:

- *Signal-to-noise ratio* (SNR) is a term for the power ratio between the signal of the useful information and the noise power and consists of a deterministic part and a stochastic part (large-scale fading and multipath).
- *Signal-to-noise plus interference ratio* (SNIR) is the ratio of the wanted signal to the total power of the interfering signals and noise that is evaluated at a specific point of the transmission channel. If we have dual polarization channels, the depolarization powers may be considered as extra interfering channels, yielding the SNIDR, signal-to-noise plus depolarization and interference ratio.
- *Bit-error-ratio (BER)*, for example, the transmission BER, the number of erroneous bits received, divided by the total number of bits transmitted, and the information BER, the number of erroneous (decoded) corrected divided by the total number of decoded bits. Its instantaneous value depends on the channel state in information (CSI).
- *Ergodic capacity* is mostly used for fast varying fading channels, either flat or frequency selective, and is given as the average Shannon capacity over the fading wireless channel.
- *Outage capacity* is used for slowly varying channel where the instantaneous SNR is assumed to be constant for a large number of transmitting symbols. Outage capacity is characterized by an achieving threshold for a given outage probability and in most of the time is more practical than the average capacity.

Moreover, in order to evaluate the general performance of mobile computing networks, some key performance indicators (KPIs) that classified into two types, whether they describe the whole network resources or the QoS provisioned, have been proposed. The main KPIs that can be simulated in any data packet switched network and their performance depends on the channel model are as follows:

- *Latency or delay*: It is the required time for a data packet to get from a specific point to another of the mobile computing system.
- *Jitter*: It represents the delay variation of the received packets over time. Packets that are transmitted at a constant rate are not received necessarily at a constant rate due to the congestion of the network and other transmission techniques (retransmission protocol/time diversity due to the best selection of channel).
- *Peak data throughput*: It is the maximum rate that is achieved during the data transmission in the mobile computing network. This KPI is referred to a single wireless link.
- *Mean user throughput*: It sis the measure of the average rate that is achieved during the data transmission in the mobile computing network. The calculation is usually made by comparing the size of the transmitted data with the time of the transmission for both downlink and uplink cases.
- *Coverage/connectivity*: A mobile computing network is defined as connected when every random node is able to communicate with every other node. This KPI strongly depends on the choice of the channel model.

Finally, the future trends are to the design of *Green* mobile computing networks. The objective of Green Communications [21] is to decrease the energy consumption without deteriorating the QoS provision in the mobile computing systems. Consequently, these lead to the following optimization problem:

$$\min \left\{ Total\ Energy\ Consumption \right\} and \max \left\{ Throughput \right\}$$

$$\text{s.t.} \quad \text{QoS constraints}$$

(8.1)

that are required accurate channel model synthesizer in order to realistically find the optimal solution of the problem.

8.4 STOCHASTIC DIFFERENTIAL EQUATIONS RADIO COMMUNICATIONS FRAMEWORK

The accurate modeling of radio channel's time evolution is of high importance in current and future mobile computing systems. Moreover, the statistical approach of the communication problems is dominant since real radio signals are not deterministic and have a stochastic nature. Single and multidimensional long-term probability distributions representing the first-order statistics are not sufficient for the accurate radio channel characterization and modeling. The second-order statistics (such as covariance matrices, cross-correlation functions, and cross power spectrum) of more than one radio channels are very important and necessary in order to efficiently design the radio system's countermeasures such as time diversity, power control, and adaptive coding and modulation. A mathematical method to predict and generate time series of a stochastic process is the SDEs [3–5]. Multidimensional SDEs (multi-SDEs) of correlated stochastic processes are described with the following general expression of the system of SDEs [15]:

$$dX_t = b(t, X_t)dt + \sigma(t, X_t)dW_t, \tag{8.2}$$

where
  X_t is an $N \times 1$ vector of the N correlated stochastic processes whose time evolution and dynamic properties are needed
  $b(t, X_t)$ is a $N \times 1$ vector named as drift vector
  $\sigma(t, X_t)$ is the $N \times r$ dispersion matrix
  W_t represents the r-dimensional Brownian motion

The components of the drift vector and the dispersion matrix are very important for the accurate modeling, due to the fact that they can be computed so that the resulted time series reproduce the first- and second-order statistics of the modeled stochastic processes, and in overall, they capture all the continuous-time statistical properties. The drift vector and the dispersion matrix can be computed either from experimental data or from analytical physical mathematical models. The SDEs can be solved either analytically and numerically or only numerically depending on the expressions of the drift vector and dispersion matrix. For the derivation of analytical solutions, various methods have been developed based on the expressions of the elements of the vector $\mathbf{b}(t,\mathbf{X}_t)$ and the matrix $\sigma(t,\mathbf{X}_t)$ using Itô's stochastic integrals and Itô's formulas [4,5].

At this point, we will briefly present a general framework that can be applied for synthesizing stochastic composite channels (narrowband and wideband) using multidimensional SDEs for the SNR of the links or the signal attenuations themselves. More specifically, we present a general framework for large-scale fading channels, and we consider the received SNR from n base stations (BSs) at k specific points. We denote by $\gamma_i(t)$, $i = 1,2, ...,m = n \cdot k$, the received SNR in linear scale, that is, every index corresponds to a specific location for a specific BS. The SNRs in linear scale $\gamma_i(t)$ can be modeled as a lognormal r.v. for each time instant. Its long-term statistical parameters are σ_i and $\ln\left(\gamma_{med_i}\right)$. We assume that the resulting shadowing vector process $\mathbf{X}(t) = [X_1(t), ...,X_m(t)]$ after applying the nonlinear transformation

$$X_i(t) = \xi \ln\left(\frac{\gamma_i(t)}{\gamma_{m_i}}\right), \quad 1 \le i \le m \tag{8.3}$$

to each component process $\gamma_i(t)$, is a solution to a linear m-dimensional SDE of Ornstein–Uhlenbeck type [4,5]:

$$d\mathbf{X}_t = \mathbf{A} \cdot \mathbf{X}_t dt + \mathbf{B} \cdot d\mathbf{W}_t, \quad \mathbf{X}_0 = \mathbf{x}_0, \tag{8.4}$$

where $\mathbf{A}$ is the diagonal matrix $\mathbf{A} = [a_{ij}]_{1 \le i,j \le m}$ with elements:

$$a_{ij} = -a_i \cdot \delta_{ij}, \tag{8.5}$$

δ_{ij} is the Kronecker delta function
$a_i, 1 \le i \le m$ are the dynamic parameters of shadowing, in principle different for each link
$\mathbf{X}(0) = \mathbf{x}_0$ is the initial condition of the SDE

The solution to the m-dimensional SDE is straightforward and is given [4,15] as follows:

$$\mathbf{X}_t = e^{t \cdot \mathbf{A}} \cdot \mathbf{X}_0 + e^{t \cdot \mathbf{A}} \cdot \int_0^t e^{-s \cdot \mathbf{A}} \cdot \mathbf{B} \cdot d\mathbf{W}_s, \tag{8.6}$$

where

$$e^{t \cdot \mathbf{A}} = \sum_{m=0}^{\infty} \frac{t^m}{m!} \mathbf{A}^m. \tag{8.7}$$

Now due to the fact that the matrix $\mathbf{A}$ is diagonal (see [15]), it can be easily verified that

$$\left[e^{t \cdot \mathbf{A}} \right]_{ij} = e^{-a_i \cdot t} \delta_{ij}. \tag{8.8}$$

The solution stochastic process $\mathbf{X}_t$ as given in (8.6) is a Gaussian process if $\mathbf{X}_0$ follows an m-variate normal distribution, including the degenerate case. The mean vector of $\mathbf{X}_t$, $\mathbf{MX}(t)$ is given by

$$\mathbf{M}_{\mathbf{X}}(t) = e^{t \cdot \mathbf{A}} \cdot \mathbf{M}_{\mathbf{X}}(0). \tag{8.9}$$

The covariance matrix $\mathbf{CX}(t)$ of the vector process $\mathbf{X}_t$ for each time instant t is given as follows:

$$\mathbf{C}_{\mathbf{X}}(t) = e^{t \cdot \mathbf{A}} \left[\mathbf{C}_{\mathbf{X}}(0) + \int_0^t e^{-s \cdot \mathbf{A}} \mathbf{B} \mathbf{B}^T e^{-s \cdot \mathbf{A}^T} ds \right] e^{t \cdot \mathbf{A}^T}, \tag{8.10}$$

which is the solution to the following linear differential equation:

$$\dot{\mathbf{C}}_{\mathbf{X}}(t) = \mathbf{A}\mathbf{C}_{\mathbf{X}}(t) + \mathbf{C}_{\mathbf{X}}(t)\mathbf{A}^T + \mathbf{B}\mathbf{B}^T. \tag{8.11}$$

The matrix $\mathbf{A}$ has by definition all of its eigenvalues real and negative (equal to $-a_i$), so the convergence of the following integral is assured:

$$\mathbf{C}_{\mathbf{X}} = \int_0^{\infty} e^{s \cdot \mathbf{A}} \mathbf{B} \mathbf{B}^T e^{s \cdot \mathbf{A}^T} ds. \tag{8.12}$$

It is easy to verify that if $\mathbf{CX}(0) = \mathbf{CX}$, then $\mathbf{CX}(t) = \mathbf{CX}$ is a solution too. This means that a stationary solution to (8.11) exists, and in this case for (8.12), we are leading to the following algebraic matrix equation:

$$\mathbf{A}\mathbf{C}_{\mathbf{X}} + \mathbf{C}_{\mathbf{X}}\mathbf{A}^T = -\mathbf{B}\mathbf{B}^T. \tag{8.13}$$

This equation forms the physical basis of the model. The stationary covariance matrix $\mathbf{CX}$ of $\mathbf{X}_t$ is equal to the covariance matrix of the shadowing process $\mathbf{S}(t)$. Thus, existing models taken from measurements or ray tracing models for the shadowing process $\mathbf{S}(t)$ permit the derivation of the covariance matrix $\mathbf{CX}$ of $\mathbf{X}_t$. Then, Equation 8.13 can be used for the determination of the transformation matrix $\mathbf{B}$, which is required for the dynamic modeling of the process $\mathbf{X}_t$ in Equation 8.6, given the matrix $\mathbf{A}$ and the stationary covariance matrix $\mathbf{CX}$.

If we denote $\mathbf{G} = \mathbf{BB}^T$, then we can determine that $\mathbf{G}$ is given by

$$\left[\mathbf{G}\right]_{ij} = \left(a_i + a_j\right)[\mathbf{C_X}]_{ij}. \tag{8.14}$$

The matrix $\mathbf{CX}$ as a covariance matrix is real and symmetric and from the same is true also for $\mathbf{G}$. The decomposition of $\mathbf{G}$ as the product $\mathbf{BB}^T$ is straightforward and can be realized via Cholesky decomposition. By use of the solution process, a straightforward calculation based on the properties of the stochastic integral leads to the cross-covariance for two time instants $t_1 < t_2$

$$\mathbf{C_X}(t_1, t_2) = \mathbf{C_X}e^{(t_2 - t_1)\cdot\mathbf{A}^T} \tag{8.15}$$

in direct relation to (8.10). This approximation may be considered valid as long as the movements are not of large scale and a stationary covariance matrix can approximate the shadowing correlations and cross-correlations, irrespective of the user's mobility. In this case, the correlations and cross-correlations have to be calculated for a representative separation angle and distance, while the dynamic parameters account also for the user's mobility.

The earlier-described general simulation framework of large-scale fading channels can be easily implemented using MATLAB® software. The suggested radio channel framework suits very well for computer simulations, and it is universal since a single structure allows the modeling of a great variety of different processes by a simple variation of the system parameters and excitations.

8.5 SIMULATION OF COMPLEX MOBILE COMPUTING NETWORKS PARADIGMS

The performance of complex modern radio communication systems taking into account channel fading is of the utmost importance in current wireless networks. The earlier-described general methodology may be used for the accurate simulation of cooperative systems, cognitive radio systems, multiple-antenna systems, multiple relays configurations, etc. More specifically, in the first part of this section, we calculate the outage probability of general multihop configurations that operate under correlated shadowing channels. In the second part of this section, we present the numerical results of generalized conditional probability vs. CIR threshold for an interference scenario between a satellite network and a fixed wireless access network that employ fade mitigation techniques (FMTs).

Summing up, the proposed radio communication tool that accurately simulates both the fading environment can be incorporated in communication protocol platforms in order to optimize the radio mobile computing system.

8.5.1 MULTIHOP TERRESTRIAL SYSTEMS

Multihop terrestrial systems are very important for the provision of multimedia services to the end users. The calculation of the outage probability of a multihop system is not an easy task, and accurate simulation tools are required for its calculation. The general geometry of the system with N relay nodes can be found in Figure 8.1. In this figure, $\mathbf{T}$ is the transmitter node, $\mathbf{RN}_i$ is the relay node $\mathbf{I}$, and $\mathbf{R}$ is the final receiver.

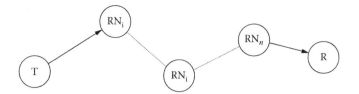

FIGURE 8.1   General geometry of a multihop mobile computing system.

Different relay types can be used in multihop system, such as regenerative and nonregenerative relays [22]. The regenerative relays use the decode-and-forward (DF) technique. In the DF technique, the signal that the relay node receives from the previous hop is regenerated using the full receiver–transmitter processing chain, which contains the sequence of demodulation, channel decoding, encoding, and modulation. Finally, the information signal is forwarded to the next hop. For N serial relays and $N + 1$-hops, it is possible to formulate the equivalent end-to-end SNR as given:

$$SNR_{reg\_eq} = \min\left(SNR_1, SNR_2, \ldots, SNR_N, SNR_{N+1}\right). \tag{8.16}$$

In the previous expression, SNR_j is the received SNR of the $j(j = 1, \ldots, N + 1)$ hop. It can be seen by the expression given earlier that the minimum SNR characterizes the overall (end-to-end) multihop performance of regenerative senor relays against outages (in other words, SNR regions below a threshold that yield system outages).

The nonregenerative relays use the amplify-and-forward (AF) technique. In this case, with either fixed or adaptive gain, the relay amplifies the received signal from the previous hop and forwards it to the next hop without decoding it. As a result, the additive noise from the previous hop is also amplified and forwarded to the next hop. More specifically, there are two types of nonregenerative relays: the adaptive gain relays and the fixed gain relays. The adaptive gain relays amplify the received signal with an adaptive gain in order that the transmitted power remains fixed. On the other hand, the fixed gain relays amplify the received signal with a fixed gain, and as a result, the transmitted power varies. These relays are less complex, but they cannot compensate for the fading of the first hop that affects the transmitted power (amplifier's back-off effect). In the following, SNR_j is defined as the received information SNR and TNR_j is defined as the total received power-to-noise ratio of hop j including the received information signal power plus the noise power of the previous hops that are also transmitted by the nonregenerative relay (as calculated in [22] for a similar dual-hop problem). The total end-to-end SNR of an adaptive gain system with N serial relays and $N + 1$ hops can be calculated as [22] follows:

$$SNR_{non\text{-}reg\_eq} = \left[\prod_{j=1}^{N+1}\left(1 + \frac{1}{TNR_j}\right) - 1\right]^{-1}. \tag{8.17}$$

In expression (18.7), each relay estimates the combination of fading and noise, and it amplifies the received signal without separating the noise from the faded transmitted signal. If a more complex relay is used that is able to estimate the exact fading of the previous hop, this fading can be compensated. As a result, the received information SNR_j remains fixed instead of the total received TNR_j. Therefore, expression (8.17) may be simplified as follows:

$$SNR_{non\text{-}reg\_eq}^{-1} = SNR_1^{-1} + SNR_2^{-1} + \cdots + SNR_N^{-1} + SNR_{N+1}^{-1}. \tag{8.18}$$

In the fixed gain relay case, the SNR_j and the TNR_j also depend on the fading of the previous hops, and expressions (8.17) and (8.18) need further calculation.

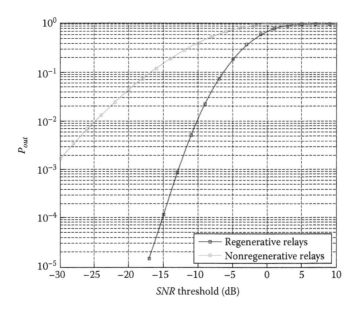

FIGURE 8.2   Outage probability for two relays with standard deviation 3.8 dB and correlation coefficient 0.5.

Employing the general simulation model in Section 8.3, the outage probability of the equivalent SNR can be calculated as

$$P_{outage} = P\left[SNR_{eq} \leq SNR_{th}\right]. \tag{8.19}$$

We derive simulated time series using the model in Section 8.3, and by employing the corresponding formulas (8.16) through (8.18) and for the calculation of equivalent SNR, we can evaluate the outage probability of a multihop system.

From Figure 8.2, we can see that the regenerative relays have much better performance than the corresponding transparent ones as it is expected. Moreover, in Figure 8.3, it is obvious that as the number of relays is increased, the outage probability deteriorates, but sometime due to the complexity of the mobile computing system, the radio relays cannot be avoided in order to establish the communication between two mobile nodes.

8.5.2 INTERSYSTEM INTERFERENCE: COEXISTENCE OF SATELLITE AND TERRESTRIAL SYSTEMS OPERATING ABOVE 10 GHz

In this subsection, we will demonstrate the usefulness of the simulation framework for the calculation of intersystem interference statistics between satellite and terrestrial systems operating above 10 GHz. The described problem is difficult, and an accurate channel model is required in order to obtain the carrier-to-interference statistics.

The effective use of the electromagnetic spectrum has necessitated the assignment of certain frequency bands above 10 GHz to both satellite and terrestrial services. Specifically, commercial geostationary satellite communication (SatCom) systems are currently making aggressive use of the *Ku* (12/14 GHz) frequency band, and operators are planning future services in the *Ka* (20/30 GHz) and *V* (40/50 GHz) bands to benefit from the extensive bandwidth available. On the other hand, recent activities from standardization institutes such as the IEEE (802.16 WirelessMAN) and ETSI (BRAN HIPERACCESS) have led to the specification of fixed broadband wireless access (FBWA) networks operating at MMWs in the range of 10–66 GHz under the coordination of the WiMax

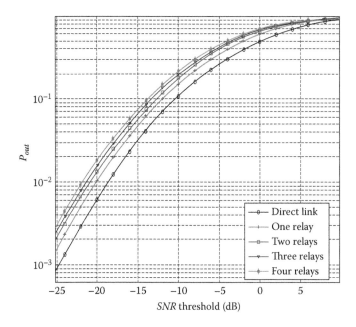

FIGURE 8.3   Outage probability for multiple relays with standard deviation 8 dB and correlation coefficient 0.5.

(worldwide interoperability for microwave access) forum. This situation clearly raises the issue of coexistence whenever the two types of services are collocated and no coordination agreement is effected.

Until recently, frequency sharing between the fixed/broadcasting satellite service (FSS/BSS) and the fixed service (FS) on a primary basis presented no serious restrictions because fixed terrestrial systems and satellite earth stations were small in number. The massive application of FSS/BSS systems oriented to deliver broadcast or direct-to-user services via the very successful Digital Video Broadcasting-Satellite (DVB-S) standard, along with the expected market penetration of FBWA networks, changed the situation dramatically and forced the European Radiocommunications Committee (ERC) of the European Conference of Postal and Telecommunications Administrations (CEPT) to determine a regulatory framework for the future deployment of FSS/BSS and FS systems. Hence, the reliable design of both SatCom and FBWA networks necessitates the exact evaluation of the potential impact of intersystem interference on the system availability and QoS. Figure 8.4 depicts an example of such an interference scenario: A beam belonging to a multibeam geostationary satellite system is located in the proximity of an FBWA cell. It is assumed that both systems operate at the same frequency and polarization. While communicating with its fixed subscriber stations, the BS interferes with the nearby satellite terminals.

For the accurate estimation of the corresponding performance degradation due to co-channel interference (CCI), it is imperative to take into account the effect of precipitation, that is, the predominant atmospheric phenomenon affecting radio wave propagation in the 10–50 GHz spectral region. With modern satellite systems migrating to higher frequency bands, rain attenuation impacts the physical layer performance to an even larger extent (attenuation in dB is proportional to the square of the frequency of operation). Note that the previous statement concerns both SatCom and FBWA, since operation in a common frequency band implies aggravation by the same atmospheric phenomena. For this reason, the new generation air interface standards for SatCom (e.g., DVB-S2) and for FBWA employ a variety of FMTs to counteract physical layer deterioration. This is done either by continuously measuring link quality and responding in real time to the change in the propagation conditions (*adaptive* FMT) or by enhancing system performance through an additional fixed margin (*static* FMT).

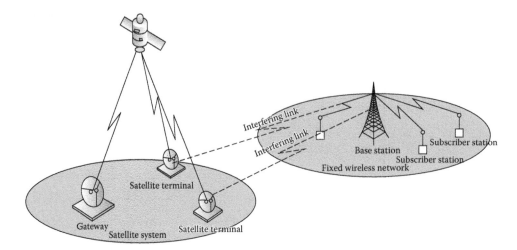

FIGURE 8.4   Intersystem interference scenario: base station of an FBWA cell interfering with a nearby satellite terminal within a beam of a SatCom system.

Generally, FMTs may be distinguished into three major categories: (1) effective isotropically radiated power control techniques, which consists in varying either the carrier power or the antenna beam shape in order to compensate for the power losses due to propagation effects; (2) adaptive transmission techniques, which switch to a more robust modulation scheme, coding rate, and/or data rate, whenever the link quality is degraded; and (3) diversity protection schemes, which offer an alternative channel by taking advantage of the spatial, spectral, and temporal properties of the rainfall medium.

Here, we employ the general simulation framework from Section 8.3 and presented in [15] to incorporate the impact of typical FMTs on the given interference scenario. A snapshot of the correlated rain attenuation time series of the wanted and the interfering links is shown in Figure 8.5. The current trend for the new generation of standards mentioned earlier is to use a combination of FMTs only when needed. In this course, DVB-S2 and WiMax standards specify the use of adaptive

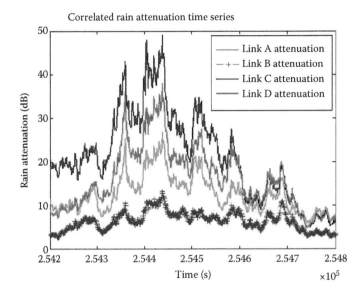

FIGURE 8.5   A snapshot of correlated rain attenuation time series of converging and parallel terrestrial and satellite links.

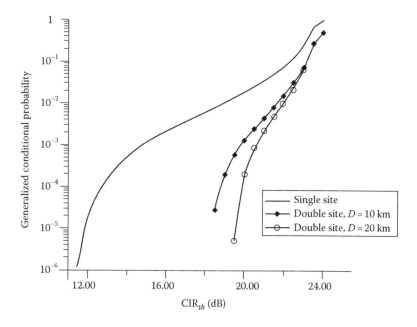

FIGURE 8.6 CIR statistics of a single- and a double-site satellite communication system.

power control, coding, and modulation techniques. Nevertheless, for the purposes of our analysis, it is assumed that the interfering terrestrial station is equipped with downstream power control (DPC), whereas the receiving satellite terminal is linked with a second earth station to form a site diversity (SD) configuration. DPC is typically employed by LMDS BSs and, under certain propagation conditions, gives rises to a significantly higher level of intersystem interference in comparison to the static output power case.

On the other hand, SD is considered a highly efficient FMT that has been shown to offer an increased tolerance to intersystem satellite interference, although very expensive due to redundant equipment. SD takes advantage of the spatial characteristics of the rainfall medium by using two earth stations to exploit the fact that the probability of attenuation due to rain occurring simultaneously on the alternative Earth–space paths is significantly less than the relevant probability occurring on either individual path.

Employing the rain attenuation time series on all the links under consideration, we can calculate the statistics of the carrier-to-interference ratio (CIR) and the carrier-to-noise plus interference ratio. In Figure 8.6, we present the generalized conditional probability vs. CIR threshold for a single satellite link and two dual reception schemes for 10 and 20 km, respectively.

It can be seen from Figure 8.6 that there is a significant improvement in the interference statistics using dual diversity reception, and there is further enhancement of the performance employing greater separation distances.

8.6 FUTURE RESEARCH DIRECTIONS AND CONCLUSIONS

A further research direction on this subject is the development of a stochastic radio channel modeling framework that deals with the problem of space–time variations of the channel in amplitude and phase that is a very critical point in current multiantenna radio systems, MIMO and virtual MIMO. New multidimensional SDEs are necessary to be developed in order to generate time series of the MIMO channels for correlated or uncorrelated paths.

Another research direction is the stochastic dynamic modeling of the aggregate radio interference in mobile computing systems. This leads to the fact that radio interference will be seen as a stochastic process and a solution of an SDE [23–25]. To be more specific, radio interference may be modeled as space–time random field, and the solution of SDE will lead to an accurate time series synthesizer. In the beginning, the long-term statistical distributions and properties of the radio network interference will be generally modeled, starting from a novel mathematical framework as presented in Section 8.3. For modeling the radio interference, the network geometry and the propagation effects will be needed. Therefore, the identification of the spatial distribution of the transmitters (e.g., Poisson process and random walk mobility model) as well as the modeling of propagation effects such as path loss, shadowing, and multipath fading is needed. Using the concept of correlated or uncorrelated scattered nodes, the probability density function of the aggregated interference may be derived and categorized in different classes for the stochastic network geometries and long-term statistics of propagation effects. After the calculation of the probability density function of the network interference and its categories, the problem of the space–time dependence of interference needs to be addressed with the adoption of space–time correlation functions for two cases: (1) various individual interfering nodes and (2) a unique space–time correlation function for the aggregate interference distribution. The necessity of these two different approaches that lead to two different synthesizers comes from the fact that the first one is more accurate but more complex than the second. Therefore, depending on the trade-off between the accuracy and the calculation time, different applications may use different synthesizers. In the first case, models examined in the previous task for the autocorrelation and the generation of time series for shadowing and multipath fading will also be used in this task. Since the interference is the sum of the radio signal from all the interfering nodes, multidimensional SDEs will also be used for the generation of the time series of every single interfering signal. For the second case, the model for generating time series of the interference will also be developed using SDEs to model the network interference as a unique stochastic process. The stochastic dynamic model of radio aggregate interference will have numerous applications in cognitive radio networks [26–29], wireless packet networks, and the coexistence between ultra wideband and narrowband radio communications systems.

This chapter has presented a general radio communication framework for the simulation modeling of wireless channels. The most important parameters for the characterization of the channel and some important performance metrics have been presented. Moreover, the presented simulation framework based on SDEs has been applied for the performance of multihop terrestrial systems and interference statistics for terrestrial and satellite systems operating above 10 GHz and employing FMTs.

ACKNOWLEDGMENTS

This work was supported by the Operational Program *Education and Lifelong Learning* under the project ARISTEIA II «FLexible RAdio CoMmunication Optimization FramEwork and Applications» (FLAME) in National Technical University of Athens.

REFERENCES

1. M. Ilyas and I. Mahgoub, *Mobile Computing Handbook*, Auerbach Publications, Boca Raton, FL, 2004.
2. D. Middleton, *An Introduction to Statistical Communication Theory*, McGraw-Hill, New York, 1960.
3. A. Goldsmith, *Wireless Communications*, Cambridge University Press, Cambridge, U.K., August 2005.
4. S. Primak, V. Kontorovich, and V. Lyandres, *Stochastic Methods and Their Applications to Communications: Stochastic Differential Equations Approach*, John Wiley & Sons, Chichester, U.K., 2004.
5. I. Karatzas and S. E. Shreve, *Brownian Motion and Stochastic Calculus*, Springer-Verlag, New York, 2005.
6. S. Karlin and H. Taylor, *A Second Course in Stochastic Processes*, Academic Press, New York, 1981.
7. S. Karlin and H. Taylor, *A First Course in Stochastic Processes*, Academic Press, New York, 1975.
8. B. Øskendal, *Stochastic Differential Equations: An Introduction with Applications*, Springer-Verlag, Berlin, Germany, 2003.

9. K. Ito, Stochastic integrals, *Proceedings of the Imperial Academy*, Tokyo, Japan, pp. 519–524, 1944.
10. T. Feng, T. R. Field, and S. Haykin, Stochastic differential equation theory applied to wireless channels, *IEEE Transactions on Communications*, 55 (8), 1478–1483, August 2007.
11. C. D. Charlambous, R. J. C. Bultitude, and X. Li, J. Zhan, Modeling wireless fading channels via stochastic differential equations: Identification and estimation based on noisy measurements, *IEEE Transactions on Wireless Communications*, 7 (2), 434–439, 2008.
12. M. M. Olama, S. M. Djouadi, and C. D. Charalambous, Stochastic differential equations for modeling, estimation and identification of mobile-to-mobile communications channels, *IEEE Transactions on Wireless Communications*, 8 (4), 1754–1763, 2009.
13. A. D. Panagopoulos, P. M. Arapoglou, and P. G. Cottis, Satellite communications at Ku, Ka and V bands: Propagation impairments and mitigation techniques, *IEEE Communication Surveys and Tutorials*, 3rd Quarter, October 1–13, 2004.
14. S. A. Kanellopoulos, G. Fikioris, A. D. Panagopoulos, and J. D. Kanellopoulos, A modified synthesis procedure for 1st order stochastic differential equations for the simulation of base band random processes signal processing, *Elsevier Signal Processing*, 87 (12), 3063–3074, December 2007.
15. G. Karagiannis, A. D. Panagopoulos, and J. D. Kanellopoulos, Multi-dimensional rain attenuation stochastic dynamic modeling: Application to earth-space diversity systems, *IEEE Transactions on Antennas and Propagation*, 60 (11), 5400–5411, November 2012.
16. D. Charilas and A. D. Panagopoulos, Network selection problem in multi-access radio network environments, *IEEE Vehicular Technology Magazine*, December 2010.
17. T. S. Rappaport, Gutierrez, F. Jr., Ben-Dor, E., Murdock, J. N., Qiao, Y., and Tamir, J. I., Broadband millimeter-wave propagation measurements and models using adaptive-beam antennas for outdoor urban cellular communications, *IEEE Transactions on Antennas and Propagation*, 61 (4), 1850–1859, April 2013.
18. K. P. Liolis, A. D. Panagopoulos, and S. Scalise, On the combination of tropospheric and local environment propagation effects for mobile satellite systems above 10 GHz, *IEEE Transactions on Vehicular Technology*, 59 (3), 1109–1120, March 2010.
19. A. D. Panagopoulos, K. P. Liolis, and P. G. Cottis, Rician K-factor distribution in broadband fixed wireless access channels under rain fades, *IEEE Communications Letters*, 11 (4), 301–303, April 2007.
20. V. K. Sakarellos, D. Skraparlis, A. D. Panagopoulos, and J. D. Kanellopoulos, Outage performance analysis of a dual-hop radio relay systems operating at frequencies above 10 GHz, *IEEE Transactions on Communications*, 58 (11), 3104–3109, November 2010.
21. K. Mantzoukas, S. Sagkriotis, and A. D. Panagopoulos, On the design of energy-efficient wireless access networks a cross-layer approach, in S. Khan and J. L. Mauri, eds., *Green Networking and Communications: ICT for Sustainability*, CRC Press, Boca Raton, FL, 2013.
22. V. K. Sakarellos, D. Skraparlis, and A. D. Panagopoulos, Cooperative transmission techniques and protocols in wireless sensor networks, in J. Loo, J. L. Mauri, and J. H. Ortiz, eds., *Wireless Sensor Networks: Current Status and Future Trends*, CRC Press, Boca Raton, FL, November 16, 2012.
23. P. C. Pinto and M. Z. Win, Communication in a Poisson field of interferers—Part I: Interference distribution and error probability, *IEEE Transactions on Wireless Communications*, 9 (7), 2176–2186, 2010.
24. P. C. Pinto and M. Z. Win, Communication in a Poisson field of interferers—Part II: Channel capacity and interference spectrum, *IEEE Transactions on Wireless Communications*, 9 (7), 2187–2195, 2010.
25. M. Z. Win, P. C. Pinto, and L. A. Shepp, A mathematical theory of network interference and its applications, *Proceedings of the IEEE*, 97 (2), 205–230, February 2009, Invited Paper in Special Issue on Ultra-Wide Bandwidth (UWB) Technology and Engineering Applications.
26. D. Niyato and E. Hossain, Competitive spectrum sharing in cognitive radio networks: A dynamic game approach, *Wireless Communications, IEEE Transactions on*, 7 (7), 2651–2660, July 2008.
27. M. van der Schaar and F. Fu, Spectrum access games and strategic learning in cognitive radio networks for delay-critical applications, *Proceedings of the IEEE*, 97 (4), 720–740, April 2009.
28. M. Poulakis, S. Vassaki, A. D. Panagopoulos, and Ph. Constantinou, Effects of spatial correlation on QoS-driven power allocation over Nakagami-m fading channels in cognitive radio systems, *Transactions on Emerging Telecommunications Technologies*, accepted for publication, published online, November 2013.
29. S. Vassaki, M. Poulakis, A. D. Panagopoulos, and Ph. Constantinou, Power allocation in cognitive satellite terrestrial networks with QoS constraints, *IEEE Communication Letters*, 17 (7), 1344–1347, July 2013.

9 How to Simulate and Evaluate Multicast Routing Algorithms

Maciej Piechowiak and Piotr Zwierzykowski

CONTENTS

ABSTRACT

This chapter describes routing algorithms for multicast connections in packet networks and analyzes their efficiency with computer simulations.

The first part presents an analysis of the methods for network topology modelling that are to represent real telecommunications and computer networks with the application of graphs. The most important parameters for the description of the network that allow the selected models to be compared qualitatively are presented and discussed. The adopted methods for modelling network topology make it possible to formulate a methodology for a comparative research study of multicast routing algorithms. The efficacy of the proposed algorithms is examined in the second part of the chapter.

The criteria for the evaluation of the effectiveness are (1) the average cost of trees constructed by the algorithms under study and (2) the average cost of paths determined in the constructed trees. The comparative analysis of the algorithm includes the network parameters and the methods for network topologies described in the second part of the chapter.

9.1 INTRODUCTION

Present-day computer networks operate under the paradigm of sustainability of distributed service delivery of multimedia data [1], videoconferences [2], and software distribution [3]. Data are transferred not just between a pair of users any more but also between single users and groups of users. It thus follows that the optimization of data transmission has never been as important as it is today and, what is even more, is becoming increasingly crucial. In issues concerning multicast routing in general, considerations on the effective method for transmission between a given node of the network (traffic source) and a given group of receivers are therefore of utmost importance. Effectiveness means in this particular case a minimization of network resources during data transmission with simultaneous provision of required quality parameters (e.g., bandwidth, delay, and cost of connection).

The communication model for multicast connections provides an opportunity to reduce traffic by transmitting single packets through routers from the sender to the locations where hosts interested in receiving the data are located. Such a communication model requires special routing algorithms to be applied. These algorithms construct distribution trees (also known as multicast trees) so that packet transmission in the network can be executed. The concept of multicast routing distinguishes between multicast core-based trees [4] and multicast shared trees [5].

The analysis of routing algorithms for multicast connections involves a concomitant definition of the way the network in which the algorithms are to be implemented will be represented. The problem of the appropriate representation of the network and its influence upon the efficiency and effectiveness of the algorithms under scrutiny is analyzed in [6,7]. Reference [8] proves that in networks in which nodes are arranged and connected randomly, the effectiveness of multicast algorithms is at least twofold lower than that in hierarchical networks that reflect the properties of the internet network.

It has then been proved that the topology of telecommunications and computer networks, understood as the method for a deployment of nodes in a given area and the way of interconnecting the nodes involved in the network, represents a significant aspect of the analysis of multicast routing algorithms [9]. In [10], such a model is proposed with the application of random graphs that reflect the real properties of the network. Barabási and Albert [11] consider scale-free networks that are characterized by preferential addition of new nodes and by incremental methods for the construction of the network. On the basis of the proposed models in [7,12,13], a number of network topology generators have been developed.

Some authors [14,15] prove that the topology parameters of the present-day Internet are characterized by power law dependencies [16]. An analysis of the Border Gateway Protocol (BGP) routing tables confirms that this functional relationship occurs at the level of autonomous systems (ASs level) [16]. This fact has been instrumental in constructing effective heuristic Internet network topology generators [17–19].

Research studies conducted in numerous research centers over the past few years clearly indicate the importance of the problem of appropriate modelling of the topology of telecommunications and computer networks. These investigations involve in the first place methods for constructing appropriate models that would reflect real phenomena taking part in networks [20–22] as well as the analysis of the changeable structure of the Internet network [23]. The latter problem is particularly important for studies on protocols and algorithms for multicast connections in which, besides the method for arranging nodes and links, the way of arranging destination nodes against one another in the network is of significant importance.

This chapter will discuss the effectiveness of the most commonly used heuristic algorithms for multicast connections, as well as their comparative usefulness. The network model, multicast tree, and representative heuristic algorithms are introduced. Methods that generate network structures representing the topology are placed under scrutiny and basic network parameters are also given attention. The most important part of the chapter is the presentation of tools forming a coherent simulation environment. The chapter will also include results of the simulation of the implemented algorithms along with their interpretation.

9.2 MULTICAST COMMUNICATIONS

Optimization of the multicast communication is based on a construction of effective routing algorithms that are to create a distribution tree with the minimum cost between the node sender and a group of receiving nodes. Such a method for communication prevents excessive multiplication of packets in the network—some defined portion of sent data reach only those nodes (routers) that are directly connected with a defined group of receivers, the so-called members of a multicast group. A construction of a tree connecting the sender with a group of receivers in the network that would take into account appropriate quality parameters of transmission is presented as the optimization problem in the graph, called the Steiner problem. The tree itself is defined as the minimum Steiner tree [24]. Due to the time and computational complexity of the problem, which is an *NP*-complete problem, the literature of the subject offers a number of heuristic solutions, for example, those presented in [10,25–38]. On the basis of these algorithms, multicast trees are constructed by way of optimization of one criterion (without any constraints) and by taking into account another limiting criterion (most frequently delay that represents the quality requirements of transmission).

Beside heuristic algorithms for optimization of multicast routing, the literature also considers a possibility of the implementation of genetic algorithms [39–43], the use of neural networks [44,45], or ant colony optimization algorithms [46,47]. These solutions, however, are not as effective as heuristic algorithms and exceed the scope of this chapter.

9.2.1 MULTICAST ROUTING ALGORITHMS

The simplest heuristic approach solves the *minimum Steiner tree* problem with delay constraints called the *constrained shortest path tree* (CSPT) and relies on computation of the shortest paths between the source and receivers. Individual paths have the minimum length, but multicast distribution tree constructed in this way is not optimal. Wang and Crowcroft prove that if network links contain at least two additive metrics, then *QoS* routing is an *NP*-complete problem [35].

The KMB heuristics (*Kou, Markowsky, Berman*) [27] is one of the best known heuristics solving the problem of the minimal Steiner tree. It is also very effective as far as the accuracy of the solution is concerned [8] and its computational complexity is $O(|M||V|^2)$. This heuristics is the basis for the KPP algorithm (*Kompella, Pasquale, Polyzos*) [26] that, additionally, takes into consideration the delay constraint. During the first phase of the KPP, a complete graph is constructed whose all vertices are the source node s and the destination nodes $m_i \in M$, while the edges represent the least cost paths connecting any two nodes a and b in the original graph $G = (V, E)$, where $a,b \in \{M \cup s\}$. Then, the minimal spanning tree is determined in this graph taking the delay constraint Δ into consideration, and then the edges of the obtained tree are converted into the paths of the original graph G. Any loops that appear in thus formed structure are removed with the help of the shortest path algorithm, for instance, by Dijkstra algorithm [48]. The computational complexity of the algorithm is $O(\Delta|V|^3)$.

The operation performed by *multicast Lagrange relaxation algorithm* (MLRA), proposed by authors in [49], consists in determining the shortest path tree between the source node s and each destination node m_i along which the maximum delay value Δ cannot be exceeded. The path calculated with an application of Lagrange relaxation algorithm refers to idea proposed by [50]. This algorithm relies on minimizing the aggregated cost function: $c\lambda = c + \lambda d$. In each iteration of the algorithm, the current value of λ parameter is calculated in order to increase the dominance of delay in the aggregated cost function, if the optimum solution of $c\lambda$ meets the delay requirements (Δ).

The paths are determined one by one and are then added to the multicast tree. If there is at least one path that does not meet the specified requirements, a multicast tree cannot be constructed. Since the network structure created in this way may contain cycles, in order to avoid them, Prim's algorithm is used [51].

The main goal of the *K-shortest path multicast algorithm* (KSPMA) proposed by the authors in [52] is to build a CSPT that calculate paths between the source and each destination node using the *K*-shortest path algorithm. The modification of the original algorithm is based on solving the delay among each of *K*-shortest paths and choosing the path with minimum cost with the delay value not exceeding the maximum delay Δ along the path.

The operation performed by the proposed KSPMA consists in determining the shortest path tree between the source node s and each destination node m_i along which the maximum delay value (Δ) cannot be exceeded. The paths are determined one by one and then added to the multicast tree. If there is at least one path that does not meet the requirements, a multicast tree cannot be constructed. Since the network structure created in this way may contain cycles, in order to avoid them, Prim's algorithm is used.

9.2.2 RESEARCH METHODOLOGY

A reliable evaluation of multicast routing algorithms requires a proper methodology of simulation research. The methodology should regard the following aspects:

- A set of results of heuristic algorithm (i.e., total cost of multicast tree, cost of path between source node, and destination node in a tree)—they specify the quality parameters of multicast trees and are the main criterion for the evaluation of an algorithm.
- Input parameters of algorithm—the algorithm retrieves data that represent a communication network in fixed format (i.e., the representation of a graph as an adjacency matrix). Then, it constructs the multicast tree and returns a result set. The algorithm also takes the set of destination (multicast) nodes (M) and abundance of this set (m). In the case of constrained algorithms, the maximum delay (Δ) is then an additional parameter.
- The way of generating network topology—network generation method determines the placement of the nodes on the plane and the way of network nodes connection in a coherent structure. It also contains a method of generating link parameters. The key topology parameter is the number of nodes (n). Each method has its own parameters, for example, the Waxman model defines parameters *alpha* and *beta* that affect the number of edges in the graph representing the network.
- The way of placement of multicast nodes in the network—the appropriate method of deployment destination nodes can construct a set of M that contains indexes of nodes constituting a multicast group in the network. Nodes can be selected randomly or based upon other network parameters, that is, node degree.
- Simulation parameters—the main parameter is the number of simulated network N (the analyzed algorithm constructs a multicast tree in N different networks).

Using the aforementioned reasons, the authors have developed a simulation environment to evaluate heuristic algorithms for multicast connections. The use of a relational database to collect information about networks and web technologies for the presentation of the results will speed up research investigations on new algorithms.

Literature confirms dependencies between methods of network topology generation and the efficiency of routing algorithms. The authors made a considerable number of experiments for different topologies to develop a research methodology and confirm the aforementioned thesis. A computer simulation is the only way to evaluate the performance of routing algorithms.

9.3 SIMULATION METHODOLOGY OF MULTICAST ROUTING ALGORITHMS

The present-day Internet is a set of connected computer domains, that is, grouped network nodes (routers). They are under joint administration and co-share routing information. The Internet is composed of thousands of administrative domains, each of which includes at least one AS.

The rapid increase in the Internet topology is accompanied with specific problems and challenges related to routing, resource reservation, and the administration of appropriate segments of the network. Designing of efficient routing algorithms is based on numerical simulations that make use of appropriate network models.

An adequate choice of a model of the real transport network has a significant influence upon the quality of the results obtained in the process. Most frequently, studies and investigations on modelling the network that are carried out in research centers focus on the four following models:

1. Regular models (ring, star, grid, lattice, etc.)
2. Real models (ARPANET, NSFNet backbone network)
3. Synthetically (randomly) generated models
4. Models that reflect the structure of wireless (ad hoc) networks

There is a limitation to the aforementioned models, which is the substantially narrow research area of algorithms under study, most frequently limited to just one structure (regular and real topologies), as well as the absence of a mock representation of a real configuration of connections (synthetic topologies).

As a result of the investigations related to the analysis of the structure of the Internet [16], a number of new parameters for the description of the network have been proposed. An adequate choice of synthetic topologies makes it possible to obtain structures with parameters that are similar to those of a real network. Because of this particular feature, synthetic topologies are widely used to model the Internet topology. The architecture of the Internet network provides an opportunity to represent its structure, with certain simplifications, by a graph. The following methods for generating graphs are used in modelling the Internet topology:

- Regular methods
- Random graphs
- Heuristic methods

Regular structures can be applied in simplified network models (e.g., by modelling particular configurations of real LAN networks) [53]. The application of regular methods in modelling Internet topologies is limited due to the irregularity of the real network that can be manifested both within domains and at the level of ASs. Hence, simulation experiments of new algorithms should be carried out according to the two different scenarios:

1. Inside a domain (or an AS)
2. At the level of ASs

The methods for generating the network topology proposed in the literature provide an opportunity to model the topology with the presented division taken into consideration. The next sections will discuss methods for generating random graphs and methods that use power law relationships in graph generation.

9.3.1 RANDOM GRAPHS

Random graphs are constructed by a random addition of edges to a given set of vertices. These models do not reflect the real structure of the network, but are widely used, due to their simple execution, in the analysis of network phenomena.

In the $G(n, p)$ model [54], each edge of the full graph K_n is considered independently and, with the probability p, added to a random graph G with n vertices. The expected value of the number of edges for graph G is $(pn(n-1))/2$.

In the $G(n, k)$ model [54], the input data are the number of vertices of graph n and the number of its edges k. The algorithm adds random edges sequentially (until the required k number of edges is

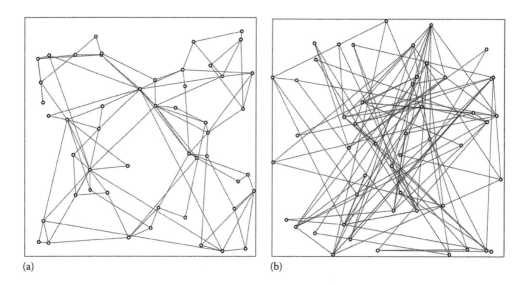

(a)   (b)

FIGURE 9.1 Network topologies generated with an application of Waxman model: (a) $\beta = 0.05$, (b) $\beta = 0.95$ (network parameters: $n = 50$, $k = 100$, $\alpha = 0.15$).

reached). In each of the algorithm's steps, the condition of the equal probability of choosing edges from the set of all possible edges in the graph is satisfied.

Other methods add edges with the probability that is the function of the distance between the vertices. The Waxman method [10] assumes the arrangement of the graph within the square plane with the length of the side p (Figure 9.1). This method defines the probability of the existence of the edge between the vertices u and v as

$$P(u,v) = \alpha e^{\frac{-d(u,v)}{\beta L}}$$

where d is the Euclidean distance between vertices u and v, whereas $L = \sqrt{2}p$ is the maximum Euclidean distance between two randomly chosen vertices in the graph. α and β are the parameters of the method, where $\alpha, \beta \in (0,1)$. An increase in the parameter α is followed by an increase in the number of edges in the graph, while an increase in the β parameter increases the ratio between the long edges and the short ones. The literature offers a number of modifications of the Waxman method [6]. One of the modifications is the *exponential* method in which the probability of existence of the edge in a graph is made dependent on the distance between vertices (the probability of existence of edge approaches 0 when the distance reaches L):

$$P(u,v) = \alpha e^{\frac{-d(u,v)}{L-d(u,v)}}.$$

Another modification of the Waxman method is the *locality* method that divides the set of edges into two length intervals and determines the probability on the basis of belonging to a given interval (in this case, r is the boundary):

$$P(u,v) = \begin{cases} \alpha \text{ if } d(u,v) < r \\ \beta \text{ if } d(u,v) \geq r \end{cases}.$$

Yet another approach is proposed in [11] (the Barabási–Albert model). This model assumes two reasons for the existence of power laws in the distribution of the number of edges starting in a given node: a gradual increase in the network size and preferential addition (of nodes). The increase in the network size results from the addition of new nodes to the existing structure, which causes the network size to be expanded. The addition of new edges is executed preferentially, that is, a new node will be added to the existing nodes with a high node degree (preferred nodes) with higher probability. If node u is added to the network, then the probability that it will be connected with node v (that already belongs to the network) is defined by the following dependence:

$$P(u,v) = \frac{d_v}{\sum_{k \in V} d_k}$$

where

d_v is the destination node degree

V is a set of nodes added to the network

$\sum_{k \in V} d_k$ is the sum of all edges that start in the nodes that have already been added to the network

9.3.2 NETWORK TOPOLOGY GENERATORS

One of the most commonly used topology generators is the simulator that has been created within the framework of the GT-ITM project [6]. The simulator introduces *transit–stub* domain structure (division of the network into transition and residual domains) to reflect the real, hierarchical topology of the Internet. The generator uses a modification of the Waxman method to construct random graphs at any level of the hierarchical structure.

A quite different approach is to be found in the generator built within the framework of the *TIERS* project [12]. Here, a three-tier hierarchical structure of LAN, MAN, and WAN is used as a base for the simulation model.

Another application for network modelling has been created within the framework of the *Boston University representative Internet topology generator* (BRITE) project [13]. In this simulator, a generator with a number of degrees of freedom that takes into account the way nodes are arranged on the plane is implemented. Depending on a selection of parameters, the output structures are similar to those that are obtained as a result of the operation of the Waxman model [10] and the Barabási–Albert model [11]. The BRITE generator also provides an opportunity to construct hierarchical networks that are two-tier structures.

Generation of a hierarchical network is based on the top-down model [6,7]. The configuration file for the generator includes parameters that control the distribution of the cost value for individual links in domains (inside ASs) and among ASs.

Still, another development of topology generators involves the use of heuristic algorithms that are based on an appropriate analysis of the properties of the Internet network. The analysis of the real network has been conducted in the *National Laboratory for Applied Network Research* (United States), collecting information from the routing tables of BGP routers [55] by a specially dedicated server (the BGP is used in the Internet network as a routing protocol between ASs). Thus, collected data made it possible to determine the so-called power laws [16] that occur between certain network parameters. These power laws have been also used in the *Inet* project [18] and the PLRG project [17].

In its first stage, the *Inet* generator determines, using the so-called power law distribution [16], the node degree of the graph (the number of edges that start in a given node). In its successive steps, connections between nodes are created. Then, the *Inet* generator checks the coherence of thus created network by setting up a spanning tree for the nodes with the degree exceeding two. To such, tree nodes with the degree equal to one are then added. In the power law random graph (PLRG) generator, input data are composed of the number of nodes in the generated network and the α parameter that is the parameter of the so-called power law distribution [16] and serves to assign a node degree to all nodes of the generated network.

9.3.3 PARAMETERS OF THE NETWORK TOPOLOGY

In order to make the results of the studied multicast connection algorithms dependent on the network topology, it is necessary to first define the basic topology parameters of the network structures that are to be represented by the graph. The most important parameters for the network description, as adopted by the literature [6,7], include

- Average node degree:

$$D_{av} = \frac{2k}{n}$$

    where
    n is the number of nodes
    k is the number of edges

- Diameter—the length of the longest path, chosen from the set of the shortest paths, where the set of the shortest paths includes the shortest paths between each pair of vertices in the graph.
  - Hop-diameter—the number of edges (unit steps, hops) that the longest path, chosen from the set of paths with the least number of edges, includes.
  - Length-diameter—the sum of Euclidean distances in the edges included in the longest path that has been chosen from the set of paths with the least sum of Euclidean distances of the edge.
- Clustering coefficient γ_v of node v is the ratio between the number of links between node v and adjacent nodes and the number of possible links that can exist between adjacent nodes [15].
- If we denote the adjacency of node v (understood as a set of all possible edges between vertices that adjoin vertex v) by the symbol $\Gamma(v)$, the count of the set $\Gamma(v)$ by the symbol $E(\Gamma(v))$, and the node degree of vertex v by D_v, then we can write

$$\gamma_v = \frac{\left|E(\Gamma(v))\right|}{\binom{D_v}{2}} = \frac{2\left|E(\Gamma(v))\right|}{D_v(D_v - 1)}.$$

The value of the clustering coefficient determined for each of the vertices in the graph can be defined as follows:

$$\hat{\gamma} = \frac{1}{|V|}\sum_{v \in V}\gamma_v.$$

The dependence is satisfied when $D_v > 2$. Let $V^{(1)} \in V$ denotes a set of nodes with the degree equal to 1. Taking this condition into consideration, we get [15]

$$\hat{\gamma} = \frac{1}{|V| - |V^{(1)}|}\sum_{v \in V}\gamma_v.$$

The parameters widely presented in the literature of the subject make it possible to describe a network with any topology. After taking into consideration the specificity of the optimization problem, an introduction of new parameters for the network topology have been proposed. These parameters take into account the way the construction of the multicast tree in a given network and the arrangement of receiving nodes are done.

Such parameters include the number of multicast nodes (group members), denoted by the symbol m. This parameter allows us to evaluate the scalability problem in analyzed solutions and can be

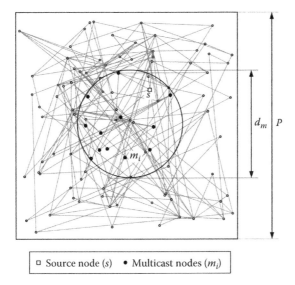

□ Source node (s)   • Multicast nodes (m_i)

FIGURE 9.2 Graphical interpretation of group scattering coefficient (network parameters: $n = 100$, $m = 12$, $D_{av} = 4$).

formulated as follows: will an increase in the number of receiving nodes cause an increase in the cost of constructed multicast tree or not?

This section also introduces the notion of the group density, defined as the ratio between the number of receiving nodes m to the number of nodes in the network n:

$$\rho_m = \frac{m}{n}.$$

The parameter described by this equation does not take into account, however, the method for arranging nodes that are already members of the group. Therefore, the so-called *group scattering coefficient* has been defined [56]. The coefficient restricts the arrangement of receiving nodes to a defined area limited by a circle. This area is centrally situated within the area confined by a square, inside of which a grid has been determined. The *group scattering coefficient* is defined as the ratio of the diameter of the area limited by the circle d_m to the length of the side of the square area P:

$$\varepsilon_m = \frac{d_m}{P}.$$

Figure 9.2 shows an exemplary network with 100 nodes and 200 edges, generated on the basis of the Waxman model in which 12 multicast nodes from the area d_m limited by the circle have been selected.

9.3.4 METHODS FOR GENERATING NETWORK TOPOLOGY

The present study on algorithms for multicast connections has been preceded by an extensive analysis of available relevant publications on the subject that have been published in the past 15 years. As a result of the preliminary literature study, simulation researches can be divided into the following groups:

- Research studies that make use of one model of the real network (e.g., the ARPANET network model) [5]
- Research studies that make use of networks generated by random graphs [26,39]
- Research studies that make use of networks generated by the Waxman model [5,37,44,57–60]
- Research studies that make use of networks with a low number of nodes (up to 100 nodes) [5,33]

The introduction of new parameters for the description of the network, such as the clustering coefficient [15] and the demonstration of power law relationships between the appropriate parameters of the Internet network at the level of ASs [16], have made it possible to intensify works on modelling of real networks [61,62].

The conclusions that come from the available literature indicate unequivocally that the Waxman model does not reflect many important parameters of the Internet network—the clustering coefficient is in generated networks twofold lower and power laws do not apply [18].

Another important network parameter is the average graph degree D_{av}. Noronha and Tobagi [63] prove that the effectiveness of the multicast routing algorithm applied to a real network is similar to the effectiveness of the same algorithm in the so-called random two-connected graph (with graph degree $D_{av} = 4$). Hence, in generated networks, the assumption is that the values of the graph degree are within the interval from 3 to 5. It is then an important issue to maintain a constant average graph degree throughout the generation of a network.

The following assumptions have been adopted in the simulation study:

- In the process of generating network topology, nodes are arranged on a square plane with the length of the side $P = 1000$ and with the grid hop equal to 1.
- Nodes of the network are located randomly according to the uniform distribution, which secures uniform arrangement of nodes within the plane.
- The method for interconnecting nodes of the network results from the variation of the adopted models used in the network topology generators.
- Each link has been assigned two metrics: cost c_{ij} and delay d_{ij}.
- The cost metric is a random value from the interval 10 to 1000 generated on the basis of uniform distribution—the difference of two orders of magnitude makes it possible to model real loads in links.
- Delay is directly proportional to the Euclidean distance $|i,j|$ between nodes within the plane ($d_{ij} = |i,j|$).
- Waxman and Barabási–Albert models are used in the study in the network with several hundred nodes.
- The adopted parameters for the Waxman model ($\alpha = 0.15$, $\beta = 0.2$) provide an easy way to generate networks with the distribution of link lengths similar to the distribution obtained in the Barabási–Albert model.

In order to increase the adequacy of the results (credibility of measurements) of any study on multicast algorithms, it is important to compare the results of the performance of the algorithms in networks generated on the basis of a variety of network topology models: Waxman, Barabási–Albert (BRITE generator [13]), and the heuristic model implemented in the *Inet* generator.

A substantial limitation of the *Inet* generator is the accurate representation of the network at the level of ASs. This means that the application provides an opportunity to generate networks with the minimum number of nodes equal to 3037 (this is the number of ASs in the Internet network in November 1997). This value, however, limits the application of the *Inet* generator in simulations of small networks.

Due to the aforementioned, in the investigation carried out by the authors, only networks generated by the Waxman method and the Barabási–Albert method were analyzed.

9.3.5 Methods for the Arrangement of Members of the Multicast Group

A significant element of the research process in a study on multicast routing algorithm is to take into account the arrangement of members of the multicast group in the network. Therefore, special methods have been developed to locate destination nodes in the network according to preestablished criteria, such as the node degree or required area.

In works related to simulations of multicast routing algorithms (e.g., [5,44,59,60]), the group of receiving nodes M is determined by a random selection of m nodes in the network from among all n nodes $n = |N|$. The source node s is also selected randomly from among n nodes of the network. Such an approach is called in the present section the *GroupRandom* method.

The authors propose the following methods for receiving nodes arrangement [64,65]:

- *GroupRadius* method
  (Nodes that create the multicast group M and the source node s are chosen from among N_r nodes that are located inside the circle with the radius $r = d_m/2$. The method reflects the arrangement of nodes in a real network.)
- *GroupHighDegree* method
  (The applied algorithm determines the node degree, i.e., the number of links outgoing from each node of the network. Then, nodes are sorted according to the decreasing number of outgoing links. The set $M \cup s$ is composed of $m + 1$ nodes that are most preferred, i.e., with the highest number of links. The transmitting node is the node with the highest degree.)
  Figure 9.3 shows the histograms of the distribution the node degree for some exemplary networks generated by the Waxman and the Barabási–Albert methods. The histograms show that in the Barabási–Albert network, there are nodes with the node degree that is twofold higher than in the network constructed on the basis of the Waxman model.

The application of the presented method to network generation will make it possible to study whether a particular arrangement of receiving nodes has any influence upon the quality of trees constructed by multicast routing algorithms. The way receiving nodes M are arranged in the network has not been previously analyzed in the literature of the subject.

9.3.6 QUALITY PARAMETERS OF MULTICAST ROUTING ALGORITHMS

The quality of group transmission trees constructed on the basis of heuristic algorithms can be determined on the basis of the following parameters:
- Cost of constructed multicast tree c_T:

$$c_T = \sum_{(i,j)\in T} c_{ij}e_{ij}.$$

- Average path cost between the source node and the receiving node in the multicast tree (c_p) that includes m receiving nodes:

$$c_p = \frac{1}{m}\sum_{m_i \in M}\sum_{(i,j)\in p} c_{ij}e_{ij}.$$

- Normalized multicast tree cost constructed by a given heuristic algorithm (c_{AH}) in relation to the cost of a tree resulted from the operation of a precise algorithm (c_{MST}) that constructs a minimal Steiner tree; the normalized tree cost is averaged for N networks in which a given heuristic algorithm has been applied:

$$\delta = \frac{1}{N}\sum_{i=1}^{N} \frac{c_{AH} - c_{MST}}{c_{MST}}.$$

- Diameter of the multicast tree.
- Operation time of a heuristic algorithm (AH)—t_{AH}.

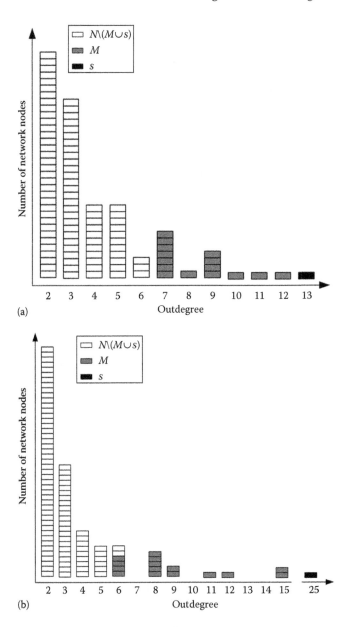

FIGURE 9.3 The histogram for the distribution of node degrees with a presentation of the operation of the *GroupHighDegree* algorithm for the exemplary networks: Waxman (a) and Barabási–Albert network (b) ($n = 100$, $m = 15$, $D_{av} = 4$).

A construction of the minimum Steiner tree is a classic optimization problem, a solution to which can be found in most of the publications devoted to routing algorithms for multicast connections [26]. However, due to the quality of data transmission between the sender of the content and the receiver, it is important to construct the tree by intermediary devices (routers) in such a way as not to minimize the total cost of the tree but also to assure quality parameters of transmission between the sender and the receiver. In addition, another important problem is to construct the shortest path trees between the sender and each of the members of a multicast group.

In the case of constrained algorithms (Section 9.2), the process of the path cost minimization takes into account link delays, the sum of which cannot exceed the required level. The proposed parameter, the average path cost in the tree, denoted as c_p, is the average of path costs in the tree between the sending node s and each of the receiving nodes m_i [31].

An accurate measurement of the time of operation for the heuristic algorithm t_{AH} with the accuracy of nanoseconds is made possible by the application of the function Win32API in the Windows system: *QueryPerformanceFrequency* (the function retrieves the frequency value at which the system efficiency counter operates) and *QueryPerformanceCounter* (retrieves the current value of the standard hardware efficiency counter) [66]. The measurement of the operation time of the algorithm does not take into account initiation procedures and copying auxiliary tables required for the correct operation of the algorithm.

9.3.7 STATISTICAL ANALYSIS OF STUDY RESULTS

Taking into account the criteria presented earlier in this section, a comparative study for the heuristic algorithms in relation to the existing and proposed network parameters was performed. The study involved an analysis of the dependence between the cost of multicast tree (c_T), the average path cost in the tree (c_p), and the diameter of the tree on the following parameters:

- The number of network nodes n
- The number of receiving nodes m
- Average graph degree D_{av}
- Maximum delay value (Δ) along the path between the source s and each of the receiving nodes $m_i \in M$
- The clustering coefficient $\hat{\gamma}$
- Method for generating the network topology and its parameters
- Method for arranging receiving nodes in the network

The simulation experiments were performed for an established sets of parameters. For individual items of the simulations, the results of the operation of each of the heuristic algorithms in 5000 networks (1000 network structures in 5 series) were collected. A simulation item is understood as an execution of a given multicast algorithm and a determination of a distribution tree with the assumed parameters of the algorithm (e.g., the number of receiving nodes m, maximum value of delay Δ, method for the selection of receiving nodes in the network) and the network parameters (e.g., the number of nodes in the network n, generation method, the average network degree D_{av}).

The rationale for the aforementioned approach (and the guarantee for creating different initial conditions for each of the 5 series) is a generation of networks with identical parameters but with a different value of the *seed* that is initiated by the pseudorandom number generator used in the applications that generate network structures.

For the determination of the confidence interval, the *t-Student* distribution with $n - 1$ degrees of freedom was used [67]:

$$\overline{c(i)} - t_{\frac{\alpha}{2}} \frac{s_n}{\sqrt{n}} \leq c_j \leq \overline{c(i)} + t_{\frac{\alpha}{2}} \frac{s_n}{\sqrt{n}}$$

where
  $c(i)$ is the average value of costs (of trees or paths) determined in each series
  s_n is the standard deviation
  $t_{\frac{\alpha}{2}}$ is the quantile read from the distribution table for $n - 1$ degrees of freedom—$\alpha/2 = 0.025$
  (for 5 series $t\alpha_{/2} = 2312$)

The results of the simulation experiments are presented in Section 9.4. In most cases, the 95% confidence intervals are so much narrow that they are included inside the points in the graph and, for better readability of the graphs, have not been marked.

Concurrently, a statistical analysis of the results within a series was performed. For that purpose, basic description statistics were established by considering the results of each of the heuristic algorithms as single random variables. For each variable, the following were determined: the average value, minimum and maximum value, standard deviation, the coefficient of variation, skewness, and kurtosis.

9.3.8 SIMULATION ENVIRONMENT

While conducting research, it became clear that there was a need to create a number of original tools and applications that would support the program implementation of new and existing routing algorithms, modelling and visualization of network topologies, and processing and storing large amount of data (describing the networks and those containing the results of the operations of the algorithms) in a relational database. All the mentioned tools form a coherent simulation environment that is not limited to the study on multicast algorithms only. This environment is presented schematically in Figure 9.4.

Due to a large number of networks used in the research process (over 300,000 generated networks with the number of nodes within the range of 100–10,000), as well as a large number of obtained results, a relational MySQL database was used [68]. Such an approach made it possible to execute a data repository that supported other simulation experiments in the networks.

The network structures obtained through the application of external applications (topology generators) are processed by a converter and then placed in a normalized way in the database. The converting module makes it also possible to import networks from other generators.

The simulator makes use of external functions (with normalized parameters) that implement multicast algorithms. It retrieves configuration data that include the simulation parameters and networks with defined parameters (through SQL queries). At the same time, it returns data that describe the constructed multicast trees to the base (e.g., tree cost, average path cost in a tree, tree diameter, operation time of the algorithm, and a set of nodes and edges that are included in a structure).

The simulator can be controlled by an Internet browser. Such an approach makes it possible to visualize a network, single multicast trees, and graphs presenting results of a simulation (Section 9.4).

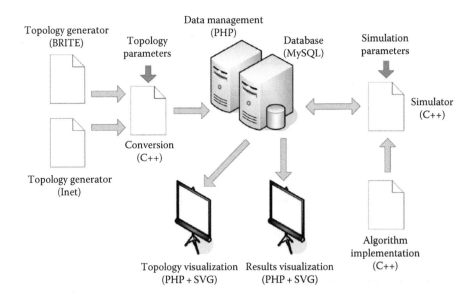

FIGURE 9.4 Programming environment used in the study on the algorithms.

A visualization makes use of the *scalable vector graphics* (SVG) language [69] that is used for descriptions of vector graphics.

The simulation experiments were performed on a PC—MAXDATA Platinum 3200, equipped with two dual-core Intel XEON 2 GHz processors and with 8 GB RAM memory.

9.4 SIMULATION STUDY ON THE EFFECTIVENESS OF THE ALGORITHMS

Simulation experiments are primarily designed to verify the effectiveness of multicast algorithms. The present article focuses on two-criteria algorithms that, additionally, take into account the delay parameter. Each of the proposed algorithms has been compared with the already existing solutions that are most frequently found (or cited) in the literature of the subject.

Algorithms for multicast constrained connections make it possible to model problems that occur in real operator networks. From the point of view of guaranteeing the quality of transmission to the receiver, the essential problem is to construct a distribution tree that would take into account the constraint of the maximum value of delay between the sender and the receiver. Within this context, the distribution tree should be characterized by the minimum total cost, while, at the same time, the paths that would connect the receiving nodes with the source should be the ones with the minimum cost.

9.4.1 ANALYSIS OF TREE COSTS CONSTRUCTED BY ALGORITHMS WITH DELAY CONSTRAINTS

The heuristic algorithms presented in Section 9.2 were put to test in wide and extensive simulation experiments. The assumption was that the average cost of constructed multicast trees would be the basic comparison criterion. The cost of a link is established as an integer number from the interval from 10 to 1000. The study was carried out for networks generated on the basis of the Waxman model ($\alpha = 0.15$, $\beta = 0.2$) and the Barabási–Albert model. The simulation experiments were conducted for the representative and well-known set of algorithms: KPP, CSPT (both described in Section 9.2), *bounded shortest multicast algorithm* (BSMA) [38], *fast delay-constrained multicast routing algorithm* (DCMA) [37], and *delay-constrained shortest path multicast algorithm* (DCSP) [28].

The results of the operation of the algorithms were also compared with the results obtained on the basis of a modified DCMA, that is, the reverse DCMA (RDCMA) and its derivative: the heuristic optimal DCMA (OPTDCMA) [70].

In the conducted study, an analysis of the dependence between the average cost of a multicast tree and the number of nodes in the network n was carried out. An increase in the number of nodes in the network is followed by a slight increase in the average cost of trees constructed by the algorithms: BSMA, DCMA, RDCMA, OPTDCMA, and CSPT (optimization of paths) and by a decrease in the costs of trees constructed by the KPP algorithm (optimization of the total cost of tree) for both methods for generating the network (Figures 9.5 and 9.6). The KPP algorithm is the most effective. The OPTDCMA proposed in this work constructs trees whose costs are 46% higher in relation to the costs of trees constructed by the KPP algorithm and 44% in relation to the costs of trees constructed by the BSMA for $n = 400$ (Figure 9.6). These trees are characterized by the average cost that is by 9.5% lower than the comparable costs of trees executed by the DCMA algorithm and by 13% lower than the results yielded by the DCSP algorithm.

The dependence of the average cost of multicast tree on the number of multicast nodes m is linear for each of the algorithms under scrutiny (Figures 9.7 and 9.8). The differences in costs of the generated trees continue at the same level throughout the confidence interval. While in the case of the cost of the link represented by the Euclidean distance between nodes the influence of the method for generating the network topology was significant, in the case of the allocation of a random value to the cost parameter, the influence of a model for generating the network on the total cost of tree is about 1%–3% (OPTDCMA, $n = 200$, $m = 180$).

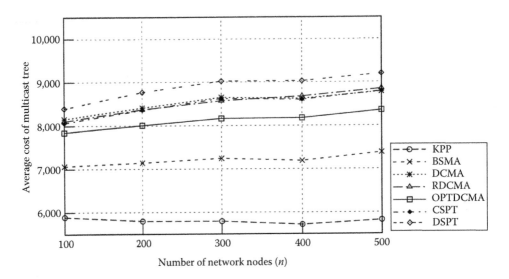

FIGURE 9.5 Average cost of multicast tree versus the number of network nodes n in graphs generated with Waxman model ($m = 20$, $D_{av} = 4$, $\Delta = 2000$).

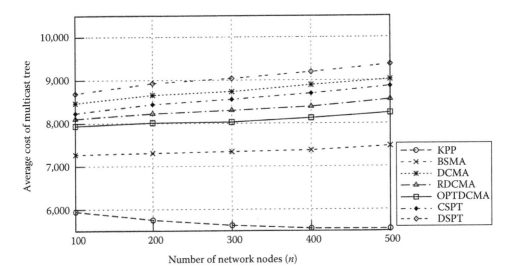

FIGURE 9.6 Average cost of multicast tree versus the number of network nodes n in graphs generated with Barabási–Albert model ($m = 20$, $D_{av} = 4$, $\Delta = 2000$).

In the next phase of the study, effectiveness of heuristic algorithms depending on the value of the maximum delay along the path was studied. The analysis of the graphs shown in Figures 9.9 and 9.10 leads to a conclusion that an increase in the Δ parameter is followed by an increase in the average cost of trees generated by the studied algorithms. A decrease in the value of the Δ parameter is followed by addition of paths that satisfy the condition of the maximum value of delay to the tree:

$$\sum_{(i,j)\in p} d_{ij}e_{ij} \leq \Delta,$$

but these are not the minimum cost paths. Eventually, this leads to an increase in the total cost of tree.

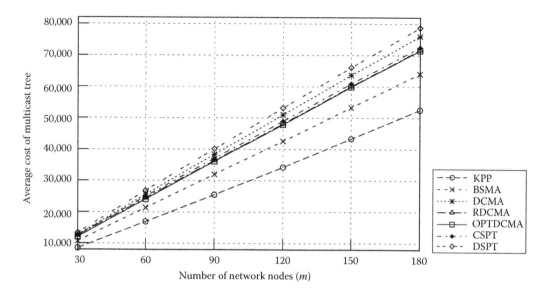

FIGURE 9.7 Average cost of multicast tree versus the number of multicast nodes m in graphs generated with Waxman model ($n = 200$, $D_{av} = 4$, $\Delta = 2000$).

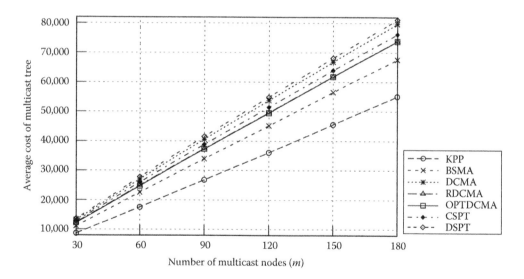

FIGURE 9.8 Average cost of multicast tree versus the number of multicast nodes m in graphs generated with Barabási–Albert model ($n = 200$, $D_{av} = 4$, $\Delta = 2000$).

Any further increase in the maximum delay ($\Delta \rightarrow \infty$) brings the optimization problem in its essence to a single-criterion problem (with no constraints). Algorithms that construct the shortest path trees converge their results for the value $\Delta > 3500$ in the networks generated on the basis of the Waxman model and for the value $\Delta > 4000$, in the networks generated on the basis of the Barabási–Albert model. The costs of trees constructed by the OPTDCMA are lower than the costs of trees constructed by the algorithms DCMA, RDCMA, DCSP, and CSPT.

The KPP algorithm is characterized by the lowest susceptibility to changes in the Δ parameter. The difference between the minimum and the maximum value of the average cost of tree is 10%, whereas for the BSMA, this difference is 63%.

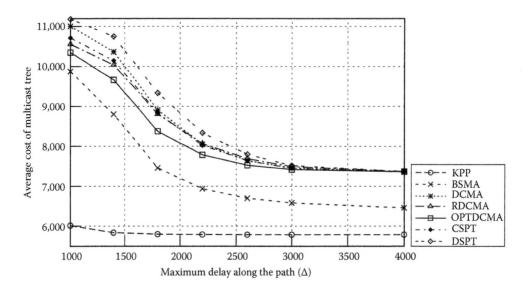

FIGURE 9.9 Average cost of multicast tree versus the maximum delay along the path Δ in graphs generated with Waxman model ($n = 200$, $m = 20$, $D_{av} = 4$).

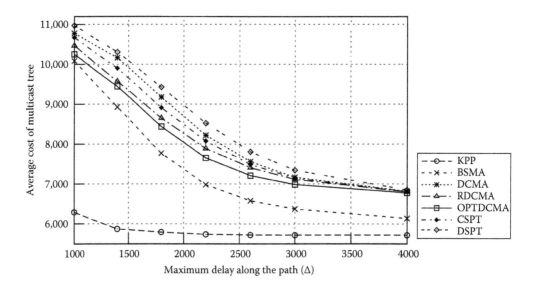

FIGURE 9.10 Average cost of multicast tree versus the maximum delay along the path Δ in graphs generated with Barabási–Albert model ($n = 200$, $m = 20$, $D_{av} = 4$).

The simulation experiments also involved a study on the dependence of the results obtained on the basis of the heuristic algorithms on the density of the network, measured by the average D_{av} graph degree parameter. Such an approach provides an opportunity to evaluate the effectiveness of the algorithms in networks with the structure different from that of the Internet network, for which D_{av} is included within the interval from 3 to 5. The study was conducted for the value D_{av} from the interval from 4 to 20. For $n = 200$, it follows from this observation that there is an increase in the number of links in the network from 400 to 2,000.

A reduction of costs of trees constructed by the algorithms under study results from the increase in the size of the set of all available minimum cost paths that can be added to the distribution tree.

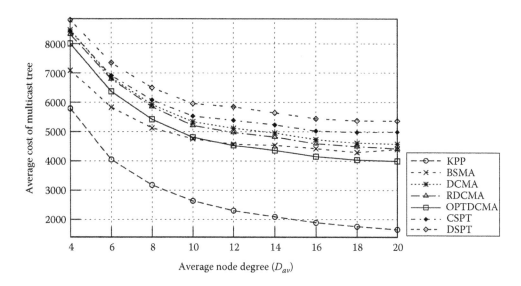

FIGURE 9.11 Average cost of multicast tree versus the average node degree D_{av} in graphs generated with Waxman model ($n = 200$, $m = 20$, $\Delta = 2000$).

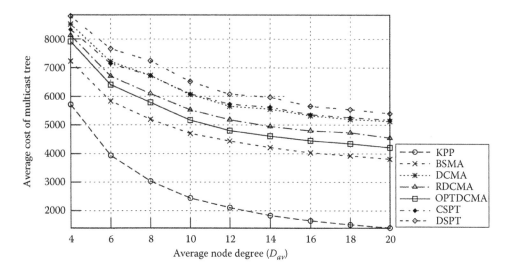

FIGURE 9.12 Average cost of multicast tree versus the average node degree D_{av} in graphs generated with Barabási–Alert model ($n = 200$, $m = 20$, $\Delta = 2000$).

Above the value $D_{av} = 12$, the proposed OPTDCMA is characterized by costs of constructed trees that exceed only those provided on the basis of the KPP algorithm. The difference between the minimum and the maximum value of the average cost of tree constructed by the KPP algorithm ($D_{av} = 4$ and $D_{av} = 20$) is 306%, while for the OPTDCMA algorithm, it is 87% (Figures 9.11 and 9.12).

9.4.2 ANALYSIS OF THE AVERAGE PATH COST IN A GROUP TRANSMISSION TREE

The next experiments involved an analysis of the average cost of paths in distribution trees constructed by appropriate algorithms. The studied heuristic algorithms also determine the shortest paths between the source node and each of the receiving nodes in the multicast tree that has been earlier constructed. The average path cost parameter c_p in a given tree is the sum of costs of all such paths divided by the number of receiving nodes. It can thus be the measure of quality of a connection between the sender and the receiver.

Figures 9.13 and 9.14 show the dependence between the average path cost in a tree and the number of nodes in the network n. The MLRA and the KSPMA, proposed in this work, are characterized by high effectiveness in terms of constructing trees in relation to all algorithms under investigation. *Least delay cost* (LDC) algorithm has the worst performance (it only minimizes delay along paths in a tree). The lowest costs of trees were obtained based on those algorithms that used networks generated by the Barabási–Albert model. The KPP constructs trees on average more costly by 19% in relation to trees obtained by the MLRA ($n = 400$, $m = 20$) and by 25% more costly

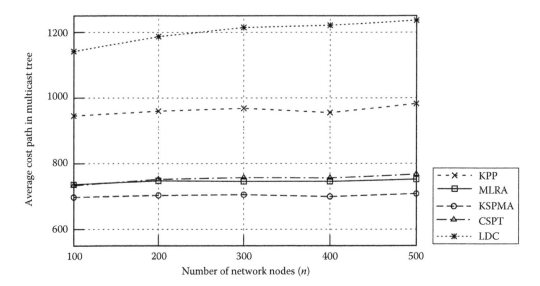

FIGURE 9.13 Average cost of path in multicast tree versus the number of network nodes n in graphs generated with Waxman model ($m = 20$, $D_{av} = 4$, $\Delta = 2000$).

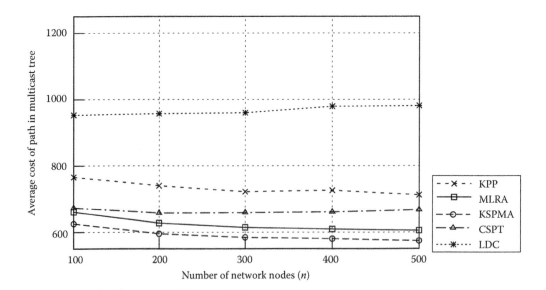

FIGURE 9.14 Average cost of path in multicast tree versus the number of network nodes n in graphs generated with Barabási–Albert model ($m = 20$, $D_{av} = 4$, $\Delta = 2000$).

in relation to trees obtained by the KSPMA. For the Waxman model, these differences are 28% and 37%, respectively. The differences under consideration result from the methods of constructing trees adopted in the algorithms. The MLRA, KSPMA, and CSPT algorithms are algorithms that construct the least cost paths, whereas the KPP algorithm optimizes the total cost of tree. Note that in the networks generated on the basis of the Barabási–Albert model, paths with lower costs are obtained than those in the networks generated on the basis of the Waxman model. It is so because networks generated on the basis of the Barabási–Albert model have nodes with high values of the node degree and are characterized by lower values of the diameter in relation to networks obtained on the basis of the Waxman model.

An increase in the number of receiving nodes m with the constant number of nodes in the network n is followed by an increase in the average cost of paths in trees (Figures 9.15 and 9.16). Within the interval m from 30 to 180, the costs of trees constructed by the KPP algorithm rise by 64%, whereas in the case of the MLRA and KSPMA, by about 40% only (for the Waxman model). The KSPMA is characterized by the best effectiveness in terms of the costs of constructed trees. The differences between KSPMA and the classic CSPT algorithm are about 5%–10% throughout the variation interval for the m parameter (with networks obtained after the application of the Barabási–Albert model).

The MLRA and the KSPMA are characterized by very low costs of paths in distribution trees, including networks that differ considerably in their structures from that of the Internet network. An increase in the density of the network (understood as an increase in the average degree of the graph D_{av}) constructed with the application of both methods for generating the network topology leads to the observation that the most effective are the MLRA and KSPMA (the differences between the results for the Waxman model and the Barabási–Albert model are 3% at the maximum). A characteristic feature of the KPP algorithm in dense networks ($D_{av} > 7$) is its better effectiveness in relation to CSPT with the application of the Barabási–Albert model (Figures 9.17 and 9.18).

A significant advantage of the MLRA and the KSPMA proposed in the work is their effective operation with rigorously maintained constraints concerning the values of the maximum delay along the path ($\Delta = 1000$)—Figures 9.19 and 9.20.

An increase in the maximum value of delay $\Delta \to \infty$ brings the problem in its essence to a single-criterion problem (without constraints). Algorithms that construct the shortest path trees (MLRA,

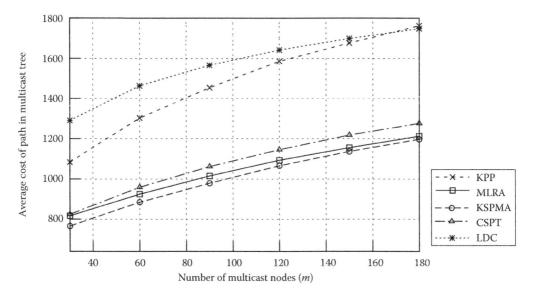

FIGURE 9.15 Average cost of path in multicast tree versus the number of multicast nodes m in graphs generated with Waxman model ($n = 200$, $D_{av} = 4$, $\Delta = 2000$).

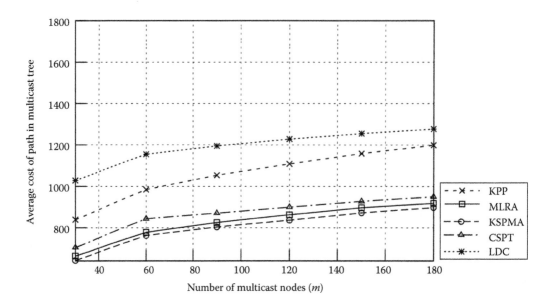

FIGURE 9.16　Average cost of path in multicast tree versus the number of multicast nodes m in graphs generated with Barabási–Albert model ($n = 200$, $D_{av} = 4$, $\Delta = 2000$).

KSPMA, and CSPT) achieve convergence of results for the value $\Delta > 3000$ in networks based on the Waxman model and $\Delta > 4000$ in Barabási–Albert networks. The KSPMA obtain the lowest values of the average path costs in group transmission trees.

9.4.2.1　Analysis of Diameters of Group in Transmission Trees

The present study on multicast algorithms has brought to attention the need to describe the span of constructed group transmission trees with the help of the number of links in a path that is included in the tree. In its essence, the problem comes to a determination of the tree diameter (hop-diameter tree), that is, the path with the longest length, chosen from the set of all shortest paths determined

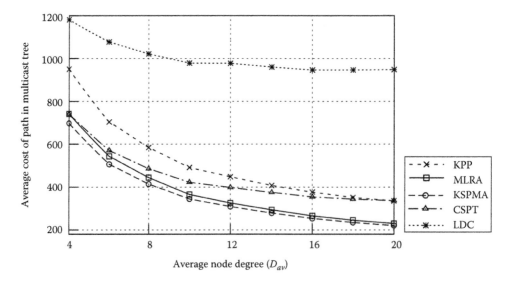

FIGURE 9.17　Average cost of path in multicast tree versus the average node degree D_{av} in graphs generated with Waxman model ($n = 200$, $m = 20$, $\Delta = 2000$).

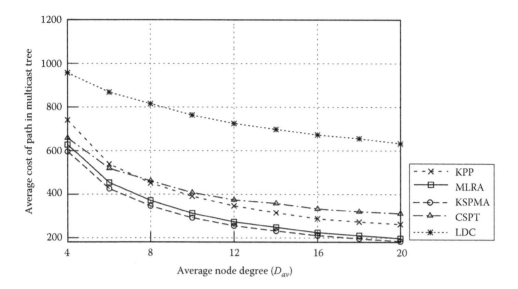

FIGURE 9.18 Average cost of path in multicast tree versus the average node degree D_{av} in graphs generated with Barabási–Albert model ($n = 200$, $m = 20$, $\Delta = 2000$).

between all possible pairs of vertices in the group transmission tree. The study was conducted for a variable number of nodes in the network n of a variable number of receiving nodes m.

The results presented in Table 9.1 show the average value of the tree diameter determined for 1000 networks (Waxman model). The biggest values of the tree diameter are those for the KPP algorithm, the lowest are for the LDC algorithm. It follows from the analyses presented in earlier sections that this algorithm is the least effective, both in terms of the cost of constructed distribution trees as well as the average value for the path cost in these trees. In addition, there is no straight dependence between the cost of tree and its diameter.

The analysis of the diameters of the trees presented in Tables 9.1 and 9.2 leads to a conclusion that the shortest path algorithms (MLRA, KSPMA, and CSPT) construct trees with similar values

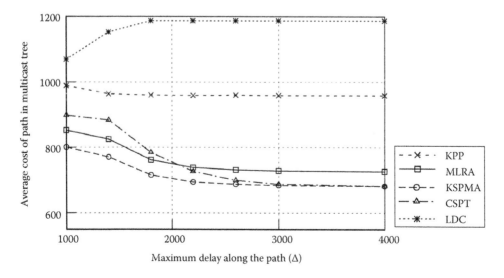

FIGURE 9.19 Average cost of path in multicast tree versus the maximum delay along the path Δ in graphs generated with Waxman model ($n = 200$, $m = 20$, $D_{av} = 4$).

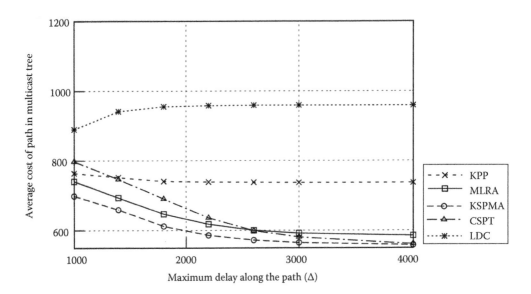

FIGURE 9.20 Average cost of path in multicast tree versus the maximum delay along the path Δ in graphs generated with Barabási–Albert model ($n = 200$, $m = 20$, $D_{av} = 4$).

TABLE 9.1

Hop-Diameter of Multicast Tree versus the Number of Network Nodes n in Graphs Generated with Waxman Model ($m = 20$, $D_{av} = 4$, $\Delta = 2000$)

n	KPP	MLRA	KSPMA	CSPT	LDC
100	11.52	9.19	8.79	8.73	7.99
200	12.02	9.59	9.21	9.11	8.43
300	12.44	9.79	9.39	9.31	8.72
400	12.20	9.95	9.61	9.48	8.89
500	12.63	10.06	9.73	9.59	9.01

TABLE 9.2

Hop-Diameter of Multicast Tree versus the Number of Network Nodes n in Graphs Generated with Barabási–Albert Model ($m = 20$, $D_{av} = 4$, $\Delta = 2000$)

n	KPP	MLRA	KSPMA	CSPT	LDC
100	10.22	7.83	7.47	7.29	6.34
200	10.94	8.06	7.83	7.62	6.59
300	11.39	8.31	7.98	7.96	6.77
400	11.61	8.39	8.09	7.93	6.89
500	11.68	8.51	8.23	7.99	6.95

TABLE 9.3

Hop-Diameter of Multicast Tree versus the Number of Multicast Nodes *m* in Graphs Generated with Waxman Model ($n = 200$, $D_{av} = 4$, $\Delta = 2000$)

m	KPP	MLRA	KSPMA	CSPT	LDC
30	14.56	10.76	10.36	10.26	9.47
60	19.68	12.84	12.56	12.49	11.13
90	22.82	14.26	13.85	13.96	12.03
120	24.20	15.65	14.95	15.01	12.54
150	26.97	16.57	16.21	16.48	13.18
180	28.41	17.55	17.23	17.42	13.54

TABLE 9.4

Hop-Diameter of Multicast Tree versus the Number of Multicast Nodes *m* in Graphs Generated with Barabási–Albert Model ($n = 200$, $D_{av} = 4$, $\Delta = 2000$)

m	KPP	MLRA	KSPMA	CSPT	LDC
30	12.77	8.92	8.63	8.43	7.43
60	16.21	10.52	10.31	9.92	8.62
90	18.22	11.57	11.22	10.82	9.24
120	19.77	12.26	12.01	11.54	9.62
150	20.71	12.92	12.69	12.05	9.93
180	21.63	13.47	13.18	12.52	10.18

of diameters, regardless of the implemented network model. In networks generated on the basis of the Barabási–Albert model, generally lower values of diameters are obtained for all analyzed algorithms.

The influence of the size of a multicast group *m* on the diameters of trees is presented in Tables 9.3 and 9.4. The lowest value of the average tree diameter value has the LDC algorithm, the highest—the KPP algorithm. The heuristic algorithms that construct the shortest path trees obtain their average values for the tree diameters at the same level, no matter what network model is applied. The increase in the number of receiving nodes is followed by an increase in the average value of the diameter (the highest for the KPP algorithm).

9.4.3 Investigations of Algorithms in Large Networks

As a complement to the study results presented in this work, an analysis of the results of the operation of the algorithms in large networks was carried out. The study employed the *Inet* network topology generator that models the real Internet network at the level of ASs. Additionally, the Waxman and the Barabási–Albert models were implemented. Until this particular implementation, these models were used to model connections between routers inside domains [71,72].

The study was conducted for networks in which the number of nodes was not less than 3037. This is the minimum value that is required by the *Inet* generator because this is the exact number of ASs the Internet network amounted to in November 1997. The time and memory complexity of the algorithms, as well as the constraints imposed by the configuration of the computer used for the simulation experiments, made it possible to perform the study in networks with up to 10,000 nodes.

The results of the investigations, presented in Figure 9.21, indicate the dependence of the average costs of constructed multicast trees on the number of nodes in the network n. The differences between the results yielded by the heuristic algorithms that construct trees in the networks obtained by the Waxman and the Barabási–Albert methods are similar to the differences resulting from the operation of these algorithms in networks with several hundred nodes (cf. Figures 9.5 and 9.6). Such dependencies do not occur though in the case of the networks obtained on the basis of the *Inet* generator.

An increase in the number of nodes in the network is followed by a decrease in the average cost of trees constructed by all heuristic algorithms. For example, the percentage difference

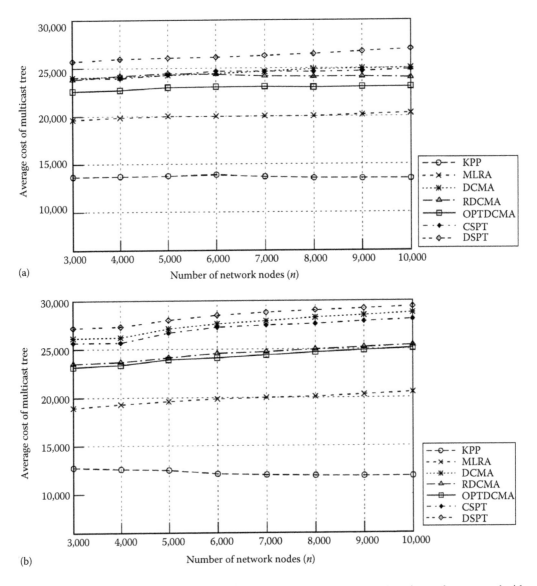

FIGURE 9.21 Average cost of multicast tree versus the number of network nodes n in graphs generated with (a) Waxman model and (b) Barabási–Albert model. (*Continued*)

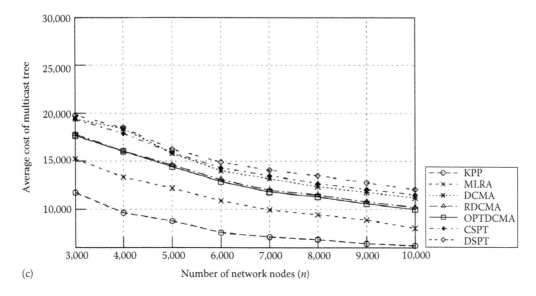

(c)

Number of network nodes (*n*)

FIGURE 9.21 (*Continued*)   Average cost of multicast tree versus the number of network nodes *n* in graphs generated with (c) *Inet* generator ($m = 50$, $D_{av} = 4$, $\Delta = 2000$).

in the results for a network composed of 3,037 and 10,000 nodes is 89% for KPP, 90% for MLRA, and 64% for DCSP, while the differences in the average costs of trees obtained with the application of the Barabási–Albert model and the *Inet* generator for a network composed of a 8,000 nodes are, respectively, 74% for KPP, 113% for MLRA, and 115% for DCSP. Such huge differences indicate strong dependence of the costs of constructed trees on a network model adopted in the study (in all the cases under investigation, the networks had the same number of nodes and links).

The drop in the average cost of trees constructed by the *Inet* application is mainly the result of the structure of connections in the networks. In order to analyze this phenomenon more thoroughly, the distribution of the node degree in the nodes of the network was investigated. A number of exemplary networks composed of $n = 3050$ nodes and generated according to the Waxman, Barabási–Albert and *Inet* models were analyzed. Because of the wide range of values of such parameters, such as the node degree and the number of nodes in the network that had a given degree, the histograms are presented in the *log–log* scale.

Figure 9.22a shows a histogram for an exemplary network generated on the basis of the Waxman model. No leaves, that is, nodes with the degree equal to 1, exist in such a network, whereas 35% of nodes have the degree equal to 2. What is also characteristic is the low value of the maximum node degree (there are two nodes in the network, each has 22 outgoing links and one node with the degree 24).

In the network generated according to the Barabási–Albert model (Figure 9.22b), 50% nodes have the degree equal to 2, whereas in the case of 30 nodes (10%), the number of outgoing links is included within the interval from 26 to 129. Despite the maximum value of the node degree being fivefold higher than in the Waxman network, the differences in costs of constructed trees are only slight (maximum 1.5% for the MLRA and 9% for OPTDCMA).

The histogram for the network constructed on the basis of the *Inet* generator (Figure 9.22c) shows that 1% of nodes are leaves, while 78% are nodes that have two adjacent nodes each. As in the Barabási–Albert model, there are 30 nodes with the degree higher than 24, whereas 5 of them have the following degrees: 128, 162, 264, 375, and 686, respectively. These nodes influence the network characteristics and cause the average value of the clustering coefficient in the network to get higher.

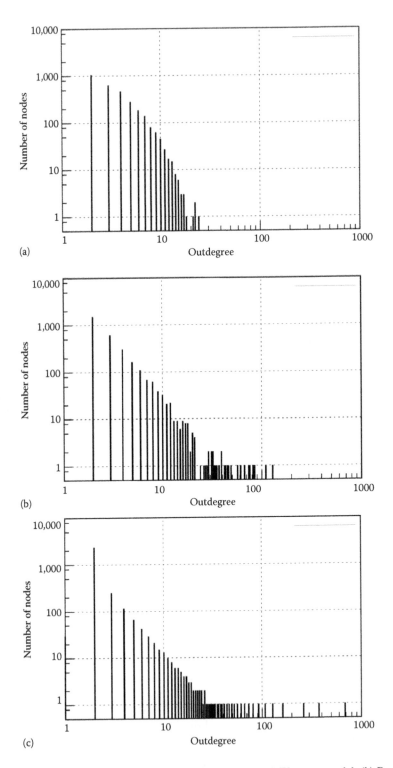

FIGURE 9.22 Distribution of outdegree in graphs generated with (a) Waxman model, (b) Barabási–Albert model, and (c) *Inet* generator ($m = 50$, $D_{av} = 4$, $\Delta = 2000$).

The average values of the clustering coefficient averaged for the number of 1000 networks ($n = 3050$) are as follows:

- With the Waxman model,

$$\hat{\gamma} = (2276.2 \pm 43.6) \times 10^{-6}.$$

- With the Barabási–Albert model,

$$\hat{\gamma} = (11.71 \pm 0.14) \times 10^{-3}.$$

- With the *Inet* generator,

$$\hat{\gamma} = (102.86 \pm 0.35) \times 10^{-3}.$$

The clustering coefficient determined in the networks constructed on the basis of the *Inet* generator is 45 times higher than the corresponding coefficient in the networks generated on the basis of the Waxman model and nearly 9 times higher in the networks generated with the Barabási–Albert model. The fact that there are nodes in the network with such a high degree, having links even up to 20% of nodes, makes it possible to construct paths and trees with lower costs in relation to the group of remaining methods for generating network topology.

9.4.4 ANALYSIS OF THE OPERATION TIME OF CONSTRAINED ALGORITHMS

A complement of the study on the effectiveness of algorithms with the constraint criterion, besides the analysis of the parameters of constructed trees, is the measurement of the operation time. The theoretical computational complexity of the studied algorithms is determined in Section 9.2. From the practical point of view, however, it is the execution time that is important. The execution time mainly depends on network parameters (the number of nodes), the parameters of the algorithm (the number of receiving nodes), and the method for the representation of data. The results presented in Table 9.5 have been averaged for the number of 1000 networks.

In order to ensure fast operation of the algorithms, the studied networks are represented in the form of adjacency matrix. The intermediate objects (trees, subgraphs) are also represented in the

TABLE 9.5

Average Execution Time of an Algorithm [*ms*] in Relation to the Number of Network Nodes *n*, Number of Multicast Nodes *m*, and Network Topology Model ($D_{av} = 4$, $\Delta = 2000$)

	Waxman			Barabási–Albert		
Algorithm	$n = 100$ $m = 20$	$n = 100$ $m = 60$	$n = 500$ $m = 20$	$n = 100$ $m = 20$	$n = 100$ $m = 60$	$n = 500$ $m = 20$
KPP	11.6	131.1	115.8	12.7	134.5	153.2
BSMA	45.3	339.8	651.5	46.2	306.2	800.7
MLRA	7.1	24.4	139.7	7.9	27.5	168.9
KSPMA	112.2	592.2	1426.8	128.4	649.7	1675.4
CSPT	1.4	4.8	20.6	1.8	5.5	25.6
DCMA	0.9	1.0	24.9	0.9	0.9	23.7
RDCMA	1.0	1.1	24.8	1.0	1.0	23.9
DCSP	6.6	6.6	52.8	4.2	4.1	31.4
LDC	1.5	4.3	20.7	1.4	4.2	19.7

form of matrices. The only exception in the proposed comparison of algorithms is the KSPMA that is characterized by a prolonged operation time as compared to the other studied algorithms. This phenomenon results from the complex operation of the algorithm (each node maintains a list of labels for each path determined in the network).

In the study, the adopted value of the K parameter in KSPMA was $K = 10$. In addition, the algorithm adopts the object-oriented method for the representation of data (one-way lists created for each of the nodes in the network), which significantly prolongs the simulation process. Despite this evident shortcoming, the KSPMA constructs trees with the least average path cost.

The measurements of the average operation times show that for a network with a low number of nodes $n = 100$ and a variable number of receiving nodes m the most effective algorithms are DCMA and RDCMA. The results of the operation of these algorithms are independent of the number of receiving nodes (for m from the interval from 20 to 60), in contrast to the KPP algorithm (12-fold increase in the operation time) and BSMA (7.5-fold increase).

In a network with the fivefold higher number of nodes, the most effective in terms of time is the CSPT algorithm. In networks with identical parameters, the proposed MLRA is nearly fivefold faster than the BSMA that constructs trees with similar costs.

The differences in average times in the operation of the algorithms in the networks constructed on the basis of the Waxman and the Barabási–Albert models are slight and thus cannot be treated as a comparative criterion for the algorithms under study. An increase in the average operation time can be though observed in the networks constructed on the basis of the Barabási–Albert model as compared to the networks based on the Waxman model, for example, for $n = 500$ and $m = 20$, KPP—33%, MLRA—21%, and CSPT—24%.

9.4.5 Descriptive Statistics for the Results of the Study

The results of the investigations presented in Sections 9.4.1 through 9.4.4 are based mainly on the mean values of the results provided by the operation of individual algorithms. To evaluate the accuracy of such an approach, basic descriptive statistics for the results obtained in 1000 different networks were determined. The results are presented in Table 9.6.

A comparison of the average values of tree costs brings a conclusion that the KPP algorithm is characterized by the lowest costs of constructed trees, while the LDC algorithm by the highest. The same dependencies are to be observed in the analysis of the maximum and minimum values of the results. The least cost trees were constructed by the KPP algorithm and were followed by the MLRA and KSPMA, whereas the highest cost trees by the LDC algorithm. The proposed MLRA heuristic constructs the minimum cost trees in networks created on the basis of the Barabási–Albert model. It is only the operation of the KPP algorithm that leads in effect to lower costs.

Standard deviation, defining the scattering of obtained results around the average value, is the least for the KPP algorithm, followed by BSMA and MLRA. The highest standard deviation has the results of the operation of the LDC algorithm in the networks that have been obtained on the basis of the Waxman model. The latter algorithm constructs multicast tree, minimizing only the delay between the source and the receivers. Due to the aforementioned, costs of trees are the highest and, as a result, their standard deviation is higher. Finally, with the networks obtained on the basis of the Barabási–Albert model, the DCSP algorithm has the highest value of standard deviation.

The variation coefficient defines relative differentiation in costs of trees constructed by a given algorithm. The values of the variation coefficient show that the least differentiated are the results for the operation of the KPP algorithm, while the most differentiated are those for the DCSP algorithm (in networks generated on the basis of the Waxman model). The variation coefficient is slightly lower for the results obtained as a result of the operation of the algorithms in networks constructed on the basis of the Barabási–Albert model.

The analysis of the skewness coefficient indicates the right-side asymmetry of the distribution of the results. The asymmetry is the highest for the algorithms: BSMA and RDCMA. The skewness

TABLE 9.6

Descriptive Statistics for the Results of Multicast Algorithms with Constraints for 1000 Networks ($n = 200$, $m = 20$, $D_{av} = 4$, $\Delta = 2000$)

Model	Parameter	KPP	BSMA	MLRA	KSPMA	OPTDCMA	CSPT	RDCMA	DCMA	DCSP	LDC
Waxman	Average	5598.9	7144.1	7453.8	7608.3	8009.3	8360.9	8372.5	8411.4	8772.4	13272.9
	Min. value	2775.0	3468.0	3280.0	3280.0	3401.0	3804.0	3401.0	3804.0	3804.0	7544.0
	Max. value	9386.0	13900.0	12900.0	13607.0	16216.0	16442.0	17949.0	16560.0	16938.0	22980.0
	Standard deviation	965.5	1387.7	1532.0	1571.6	1748.7	1911.0	1991.8	1917.1	2071.9	2379.1
	Variation coefficient	0.172	0.194	0.205	0.206	0.218	0.228	0.237	0.227	0.236	0.179
	Skewness coefficient	0.164	0.641	0.371	0.355	0.502	0.631	0.682	0.604	0.582	0.542
	Kurtosis	0.021	1.306	0.237	0.274	0.598	0.971	1.117	0.863	0.696	0.561
Barabási	Average	5423.0	7163.7	7075.7	7242.7	7821.2	8276.4	8036.8	8504.6	8927.6	11970.4
	Min. value	2635.0	3179.0	2898.0	2936.0	3159.0	3159.0	3189.0	3159.0	3765.0	6693.0
	Max. value	8234.0	10948.0	12091.0	11948.0	12904.0	13930.0	13842.0	13930.0	15137.0	17869.0
	Standard deviation	877.0	1327.3	1325.6	1265.0	1491.1	1785.3	1591.3	1835.8	1973.8	1808.0
	Variation coefficient	0.161	0.185	0.187	0.174	0.190	0.215	0.198	0.215	0.221	0.151
	Skewness coefficient	0.232	0.355	0.321	0.225	0.338	0.384	0.422	0.342	0.353	0.158
	Kurtosis	0.597	0.133	0.798	0.137	0.422	-0.012	0.705	-0.037	0.109	-0.186

coefficient takes on higher values for the results of the operation of algorithms in the networks constructed on the basis of the Waxman model than in the networks constructed on the basis of the Barabási–Albert model (in the case of the BSMA, this difference is higher than 80%).

The values of the kurtosis of the results of the operation of the algorithms in the networks generated with the application of both models are differentiated. In the networks generated on the basis of the Waxman model, only the results of the operation of the KPP algorithm have the distribution that resembles the mesokurtic distribution. The highest value of the parameter has the BSMA (10 times higher kurtosis than in the Barabási–Albert networks) and RDCMA. The results of the operation of these algorithms are characterized by the leptokurtic distribution in which the values of tree costs are more concentrated than in the normal distribution (mesokurtic). The use of networks generated on the basis of the Barabási–Albert model effects in values of kurtosis that are similar to distributions with zero excess for the algorithms: DCMA and CSPT. The results of the remaining algorithms have the leptokurtic distribution.

The confidence interval as determined in the experiment indicates a possibility for a comparison of the results of the operation of the algorithms for a sample equal to 1000. The data in Table 9.7 are ordered according to the increasing value of the average cost of trees c_T.

The costs of trees in the networks generated on the basis of the Waxman model can be ordered as follows:

$$c_{KPP} < c_{BSMA} < c_{MLRA} < c_{KSPMA} < c_{OPTDCMA} < c_{CSPT} \approx c_{RDCMA} \approx c_{DCMA} < c_{DCSP} < c_{LDC}.$$

Next, the use of networks generated on the basis of the Barabási–Albert model leads to the following dependencies:

$$c_{KPP} < c_{BSMA} \approx c_{MLRA} < c_{KSPMA} < c_{OPTDCMA} < c_{CSPT} < c_{RDCMA} < c_{DCMA} < c_{DCSP} < c_{LDC}.$$

It is adopted in the notations (X and Y) that the relations $c_X < c_Y$ correspond to higher costs of constructed trees in the algorithm c_Y than in the algorithm c_X (confidence intervals are separable). The notation $c_X \approx c_Y$ in the adopted notation stands for comparable costs of constructed trees (confidence intervals overlap). When this is the case, the results of the operation of the algorithms cannot be differentiated (distinguished).

The analysis of the ordered data available in Table 9.7 leads to a conclusion that in networks generated on the basis of the Waxman model, the confidence intervals of the results of the

TABLE 9.7

Confidence Intervals for the Average Costs of Multicast Trees for 1000 Networks ($n = 200$, $m = 20$, $D_{av} = 4$, $\Delta = 2000$)

	Waxman			Barabási–Albert		
Algorithm	$\overline{c_T} - u_\alpha \dfrac{s_N}{\sqrt{N}}$	$\overline{c_T}$	$\overline{c_T} + u_\alpha \dfrac{s_N}{\sqrt{N}}$	$\overline{c_T} - u_\alpha \dfrac{s_N}{\sqrt{N}}$	$\overline{c_T}$	$\overline{c_T} + u_\alpha \dfrac{s_N}{\sqrt{N}}$
KPP	5539.03	5598.81	5658.71	5368.62	5422.98	547.33
BSMA	7058.07	7144.09	7230.10	7081.41	7163.68	7245.94
MLRA	7358.87	7453.82	7548.78	6993.52	7075.68	7157.84
KSPMA	7510.90	7608.31	7705.72	7164.32	7242.73	7321.13
OPTDCMA	7900.92	8009.31	8117.69	7728.76	7821.17	7913.59
CSPT	8242.48	8360.93	8479.37	8165.79	8276.44	8387.10
RDCMA	8249.00	8372.45	8495.91	7938.19	8036.81	8135.44
DCMA	8292.57	8411.39	8530.21	8390.78	8504.56	8618.34
DCSP	8644.45	8772.39	8901.28	8805.22	8927.56	9041.91
LDC	13419.35	13271.89	13419.35	11858.39	11970.45	12082.51

operation of the algorithms: CSPT, DCMA, and RDCMA overlap. For the algorithms MLRA and KSPMA, one can adopt the separation of these results.

9.5 SUMMARY

This chapter presents the methodology for the conducted study on algorithms for multicast connections. Methods for modelling the network topology of the telecommunications and computer networks that form the main element of the comparative analysis of the results obtained in the course of the operation of the studied algorithms are presented. A review of the existing parameters for the network topology is included and new parameters for the network topology are defined. In addition, new methods for the arrangement of receiving nodes in the network are proposed. The method for the statistical analysis of the results is presented as well as the simulation environment in which the study on multicast routing algorithms was performed is introduced.

The chapter extends the existing methodology for the evaluation of multicast routing algorithms. The analysis of the arrangement of group members constitutes an essential input in the research methodology. This kind of approach has not been considered in relevant literature as yet. The chapter examines unconstrained routing algorithms for multicast connections emphasizing the quality of the network model (the accuracy of the illustration of a real Internet topology) and presents a new proposal.

The research has been conducted by the authors for several years. Initially, the studies focused on the accuracy of multicast routing algorithms in relation to the exact algorithm (MST) and were provided for networks that were composed of several nodes. The next stage of research work involved an evaluation of the influence of the parameters describing graphs (that represents real topologies) on the costs of trees constructed by examined algorithms. The studies are unique because they analyze the algorithms in a wide range of network sizes (from several to ten thousand nodes).

Analyses are conducted for networks generated as random graphs with an implementation of Waxman and Barabási–Albert method and with an application of *Inet* heuristic generator. Separate track of the research analyzed the influence of network topology parameters on the cost of multicast tree constructed by selected genetic algorithms.

The research results also show that obtaining of a tree with the lower cost can be the result of the application of the generative methods and not only the result of the application of a more efficient routing algorithm.

REFERENCES

1. J. Pasquale, G. Polyzos, and G. Xylomenos. The multimedia multicasting problem. *ACM Multimedia Systems Journal*, 6:43–59, 1998.
2. H. Eriksson. MBONE: The multicast backbone. *Communications of ACM*, 37:54–60, 1994.
3. L. Han and N. Shahmehri. Secure multicast software delivery. *IEEE Ninth International Workshops on Enabling Technologies: Infrastructure for Collaborative Enterprises (WET ICE'00)*, Gaithersburg, MD, pp. 207–212, 2000.
4. A. Ballardie, P. Francis, and J. Crowcroft. Core-based trees (CBT)—An architecture for scalable inter-domain multicast routing. *Computer Communication Review*, 23:85–95, 1993.
5. L. Wei and D. Estrin. The trade-offs of multicast trees and algorithms. *Proceedings of ICCCN'94*, San Francisco, CA, pp. 55–64. IEEE, 1994.
6. E. W. Zegura, K. L. Calvert, and S. Bhattacharjee. How to model an internetwork. *IEEE INFOCOM'96*, San Francisco, CA, 1996.
7. E. W. Zegura, M. J. Donahoo, and K. L. Calvert. A quantitative comparison of graph-based models for internet topology. *IEEE/ACM Transactions on Networking*, 5:770–783, 1997.
8. M. Doar and I. Leslie. How bad is naive multicast routing? *IEEE INFOCOM'1993*, San Francisco, CA, pp. 82–89, 1993.

9. M. Newman, A. L. Barabási, and D. J. Watts. *The Structure and Dynamics of Networks*. Princeton Studies in Complexity. Princeton University Press, Princeton, NJ, 2007.

10. B. Waxmann. Routing of multipoint connections. *IEEE Journal on Selected Area in Communications*, 6:1617–1622, 1988.

11. A. L. Barabási and R. Albert. Emergence of scaling in random networks. *Science*, 286:509–512, 1999.

12. M. Doar. A better model for generating test networks. *Proceedings of IEEE GLOBECOM*, San Francisco, CA, 1996.

13. A. Medina, A. Lakhina, I. Matta, and J. Byers. BRITE: An approach to universal topology generation. *9th International Symposium on Modeling, Analysis and Simulation of Computer and Telecommunication Systems (IEEE/ACM MASCOTS 2001)*, pp. 346–356, Cincinnati, OH, 2001.

14. T. Bu and D. Towsley. On distinguishing between Internet power law topology generators. *Proceedings of INFOCOM*, New York, 2002.

15. D. J. Watts and S. H. Strogatz. Collective dynamics of 'small-world' networks. *Nature*, 12(393):440–442, 1998.

16. M. Faloutsos, P. Faloutsos, and C. Faloutsos. On power-law relationships of the internet topology. *ACM Computer Communication Review*, Cambridge, MA, pp. 111–122, 1999.

17. W. Aiello, F. Chung, and L. Lu. A random graph model for massive graphs. *32nd Annual Symposium in Theory of Computing*, Portland, OR, 2000.

18. C. Jin, Q. Chen, and S. Jamin. Inet: Internet topology generator. Technical Research Report CSE-TR-433-00, University of Michigan, Ann Arbor, MI, 2000.

19. J. Winick and S. Jamin. Inet 3.0: Internet topology generator. Technical Research Report CSE-TR-456-02, University of Michigan, Ann Arbor, MI, 2003.

20. D. Achlioptas, A. Clauset, D. Kempe, and C. Moore. On the bias of traceroute sampling; or, power-law degree distributions in regular graphs. *STOC'05: Proceedings of the 37th Annual ACM Symposium on Theory of Computing*, Baltimore, MD, pp. 694–703. ACM, 2005.

21. A. Clauset, C. Moore, and M. E. Newman. Hierarchical structure and the prediction of missing links in networks. *Nature*, 453:98–101, 2008.

22. C. Moore, G. Ghoshal, and M. E. Newman. Exact solutions for models of evolving networks with addition and deletion of nodes. *Physical Review (E), Statistical, Nonlinear, and Soft Matter Physics*, 74:036121, 2006.

23. R. Pastor-Satorras and A. Vespignani. *Evolution and Structure of the Internet: A Statistical Physics Approach*. Cambridge University Press, Cambridge, U.K., 2004.

24. F. K. Hwang, D. S. Richards, and P. Winter. *The Steiner Tree Problem*. Annals of Discrete Mathematics, Vol. 53. North-Holland, Amsterdam, the Netherlands, 1992.

25. J. S. Crawford and A. G. Waters. Heuristics for ATM multicast routing. *Proceedings of Sixth IFIP Workshop on Performance Modeling and Evaluation of ATM Networks*, Ilkley, U.K., pp. 5/1–5/18, July 1998.

26. V. P. Kompella, J. Pasquale, and G. C. Polyzos. Multicasting for multimedia applications. *INFOCOM*, Florence, Italy, pp. 2078–2085, 1992.

27. L. Kou, G. Markowsky, and L. Berman. A fast algorithm for Steiner trees. *Acta Informatica*, 15:141–145, 1981.

28. M. F. Mokbel, W. A. El-Haweet, and M. N. El-Derini. A delay constrained shortest path algorithm for multicast routing in multimedia applications. *Proceedings of IEEE Middle East Workshop on Networking*, San Francisco, CA. IEEE Computer Society, 1999.

29. C. A. S. Oliveira and P. M. Pardalos. A survey of combinatorial optimization problems in multicast routing. *Computers & Operations Research*, 32:1953–1981, 2005.

30. M. Piechowiak and P. Zwierzykowski. A new multicast routing algorithm with Lagrange relaxation. Poznanskie Warsztaty Telekomunikacyjne. Politechnika Poznanska, Grudzien, 2007 (in Polish).

31. M. Piechowiak, P. Zwierzykowski, and T. Bartczak. Efficiency of QoS-oriented multicast routing algorithms. *Information Systems Architecture and Technology, Information Systems and Computer Communication Networks*, pp. 113–124, 2008.

32. H. F. Salama, D. S. Reeves, and Y. Viniotis. Evaluation of multicast routing algorithms for real-time communication on high-speed networks. *IEEE Journal on Selected Areas in Communications*, 15(3):332–345, 1997.

33. A. Shaikh and K. G. Shin. Destination-driven routing for low-cost multicast. *IEEE Journal on Selected Areas in Communications*, 15(3):373–381, 1997.

34. Q. Sun and H. Langendoerfer. Efficient multicast routing for delay-sensitive applications. *Proceedings of the Second Workshop on Protocols for Multimedia Systems (PROMS'95)*, pp. 452–458, October 1995.

35. Z. Wang and J. Crowcroft. Quality-of-service routing for supporting multimedia applications. *IEEE Journal on Selected Area in Communications*, 14(7):1228–1234, 1996.

36. R. Widyono. The design and evaluation of routing algorithms for real-time channels. *Technical Report TR-94-024*, University of California, Berkeley, CA, 1994.
37. B. Zhang, M. Krunz, and C. Chen. A fast delay-constrained multicast routing algorithm. *IEEE ICC 2001 Conference*, Helsinki, Finland, 2001.
38. Q. Zhu, M. Parsa, and J. J. Garcia-Luna-Aceves. A source-based algorithm for delay-constrained minimum-cost multicasting. *INFOCOM'95: Proceedings of the 14th Annual Joint Conference of the IEEE Computer and Communication Societies*, Boston, MA, pp. 377. IEEE Computer Society, 1995.
39. N. Banerjee and S. K. Das. Fast determination of QoS-based multicast routes in wireless networks using genetic algorithm. *ICC 2001 IEEE International Conference on Communications*, pp. 2588–2592, 2001.
40. A. T. Haghighat, K. Faez, M. Dehghan, A. Mowlaei, and Y. Ghahremani. A genetic algorithm for Steiner tree optimization with multiple constraints using Prüfer number. *EurAsia-ICT'02: Proceedings of the First EurAsian Conference on Information and Communication Technology*, Shiraz, Iran, pp. 272–280. Springer-Verlag, 2002.
41. M. Hamdan and M. E. El-Hawary. Multicast routing with delay variation constraints using genetic algorithm. *Canadian Conference on Electrical and Computer Engineering*, Niagara Fall, Ontario, Canada, pp. 2363–2366, 2004.
42. N. Shimamoto, A. Hiramatsu, and K. Yamasaki. A dynamic routing control based on a genetic algorithm. *IEEE International Conference on Neural Networks*, Nagoya, Japan, pp. 1123–1128, 1993.
43. G. Zhou and M. Gen. An effective genetic algorithm approach to the quadratic minimum spanning tree problem. *Computers & Operations Research*, 25(3):229–237, 1998.
44. E. Gelenbe, A. Ghanwani, and V. Srinivasan. Improved neural heuristics for multicast routing. *IEEE Journal on Selected Areas in Communications*, 15:147–155, 1997.
45. T. L. Hemminger and C. A. Pomalaza-Raez. Using neural networks to solve the multicast routing problem in packet radio networks. *International Journal of Neural Systems*, 7(5):617–626, 1996.
46. G. Lu and Z. Liu. Multicast routing based on ant-algorithm with delay and delay variation constraints. *Proceedings of IEEE Asia-Pacific Conference on Circuits and Systems*, Tianjin, China, pp. 243–246, 2000.
47. M. Głąbowski, B. Musznicki, P. Nowak, and P. Zwierzykowski. An algorithm for finding shortest path tree using ant colony optimization metaheuristic. In *Image Processing and Communications Challenges 5*, R. Choras (ed.), Advances in Intelligent Systems and Computing, vol. 233, pp. 317–326. Springer, New York, 2014.
48. E. Dijkstra. A note on two problems in connexion with graphs. *Numerische Mathematik*, 1:269–271, 1959.
49. M. Piechowiak, P. Zwierzykowski, and M. Stasiak. Multicast routing algorithm for packet networks with the application of the Lagrange relaxation. *Proceedings of NETWORKS 2010—14th International Telecommunications Network Strategy and Planning Symposium*, Warsaw, Poland, pp. 197–202, September 2010.
50. A. Juttner, B. Szviatovszki, I. Mecs, and Z. Rajko. Lagrange relaxation based method for the QoS routing problem. *IEEE INFOCOM'2001*, Anchorage, AK, 2001.
51. R. Prim. Shortest connection networks and some generalizations. *Bell Systems Technical Journal*, 36:1389–1401, 1957.
52. M. Piechowiak, M. Stasiak, and P. Zwierzykowski. The application of k-shortest path algorithm in multicast routing. *Theoretical and Applied Informatics*, 21(2):69–82, 2009.
53. M. Glabowski, B. Musznicki, P. Nowak, and P. Zwierzykowski. Efficiency evaluation of the shortest path algorithms. *Proceedings of the Ninth Advanced International Conference on Telecommunications (AICT 2013)*, M. D. Logothetis, M. Glabowski, D. Krstic, eds., IARIA, ThinkMind, Rome, Italy, pp. 154–160, June 2013.
54. K. Balinska. Projektowanie algorytmów i struktur danych. Wydawnictwo Politechniki Poznanskiej, 2003 (in Polish).
55. Y. Rekhter and T. Li. A Border Gateway Protocol 4 (BGP-4). RFC 1654 (Proposed Standard), 1994. Obsoleted by RFC 1771.
56. M. Piechowiak, P. Zwierzykowski, and T. Bartczak. An application of the switched tree mechanism in the multicast routing algorithms. *First Interdisciplinary Technical Conference of Young Scientists InterTech 2008*, Poznań, Poland, pp. 282–286, 2008.
57. S. Hong, H. Lee, and B. H. Park. An efficient multicast routing algorithm for delay-sensitive applications with dynamic membership. *Proceedings of IEEE INFOCOM'98*, San Francisco, CA, pp. 1433–1440, 1998.
58. C. P. Low and N. Wang. On group multicast routing with bandwidth constraint: A lower bound and performance evaluation. *IEICE Transactions on Communications*, 1:124–131, 2004.

59. G. N. Rouskas and I. Baldine. Multicast routing with end-to-end delay and delay variation constraints. *IEEE Journal on Selected Areas in Communications*, 15:346–356, 1997.
60. D. G. Thaler and C. V. Ravishankar. Distributed center-location algorithms. *IEEE Journal on Selected Areas in Communications*, 15:291–303, 1997.
61. K. Stachowiak and P. Zwierzykowski. Innovative method of the evaluation of multicriterial multicast routing algorithms. *Journal of Telecommunications and Information Technology*, 13(1):49–55, 2013.
62. T. Bartczak and P. Zwierzykowski. Lightweight PIM—A new multicast routing protocol. *International Journal of Communication Systems*, Wiley, July 2012, DOI: 10.1002/dac.2407.
63. C. A. Noronha and F. A. Tobagi. Evaluation of multicast routing algorithms for multimedia streams. *Proceedings of IEEE International Telecommunications Symposium*, Rio de Janeiro, Brazil, p. 94, 1994.
64. M. Piechowiak, M. Stasiak, and P. Zwierzykowski. Analysis of the influence of group members arrangement on the multicast tree cost. *Proceedings of the Fifth Advanced International Conference on Telecommunications AICT*, Venice, Italy, 2009.
65. M. Piechowiak, M. Stasiak, and P. Zwierzykowski. The influence of group members arrangement on the multicast tree cost. *International Journal on Advances in Systems and Measurements*, 2(2–3):248–257, 2009.
66. Microsoft. MSDN Library Documentation.
67. A. Plucinska and E. Plucinski. Rachunek prawdopodobienstwa. Statystyka matematyczna. Procesy stochastyczne. Wydawnictwa Naukowo-Techniczne, Warszawa, 2000 (in Polish).
68. MySQL Documentation. http://dev.mysql.com/doc/ (accessed January 14, 2014).
69. J. Eisenberg. *SVG Essentials*. O'Reilly Media, Inc., Sebastopol, CA, 2002.
70. M. Piechowiak and P. Zwierzykowski. Heuristic algorithm for multicast connections in packet networks. *Proceedings of EUROCON 2007 the International Conference on: Computer as a Tool*, Warsaw, Poland, pp. 948–955. IEEE, September 2007.
71. M. Piechowiak and P. Zwierzykowski. Efficiency analysis of multicast routing algorithms in large networks. *Proceedings of the Third International Conference on Networking and Services ICNS 2007*, Athens, Greece, pp. 101–106. IEEE, June 2007.
72. M. Piechowiak and P. Zwierzykowski. Performance of fast multicast algorithms in real networks. *Proceedings of EUROCON 2007 the International Conference on: Computer as a Tool*, Warsaw, Poland, pp. 956–961. IEEE, September 2007.

10 Statistical Simulation of Multipath Fading Channels for Mobile Wireless Digital Communication Systems

David Luengo and Luca Martino

CONTENTS

ABSTRACT

The accurate simulation of multipath fading channels is a crucial issue in the development and evaluation of modern wireless communication networks. Since the received signal depends on many fast-changing factors, statistical models are typically used to simulate fading. Many fading models have been developed over the last four decades, making use of several statistical distributions (Rice, Rayleigh, Nakagami-m, etc.) and with different degrees of complexity. In this chapter, we describe some of the most important statistical models developed for the simulation of wireless fading channels, paying special attention to the accuracy–complexity trade-off and discussing the application domain where each of them is more appropriate. Indeed, the main purpose of this chapter is providing clear algorithms for the efficient simulation of each of the channels described (which are often hard to find in the literature), rather than focusing only on the theoretical aspects (which are also covered). In order to achieve this purpose, several case studies are introduced throughout the chapter, showing how to simulate real-world channels (at different degrees of detail and complexity) for practical applications.

The chapter is structured as follows. Firstly, we provide an introduction to mobile wireless communication networks, motivating the use of statistical tools to simulate multipath fading channels in Section 10.1. In Section 10.2, we describe the general framework of the statistical models used for the simulation of wireless fading channels, emphasizing the different types of channels that can be encountered and the key elements that have to be modelled. A brief description of a digital communications transmitter and receiver is also provided in this section. Then, Section 10.3 focuses on large-scale fading, which is typically simulated using a fixed average loss plus a log-normal distribution to describe local variations. Section 10.4, which is the main section of the chapter, concentrates on the simulation of small-scale fading for the two most common situations found in modern communication systems: slow flat fading and slow frequency-selective fading. Finally, Section 10.5 closes the chapter.

10.1   INTRODUCTION TO MOBILE WIRELESS COMMUNICATION NETWORKS

This chapter deals with the discrete-time computer simulation of mobile wireless digital communication systems. More precisely, we focus on the important and challenging issue of the generation of accurate and computable statistical models for fading channels, which are always encountered in the simulation of the physical-link layer of mobile wireless networks. Indeed, a great deal of effort has been devoted to developing statistical fading models, since fading is often one of the main limiting factors for reliable transmission throughout these networks [1–3]. However, before we address this issue in the following sections, we first have to answer one important question: what are mobile wireless digital communication systems?

First of all, let us remark that there is not a simple and widely accepted answer to this question. Indeed, mobile wireless digital communication systems comprise a wide variety of different techniques and standards that have some broad common characteristics:

- At least one of the users in the communications link is assumed to be able to move around the network, although many degrees of mobility are possible: nomadic (i.e., fixed users with occasional and reduced mobility), low mobility (i.e., local and/or low speed), high mobility (i.e., wide range and/or high speed), etc. Often, in a two-way communication, only one of the users is assumed to move, while the other remains fixed (e.g., base stations [BSs] in cellular communications or access points in wireless data networks).
- At least the last stage of the network is based on wireless (i.e., radio frequency [RF]) transmission. However, wireless networks often make use of wired transmission at higher levels of the network (e.g., as backhaul technology among BSs in cellular networks) or at least are interconnected to other wired networks (e.g., access points in wireless data networks are typically connected to the Internet).
- The physical-link layer is based on digital communication techniques to allow an efficient and reliable transmission throughout the hostile RF environment.

Mobile wireless communication systems were introduced more than 40 years ago to allow users to get rid of the need to be attached to a particular place in order to communicate with other users (i.e., one of their main goals was introducing user mobility) [4]. Since then, mobile wireless communications have revolutionized the concept of communications, implying a change of paradigm: from fixed, wired, place-to-place communications to mobile, wireless, person-to-person communications. One of the first wireless mobile technologies was cordless telephones, which were introduced as an in-house replacement of wireline telephones in the 1970s. The aim of cordless telephony is mainly providing low-cost and low-power voice communication to low-mobility users (in terms of both speed and range) and has evolved from its first analog versions to the Digital Enhanced Cordless Telecommunications (DECT) standards [5–7], which are ubiquitously used worldwide nowadays.

However, the best known example of mobile wireless digital communication systems is probably cellular telephony systems. The first analog cellular telephony system, the Advanced Mobile Phone System (AMPS), was introduced in the United States at the end of the 1970s [8]. From the beginning, these systems were designed for high-mobility users (up to vehicular speeds) with a widespread coverage (at least nationwide). Cellular systems have evolved much from these analog first-generation devices, becoming digital already in the second generation [9,10]: the Global System for Mobile Communications (GSM) in Europe [11–13], the IS-54 [14,15] and IS-95 [16,17] systems in the United States, the Japanese Digital Cellular (JDC) system [18], etc. Although these systems allowed for limited data transmission (through short text messages), they were still primarily developed for voice transmission. However, third- and fourth-generation (and beyond) cellular networks are evolving to become truly worldwide multimedia networks (integrating voice, text, images, and video) with an ever-increasing capacity [19–21].

A third important class of mobile wireless digital communication systems is wireless data networks, which can be divided primarily into wireless local area networks (WLANs) and wireless metropolitan area networks (WMANs). On the one hand, WLANs were designed from the start to provide high-speed data communications to low-mobility users (i.e., typically nomadic user moving only occasionally or users moving at walking speed) confined in a reduced local area (a large building or a campus at most). The best known example of WLANs is the IEEE 802.11 family of standards (known popularly as Wi-Fi), which are integrated nowadays in most portable devices (from laptops to tablets and mobile phones) [22,23]. On the other hand, WMANs were initially designed for high-mobility, wide-range (i.e., within a city or even a larger region), low-data-rate communications to both vehicles and pedestrians. The best example of WMANs is the IEEE 802.16 family of standards (known popularly as WiMAX) [24].

As a summary, let us remark again that wireless networks were initially introduced to provide mobility to the users. The first wireless networks were very specialized, being roughly divided into analog voice (cordless telephony and first-generation cellular networks) and diffusion networks (satellite and land based) and digital data networks (WLANs and WMANs). In a second stage, analog networks quickly evolved into digital networks (digital cordless telephony and second-generation cellular networks) and started integrating some (very limited) data capabilities (e.g., text messages). In a third stage, different networks have integrated different types of data and services, thus becoming truly multimedia networks. Nowadays, we are in the middle of the fourth stage, where faster and more efficient communications are being sought and interconnections between networks are becoming more and more ubiquitous.

However, regardless of the particular communications network being considered, the need for developing simulators of mobile wireless digital communication systems has always existed. The discrete-time simulation of communication systems on a digital computer attempts to extract the key aspects of the system under study, with the goal of replicating some desired aspect of its behavior in an accurate, yet computable, manner [25]. In the case of a digital communications system, the simulation of the physical-link layer provides important performance parameters about the quality of the network: bit or symbol error rate (BER or SER), outage probability and duration, acceptable level of interference, etc. These simulations are crucial during the network's design stage (a lot of time and money can be saved by not requiring the development of a prototype network from the start) but also during its operation (e.g., to identify potential network problems before they occur) and when planning the development and introduction of new physical-layer techniques. Furthermore, physical-layer simulations are also frequently used by standardization bodies for deciding among several competing technologies.

In this chapter, we focus on the statistical simulation of multipath fading channels, which are often the limiting factor for reliable transmission through mobile wireless digital communication systems, as stated before. Finally, it is important to remark that there are many other mobile wireless digital communication systems that we have not discussed in this introduction: satellite networks, trunking networks, wireless sensor networks, cognitive radio networks, wireless personal area networks (WPANs), etc. However, all of these systems may also benefit from the use of the techniques described in the sequel to simulate fading channels, since they have been developed for different classes of channels and are often not tailored to a specific digital communications system.

10.2   STATISTICAL SIMULATION OF MULTIPATH FADING CHANNELS

From a statistical point of view, the discrete-time simulation of a mobile wireless communication channel may be performed according to the model shown in Figure 10.1.

This model consists of three large blocks: transmitter, channel, and receiver. In the following subsections, we provide a high-level view of each of these blocks, before focusing on the simulation of the fading channel in Sections 10.3 and 10.4.

10.2.1   Transmitter

The first block in a mobile wireless digital communications system is the transmitter, which generates a wave form containing the desired information to be sent to the receiver at a carrier frequency $\omega_c = 2\pi f_c$ rad/s, which can be typically expressed as [26,27]

$$\tilde{x}(t) = \sum_k \left[A_I[k] \cos(\omega_c t + \phi) - A_Q[k] \sin(\omega_c t + \phi) \right] p(t - kT),$$

where
  $k = 0,1,2,\ldots$ indicates the transmitted symbol's number
  $\phi \in [0,2\pi)$ is an arbitrary phase (unknown in practice, but assumed to be fixed)
  T is the transmitted symbol's period
  $A_I[k]$ and $A_Q[k]$ denote the in-phase and quadrature components
  $p(t)$ is the impulse response of the baseband shaping filter

Regarding the in-phase and quadrature components, their values depend on the *constellation* used. For instance, for two of the most widely used constellations in mobile wireless communications,

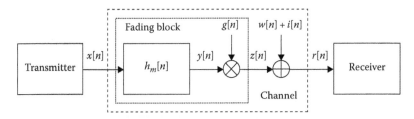

FIGURE 10.1   Basic statistical model for the discrete-time simulation of a mobile wireless communications channel.

binary phase shift keying (BPSK) and quadrature phase shift keying (QPSK), their values are $A_I[k] = \pm\sqrt{2E_b/T_b}$ and $A_Q[k] = 0$ for BPSK (with E_b denoting the desired energy per bit and $T_b = 1/R_b$ the bit's period given a required bit rate R_b), whereas for QPSK, either $A_I[k] = \pm\sqrt{2E_s/T}$ and $A_Q[k] = 0$ or $A_Q[k] = \pm\sqrt{2E_s/T}$ and $A_I[k] = 0$ (with E_s denoting the desired energy per symbol and $T = 1/R_s$ the symbol's period for a transmitted symbol's rate $R_s = R_b/2$).* More complex modern modulation formats, such as orthogonal frequency division multiplexing (OFDM), can also be described in terms of their phase and quadrature components, which are obtained applying the inverse fast Fourier transform (IFFT) to the transmitted symbols per carrier [28,29]. With respect to (w.r.t.) the shaping pulse, the simplest option is using a rectangular pulse centered at $t = 0$ and a width equal to the symbol's period:

$$p(t) = \begin{cases} 1, & -T/2 < t < T/2 \\ 0, & \text{otherwise} \end{cases}.$$

Although this option is never used in practice in mobile wireless communication systems (due to the infinite bandwidth occupied by this pulse), it is the easiest option for the simulation and is often used in this case, especially when ideal synchronization is considered. In practice, the most common shaping filter is the so-called *raised cosine* pulse:[†]

$$p(t) = \frac{\sin(\pi t/T)}{\pi t}\frac{\cos(\alpha\pi t/T)}{1-(2\alpha t/T)^2},$$

where $0 \leq \alpha \leq 1$ is the so-called roll-off factor, which controls the height of the secondary lobes of the filter in the time domain (the smaller the value of α, the higher the lobes) and accordingly the bandwidth of the filter, since the Fourier transform of $p(t)$ is

$$P(f) = \begin{cases} T, & |f| \leq \frac{1-\alpha}{2T}, \\ \frac{T}{2}\left[1+\cos\left(\frac{\pi T}{\alpha}\left[|f|-\frac{1-\alpha}{2T}\right]\right)\right], & \frac{1-\alpha}{2T} < |f| \leq \frac{1+\alpha}{2T}, \\ 0, & |f| > \frac{1+\alpha}{2T}. \end{cases}$$

Hence, the bandwidth of the filter is $W = (1 + \alpha)/2T$, and it can be seen that the higher the value of α, the larger the occupied bandwidth, which has a range $\frac{1}{2T} \leq W \leq \frac{1}{T}$. The value $\alpha = 0$ corresponds to the well-known *Nyquist shaping filter*:

$$p(t) = \frac{\sin(\pi t/T)}{\pi t},$$

---

* Some authors consider the following alternative constellation for QPSK: $A_I[k] = \pm\sqrt{E_s/T}$ and $A_I[k] = \pm\sqrt{E_s/T}$. This corresponds simply to a $\pi/4$ rotation of the constellation considered here and does not affect the system's performance.
† Note that a raised cosine filter is noncausal and has an infinite length. Thus, in practice, a truncated and time-shifted version of the filter is typically used [26,27].

whose Fourier transform is a rectangular pulse:

$$
P(f) = \begin{cases} T, & |f| \le \dfrac{1}{2T} \\[2mm] 0, & |f| > \dfrac{1}{2T} \end{cases}
$$

and which occupies the smallest possible bandwidth, $W = 1/2T$, but has the highest secondary lobes (thus being very sensitive to synchronization errors and requiring long filters to achieve a good performance). Good compromise solutions typically use $0.25 \le \alpha \le 0.5$. Figure 10.2 shows several examples of the aforementioned shaping filters, both in the time and the frequency domains. For the raised cosine filters, note the decrease in the amplitude of the side lobes as α increases and the

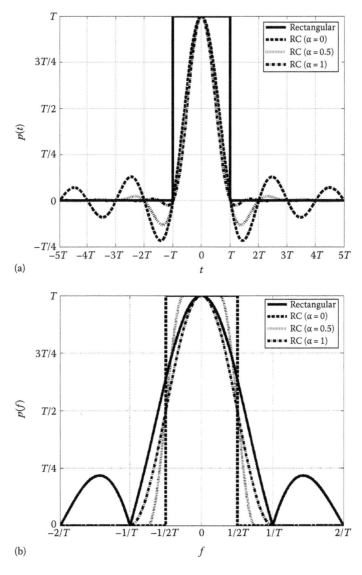

(a)

(b)

FIGURE 10.2 Examples of typical shaping filters used in digital communications: (a) time domain and (b) frequency domain.

corresponding increase in the occupied bandwidth. For the rectangular shaping filter, note the finite duration in the time domain and the corresponding infinite occupied bandwidth, with large side lobes on both sides of the spectrum.

Regardless of the constellation and shaping filter used, any computer simulation of a digital communications system has to be performed in the discrete-time domain, typically using equispaced samples. According to the *sampling theorem* [30], a sampling frequency $\omega_s > 2\omega_c$ would have to be used in order to avoid aliasing,* with the subsequent computational burden given the high carrier frequencies used in wireless communications (typically in the range of hundreds of MHz up to a few GHz and even more). Hence, computer simulators of digital communications systems usually work with the *complex low-pass equivalent signal*,

$$x(t) = \sum_k \left(A_I[k] + jA_Q[k] \right) p(t - kT),$$

which contains the baseband portion of the communications signal and which allows us to recover the original band-pass signal as

$$\tilde{x}(t) = \mathrm{Re}\left\{ x(t) e^{j\omega_c t} \right\}.$$

Using a raised cosine shaping filter, the complex low-pass equivalent signal has a bandwidth equal to $W = \dfrac{1+\alpha}{2T} \ll f_c$, thus allowing us to perform a discrete-time simulation using a sampling frequency $f_s = R/T \ll 2f_c$, where $R = T/T_s \geq 1 + \alpha$ is the *oversampling ratio* and T_s is the sampling period.† The discrete-time low-pass equivalent signal is finally given by

$$x[n] = \sum_k \left(A_I[k] + jA_Q[k] \right) p[n - kR],$$

where
$p[n - kR] = p\left((n - kR)T_s \right)$
$T_s = T/R$ is the sampling rate

10.2.2 Channel: Fading, Noise, and Interference

The central block in any mobile wireless digital communications simulator is the channel. For a mobile wireless system, it is typically composed of a fading generator, a noise generator, and possibly also an interference generator. The signal at the output of the transmitter is the input of the fading generator (as shown in Figure 10.1), which returns an output

$$z[n] = g[n] \times \left(h_m[n] * x[n] \right),$$

where the first term, $g[n]$, indicates the variable loss due to large-scale fading and the second term, $y[n] = h_m[n] * x[n]$, corresponds to the distortion introduced by small-scale fading, with $h_m[n]$ being

---

* *Band-pass sampling* approaches could be applied to reduce the required sampling frequency [96], but this is usually more complex than the low-pass equivalent signal approach.
† A *symbol rate simulation*, performed when $\alpha = 0$ and $R = 1$ (the minimum sampling rate required), is often enough (e.g., when working only with additive white Gaussian noise or simulating a flat fading channel). When more precise signal representations are required (e.g., for the simulation of frequency selective fading), higher oversampling ratios (e.g., $R = 4$, $R = 8$ or even $R > 8$) have to be used.

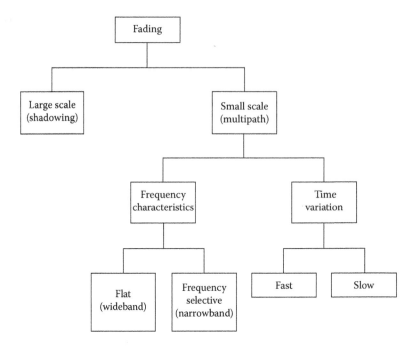

FIGURE 10.3 Block diagram summarizing the different types of fading in mobile wireless communications systems.

the *complex baseband equivalent channel*. In this equation, the large-scale fading term models the average signal attenuation or path loss due to motion over large areas, whereas the small-scale fading term takes into account the dramatic changes in amplitude that can occur even for very small displacements (even on the order of a fraction of a wavelength). Figure 10.3 shows a tree diagram with the different types of fading considered in the sequel [2,3,31]. A more precise definition of the different types of fading shown in Figure 10.3 is provided in Section 10.3 for large-scale fading and Section 10.4 for small-scale fading. The difference between the four classes of small-scale fading is also provided in this section, once the required statistical concepts (i.e., coherence bandwidth and coherence time) have been introduced.

The signal is then contaminated by the additive white Gaussian noise (AWGN) present in all digital communications systems, $w[n]$, and possibly also by interference from other users of the spectrum (either from the same system, as it happens in CDMA applications or from other systems), $i[n]$.* Therefore, the received signal is given by

$$r[n] = z[n] + w[n] + i[n]$$

$$= g[n] \times \left(h_m[n] * x[n] \right) + w[n] + i[n]$$

$$= g[n] \times \left(h_m[n] * \sum_k \left(A_I[k] + jA_Q[k] \right) p[n - kR] \right) + w[n] + i[n].$$

Table 10.1 shows the steps required for simulating the block diagram of Figure 10.1. In the following sections, we consider only AWGN, $w[n] \sim N\left(0, \sigma_w^2\right)$ with $x \sim N(\mu, \sigma^2)$ indicating that x is a Gaussian

---

* Interference from other systems can be simulated either by generating full valid signals from the interfering systems or by drawing $i[n]$ from a statistical model that replicates the typical characteristics of the interference. However, this issue is out of the scope of this chapter.

TABLE 10.1

High-Level Simulation of a Mobile Wireless Digital Communications System

1. Generate the sequence of digital symbols to be transmitted, $x[n]$ for $n = 0,1,...,N_s-1$. These symbols may be generated randomly according to the desired constellation (typically by assuming them to be equiprobable) or replicating the transmitter's structure if a more detailed simulation is required.

2. Determine the most appropriate fading model for the system under study (i.e., the large-scale fading and small-scale fading channel characteristics) and use it to generate samples from $h_m[n]$ and $g[n]$, obtaining $z[n] = g[n] \times (h_m[n] * x[n])$ for $n = 0,1,...,N_s-1$.

3. Add the appropriate level of noise for the signal-to-noise ratio (SNR) required by the simulation, as well as the interference from other users or systems (optional), so that the received signal is $r[n] = z[n] + w[n] + i[n]$.

4. Estimate the transmitted symbols in the receiver, $\hat{x}[n]$ for $n = 0,1,...,N_s-1$, and obtain the desired performance parameters (e.g., the BER or SER).

random variable (RV) with mean μ and variance σ^2, and focus on the implementation of the different models developed for the simulation of large-scale and especially small-scale fading.

10.2.3 RECEIVER

The receiver in a digital communications system tries to provide a faithful estimate of the transmitted symbols from the noisy received signal. The basic structure of a digital communications receiver is shown in Figure 10.4.

The first element in the receiver (after the analog front end [not shown], which typically includes the antenna and a low-noise amplifier) is usually an automatic gain control (AGC) block. The AGC tries to compensate the variable attenuation introduced by the large-scale fading (see Section 10.3).* Essentially, the AGC estimates the variable channel's gain (e.g., through a moving average [MA] filter) and compensates it, trying to obtain a fixed energy per symbol at its output. Assuming that the estimated channel gain at the nth sample is $\hat{g}[n] \approx g[n]$ (with $g[n]$ denoting the path loss due to large-scale fading), the desired gain at the output is equal to one, and no interference is present in the system, the output of the AGC will be

$$\hat{y}[n] = \frac{r[n]}{\hat{g}[n]} = \left(x[n] * h_m[n] \right) \times \frac{g[n]}{\hat{g}[n]} + \frac{w[n]}{\hat{g}[n]} \approx x[n] * h_m[n] + \tilde{w}[n].$$

This constant gain signal is then passed to the second element in the receiver: the equalizer. This block tries to compensate the effect of the channel, modelled through its time-varying impulse response, $h_m[n]$. The equalizer has been traditionally implemented as an adaptive filter [32–34], which tries to approximate the inverse of the time-varying channel's impulse response (i.e., $h_{eq}[n] \approx h_t^{-1}[n]$), so that its output is given by

$$\hat{x}[n] = \hat{y}[n] * h_{eq}[n] \approx x[n] * h_t[n] * h_{eq}[n] + \tilde{w}[n] * h_{eq}[n] \approx x[n] + \tilde{w}_{eq}[n].$$

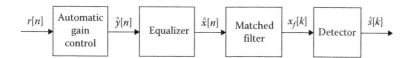

FIGURE 10.4 Basic structure of a digital communications receiver.

* In practice, the AGC may also be able to compensate (at least partially) the small-scale slow flat fading (see Section 10.4.2).

More recently, equalization in the frequency domain has been shown to be advantageous in many cases. This is particularly true for multicarrier modulation schemes (such as those based on OFDM or filter banks) [35,36], but this approach has been considered even for single-carrier modulations [37,38]. In any case, the goal of the equalizer is producing an output like the one shown earlier, where the effect of the channel's impulse response has been removed. Finally, let us remark that this block is only required when we have a frequency-selective channel (see Section 10.4.3). Otherwise, the channel is simply given by $h_k[n] = A_k\delta[n]$, and this gain can either be compensated by the AGC block or using a simple single-tap equalizer.

The output of the equalizer (or the output of the AGC if the equalizer is not required) is used as the input of the matched filter. This is a filter adapted to the shaping filter used in the transmitter in the sense that its impulse response is $p^*[-n]$, that is, it is the time-reversed complex conjugate of the transmitter's shaping pulse $p[n]$.* Hence, the output of the matched filter will be given by

$$x_f[n] = \hat{x}[n] * p^*[-n] \approx \sum_k \left(A_I[k] + jA_Q[k] \right) p[n - kR] * p^*[-n] + \tilde{w}_{eq}[n] * p^*[-n].$$

It can be shown that the output of this filter maximizes the signal-to-noise ratio (SNR) for samples taken at the appropriate time instants, which are the times where $p[n-kR]$ and $p^*[-n]$ are aligned (i.e., $n = kR$ for $k = 0,1,\ldots,N_s-1$) and the result of $p[n-kR]^* p^*[-n]$ corresponds to the energy of the shaping pulse [26,27,32]. Considering only the optimal sampling instants and assuming that the shaping pulse's energy is normalized to unity, we have[†]

$$x_f[k] = x_f[n]\big|_{n=kR} \approx \sum_k \left(A_I[k] + jA_Q[k] \right) + \tilde{w}_f[k].$$

The last block, the detector, tries to estimate the transmitted symbols from the noisy symbols at the output of the matched filter, that is, it performs a mapping in such a way that the SER is minimized:

$$\left(A_I[k] + \mathrm{Re}\{\tilde{w}_f[k]\} \right) + j\left(A_Q[k] + \mathrm{Im}\{\tilde{w}_f[k]\} \right) \rightarrow \hat{A}_I[k] + j\,\hat{A}_Q[k]$$

where
  $\mathrm{Re}\{\tilde{w}_f[k]\}$ and $\mathrm{Im}\{\tilde{w}_f[k]\}$ denote the real and imaginary parts of $\tilde{w}_f[k]$, respectively
  $\hat{A}_I[k] + j\hat{A}_Q[k]$ is the valid symbol from the transmitter's constellation (see Section 10.2.1) that minimizes the SER

For example, for BPSK, $A_I[k] = \pm\sqrt{2E_b/T_b}$ and $A_I[k] = \pm\sqrt{2E_b/T_b}$, and the optimum detector is obtained by taking the sign of the output of the matched filter, that is, $\hat{A}_I[k] = \sqrt{2E_b/T_b} \times sign\left(x_f[k] \right)$.

Finally, let us remark that many other blocks are usually present in the simulation of a real-world receiver (e.g., synchronization). However, the detailed description of these blocks is out of the scope of this chapter. A detailed description of these blocks can be found in any digital communications book [26,27,32].

---

* Note that all the shaping filters considered in Section 10.2.1 (rectangular pulse, Nyquist filter, and raised cosine filter) are real, so the matched filter becomes simply $p[-n]$.
[†] Note the change in data rate after the matched filter: until then, $R \geq 1 + \alpha$ samples/symbol have to be used. Afterwards, only one sample/symbol is typically used to estimate the transmitted symbols.

10.3 LARGE-SCALE FADING MODELS

10.3.1 GENERAL EXPRESSION FOR THE LARGE-SCALE FADING ATTENUATION

Large-scale fading or shadowing refers to the average signal attenuation or path loss due to motion over large areas and depends on the distance between the transmitter and the receiver, as well as on the obstructions present in the communications channel (prominent terrain contours, buildings, etc.). Large-scale fading is typically modelled as a mean path loss, which represents the average loss as a function of distance and the type of environment where transmission takes place (urban, suburban, rural, etc.), and a log-normally distributed variation about the mean describing local variations w.r.t. the typical conditions [2,3,31]. Mathematically,

$$L_{total}(t) = \bar{L}(d_0) + 10\ell \log_{10}\left(\frac{d(t)}{d_0}\right) + L_{LS}(t),$$

where the first two terms represent the average loss following the so-called ℓth power law,* which consists of the reference average loss (first term, $\bar{L}(d_0)$) and the excess average loss w.r.t. the reference distance (second term, $\overline{\Delta L_d}(t)$), whereas the third term, $L_{LS}(t) \sim N(0, \sigma_{LS}^2)$, is a Gaussian-distributed RV power variation in decibels (log-normally distributed in natural units) [2]. Here, d_0 is the *reference distance* corresponding to a point located in the far field of the antenna (typically, $d_0 = 1$ km for large cells, $d_0 = 100$ m for microcells and $d_0 = 1$ m for indoor channels), $d(t)$ is the distance between the transmitter and receiver (which varies with time as one or both of them move), and $\bar{L}(d_0)$ is the *reference path loss*, found through measurements or calculated using the classical *free-space path loss rule*:

$$\bar{L}(d_0) = 10\ell \log_{10}\left(\frac{P_t G_t G_r \lambda^2}{(4\pi d_0)^2}\right),$$

where
  P_t is the transmitted power
  G_t is the transmitter's antenna gain
  G_r is the receiver's antenna gain
  ℓ is the *propagation* or *path loss exponent* ($\ell = 2$ for free space, $\ell < 2$ for environments where strong guided wave phenomena appear, like tunnels or urban streets with very high buildings, and $\ell > 2$ for most urban environments, where obstructions between the transmitter and receiver are present)

Simulating this model for a given trajectory, $d[n] = d(t)\big|_{t=nT_s}$ where T_s is the sampling period, is as easy as calculating the fixed part for each time instant, adding to it the random part drawn from a zero-mean RV with variance σ_{LS}^2, and setting the gain of the transmitted signal in the discrete-time simulation as

$$g[n] = \sqrt{10^{L_{total}[n]/10}},$$

where $L_{total}[n] = L_{total}(t)\big|_{t=nT_s}$. Table 10.2 provides the steps required for simulating this model, whereas the block diagram corresponding to its implementation is shown in Figure 10.5. Note that

---

* Let us remark that many models for the average loss have been developed, but they all follow similar expressions to the one provided here [97–102].

TABLE 10.2

Procedure Used to Simulate the nth Sample of the Loss due to Large-Scale Fading When No Correlation among Samples Is Considered

1. Obtain the local variation of the large-scale fading as $L_{LS}[n] \sim N\left(0, \sigma_{LS}^2\right)$.

2. Obtain the mean path loss as $\bar{L}_d[n] = \bar{L}\left(d_0\right) + \overline{\Delta L_d}[n]$, where $\bar{L}\left(d_0\right)$ is the average loss at the reference distance d_0, which is typically calculated either using the free-space loss rule or through channel measurements, and $\overline{\Delta L_d}[n] = 10\ell \log_{10}\left(\dfrac{d[n]}{d_0}\right)$ is the excess average loss w.r.t. the reference distance for a given path loss exponent ℓ.

3. The scale factor due to large-scale fading for the nth sample is finally given by $g[n] = \sqrt{10^{L_{total}[n]/10}}$, with $L_{total}[n] = \bar{L}_d[n] + L_{LS}[n]$.

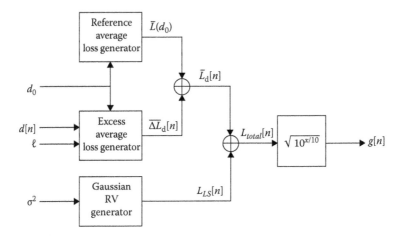

FIGURE 10.5 Block diagram for the simulation of the large-scale fading loss without taking into account the temporal correlation.

the simulation of this model simply requires the precalculation of $\bar{L}\left(d_0\right)$, updating $\overline{\Delta L_d}[n]$ if the distance between the transmitter and receiver has changed, and one sample from a Gaussian RV per output sample.

10.3.2 INTRODUCING CORRELATION IN THE LARGE-SCALE FADING LOSS

The model described in the previous subsection does not take into account the correlation among consecutive samples, thus providing unrealistically large variations in the large-scale fading loss. In this section, we describe in detail the most widely used approach to introduce the temporal correlation (through an exponential autocorrelation function [ACF]) and then briefly mention other alternatives considered in the literature.

10.3.2.1 Exponential Autocorrelation Function

A simple correlation model that has been widely used for computer simulations (especially in the field of cellular communications) was proposed in [39]. According to this model, the discrete-time ACF among consecutive samples decays exponentially:

$$R_{gg}[m] = \sigma_{LS}^2 \times \exp\left(-\beta|k|\right),$$

where uniform sampling is assumed (i.e., $R_{gg}[m] = R_{gg}(\tau)\big|_{\tau=mT_s}$ for $m = 0, \pm1, \pm2,\ldots$), and

$$\beta = -\frac{v(\text{m/s})T_s(\text{s})}{d(\text{m})}\ln\varepsilon_D,$$

where
    $v(\text{m/s})$ is the mobile's speed (assuming a fixed BS's position)
    $T_s(\text{s})$ is the sampling period
    $d(\text{m})$ is the distance between the mobile and the BS
    $0 < \varepsilon_D < 1$ is a constant that controls the strength of the correlation (the larger the value of ε_D, the smaller the value of β, the stronger the correlation, and thus the slower the exponential decay)

Figure 10.6 shows this ACF for $\sigma_{LS}^2 = 1$ and three values of β (0.1, 0.2, and 0.4), as well as the corresponding power spectral density (PSD), $S_{gg}(\omega)$.

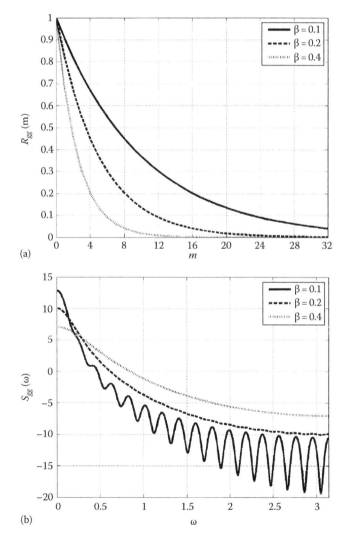

FIGURE 10.6 Theoretical (a) ACFs and (b) PSDs for simulating large-scale fading.

This model is able to provide realistic correlations when compared to measured data (especially for moderate and large cells in mobile cellular communications), is easy to analyze, and is extremely easy to simulate, since the desired samples can be obtained as the output of a first-degree infinite impulse response (IIR) filter with transfer function [39,40]:

$$H(z) = \frac{z}{z-a} = \frac{1}{1-az^{-1}}, \quad |z| < |a| < 1,$$

where $a = \exp(-\beta)$. Hence, the difference equation associated to this filter is simply

$$y[n] = x[n] + ay[n-1],$$

and the desired noise samples are obtained by using a zero-mean AWGN input with a variance $\sigma_a^2 = \sigma_{LS}^2 (1-a^2)$ to drive the filter.

Figure 10.7 shows the block diagram required for the complete simulation of the large-scale fading model described earlier, whereas Table 10.3 details the corresponding stages in the simulation. Then, Figure 10.8 shows the empirical ACFs and PSDs obtained applying this simulation model for the same values of β used before (0.1, 0.2 and 0.4). Comparing Figures 10.8 and 10.6, we notice the good match among the simulated ACFs and the desired theoretical ACFs. Finally, a case study

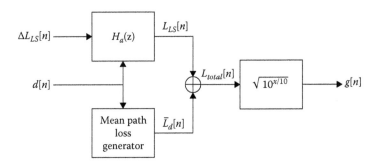

FIGURE 10.7 Block diagram for the simulation of the large-scale fading loss, taking into account the temporal correlation of the channel.

TABLE 10.3

Procedure Used to Simulate the *n*th Sample of the Loss due to Large-Scale Fading

1. Draw $\Delta_{LS}[n] \sim N(0, \sigma_a^2)$, with $\sigma_a^2 = \sigma_{LS}^2 (1-a^2)$ and $a = \exp(-\beta) = \varepsilon_D^{v[n]T_s / d[n]}$.

2. Obtain the local variation of the large-scale fading as $L_{LS}[n] = \Delta_{LS}[n] + aL_{LS}[n-1]$, using $L_{LS}[-1] \sim N(0, \sigma_{LS}^2)$, where σ_{LS}^2 is the large-scale fading variance.

3. Obtain the mean path loss as $\bar{L}_d[n] = \bar{L}(d_0) + \overline{\Delta L_d}[n]$, where $\bar{L}(d_0)$ is the average loss at the reference distance d_0, which is typically calculated either using the free-space loss rule or through channel measurements, and

   $$\overline{\Delta L_d}[n] = 10\ell \log_{10}\left(\frac{d[n]}{d_0}\right)$$ is the excess average loss w.r.t. the reference distance for a given path loss exponent ℓ.

4. The scale factor due to large-scale fading for the *n*th sample is finally

   $$g[n] = \sqrt{10^{L_n(\text{dB})/10}},$$

   with $L_n(\text{dB}) = \bar{L}_d[n] + L_{LS}[n]$.

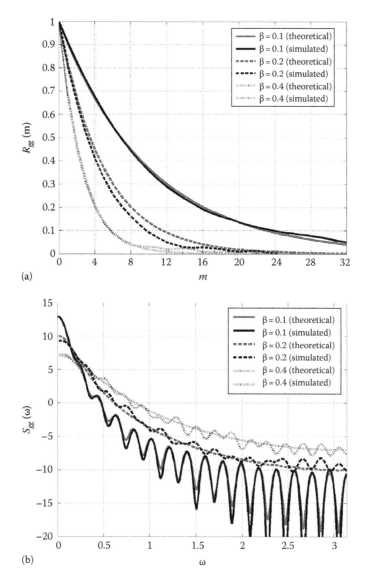

FIGURE 10.8 Theoretical and simulated large-scale fading (a) ACFs and (b) PSDs.

showing an example of the use of this model to obtain the total channel gain in the simulation of two GSM scenarios is provided in Section 10.3.3.

10.3.2.2 Other Autocorrelation Functions

The ACF described in the previous section is not differentiable for a zero lag. Hence, in order to avoid problems in the calculation of several performance parameters (e.g., outage duration), some authors consider the following modified ACF that avoids the discontinuity in the derivative at the origin [41]:

$$R_{gg}[k] \propto \exp\left(-\beta k^2\right).$$

Furthermore, these two ACFs consider only the autocorrelation within an isolated link between a mobile user and a single BS. More recently, extended correlation functions that consider the cross-correlation between the uplink (UL) and the downlink (DL) [42] or among the links between a mobile station and two BSs [43] have been developed. However, the simulation of these channel models (although more involved from a computational point of view) is very similar to that of the single link described in Table 10.3 and simply requires an additional effort for the implementation of the appropriate cross-correlation filters.

10.3.3 CASE STUDY: SIMULATION OF LARGE-SCALE FADING IN CELLULAR NETWORKS

Let us consider the two scenarios shown in Figure 10.9, where a GSM BS is located in the center and a mobile user's station (MS) is moving around the BS (scenario S1) and moving away from it radially (scenario S2).

In both cases, we consider a sampling period $T_s = 1$ ms; a simulation interval $\Delta T = 10$ s (i.e., $\Delta T/T = 10,000$ samples); a typical suburban vehicular moving speed $v = 45$ km/h (i.e., $v = 12.5$ m/s); an initial distance between the mobile user and the BS $d_0 = 100$ m, $P_t = 1$ W, $G_t = G_r = 1$, $f_c = 900$ MHz (i.e., $\lambda = 1/3$ m) as in the P-GSM-900 frequency band; and the exponential correlation function shown before with parameters $\varepsilon_D = 0.82$ and $\sigma_{LS} = 7.5$ dB [39]. Using the simulation process described in Table 10.3 and illustrated in Figure 10.7 with $\ell = 2$, we obtain the curves shown in Figures 10.10 and 10.11. Figure 10.10 illustrates four examples of the local variation in the channel's gain, $L_{LS}(t)$, whereas Figure 10.11 displays the total channel gain, $L_t(\text{dB})$, for 100 examples in both scenarios. Altogether, both figures show that, although the local variation in the channel's gain can occasionally be high, it only suffers very small modifications from one sample to the next and overall the mean channel gain is obtained as expected.

Finally, this case study is concluded by analyzing the effect of the large-scale fading on a GSM signal. At the physical layer, GSM uses a modulation known as Gaussian minimum shift keying (GMSK), which consists of an MSK-modulated signal smoothed by using a Gaussian filter [31,44]. The GMSK modulation was selected in GSM due to its constant envelope (which allows using efficient nonlinear power amplifiers in the transmitter) and its excellent power efficiency [13,31]. Figure 10.12 shows the transmitter (MSK modulator plus Gaussian filter) and the channel (LS fading and AWGN).

First of all, a sequence of equiprobable independent bits, $b[n]$, was generated to feed the MSK modulator, which was implemented using the standard function provided by MATLAB's communications toolbox. This output is then passed to the discretized Gaussian filter, $h_G[n] = h_G(t)\big|_{t=nT_s}$, with

$$h_G(t) = \frac{1}{\sqrt{2\pi}\delta T} \exp\left(-\frac{t^2}{2\delta^2 T^2}\right),$$

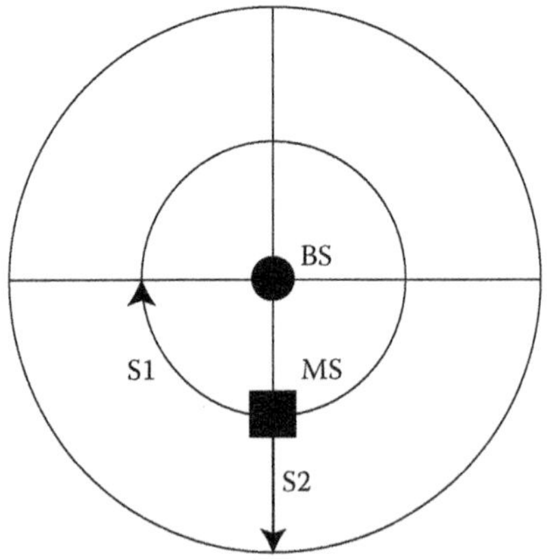

FIGURE 10.9 Two scenarios considered for the large-scale GSM fading simulation example.

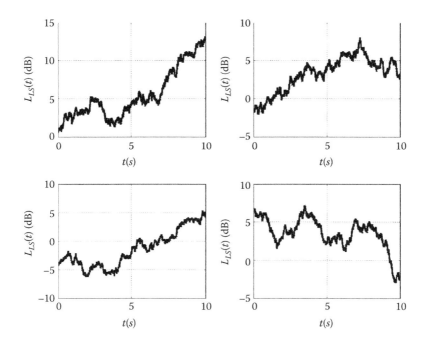

FIGURE 10.10   Four examples of the local variation in the channel's gain.

where $T = T_b = 1/R_b$ (with $R_b = 1625/6$ kbps ≈ 270.833 kbps for GSM), $T_s = T/R$ ($R = 8$), and $\delta = \sqrt{\ln 2}/2\pi WT$ with the product of the bandwidth times the symbol's period, WT, determining the width of the pulse (for GSM, $WT = 0.3$, implying that $\delta \approx 0.44$ and the width of the pulse is approximately equal to four symbol periods). The output of the Gaussian filter, $x[n] = s[n]* h_G[n]$, goes through the channel (after an energy normalization to ensure that $\mathbb{E}\left\{|x[n]|^2\right\} = 1$), composed of the large-scale fading (generated using the approach described in Table 10.3 and the same parameters as before) and AWGN with $\sigma_w = 10^{-5}$ for both the in-phase and quadrature components. Thus, the received signal is

$$r[n] = g[n]\times\left(s[n]* h_G[n]\right) + w[n].$$

Figure 10.13 shows the real part of the output of the MSK modulator, the Gaussian-filtered signal, and the received signal (the imaginary parts, not shown, are similar). Note how LS fading essentially causes a time-variant attenuation of the received signal, but no distortion: the signal can be perfectly recognized in spite of the attenuation and the small amount of noise.

As mentioned before, the effect of LS fading can be removed (at least partially) using an AGC block. One simple option for estimating the LS path loss is

$$\hat{g}[n] = \sqrt{\frac{1}{M}\sum_{m=0}^{M-1}|r[n-m]|^2 - 2\sigma_w^2},$$

where the AWGN variance, σ_w^2, must either be known or estimated somehow. Figure 10.14 shows the LS path loss, its estimated value (assuming σ_w^2 known), and the received signal's envelope before and after compensation.

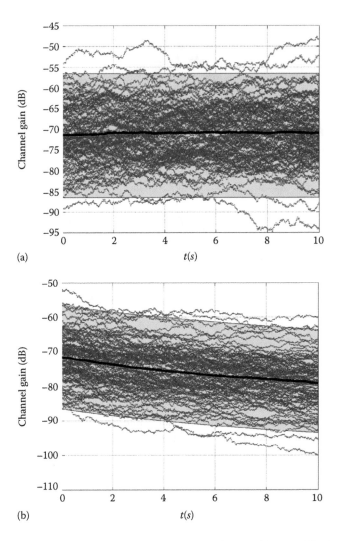

(a)

(b)

FIGURE 10.11 Total channel gain, L_t(dB), for 100 examples (shown with dotted lines). Also shown is the average channel gain (thick black line), as well as L_t(dB) $\pm 2\sigma_{LS}$ (gray-filled area).

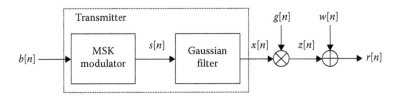

FIGURE 10.12 Block diagram for the simulation of large-scale fading in a GSM system with GMSK modulation.

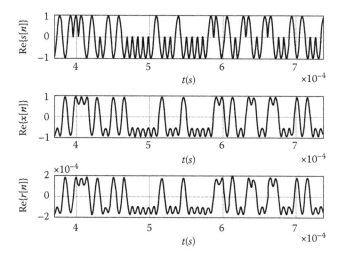

FIGURE 10.13 Example of the GMSK modulation used in GSM:MSK-modulated signal ($x[n]$), signal at the output of the Gaussian filter ($s[n]$), and received signal after LS fading and addition of white Gaussian noise ($r[n]$).

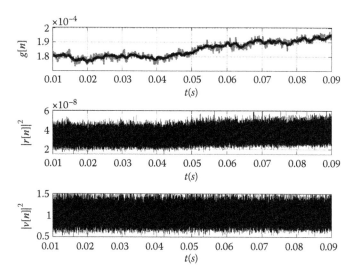

FIGURE 10.14 From top to bottom: LS path loss, $g[n]$ (black line), and estimated value, $\hat{g}[n]$ (gray line); envelope of the received signal, $|r[n]|^2$, where the trend (increase) due to the time-variant LS path loss can be appreciated; and envelope after compensation, $|v[n]|^2 = |r[n]|^2/\hat{g}^2[n]$, centered around $|v[n]|^2 = 1$ and without trend.

10.4 SMALL-SCALE FADING MODELS

Small-scale fading refers to the dramatic changes in amplitude that the received signal can suffer as a result of very small changes in location (even on the order of a fraction of a wavelength). These changes are due to small changes in the path between the transmitter and the receiver that modify radically the scattering environment and thus the amplitude (and also the phase) of the received signal. Looking back at Figure 10.1, we notice that small-scale fading may be classified either according to its frequency characteristics or to its time variation. On the one hand, fading may be *flat* (also known as *wideband*) or *frequency selective* (i.e., *narrowband*) depending on the channel's bandwidth w.r.t. the bandwidth occupied by the transmitted signal. On the other hand, fading may be *fast* or *slow* depending on the speed of variation of the channel w.r.t. the symbol's period. Both classifications are independent of each other, so we may have four types of small-scale fading [2,3,31]:

- Slow flat fading
- Slow frequency-selective fading
- Fast flat fading
- Fast frequency-selective fading

However, in order to define more precisely all of these categories and discuss their simulation, we need to define first several parameters that are typically used to characterize a mobile communications channel from a statistical point of view.

10.4.1 STATISTICAL CHARACTERIZATION OF WIRELESS COMMUNICATION CHANNELS

10.4.1.1 Power-Delay Profile

The basic measurement used to characterize a mobile wireless communications channel is the so-called power-delay profile (PDP) [2,3,31]. An ensemble PDP is obtained simply as a spatial average of the channel's impulse response, $|h_t(\tau)|^2$, over a local area by making many measurements in several close locations. The impulse response of a mobile wireless communications channel is typically described by a collection of impulses:

$$h_t(\tau) = \sum_{i=0}^{N_t-1} a_{t,i} e^{j\theta_{t,i}} \delta(\tau - \tau_{t,i}).$$

Hence, the PDP may be expressed as

$$P_h[i] = \mathbb{E}\left\{|h_t(\tau)|^2 \mid \tau = iT_s\right\} = \sum_{i=0}^{N_t-1} |a_i|^2 \delta(t - iT_s),$$

where

$\mathbb{E}\left\{|h_t(\tau)|^2 \mid \tau = iT_s\right\}$ denotes the expected value of $|h_t(\tau)|^2$ at $\tau = iT_s$

T_s is the sampling period used to obtain the PDP

$|a_i|^2$ is the average power found at the ith sampled delay, that is, $|a_i|^2 = \mathbb{E}\left\{|a_{t,i}|^2\right\}\Big|_{t=iT_s}$

Given the importance of the PDP in the characterization of mobile communications channels, extensive measurement campaigns have been developed since the early work of [45], and many PDPs are available for all types of environments and frequency bands: urban and suburban environments in the P-GSM-900 band [46,47], houses and office buildings in the 800–900 MHz bands [48,49], open-plan factories in the 1300 MHz band [50,51], the wideband UMTS channel [52,53], over-the-sea radio channels at 2 GHz [54,55], intra- and intervehicular communications at 5 GHz [56,57], etc.

10.4.1.2 Excess Delay and Coherence Bandwidth: Flat vs. Frequency-Selective Fading

Several useful definitions for the statistical characterization of the channel may be derived from the PDP [2,3,31]. First of all, the *mean excess delay* is defined as the first moment of the power-delay profile:

$$\mu_\tau = \frac{\sum_i |a_i|^2 \tau_i}{\sum_i |a_i|^2}.$$

A second measure of interest is the *RMS delay spread*, which is defined as the square root of the second central moment of the PDP:

$$\sigma_\tau = \sqrt{\mathbb{E}\{\tau^2\} - \mu_\tau^2},$$

with the mean squared delay, $\mathbb{E}\{\tau^2\}$, defined as

$$\mathbb{E}\{\tau^2\} = \frac{\sum_i |a_i|^2 \tau_i^2}{\sum_i |a_i|^2}.$$

The RMS delay spread, σ_τ, is a measure of the time dispersion of the channel's impulse response and serves to define the key parameter for the frequency characterization of the channel: the *coherence bandwidth*. The coherence bandwidth is a statistical measure of the range of frequencies over which the channel can be considered *flat*. The coherence bandwidth is inversely proportional to the RMS delay spread (i.e., $B_c \propto 1/\sigma_\tau$), but a universally accepted and precise relationship does not exist. However, one widely accepted definition considers the bandwidth over which the frequency correlation is above 0.9:

$$B_c(0.9) = \frac{1}{50\sigma_\tau}.$$

Alternatively, a more relaxed definition considers the bandwidth over which the frequency correlation is above 0.5:

$$B_c(0.5) = \frac{1}{5\sigma_\tau}.$$

The RMS delay spread and the coherence bandwidth allow us to provide a precise definition of whether a channel is flat or frequency selective for a given digital communications network:

- *Flat fading* occurs when the channel's coherence bandwidth, B_c, is larger than the bandwidth occupied by the transmitted signal, that is, $B_c > W$ with $W \propto 1/T$ and T denoting the symbol's period. In the time domain, this is equivalent to stating that the RMS delay spread of the channel's impulse response is smaller than the symbol's period, that is, $\sigma_\tau < T$.
- *Frequency-selective fading* occurs when the channel's coherence bandwidth is smaller than the bandwidth occupied by the transmitted signal, that is, $B_c > W$. In the time domain, this is equivalent to stating that the RMS delay spread of the channel's impulse response is larger than the symbol's period, that is, $\sigma_\tau < T$.

10.4.1.3 Doppler Spectrum and Coherence Time: Slow vs. Flat Fading

The RMS delay spread and the coherence bandwidth characterize the frequency behavior of the channel (or equivalently, its time-dispersive nature), but do not provide any information about the speed of variation of the channel. In order to describe precisely the local time-varying nature of a mobile communications channel, we need to introduce two additional measures: the Doppler spectrum and the coherence time.

Whenever the transmitter, the receiver, or both are moving, a change in the received frequency w.r.t. the transmitted frequency occurs. This is the so-called Doppler shift, and the frequency displacement is given by [31]

$$\Delta f = \frac{v}{\lambda} \cos \theta,$$

where
  v(m/s) is the relative motion speed between the transmitter and the receiver
  λ(m) = c/f_c is the wavelength (with c(m/s) $\approx 3 \cdot 10^8$ the free-space speed of light and f_c(Hz) the transmitted [carrier] frequency)
  $\theta(t)$ the relative angle between the transmitter and the receiver at the time instant t

In a multipath propagation environment, many signals are received simultaneously with different amplitudes and incidence angles, corresponding to different delayed versions of the original transmitted signal. Hence, it does not make sense to consider a single Doppler shift. Instead, we have to perform a statistical characterization of the Doppler shift through the *Doppler PSD* or *Doppler spectrum* [31], which provides us with the statistical distribution of the frequencies of the received signals w.r.t. the central or carrier frequency. Regarding the Doppler spectrum, the most widely used expression is the so-called classical or *Jakes Doppler spectrum* [31,58], which stems from *Clarke's fading model* [59], one of the first fading models that has been widely used since then, and Gans analysis for a $\lambda/4$ vertical antenna [60]*:

$$S_{RR}\left(f\right) = \begin{cases} \dfrac{3}{2\pi f_m \sqrt{1 - \left(\dfrac{f - f_c}{f_m}\right)^2}}, & \left|f - f_c\right| < f_m, \\ 0, & \left|f - f_c\right| > f_m, \end{cases}$$

where
  $f_m = v_{\max}/\lambda$ is the maximum Doppler shift
  f_c is the carrier frequency

A simpler alternative, considered by some authors, is the so-called flat or *uniform Doppler spectrum* [61][†]:

$$S_{RR}\left(f\right) = \begin{cases} \dfrac{1}{2\left|f - f_c\right|}, & \left|f - f_c\right| < f_m, \\ 0, & \left|f - f_c\right| > f_m. \end{cases}$$

From the Doppler spectrum, and the maximum Doppler shift, we may derive the fundamental statistical measure used to describe the speed of the time variation of the channel: the coherence time (T_c). The *coherence time* is a statistical measure of the time duration over which the channel's impulse response can be considered to be time invariant (i.e., the time duration over which two signals have a strong potential for amplitude correlation). As it happens with the coherence bandwidth,

---

* The name Jakes Doppler spectrum comes from the fact that Jakes derived the final baseband expression for the spectrum and helped to popularize its use through his 1974 book [58].
† Interestingly, it was shown in [103] that both the classical and the uniform Doppler spectra are equivalent w.r.t. the first- and second-order moments when using a maximum Doppler frequency $\tilde{f}_m$ in the uniform Doppler spectrum and $f_m = \sqrt{2/3}\,\tilde{f}_m$ in the classical Doppler spectrum.

there is not a single universally accepted definition of coherence time. The simplest definition considers it to be directly the inverse of the maximum Doppler shift:

$$T_c(0) = \frac{1}{f_m} = \frac{\lambda}{v_{max}}.$$

Unfortunately, this definition provides an overly optimistic value of T_c. A more principled and restrictive definition considers the time lag during which the time correlation is above 0.5, resulting in

$$T_c(0.5) = \frac{9}{16\pi f_m} = \frac{9\lambda}{16\pi v_{max}}.$$

However, this second definition tends to be too conservative. Thus, a heuristic rule of thumb (widely used for the design of communication systems) is taking the coherence time as the geometric mean of $T_c(0)$ and $T_c(0.5)$:

$$\overline{T}_c = \sqrt{T_c(0)T_c(0.5)} = \frac{3}{4\sqrt{\pi}f_m} \approx \frac{0.423}{f_m}.$$

The coherence time allows us to provide a precise definition of the concepts of slow and fast fading:

- *Slow fading* occurs when the coherence time of the channel is larger than the transmitted symbol's period, that is, $\overline{T}_c > T$.
- *Fast fading* occurs when the coherence time of the channel is smaller than the transmitted symbol's period, that is, $\overline{T}_c < T$.

10.4.1.4   Case Study: Slow vs. Fast Fading
In modern digital communications systems, fast fading rarely occurs, since this requires $\overline{T}_c < T$, or equivalently,

$$\overline{T}_c = \frac{3}{4\sqrt{\pi}f_m} = \frac{3\lambda}{4\sqrt{\pi}v_{max}} < T = \frac{1}{R_s},$$

which results in the following condition for the data rate:

$$R_s < \frac{4\sqrt{\pi}v_{max}}{3\lambda}.$$

This condition is only fulfilled for very slow data rates or extremely high transmission frequencies (or both), as shown in the following examples.

Let us consider first a GSM connection in the P-GSM-900 band ($f_c \approx 900$ MHz) from a car travelling at a typical highway speed ($v_{max} = 120$ km/h). Applying the previous equation, the data rate must be $R_s < 236.33$ bauds (symbols/second) to have fast fading. An equivalent UMTS connection in band 1 ($f_c \approx 2100$ MHz) would require $R_s < 551.43$ bauds for fast fading. Both data rates are far below the transmission rates achieved by modern mobile communications systems. Thus, in both cases, the scenario that has to be considered is slow (either flat or frequency-selective) fading. The same situation occurs in two other popular wireless communications standards: Wi-Fi and WiMAX. In the first case, considering the highest frequency band ($f_c \approx 5$ GHz), which is used by 802.11a and optionally by 802.11n, and a walking speed ($v_{max} = 6$ km/h), we obtain $R_s < 65.65$ bauds. Similarly, for a WiMAX connection in the 2.5 GHz band from a high-speed train ($v_{max} = 300$ km/h), we have $R_s < 1641.16$ bauds, which is still much less than the typical data rate.

As an example of applications where we may have to consider fast fading scenarios, let us consider the following case studies. First of all, consider a commercial aeronautical RF communication in the 5 GHz band at a maximum speed $v_{max} = 1000$ km/h. In this case, we have $R_s < 10941.07$ bauds for fast fading, which is closer to the limit between slow and fast fading than in the systems analyzed before. Indeed, if we consider an RF link to a military aircraft, which may travel at Mach 2 or Mach 3 speeds (and even beyond),* in the 5 GHz band, we have $R_s < 26783.75$ bauds and $R_s < 40175.62$ bauds for Mach 2 and Mach 3, respectively. This situation becomes worse when we consider spacecrafts or satellites, which can move at speeds well beyond Mach 3. For example, a low earth orbit (LEO) satellite, such as the ones in the Iridium satellite communications system, can move at speeds ranging from 6.5 to 8.2 km/s (i.e., 23,400 to 29,520 km/h). In these circumstances, and taking into account that Iridium uses the 1600 MHz band [62], we have $R_s < 94530.87$ bauds for fast fading at a speed of 7.5 km/s. Finally, communications at higher and higher carrier frequencies are being considered lately, given the saturation of the RF spectrum. Indeed, communications in the 60 GHz band has been gaining momentum during the last decade [63]. In this case, an RF link at 60 GHz to a car moving at a typical highway speed of 120 km/h would require $R_s < 15755.14$ bauds to be in a slow fading scenario. If we try to establish a link to a high-speed train moving at 300 km/h or to an aircraft moving at 1000 km/h, we need $R_s > 39387.86$ bauds and $R_s > 131292.88$ bauds, respectively.

Hence, we see that most modern digital communications systems typically operate under a slow fading scenario. Indeed, only very specialized applications (working at high speeds and/or frequencies) may need to consider fast fading channels. Figure 10.15 shows the frontier between slow and fast fading for three speeds (6, 60, and 600 km/h): systems operating above these curve can be considered to operate in a slow fading regime, whereas systems below the curve are subject to fast fading.[†] Almost all of the commercially available systems nowadays (UMTS, Wi-Fi, WiMAX, etc.)

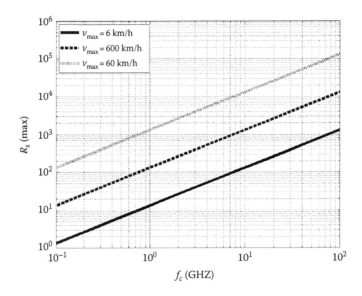

FIGURE 10.15 Slow fading–fast fading frontier for three moving speeds. Systems operating above the curves are in a slow fading regime, whereas systems below the curve suffer fast fading.

---

* The Mach number is a dimensionless quantity representing the ratio of speed of an object (typically an aircraft or a spacecraft) w.r.t. the speed of sound, 340.3 m/s (i.e., 1225.08 km/h), in the Earth's atmosphere.

† Let us remark that many authors use slow and fast fading to refer to large-scale and small-scale fading, as the first one changes slowly in time, whereas the changes of the second one are much faster. This should not be confused with the use of slow and fast fading here (and many references in the literature), which refers to the speed of variation in the characteristics of the channel w.r.t. the symbol's period.

are above these curves and thus can be considered to operate under slow fading conditions. Hence, in the sequel, we will focus exclusively on the simulation of slow fading channels. We will first describe the simulation of slow flat fading channels in Section 10.4.2, and then we will consider the simulation of slow frequency-selective channels, which involves a greater computational effort, in the following section.

10.4.2 SIMULATION OF SLOW FLAT FADING CHANNELS

10.4.2.1 Monte Carlo Simulation of Fading Channels: Static vs. Dynamic Simulation

The usual model for the simulation of a slow flat fading channel considers all the energy concentrated on a single ray, that is, considering uniform sampling with a period T_s, the channel at the nth time instant, $t_n = nT_s$, is given by

$$h_n[n] = \alpha_n e^{j\theta_n} \delta[n],$$

where
    the amplitude α_n typically follows either a Rice, Rayleigh, or Nakagami-m distribution (for other distributions, see Section 10.4.2.2.4)
    the phase θ_n is usually distributed uniformly (i.e., $\theta_n \sim U\left(\left[0, 2\pi\right]\right)$)*

In order to simulate this channel, we have three options. On the one hand, when we are only interested in analyzing the expected system's behavior (e.g., the expected BER in a flat fading situation), the channel may change drastically among two consecutive uses, or the coherence time is larger than the simulation's duration (i.e., $T_c > N_s T$, where N_s is the number of symbols to be simulated and T is the symbol's period), we can simply perform a *static* or *time-invariant Monte Carlo channel simulation*: generate a large enough number of channels (N_c) according to the previous equation, simulate the system using each of those channels, and obtain the desired system performance parameters (e.g., the SER or the BER) by averaging the results. This approach is summarized in Table 10.4. In this case, the main concern is the generation of an amplitude $\alpha^{(i)}$ for the ith Monte Carlo simulation that follows the desired distribution (Rice, Rayleigh, or Nakagami typically) in a computationally efficient way (see Section 10.4.2.2).

TABLE 10.4
Static or Time-Invariant Monte Carlo Simulation of a Slow Flat Fading Channel

For $i = 1, \ldots, N_c$,

1. Draw $\theta^{(i)} \sim U\left(\left[0, 2\pi\right]\right)$.

2. Draw $\alpha^{(i)}$ from the desired amplitude distribution (typically a Rice, Rayleigh, or Nakagami-m distribution).

3. Simulate the desired digital communications system under study using a channel $h^{(i)}[n] = \alpha^{(i)} e^{j\theta^{(i)}} \delta[n]$, and the specified level of additive white Gaussian noise, i.e., the received signal (for $n = 0,1,\ldots,N_s-1$) is
   $r[n] = \alpha^{(i)} e^{j\theta^{(i)}} x[n] + w[n]$.

4. Calculate and store the desired performance parameters (e.g., BER$^{(i)}$).
   Return the average performance metric, e.g., BER $= \dfrac{1}{N_c} \displaystyle\sum_{i=1}^{N_c} \text{BER}^{(i)}$.

---

* For Rice or Rayleigh fading, the uniform distribution of the phase comes naturally from the physical model from which these two amplitude distributions arise. The Nakagami-m amplitude distribution does not have a physical interpretation, so any phase distribution could be used. However, although some authors have considered the generation of Nakagami fading with a nonuniform phase distribution [104], the uniform phase distribution is still the one more commonly used in this case.

A more sophisticated approach has to be followed when we intend to perform a more realistic and accurate simulation of the digital communications system. This so-called *dynamic* or *time-variant* approach requires simulating the time variation of the channel. This is done by taking into account the Doppler spectrum of the channel, which is introduced through an additional filter that shapes the originally white signals according to either the classical or the uniform Doppler spectrum. This approach is summarized in Table 10.5 (note that, although Table 10.5 looks very similar to Table 10.4, the FOR loop here is not on the channels generated but on the samples, which corresponds to the inner loop (not shown) in Table 10.4) and described graphically in Figure 10.16. In this case, the main concern is the construction of the proper Doppler spectrum filter, $h_{DS}[n]$, in such a way that the ACF at its output is $R_{DS}(\tau) = F^{-1}\{S_{RR}(f)\}$, with $S_{RR}(f)$ being the desired Doppler spectrum and $F^{-1}\{\cdot\}$ denoting the inverse Fourier transform. As shown in Section 10.4.2.3, given the complexity of the filter required in the time domain, this filter is often implemented in the frequency domain using $H_{DS}(f) = \sqrt{S_{RR}(f)}$, which guarantees the desired correlation at the output when the input is white.*

Finally, we have to consider also a third intermediate approach (which may be called a *block time-variant simulation*), where the channel is fixed for $\lfloor T_c/T \rfloor$ symbols at most and a completely new channel is generated afterwards. This approach, summarized in Table 10.6 for N_B simulated blocks, is particularly interesting for many *block transmission* approaches (frequently used in modern digital communications), such as time-division multiple access (TDMA) communications, block single-carrier transmission, or OFDM systems.

TABLE 10.5

Dynamic or Time-Variant Simulation of a Slow Flat Fading Channel

For $n = 0,1,\ldots,N_s-1$,

1. Draw $\theta_n \sim U([0,2\pi))$.

2. Draw α_n from the desired amplitude distribution (typically a Rice, Rayleigh, or Nakagami-*m* distribution).

3. Generate $h_n[n] = \alpha_n e^{j\theta_n}\delta[n]$ and pass it through the Doppler spectrum filter to obtain a stochastic channel with the desired autocorrelation.

4. Convolve the transmitted signal with $h_n[n]$, adding the specified level of additive white Gaussian noise to obtain the received signal:

$r[n] = \alpha_n e^{j\theta_n}x[n] + w[n]$.

Calculate and store the desired performance parameters (e.g., the BER).

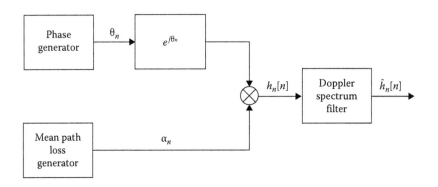

FIGURE 10.16 Block diagram for the time-variant simulation of a slow flat fading channel.

---

* The frequency domain approach is particularly appropriate for block transmission schemes (such as block single-carrier, orthogonal frequency division multiplexing [OFDM], or time-division multiplexing [TDM] systems), since the number of samples can be set equal to the length of the block.

TABLE 10.6

Block Time-Variant Monte Carlo Simulation of a Slow Flat Fading Channel

For $i = 1,\ldots,N_B$,

1. Draw $\theta_n^{(i)} \sim U\left([0,2\pi)\right)$.

2. Draw $\alpha_n^{(i)}$ from the desired amplitude distribution (typically a Rice, Rayleigh, or Nakagami-m distribution).

3. Simulate the nth block (containing up to $\lfloor T_c/T \rfloor$ symbols) of the digital communications system under study, using $h_n^{(i)}[n] = \alpha_n^{(i)} e^{j\theta_n^{(i)}} \delta[n]$, and the specified level of additive white Gaussian noise.

Calculate and store the desired performance parameters (e.g., the BER).

10.4.2.2   Generation of the Amplitudes

Assuming a uniform phase, the time-invariant and block time-variant simulation of a slow flat fading channel simply requires generating RVs according to the appropriate statistical distribution. For the time-variant simulation, these RVs are also required to generate the channel at the input of the Doppler spectrum filter, as shown in Figure 10.16. However, the amplitude for a flat fading channel typically follows either a Rice, Rayleigh, or Nakagami-m distribution in all cases. In the sequel, we will briefly discuss each of these distributions in Sections 10.4.2.2.1 through 10.4.2.2.3, emphasizing which one is more appropriate in each situation and showing how to draw samples from each of those distributions. Finally, other distributions proposed in the literature will be briefly mentioned in Section 10.4.2.2.4.

10.4.2.2.1   Rayleigh Distribution

The so-called Rayleigh distribution was developed by Lord Rayleigh when analyzing the distribution of the amplitude of a large number of vibrations of the same pitch and arbitrary phase [64]. The Rayleigh distribution has been traditionally used to model fading when there is no dominant path between the transmitter and the receiver, that is, when there is no line of sight (NLoS) between them and the transmitted signal is received scattered from many sources, as in Clarke's model [59]. The Rayleigh probability density function (PDF) is given by [27,31,40]

$$p(x) = \frac{x}{\sigma^2} \exp\left(-\frac{x^2}{2\sigma^2}\right) u(x),$$

where $u(x)$ is Heaviside's unit step function:

$$u(x) = \begin{cases} 0, & x < 0, \\ 1, & x > 0. \end{cases}$$

We will denote a Rayleigh-distributed RV as $x \sim R(\sigma^2)$, where σ^2 is the only parameter of the Rayleigh distribution, which has an expected value $\mathbb{E}\{x\} = \sigma\sqrt{\pi/2}$, a mean squared value $\mathbb{E}\{x^2\} = 2\sigma^2$, and a variance $\text{Var}\{x\} = (4-\pi)\sigma^2/2$.*

Drawing samples from a Rayleigh distribution is straightforward: given two independent zero-mean Gaussian RVs with variance σ^2, $v,w \sim N(0,\sigma^2)$, the RV $z = \sqrt{v^2 + w^2}$ follows the desired

---

* Sometimes the Rayleigh density is alternatively expressed as $p(x) = \frac{2x}{\Omega} \exp\left(-\frac{x^2}{\Omega}\right) u(x)$,

  where $\Omega = \mathbb{E}\{x^2\} = 2\sigma^2$ represents the average path energy, and we have $\mathbb{E}\{x\} = \sqrt{\pi\Omega}/2$ and $\text{Var}\{x\} = (1-\pi/4)\Omega$.

Rayleigh distribution with the same variance σ^2. Thus, simulating a time-invariant Rayleigh slow flat fading channel simply requires, for each simulation, a uniform RV (for the phase), as well as two independent Gaussians and a simple mathematical operation (for the amplitude). Figure 10.17 shows the good match between the empirical density obtained following this approach (represented through a histogram) and the theoretical expression for two values of σ^2 using 10^6 samples for the simulation.

10.4.2.2.2 Rice Distribution

A second commonly used distribution in the simulation of wireless communication channels is the so-called Rice distribution, which was obtained by Stephen Rice when analyzing the statistical properties of a current consisting of a sinusoidal component plus a random noise component [65–67]. The Rice distribution is a generalization of the Rayleigh distribution, which is used when one path is

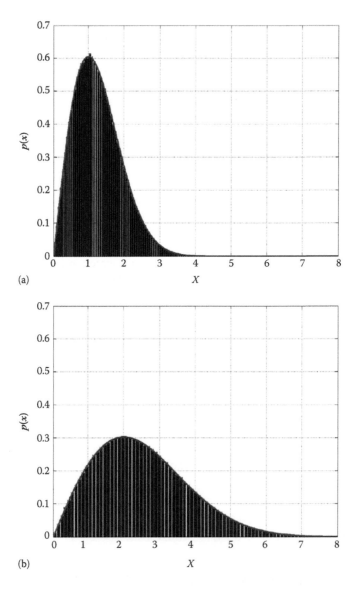

FIGURE 10.17 Example of the simulated (using 10^6 samples) and theoretical Rayleigh densities: (a) $\sigma^2 = 1$ and (b) $\sigma^2 = 4$.

much stronger than the rest, that is, when there is a direct line of sight (LoS) between the transmitter and the receiver. The Rice PDF is given by

$$p(x) = \frac{x}{\sigma^2} \exp\left(-\frac{x^2 + \mu^2}{2\sigma^2}\right) I_0\left(\frac{\mu x}{\sigma^2}\right) u(x),$$

with $I_0(x)$ being the *modified Bessel function* of order zero [68]. A Rice distributed RV will be denoted as $x \sim \text{Rice}(\mu, \sigma^2)$, where μ and σ^2 are the two parameters of the Rice distribution. The expected value is now $\mathbb{E}\{x\} = \sigma\sqrt{\pi/2} L_{1/2}\left(-\mu^2/(2\sigma^2)\right)$, where $L_q(x)$ denotes the qth-order *Laguerre polynomial* [68], the mean squared value is $\mathbb{E}\{x^2\} = 2\sigma^2 + \mu^2$, and the variance is $\text{Var}\{x\} = 2\sigma^2 + \mu^2 - \frac{\pi\sigma^2}{2} L_{1/2}^2\left(-\mu^2/(2\sigma^2)\right)$.

The Rice PDF can be expressed in an alternative form as

$$p(x) = \frac{2(K+1)x}{\Omega} \exp\left(-K - \frac{(K+1)x^2}{\Omega}\right) I_0\left(2\sqrt{\frac{K(K+1)}{\Omega}} x\right) u(x),$$

where $K = \mu^2/(2\sigma^2)$ is the so-called Rice factor, which provides the ratio of the power received via the LoS path to the power contribution of the NLoS paths [69], and $\Omega = \mathbb{E}\{x^2\} = 2\sigma^2 + \mu^2$ is the average path energy (i.e., the mean squared value of the PDF).* Although this expression looks more complicated, it is usually preferred in mobile wireless communications (and it is much more commonly used actually), given the interpretability of the parameters. Indeed, all of the statistics of the PDF can be rewritten now as a function of K and Ω: $\mathbb{E}\{x\} = \frac{1}{2}\sqrt{\pi\Omega/(K+1)} L_{1/2}(-K)$, $\mathbb{E}\{x^2\} = \Omega$, and $\text{Var}\{x\} = \Omega - \frac{\pi\Omega}{4(K+1)} L_{1/2}^2(-K)$. Using this alternative formulation, a Rice distributed RV can also be denoted as $x \sim \text{Rice}(K, \Omega)$.

Drawing samples from a Rice distribution is also straightforward, since, given two independent Gaussian RVs with variance σ^2 and means $\mu\cos\theta$ and $\mu\sin\theta$, respectively, $v \sim N(\mu\cos\theta, \sigma^2)$ and $w \sim N(\mu\sin\theta, \sigma^2)$, the RV $z = \sqrt{v^2 + w^2}$ follows the desired Rice distribution, that is, $z \sim \text{Rice}(\mu, \sigma^2)$.† Thus, simulating a time-invariant Rice slow flat fading channel has a similar complexity to a Rayleigh fading channel, as it simply requires, for each simulation, a uniform RV (for the phase), as well as two independent Gaussians, an angle (which can be fixed or selected randomly), and a simple mathematical operation (for the amplitude). Figure 10.18 shows the good match between the empirical density obtained following this approach (represented through a histogram) and the theoretical expression for $\Omega = 1$ and two values of K (corresponding to weak and strong LoS paths, respectively) using 10^6 samples for the simulation. Note also the similarity between the Rayleigh and Rice PDFs by comparing Figures 10.17 and 10.18.

---

* Note that, given K and Ω, we have $\mu = \sqrt{\frac{K\Omega}{K+1}}$ and $\sigma^2 = \frac{\Omega}{2(K+1)}$.

† Given K and Ω, the two Gaussian RVs that we need to generate $z = \sqrt{v^2 + w^2} \sim \text{Rice}(K, \Omega)$ are $v \sim N\left(\sqrt{\frac{K\Omega}{K+1}}\cos\theta, \frac{\Omega}{2(K+1)}\right)$ and $w \sim N\left(\sqrt{\frac{K\Omega}{K+1}}\sin\theta, \frac{\Omega}{2(K+1)}\right)$.

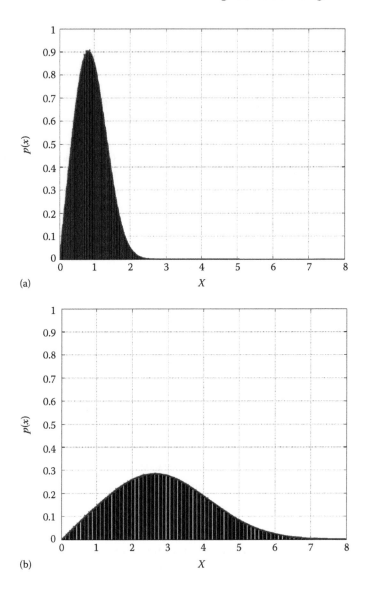

(a)

(b)

FIGURE 10.18 Example of the simulated (using 10^6 samples) and theoretical Rice densities for $\Omega = 1$: (a) $K = 1$ and (b) $K = 10$.

Finally, let us remark that setting $K = 0$ (or equivalently $\mu = 0$), there is no power received through the LoS path, and the Rice density degenerates into the Rayleigh PDF. Thus, the Rice fading model can be seen as a generalization of the Rayleigh fading model. Furthermore, regarding the parameter values commonly used in Rice fading models, Ω is usually set to one (i.e., a normalized average path energy equal to one is considered), and $0 \leq K \leq 20$, with $K = 0$ corresponding to the Rayleigh fading model (no LoS path, as stated before) and $K = 20$ corresponding to a strong direct path that clearly predominates over the reflected paths [69–71].

10.4.2.2.3 Nakagami-m Distribution
Finally, a third distribution that has been widely used over the last two decades is the so-called Nakagami-m distribution, developed by Minoru Nakagami as a mathematical model for small-scale fading [72]. Unlike the Rayleigh and Rice distributions, the Nakagami-m distribution does not have

a physical interpretation, but it has been adopted by many authors lately, due to its good agreement (better than the Rayleigh or Rice distributions) with empirical channel measurements for some urban multipath environments [73–75]. The Nakagami-m PDF is given by

$$p(x) = \frac{2}{\Gamma(\mu)} \left(\frac{\mu}{\Omega}\right)^{\mu} x^{2\mu-1} \exp\left(-\frac{\mu}{\Omega} x^2\right) u(x),$$

where $\Gamma(\mu)$ is the *gamma function* [68]. A Nakagami-m distributed RV will be denoted as $x \sim$ Nakagami (μ,Ω), where μ and Ω are the two parameters of the Nakagami distribution. The expected value is now $\mathbb{E}\{x\} = \frac{\Gamma(\mu+1/2)}{\Gamma(\mu)} \sqrt{\frac{\Omega}{\mu}}$, the mean squared value is $\mathbb{E}\{x^2\} = \Omega$, and the variance is $\mathrm{Var}\{x\} = \Omega \left[1 - \frac{1}{\mu}\left(\frac{\Gamma(\mu+1/2)}{\Gamma(\mu)}\right)^2\right]$.

With respect to the interpretation of the parameters of the PDF, the parameter Ω has the same meaning as in the Rayleigh and Rice densities: it represents the average path energy, that is, $\Omega = \mathbb{E}\{x^2\} > 0$. The other parameter, $\mu \geq 0.5$, is a fading parameter that indicates the *fading depth* of the channel (the smaller the value of μ, the greater the fading depth), and we have two possible situations:

- $0.5 \leq \mu \leq 1$: The Nakagami PDF corresponds to the particular case where no LoS path exists, with the Nakagami PDF degenerating into a *half-Gaussian* PDF for $\mu = 0.5$:

$$p(x) = \frac{1}{\Gamma(0.5)} \sqrt{\frac{2}{\Omega}} \exp\left(-\frac{x^2}{2\Omega}\right) u(x),$$

  where $\dfrac{1}{\Gamma(0.5)} \sqrt{\dfrac{2}{\Omega}} = \dfrac{2}{\sqrt{2\pi\Omega}}$ is the right scaling factor for the half-Gaussian PDF and into a Rayleigh PDF for $\mu = 1$. In general, for $0.5 \leq \mu < 1$, the channel is in a so-called worse-than-Rayleigh fading situation, where the probability of having a large fading is higher than for the Rayleigh fading model.
- $\mu > 1$: The Nakagami PDF corresponds to cases where an LoS path or a specular component exists, and for $\mu \to \infty$, the channel becomes the classical and well-known AWGN channel (i.e., we are in a no fading situation) [76].

Additionally, for $\mu > 1$, the Nakagami PDF is a very good approximation of the Rice density, with a ratio between the power received via the LoS path to the power contribution of NLoS paths given by a Rice factor $K = \dfrac{\mu - \sqrt{\mu(\mu-1)}}{\sqrt{\mu(\mu-1)}}$.* Furthermore, it can even be used to approximate the log-normal PDF (widely used to simulate large-scale fading or shadowing, as discussed in Section 10.3) with a small value of σ over a specific domain. Regarding the parameter values commonly used in Nakagami fading models, Ω is usually set to one (thus considering a normalized average path energy equal to one, as before) and $0.5 \leq \mu \leq 8$ [76–79].

---

* Hence, in order to emulate a given Rice factor K, we have to choose $\mu = \dfrac{(K+1)^2}{K(K+2)}$.

The generation of Nakagami RVs is more complicated than drawing samples from Rayleigh or Rice RVs, since no direct approach can be applied, except when $\mu = m/2$ and m is an integer value. In this case, a Nakagami RV can be generated as the squared root of the sum of squares of $m = 2\mu$ zero-mean independent and identically distributed (i.i.d.) Gaussian RVs, that is, if we have

$v_i \sim N(0,\sigma^2)$ i.i.d. for $1 \le i \le m = 2\mu$, then $z = \sqrt{\dfrac{\Omega}{2\mu}\displaystyle\sum_{i=1}^{m} v_i^2} \sim$ Nakagami (μ,Ω). However, this so-called

brute-force approach is only valid when μ is a half-integer value and requires drawing 2μ Gaussian RVs for each Nakagami RV (e.g., for $\mu = 8$ each Nakagami RV requires generating 16 Gaussians). Hence, several alternative approaches have been developed to allow the generation of Nakagami RVs with arbitrary values of μ efficiently. One of the most efficient and interesting approaches is based on the *rejection sampling* (RS) technique [80–82], which consists of drawing samples from a simpler *proposal density* and accepting or rejecting them according to the ratio of the target and proposal densities (i.e., between the desired density and the density used to draw samples). Let us consider a proposal density, $\pi(x)$, from which samples can be easily and efficiently generated, such that $L\pi(x) \ge p(x)$ for all $x \in \mathcal{X} \subseteq \mathbb{R}$ (i.e., $L\pi(x)$ is an envelope function for $p(x)$). Then, the RS approach consists of the two steps provided in Table 10.7.

It can be shown that this algorithm always provides samples distributed according to $p(x)$ regardless of the proposal density used (as long as $L\pi(x) \ge p(x)$ for all $x \in \mathcal{X} \subseteq \mathbb{R}$) [80]. However, the efficiency of an RS algorithm depends critically on the choice of a good proposal density. Indeed, the average number of samples required to draw a sample from $p(x)$ using the RS algorithm is $1/(1-P_r)$, where P_r is the rejection probability (i.e., the probability of rejecting a sample in Step 2 of the algorithm), which can be shown to be directly proportional to the L_1 distance between the proposal and the target densities, that is,

$$P_r \propto D_1 = \int_{x \in \mathcal{X} \subseteq \mathbb{R}} \left[Lp(x) - \pi(x) \right] dx.$$

Hence, $P_r \to 1$ when D_1 is large, and the number of samples required to generate a single valid sample from the desired distribution can be very large (indeed, when $P_r \to 1$ the number of samples tends to infinity). On the other hand, $P_r \to 0$ when D_1 is small, and the RS algorithm becomes extremely efficient, since we only require a single RV from a simple distribution and a uniform RV to generate a valid sample from the desired distribution.

Thus, the key for the good performance of an RS algorithm is the choice of a proposal density that is always above the target but as close as possible to it. For the Nakagami distribution, several efficient proposal distributions have been developed over the last decade [83–87]. Currently, the most efficient approach for $\mu \ge 1$ is obtained by using an unnormalized Nakagami density for the proposal function [86]

$$\pi(x) = \gamma_p x^{2\mu_p - 1} \exp\left(-\frac{\mu_p}{\Omega_p} x^2 \right) u(x),$$

TABLE 10.7

Rejection Sampling Approach for Drawing Samples from an Arbitrary PDF $p(x)$ by Using a Simpler Proposal $\pi(x)$

1. Draw a sample from the proposal density, $x \sim \pi(x)$, and a uniform random variable, $u \sim U\left([0,1]\right)$.

2. If $u \le \dfrac{p(x)}{L\pi(x)}$, then accept x (which will be distributed according to the desired target density, $p(x)$). Otherwise, discard x and return to Step 1.

---

where $\mu_p = 2\mu/2$ (i.e., it is the closest half-integer value of the fading parameter to the desired one [e.g., if $\mu = 1.73$, then $\mu_p = 1.5$]), $\Omega_p = \dfrac{\mu_p(2\mu-1)}{\mu(2\mu_p-1)}$, and

$$\gamma_p = \exp(\mu_p - \mu)\left(\frac{\Omega(2\mu-1)}{2\mu}\right)^{\mu-\mu_p}.$$

This approach is extremely efficient, attaining acceptance rates that are above 0.9 for most values of μ. Furthermore, since μ_p is a half-integer value, samples can be easily drawn from $\pi(x)$ by using the brute-force approach described earlier. Unfortunately, this approach is not valid for $\mu < 1$ and thus cannot be used to model worse-than-Rayleigh channels, which have attracted much interest recently. In this case, a very efficient RS algorithm is obtained using a proposal function composed of three truncated Gaussians [87]:

$$\pi(x) = \pi_1(x)\mathbb{I}_1(x) + \pi_2(x)\mathbb{I}_2(x) + \pi_3(x)\mathbb{I}_3(x),$$

where
$\pi_i(x)$ denotes each of the three Gaussian densities (for $i \in \{1,2,3\}$)
$\mathbb{I}_i(x)$ their domain of definition (i.e., the interval where each one is used)

See [87] for further details on this proposal function. Figure 10.19 shows the good match between the empirical density obtained using this last proposal (represented through a histogram) and the theoretical expression for $\Omega = 1$ and two values of μ (corresponding to an NLoS worse-than-Rayleigh and an LoS case, respectively) using 10^6 samples for the simulation.

10.4.2.2.4 Other Distributions
Apart from these three distributions, many other distributions have been considered for particular cases. For example, the gamma distribution has been recently applied in the simulation of fading/shadowing channels using the Weibull–gamma model [88,89] or the effects of the turbulent atmosphere in free-space optical links with the gamma–gamma approach, which requires two independent gamma RVs [90,91].* Another distribution that has been proposed to model small-scale fading for indoor peer-to-peer scenarios is the mixture of Gaussians [92]. Even the log-normal distribution, widely used for large-scale fading, has also been proposed to model small-scale fading.

10.4.2.3 Doppler Spectrum for Time-Variant Simulations
In this section, we show how to generate the desired Doppler spectrum following the frequency domain approach already proposed by Smith in 1975 [93]. This approach is based on two independent, complex, and white Gaussian noise processes in the frequency domain, $V_I(f)$ and $V_Q(f)$. These two white processes are then shaped independently in the frequency domain using the squared root of the desired Doppler spectrum, $\sqrt{S_{gg}(f)}$, thus obtaining $W_I(f) = V_I(f)\sqrt{S_{gg}(f)}$ and $W_Q(f) = V_Q(f)\sqrt{S_{gg}(f)}$. The next step is obtaining their corresponding time-domain wave forms through an inverse Fourier transform, that is, $w_I[n] = TF^{-1}\{W_I(f)\}$ and $w_Q[n] = TF^{-1}\{W_Q(f)\}$. Finally, the desired amplitudes are obtained as

$$\alpha[n] = \sqrt{|w_I[n]|^2 + |w_Q[n]|^2},$$

---

* Note that Gamma random variables are closely related to Nakagami RVs (since a Gamma RV can be used to draw samples from a Nakagami density) and are also challenging to generate [81,105].

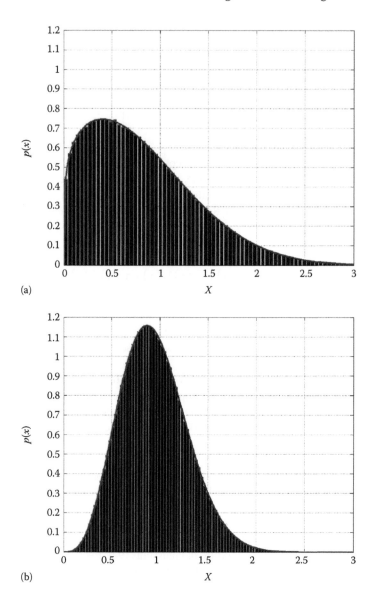

(a)

(b)

FIGURE 10.19 Example of the simulated (using 10^6 samples) and theoretical Nakagami densities for $\Omega = 1$: (a) $\mu = 0.6$ and (b) $\mu = 2$.

whereas the phases are given by

$$\theta[n] = \tan^{-1}\left(\frac{w_Q[n]}{w_I[n]}\right).$$

It can be shown that $\alpha[n]$ follows a Rayleigh distribution and $\theta[n]$ is uniformly distributed.

The steps required by the discrete-time version of this approach (implemented using discrete Fourier transforms of length N for a block of N samples) are described in Table 10.8, whereas the corresponding block diagram is shown in Figure 10.20. Finally, Figure 10.21 displays an example of the path loss ($\alpha[n]$) and the phase ($\theta[n]$) for $N = 1024$, $\sigma_w^2 = 1$, and $R_m = 0.01$. Notice the deep fades in $\alpha[n]$ and the sudden changes in $\theta[n]$, both of which are characteristics of Rayleigh fading.

TABLE 10.8

Steps Required by Smith's Frequency Domain Doppler Spectrum Generator

1. Generate two independent, complex, white Gaussian-distributed spectra for first $N/2-1$ samples of both the real and the imaginary parts, i.e., $V_I[k] \sim N\left(0, \sigma_w^2\right) + jN\left(0, \sigma_w^2\right)$ and $V_Q[k] \sim N\left(0, \sigma_w^2\right) + jN\left(0, \sigma_w^2\right)$ for $k = 1, \dots, N/2-1$.

2. For $k = 0$ and $k = N/2$, $V_I[0] \sim N\left(0, \sigma_w^2\right)$, $V_Q[0] \sim N\left(0, \sigma_w^2\right)$, $V_I\left[N/2\right] \sim N\left(0, \sigma_w^2\right)$, and $V_Q\left[N/2\right] \sim N\left(0, \sigma_w^2\right)$.

3. Obtain the remaining samples of the spectrum through the conjugate symmetry property of the Fourier transform of real signals, i.e., by conjugating the corresponding processes to ensure that the resulting signals in the time domain are real valued: $V_I[N/2 + k] = V_I[N/2-k]$ and $V_Q[N/2 + k] = -V_Q[N/2-k]$ for $k = 1, \dots, N/2-1$.

4. Multiply the in-phase and quadrature noise processes separately by the squared root of the desired Doppler spectrum, i.e., $W_I[k] = V_I[k]\sqrt{S_{gg}[k]}$ and $W_Q[k] = V_Q[k]\sqrt{S_{gg}[k]}$ for $k = 0, 1, \dots, N-1$. Note that $S_{gg}[k]$ is the discretized baseband version of the desired Doppler spectrum, which is given (for Jakes fading spectrum) by

$$S_{RR}[k] = \begin{cases} \dfrac{3}{2\pi R_m \sqrt{1 - \left(\dfrac{k}{NR_m}\right)^2}}, & |k| < NR_m, \\ 0, & |k| > NR_m, \end{cases}$$

where the maximum Doppler frequency is $f_m = R_m \times f_s$ with $0 \le R_m < 0.5$ and f_s denoting the sampling frequency used in the simulation.

5. Perform an inverse fast Fourier transform (IFFT) of $W_I[k]$ and $W_Q[k]$ separately to obtain the corresponding time-domain signals, $w_I[n]$ and $w_Q[n]$.

6. Compute the square of these two signals, add them, and compute the squared root to obtain the desired Rayleigh envelope:

$$\alpha[n] = \sqrt{\left|w_I[n]\right|^2 + \left|w_Q[n]\right|^2}.$$

7. Compute the desired phase as the arc whose tangent is given by the quotient between $w_Q[n]$ and $w_I[n]$:

$$\theta[n] = \tan^{-1}\left(\frac{w_Q[n]}{w_I[n]}\right).$$

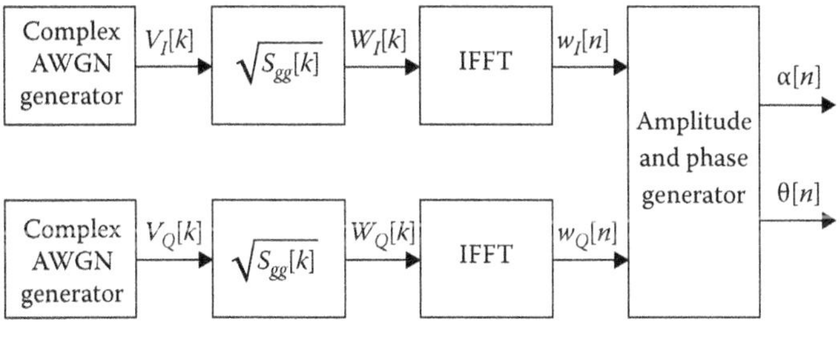

FIGURE 10.20 Block diagram for the generation of a Rayleigh fading amplitude and phase with the desired Doppler spectrum in the frequency domain.

Note also the smoothness in the envelope and phase processes (when compared to a white noise process) provided by the time correlation introduced by the Doppler spectrum filter.

10.4.3 SIMULATION OF SLOW FREQUENCY-SELECTIVE CHANNELS

The simulation of slow frequency-selective channels can be accomplished in a similar way to slow flat fading channels: following either a simple time-invariant approach (or the equivalent block time-variant scheme if deemed more appropriate) or the more realistic time-variant approach.

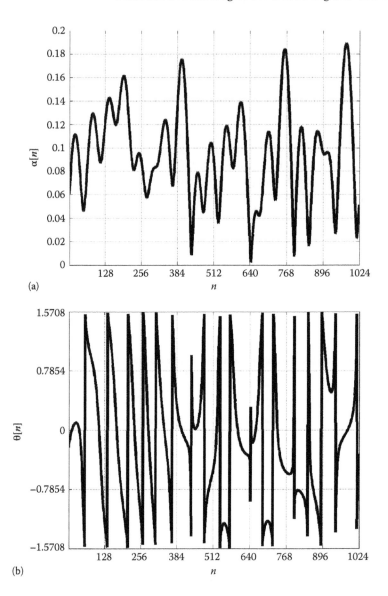

(a)

(b)

FIGURE 10.21 Example of a Rayleigh fading process according to Jakes Doppler spectrum: (a) path loss $\alpha[n]$ and (b) phase $\theta[n]$.

In any case, the full multipath channel model introduced previously has to be used. The discretized version (assuming a sampling period T_s) is given by

$$h_k[n] = \sum_{i=1}^{N_t-1} \kappa_i \alpha_{i,k} e^{j\theta_{i,k}} \delta\left[n - m_{i,k}\right],$$

where $h_k[n] = h_{kT_s}\left(nT_s\right)$. In order to simulate this channel, we have to take into account the statistical distribution of a larger number of parameters than before:

- The number of paths, N_t, which is fixed in many simple models, but not in other more sophisticated models (particularly in models derived from empirical measurements).
- The path gain factor, κ_i, which indicates the relative importance of the ith path. Usually this factor is either equal to one or fixed to some prespecified value.

- The amplitudes of the signals received through the different paths, $\alpha_{i,k}$, which typically follow one of the distributions described earlier for slow flat fading models (i.e., either a Rice, Rayleigh, or Nakagami distribution).
- The phases of the signals received through the different paths, $\theta_{i,k}$, which typically follow a uniform distribution.
- The delays, $\tau_{i,k} = m_{i,k}T_s$, which are implicitly assumed to be integer multiples of the sampling period in this model. Typically, $m_{i,0} = 0$, and the remaining delays are often fixed, but some sophisticated models (again, particularly models derived from empirical measurements) consider a distribution over $m_{i,k}$ to simulate the delays.

Besides, for the more realistic time-variant simulation of the channel, we have to address the time correlation of the channel (as before), as well as the correlation among the different rays in the most complex models.

Many models have been proposed for frequency-selective fading. In the sequel, we review two widely used frequency-selective models: the simple well-known two-ray Rayleigh fading model and an example of a model derived from empirical measurements (simulation of indoor radio channel impulse models [SIRCIM]).

10.4.3.1 Two-Ray Rayleigh Fading Model

This is probably the simplest frequency-selective fading model and one of the most widely used in practice for the simulation of frequency-selective fading. This model considers two paths (one main path corresponding to a zero lag and one secondary path that typically has a fixed prespecified delay) with independently distributed Rayleigh amplitudes and phases,* that is,

$$h_k[n] = \alpha_{0,k}e^{j\theta_{0,k}}\delta[n] + \kappa_1\alpha_{1,k}e^{j\theta_{1,k}}\delta\left[n - m_{1,k}\right],$$

with $\theta_{0,k}, \theta_{1,k} \sim U\left(\left[0, 2\pi\right)\right)$, $\alpha_{0,k} \sim R\left(\sigma_0^2\right)$, $\alpha_{1,k} \sim R\left(\sigma_1^2\right)$, and the delay $(m_{1,k})$ and the relative importance of the reflected ray w.r.t. the main path (κ_1 with $|\kappa_1| < 1$ usually) typically fixed to a prespecified value. Furthermore, the two paths are usually considered to be independent, and their time correlation is typically provided by the classical or Jakes Doppler spectrum. Hence, the approach described in Section 10.4.2.3 can be applied independently to introduce time correlation in each of the two rays. Indeed, this approach can also be easily extended to include $3,4,\ldots,N_t-1$ independent rays.

10.4.3.2 Models Derived from Empirical Measurements: SIRCIM

As an example of a model derived from empirical measurements, we will consider SIRCIM, a model widely used for the characterization and simulation of indoor wireless communication systems [94,95]. This model is extremely complete, as it considers the statistical distribution of all the parameters involved in the model, as well as the time and inter-ray correlation functions. More precisely, SIRCIM considers the following channel model:

$$h_k[n] = \sum_{i=1}^{N_t-1}\alpha_{i,k}e^{j\theta_{i,k}}\delta\left[n - m_{i,k}\right],$$

where all of the parameters are characterized according to their probabilistic distributions, obtained from real-world measurements and an exhaustive statistical analysis. For instance, the number of

---

* This model can be easily generalized by considering Rice or Nakagami distributed amplitudes for each of the paths or even a hybrid Rice–Rayleigh model, where the main (direct) path follows a Rice distribution and the secondary (reflected) path follows a Rayleigh distribution.

paths is initially chosen from a Gaussian distribution, that is, $N_t \sim N\left(\mu_N, \sigma_N^2\right)$ with μ_N and σ_N^2 depending on whether we are in an LoS or NLoS situation [94]. Once the number of paths has been determined, for each path, we still need to determine the following:

1. A path delay, drawn from a multinomial density
2. A uniformly distributed phase, $\theta_{i,k} \sim U\left(\left[0, 2\pi\right)\right)$
3. A Gaussian amplitude, $\alpha_{i,k} \sim N\left(\mu_\alpha, \sigma_\alpha^2\right)$, where μ_α and σ_α^2 are themselves Gaussian RVs

Additionally, SIRCIM considers both time correlation and inter-ray correlation (see [94] for further details). Hence, SIRCIM is a very sophisticated model, which can provide an extremely realistic channel simulation when properly parameterized but which can be very difficult to parameterize and to simulate.

10.4.4 CASE STUDY: RAYLEIGH FADING IN WI-FI NETWORKS

As an example of the simulation of small-scale fading, we consider an IEEE 802.11 Wi-Fi network transmitting in the 2.4 GHz frequency band using an OFDM modulation with 64 subcarriers and PSK modulation in each subcarrier. We consider three channel conditions for this signal:

1. Signal corrupted only by AWGN. Under these circumstances, the BER of the system is identical to that of a BPSK signal contaminated by AWGN:

$$BER = Q\left(\sqrt{\frac{2E_b}{N_0}}\right),$$

where
    E_b denotes the average energy per bit
    N_0 is the one-sided noise PSD
    $Q(x)$ is the Q function [26,68]

2. Signal corrupted by AWGN as before, plus a flat fading Rayleigh component without time correlation. The block time-variant simulation is used for this approach, with a new path loss generated for each OFDM symbol (i.e., each block of 64 samples). No phase is generated (i.e., we consider $\theta[n] = 0$), as any phase rotation would force us to introduce a phase synchronization stage in the receiver, thus complicating a lot the simulation scenario.*
3. Signal corrupted by AWGN as before, plus a flat fading Rayleigh component with time correlation introduced using the approach described in Table 10.8 with $N = 64$, $\sigma_w^2 = 1$, and $R_m = 0.1$. The phase is again set to zero to avoid having to introduce a phase synchronization stage in the receiver.

Figure 10.22 shows the results as a function of the E_b/N_0 ratio, which is the usual simulation parameter in digital communications. Note the much slower decay in BER w.r.t. E_b/N_0 when the Rayleigh fading is introduced and the error floor when we introduce the Doppler spectrum.

---

* Note that the results should be roughly independent of the phase in this case (i.e., flat fading and block time-variant simulation), provided that a good phase synchronizer is implemented.

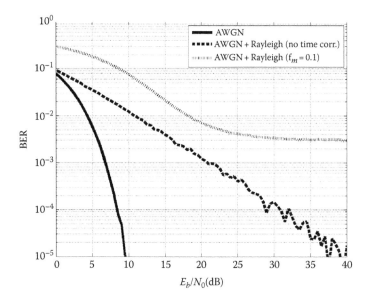

FIGURE 10.22 Performance of a 64-OFDM system under AWGN conditions (solid line), Rayleigh fading without time correlation (dashed line), and time-correlated Rayleigh fading (dotted line).

10.5 CONCLUSIONS AND DISCUSSION

The efficient and accurate simulation of mobile wireless communication systems is a challenging issue that has evolved much during the last decades: from simple mathematical models based on simplistic assumptions to more complex models based on more realistic hypotheses and even real-data obtained through extensive measurement campaigns. In this chapter, we have considered the statistical simulation of wireless communication models, focusing on multipath fading, which is the main impairment for many real-world channels. We have introduced the main parameters used to describe a mobile wireless communication channel and used them to classify fading into four different types according to the frequency characteristics and the time variation of the channel. We have finally concentrated on the two most common situations found in modern wireless communication systems: slow flat and frequency-selective fading. Several statistical channel models have been considered for these two cases, describing in detail how they are simulated in practice through block diagrams and tables providing step-by-step algorithms for their simulation. Many examples of the results of these algorithms have been provided and three case studies have been considered in more detail: the simulation of large-scale fading in a cellular communications network, the issue of fast versus slow fading, and the simulation of small-scale fading in a Wi-Fi network.

Finally, it is important to remark that this is still a very active and attractive research field. Indeed, although the field has witnessed very important advances over the last decades, there are still many open issues that deserve further attention. First of all, let us remark that the efficient generation of some complicated RVs used in the simulation (e.g., Nakagami or gamma) is still an open issue, with new approaches appearing every year. In this area, emphasis is placed on the generation of variables corresponding to nonstandard settings (e.g., with nonuniform phase) as well as on the very challenging issue of the efficient generation of correlated multivariate RVs. A second field of interest is on the simulation of novel fading channels developed more recently, such as those corresponding to the so-called worse-than-Rayleigh fading scenarios. Finally, another important field is in the development of fading models for new communication scenarios, such as vehicle-to-vehicle communications or communications in the 60 GHz frequency band, which is attracting more and more attention recently.

ACKNOWLEDGMENTS

This work has been partly financed by the Spanish government through the CONSOLIDER-INGENIO 2010 Program (Project CSD2008-00010) and the DISSECT (TEC2012-38058-C03-01) project.

REFERENCES

1. S. Stein, Fading channel issues in system engineering, *IEEE Journal on Selected Areas in Communications*, 5 (2), 68–89, February 1987.
2. B. Sklar, Rayleigh fading channels in mobile digital communication systems. Part I: Characterization, *IEEE Communications Magazine*, 35 (7), 90–100, July 1997.
3. T. K. Sarkar, Z. Ji, K. Kim, A. Medouri, and M. Salazar-Palma, A survey of various propagation models for mobile communication, *IEEE Antennas and Propagation Magazine*, 45 (3), 51–82, June 2003.
4. D. C. Cox, Wireless personal communications: A perspective, *The Mobile Communications Handbook*, J. D. Gibson, ed., CRC Press, Boca Raton, FL, 1996, pp. 209–241.
5. S. Asghar, Digital European cordless telephone, *The Mobile Communications Handbook*, J. D. Gibson, ed., CRC Press, Boca Raton, FL, 1996, pp. 478–499.
6. ETSI TR 101 178 V1.5.1, *Digital Enhanced Cordless Telecommunications (DECT); A High Level Guide to the DECT Standardization*, Valbonne, France, 2005.
7. ETSI TR 102 570 V1.1.1, *Digital Enhanced Cordless Telecommunications (DECT); New Generation DECT; Overview and Requirements*, Valbonne, France, 2007.
8. R. H. Frenkiel, Creating cellular: A history of the AMPS project (1971–1983), *IEEE Communications Magazine*, 48 (9), 14–24, 2010.
9. M. Delprat and V. Kumar, Second generation systems, *The Mobile Communications Handbook*, J. D. Gibson, ed., CRC Press, Boca Raton, FL, 1996, pp. 381–398.
10. T. Farley, Mobile telephone history, *Telektronikk*, 101 (3/4), 22–34, 2005.
11. M. Mouly and M.-B. Pautet, *The GSM System for Mobile Communications*, Telecom Publishing, Palaiseau, France, 1992.
12. M. Rahnema, Overview of the GSM system and protocol architecture, *IEEE Communications Magazine*, 31 (4), 92–100, 1993.
13. L. Hanzo, The pan-European cellular system, *The Mobile Communications Handbook*, J. D. Gibson, ed., CRC Press, Boca Raton, FL, 1996, pp. 399–418.
14. EIA/TIA IS-54, *Cellular System Dual-Mode Mobile Station—Base Station Compatibility Standard*, Washington, DC, 1989.
15. P. Mermelstein, The IS-54 digital cellular standard, *The Mobile Communications Handbook*, J. D. Gibson, ed., CRC Press, Boca Raton, FL, 1996, pp. 419–429.
16. J. S. Lee, Overview of the technical basis of Qualcomm's CDMA cellular telephone system design— A view North American TIA/EIA IS-95, *Proceedings of Singapore ICCS'94*, Singapore, 1994.
17. A. H. M. Ross and K. S. Gilhousen, CDMA technology and the IS-95 North American standard, *The Mobile Communications Handbook*, J. D. Gibson, ed., CRC Press, Boca Raton, FL, 1996, pp. 430–448.
18. K. Kinoshita and M. Nakagawa, Japanese cellular standard, *The Mobile Communications Handbook*, J. D. Gibson, ed., CRC Press, Boca Raton, FL, 1996, pp. 449–461.
19. Q. Bi, G. I. Zysman, and H. Menkes, Wireless mobile communications at the start of the 21st century, *IEEE Communications Magazine*, 39 (1), 110–116, January 2001.
20. M. Frodigh, S. Parkvall, C. Roobol, P. Johansson, and P. Larsson, Future-generation wireless networks, *IEEE Personal Communications*, 8 (5), pp. 10–17, 2001.
21. A. Akan and Ç. Edemen, Path to 4G wireless networks, *Proc. of the EEE 21st International Symposium on Personal, Indoor and Mobile Radio Communications (PIMRC)*, Istanbul, Turkey, 2009.
22. P. S. Henry and H. Luo, WiFi: What's next?, *IEEE Communications Magazine*, 40 (12), 66–72, December 2002.
23. L. Verma, M. Fakharzadeh, and S. Choi, WiFi on steroids: 802.11AC and 802.11AD, *IEEE Wireless Communications*, 20 (6), 30–35, December 2013.
24. K. Etemad, Overview of mobile WiMAX technology and evolution, *IEEE Communications Magazine*, 46 (10), October 2008.
25. M. C. Jeruchim, P. Balaban, and K. S. Shanmugan, *Simulation of Communication Systems*, 2nd edn., Springer, Boston, MA, 2000.

26. B. Sklar, *Digital Communications. Fundamentals and Applications*, Prentice Hall, Englewood Cliffs, NJ, 1988.
27. J. G. Proakis and M. Salehi, *Digital Communications*, McGraw-Hill, Boston, MA, 2007.
28. R. Van Nee and R. Prasad, *ODFM for Wireless Multimedia Communications*, Artech House, Norwood, MA, 2000.
29. A. R. S. Bahai and B. Saltzberg, *Multi-Carrier Digital Communications. Theory and Applications of ODFM*, Kluwer Academic/Plenum Publishers, New York, 2004.
30. A. V. Oppenheim, A. S. Willsky, and S. Hamid, *Signals and Systems*, 2nd edn., Prentice Hall, Upper Saddle River, NJ, 1996.
31. T. S. Rappaport, *Wireless Communications. Principles and Practice*, 2nd edn., Prentice Hall, Upper Saddle River, NJ, 2002.
32. H. Meyr, M. Moeneclaey, and S. A. Fechtel, *Digital Communication Receivers. Synchronization, Channel Estimation, and Signal Processing*, John Wiley & Sons, New York, 1998.
33. Z. Ding and Y. Li, *Blind Equalization and Identification*, Marcel Dekker, New York, 2001.
34. C.-Y. Chi, C.-C. Feng, C.-H. Chen, and C.-Y. Chen, *Blind Equalization and System Identification. Batch Processing Algorithms, Performance and Applications*, Springer-Verlag, London, U.K., 2006.
35. K. Van Acker, G. Leus, M. Moonen, O. van de Wiel, and T. Pollet, Per tone equalization for DMT-based systems, *IEEE Transactions on Communications*, 49 (1), 109–119, January 2001.
36. T. H. Stitz, T. Ihalainen, A. Viholainen, and M. Renfors, Pilot-Based synchronization and equalization in filter bank multicarrier communications, *EURASIP Journal on Advances in Signal Processing*, 2010, 1–18, 2010.
37. D. Falconer, S. L. Ariyavisitakul, A. Benyamin-Seeyar, and B. Eidson, Frequency domain equalization for single-carrier broadband wireless systems, *IEEE Communications Magazine*, 40 (4), 58–66, April 2002.
38. Y. Zeng and T. S. Ng, Pilot cyclic prefixed single carrier communication: Channel estimation and equalization, *IEEE Signal Processing Letters*, 12 (1), 56–59, January 2005.
39. M. Gudmunson, Correlation model for shadow fading in mobile radio systems, *Electronics Letters*, 27 (23), 2145–2146, November 1991.
40. A. Papoulis, *Probability and Statistics*, 1st edn., Prentice Hall, New York, 1989.
41. N. B. Mandayam, P.-C. Chen, and J. M. Holtzman, Minimum duration outage for cellular systems: A level crossing analysis, *Proceedings of the 46th IEEE Int. Vehicular Technology Conference*, Atlanta, GA, 1996.
42. H. Kim and Y. Han, Enhanced correlated shadowing generation in channel simulation, *IEEE Communications Letters*, 6 (7), 279–281, July 2002.
43. F. Graziosi, A general correlation model for shadow fading in mobile radio systems, *IEEE Communications Letters*, 6 (3), 102–104, March 2002.
44. G. Stüber, Modulation methods, *The Mobile Communications Handbook*, J. D. Gibson, ed., CRC Press, Boca Raton, FL, 1996, pp. 526–539.
45. D. C. Cox, Delay Doppler characteristics of multipath propagation at 910 MHz in a suburban mobile radio environment, *IEEE Transactions on Antennas and Propagation*, 20 (5), 625–635, September 1972.
46. T. S. Rappaport, S. Y. Seidel, and R. Singh, 900-MHz multipath propagation measurements for U.S. digital cellular radiotelephone, *IEEE Transactions on Vehicular Technology*, 39 (2), 132–139, May 1990.
47. S. Y. Seidel, T. S. Rappaport, S. Jain, M. L. Lord, and R. Singh, Path loss, scattering, and multipath delay statistics in four European cities for digital cellular and microcellular radiotelephone, *IEEE Transactions on Vehicular Technology*, 40 (4), 721–730, November 1991.
48. D. M. J. Devasirvatham, Time delay spread measurements of wideband radio signals within a building, *Electronics Letters*, 20 (23), 950–951, November 1984.
49. A. A. M. Saleh and R. A. Valenzuela, A statistical model for indoor multipath propagation, *IEEE Journal of Selected Areas in Communications*, 5 (2), 138–146, February 1987.
50. T. S. Rappaport and C. D. McGillem, UHF fading in factories, *IEEE Journal on Selected Areas in Communications*, 7 (1), 40–48, January 1989.
51. T. S. Rappaport, Characterization of UHF multipath radio channels in factory buildings, *IEEE Transactions on Antennas and Propagation*, 37 (8), 1058–1069, August 1989.
52. J. Kivinen, T. O. Korhonen, P. Aikio, R. Gruber, P. Vainikainen, and S.-G. Haggman, Wideband radio channel measurement system at 2 GHz, *IEEE Transactions on Instrumentation and Measurement*, 48 (1), 39–44, February 1999.
53. S. Salous and H. Gokalp, Propagation measurements in FDD UMT bands, *Proceedings of the 11th International Conference on Antennas and Propagation (ICAP)*, Manchester, U.K., 2001.

54. K. Yang, T. Roste, F. Bekkadal, and T. Ekman, Land-to-ship radio channel measurements over sea at 2 GHz, *Proceedings of the Sixth International Conference on Wireless Communications, Networking and Mobile Computing (WiCOM)*, Chengdu, China, 2010.

55. K. Yang, T. Roste, F. Bekkadal, and T. Ekman, Experimental multipath delay profile of mobile radio channels over sea at 2 GHz, *Proceedings of the Loughborough Antennas and Propagation Conference (LAPC)*, Loughborough, U.K., 2012.

56. A. Paier, J. Karedal, N. Czink, and H. Hofstetter, Car-to-car radio channel measurements at 5 GHz: Pathloss, power-delay profile, and delay-Doppler spectrum, *Proceedings of the Fourth International Symposium on Wireless Communication Systems (ISWCS)*, Trondheim, Sweden, 2007.

57. D. W. Matolak and A. Chandrasekaran, 5 GHz intra-vehicle channel characterization, de *Proc. of the IEEE Vehicular Technology Conference (VTC Fall)*, Quebec City, Quebec, Canada, 2012.

58. W. C. Jakes, *Microwave Mobile Communications*, John Wiley & Sons, New York, 1974.

59. R. H. Clarke, A statistical theory of mobile-radio reception, *Bell Systems Technical Journal*, 47, 957–1000, 1968.

60. M. J. Gans, A power spectral theory of propagation in the mobile radio environment, *IEEE Transactions on Vehicular Technology*, 21 (1),27–38, February 1972.

61. P. Robertson and S. Kaiser, Doppler spread analysis and simulation for multi-carrier mobile radio and broadcast systems, *Multi-Carrier Spread Spectrum & Related Topics*, K. Fazel and S. Kaiser, eds., Springer, Dordrecht, the Netherlands, 2000, pp. 279–286.

62. P. W. Lemme, S. M. Glenister, and A. W. Miller, Iridium aeronautical satellite communications, *IEEE Aerospace and Electronic Systems Magazine*, 14 (11), 11–16, November 1999.

63. T. S. Rappaport, J. N. Murdock, and F. Gutiérrez, State of the art in 60-GHz integrated circuits and systems for wireless communications, *Proceedings of the IEEE*, 99 (8), 1390–1436, August 2011.

64. L. Rayleigh, On the resultant of a large number of vibrations of the same pitch and of arbitrary phases, *Philosophical Magazine and Journal of Science, Fifth Series*, 10 (60), 73–78, August 1880.

65. S. O. Rice, Mathematical analysis of random noise. Parts I and II, *Bell System Technical Journal*, 23 (3), 282–332, July 1944.

66. S. O. Rice, Mathematical analysis of random noise. Parts III and IV, *Bell System Technical Journal*, 24 (1), 46–156, January 1945.

67. S. O. Rice, Statistical properties of a sine wave plus random noise, *Bell System Technical Journal*, 27, 109–157, January 1948.

68. M. Abramowitz and I. A. Stegun, eds., *Handbook of Mathematical Functions: With Formulas, Graphs and Mathematical Tables*, Dover, New York, 1972.

69. A. Abdi, C. Tependelenlioglu, M. Kaveh, and G. B. Giannakis, On the estimation of the K parameter for the rice fading distribution, *IEEE Communications Letters*, 5 (3), 92–94, March 2001.

70. C. Chayawan and V. A. Aalo, On the outage probability of optimum combining and maximal ratio combining schemes in an interference-limited rice fading channel, *IEEE Transactions on Communications*, 50 (4), 532–535, April 2002.

71. A. Doukas and G. Kalivas, Rician K factor estimation for wireless communication systems, *Proceedings of the IEEE International Conference on Wireless and Mobile Communications (ICWMC)*, Bucharest, Romania, 2006.

72. M. Nakagami, The m-distribution: A general formula of intensity distribution of rapid fading, *Statistical Models in Radio Wave Propagation*, W. C. Hoffman, ed., Pergamon Press, New York, 1960, pp. 3–36.

73. H. Suzuki, A statistical model for urban radio propagation, *IEEE Transactions on Communications*, 25 (7), 673–680, July 1977.

74. W. Braun and U. Dersch, A physical mobile radio channel model, *IEEE Transactions on Vehicular Technology*, 40 (2), 472–482, May 1991.

75. L. Rubio, J. Reig, and N. Cardona, Evaluation of Nakagami fading behaviour based on measurements in urban scenarios, *International Journal of Electronics and Communications*, 61, 135–138, 2007.

76. C.-J. Chen and L.-C. Wang, A unified capacity analysis for wireless systems with joint multiuser scheduling and antenna diversity in Nakagami fading channels, *IEEE Transactions on Communications*, 54 (3), 469–478, March 2006.

77. M.-S. Alaouini and A. J. Goldsmith, Adaptive modulation over Nakagami fading channels, *Wireless Personal Communications*, 13, 119–143, 2000.

78. T. Eng and L. B. Milstein, Coherent DS-CDMA performance in Nakagami multipath fading, *IEEE Transactions on Communications*, 43 (2/3/4), 1134–1143, February/March/April 1995.

79. S. Ikki and M. H. Ahmed, Performance analysis of cooperative diversity wireless networks over Nakagami-*m* fading channel, *IEEE Communications Letters*, 11 (4), 334–336, April 2007.

80. L. Devroye, *Non-Uniform Random Variate Generation*, Springer-Verlag, New York, 1986.
81. J. Dagpunar, *Principles of Random Variate Generation*, Clarendon Press, Oxford, U.K., 1988.
82. W. Hörmann, J. Leydold, and G. Derflinger, *Automatic Nonuniform Random Variate Generation*, Springer, Berlin, Germany, 2003.
83. L. Cao and N. C. Beaulieu, Simple efficient methods for generating independent and bivariate Nakagami-*m* fading envelope samples, *IEEE Transactions on Vehicular Technology*, 56 (4), 1573–1579, July 2007.
84. M. Matthaiou and D. Laurenson, Rejection method for generating Nakagami-*m* independent deviates, *Electronics Letters*, 43 (25), 1474–1475, December 2007.
85. Q. Zhu, X. Dang, D. Xu, and X. Chen, Highly efficient rejection method for generating Nakagami-*m* sequences, *Electronics Letters*, 47 (19), 1100–1101, September 2011.
86. D. Luengo and L. Martino, Almost rejectionless sampling from Nakagami-*m* distributions (m > 1), *Electronics Letters*, 48 (24), 1559–1561, November 2012.
87. D. Luengo and L. Martino, Extremely efficient acceptance–rejection method for simulating uncorrelated Nakagami fading channels, *viXra*, 1407.0133, July 2014.
88. P. S. Bithas, Weibull-gamma composite distribution: alternative multipath/shadowing fading model, *Electronics Letters*, 45 (14), 749–751, July 2009.
89. J. A. Anastasov, G. T. Djordjevic, and M. C. Stefanovic, Outage probability of interference limited system over Weibull-gamma fading channel, *Electronics Letters*, 48 (7), 408–410, March 2012.
90. S. Gappmair and S. S. Muhamad, Error performance of PPM/Poisson channels in turbulent atmosphere with gamma-gamma distribution, *Electronics Letters*, 43 (16), 880–882, August 2007.
91. C. Liu, Y. Yao, X. Zhao, Average capacity for heterodyne FSO communication systems over gamma-gamma turbulence channels with pointing errors, *Electronics Letters*, 46 (12), 851–853, 2010.
92. B. Bandemer, C. Oestges, N. Czink, and A. Paulraj, Physically motivated fast-fading model for indoor peer-to-peer channels, *Electronics Letters*, 45 (10), 515–517, May 2009.
93. J. L. Smith, A computer generated multipath fading simulation for mobile radio, *IEEE Transactions on Vehicular Technology*, 24 (3), 39–40, August 1975.
94. T. S. Rappaport and S. Y. Seidel, Statistical channel impulse response models for factory and open plan building radio communications system design, *IEEE Transactions on Communications*, 39 (5), 794–807, May 1991.
95. V. Fung, T. S. Rappaport, and B. Thoma, Bit error simulation for pi/4 DQPSK mobile radio communications using two-ray and measurement-based impulse response models, *IEEE Journal on Selected Areas in Communications*, 11 (3), 393–405, April 1993.
96. R. G. Vaughan, N. L. Scott, and D. R. White, The theory of bandpass sampling, *IEEE Transactions on Signal Processing*, 39 (9), 1973–1984, September 1991.
97. M. Hata, Empirical formula for propagation loss in land mobile radio services, *IEEE Transactions on Vehicular Technology*, 29 (3), 317–325, August 1980.
98. Y. Okumura, E. Ohmori, and K. Fukuda, Field strength and its variability in VHF and UHF land mobile radio service, *Review of the Electrical Communications Laboratory*, 16 (9–10), 825–873, 1968.
99. M. Jeong and B. Lee, Comparison between path-loss models for wireless telecommunication system design, *Proc. of the IEEE Antennas and Propagation Society International Symposium*, Boston, MA, 2001.
100. D. Har, A. M. Watson, and A. G. Chadney, Comment on diffraction loss of rooftop-to-street in COST 231-Walfisch-Ikegami model, *IEEE Transactions on Vehicular Technology*, 48 (5), 1451–1452, September 1999.
101. F. Ikegami, S. Yoshida, T. Takeuchi, and M. Umehira, Propagation factors controlling mean field strength on urban streets, *IEEE Transactions on Antennas and Propagation*, 32 (8), 822–829, August 1984.
102. M. Hata, Propagation loss prediction models for land mobile communications, *Proceedings of the IEEE International Conference on Microwave and Millimeter Wave Technology (ICMMT)*, Beijing, China, 1998.
103. Y. Rosmansyah, S. R. Saunders, P. Sweeney, and R. Tafazolli, Equivalence of flat and classical Doppler sample generators, *Electronics Letters*, 37 (4), 243–244, February 2001.
104. Y. Ma and D. Zhang, A method for simulating complex Nakagami fading time series with nonuniform phase and prescribed autocorrelation characteristics, *IEEE Transactions on Vehicular Technology*, 59 (1), 29–35, January 2010.
105. L. Martino and D. Luengo, Extremely efficient generation of gamma random variables for alpha > 1, Arxiv:1304.3800v3, 2013.

Section III

Section III

11 Simulation Tools for Cloud Computing

*Bernardi Pranggono, Dabiah Alboaneen,
and Huaglory Tianfield*

CONTENTS

ABSTRACT

Simulation tools are essential to evaluate cloud computing services and applications. This chapter surveys the various simulation platforms and tools of cloud computing through a literature review and classification of journal articles, conference papers, and technical reports. The review and discussion on the cloud computing simulation tools also provides insights on modeling and evaluation of cloud computing services and applications. Various cloud computing simulation tools are reviewed and characteristics of the tools are identified. Finally, the chapter notes the challenges and the future research for next-generation cloud computing simulation tools to obtain more accurate results and keep in pace with the advance of technologies.

11.1   INTRODUCTION

A cloud platform can be defined as a type of distributed data center that provides infrastructure, platform, or software as services that are subscription-based, delivered in a pay-as-you-go model to clients. Typically, cloud computing systems are classified into three types (see Figure 11.1):

1. On-demand virtual machine (VM) images, such as Amazon's Elastic Compute Cloud (EC2) [1] or Eucalyptus [2], which are known as infrastructure as a service (IaaS)
2. Storage, data processing, and compute services such as provided by S3, Hadoop [3], CloudStore, and Sector/Sphere [4,5] that can be used to build applications, which are also known as platform as a service (PaaS)
3. Software applications delivered as a service such as Google Apps, which are known as software as a service (SaaS)

Recent years have seen a significant increase in cloud computing simulation tools. It involves considerable complexity and the scale of shared resources to evaluate new scheduling and provisioning algorithms on actual cloud platforms. Simulation tools are becoming more and more important in the evaluation of the cloud computing model. The performance, efficiency, and reliability of new algorithms for a cloud platform of a large heterogeneous infrastructure can be evaluated more easily by means of suitable simulation tools.

The validity and the accuracy of a simulator are greatly reliant on how much detail each component of the simulator can mimic the behavior of the physical platform. The simulation tools of cloud computing have many advantages to cloud clients or potential customers. For example, the simulator can help to test, evaluate, and replicate the assumptions cost-effectively before a cloud platform is deployed in a real environment. It also provides an option to change and control the bottlenecks of the performance prior to implementing on an actual cloud platform [6].

The vast amount of cloud computing services and the lack of standards make it hard for customers to choose the best services and applications that meet their needs. The simulation tools for cloud computing are also important to help users choose which cloud computing service is suitable

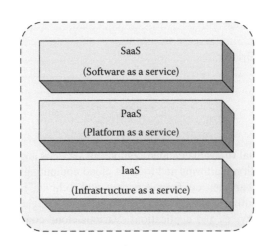

FIGURE 11.1   Service delivery models of cloud computing.

for their requirements as deploying and operating a cloud platform in a real environment are often expensive, time consuming, or even impractical due to multiple test runs in different conditions and requirements [7].

The remainder of the chapter is organized as follows: Section 11.2 presents general background in simulation tools. Section 1.3 discusses simulation tools for cloud computing in detail. Finally, open issues are presented in Section 11.4.

11.2 SYSTEMS EVALUATION

Typically, there are three approaches to evaluate systems [8]. First, performance evaluation can be conducted by emulation through building a testbed or miniscale of the system. Emulation basically can be defined as the series of activities that implement the real hardware and software that is going to be used in a real-world scenario in a smaller scale. This needs a relatively realistic amount of hardware and software. The real advantage of this method is that the evaluation results are expected to be very close to the real system's performance as we use very similar hardware and software configurations when evaluating it. On the other hand, building a miniscale of the proposed network architecture is often costly and time consuming. Probably, the time needed to build and develop the testbed can be close to or even the same as the time needed to implement the real system. In terms of facilities, in many large enterprises, maybe it is not a major issue as many blue chip companies usually have test-bed facilities in place. However, it is a serious issue for small and medium companies (SME), which tend to have a limited budget and therefore limited resources.

Second, mathematical analysis or modeling with numerical analysis can be used to evaluate the system in question. Compared to emulation through building a miniscale of the system, performing the tests and performance evaluation on the miniscale system can theoretically design the system and evaluate analytically (usually by creating a model and a set of computer codes) its performance. Obviously, the main advantage of the mathematical analysis is the engineering cost, as it does not require implementing hardware and software to evaluate the performance of the system in question. However, as many assumptions need to be made to build a reasonable tractable analytical model, the output may not be very close to the real system. Often, numerical analysis shows the borderline behavior of system characteristics or offer upper and lower bounds for specific research questions. However, more fine-grained analysis often leads to an unacceptable complexity of the analytical models.

Third, performance evaluation can be carried out through simulations. Simulation is a compromise between emulation and analytical modeling, or even to get the best of both worlds: emulation and analytical modeling. Simulation offers a controlled environment in which a system can be investigated in more detail. Different parameter settings and scenarios can be analyzed with comparably limited effort. Through proper design and implementation, relatively accurate results can be obtained from the simulations. Modest cost, effectiveness, and the short time needed are the main advantages of the simulation. Simulation results are also easier to analyze than experimental results because it is relatively easy to log and trace important information at critical points in order to diagnose system's behavior. Moreover, simulation modeling techniques offer more flexibility and accuracy compared with analytical modeling. Therefore, simulation modeling is the choice by researchers. In particular, simulation is used to analyze systems that are highly dynamic and whose properties are difficult to capture in a mathematical way.

In a computer simulation, a real-world system is *imitated* over time. Computer simulation is applied in many different fields. Among different types of computer simulation, like discrete-event simulation (DES), event-based simulation, continuous simulation, Monte Carlo simulation, spreadsheet simulation, and trace-driven simulation, the dominant simulation technique is discrete-event or event-driven simulation. The main reason behind the popularity of discrete-event-based simulation is that its simulation paradigm fits well to the systems concerned and is easily applied. Hence, DES provides a simple and flexible way to evaluate the systems behavior under different conditions.

11.3 VIRTUALIZATION AND CLOUD COMPUTING

The main underlying technologies of cloud computing are as follows: virtualization technologies, web services and service-oriented architecture (SOA), and distributed storage [9]. Virtualization is considered a key enabling technology for cloud computing as it enables scalable and flexible computing platforms.

At a fundamental level, virtualization technology enables the abstraction or decoupling of the application payload from the underlying physical resource [10]. This means that the physical resource can then be carved up into logical or virtual resources as needed. This is known as provisioning. By introducing a suitable management infrastructure on top of this virtualization functionality, the provisioning of these logical resources could be made dynamic, that is, the logical resource could be made bigger or smaller in accordance with the demand. This is known as dynamic provisioning. To enable a true *cloud* computer, every single resource should be capable of being dynamically provisioned and managed in real time [11].

Virtualization takes many forms. System virtualization [12], also commonly referred to as server virtualization, is the ability to run multiple heterogeneous operating systems on the same physical server [13]. In server virtualization, a *hypervisor* or *VM monitor* is run on the hardware platform, simulating one or more other computer environments (VMs). Each VM, in turn, runs its respective *guest* software, typically an operating system, which runs just as if it were installed on the stand-alone hardware. Server virtualization can be achieved through paravirtualization or full virtualization [14]. Virtualization includes storage virtualization and network virtualization, namely, logical representations of the physical storage and network resources [15].

Amazon Web Services [1] is considered a pioneer in virtualization. Amazon's EC2 is one of the most widely used infrastructure platforms [16].

Consolidating underutilized servers into a single server can significantly reduce energy consumption. Virtualization technology increases the utilization of resources by loading more than one VM on a physical server. According to Gartner, virtualization can increase hardware utilization by 5–20 times and allows organizations to reduce the number of power-consuming servers. By implementing virtualization technology, it is possible to combine and consolidate different applications and execute them in a smaller number of machines [17,18]. This implies less hardware and reduces the amount of energy consumption.

The combination of virtualization and consolidation approaches can significantly reduce the power consumption of virtualized and consolidated data centers as pointed out in [19]. Energy efficiency can be increased by modifying the consolidation strategy to allow an important reduction in the number of active nodes required to process a workload without degrading the level of service. It is shown in [19] that introducing memory compression and request discrimination in the consolidation strategy can dramatically decrease the power consumption in data centers.

Cloud computing is defined by National Institute of Standards and Technology (NIST) in [20] as "a model for enabling ubiquitous, convenient, on-demand network access to a shared pool of configurable computing resources (e.g., networks, servers, storage, application and services) that can be rapidly provisioned and released with minimal management effort or service provider interaction." Five essential characteristics of cloud computing are as follows [20]:

1. *Broad network access*: Resources are available over the network through standard mechanisms.
2. *Rapid elasticity*: Capabilities can be rapidly and elastically provisioned, preferably automatically.
3. *Measured service*: Resource usage is monitored and automatically controlled and optimized. The organization provides transparency for both itself and the customer of the utilized service.

4. *On-demand self-service*: Customers can provision computing capabilities.
5. *Resource pooling*: The provider's computing resources are pooled to serve multiple consumers using a multitenant model.

Cloud computing is the convergence of several concepts from resource pooling, virtualization, dynamic provisioning, utility computing, on-demand deployment, and Internet delivery of services to enable a more flexible approach to deploying and scaling applications. There are three cloud computing service models—SaaS, PaaS, and IaaS—as shown in Figure 11.1. In contrast to traditional IT services, cloud services have attributes such as a pay-per-use model, self-service usage, flexible scaling, and shared underlying IT resources.

There are four deployment models for cloud computing:

1. *Public cloud*: The cloud infrastructure is available to a general public.
2. *Private cloud*: The cloud infrastructure is operated solely for an organization.
3. *Community cloud*: The cloud infrastructure is shared by multiple organizations that share the same goal.
4. *Hybrid cloud*: The cloud infrastructure is a composed by two or more cloud deployment models (private, community, or public) that are bound together to enable data and application portability. Most common applications consist of cloud bursting (handling temporary peaks) and load balancing between clouds.

In summary, the visual model of cloud computing as defined by NIST is shown in Figure 11.2.

Cloud computing services and applications are nowadays increasing rapidly. The wide variety of cloud computing services and applications makes it difficult for users to compare and find the right service. Cloud computing simulation tools are important for users to evaluate the services offered by different cloud service providers. On the other hand, researchers are often facing difficulty to analyze the performance of new scheduling and provisioning on actual cloud computing platform. Also, researchers need a way to understand and visualize the micro and macro behavior of the

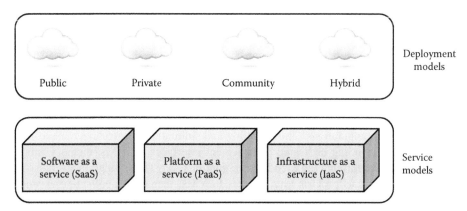

FIGURE 11.2   NIST cloud computing visual model.

systems they are developing. When building a cloud platform is impractical and costly, simulation tools are becoming important for the evaluation of cloud services and applications.

11.4 SIMULATION TOOLS FOR CLOUD COMPUTING

In general, there is no single simulation tool that can solve all networking and communications problems. There is no silver bullet. This also applies for cloud computing simulation tools. Every simulation tool has strengths and weaknesses. We need to understand which simulation tools are suitable for which problem(s).

In this section, several cloud computing simulator tools are discussed in detail. Interested readers can study further the references provided for each tool for more detailed information.

11.4.1 CloudSim and Variants

11.4.1.1 CloudSim

Proposed in 2009, CloudSim is a simulation toolkit for enabling continuous modeling, simulation, and experimentation of cloud computing and application services [21]. CloudSim does not provide a ready-to-use environment for execution of a complete scenario with a specific input. Instead, users of CloudSim need to develop their own cloud scenario to evaluate, define the required output, and provide the input parameters. CloudSim scales well for cloud computing and has a low simulation overhead but is arguably the most sophisticated DES. CloudSim enables fast evaluation of scheduling and resource allocation mechanisms within cloud data centers. It represents data center topology in the form of direct graph but no specific network topology is involved for modeling an internal data center network [6]. CloudSim allows researchers to study specific system problems without worrying low-level details related to cloud-based infrastructures and services.

CloudSim has the following advantages [6]:

- It helps in developing new resource distribution policies and scheduling algorithms rapidly.
- It provides modeling and simulation of large-scale infrastructure that involves data centers of multiple physical computing nodes.
- It creates and manages various, independent, and cohosted virtualized services on a data center node by using a virtualization engine.
- It offers great flexibility due to the switch between space-shared and time-shared allocation of processing cores to virtualized services.

CloudSim is based on DES where the state of the simulation model can only change at discrete points in time, which are referred to as events. This means the model is time based and takes into account resources, constraints, and interactions between the events as time passes. Central to DES are a clock and an event list that tells what steps have to be executed. Therefore, DES in general is able to cut the simulation time significantly compared to packet-level simulation at the expense of simulation accuracy.

The high-level multilayer architecture of CloudSim software framework is illustrated in Figure 11.3. CloudSim simulates cloud systems following a cloud provider perspective and uses so-called cloudlets to model abstract computations. The modeling and simulation of virtualized cloud-based data center environments including dedicated management interfaces for VMs, memory, storage, and bandwidth are supported through the CloudSim simulation layer. The simulation layer also manages fundamental issues, such as provisioning of hosts to VMs, managing application execution, and monitoring dynamic system state. The user code is situated at the topmost layer in the CloudSim. It exposes basic entities for hosts (number of machines, their specification, and so on), applications (number of tasks and their requirements), VMs, number of users and their application types, and broker scheduling policies [6].

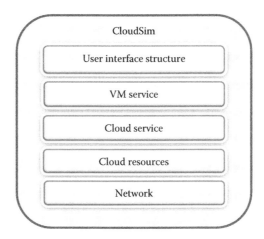

FIGURE 11.3 CloudSim architecture. (Adapted from Calheiros, R.N. et al., *Software Prac. Ex.*, 41, 23, 2011.)

CloudSim is very popular tool to simulate cloud computing environment. Many researchers developed other simulation tools based on CloudSim. For example, CDOSim [22], CloudAnalyst [23], NetworkCloudSim [24], and TechCloud [25] were all built or developed on top of the CloudSim. Werner et al. also proposed an organization theory model for resource management for green cloud computing based on CloudSim [26].

11.4.1.2 Cloud Analyst

CloudAnalyst is developed based on the SimJava and CloudSim toolkit by extending its functionality with the introduction of concepts that model the Internet and Internet application behavior [23]. It is developed on Java platform to simulate large-scale cloud applications with the purpose to study and analyze the behavior of such applications under various deployment scenarios. CloudAnalyst allows to set up repeat simulations easily and displays simulation experiments sequentially with slight parameters variations in a quick and easy way. It also provides many user-friendly capabilities that make CloudAnalyst simulator more flexible and configurable to use. The graphical user interface (GUI) facilitates users to set up and design their experiments easily. The GUI is also convenient for users to combine technology with the easing of extension. Making a simulation implementation separated from a programming implementation is considered one of the key benefits of CloudAnalyst. Therefore, by using CloudAnalyst users can focus on the simulation problems without being heavily involved in programming. CloudAnalyst allows users to model scenarios where SaaS data centers and users are in different geographic locations. By using CloudAnalyst, application developers or designers are also able to choose the optimum strategy for allocating the limited resources among data centers, strategies for selecting data centers to serve specific requests, and reducing costs related to such operations [23].

There are three algorithms that are implemented in CloudAnalyst including VM load balancing, throttled load balancer, and active monitoring load balancer. The design of CloudAnalyst architecture is illustrated in Figure 11.4. The main components of CloudAnalyst are as follows [23]:

- *GUI Package* to manage GUI
- *Simulation* to manage creation and execution of the simulation
- *UserBase* for generating the traffic that represent users
- *DataCenterController* to control all activities of users
- *InternetCharacteristics* for defining the Internet characteristics that applied during the simulation

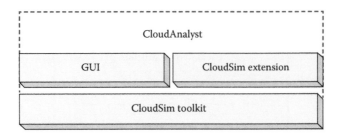

FIGURE 11.4 CloudAnalyst architecture. (Adapted from Wickremasinghe, B. et al., CloudAnalyst: A CloudSim-based visual modeller for analysing cloud computing environments and applications, in *Advanced Information Networking and Applications (AINA), 2010 24th IEEE International Conference on*, Perth, Western Australia, Australia, 2010, pp. 446–452.)

- *VmLoadBalancer* for displaying the policy of load balance that are used by data centers when serving the requests of allocation
- *CloudAppServiceBroker* for displaying the service brokers that control traffic routing between user bases and data centers

The main characteristics of CloudAnalyst are as follows [23]:

- Easy to use GUI
- Provides a high degree of configurability and flexibility
- Repeatability of experiments
- Easy to read output (graphical output)
- Use of consolidated technology and ease of extension

11.4.1.3 CDOSim

CDOSim [22], cloud deployment option (CDO) simulator, is built based on CloudSim and extends it to combine with cloud migration framework CloudMIG [27]. CDOSim is able to simulate various scenarios, such as [22]

- *The response times*: The total rating for the response times will be calculated.
- *Service-level agreement (SLA) violations*: Counts the number of each SLA violations.
- *Rating*: Making ratings for each run and the scale ranges for rating from 1 to 5 (where 1 is the best performance).
- *Costs of a CDO*: Presents the overall cost that is the total costs for VM instances and the used bandwidth.

CDOSim is developed with objective to overcome the limitations of other existing cloud simulator that related to CDOs. The CDOSim has the following characteristics [22]:

- It does not expose fine-grained internals of a cloud platform when following the cloud provider perspective; therefore, it focuses on client perspective.
- It keeps a balance between the lack of knowledge of cloud user's and control regarding an underlying cloud platform structure and its effect on a total performance of applications.
- Workload profiles from production monitoring data are used to rerun actual user behavior to simulate CDOs.

Based on these characteristics, CDOSim can be considered as the best simulator tool for evaluating the trade-off between costs and performance or for comparing runtime reconfiguration plans. Hence, the significance of using CDOSim is to obtain an accurate prediction for the cost and performance features of CDOs. CDOSim requires a mapping model that contains the value of mega integer plus instructions per second (MIPIPS), weights per statement, and the actual knowledge discovery metamodel (KDM). CDOSim introduce MIPIPS as the measure for describing the computing performance and express instructions like double plus as integer plus instructions through a conversion [22]. CDOSim also needs the instruction count (IC) for each method call. The method for the IC is dynamic, static, and hybrid approaches. However, TCP/IP protocol is not supported in CDOSim.

The main features of CDOSim include the following [22]:

- *IC derivation*: The derivation is implemented with one of the IC mechanisms.
- The size of included types is derived to approximate the bandwidth that is used when there are distributed calls to other VM instances.
- The transformation from the mapping model provided by CloudMIG Xpress to the extended CloudSim metamodel.
- The actual simulation with CloudSim.
- The new simulation result is rated relative to the other runs.

We understand that CloudSim simulates cloud infrastructure from a cloud service provider perspective. CDOSim extends this capability to include the cloud user perspective to enable simulation of CDOs. These range from the implementation of a CPU utilization model that is not based on randomness to a new cloudlet scheduler that allows cloudlets to call other cloudlets. In addition, VMs in CDOSim can be started and stopped during runtime. Unlike CloudSim, which utilize a process-based approach that runs a separate thread for each entity, CDOSim employs an event-based simulator that requires only one thread per simulation [22].

11.4.1.4 EMUSIM

A key question raised by potential cloud customers before renting resources from any cloud service provider is how to know the services received will behave in a set of resources, and how much costs are involved when their resource pool has changed. EMUSIM (emulation and simulation) [7] is developed with the objective to answer these questions. It is designed for both emulation and simulation objectives in order to obtain better understanding of service's behavior on cloud platforms. Emulation in EMUSIM allows the simulator to automatically extract information from the application behavior and then uses this information to generate the corresponding simulation model [7].

EMUSIM is illustrated in Figure 11.5. EMUSIM is built on top of two software systems: automated emulation framework (AEF) for emulation purpose and CloudSim for simulation purpose. As it employs both emulation and simulation, in theory, EMUSIM should improve the accuracy of the simulation. The main advantages of this integration are as follows: Emulation allows execution of the actual application in a small-scale environment that models the actual production infrastructure, and simulation allows assessment of how a system or an application performs in response to various conditions, for example, request arrival times and amount of work, in a controlled manner.

By using EMUSIM, cloud service provider, we are able to know how many resources are required to have a given response time considering a specific user arrival distribution and the effects of the changes in the request arrival rate. There are important requirements as an input when

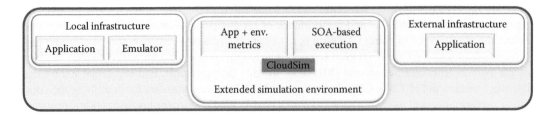

FIGURE 11.5   EMUSIM architecture. (Adapted from Calheiros, R.N. et al., *Software Prac. Ex.*, 00, 1, 2012.)

applying EMUSIM: description of the physical environment hosting the emulation, description of the emulated environment, application configuration, and simulation configuration.

11.4.1.5   NetworkCloudSim

NetworkCloudSim [24] is an extension of CloudSim that supports a scalable network model of data center and generalized applications such as high-performance computing (HPC), e-commerce, social network, and web applications. NetworkCloudSim claims to have more realistic application models than any other cloud computing simulator. The main components of NetworkCloudSim are as follows: classes to model a network topology and classes for application modeling. NetworkCloudSim has three main entities: Switch, NetworkDatacenter, and NetworkDatacenterBroker. Switch represents a network entity that can be configured as a router or switch. NetworkCloudSim has two levels of scheduling, one at the host level (i.e., scheduling of tasks on VMs) and another at the VM level where real applications are executed.

The CloudSim architecture with the key elements of NetworkCloudSim (shown in dark boxes) is depicted in Figure 11.6. The bottom layer of the CloudSim architecture handles the interaction

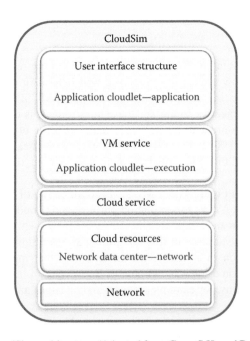

FIGURE 11.6   NetworkCloudSim architecture. (Adapted from Garg, S.K. and Buyya, R., NetworkCloudSim: Modelling parallel applications in cloud simulations, in *Fourth IEEE International Conference on Utility and Cloud Computing (UCC), 2011*, Melbourne, Victoria, Australia, 2011, pp. 105–113.)

between CloudSim entities and components. All components in CloudSim communicate through message passing operations. The second layer consists of several sublayers that model the core elements of cloud computing. The cloud resources sublayers model datacenter, cloud coordinator, and network topology between different datacenters.

A network flow model for cloud data center utilizing bandwidth sharing and latencies is proposed to enable scalable and fast simulations. While the network flow model is sufficient for most network calculations, it is less accurate compared to the packet-level model. NetworkCloudSim implements network flow model design with low computational overhead and good approximation to real network models such as TCP/IP [28].

NetworkCloudSim is capable of simulating cloud data center network and applications with communicating tasks such as MPI with a high degree of accuracy [24].

11.4.1.6 TeachCloud

TeachCloud is an extension of CloudSim, particularly designed for teaching purposes. TeachCloud is equipped with a simple GUI, which enables students and scholars to modify a cloud's configuration and perform simple experiments easily [25]. Students can use TeachCloud to experiment with different cloud components such as processing elements, data centers, networking, SLA constraints, web-based applications, SOA, virtualization, management and automation, and business process management (BPM). TeachCloud introduces many new enhancements on top of CloudSim basic platform, such as providing a GUI toolkit, adding the cloud workload generator to the CloudSim simulator, adding new modules related to SLA and BPM, and adding new cloud network models such as VL2, Bcube, Portland, and Dcell. TeachCloud enables students to reconfigure the cloud system any time to investigate the effects of changes in the overall system performance [25]. It is an open source for students and researchers.

11.4.2 SIMULATION OF VIRTUAL DATA CENTERS

11.4.2.1 DCSim

DCSim or data center simulator is an extensible simulation framework for simulating a multitenant, virtualized data center hosting an IaaS cloud to any clients [29,30]. DCSim is specifically designed and developed for transactional and continuous workloads such as a web server (WS). Hence, by using DCSim, there is a further capability of modeling that replicates the VMs sharing incoming workload and the dependencies between VMs that are part of a multitiered application. DCSim is especially focused on a virtualized data center providing IaaS to multiple tenants. This characteristic makes it differ from other simulators such as GreenCloud, MDCSim, and GDCSim, which are not equipped with any modeling of virtualization.

DCSim is classified as an event-driven simulator and is developed in Java to allow rapid development and evaluation of data center management techniques. In addition, the implementation of SLA can be easily measured and directed. The SLA is also available to manage components in the simulation. The main component of DCSim is the DataCenter, which contains hosts, VMs, and different management components and policies. A data center has a set of interconnected hosts that are managed by a set of management policies. Each host consists of a set of resource managers that manage local resource allocation, a CPU scheduler to decide how to run VMs, and a power model that models how much power is being consumed by the host at any point in time. The design of DCSim is illustrated in Figure 11.7. There are different metrics to be computed by DCSim for each simulation including SLA violation, active hosts, host hours, active host utilization, number of migrations, power consumption and simulation, and algorithm running time [29,30].

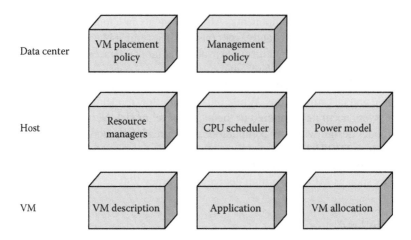

FIGURE 11.7 DCSim architecture. (Adapted from Tighe, M. et al., DCSim: A data centre simulation tool for evaluating dynamic virtualized resource management, in *Network and Service Management (CNSM), 2012 Eighth International Conference and 2012 Workshop on Systems Virtualization Management (SVM)*, Las Vegas, NV, 2012, pp. 385–392.)

11.4.2.2 GDCSim

GDCSim [31] stands for green data center simulator. It combines both the simulation of management techniques and the simulation of the physical behavior of a data center. GDCSim is developed as a part of the BlueTool infrastructure project at Impact Lab. It has the following design features [31]:

- Processing will be automatic without any intervention from the user.
- Capability to analyze online, which means changes in the physical environment in the data center will be taken into consideration to make decisions for the real-time simulation of management.
- Capability to repeat design analysis, which helps to design time testing and analyze different data center configurations before deployment.
- Capability to conduct thermal analysis.
- Workload and power management, which supports scheduling of workload and controlling of power modes of servers in the data center.
- It aims to save energy consumption by providing feedback of information on temperature and airflow designs in the data center to the management algorithms and the closed-loop operation of the servers and cooling units (CRAC).

The main components of GDCSim are as follows [31]:

- *BlueSim*: Which takes XML-based specification as input and executes computational fluid dynamics (CFD) simulations to output a heat recirculation profile of the data center for fast thermal evaluation.
- *Resource management*: To build informed decisions about workload and cooling management based on physical behavior of data center.
- *Simulator*: To capture the physical behavior of the data center in response to the management decisions that then provides feedback to the resource management in order to manage the changes in data center physical behavior.

As shown in Figure 11.8, GDCSim consist of four modules: BlueSim, Input/Output Management (IOM), Resource Management (RM), and Simulator.

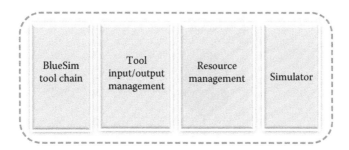

FIGURE 11.8 GDCSim architecture. (Adapted from Umeno, H. et al., Performance evaluation on server consolidation using virtual machines, in *International Joint Conference SICE-ICASE*, Busan, Korea, 2006, pp. 2730–2734.)

11.4.2.3 GreenCloud

GreenCloud is an open-source packet-level simulator for cloud computing that is specifically designed for data center simulation by implementing detailed modeling of communication aspects of the data center [32,33]. Along with the workload distribution, the simulator is designed to capture details of the energy consumed by data center components (servers, switches, routers, and connection links between them) as well as packet-level communication patterns in realistic setups. GreenCloud also allows analysis of the load distribution through the network, as well as communication with high accuracy (TCP packet level). GreenCloud views data center resources as a collection of VMs. It implements a simplistic applications model without any communicating tasks or limited network model within the data center.

GreenCloud simulator is an extension of the well-known simulator in computer networking, ns-2 [34]. Using ns-2 as the foundation, GreenCloud implements a full TCP/IP protocol reference model, which allows seamless integration of a wide variety of communication protocols, such as IP, TCP, and UDP with the simulation. Similar to ns-2, GreenCloud is also coded in C++ with OTcl. The main drawback of packet-level simulation is on the simulation duration, which is usually longer than DES, but with more accurate results. Typical GreenCloud simulation is about tens of minutes.

In order to accurately assess and evaluate the energy consumption of data center, GreenCloud defines three types of energy: calculation (CPU), communications, and physical computing center (cooling system). It also includes two built-in mechanisms of energy reduction: dynamic voltage scaling (DVS) that allows to reduce the voltage switches and dynamic network shutdown (DNS) to shutdown switches when possible. As GreenCloud is based on ns-2, it is possible to add more modules to the simulator, for example, to add different energy reduction mechanisms such as dynamic voltage and frequency scaling (DVFS).

A typical data center scenario is composed of three-tier architecture: access network, aggregation network, and core network. "The availability of the aggregation layer facilitates the increase in the number of server nodes (to over 10,000 servers) while keeping inexpensive Layer-2 (L2) switches in the access network, which provides a loop-free topology" [32]. The main reason behind the three-tier high-speed data center architectures is to obtain the optimum number of nodes, capacity of core, and aggregation networks that are typical bottleneck in data center and in turn limit the maximum number of nodes in a data center or a per-node bandwidth.

The structure of GreenCloud simulator mapped onto the three-tier data center architecture is presented in Figure 11.9. The three main components are identified in the design: servers (S), switches and links, and workloads.

Servers in a data center are responsible for task execution. Single-core nodes, associated size of the storage resources, and different task scheduling mechanisms ranging from the simple round robin to the more sophisticated DVFS- [35] and DNS-enabled are implemented in the servers. Switches and links are responsible to deliver workload to any of the computing servers for execution in a

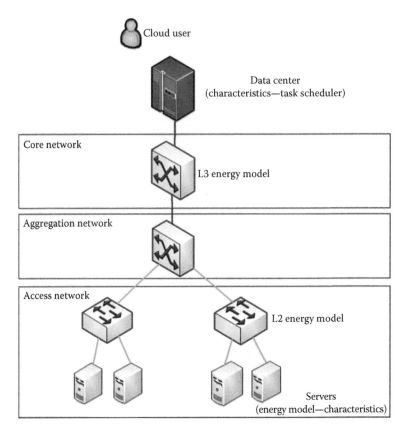

Cloud user

Data center
(characteristics—task scheduler)

Core network

L3 energy model

Aggregation network

Access network

L2 energy model

Servers
(energy model—characteristics)

FIGURE 11.9 Architecture of the GreenCloud simulation environment. (Adapted from Kliazovich, D. et al., *J. Supercomput.*, 62, 1263, 2012.)

timely fashion. The number of switches installed depends on the data center scenario. Workloads are the objects designed for universal modeling of cloud user services, such as social networking, instant messaging, and content delivery. In cloud computing, incoming requests are typically generated applications like web browsing, instant messaging, or various content delivery applications. In GreenCloud, three different types of jobs are considered:

- *Computationally intensive workloads (CIWs)*: To model HPC applications aiming at solving advanced computational problems
- *Data-intensive workloads (DIWs)*: To produce almost no load at the computing servers but require heavy data transfers
- *Balanced workloads (BWs)*: To model the applications having both computing and data transfer requirements

The execution of each workload object in GreenCloud requires a successful completion of its two components: (1) computing and (2) communication.

11.4.2.4 iCanCloud

iCanCloud is a simulation platform for large storage networks cloud infrastructures (thousands of nodes) that has special characteristics such as usability, flexibility, performance, and scalability [36–38]. iCanCloud is initiated in 2010 as an open source and is developed based on SIMCAN [39], a modular simulation platform that can be configured for modeling a wide range of HPC architectures. The main objective of iCanCloud is to predict the trade-offs between cost and performance of a given set of applications executed in a specific hardware and then provide users useful information about such costs. iCanCloud can be used by a wide range of users, from basic active users to developers of large distributed applications.

The main features of iCanCloud are as follows [36–38]:

- It supports a wide variety of cloud computing architectures.
- It supports GUI for designing and running the experiments but not for modeling the software system.
- It implements a pay-as-you-go method as a policy of charging users.
- It supports parallel processing: run one experiment is run over various machines in parallel.
- Customizable VMs can be used to quickly simulate unicore/multicore systems.
- A flexible cloud hypervisor module provides an easy method for integrating and testing both new and existent cloud brokering policies.
- It provides a wide range of configurations for storage systems, which include models for local storage systems, remote storage systems, like NFS, and parallel storage systems, like parallel file systems and RAID systems.
- New components can be added to the repository of iCanCloud to increase the functionality of the simulation platform.

iCanCloud is developed based on various platforms: OMNeT++, MPI, and C++. iCanCloud provides a set of modules that are grouped by functionality to simulate the performance of specific components. The basic design of iCanCloud is illustrated in Figure 11.10. iCanCloud comprises four main layers: VMs repository, application repository, cloud hypervisor, and cloud system. The bottom of the architecture consists of the hardware models layer. Above the hardware layer sits a VM repository. This repository contains a collection of VMs previously defined by the user. A module to deal with all incoming jobs and the instances of VMs where those jobs are executed is located at cloud hypervisor layer. Finally, at the top of the architecture is the cloud system module to define the entire cloud system, which basically consists of the definition of hypervisor, and the definition of each VM that comprises the system.

iCanCloud models the simulated cloud computing scenarios using a set of existent components to simulate the behavior of real components such as storages, memories, networks, and file systems. Those components are hierarchically organized within the repository of iCanCloud, which form the

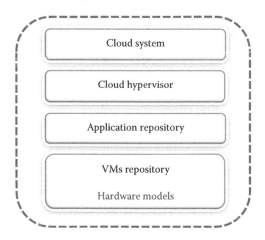

FIGURE 11.10 iCanCloud architecture. (Adapted from Nuñez, A. et al., Design of a new cloud computing simulation platform, in *Computational Science and Its Applications—ICCSA 2011*, vol. 6784, B. Murgante, O. Gervasi, A. Iglesias, D. Taniar, and B. Apduhan, eds., Springer, Berlin, Germany, 2011, pp. 582–593.)

core simulation engine. New components can be added to its repository if necessary. Furthermore, iCanCloud also allows an easy substitution of components for a particular component. Those interchangeable components can differ in the level of detail (to make performance versus accuracy trade-offs), in the functional behavior of the component, or both.

11.4.2.5 MDCSim

Developed in 2009, MDCSim is a comprehensive, flexible, and scalable simulation platform for in-depth analysis of multitier data centers [40]. It is equipped with specific hardware characteristics of data center components such as servers, switches from various vendors, and communication links in order to mimic real-world data centers. Furthermore, MDCSim allows capturing all the important design details of the underlying communication paradigm, kernel level scheduling artifacts, and the application level interactions among the tiers of a three-tier data center [40]. MDCSim is flexible in implementing different design scenarios in the three layers and analyzing the performance and power consumption of the design setup. It is also scalable in analyzing various design scenarios of data centers.

MDCSim has been validated by means of building a mini testbed of three-tier Linux-cluster-based data center connected with InfiniBand Architecture (IBA). MDCSim shows very close scores in estimating throughput, response time, and power consumption [40]. Three further studies are considered:

- To validate the advantages of cluster-based design servers by comparing the IBA and 10-gigabit Ethernet (10 GbE) under different traffic conditions and cluster specifications
- To measure and identify the characterization of power consumption across the servers of a three-tier data center
- To obtain the best optimization for different servers: WS, application server (AS), and database server (DB)

The three-layer architecture of MDCSim is shown in Figure 11.11. The simulation is configured into three layers (a communication layer, a kernel layer, and a user-level layer) in order to model and mimic real-world stack from the communication protocols to the applications [40].

11.4.2.6 SPECI

Simulation Program for Elastic Cloud Infrastructures (SPECI) is a DES. SPECI is developed on SimKit [41] and uses an existing package for DES in Java in order not to reimplement common features of DES. In order to realize DES, the clock with a queue of events is employed by the simulator, each of which is associated with a start time. The computation of the event then takes place with

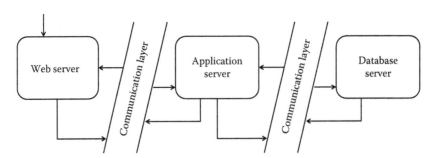

FIGURE 11.11 Three-layer architecture of the MDCSim platform. (Adapted from Seung-Hwan, L. et al., MDCSim: A multi-tier data center simulation, platform, in *Cluster Computing and Workshops, 2009. CLUSTER'09. IEEE International Conference on*, New Orleans, LA, 2009, pp. 1–9.)

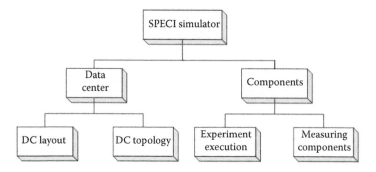

FIGURE 11.12   SPECI architecture.

duration of zero time interval. When the computation of the event is completed, the clock moves forward to the time of the next event in the schedule. Another benefit of using SimKit is it also supports various distributions for random-number generation [42].

SPECI is a simulation tool that allows exploration and analysis of aspects of scaling as well as performance properties of large data centers. SPECI is composed of two packages, that is, the data center layout and topology and the experiment execution and measuring components, which are stored in other package [42]. Event scheduling and random distribution drawing are managed in the experiment part of the simulator that is based on SimKit. The architecture of SPECI is illustrated in Figure 11.12.

11.4.3   OTHER SIMULATION TOOLS

11.4.3.1   GSSIM

The GSSIM [43], grid scheduling simulator, is based on GridSim and SimJava2 simulators. It provides a layer on top of the GridSim that enables the investigation of scheduling policies with a flexible description of the architecture and interactions between modules. GSSIM also offers an advanced generator module using real and unreal workloads. One of the main characteristics of GSSIM is the flexibility to simulate a wide range of scheduling algorithms in varied grid infrastructures and at multilevel by providing an easy-to-use grid scheduling framework. GSSIM is developed with two main objectives: to offer capabilities to generate workloads, resources, and events and to model different grid levels, that is, resource brokers, and local level scheduling systems. The workloads used are compliant with known workload formats such as standard workload format (SWF) and grid workload format (GWF). The overall architecture of GSSIM is presented in Figure 11.13.

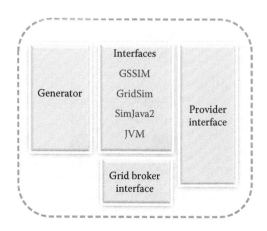

FIGURE 11.13   Overall architecture of GSSIM. (Adapted from Bąk, S. et al., *Sci. Program.*, 19, 231, 2011.)

11.4.3.2 Open Cirrus

Open Cirrus is a cloud computing testbed for the research community that, unlike existing alternatives, federates heterogeneous distributed data centers. It is a platform for real-world applications and services that aims to spur innovation in systems and applications research and to analyze the development of an open-source service stack for cloud computing [44]. Open Cirrus testbed is a joint project sponsored by HP Labs, Intel, and Yahoo! in collaboration with many institutions. Open Cirrus provides a cloud stack consisting of physical machines and VMs and global services, such as single sign-on, monitoring, storage, and job submission [45].

Open Cirrus has four objectives: foster system-level research in cloud computing, encourage new cloud computing applications and applications-level research, offer a collection of experimental data, and develop an open-source stacks and APIs for the cloud [44].

11.4.3.3 Open Cloud Testbed

In 2009 [46], Grossman et al. proposed Open Cloud Testbed (OCT) as a smaller-scale testbed with main objectives to benchmark different cloud computing systems and to study their interoperability. The main characteristics of OCT are as follows:

- Different cloud systems and services are installed, including Eucalyptus, Hadoop, CloudStore (KosmosFS), Sector/Sphere, and Thrift in order to make OCT easier to study interoperability and benchmarking, to develop network libraries, monitoring systems, and benchmark suites to support the development and experimental studies of cloud computing stacks.
- The architecture of OCT includes high-performance protocols, services, and infrastructure at all levels. It uses a high-performance 10 Gb/s network based on extremely fast transport protocols supported by dedicated light paths instead of using the Internet.

The architecture of OCT is presented in Figure 11.14. It can be seen from the figure that there are four racks of servers in the OCT located in four data centers locations: Johns Hopkins University (Baltimore), StarLight (Chicago), the University of Illinois (Chicago), and the University of California (San Diego). Each rack has 32 nodes. Each node has dual dual-core AMD 2.4 GHz CPU,

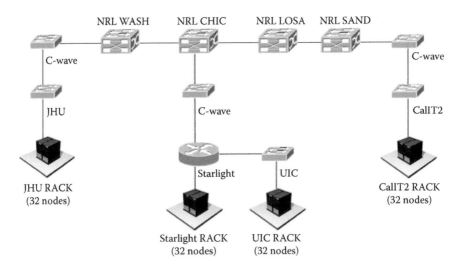

FIGURE 11.14 OCT architecture. (Adapted from Grossman, R. et al., The open cloud testbed: A wide area testbed for cloud computing utilizing high performance network services, in *GridNets*, Athens, Greece, 2009.)

12 GB memory, 1 TB single SATA disk, and dual 1 GE NICs. Two Cisco 3750E switches connect the 32 nodes, which then connects to the outside by a 10 Gb/s uplink.

11.4.3.4 PerfCloud

PerfCloud is a cloud environment built on the top of a GRID system, which allows the user to instantiate virtual clusters (VCs) that become part of the starting GRID. It is a cloud computing simulation framework that provides performance prediction services in an e-science cloud. PerfCloud relies on a set of grid services that are able to create a VC and to predict the performance of a given target application on that VC taking into account the actual amount of computing and communication resources allocated to the VC, as well as the presence of the virtualization layer. The obtained predictions allow the user to evaluate on-the-fly whether the created VC is compatible with the user's performance expectations or not [47].

PerfCloud provides three subsets of services, which manage and obtain the parameters for the time needed for simulation, to predict its performance and to execute the actual application. They are (1) Configuration services, which are used to design the characteristics of VC; (2) VC Management and Evaluation services, which are used to let users create a VC; and (3) Performance Prediction and Application Execution services, which are used to run the simulation [47]. The PerfCloud architecture is illustrated in Figure 11.15. PerfCloud client resides on a user machine (which has access to the GRID environment) and interacts with the PerfCloud system through invocation of GRID services, which enable the user to obtain a new cluster to which he has full access.

11.4.3.5 GroudSim

GroudSim [48] is a grid and cloud simulation toolkit for scientific applications based on a scalable simulation-independent discrete-event core. GroudSim is developed in Java environment and

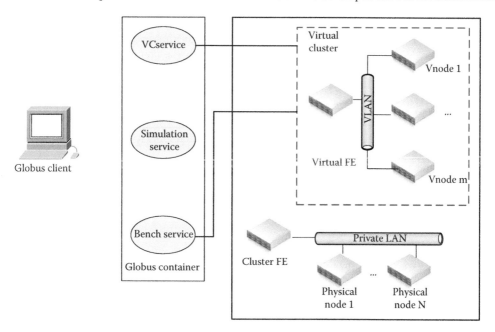

FIGURE 11.15 PerfCloud architecture. (Adapted from Mancini, E.P. et al., PerfCloud: GRID services for performance-oriented development of cloud computing applications, in *Enabling Technologies: Infrastructures for Collaborative Enterprises, 2009. WETICE'09. 18th IEEE International Workshops on*, Groningen, the Netherlands, 2009, pp. 201–206.)

can simulate both cloud and grid computing. GroudSim provides a comprehensive set of features for complex simulation scenarios from simple job execution on leased computing resources to calculation of costs and background load on resources. Simulations can be parameterized and is easily extended by probability distribution packages for failures that normally occur in complex environments. GroudSim focuses on IaaS but is easily extendable to support additional models like cloud storage or platform as a service [48].

One of the main characteristics of GroudSim is the definition of failures, that is, failures can be produced in a defined interval for a specific registered resource. Like CloudSim, GroudSim is also a DES. GroudSim also has GroudJobs, which is functionally similar to cloudlets in CloudSim. However, unlike CloudSim, which employs process-based approach that runs a separate thread for each entity, GroudSim requires one simulation thread only (instead of one thread per entity). The GroudSim framework is integrated as a back end in the ASKALON Grid computing environment, which performs both real and simulated execution of real-world applications using the same integrated development, monitoring, and analysis interface [48].

11.4.3.6 SimGrid

SimGrid was initiated in 1999 for the study of scheduling algorithms on varied platforms. SimGrid framework provides basic functionality for the simulation of distributed applications in varied distributed environments [49]. SimGrid is a simulation-based framework developed for evaluating cluster, grid, and peer-to-peer (P2P) algorithms and heuristics. It can be used to evaluate heuristics, to prototype applications, or even to assess legacy MPI applications. SimGrid has many versions with various new features in each. In SimGrid v3.10 release, preliminary DVFS support to track the energy consumption is included, which is very useful to evaluate the energy consumption in a data center.

SimGrid is considered an active simulator in terms of research and development and is supported by a large number of user communities. SimGrid provides ready-to-use models and API to simulate many different distributed systems: clusters, wide-area and local-area networks, peers over DSL connections, data centers, etc. The main components of SimGrid are two APIs: grid reality and simulation (GRAS) and SMPI for researchers who study algorithm and need to model simulations quickly and for developers who can develop applications in the comfort of the simulated world before deploying them seamlessly into the real world.

11.5 SUMMARY

The chapter surveys significant works in the field of simulation tools for cloud computing systems. Table 11.1 gives a summary.

Simulation of cloud computing may develop in three strands:

1. Cloud network connections
2. Cloud dynamic resource allocations
3. Cloud service interactions

CloudSim and variants are mostly concerned with resource allocations, while some virtual data center simulators are concerned with network connections; however, there seem to be very few concerned with cloud service interactions.

There are still a lot of rooms for further research in this hot field. While most of the approaches build their simulation based on DES, the use of packet-level simulation is encouraged to achieve more accurate results. The rapid development in CPU technology, processor speed, etc., will reduce time in packet-level simulation significantly.

TABLE 11.1

Summary of Cloud Computing Simulation Tools

Simulator	Characteristics and Limitations	Programming Language	Platform	Open Source/ Commercial
CloudSim [6,21]	• Discrete-event simulation • Comprehensive • Complex scheduling support • No TCP/IP support • Limited communication model (only transmission delay and bandwidth) • Limited GUI support (via CloudAnalyst) • No MPI and workflow applications support • No energy model	Java	GridSim	Open source (http://www. cloudbus.org/ cloudsim/)
CloudAnalyst [23]	• Simulation of large-scale cloud applications • Easy to repeat simulations and to change different parameters • Easy to read output (graphical output) • GUI support • Flexible: easy to combine use with other technologies	Java	CloudSim, SimJava	Open source (http://www. cloudbus.org/ cloudsim/ CloudAnalyst. zip)
CDOSim [22]	• Simulation of response times, SLA violations, rating, and costs of a CDO • No TCP/IP support	Java	CloudSim	
DCSim [50]	• Focused on a virtualized data center providing IaaS to multiple tenants • Event-driven simulation • No TCP/IP support • Developed for transactional, continuous workloads • Model of replicated VMs	Java		Open source (https://github. com/digs-uwo/ dcsim)
EMUSIM [7]	• Implementation of both emulation and simulation • Assessment of system and application • Improve simulation's accuracy	Java	AEF, CloudSim	Open source (http://www. cloudbus.org/ cloudsim/ emusim/)
GDCSim [31]	• Integrated simulation: physical data center description and a management technique description ∘ Online analysis • Modular and extensible	C++, XML	BlueTool	Open source (https:// impact.asu. edu/BlueTool/ wiki/index. php/BlueTool)
GreenCloud [32,33]	• Packet-level simulation • Suitable for modeling data center • Implementation of a full TCP/IP protocol reference model • Implementation of full communication model • Limited GUI support (via Nam) • Precise energy model (servers + network)	C++ and OTcl	ns-2	Open source (http:// greencloud. gforge.uni.lu/)
GroudSim [48]	• Discrete-event simulation • Simulation of clouds and grids • Provides a comprehensive set of features for complex simulation scenarios	Java		

(Continued)

TABLE 11.1 (*Continued*)
Summary of Cloud Computing Simulation Tools

Simulator	Characteristics and Limitations	Programming Language	Platform	Open Source/ Commercial
GSSIM [43]	• An advanced generator module using real and unreal workloads • Workloads compliant with known workload formats such as standard workload format (SWF) and grid workload format (GWF) • Easy-to-use grid scheduler • Support to generation of resource failures	Java	GridSim and SimJava2	Open source (http://www. gssim.org/)
iCanCloud [37,38]	• Design for simulate large storage networks • Implementation of detailed modeling of communication aspects of data center • GUI supports • Support to parallel processing • Easy to add new modules	OMNeT, MPI, and C++	SIMCAN [39]	Open source (http://www. arcos.inf. uc3m. es/~icancloud/ downloads. html)
MDCSim [40]	• No GUI support • Limited communication model (only transmission delay and bandwidth) • No TCP/IP support • Rough energy model (servers only)	C++/Java	CSIM	Commercial
NetworkCloudSim [24]	• Discrete-event simulation • Scalability, flexibility, and reliability • Limited support for MPI and workflow applications	Java	CloudSim	Open source (http://www. cloudbus.org/ cloudsim/)
Open Cloud Testbed (OCT) [51]	• Geographically distributed cloud testbed spanning four data centers • Connected with 10G and 100G network connections • Different cloud systems and services installed • Use to develop new cloud computing software and infrastructure • Focused: developing an OpenFlow-enabled version of Hadoop			Limited (registration is needed)
Open Cirrus [44]	• A platform for real-world applications and services • To encourage new cloud computing applications and applications-level research			Open source (http:// opencirrus. intel-research. net/)
PerfCloud [47]	• Virtual clusters (VCs) created on user request • Performance prediction of user applications on the newly generated VC in real time	MetaPL	GRID system	
SimGrid [49]	• Evaluate cluster, grid, and P2P algorithms and heuristics • Employs a modular simulation kernel • Provides ready-to-use models and API to simulate many different distributed systems • Provides a *libvirt-like* interface for cloud simulation	C or Java		Open source (http:// simgrid. gforge.inria. fr/)

(*Continued*)

TABLE 11.1 (*Continued*)
Summary of Cloud Computing Simulation Tools

Simulator	Characteristics and Limitations	Programming Language	Platform	Open Source/ Commercial
SPECI [42]	• Discrete-event simulation • Analyze the performance properties of large data centers	Java	SimKit	
TeachCloud [25]	• GUI support • Cloud system is reconfigurable many times • Easy to add new modules • No TCP/IP support	Java	CloudSim	Open source (https://code. google.com/p/ teachcloud/)

Often issues are too complex and too sophisticated to be investigated by a single simulation tool. Therefore, co-simulation or hybrid simulation is sometimes a necessity. Co-simulation (or cooperative simulation) is a simulation mechanism that allows individual components to be simulated by different simulation tools running simultaneously and exchanging information in a collaborative fashion.

REFERENCES

1. Amazon. (2013). Amazon Web Services EC2 and S3. http://aws.amazon.com/ (accessed January 31, 2014).
2. D. Nurmi, R. Wolski, C. Grzegorczyk, G. Obertelli, S. Soman, L. Youseff et al., The Eucalyptus open-source cloud-computing system, in *Cluster Computing and the Grid, 2009. CCGRID'09. Ninth IEEE/ACM International Symposium on*, Shanghai, China, 2009, pp. 124–131.
3. Hadoop. (2013). Hadoop. http://hadoop.apache.org/core (accessed January 31, 2014).
4. Sector/Sphere. (2013). http://sector.sourceforge.net/ (accessed January 31, 2014).
5. Y. Gu and R. L. Grossman, Sector and Sphere: The design and implementation of a high-performance data cloud, *Philosophical Transactions of the Royal Society A: Mathematical, Physical and Engineering Sciences*, 367, 2429–2445, 2009.
6. R. N. Calheiros, R. Ranjan, A. Beloglazov, C. A. F. De Rose, and R. Buyya, CloudSim: A toolkit for modeling and simulation of cloud computing environments and evaluation of resource provisioning algorithms, *Software: Practice and Experience*, 41, 23–50, 2011.
7. R. N. Calheiros, M. A. Netto, C. A. De Rose, and R. Buyya, EMUSIM: An integrated emulation and simulation environment for modeling, evaluation, and validation of performance of cloud computing applications, *Software: Practice and Experience*, 00, 1–18, 2012.
8. R. Jain, *The Art of Computer Systems Performance Analysis*, John Wiley & Sons, New York, 1991.
9. L. Wang, G. Laszewski, A. Younge, X. He, M. Kunze, J. Tao et al., Cloud computing: A perspective study, *New Generation Computing*, 28, 137–146, 2010.
10. R. Buyya, C. S. Yeo, S. Venugopal, J. Broberg, and I. Brandic, Cloud computing and emerging IT platforms: Vision, hype, and reality for delivering computing as the 5th utility, *Future Generation Computer Systems*, 25, 599–616, 2009.
11. V. Sarathy, P. Narayan, and R. Mikkilineni, Next generation cloud computing architecture: Enabling real-time dynamism for shared distributed physical infrastructure, in *Enabling Technologies: Infrastructures for Collaborative Enterprises (WETICE), 2010 19th IEEE International Workshop on*, Larissa, Greece, 2010, pp. 48–53.
12. G. J. Popek and R. P. Goldberg, Formal requirements for virtualizable third generation architectures, *Communications of the ACM*, 17, 412–421, 1974.
13. P. Barham, B. Dragovic, K. Fraser, S. Hand, T. Harris, A. Ho et al., Xen and the art of virtualization, *Operating Systems Review (ACM)*, 37, 164–177, 2003.
14. H. Umeno, M. L. C. Parayno, K. Teramoto, M. Kawano, H. Inamasu, S. Enoki et al., Performance evaluation on server consolidation using virtual machines, in *International Joint Conference SICE-ICASE*, Busan, Korea, 2006, pp. 2730–2734.

15. B. Rochwerger, D. Breitgand, E. Levy, A. Galis, K. Nagin, I. M. Llorente et al., The reservoir model and architecture for open federated cloud computing, *IBM Journal of Research and Development*, 53, 4:1–4:11, 2009.

16. D. Hilley, Cloud computing: A taxonomy of platform and infrastructure-level offerings. Technical Report GIT-CERCS-09-13, CERCS, Georgia Institute of Technology, Atlanta, GA, 2009.

17. L. George, Powering down the computing infrastructure, *Computer*, 40, 16–19, 2007.

18. J. Torres, D. Carrera, K. Hogan, R. Gavalda, V. Beltran, and N. Poggi, Reducing wasted resources to help achieve green data centers, in *Parallel and Distributed Processing, 2008. IPDPS 2008. IEEE International Symposium on*, Miami, FL, 2008, pp. 1–8.

19. J. Torres, D. Carrera, V. Beltran, N. Poggi, K. Hogan, J. L. Berral et al., Tailoring resources: The energy efficient consolidation strategy goes beyond virtualization, in *Autonomic Computing, 2008. ICAC'08. International Conference on*, Chicago, IL, 2008, pp. 197–198.

20. NIST, NIST SP—500-292: NIST cloud computing reference architecture, The National Institute of Standards and Technology (NIST), Gaithersburg, MD, 2011.

21. R. Buyya, R. Ranjan, and R. N. Calheiros, Modeling and simulation of scalable cloud computing environments and the CloudSim toolkit: Challenges and opportunities, in *High Performance Computing & Simulation, 2009. HPCS'09. International Conference on*, Leipzig, Germany, 2009, pp. 1–11.

22. F. Fittkau, S. Frey, and W. Hasselbring, CDOSim: Simulating cloud deployment options for software migration support, in *Maintenance and Evolution of Service-Oriented and Cloud-Based Systems (MESOCA), 2012 IEEE Sixth International Workshop on the*, Trento, Italy, 2012, pp. 37–46.

23. B. Wickremasinghe, R. N. Calheiros, and R. Buyya, CloudAnalyst: A CloudSim-based visual modeller for analysing cloud computing environments and applications, in *Advanced Information Networking and Applications (AINA), 2010 24th IEEE International Conference on*, Perth, Western Australia, Australia, 2010, pp. 446–452.

24. S. K. Garg and R. Buyya, NetworkCloudSim: Modelling parallel applications in cloud simulations, in *Fourth IEEE International Conference on Utility and Cloud Computing (UCC), 2011*, Melbourne, Victoria, Australia, 2011, pp. 105–113.

25. Y. Jararweh, Z. Alshara, M. Jarrah, M. Kharbutli, and M. Alsaleh, Teachcloud: A cloud computing educational toolkit, in *Proceedings of the First International IBM Cloud Academy Conference (ICA CON 2012), IBM*, Research Triangle Park, NC, 2012.

26. J. Werner, G. Geronimo, C. Westphall, F. Koch, and R. Freitas, Simulator improvements to validate the Green Cloud Computing approach, in *Network Operations and Management Symposium (LANOMS), 2011 Seventh Latin American*, Quito, Ecuador, 2011, pp. 1–8.

27. S. Frey, W. Hasselbring, and B. Schnoor, Automatic conformance checking for migrating software systems to cloud infrastructures and platforms, *Journal of Software: Evolution and Process*, 25, 1089–1115, 2013.

28. H. Casanova, Network modeling issues for grid application scheduling, *International Journal of Foundations of Computer Science*, 16, 145–162, 2005.

29. M. Tighe, G. Keller, M. Bauer, and H. Lutfiyya, DCSim: A data centre simulation tool for evaluating dynamic virtualized resource management, in *Network and Service Management (CNSM), 2012 Eighth International Conference and 2012 Workshop on Systems Virtualization Management (SVM)*, Las Vegas, NV, 2012, pp. 385–392.

30. DCSim. (December 13, 2013). *Distributed and Grid Systems (DiGS) Research Group*, Western University, London, U.K. Available: http://digs.csd.uwo.ca/

31. S. K. S. Gupta, R. R. Gilbert, A. Banerjee, Z. Abbasi, T. Mukherjee, and G. Varsamopoulos, GDCSim: A tool for analyzing Green Data Center design and resource management techniques, in *Green Computing Conference and Workshops (IGCC), 2011 International*, Orlando, FL, 2011, pp. 1–8.

32. D. Kliazovich, P. Bouvry, and S. Khan, GreenCloud: A packet-level simulator of energy-aware cloud computing data centers, *The Journal of Supercomputing*, 62, 1263–1283, 2012.

33. D. Kliazovich, P. Bouvry, Y. Audzevich, and S. U. Khan, GreenCloud: A packet-level simulator of energy-aware cloud computing data centers, in *GLOBECOM 2010, 2010 IEEE Global Telecommunications Conference*, Miami, FL, 2010, pp. 1–5.

34. ns-2. (October 12, 2013). The Network Simulator—ns-2. Available: http://www.isi.edu/nsnam/ns/.

35. S. Li, P. Li-Shiuan, and N. K. Jha, Dynamic voltage scaling with links for power optimization of interconnection networks, in *High-Performance Computer Architecture, 2003. HPCA-9 2003. Proceedings. The Ninth International Symposium on*, Anaheim, CA, 2003, pp. 91–102.

36. A. Nuñez, J. L. Vázquez-Poletti, A. C. Caminero, J. Carretero, and I. M. Llorente, Design of a new cloud computing simulation platform, in *Computational Science and Its Applications—ICCSA 2011*, vol. 6784, B. Murgante, O. Gervasi, A. Iglesias, D. Taniar, and B. Apduhan, eds., Springer, Berlin, Germany, 2011, pp. 582–593.

37. A. Nunez, G. G. Castane, J. L. Vazquez-Poletti, A. C. Caminero, J. Carretero, and I. M. Llorente, Design of a flexible and scalable hypervisor module for simulating cloud computing environments, in *Performance Evaluation of Computer & Telecommunication Systems (SPECTS), 2011 International Symposium on*, The Hague, the Netherlands, 2011, pp. 265–270.

38. A. Núñez, J. L. Vázquez-Poletti, A. C. Caminero, G. G. Castañé, J. Carretero, and I. M. Llorente, iCanCloud: A flexible and scalable cloud infrastructure simulator, *Journal of Grid Computing*, 10, 185–209, 2012.

39. SIMCAN. (December 13, 2013). SIMCAN. Available: http://www.arcos.inf.uc3m.es/~simcan/.

40. L. Seung-Hwan, B. Sharma, N. Gunwoo, K. Eun Kyoung, and C. R. Das, MDCSim: A multi-tier data center simulation, platform, in *Cluster Computing and Workshops, 2009. CLUSTER'09. IEEE International Conference on*, New Orleans, LA, 2009, pp. 1–9.

41. A. Buss, Component based simulation modeling with Simkit, in *Simulation Conference, 2002. Proceedings of the Winter*, San Diego, CA, vol. 1, 2002, pp. 243–249.

42. I. Sriram, SPECI, a simulation tool exploring cloud-scale data centres, in *Cloud Computing*, vol. 5931, M. Jaatun, G. Zhao, and C. Rong, eds., Springer, Berlin, Germany, 2009, pp. 381–392.

43. S. Bąk, M. Krystek, K. Kurowski, A. Oleksiak, W. Piątek, and J. Wąglarz, GSSIM—A tool for distributed computing experiments, *Scientific Programming*, 19, 231–251, 2011.

44. A. I. Avetisyan, R. Campbell, I. Gupta, M. T. Heath, S. Y. Ko, G. R. Ganger et al., Open Cirrus: A global cloud computing testbed, *Computer*, 43, 35–43, 2010.

45. R. Campbell, I. Gupta, M. Heath, S. Y. Ko, M. Kozuch, M. Kunze et al., Open Cirrus™ cloud computing testbed: Federated data centers for open source systems and services research, in *Proceedings of the 2009 Conference on Hot Topics in Cloud Computing*, San Diego, CA, 2009, pp. 1–1.

46. R. Grossman, Y. Gu, M. Sabala, C. Bennet, J. Seidman, and J. Mambratti, The open cloud testbed: A wide area testbed for cloud computing utilizing high performance network services, in *GridNets*, Athens, Greece, 2009.

47. E. P. Mancini, M. Rak, and U. Villano, PerfCloud: GRID services for performance-oriented development of cloud computing applications, in *Enabling Technologies: Infrastructures for Collaborative Enterprises, 2009. WETICE'09. 18th IEEE International Workshops on*, Groningen, the Netherlands, 2009, pp. 201–206.

48. S. Ostermann, K. Plankensteiner, R. Prodan, and T. Fahringer, GroudSim: An event-based simulation framework for computational grids and clouds, in *Euro-Par 2010 Parallel Processing Workshops*, vol. 6586, M. Guarracino, F. Vivien, J. Träff, M. Cannatoro, M. Danelutto, A. Hast et al., eds., Springer, Berlin, Germany, 2011, pp. 305–313.

49. H. Casanova, A. Legrand, and M. Quinson, SimGrid: A generic framework for large-scale distributed experiments, in *Computer Modeling and Simulation, 2008. UKSIM 2008. 10th International Conference on*, Cambridge, U.K., 2008, pp. 126–131.

50. G. Keller, M. Tighe, H. Lutfiyya, and M. Bauer, DCSim: A data centre simulation tool, in *Integrated Network Management (IM 2013), 2013 IFIP/IEEE International Symposium on*, Ghent, Belgium, 2013, pp. 1090–1091.

51. The Open Cloud Testbed. (2013). http://www.opencloudconsortium.org (accessed January 31, 2014).

12 Simulation Tools for Broadband Passive Optical Networks

Rastislav Róka and Filip Čertík

CONTENTS

ABSTRACT

With the emerging applications and needs of ever-increasing bandwidth, it is anticipated that the next-generation broadband passive optical network with much higher bandwidth is a natural path forward to satisfy these demands. For network operators, it is therefore interesting and profitable to develop this scenario of valuable access networks. Broadband passive optical networks utilize the optical transmission medium, and for increasing transmission rates of broadband applications and services, it is necessary to characterize and investigate some specific features of new optical signal processing techniques at the signal transmission. Furthermore, there also exist various architectures and optical infrastructures for possible developing of broadband passive optical networks. Also in this case, they must be specified with emphasis on their characteristics and possible implementations in real access networks.

In this chapter, simulation tools for investigating new optical signal processing techniques in the optical transmission medium and for analyzing considered optical infrastructures in broadband passive optical networks are presented together with possibilities of their practical implementations in real passive optical networks.

12.1　INTRODUCTION

For new multimedia broadband services and advanced modern applications provisioned to both residential and business customers, a reliable and safety transmission medium—the optical fiber—must be used. One of the prominent access technologies is the broadband passive optical network (PON). Architectures of broadband PONs must be reliable and cost effective; therefore, no switching and control elements in the optical distribution network are preferable against active ones. Moreover, optical network terminals on the subscriber side must be cheap and simple. The optical line terminal on the central office side can be more sophisticated because it represents the main managing equipment of broadband PONs.

The general PON is a bidirectional point-to-multipoint system that contains passive (optical fibers, passive optical splitters, couplers, connectors) optical elements in the distribution part of the access network and active (the OLT terminal, multiple ONT terminals) optical elements placed in end terminating points of the access network. The transmission path between optical transmitters and receivers is presented by single-mode optical fibers. It can be divided into the three main parts:

1. A transmitting part is responsible for generating optical signals using the laser and for modulating generated signals according to relevant input data into a form suitable for transmitting in the environment of the optical fiber.
2. A transmission channel represents the environment of optical transmission media with linear and nonlinear effects with negative influences on the transmitted signal. Main attention is focused on the attenuation, various types of the dispersion, and different nonlinearities. Because these negative influences expressively interfere into the communication and represent its main limiting factors, they present a critical part of the simulation model and, therefore, it is necessary exactly to recognize and express their characteristics by correct transmission parameters.
3. A receiving part is responsible for demodulating received optical signals and for calculating of the bit error rate/ratio (BER). Moreover, graphical presentations (eye diagrams, signal constellation diagrams) of optical signals in any part of the transmission path can be utilized.

12.2　SIMULATION TOOL FOR NEW OPTICAL SIGNAL PROCESSING TECHNIQUES

For successful understanding of the signal transmission in optical access networks that utilized fixed transmission media, it is necessary exactly to recognize essential negative influences in the real environment of optical fibers [1,2]. Main attention of the optical transmission media's part is focused on the description of the proposed optical fiber's simulation model and on the explanation of simulation methods for its substantial linear and nonlinear effects. The simulation model represents a reach enough knowledgebase that can be helpful for various tests and performance comparisons of various advanced modulation and encoding techniques.

For modeling of the optical transmission path, we can use the software program MATLAB® 2010 Simulink® together with additional libraries. The realized model represents the signal transmission in the environment utilizing optical fibers for very high-speed data signals in both directions. The whole program is controlled by a main screen, where a user is able to perform adequate operations with only basic knowledge about optical fibers. A program has two main functions (calculation and simulation) representing two independent systems. The calculation part is used for calculating linear and nonlinear effects with inserted optical fiber parameters. The simulation part simulates signal transmissions in the optical transmission path under different modulation and encoding techniques using the Communication Blockset and Communication Toolbox tools. In these tools, all necessary blocks as modulators/demodulators, generators, blocks with operation functions, and scope blocks are created and utilized.

12.2.1 Characteristics of the Optical Transmission Medium

The optical fiber is a complex transmission medium intended for transmission systems that are frequency dependent. To design and simulate such an optical system, we need to describe pulse propagation and environmental influences inside this transmission medium. This pulse propagation can be described by the nonlinear Schrödinger equation (NLSE) derived from Maxwell equations. A solution of the NLSE equation is the Gaussian pulse:

$$\frac{\partial a(z,t)}{\partial z} = -\frac{\alpha}{2} a(z,t) - \beta_1 \frac{\partial a(z,t)}{\partial t} - j \frac{\beta_2}{2} \frac{\partial^2 a(z,t)}{\partial t^2} + \frac{\beta_3}{6} \frac{\partial^3 a(z,t)}{\partial t^3}$$

$$+ j\gamma |a(z,t)|^2 a(z,t) - j\gamma T_R \frac{\partial |a(z,t)|^2}{\partial t} a(z,t) - \frac{\gamma}{\omega_0} \frac{\partial |a(z,t)|^2 a(z,t)}{\partial t} \qquad (12.1)$$

where
$a(z, t)$ is the intensity process of optical signals
z is the distance
t is the time
α is the specific attenuation of optical fibers
$\beta_1, \beta_2, \beta_3$ are dispersion coefficients of the first, the second, and the third place values, respectively
γ is the nonlinear coefficient

Each term of the equation sequentially represents a linear attenuation; a dispersion of the first, the second, and the third place values; the Kerr effect; the stimulated Raman scattering (SRS); and a change of the pulse slope [3]. As we can see from the Schrödinger equation, the optical signal is changed by these effects classified as

1. Linear effects that are wavelength depended
2. Nonlinear effects that are intensity (power) dependent

12.2.1.1 Linear Effects

Major impairments of optical signals transmitted via the optical fiber are mainly caused by the dispersion and the attenuation. The attenuation represents a transmission loss that limits power levels of optical signals. The total signal attenuation a[dB] is defined for a particular wavelength that is defined by

$$a[dB] = 10 \log_{10} \frac{P_i}{P_0} \qquad (12.2)$$

where
P_i is the input power
P_0 is the output power

Nowadays, optical transmission systems are able to minimize impact of the attenuation by deploying special optical fibers with low attenuation values and/or to increase a reachable range of transmission systems with optical regenerators and amplifiers. Despite of attenuation limitations with different methods, optical transmission systems are limited by the dispersion that influences their transmission rates. Three basic types of the dispersion occur in the optical fiber:

- Modal dispersion (intermodal, multimode)
- Chromatic dispersion (CD) (intramodal)
- Polarization mode dispersion (PMD)

The modal dispersion occurs only in multimode fibers that are mainly used for very short distances; therefore, it will not be included in this chapter. The CD is caused by the fact that it is impossible to create a practically monochromatic light source. Each pulse wavelength is transmitted with a different velocity that leads to broadening and phase shifting at the optical fiber's end. The PMD is a random phenomenon that can be only statistically evaluated. The PMD occurs due to imperfections in the optical fiber shape that is not perfectly symmetrical and is changing along the fiber. Asymmetries of optical fibers can be caused by defects in the manufacture or by external effects that influence on the optical transmission medium. External effects are created by a combination of environmental effects (e.g., temperature changes, a shift in the ground) and cabling (e.g., neighboring fibers in the bundle, fiber bending) [4–8].

In general, the dispersion causes time broadening and phase shifting of optical pulses at the optical fiber's end. The relationship between time broadening and phase shifting can be described by these equations [9]:

$$t = \frac{1}{2\pi} \frac{d\varphi}{df_m} \qquad (12.3)$$

$$GVD = \frac{1}{2\pi} \frac{d^2\varphi}{df_m^2} \qquad (12.4)$$

where
t is the travel time of optical pulses through the optical fiber
φ is the signal phase shift
GVD represents the group velocity dispersion coefficient

The closer analysis of linear effects is explained in article [9].

12.2.1.2 Nonlinear Effects

The nonlinear effects play an important role at the transmission of optical signals. We can classify nonlinear effects in the following way:

- *Kerr nonlinearities* are self-induced effects, where the phase velocity of the optical pulse is depending on the pulse's own intensity. The Kerr effect describes a change in the fiber refractive index due to electrical perturbations. Due to the Kerr effect, we are able to describe the following effects:
  - Self-phase modulation (SPM)—changes the refractive index of the optical transmission medium caused by the pulse intensity.
  - Four-wave mixing (FWM)—the fourth wavelength can be arisen by mixing of three wavelengths and can appear in the same position as one from mixed wavelengths.
  - Cross-phase modulation (XPM)—the optical pulse can change the phase of another optical pulse with a different wavelength and by this way causes a spectral broadening.
- *Scattering nonlinearities* occur due to photon inelastic scattering to lower-energy photons. The pulse energy is transferred to another wave with a different wavelength. We can determine the following effects:
  - Stimulated Brillouin scattering (SBS) and SRS—change a variance of the light wave into different waves when the intensity reaches a certain threshold [4,9]

12.2.1.3 Digital Modulation Formats

The digital modulation represents the signal transformation into a form suitable for the transmission medium. We change the information signal by adding a carrier signal and therefore we get a new signal called the modulated signal that is suitable for transmission via the optical fiber. There are four basic physical attributes of optical signals that can be adapted, and therefore, we can divide digital modulations into these groups [2,10]:

- Amplitude shift keying (ASK)
- Phase shift keying (PSK)
- Polarization shift keying (PolSK)
- Duobinary modulation

The on–off keying (OOK) modulation is a two-level ASK modulation characterized by the optical emission from the optical generator for a logical "1" and the absence of the optical flow for a logical "0". The PSK modulation is characterized by a phase shift in point of changing between logical levels. The information is hidden not in the amplitude, but in the change of signal phase. The closer analysis of optical modulation techniques is explained in articles [9,11,12].

12.2.1.4 Design and Simulation of the Optical Transmission Medium

The design and simulation of the optical transmission medium is performed in MATLAB Simulink 2010 and MATLAB GUI. The simulation program starts with a main screen that control whole simulations. The main screen allows performing adequate operations to users with required only basic knowledge about characteristics of optical fibers. The MATLAB GUI is used to create a main screen with the drag and drop system. The main screen can run and control the simulation model after the calculation performed for the simulation model. As we can see in Figure 12.1, the main screen contains basic input optical fiber parameters that characterize

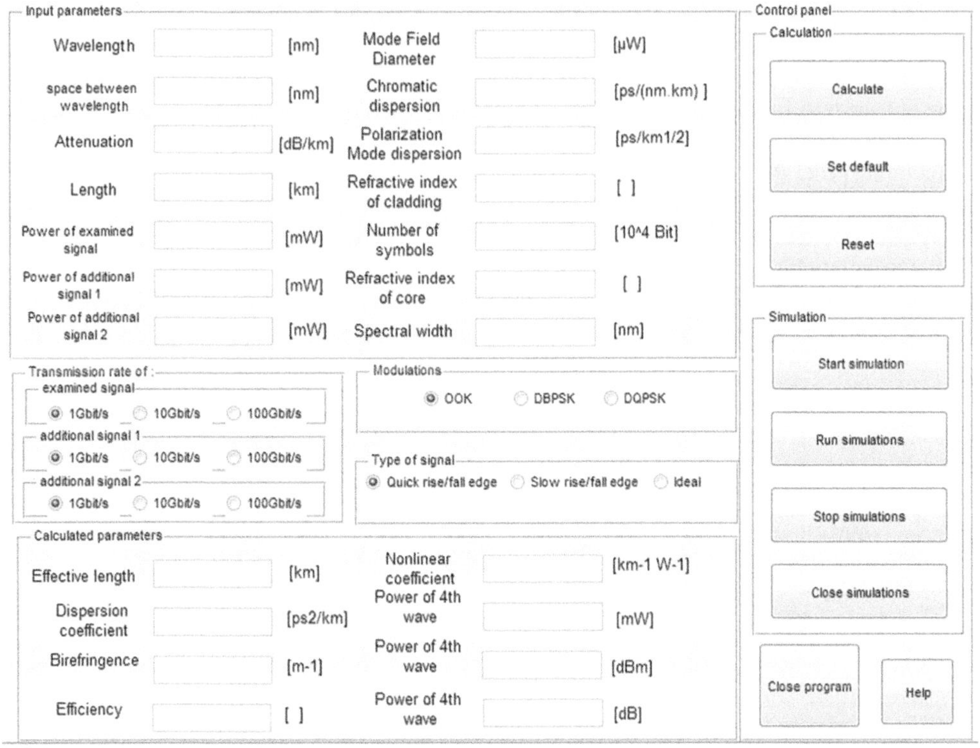

FIGURE 12.1 The main screen of the optical simulation program.

its features and can be provided by optical fiber manufactures. The transmission rate of optical signals, different modulation formats, and the shape of optical signals are included in this main screen. Our simulation program includes 1, 10, and 100 Gbit/s transmission rates. The design of optical transmission systems allows selecting from three different modulation formats and various signal types.

A simulation program has two main functions (calculation and simulation) representing two independent systems. The simulation part simulates an environment of the optical transmission medium with linear and nonlinear effects influenced on optical signals transmitted with different modulation techniques. The simulation part uses the Communication Blockset and Communication Toolbox tools to simulate transmission of optical signals through the optical transmission path. In this simulation program, except already created blocks—modulators, generators, operation functions, and scope blocks—new designed and prepared blocks to simulate some of the linear or nonlinear effects in the optical fiber are also used.

12.2.2 FEATURES OF THE OPTICAL SIMULATION PROGRAM

A design of the optical transmission system for desired digital modulation techniques is shown in Figure 12.2. Within this chapter, each block in the optical simulation program will be discussed. For demo purposes, we can consider a standard single-mode fiber (SSMF) with these system parameters: three optical source generators with the power 1 mW at wavelengths of 1550.5, 1551, and 1551.5 nm; the optical fiber length 10 km; the CD coefficient 18 ps/(nm·km); the PMD coefficient 0.4 ps/km^2; and the attenuation coefficient 0.21 dB/km.

As a bit source, we used the Bernoulli generator that generates two pulses "1" and "0". As an alternative generator, a random integer generator can be used. Additional parameters for selected generator must be inserted: a probability of zero and an initial speed that represents the system transmission rate. If the random integer generator is preferred, the number of states must be inserted. We use the interconnection between MATLAB GUI and MATLAB Simulink to insert these input parameters via the main screen. The Bernoulli-generated signal diagram with the appropriate block scheme is shown in Figure 12.3. The complete generator scheme involves the Bernoulli generator that generates signals and the discrete generator that changes samples per 1 bit.

After being generated, the ideal signal is modulated by appropriate modulations that are shown in Figures 12.4 and 12.5. Modulation blocks are included in the MATLAB Simulink Communication Blockset. As we can see from the figures, the OOK modulation changes the signal magnitude to 1 and 0. Differential binary PSK (DBPSK) and differential quadrature PSK (DQPSK) modulations have constant signal envelope and change only the signal phase; therefore, constellation diagrams

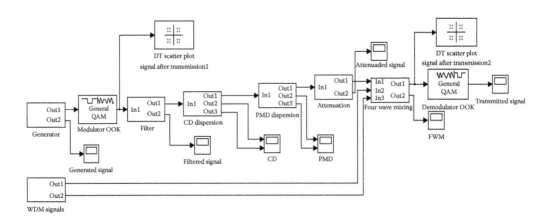

FIGURE 12.2 The block scheme of the optical transmission path with the OOK modulation.

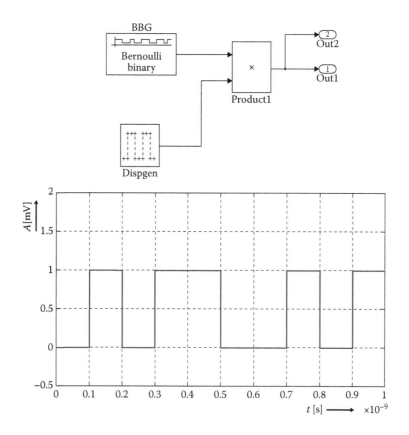

FIGURE 12.3 The ideal electrical signal generated by the Bernoulli generator and its block scheme.

are presented. The aim of this optical simulation program is to simulate a behavior of signal transmission in the environment of real optical fibers utilized in applied optical transmission systems without infinity bandwidth. Assuming a finite bandwidth, we restrict modulated signals with filtering, and the appropriate filter block scheme is shown in Figure 12.6 and the OOK filtered signal is shown in Figure 12.7. The filter block uses a discrete filter for reshaping an ideal pulse into the super-Gaussian pulse and few switch blocks for keeping optical signals in boundaries (between 0 and 1). We use super-Gaussian pulses to simulate optical pulses from directly modulated laser diodes. The magnitude of DQPSK- and DBPSK-modulated signals is exactly the same and equals to one; therefore, there is no special reason for showing it after filtering.

After signal filtering, the CD and PMD blocks are integrated. We assume that SSMFs are utilized and therefore we are not considering simulating a multimodal dispersion. CD and PMD blocks are expanding an original signal in the time domain and the phase shift occurs due to this signal broadening. The block scheme for simulating of the CD effect is shown in Figure 12.8. In both CD and PMD schemes, we use the discrete generator, where a signal magnitude is set, respectively, to the GVD value (constant for the CD effect) and the differential group delay (DGD) value (for the PMD effect). This original signal inputs to the variable delay block that delays the modulated signal by the signal magnitude values. Modulated and delayed modulated signals are brought to the disp_plane block where a new signal is created from the interference of input signals. This signal is inputted to the disp_loss block, where it is attenuated with appropriate values (the law of the energy conservation). After this block, the complex phase shift block is used, where the modulated signal is phase shifted with a signal driven from the discrete generator. The appropriate value is calculated at start of the simulation. The PMD block is very similar to the CD block and includes a dynamic generator with the normal distribution instead of a discrete

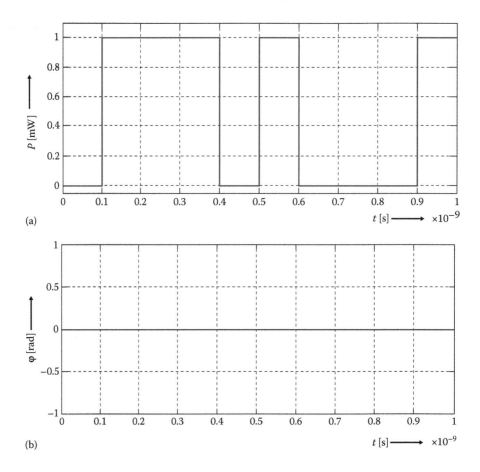

FIGURE 12.4   The ideal optical signal modulated with the OOK modulation: (a) Magnitude and (b) phase.

generator to generate the dynamic DGD parameter. Disp_plane and disp_loss blocks are programmed in s-function that uses C++ coding. First, we can demonstrate the influence of presented blocks on optical signals modulated with OOK, DBPSK, and DQPSK modulations with the 10 Gbit/s transmission system rate for the 10 km fiber length. For better signal representation of the OOK modulation, its magnitude in time is shown. For DBPSK and DQPSK modulations, constellation diagrams are presented. As we can see in Figure 12.9, the CD broadens the optical signal with the value of 180 ps and therefore causes the intersymbol interference (ISI) generation. As we mentioned before, the dispersion broadening limits must be 1/10 of a bit time; otherwise, signal boundaries are broken and a receiver is not able to recover an original form of the received signal. This is the main reason for using dispersion-shifted fibers or dispersion compensators. DBPSK and the DQPSK modulations have a constant magnitude during the whole transmission; therefore, signal broadening does not arise and only phase shift occurs as we can see in Figure 12.10.

   For demonstration of the CD compensation, we can consider an SSMF with the CD coefficient reduced to 0.1 ps/(nm·km) value and with the extended 100 km optical fiber length. If we change the fiber length from 10 km up to 100 km, the CD can broaden the optical signal by 10 ps. After CD compensation, the transmission system errors are mainly resulting from the PMD effect, where the signal broadening exceeds the limit as shown in Figure 12.11. However, DBPSK- and DQPSK-modulated signals are more influenced by the dynamic phase shifting as shown in Figure 12.12.

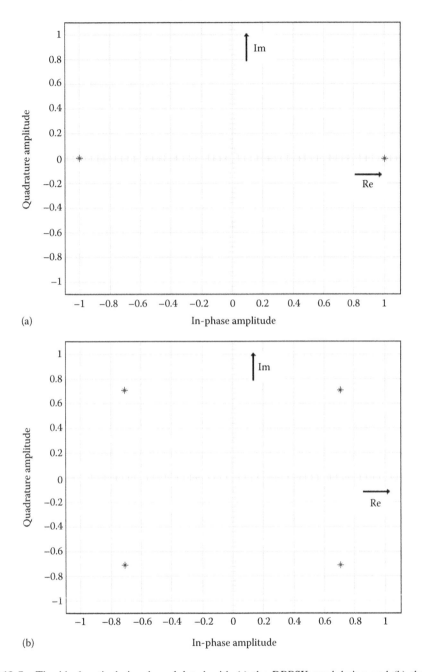

FIGURE 12.5 The ideal optical signal modulated with (a) the DBPSK modulation and (b) the DQPSK modulation.

The 10 Gbit/s transmission system with the 100 km fiber length and with the attenuation coefficient 0.21 dB/km causes the total signal attenuation on the transmission path equal to 21 dB. The attenuation block scheme and the OOK-modulated signal after attenuation is shown in Figure 12.13. We are using the dB gain block to simulate signal attenuation. First, signal attenuation is calculated in the calculation part, then an acquired value is converted into decibel units and consequently a result is inserted into the dB gain block. A negative gain value causes loss of the signal power.

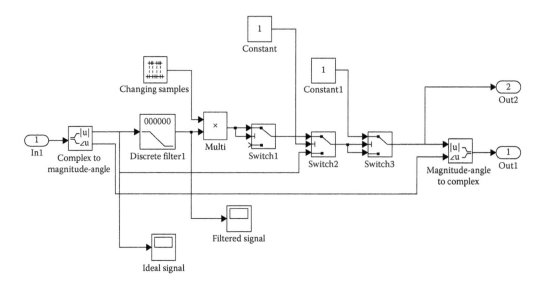

FIGURE 12.6 The filter block scheme.

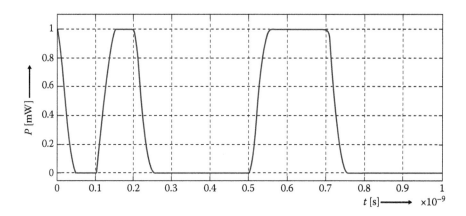

FIGURE 12.7 The OOK-modulated signal after filtering.

Another source of bit errors can be caused by the FWM effect. This FWM effect occurs only in wavelength division multiplexing (WDM) transmission systems, and therefore, we must generate additional optical signals with the same modulation technique on different wavelengths. Three originated signals are mixed (in the FWM block) with the sum block and a new generated optical signal is created with the adequate amplitude that is ensured by the dB gain block and calculated in the calculation part. This fourth wavelength signal interferes with transmitted signals. The FWM block scheme is shown in Figure 12.14. The FWM effect between various wavelengths in WDM transmission system mainly depends on optical signal powers, dispersion, and attenuation parameters, so we can present the FWM influence on the OOK-modulated signal with the 10 Gbit/s transmission rate with the CD compensation in Figure 12.15.

For analyzing and comparing different optical modulation techniques and for evaluating influences of various simulated effect of the optical transmission medium, the BER parameter can be calculated. The BER measurement can be accomplished by bit-by-bit comparing original input data sequences with received output ones. This option can be set in the main screen of the simulation program. As we can see in Figure 12.16, the BER calculation block contains displays showing the BER, a number of error bits, and a number of transmitted bits.

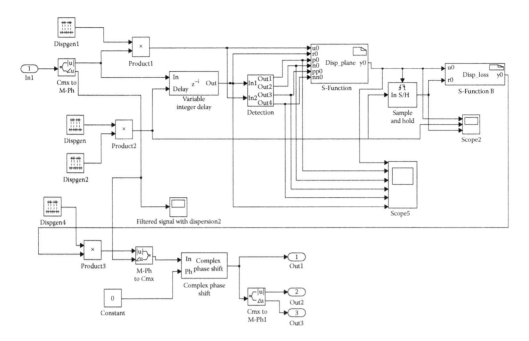

FIGURE 12.8 The CD block scheme.

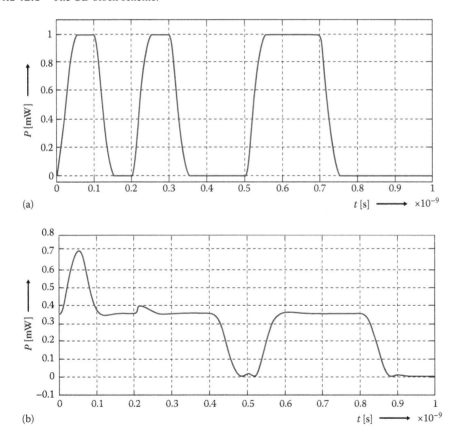

FIGURE 12.9 The OOK filtered signal (a) without CD and PMD influences and (b) with CD and PMD influences.

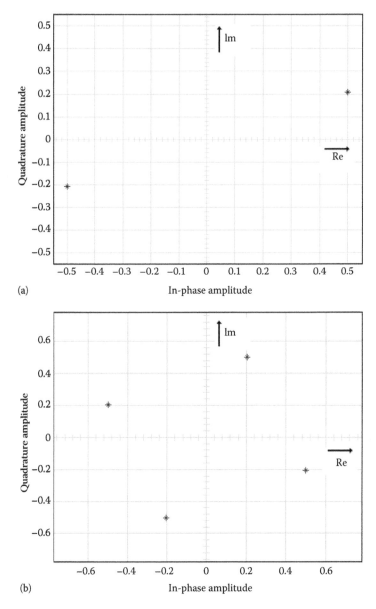

FIGURE 12.10 CD and PMD influences (a) on the DBPSK-modulated signal and (b) on DQPSK-modulated signal.

12.3 SIMULATION TOOL FOR OPTICAL INFRASTRUCTURES

Hybrid PONs present a future phase of PON classes. For developing broadband PONs, there exist various architectures and directions. They are specified with emphasis on basic characteristics and distinctions.

The HPON simulator represents real possibilities for a consistent transition from the time division multiplexing PON (TDM-PON) to the HPON based on various specific parameters. This HPON configurator as the interactive software tool allows comparing possibilities of various passive optical access networks. Its main aim is helping users, professional workers, network operators, and system analysts to design, configure, analyze, and compare various variations of possible HPONs. This HPON simulator has a main dialogue window for designing a transition from TDM-PON to HPONs.

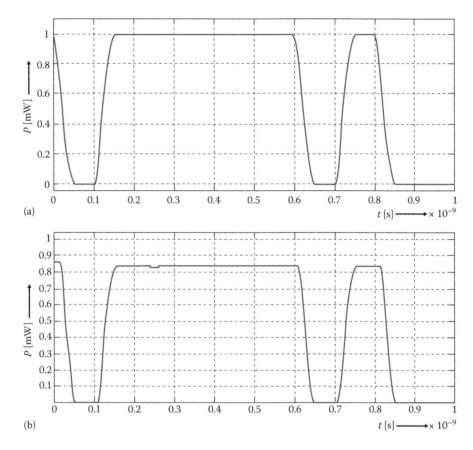

(a)

(b)

FIGURE 12.11 The OOK filtered signal (a) without CD and PMD influences and (b) with CD and PMD influences after CD compensating on the 100 km optical fiber length.

Additional dialogue windows with basic network schemes and short interactive descriptions serve for the specific HPON configuration setup.

12.3.1 Simulation Environment of the HPON Configurator

The HPON configurator allows comparing possibilities of various passive optical access networks. This HPON simulator [13,14] represents real possibilities for a consistent transition from the TDM-PON to the HPON based on various specific network parameters—the network capacity from the viewpoint of the optical layer, the number of original TDM and new WDM subscribers, the number of exploited wavelengths and multiplexing types, the growth of the channel capacity by connecting of new subscribers, and other feasibilities.

At the creation of simulated HPON configurations, it is necessary to deal with these problems:

- Parameters of the optical fiber and an availability of wavelengths
- Input parameters of the deployed TDM-PON
- Expected parameters of the selected HPON:
  - The total network capacity
  - The total number of subscribers
  - Power relationships (the transmitting power of light sources, the sensitivity of receivers, the attenuation of the transmission path, and optical components)
- Type and number of deployed optical components
- Possibilities for expanding of the network

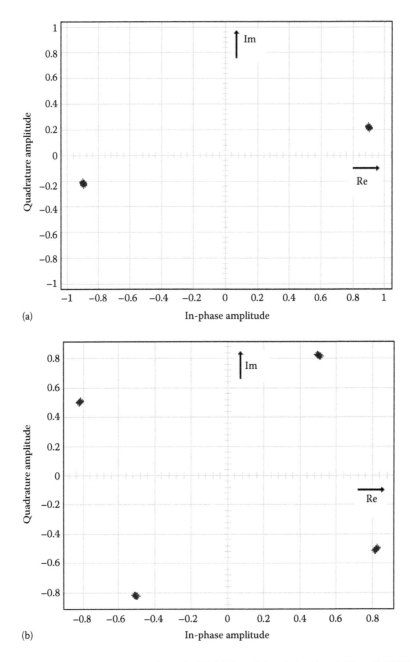

FIGURE 12.12 CD and PMD influences (a) on the DBPSK-modulated signal and (b) on DQPSK-modulated signal after CD compensating on the 100 km optical fiber length.

12.3.2 Main Interactive Window

The HPON configurator for comparing possibilities of various passive optical access networks is created by using the Microsoft Visual Studio 2008 software in the integrated development environment (IDE) [15–17]. There exist possibilities for the graphical interface created by using the Microsoft Foundation Class (MFC) library for the C++ programming language. The simulation model has one main interactive dialogue window (Figure 12.17) for inserting and presenting

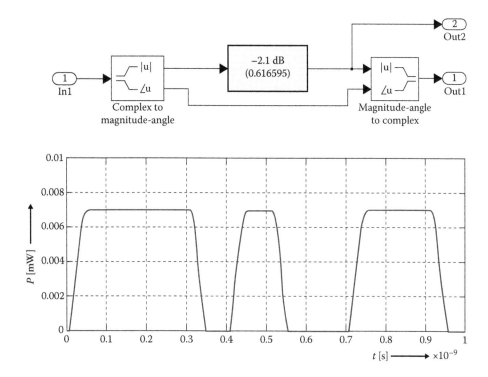

FIGURE 12.13 The attenuation block scheme and the OOK-modulated signal after attenuation.

parameters of transitions from TDM-PON to HPONs. It allows comparing and analyzing four principal approaches for designing and configuring HPONs. Therefore, additional dialogue windows with basic network infrastructures and relations can be used for the specific HPON configuration setup. For WDM/TDM-PON and SUCCESS HPONs, transitions from the original TDM-PON architecture are expressed by GIF animations. For their presentation, a free available CPictureEx class is used. For SARDANA HPON and long-reach (LR)-PONs, features of networks are presented in a simple picture.

The HPON configurator is working in several steps:

1. Setting parameters of the optical fiber—the type of the optical fiber (according to the ITU-T standards), the DWDM density.
2. Evaluating optical fibers and wavelengths—standard or inserted specific attenuation values in dB/km, the calculation of numbers of utilizable coarse wavelength division multiplexing (CWDM) and dense wavelength division multiplexing (DWDM) carriers.
3. Inserting input parameters of the deployed TDM-PON—the number of original TDM networks, the type of the network, the number of subscribers per one network, the distance between the ONT and the OLT.
4. Evaluating input parameters—the calculation of the total transmission capacity of the TDM network together with the average capacity per one subscriber, the total number of subscribers, and the maximum attenuation of the TDM network; also, the attenuation class is presented. This step is terminating with the selection of detailed HPON configuration design.
5. Setting input parameters for the HPON configuration—based on the stored TDM-PON data and selecting one from possible HPON types.

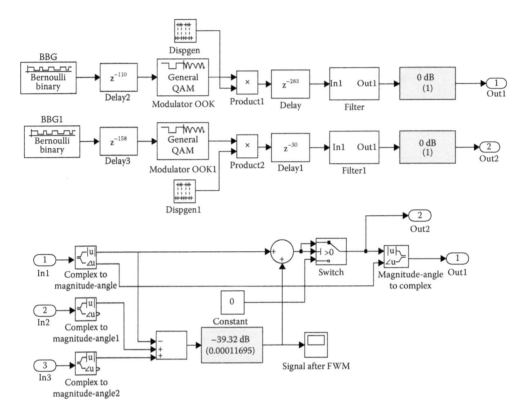

FIGURE 12.14 The FWM effect block scheme.

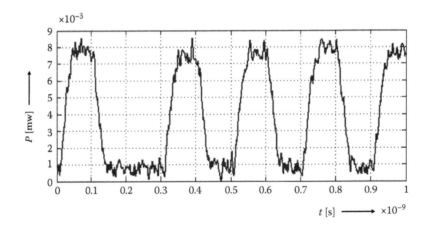

FIGURE 12.15 The OOK-modulated signal with the 10 Gbit/s transmission rate with the FWM influence.

6. Application input parameters and specific parameters of the HPON configuration (the total capacity of the hybrid network, the total number of subscribers, the average capacity per one subscriber, the maximum attenuation of the hybrid network between the OLT and the ONT, the number and type of used active and passive components), summing up the type and number of deployed optical components and presenting possibilities for future expansion of HPON types.

BER measurement—OOK (NRZ)

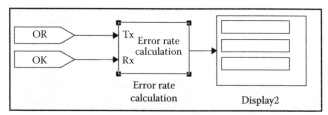

FIGURE 12.16   The BER measurement block scheme.

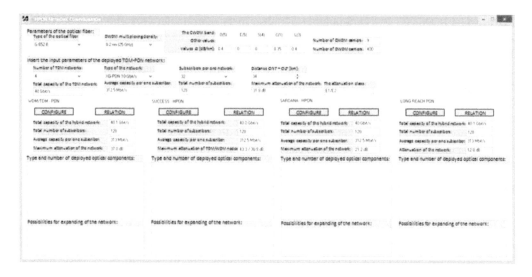

FIGURE 12.17   The main window of the HPON Network Configurator.

After setting optical fibers and wavelengths, the number of new DWDM subscribers and the required capacity per one subscriber are inserted. Based on given values, the total capacity of the hybrid network (only for the downstream direction), the growth of the capacity due to connecting of new subscribers, the number of necessary DWDM transmitters and receivers, and the number of utilized CWDM and DWDM wavelengths are evaluated (Figure 12.18). Thereafter, further possible changes of the network topology and the subsequent transition to the HPON are presented to the user. By this way, we can take a decision to regulate or to exchange primary input values for adapting to demands of the HPON.

Parameters of the optical fiber can be executed by the selection of the optical fiber's type and the DWDM density. A selected type of the optical fiber is presented by the specific attenuation

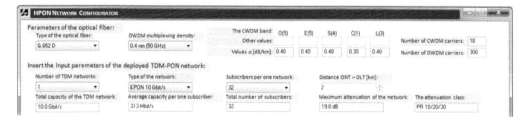

FIGURE 12.18   Setting parameters for the optical fiber and the deployed TDM-PON network.

values and by the number of transmission bands. For specifications, various ITU-T recommendations—ITU-T G.652A, G.652B, G.652C, G.652D, G.656, G.657—together with the "Other values" option can be inserted. Then, specific attenuation coefficients are concretely displayed. Also, the total number of possible CWDM and DWDM carrier wavelengths for particular bands is presented.

Also, the attenuation of optical fibers for available wavelengths is calculated in specific network configurations. In this case, we prefer attenuation coefficients for common optical fibers according to the ITU-T G.652A, B, C, D [18]. Moreover, we can incorporate attenuation coefficients for new types of optical fibers according to the ITU-T G.656 [19] and the ITU-T G.657 [20] standards and evaluate their utilization in the HPON infrastructure.

Parameters of the TDM-PON infrastructure can be executed by a selection and a listing of appropriate parameters for the deployed TDM-PON. A number and type of networks (EPON, GPON, 10G-EPON, XG-PON), a number of subscribers per one network and a network reach, respectively the distance OLT-ONT (max. 999 km) can be selected. Then, features of the selected TDM-PON configuration—the total capacity, the average capacity per one subscriber, the total number of subscribers, the maximum network attenuation, and the attenuation class—are presented.

In the case that the maximum network attenuation exceeds a specified value depending on the parameters of the optical fiber, deployed TDM network types, the number of subscribers, and the distance between OLT and ONT, options of WDM/TDM-PON, SUCCESS HPON, and SARDANA HPON configurations are turned off and a challenge for utilization of the LR-PON is appearing.

12.3.3 ADDITIONAL INTERACTIVE WINDOWS

In the next step, four HPONs are reserved. In the main dialogue window, autonomous dialogue windows for the specific hybrid network configuration are opened by using CONFIGURE push buttons. Then, a configuration of specific network parameters can proceed for the WDM/TDM-PON, SUCCESS HPON, SARDANA HPON, or LR-PON options. Finally, a short list of basic network characteristics—the total capacity of the hybrid network, the total number of subscribers, the average capacity per one subscriber, the maximum network attenuation—calculated for each specific option are displayed.

Applications of the HPON selection are possible in one from four various HPON approaches. In the first case, the WDM/TDM-PON represents a hybrid network based on the combined WDM/TDM approach. The WDM/TDM-PON configuration window can be opened (Figure 12.19). After inserting requested parameters of the hybrid WDM/TDM-PON and also the number of TDM subnetworks, the APPLY button can be pushed. Then, a complete set introducing the type and number of optical components is introduced together with their basic characteristics.

In the second case, the SUCCESS HPON allows the coexistence of TDM and WDM technologies within one network infrastructure and provides a possibility of the smooth transition from TDM subscribers to new WDM ones. The SUCCESS HPON configuration window can be opened (Figure 12.20). After inserting requested input parameters for TDM nodes and WDM nodes, the CONFIGURE button can be pushed. Then, a complete set introducing the type and number of optical components is introduced together with their basic characteristics. Moreover, a graphical presentation of the bandwidth allocation between TDM downstream (CWDM), TDM upstream (DWDM), and WDM (DWDM) wavelengths is presented.

In the third case, the SARDANA HPON architecture is created by the two-fiber ring with connected remote nodes (RNs) that ensure bidirectional signal amplification and dropping/adding of DWDM wavelengths for particular TDM trees. The SARDANA HPON configuration window can be opened (Figure 12.21). After inserting the requested input parameters for RNs,

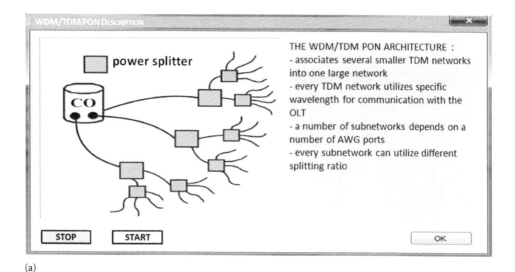

FIGURE 12.19 Windows of the WDM/TDM-PON approach.

TDM trees, and the WDM ring length, the CONFIGURE button can be pushed. Then, a complete set introducing the type and number of optical components is introduced together with their basic characteristics.

In the fourth case, the LR-PON utilizes moreover active components (optical amplifiers) that can extend a network reach or improve the splitting ratio in RNs. The LR-PON configuration window is opened (Figure 12.22). An option for selecting a higher splitting ratio (1:128 and more) is supplemented as a special feature of the LR-PON. When this higher splitting ratio of subscribers per network is selected, options for configuration of other HPONs—WDM/TDM-PON, SUCCESS HPON, and SARDANA HPON—are automatically deactivated because they don't support this selected splitting ratio. For this case, an option for the LR-PON configuration is appearing.

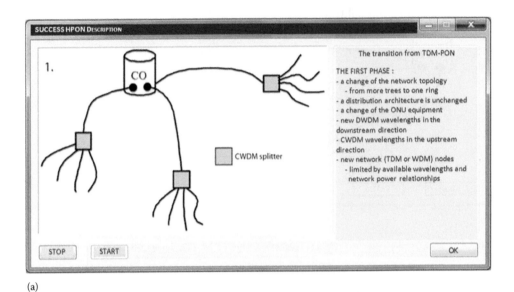

(a)

(b)

FIGURE 12.20   Windows of the SUCCESS HPON approach.

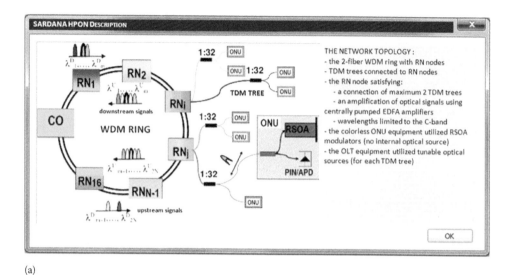

(a)

(b)

FIGURE 12.21   Windows of the SARDANA HPON approach.

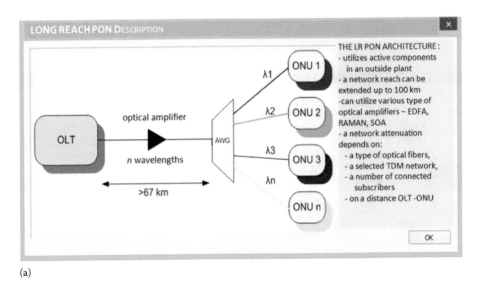

(a)

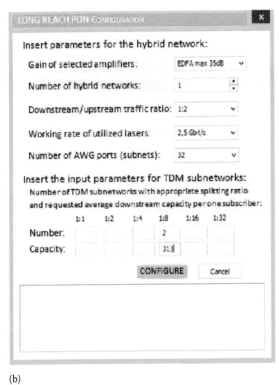

(b)

FIGURE 12.22   Windows of the LR-PON approach.

Possibilities for new configuration of this LR-PON are very similar to the hybrid WDM/TDM-PON configuration. One option for selecting the specified optical amplifier's type is supplemented as the main factor that distinguishes the LR-PON from other PONs.

12.3.4   PARAMETERS

One of the issues with any DWDM scheme is provisioning of available wavelengths. For the OLT side, there is no issue with equipment—a new OLT interface supporting all the channels in a single

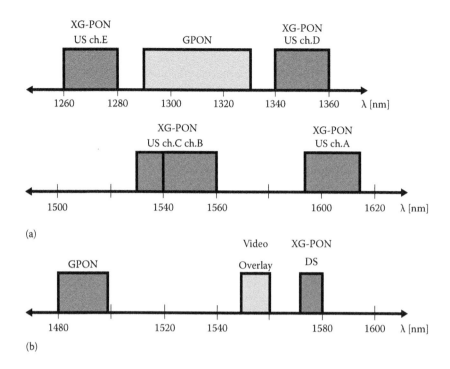

FIGURE 12.23 The NG-PON wavelength spectrum plan.

module. However, each optical network unit (ONU) terminal should support only a single wavelength and an option of wavelength selectable sources for upstream transmitters might be possible. Alternatively, ONU interfaces may be colored and this may present operational problems. Therefore, colorless ONU equipment is considered in some architecture.

The overall wavelength plan for next-generation (NG)-PONs is shown in Figure 12.23. This plan pulls together the diverse set of requirements from PON deployments and the NG-PON system concept into a minimal set of wavelength assignment [21,22]:

1. The XG-PON1: The downstream wavelength band of 1575–1580 nm is used since it is the only wavelength band that is left in the system with the video overlay. For the upstream, five channel assignments can be discussed—within the L-band (channel A), the C-band (channel B), the video-compatible C-band (channel C), the O-plus band (channel D), and the O-minus band (channel E). According to the ITU-T G.987 series [23], parameters for the upstream wavelength allocation are determined to be 1260–1280 nm.
2. The ER+WC XG-PON1: The 0.5 nm wide wavelength windows (the 200 GHz channel spacing) are specified.
3. The XG-PON2: The downstream band is the same as for the XG-PON1. The upstream band spans 1260–1280 nm that corresponds to the O-band placement and permits the use of directly modulated lasers without excessive dispersion penalties and also use of uncooled lasers.
4. The hybrid DWDM/XG-PON: The channel spacing can be selected at either 100 or 50 GHz. It requires the C-band upstream allocation (channel B) because it is sensitive to fiber losses, but this is in conflict with the video overlay. The downstream wavelengths are located from 1575 to 1582 nm.

The allocation of CWDM and DWDM carriers in particular transmission bands is presented. By reducing DWDM channel spacing, negative influences of nonlinear effects (e.g., FWM) can be increased in the real optical transmission media.

In the WDM/TDM-PON, two sets of wavelengths can be utilized—for CWDM and for DWDM systems. Because these bands are overlapped, the number of available wavelengths depends on the utilized bandwidths and on the density of the wavelength allocation (100, 50, or 25 GHz channel spacing). In the simulation program, the dependency between wavelengths is exactly scheduled [14,16].

In the SUCCESS HPON, the situation is different. In the first step of transition to the HPON, the network topology is changing from various point-to-multipoint infrastructures to one ring by means of one optical fiber. Therefore, DWDM wavelengths (one wavelength per one TDM node) are utilizing the downstream direction. In the upstream direction, each TDM node utilizes a different wavelength from the CWDM grid.

In the SARDANA HPON, transmissions in both directions are limited by the use of wavelengths in the C-band due to the erbium-doped fiber amplifier (EDFA) operational range. So the utilization of DWDM systems can be varied with different densities of the DWDM channel spacing.

In the LR-PON, a selection from three types of optical amplifiers—EDFA, Raman amplifier (RAMAN), and semiconductor optical amplifier (SOA)—is possible. These optical amplifiers have various features and characteristics and different values of optical amplification (gain). Options of appropriate amplifiers type are depending on specified network configurations. In the case of mismatched network configuration or/and parameter setting, error messages are presented at the bottom part of the LR-PON configuration window. A summary of selected parameters for the mentioned optical amplifiers is introduced in Table 12.1.

In the HPON configurator, total power relationships depend on specific network characteristics and applied optical component parameters. We prefer real values of optical components utilized in PONs (Table 12.2). For the XG-PON, there will be two loss budgets denoted normal

TABLE 12.1

Parameters of Optical Amplifiers in the LR-PONs

Features	EDFA	Raman	SOA
Gain	>35 dB	>25 dB	>30 dB
Wavelength band	1530–1560 nm	1280–1650 nm	1280–1650 nm
Noise	5 dB	5 dB	8 dB
Cost	Medium	High	Low

TABLE 12.2

Attenuation Specifications of HPON Optical Components

Symbol	Description	Value (dB)
a_{filter}	The attenuation of the WDM filter	0.4
a_{AWG}	The AWG attenuation	5
$a_{50:50}$	The 50:50 power splitter attenuation	4.4
$a_{90:10}$	The 90:10 power splitter attenuation	0.8:12
	The 1:16 splitter attenuation	14.1
$a_{SPLIT1:N}$	The 1:32 splitter attenuation	17.4
	The 1:64 splitter attenuation	21.0
a_{TDM-RN}	The TDM node attenuation (including connectors)	1.5
a_{WDM-RN}	The WDM node attenuation (including connectors)	1
$a_{add/drop}$	The attenuation of added/dropped wavelengths	1.2
$a_{isolator}$	The attenuation of the isolator	0.3
a_{CON}	The connector attenuation	0.2

and extended. The normal loss budget is defined with a Class B+ loss budget plus an insertion loss from a WDM1 filter. The link loss will be approximately 28.5–31 dB at $BER = 10^{-12}$. The extended loss budget is defined with a Class C+ loss budget plus an insertion loss from a WDM1 filter [21,22].

12.4 CONCLUSION

In the first part of this chapter, a simulation tool for investigating new optical signal processing techniques in the optical transmission medium is introduced. This simulation program analyzes pulse propagation inside optical fibers and influences of linear and nonlinear effects on information signals transmitted in the optical transmission medium. The analysis includes a detailed view of particular effects and focuses on their influences on optical pulses separately and of course in conjunction with each other. The complex simulation model is designed with an adaptability, flexibility, scalability, and extensibility for future improvements. Results of the investigation can be used for designing a robust simulation model with a smooth transition to advanced modern and future optical transmission systems.

A design of the optical transmission path offers possibilities for visual displaying of transmitted signals via the optical transmission medium representing with its negative environmental influences by means of graphical tools such as eye diagrams and BER blocks. The simulation program of the optical transmission path is used for analyzing different optical modulation techniques at the signal transmission [9,24,25]. Also, this simulation program can be utilized for presenting influences of linear and nonlinear effects on the optical pulse propagation in the optical fiber. First, effects of the CD and the attenuation on the optical signal can be found in [26] and then the FWM effect is focused [27]. Moreover, a mentioned simulation program as a research tool for different optical transmission systems is presented in [28].

In the second part of this chapter, a simulation tool for analyzing considered optical infrastructures in broadband PONs is described. The authentic HPON configurator is still developing during its evolution, from the initial phase through several extensions. Another possibility of this network simulator can be obtained by practical utilization in the design of real PONs that can utilize various multiplexing techniques on their optical layer. At the same time, this simulation tool can be employed for comparing different HPONs mutually and for analyzing their pros and cons, strong and weak features.

By this way, the HPON configurator is utilized in the analysis of SUCCESS and SARDANA HPONs with presented results in the following works [29–31].

ACKNOWLEDGMENTS

This work is a part of research activities conducted at Slovak University of Technology Bratislava, Faculty of Electrical Engineering and Information Technology, Institute of Telecommunications, within the scope of the project KEGA No. 039STU-4/2013 "Utilization of Web-based Training and Learning Systems at the Development of New Educational Programs in the Area of Optical Transmission Media".

REFERENCES

1. R. Róka. Fixed transmission media. In: *Technology and Engineering Applications of Simulink*, S. Chakravarty, editor, InTech, Rijeka, Croatia, May 2012.
2. J. Čuchran and R. Róka. *Optocommunication Systems and Networks*. STU Publishing House, Bratislava, Slovakia, 2006, pp. 6–140.
3. B.E.A. Saleh and M.C. Teich. *Fundamentals of Photonics*. Wiley-Interscience, New York, 1991, pp. 739.

4. E. Iannone, F. Matera, A. Mecozzi, and M. Settembre. *Nonlinear Optical Communication Networks.* John Wiley & Sons, New York, 1998, pp. 20–50.

5. Z. Jamaludin, A.F. Abas, A.S.M. Noor, and M.K. Abfullah. Issues in polarization mode dispersion for high speed fiber optics transmission. *Suranaree Journal of Science and Technology*, 12 (2), 98–106, April–June 2005.

6. H.A. Yasser. Polarization losses in optical fibers. In: *Recent Progress in Optical Fiber Research* M. Yasin, S.W. Harun, and H. Arof, editors, InTech, Rijeka, Croatia, January 2012.

7. Ch.T. Allen, P.K. Kondamuri, D.L. Richards, and D.C. Hague. Measured temporal and spectral PMD characteristics and their implications for network-level mitigation approaches. *Journal of Lightwave Technology*, 21 (1), 79–86, January 2003.

8. B.L. Heffner. PMD measurement techniques—A consistent comparison. In: *OFC'96 Technical Digest*, San Jose, CA, 1996.

9. R. Róka and F. Čertík. Modeling of environmental influences at the signal transmission in the optical transmission medium. *International Journal of Communication Networks and Information Security*, 4 (3), 146–162, 2012.

10. F.G. Xiong. *Digital Modulation Techniques.* Artech House, Boston, MA, 2000, pp. 23–234.

11. J. Seams. A comparison of resistive terminators for high speed digital data transmission. In: *High Frequency Electronics*, Summit Technical Media, Bedford, NH, October 2005, pp. 18–26.

12. J.M. Kahn and K.-P. Ho. Spectral efficiency limits and modulation/detection techniques for DWDM systems. *IEEE Journal on Selected Topics in Quantum Electronics*, 10 (2), 259–272, March–April 2004.

13. R. Róka. The designing of passive optical networks using the HPON network configurator. *International Journal of Research and Reviews in Computer Science*, 1 (3), 2079–2557, 2010.

14. R. Róka. The utilization of the HPON network configurator at designing of passive optical networks. In: *TSP 2010—International Conference on Telecommunication and Signal Processing*, Vienna, Austria, 2010, Vol. 33, pp. 444–448.

15. R. Rókaand and S. Khan. The modeling of hybrid passive optical networks using the network configurator. *International Journal of Research and Reviews in Computer Science*, 2, 48–54, April 2011.

16. R. Róka. The extension of the HPON network configurator at designing of NG-PON networks. In: *TSP 2011—International Conference on Telecommunication and Signal Processing*, Budapest Hungary, 2011, Vol. 34, pp. 79–84.

17. R. Róka. The designing of NG-PON networks using the HPON network configurator. *Journal of Communication and Computer*, 9 (6), 669–678, 2012.

18. ITU-T Telecommunication Standardization Sector: Recommendation G.652—Characteristics of a single-mode optical fiber and cable. 2009.

19. ITU-T Telecommunication Standardization Sector: Recommendation G.656—Characteristics of a fibre and cable with non-zero dispersion for wideband optical transport. 2010.

20. ITU-T Telecommunication Standardization Sector: Recommendation G.657—Characteristics of a bending-loss insensitive single-mode optical fiber and cable for the access network. 2009.

21. F. Effenberger et al. Next-generation PON—Part II: Candidate systems for next-generation PON. *IEEE Communications Magazine*, 47 (11), 50–57, 2009.

22. F. Effenberger et al. Next-generation PON—Part III: System specification for XG-PON. *IEEE Communications Magazine*, 47 (11), 58–64, 2009.

23. ITU-T Telecommunication Standardization Sector: Recommendation G.987—10-Gigabit-capable passive optical network systems: Definitions, abbreviations, and acronyms. Recommendation G.987.1—General requirements. Recommendation G.987.2—Physical media dependent (PMD) layer specification. 2010.

24. F. Čertík and R. Róka. Analysis of modulation techniques utilized in the optical transmission medium. In: *ELEKTRO 2012—International Conference*, Žilina, Slovakia, May 21—22, 2012.

25. F. Čertík. The propagation of higher modulated formats via single mode optical fiber. In: *Workshop RTT 2013*, Senec, Slovakia, 2013, pp. 71–82.

26. F. Čertík. Using MATLAB tools for simulation of the optical transmission medium. In: *Technical Computing*, Bratislava, Slovakia, 2012, pp. 8.

27. F. Čertíkand R. Róka. The nonlinear FWM effect and its influence on optical signals utilized different modulation techniques in the WDM transmission systems. In: *OK 2012—24th Conference*, Praha, Czech Republic, 2012.

28. F. Čertík. The propagation of 10 Gbit/s OOK modulated signal via SM optical fiber. In: *RTT 2013*, Senec, Slovakia, 2013, pp. 96–101.

29. R. Róka. The analysis of hybrid passive optical networks using the HPON network configurator. In: *IN-TECH 2013—International Conference on Innovative Technologies*, Budapest, Hungary, 2013, pp. 401–404.
30. R. Róka. The analysis of SUCCESS HPON networks using the HPON network configurator. In: *Advances in Electrical and Electronic Engineering—AEEE*, 11 (5), 420–425.
31. R. Róka. The analysis of SARDANA HPON networks using the HPON network configurator. In: *Advances in Electrical and Electronic Engineering—AEEE*, 11 (6), 522–527, December 2013.

13 Simulation Tools for the Evaluation of Radio Interface Technologies for IMT-Advanced and Beyond

Krzysztof Bąkowski, Krzysztof Wesołowski, and Marcin Rodziewicz

CONTENTS

ABSTRACT

Simulation tools for the evaluation of several radio interface technologies including International Mobile Telecommunications-Advanced (IMT-Advanced) and 5G radio interface technologies are described. First, the importance of simulation as an experimental and evaluation method and its evolution is outlined against the background of the history of development of cellular systems. Subsequently, the guidelines for evaluation of 4G radio interface technology defined by the

International Telecommunication Union—Radiocommunication Sector are shortly presented. Then, the detailed simulation model implemented by the authors and useful in the evaluation of IMT-Advanced system proposals is described. Finally, the latest development of the simulation tools needed for experiments and proving the concepts of future 5G systems is discussed.

13.1 INTRODUCTION

In the last 40 years, simulation has become a powerful tool in the modeling and construction of several types of digital communication systems. Simulation has also become an important and obligatory topic of university courses in communications. Several books have been devoted to specific simulation methods used in communications. Jeruchim et al. [1] published their best-selling classic on the simulation of communication systems presenting a wide selection of topics related to the modeling of linear and nonlinear communication systems, simulating randomness that is an inherent feature of communications, and showing methods of effective simulation time shortening. The book by Gardner and Baker [2] is also devoted to simulation techniques and the models of communication systems and processes. The book by Tranter et al. [3] is devoted to similar issues. Their authors applied MATLAB® to model illustrating examples of the simulation of communication systems with wireless applications. All books mentioned so far are related to the simulation of the physical layer in communication systems. It means that they focus on explaining the so-called link-level simulation methods in which such processes as data generation, data source and channel encoding, waveform and noise generation and filtration, digital modulations, signal propagation over a transmission channel, and signal distortion and reception (demodulation, filtration, detection, channel decoding, etc.) are elaborated. As in digital communication systems, their main performance criteria are bit, symbol, or data block error rates, and their estimators are inherent blocks of the simulated system.

Scientists and developers of digital communication systems are interested not only in the modeling and performance evaluation of single communication links. Typically, the performance of the whole network containing many links working in parallel and interfering with each other is of primary interest; therefore, the so-called system-level simulation is also necessary. This type of simulation allows the modeling of resource management issues, admission of system users, interaction and network loading by several types of traffic sources, etc. The aim of such modeling is network capacity evaluation. Many processes in communication networks occur randomly in time; therefore, they can be modeled using the event-driven simulation approach widely presented in the book by Tyszer [4].

Communication systems and networks can be modeled using different software packages and programming languages. Several specialized software packages were offered over the history of using simulation and modeling of digital communication systems. Obviously, general-purpose languages are widely used, such as Fortran (mostly of historical meaning), C, C++, and the already mentioned MATLAB. However, special software packages have been offered for simulation of communication systems as well. COSSAP, which stood for Communication System Simulation and Application Processor, was developed at the RWTH University of Aachen and distributed by Cadis. It was widely known and used in Europe in the nineties. The competing package was SPW®, which is further being offered by Synopsis [5]. SPW, similarly to COSSAP, promises fast development of a communication system model by building its scheme on a computer screen from ready-to-use blocks. However, probably the most popular simulation package is Simulink® [6], having similar capabilities as SPW and featuring a huge number of building blocks implemented in both floating- and fixed-point arithmetics. The previously mentioned packages have wide capabilities of modeling and display of interesting waveforms and partial results in the selected places of the simulated system; however, in the case of really innovative investigations supported by intensive simulations, many new blocks have to be constructed by the simulation developer. This is often not an easy and efficient task due to constraints set by the block interface applied in the simulation software.

Therefore, many engineers and scientists prefer to use their proprietary software in which particular blocks that are characteristic for communication systems of interest have already been collected in their own libraries. Such programs can also be individually and flexibly optimized with respect to the required simulation speed or used memory size. In this context, the IT++ library is worth mentioning [7]. IT++ is a C++ library of mathematical, signal processing, and communication classes and functions. Its main use is in the simulation of communication systems and in research in the area of communications. The IT++ library has originated from the former department of Information Theory at the Chalmers University of Technology, Gothenburg, Sweden; however, it is now released under the terms of the GNU General Public License and is well known among the scientific community interested in communication systems modeling.

Although the simulation of communication systems has been used for many years, its power was shown during the works on the standardization of the second-generation GSM cellular system. To our best knowledge, after the presentation of several competing prototypes [8], the system was agreed upon as a compromise among them and its performance was evaluated by simulation followed by a direct standardization by ETSI. A similar situation occurred during the development of the 3G UMTS system that conformed to ITU-2000 requirements [9,10]. Between 1991 and 1995, two European Union (EU)-funded research projects called Code Division Testbed (CODIT) and Advanced Time Division Multiple Access (ATDMA) were carried out by major European telecom manufacturers, network operators, and the academia. The CODIT and ATDMA projects investigated the suitability of wideband code division multiple access (CDMA)- and time division multiple access (TDMA)-based radio access technology for 3G systems. This work was later continued in the Future Radio Wideband Multiple Access System (FRAMES) project and became the basis of the further ETSI UMTS work until decisions were taken in 1998. Although a CDMA-based system in the form of IS-95 [11] was earlier introduced in the United States, the main concepts and details of a UMTS radio interface were checked by simulation. All following system improvements such as HSDPA, HSUPA, HSPA, HSPA+, and LTE, introduced and proposed at 3GPP meetings and finally found in 3GPP standards, were intensively investigated by simulation as well. The significance of simulation as a modeling and performance evaluation method has further increased with the introduction of new requirements and standards in wireless cellular systems [12]. Intensive works on wireless systems conforming to the requirements for 4G systems defined in [12] were based on simulations as well. As it was already mentioned, both link- and system-level simulations were applied as powerful tools for investigations and verifications of several concepts and ideas. Moreover, in order to evaluate International Mobile Telecommunications-Advanced (IMT-Advanced) radio interface technology (RIT) proposals, International Telecommunication Union—Radiocommunication Sector (ITU-R) formulated a special report [13] in which the rules of evaluation are precisely described. In Section 13.2, we will present them briefly. Section 13.3 is devoted to the detailed description of a radio wave propagation model in which a RIT should operate/work. In Section 13.4, simulator software architecture, its development, and calibration procedures are described. Finally, in Section 13.5, possible simulation tool extensions for the evaluation of 5G systems are discussed.

13.2 RULES OF EVALUATION OF RADIO INTERFACE TECHNOLOGIES FOR IMT-ADVANCED

Stating the requirements for the IMT-Advanced RIT has triggered intensive research in industrial and academic centers, aimed at developing a new radio interface. Its natural consequence was a necessity to formulate the rules according to which the proposals submitted to ITU-R will be assessed and evaluated. In the report [13], guidelines for both the procedure and the technical, spectrum, and service criteria of evaluation of the proposed IMT-Advanced RITs have been provided for a number of test environments and deployment scenarios. From our perspective, this guideline is a very instructive report explaining several rules of simulation that should be applied in order to reliably investigate a new radio interface. Let us recall that the requirements for IMT-Advanced

[12] have determined, among some other criteria, the enhanced peak data rates equal to 100 Mbps for high and 1 Gbps for low mobility to support advanced services and applications. The proposals should be evaluated through simulation, as well as through analytical and inspection procedures. We will concentrate on the simulation methodology defined in these guidelines.

According to [13], the simulation should be performed on the system and link levels. Selected topics of evaluation that need system-level simulation are cell spectral efficiency, cell edge user spectral efficiency, mobility, and voice over IP (VoIP) capacity. Additionally, mobility should be also investigated using link-level simulations.

System-level simulations should be based on two particular network layouts. The first one is a regular grid of cells in a wrap-around configuration of 19 sites, each determining 3 cells. No specific topographical details are taken into account. Such configuration should model the rural/high-speed, base coverage urban and microcell cells. The second network layout is characteristic for an indoor hotspot environment. Details characterizing both layouts are discussed in the next section.

System-level simulations are based on the concept of the so-called drops. Each drop can be understood as a single statistical event. A sufficiently high number of drops have to be drawn in order to obtain a statistical validity of the results expressed in the form of appropriate user and system metrics. An important piece of information about statistical validity of the estimated performance metrics is the width of confidence intervals. The most important principles in designing system-level simulation for each drop are described in the following.

For each drop, an assumed number of user terminals are applied and their locations are selected according to the uniform distribution within a predefined area of the network layout. Active user terminals are randomly assigned line of sight (LoS) and non–line of sight (NLoS) channel conditions. Each user terminal is assigned to a certain cell according to the algorithm proposed by the entity submitting the system to be evaluated. Fading signals and interferences are determined from each terminal to each base station (BS) and vice versa. For each type of environment in which transmission is modeled, an appropriate channel propagation formula is defined in [13]. The full-buffer traffic model (infinite queue depth) or the VoIP traffic model is assumed. In the first model, there is an infinite amount of data bits waiting for transmission. In the second model, it is assumed that the source rate is equal to 12.2 kbps, as in a typical speech codec used in cellular telephony. Voice activity factor is 50%. A two-state voice activity model similar to the Gilbert model [14] is assumed. Users are assigned a traffic class and their traffic should be modeled according to one of the previously mentioned traffic models. More details on traffic models can be found in [13]. Proprietary schedulers are used in radio resource management (RRM) ensuring appropriate packet scheduling for each traffic class. All important factors affecting the RRM quality have to be taken into account, such as channel quality feedback delay, feedback errors, realistic channel estimation and its inaccuracies, protocol data unit (PDU) errors, and possible retransmissions and overhead due to feedback and control channels. The channel models applied in simulations should have dynamic properties. Thus, for a realistic evaluation of the proposed system, the assumed channel model is of primary importance. All parameters of the channel models to be applied are precisely defined in [13]. It allows the generation of statistically equivalent results in the simulators owned by different partners of the system evaluation group, making it possible to calibrate them. Detailed description of channel models and their implementation will be the subject of considerations in the next section. It was also presented in short in [15].

Due to a high level of complexity, the fully realistic simulation of all component links and all RRM and other data link control (DLC—Layer 2) issues would be impossible or time consuming to the extent that valuable results would not be available within an acceptable time period. Therefore, in order to simplify the simulation process, link-level simulations are performed. As their results, error statistics as a function of signal-to-noise ratio as well as interference level are estimated. Then, due to an appropriately designed link-to-system interface, these estimates are used in the system simulations performed without full modeling of physical layer (PHY—Layer 1) details.

13.3 SIMULATION FRAMEWORK OF THE IMT-ADVANCED SYSTEM

13.3.1 INTRODUCTION

As it has already been stressed earlier, in order to evaluate the performance of any radio access network (RAN) technology by means of simulation, appropriate modeling of the propagation environment is required. The level of details reflected in the model of radio wave propagation environment is mostly driven by the RITs specified to be used in RAN. As the peak performance of 4G RITs is achieved when multiple transmitting and receiving antennas are assumed at both ends of the radio link, an appropriate multiple-input–multiple-output (MIMO) mobile radio channel model must be used as well. Furthermore, advanced schemes of transmission in RAN assume the so-called coordinated multiple point (CoMP) transmission applied in RAN. Let us recall that transmission quality of users located on cell edges can be improved by coordinated transmission and reception by a few surrounding base stations (CoMP). Therefore, appropriate spatial correlation of mobile radio channel parameters of deployed radio transmission capable stations must be considered to assure adequate evaluation of a given RIT. Nowadays, cellular systems are multiuser systems in which transmission time, available spectrum, orthogonal codes, transmission power, and spatial dimensions are shared among currently active users. In wireless systems, the performance of a single link depends on other links established in a given cell and its neighboring cells. This is due to a cochannel interference and the applied RRM algorithms. The simulation of such an environment is doubtlessly challenging and requires appropriate modeling of the propagation environment.

13.3.2 IMPACT OF THE WINNER PROJECTS

For the evaluation of candidate technologies for the IMT-Advanced system, the requirements for which have been specified in [12], among many other guidelines, a new channel model has been defined in the ITU-R report [13]. The defined model is an extension of the channel model described, for example, in [16,17], which has previously been developed within the EU WINNER I, WINNER II, and WINNER+ projects [18]. In the ITU-R recommended model, the concept of the spatial channel model (SCM) for MIMO channel simulations presented in the 3GPP Technical Report [19] has been adapted. The reader interested in the overview of the SCM and WINNER channel models is advised to read the relevant chapters in [20,21]. An overview of state-of-the-art channel models can be found in [22].

The goal of the WINNER projects was to develop a single ubiquitous radio access system adaptable to a comprehensive range of mobile communication scenarios from a short range to a wide area. This access system should comply with the requirements for the IMT-Advanced system. In the WINNER I project, the overall system concept was developed. The aim of the WINNER II project was to develop and optimize the previously mentioned concept and to define the radio access system. Finally, the goal of WINNER+ was to develop, optimize, and evaluate IMT-Advanced-compliant technologies by integrating innovative and cost-effective additional concepts and functions providing an evolution path toward further improved performance of the IMT-Advanced system. In all these research and development activities, the wireless channel model was of primary importance. The works on channel modeling in WINNER I focused on a wideband MIMO channel at the 5 GHz frequency range starting from the 3GPP SCM model [19] followed by its wideband extension (SCME) [23] and the creation of the generic model. This model, based on measurements performed in 2 and 5 GHz bands [24], was a ray-based double-directional [25] multilink model that was antenna independent, scalable, and capable of modeling channels for MIMO links. In WINNER II, the developed model was updated and the number of the considered propagation scenarios was increased. Further measurement campaigns were performed, leading to better determination of channel model parameters [26–29]. As a result, the channel model for the frequency range 2–6 GHz was specified in [16]. The concept of time evolution of channel parameters was presented in [30]. In the CELTIC/EUREKA WINNER+ project, the works on channel modeling resulted in the extension of the frequency range to 450 MHz–1 GHz bands, developing 3D channel models [31] and enriching the propagation scenarios. The newly developed models are based on a generic channel modeling approach that enables the variation of the number of antennas, their configurations, geometry, and antenna beam patterns without changing the basic propagation model. Applying this approach allows the use of the same channel data in different link-level and system-level simulations. The final set of the WINNER channel models has been described in [17]. The WINNER model has been compared with other reference models, such as the 3GPP SCM [19] and COST 273 [32] models in [33,34].

13.3.3 IMT-ADVANCED FRAMEWORK

As one may imagine, addressing the needs of various possible RITs and propagation and deployment scenarios in a single evaluation framework is not an easy task. To address this task in [13], a generic model with scenario-specific parameters was specified. In this section, we first describe the considered test environments together with deployment scenarios associated with them. The mapping of channel models is presented as well. Subsequently, a model structure is presented and after that, a single link model is elaborated. This description is followed by a network layout presentation. Due to the rather complex structure of the IMT-Advanced simulation-based evaluation framework, it is important for a possible user of the simulation tool to understand the way it has been constructed and the output data it produces. Therefore, we present the procedure of channel coefficient generation.

13.3.3.1 Propagation Scenarios

The purpose of the RAN is to operate the radio interface to allow users to gain access to services like the Internet or voice calls. Due to various propagation conditions that exist in the deployed networks, the performance of a given RIT must be tested in different environments. For the purpose of the evaluation of RITs in [13], representative propagation scenarios were selected. The following so-called test environments were chosen [13]:

- An indoor environment targeting isolated cells at offices and/or in hotspots assuming stationary and pedestrian users. The indoor environment is characterized by the smallest cells and high user throughput or user density in buildings.
- An urban microcellular environment with higher user density focusing on pedestrian and slow vehicular users. This test environment targets small cells and high user densities and traffic loads in city centers and dense urban areas.
- An urban macrocellular environment targeting continuous coverage for pedestrian up to fast vehicular users. This test environment focuses on large cells and continuous coverage.
- A macrocell environment with high-speed vehicles and trains. The high-speed test environment focuses on larger cells and continuous coverage.

Each test environment is associated with a deployment scenario describing the configuration of baseline evaluation of the radio network for system-level and link-level simulations. Each of the deployment scenarios is supplemented with an appropriate channel model. Although different channel models are specified, they use a generic framework that will be described in the next paragraph.

13.3.3.2 Model Structure

The recommended IMT-Advanced channel model structure is shown in Figure 13.1 [13]. The channel model consists of a primary module and an extension module. The primary module originates from the WINNER II channel model, whereas the extension module has been added to enable the tests of wireless systems operating in scenarios beyond those specified by the IMT-Advanced system requirements [12,13].

As seen in Figure 13.1, four different scenarios are modeled in the primary module: indoor hotspot (InH), urban microcell (UMi), urban macrocell (UMa), and rural macrocell (RMa).

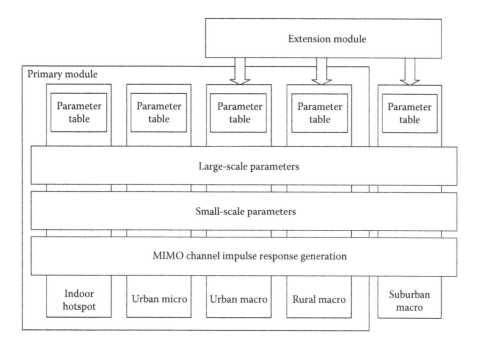

FIGURE 13.1 IMT-Advanced channel model structure. (Adapted from Report ITU-R M.2135-1, Guidelines for evaluation of radio interface technologies for IMT-Advanced, Geneva, Switzerland, 2009.)

Additionally, the optional suburban macrocell (SMa) scenario has been taken into account in the evaluation of IMT-Advanced system proposals. Each of the scenarios is accompanied by a set of parameters that are fed to the generic large- and small-scale (LS and SS) parameters and channel impulse response generation procedures. The extension module is a tool allowing the selection of modified parameters to generate LS parameters of the simulated scenarios for macrocell scenarios, such as UMa, SMa, and RMa. The extension module provides an additional level of parameter variability. In the primary module, SS and LS parameters are variables, and the extension module provides new parameter values for the primary module based on environment-specific parameters [13]. The extension module is optional for implementation.

13.3.3.3 Single Link Model

As explained in [13,14,16,17,19], the ITU-R IMT-Advanced channel model is a geometry-based stochastic model and it is often called a double-directional channel model [22]. In such a model, directions of the rays departing from the transmit antennas and arriving at the receive antennas are specified without the precise location of the scatterers. Thanks to this approach, propagation parameters and antenna parameters can be separated. Figure 13.2 illustrates the scheme of a single link between a BS and a particular mobile station.

The signal generated by the sth transmit antenna ($s = 1, ..., S$) is emitted according to the angle of departure (AoD) distribution, for which a particular angle is denoted by φ_i. It reaches the uth receive antenna ($u = 1, ..., U$) according to the angle of arrival (AoA) distribution. The AoA angle is denoted by φ_j. Local scatterers disperse the transmitted rays, which leads to the creation of the so-called clusters. Each cluster constitutes a multipath component. Thus, the impulse response of the MIMO channel characterized by S transmit antennas and U receive antennas has a matrix form and is described by the formula

$$\mathbf{H}(t;\tau) = \sum_{n=1}^{N} \mathbf{H}_n(t;\tau)$$

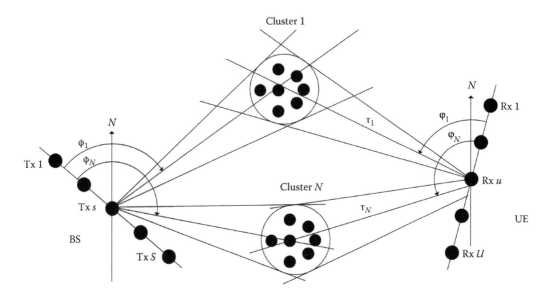

FIGURE 13.2 Geometrical model of a single-link MIMO channel recommended for IMT-Advanced testing. (Adapted from Report ITU-R M.2135-1, Guidelines for evaluation of radio interface technologies for IMT-Advanced, Geneva, Switzerland, 2009.)

Each matrix $\mathbf{H}_n(t; \tau)$ represents a single multipath component. Such a component can be expressed by the transmit and receive antenna array response matrices $\mathbf{F}_{tx}$ and $\mathbf{F}_{rx}$ and the propagation channel response matrix $\mathbf{h}_n$ resulting in the following formula for the matrix characterizing the nth cluster:

$$\mathbf{H}_n(t;\tau) = \iint \mathbf{F}_{rx}^T(\varphi)\mathbf{h}_n(t;\tau,\phi,\varphi)\mathbf{F}_{tx}(\phi)d\phi\, d\varphi$$

If we assume that the number of rays in each cluster is not higher than M, it is possible to express the nth multipath component between the sth transmit antenna and uth receive antenna, that is, the entry of the matrix $\mathbf{H}_n(t; \tau)$ in the position (s, u) in the form

$$\mathbf{H}_{u,s,n}(t;\tau) = \sum_{m=1}^{M} \begin{bmatrix} F_{rx,u}^V(\varphi_{n,m}) \\ F_{rx,u}^H(\varphi_{n,m}) \end{bmatrix}^T \begin{bmatrix} \alpha_{n,m}^{VV} & \alpha_{n,m}^{VH} \\ \alpha_{n,m}^{HV} & \alpha_{n,m}^{HH} \end{bmatrix} \begin{bmatrix} F_{tx,s}^V(\phi_{n,m}) \\ F_{tx,s}^H(\phi_{n,m}) \end{bmatrix}$$

$$\times \exp\left(\frac{j2\pi}{\lambda_0}\left(\overline{\varphi}_{n,m} \cdot \overline{r}_{rx,u}\right)\right)\exp\left(\frac{j2\pi}{\lambda_0}\left(\overline{\phi}_{n,m} \cdot \overline{r}_{tx,s}\right)\right)$$

$$\times \exp\left(j2\pi\vartheta_{n,m}t\right)\delta\left(\tau-\tau_{n,m}\right)$$

where the superscripts V and H denote the vertical and horizontal polarization, respectively; $F_{rx,u}$ and $F_{tx,s}$ with the appropriate superscripts denote the field patterns for the uth receive antenna array element and the sth transmit antenna array element; the variables $\alpha_{n,m}$ with appropriate superscripts denote the complex gains for the mth ray of the nth cluster; λ_0 is the wavelength of the carrier; and $\overline{\phi}_{n,m}$ and $\overline{\varphi}_{n,m}$ denote the AoD and AoA unit vectors, respectively. Finally, $\overline{r}_{tx,s}$ and $\overline{r}_{rx,u}$ are the location vectors of the sth transmit antenna and the uth receive antenna and $\vartheta_{n,m}$ is the Doppler frequency for the mth ray of the nth cluster. If the time variability of the channel is applied in the simulations, these listed SS parameters are functions of time t.

13.3.3.4 Network Layout

The previous paragraphs show the basic dependencies leading to the determination of the channel impulse response for the MIMO link between a given mobile terminal and the selected BS. As mentioned earlier, the simulation of a single link is insufficient for the system evaluation; therefore, the developed software channel simulator generates MIMO channel impulse responses for all links in the simulated mobile network. These links depend on the layout of the BSs and mobile terminals. For that reason, a network layout has to be assumed.

In the software simulator designed according to the guidelines contained in the ITU-R Report [13], the network layout depends on the selected scenario. For the indoor hotspot network, deployed on a single floor of a building, the simulation is performed for two sites placed in the hall of a 120 m × 20 m building floor. There are 8 rooms (15 m × 15 m) on each side of the hall (see Figure 13.3).

In other scenarios, such as the rural macrocell and urban micro- and macrocell, a basic regular hexagon layout is assumed with a central hexagon surrounded by two tiers of cells. In consequence, there are 19 sites. Each hexagon contains three 120° cells, so 57 cells are simulated. As it has been written earlier, a wrap-around technique is applied to equalize interference conditions in all cells (see Figure 13.4). It is assumed that mobile terminals are uniformly distributed over the whole modeled area.

13.3.3.5 Model Implementation

In order to obtain sufficient statistical knowledge on the system performance in the simulated environments, a sufficiently high number of statistical events have to be modeled. A statistical event describes a particular distribution of mobile terminals in the simulated network coverage area for which all transmission channels to the assigned BSs are determined. Thus, a statistical event involves a random selection of mobile terminal locations and a selection of random properties of all channels between the mobile terminals and BSs. As it has been mentioned earlier, such an approach to deriving statistical properties of the investigated system is called the *drop concept*. In a single drop, mobile station locations and LS channel parameters are fixed. Only fast fading is dynamically simulated.

Deriving performance measures for each drop requires the collection of an appropriately high volume of simulation data. Reliable statistical properties are obtained by the selection of a sufficiently high number of drops. In the assumed model [13], the properties of each drop are selected randomly and independently of the properties of the previous drops.

The coefficients of each channel are generated in a stepwise procedure visualized in Figure 13.5. First, the propagation scenario is selected and the environment parameters for this scenario are defined. After that, the network layout is set up by generating BS positions according to the network topology and by randomly generating the positions of user terminals and directions of their motion. For a given scenario, the speed of all user terminals is fixed.

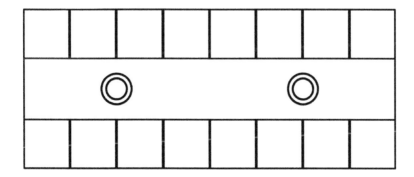

FIGURE 13.3 Network layout—single floor model.

FIGURE 13.4   Network layout—hexagonal grid model.

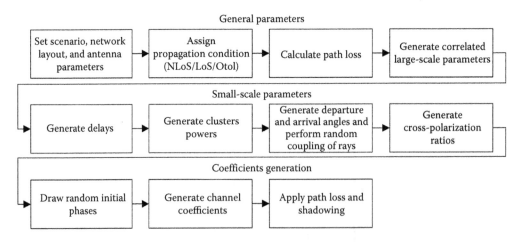

FIGURE 13.5   Generation of channel coefficients.

In the next step, LS parameters are generated. First, LoS or NLoS propagation conditions are chosen. They can be selected randomly for all terminals in the whole network; however, in the calibration mode, the same LoS or NLoS propagation can also be chosen for all terminals. After setting the LoS/NLoS conditions, an appropriate path loss formula is applied for each link.

Finally, LS parameters are generated for each link. The LS parameters are delay spread (DS), angle of arrival spread (ASA), angle of departure spread (ASD), Rician K factor (K), and shadow fading (SF). In order to receive correlated values of each LS parameter, first, random Gaussian distributed numbers for each of them are generated on a dense 2D grid covering the area of the whole simulated network. Then, each parameter surface is filtered in two dimensions by an exponential filter with parameters determined on the basis of measurements and given in [13]. Similarly to what has been done in [35], filtering in two dimensions is performed using a 1D filter, by implementing it as a two-step procedure. In the first step, the filtration along the x-axis is performed. In the second step, the resulting surface is filtered again along the y-axis. Figure 13.6 illustrates the filter impulse responses and the filtered parameters on the grids for the indoor hotspot scenario. The filtered surfaces correspond to normalized LS parameters on the dB scale (with the unity variance and the zero mean). Let us consider K mobile terminals located in specified places and linked to the same BS. The filtered values of each parameter at a specific location of each mobile terminal constitute a basis for the calculation of the final parameter values that are received due to the cross correlation between all LS parameters. This operation is performed by a matrix multiplication. The entries of the cross-correlation matrix are determined by measurements. Their values are given in [13]. In order to apply them in the simulator, inverse mapping to the log-normal distribution with scaling by the relevant standard deviation and shifting them by the selected mean value is performed.

In the exemplary simulator implemented at the Poznan University of Technology, the path losses between each mobile terminal and each BS are known. This enables the assignment of each mobile terminal to the serving BS. Such an assignment is a procedure in which a mobile station selects the serving BS from a list of candidate BSs featuring the highest received power within the 1 dB margin. In Figure 13.7, the resulting wideband signal-to-noise and interference ratio (wideband SINR) is presented, and in Figure 13.8, the selected serving cells ID is visualized. Note that due to the data generation procedure, only the cross correlation of LSPs was performed with reference to Figures 13.7 and 13.8.

After the LS parameters are fixed for each link, the generation of SS parameters is performed. For each propagation scenario, the number of clusters applied in the channel model is given in [13]. First, cluster delays in a given link are generated on the basis of the DS parameter. After that, cluster powers are generated using a particular cluster delay as a parameter. Then, each of the two clusters featuring the highest power is divided into three subclusters. The last step in the SS parameter generation is the calculation of the departure and arrival angles, first for the clusters and then for the rays of which the clusters are composed.

Figures 13.9 and 13.10 illustrate the previously mentioned operation performed for a single link. Two plots in Figure 13.9 show the link power profile normalized to unity. Plot 9a illustrates the power profile for clusters before the division of the two strongest clusters into subclusters. Plot 9b shows the power profile after performing this operation. In Figure 13.10, angular parameters of a single link are presented. Plots a and b present the distributions of the departure and arrival angles for clusters before the division, whereas plots c and d show the same parameters after subcluster division. Finally, plots e and f show the ray angle distribution. In the last row of the plots, the angles are wrapped into the $[-\pi, \pi]$ interval.

When all LS and SS parameters are set, the final MIMO channel impulse response coefficients are calculated. There is a separate impulse response for each pair of the sth transmit and uth receive antenna array elements.

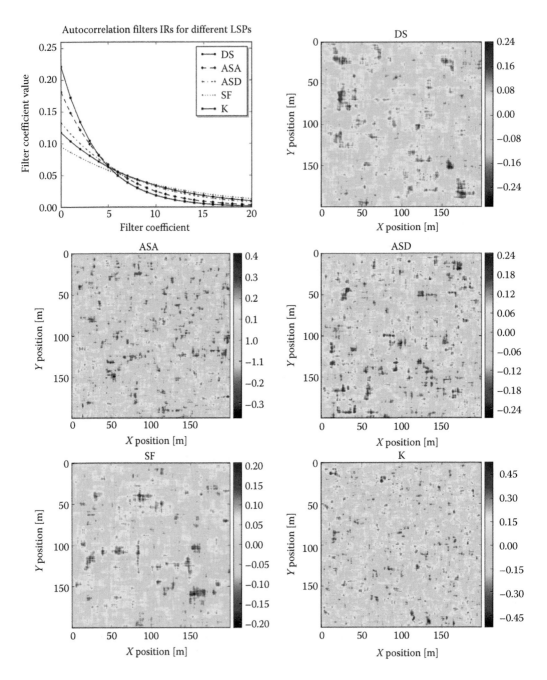

FIGURE 13.6 Filtration of the LS parameters in an indoor hotspot scenario for LoS propagation conditions (IR, impulse response; LSP, LS parameters).

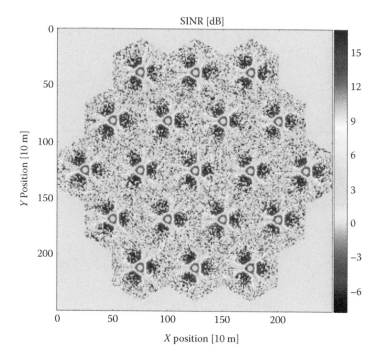

FIGURE 13.7 Wideband SINR experienced by UEs.

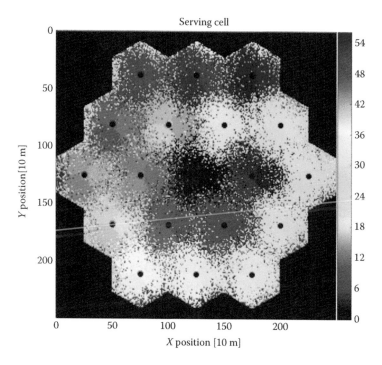

FIGURE 13.8 Serving cell ID map (ID numbers visualized by mapping denoted in the right column).

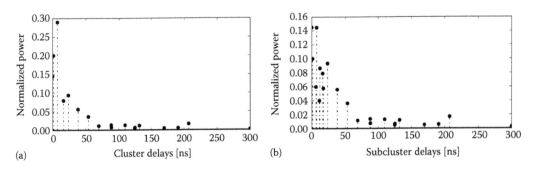

FIGURE 13.9 Example of power-delay profile (PDP) for a single link of indoor hotspot scenario (NLoS). (a) PDP for clusters and (b) PDP for subclusters.

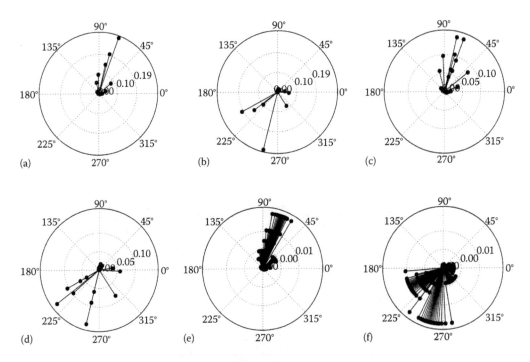

FIGURE 13.10 Example of generated angles for a single link of indoor hotspot scenario (NLoS). (a) Distribution of AoD for clusters, (b) distribution of AoA for clusters, (c) distribution of AoD for subclusters, (d) distribution of AoA for subclusters, (e) distribution of AoD for rays, and (f) distribution of AoA for rays.

13.4 SOFTWARE ARCHITECTURE, ITS DEVELOPMENT, AND CALIBRATION PROCEDURES

Implementation of the IMT-Advanced evaluation framework is a quite complex task. The initial C++ implementation of the PUT simulator was described by the authors in [15]. Since then, the tool was substantially enriched to accommodate new extensions like relays and direct links between mobile stations. The weakness of the initial implementation was caused by insufficient structuring of the variables' containers, so that it was crucial to extend the model by a new set of stations and links that interconnect them. In the current implementation, there is still one main C++ class that stores various parameters of modes and provides model instantiation, configuration, initialization, and finally execution. Besides the main class, there is a set of classes representing various nodes of wireless networks like BSs, relay stations (RSs), and user equipments (UEs). A large part of the required processing is now focused on a class representing a two-way communication link. This class stores the LS parameters and, if required, the SS parameters together with channel coefficients generated for a given communication link. As it was stated in [15], because of the model complexity and its processing power requirements, the crucial issue was the selection of an appropriate programming language. Scripting languages and all high-level languages with managed memory were excluded on account of the processing power requirements. Owing to the complexity of the model, a language with sufficient code structuring capabilities that supports high-level data management had to be used. The selection of C++ was an obvious solution because of the two main features of this language: object-oriented programming and code reusability through inheritance. Another important issue is providing a flexible and easy-to-use interface for the simulated system. This goal was accomplished by implementing the IMT-Advanced channel model features related to the previously mentioned stations as separate classes that can be easily inherited by similar classes representing features of the modeled system. This design is visualized in Figure 13.11, where the top-level architecture of IMT-Advanced evaluation tool is presented. Because the IMT-Advanced channel model is a geometry-based model, it can yield position and orientation data. Modern radio communication systems can take advantage of these location data in addition to the channel impulse response resulting from the radio propagation environment; therefore, in the simulation process, not only the channel impulse response but also some additional data derived from the channel model are taken into account. Thus, reusability through inheritance plays a significant part. Inheritance is used to create a station class of a given type composed of the class with channel model–specific information and another class containing specific information on the simulated system. As a result, two pieces of the software code may be integrated: the one corresponding to the channel model and the one modeling the simulated system. This design, together with the class representing various wireless links in the network, facilitates a clear design of the tool. Obviously, not all the tools designed to simulate a radio communication system must follow this design. Therefore, it is still possible to run the simulation tool in the "stand-alone" mode named so by the authors,

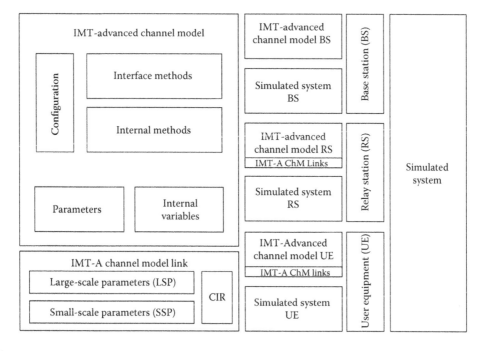

FIGURE 13.11 Top-level architecture of the simulator.

where no inheritance from IMT-Advanced evaluation tool classes is required. The tool is built as a library and is used in various link-level and system-level tools implemented in the Chair of Wireless Communications of PUT.

Parts of the program use the previously mentioned IT++ library [7]. This valuable library is used to facilitate some of the operations needed to generate channel model realizations. Data containers of the IT++ library are used to store the generated parameters to test model operation and validate its modules. In our software channel model, data visualization uses the matplotlib library that is written in Python. This language is also used for initial data processing and data visualization. Summarizing, C++ is applied for the core processing implementation, with some help of the IT++ library and Python for data postprocessing and visualization with the help of the matplotlib library. Almost all plots shown in this chapter have been generated using Python scripts.

After the implementation of the channel model, the verification of its correct operation should be the last step before software exploitation. Such work was done in the WINNER+ project by all eight partners participating in the IMT-Advanced system Evaluation Group and its results are available at the WINNER+ Evaluation Group web page [36]. The PUT implementation was also calibrated and all metrics are consistent with the available data. Metrics used for the calibration are cumulative distribution functions (CDFs) of the following parameters: path gain and WSINR for LS parameters and delay spread, AoA spread, and AoD spread for SS parameters. All metrics were obtained separately for LoS, NLoS, and outdoor-to-indoor (OtoI) propagation conditions using a sufficient number of system-level simulations. Details related to the calculation of the mentioned metrics can also be found in [36]. It is worth stressing that the implemented channel model has been verified by comparison with other channel model implementations performed by other partners to the WINNER+ project. The metrics used for the implemented model calibration were fully consistent with those obtained by the other partners. As practice shows, complex simulation packages should be extensively tested and, if only possible, calibrated before being used to produce results and draw valuable conclusions from them.

To the best knowledge of the authors, the IMT-Advanced evaluation framework (or at least the channel model) and its several implementations are publicly available. A MATLAB-based software channel simulator [23] based on a WINNER-type channel model freely available on the Internet is worth mentioning, although the network layout must be provided by the user of the tool or otherwise is generated randomly. In order to apply this MATLAB simulator in the IMT-Advanced system evaluation, the user should provide the network topology by applying the appropriate wrap-around technique. To validate the network layout implementation, the user should also implement the process of metric calibration on his or her own. In the channel simulator described in this chapter, all these features are already built in. Another implementation available at the ITU-R web page devoted to IMT-Advanced evaluation is shown in [37]. According to our analysis, the network layout is implemented in this tool, but modeling of the SS effects is only partial, because autocorrelation of LS parameters is not implemented. Also, some bugs have been identified, and this tool is not recommended to be used without proper calibration. Nevertheless, its performance measured as the execution time is much higher as compared with the previously described tool. On the other hand, the simulation tool available at [38] is worth recommendation. It is fully functional, calibrated, and designed for high performance.

13.5 DISCUSSION ON POSSIBLE SIMULATION TOOL EXTENSIONS FOR THE EVALUATION OF 5G SYSTEMS

The necessity of system and network evaluation is further enhanced with the current development of several technologies that will be applied in future 5G wireless systems. The exponential growth in mobile data traffic—typical estimates give an approximately 1000× increase over the next decade— requires an adequate response from the world's wireless communication researchers. Since the current 4G technology will not be capable of carrying this rapid increase in data consumption, the attention of the research community is shifting toward what will be the next set of innovations in wireless communication technologies referring to 5G technologies [39]. These new requirements set new challenges to the evaluation framework of 5G technologies. One of the projects currently performed with reference to 5G and, among other things, the evaluation framework is the *Mobile and wireless communications Enablers for the Twenty-twenty Information Society* (METIS) project. Compared to the IMT-Advanced evaluation framework, the approach of the METIS project is scenario driven. A set of scenarios was described in [39], with accompanying test cases (TCs). So instead of specifying a single set of requirements like 1 Gbps data rate for low mobility and 100 Mbps data rate for high-speed scenarios, the requirements are specific for each real-world scenario that is presented in detail. To enable the evaluation of technologies investigated in specific TCs, simulation guidelines were formulated [40]. In this paragraph, we will present the best evaluation framework specified so far by Work Package 6 (WP6) members of the METIS project (listed in [40]) for the so-called TC 2.

13.5.1 METIS Scenarios and Test Cases

As we have mentioned, the METIS project evaluation framework is scenario driven. The scenarios outline the scope of METIS and correspond to one specific challenge. The challenge is a fundamental technical difficulty that should be addressed within wireless communications systems of the future. Moreover, each scenario addresses at least one technical goal identified by the METIS consortium. These technical goals are

- 1000 times higher mobile data volume per area
- 10–100 times higher number of connected devices
- 10–100 times higher typical user data rate
- 10 times longer battery life for low-power massive machine communication (MMC) devices
- 5 times reduced end-to-end (E2E) latency

Within the METIS project, five scenarios have been identified by the WP1 members (listed in [39]):

1. *Amazingly fast*, where instantaneous connectivity and flash behavior of applications are assumed and the users get the feeling that they get all they need, when they need, wherever they will have the need.
2. *Great service in a crowd*, where the assumption is that the user of future wireless access systems will expect good service in very crowded places despite the increase in traffic volume. This scenario mostly includes infotainment applications in shopping malls, stadiums, open-air festivals, or other public events that attract crowds.
3. *Ubiquitous things communicating*, where the communication needs of a massive deployment of ubiquitous machine-type devices are addressed. A broad range of machine-type devices is assumed, from low-complexity devices to more advanced ones, which results in widely varying requirements in several domains, for example, in terms of energy consumption, cost, complexity, transmission power, or latency.
4. *Best experience follows you*, where the aim is to bring a similar user experience for moving end users as for the static ones. Obviously, this concerns both machine-type and human end users. Devices that are highly mobile, for example, cars and trains, are evident examples of communicating machines, but sensors or actuators used, for example, for monitoring transported goods or moving components in industries are also relevant.
5. *Super real-time and reliable connections*, where the E2E latency is tackled under the assumption that the reliability of the connection is very high, for example, 99.999%. This scenario envisions that in future systems, M2M communication with real-time constraints will enable new functionalities for traffic safety and efficiency, smart grid, e-health, or industrial communications.

Based on the identified challenges and scenarios, twelve different TCs have been selected in WP1 of the METIS project [39]. The TC is a practical aspect formulated from the end user's perspective containing a set of assumptions, constraints, and requirements. The relation between scenarios and TCs is that a scenario may cover several TCs, whereas a TC can consider several challenges and therefore belongs to several scenarios.

TC 1: Virtual reality office

The virtual reality office TC is a future indoor setting where very high data rates and challenging capacity requirements at a reasonable cost are envisioned.

TC 2: Dense urban information society

This TC is formulated for a future very dense urban environment where requirements regarding the performance of wireless communication system are very demanding.

TC 3: Shopping mall

The shopping mall TC refers to an environment with a high density of users, where high traffic volumes and user data rates with good availability are required. In this TC, both human and machine-type communications are assumed.

TC 4: Stadium

The stadium TC represents a challenging use case from the network operator's perspective, that is, a mass event where the probability of correlated demand for data is very high.

TC 5: Teleprotection in smart grid network

This TC touches the challenges of low latency and high reliability of communication within a smart grid. The elements of the smart grid network have to perform with very tight latency requirements, for example, 8 ms over 10 km distances, and data reliability of, for example, 99.999% of messages transferred correctly.

TC 6: Traffic jam

The traffic jam TC captures the challenge of providing good-quality network experience for in-vehicle users that utilize bandwidth-demanding services during traffic jam situations.

TC 7: Blind spots

The target of this TC is to inspect ubiquitous capacity demands of future users in blind spots, such as rural areas with sparse network infrastructure or in deeply shadowed urban areas.

TC 8: Real-time remote computing for mobile terminals

The real-time remote computing for mobile terminals TC focuses on providing real-time access to remote computing and cloud facilities for highly mobile terminals. The main challenge of this TC is to provide high data rates and low latency, even in the presence of high mobility.

TC 9: Open-air festival

The open-air festival TC considers an event where for a few days, a significant number of users visit a small rural area that does not require handling a high volume of data for the majority of time. Thus, the infrastructure of a legacy network is highly underdimensioned. Therefore, the network needs to be complemented in a cost-efficient way for this time period.

TC 10: Emergency communications

This TC targets the communication expectations for an event of natural disaster in a dense urban environment. Power consumption and ultrareliability of communication are the main challenges of this TC.

TC 11: Massive deployment of sensors and actuators

In this TC, the challenge of connecting a large number of connected devices is addressed. These devices, typically sensors or actuators, need to transmit data with small payloads only occasionally, with moderate latency requirements, for example, a few seconds.

TC 12: Traffic efficiency and safety

The traffic efficiency and safety TC presents challenges regarding required reliability, availability, and latency of automotive safety services.

13.5.2   METIS Test Case 2: Madrid Grid Model

The METIS project TC 2, also known as the dense urban information society scenario, is a future urban setting where the requirements regarding system performance are very demanding. In particular, it is required that the future systems are able to handle high traffic volumes and high experienced data rates under the constraint of a reasonable cost of solutions.

13.5.2.1 Propagation Scenarios

The METIS project participants involved in WP1 recognized several propagation scenarios that are relevant for various TCs. These propagation scenarios can be divided into three subclasses based on transmitter and receiver locations. The first class of propagation scenarios corresponds to the outdoor-to-outdoor (O2O) propagation that applies to the situations when both the receiver and the transmitter are located outside buildings. The second class is the outdoor-to-indoor (O2I) propagation, in which one transmitter/receiver is located in a building, while the other one is outside. Finally, the third one is the indoor-to-indoor (I2I) propagation that applies when both the receiver and the transmitter are located inside buildings. These three subclasses of propagation scenarios can be further divided depending on the type of the communication link. Three types of links are defined, that is, macro BS to UE link, micro BS to UE link, and UE to UE link. The macro BS to UE link propagation scenarios correspond to a situation where the BS is located over a building rooftop. The micro BS to UE link propagation scenarios are defined with the assumption that the transmitter is located below a building's rooftop and the dominant part of the propagation is due to the reflections between buildings. The UE to UE link is defined because the need for device-to-device (D2D) communication in a future wireless system is recognized in the METIS project; thus, there are also propagation scenarios for this type of communication. In the TC2 Madrid grid model, eight propagation scenarios are identified:

1. PS#1 urban micro O2O
2. PS#2 urban micro O2I
3. PS#3 urban macro O2O
4. PS#4 urban macro O2I
5. PS#7 indoor office I2I
6. PS#9 urban O2O for D2D (also vehicle-to-vehicle)
7. PS#10 urban O2I for D2D
8. PS#13 indoor office I2I for D2D

The propagation scenarios were defined with two important assumptions in mind. The first one is that there is a need for a realistic (not synthetic) scenario. The second one is the use of 3D models for propagation. In the current commonly used propagation models, the LoS or NLoS conditions are randomly selected. However, for scenarios that are realistic, the sight conditions between transmitter and receiver should be evaluated on a real-time basis. A more detailed description of propagation scenarios can be found in [40].

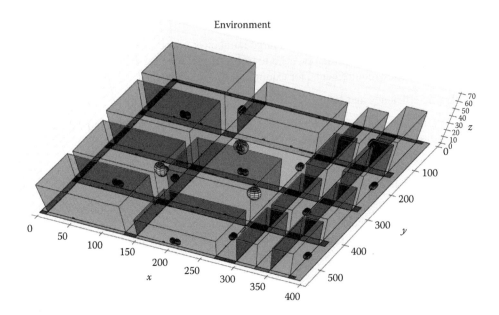

FIGURE 13.12 3D visualization of the Madrid grid.

13.5.2.2 Environment Model

The Madrid grid model is a realistic urban environment model. The realism is achieved by considering different environments of buildings, roads, parks, bus stops, metro entrances, sidewalks, and crossing lanes. The METIS consortium has developed this model on the basis of observations regarding the structure of Madrid. Madrid is an example of a typical European city environment that captures more aspects than the Manhattan grid. The TC 2 environmental model is presented in Figure 13.12, and it is described in more detail in Annex Section 9.2.1 of [40].

The Madrid grid model is a result of a compromise between the need to model realistic dense urban architecture and existing commonly used models, for example, the Manhattan grid model. The Madrid grid model consists of nonhomogeneously placed buildings, which makes it more realistic and allows better capturing of, for example, real-life behavior of users in motion, diversity of SINR distribution, or heterogeneity of cellular network deployment.

Apart from the macro- and micro-/picostation, the network can also be enhanced by a dense network of small cells (e.g., femto cells) with perfectly isotropic antennas. Outdoor small cells are positioned on the facade of a building 5 m above the ground level. Indoor small cells are positioned at room ceilings at the height of 3 m above the floor ground. Similarly to the IMT-Advanced framework, a wideband SINR experienced by UEs and serving cell ID were plotted in Figures 13.13 and 13.14, respectively. The currently proposed modeling of SS effects is based on the approach of the IMT-Advanced evaluation framework, but, as it was stated previously, the LoS/NLoS condition is not random but actually taken from the environment and determined using the ray-tracing approach. Implementation of such complex environment model is not trivial and more details on the PUT implementation are given in Section 13.5.2.4.

13.5.2.3 Outdoor Mobility Model

In the dense urban information society TC, there is a need to consider both indoor and various outdoor mobility models. They are pedestrian mobility and vehicular mobility models. These models should also take into account the behavior of traffic lights. Here, we focus on the outdoor models.

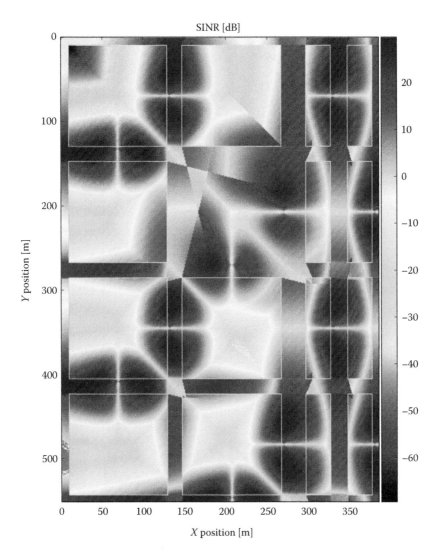

FIGURE 13.13 Wideband SINR experienced by UEs.

TC 2 defines two versions of outdoor mobility models, the simplified and detailed one. The simplified pedestrian mobility model is based on the urban mobility model considered in 3GPP, where a fixed number of pedestrians are randomly initialized at different building exits with uniformly chosen speed from the [0, 3] kmph interval. The pedestrians with nonzero speed are additionally assigned a direction of movement. More details on the simplified pedestrian mobility model can be found in [40]. The detailed version of the mobility model for pedestrians follows the initialization procedure of the simplified version, with the difference that instead of the direction of movement, a destination is chosen. The destination can be either the closest metro stop or the nearest bus stop. Pedestrians then move toward the destination along the shortest path. Depending on the type of the destination, the pedestrian either enters a metro station or waits for a bus at a bus stop. After entering the metro station, the user temporarily disappears for a random time interval to be then reinitialized at a different metro stop with a new speed and destination. The users that reach the bus stop as their destination enter a bus on a first-come–first-served basis until the bus capacity is reached. Buses disappear on the boundaries of the simulation area and each user on a bus is reinitialized at a randomly chosen bus stop.

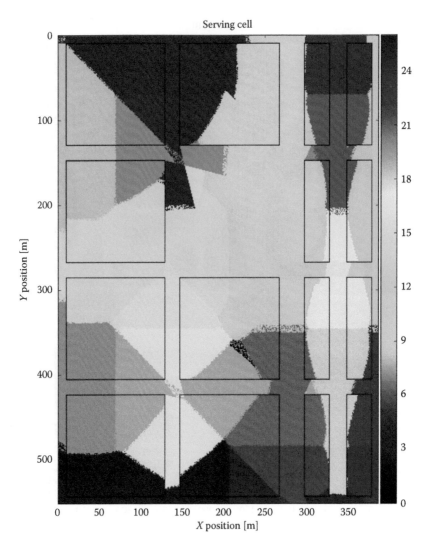

FIGURE 13.14 Serving cell ID map (ID numbers visualized by mapping denoted in the right column).

Regarding the vehicular mobility model, the TC recognizes two types of vehicles, namely, cars and buses. Car mobility is the same for both the simplified and detailed version of the model. A fixed number of cars are distributed on the streets of the scenario with a fixed velocity of 50 kmph. Each car is assigned a uniformly chosen number of users from the interval [1, 5]. Cars move on street lanes until they reach a junction where they can turn, and the decision about a turn is made with an arbitrary probability. At the junctions, cars stop at red traffic light and also when there is another vehicle less than 4 m in front of them. The simplified mobility model for buses assumes that they arrive through Poisson process with an interarrival time of 2 min on each street with the fixed speed of 50 kmph and carry a uniformly chosen number of passengers from the interval of [1, 50]. Buses move only in straight lines and stop only on red traffic lights or when there is another vehicle less than 4 m in front of it. Both cars and buses bounce back at the boundaries of the simulation environment. The detailed mobility model for buses employs similar assumptions to the simplified version; however, the buses are initialized with only one user (which is a bus driver) and can stop additionally at a bus stop, where the passengers enter the bus until either there are no pedestrians left or the bus capacity is reached. The bus can stop at a bus stop only if the bus maximum capacity is not reached.

A traffic light model that complements the mobility models is also defined. All the traffic lights in the scenario grid have two options, red and green. The switching between the lights is performed simultaneously by all traffic lights with a pattern that repeats itself every 90 s.

As we see from the previous description, the level of details, especially in the mobility model, is enormous. The model is far from that in which mobile stations move in random directions with the same speed that was used in many simulators so far. The previously listed TCs and the mobility model shown for TC 2 show the level of details to be modeled in the future evaluation framework of 5G system proposals. An important observation is that the evaluation framework will be based on carefully selected scenarios. Taking into account all the details of these scenarios presents a considerable challenge for simulation tools that have to supply statistically valuable results within a reasonable time interval.

13.5.2.4 Model Implementation

Each of the METIS project partners proposes one or more of the so-called technology components (TeCs), and if possible, they should evaluate it as proposed in [40]. Even though PUT is not a member of WP6 in which simulation guidelines have been developed, the propagation model presented earlier was implemented at PUT to enable the evaluation of the proposed TeCs. The PUT implementation is written in C++ and its architecture is shown in Figure 13.15. One of the significant parts of the simulation tool is environment modeling. The entire environment consists of various elements, that is, buildings, streets, and pavements. In order to manage these elements efficiently, an object-oriented implementation is proposed by PUT. All elements of the environment are put together in a single object. This approach allows the creation of a generic environment where the placement and dimensions of different elements can be defined manually depending on the simulation scenario. In addition, the map container provides easy access to all elements of the environment, which allows to perform an environment-based calculation, for instance, the determination of LoS conditions.

As compared to the MATLAB reference implementation [41] authored by J. F. Monserrat and J. Calabuig, the path loss calculation is based on the aforementioned map with objects representing various elements of the environment. The PUT implementation was compared to the reference one, and good alignment was achieved. The results of the current implementation are shown in Figures 13.13 and 13.14. Further development work is focused on implementing SS effects model in the TC2.

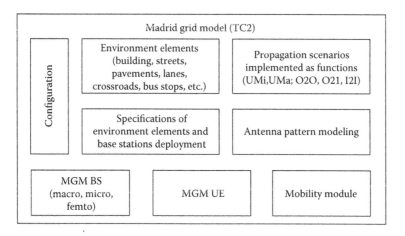

FIGURE 13.15 Architecture of the implemented Madrid grid model.

13.6 CONCLUSIONS

In this chapter, the development of simulators used in the evaluation of wireless cellular systems has been sketched. We mentioned the validity of simulation methods used in GSM and 3G system development. Then we described the main guidelines stated by ITU-R for the evaluation of wireless system proposals that should comply with IMT-Advanced requirements. These guidelines can be considered as clear hints on how to perform system- and link-level simulations not only for the IMT-Advanced system but also for any other wireless system. As the authors of the chapter have their own experiences in the construction of highly advanced wireless system simulators, their experiences from the work in the IMT-Advanced Evaluation Group performed within the WINNER+ EUREKA/CELTIC project have been reported. In particular, channel models have been described showing some exemplary plots illustrating system-level simulation results. As the works on wireless system modeling aimed at the evaluation of future 5G systems are continued, for example, within the EU Seventh Framework METIS project, some descriptions of TCs and evaluation scenarios are given. These works need further advancement and they present a substantial challenge not only to 5G system designers but also to system simulator developers.

ACKNOWLEDGMENTS

The described software IMT-Advanced channel simulator was initially developed within the EUREKA/CELTIC project WINNER+ (No. CELTIC CP5-026). The authors thank the members of the WINNER+ IMT-Advanced Evaluation Group for many helpful hints and remarks during the software development and its calibration. Also, part of the research leading to the presented results has received funding from the EU Seventh Framework program FP7-ICT-2012 under grant agreement No. 317669, also referred to as METIS.

REFERENCES

1. M. C. Jeruchim, P. Balaban, and K. S. Shanmugan, *Simulation of Communication Systems, Modeling, Methodology, and Techniques*, 2nd edn., Kluwer Academic Publishers, New York, 2000.
2. F. M. Gardner and J. D. Baker, *Simulation Techniques, Models of Communication Signals and Processes*, John Wiley & Sons, New York, 1997.
3. W. H. Tranter, K. S. Shanmugan, T. S. Rappaport, and K. L. Kosbar, *Principles of Communication Systems Simulation with Wireless Applications*, Prentice Hall, Upper Saddle River, NJ, 2004.
4. J. Tyszer, *Object-Oriented Computer Simulation of Discrete-Event Systems*, Kluwer Academic Publishers, New York, 1999.
5. Synopsys, Inc., http://www.synopsys.com/Systems/BlockDesign/DigitalSignalProcessing/Pages/Signal-Processing.aspx (accessed March 2014).
6. The MathWorks, Inc., http://www.mathworks.com/products/simulink/ (accessed March 2014).
7. Documentation of the IT++ library, GPL v3, http://itpp.sourceforge.net/4.3.1/ (accessed March 2014).
8. M. Mouly and M.-B. Paulet, *The GSM System for Mobile Communications*, Mouly et Paulet, Palaiseau, France, 1992.
9. Recommendation ITU-R M.687, International Mobile Telecommunications-2000 (IMT-2000), Geneva, Switzerland, 1997.
10. Recommendation ITU-R M.1034, Requirements for the radio interface(s) for International Mobile Telecommunications-2000 (IMT-2000), Geneva, Switzerland, 1997.
11. J. S. Lee and L. E. Miller, *CDMA Systems Engineering Handbook*, Artech House Publishers, Boston, MA, 1998.
12. Recommendation ITU-R M.1645, Framework and overall objectives of the future development of IMT-2000 and systems beyond IMT-2000, Geneva, Switzerland, 2003.
13. Report ITU-R M.2135-1, Guidelines for evaluation of radio interface technologies for IMT-Advanced, Geneva, Switzerland, 2009.
14. K. Wesołowski, *Introduction to Digital Communication Systems*, John Wiley & Sons, Chichester, U.K., 2009.

15. K. Bąkowski and K. Wesołowski, Change the channel, *IEEE Vehicular Technology Magazine*, 6 (2), 82–91, 2011.
16. IST-WINNER II Deliverable 1.1.2 v.1.2. WINNER II channel models, IST-WINNER II Technical Report 2007, http://www.ist-winner.org/deliverables.html.
17. WINNER+ Final channel models, CELTIC CP5-026 Deliverable D5.3, June 30, 2010, http://projects.celtic-initiative.org/winner+/WINNER+%20Deliverables/D5.3_v1.0.pdf.
18. The WINNER Project IST 2004-507581, WINNER II Project IST-4-027756, and WINNER+ Project CELTIC CP5-026, http://projects.celtic-initiative.org/winner+/ (accessed March 2014).
19. Spatial channel model for multiple input multiple output (MIMO) simulations, Release 6, 3GPP TR25.996 6.1.0, 2003–2009.
20. S. Sesia, I. Toufik, and M. Baker (eds.), *LTE—The UMTS Long Term Evolution. From Theory to Practice*, John Wiley & Sons, Chichester, U.K., 2009.
21. M. Döttling, W. Mohr, and A. Osseiran (eds.), *Radio Technologies and Concepts for IMT-Advanced*, John Wiley & Sons, Chichester, U.K., 2009.
22. P. Almers, E. Bonek, A. Burr et al., Survey of channel and radio propagation models for wireless MIMO systems, *EURASIP Journal on Wireless Communications and Networking*, 2007, Article ID 19070, 19, 2007.
23. D. S. Baum, J. Hansen, G. Del Galdo, M. Milojevic, J. Salo, and P. Kyösti, An interim channel model for beyond-3G systems: Extending the 3GPP spatial channel model (SCM), *Proceedings of the 61st IEEE Vehicular Technology Conference (VTC'05)*, vol. 5, pp. 3132–3136, Stockholm, Sweden, May–June 2005.
24. H. El-Sallabi, D. S. Baum, P. Zetterberg, P. Kyösti, T. Rautiainen, and C. Schneider, Wideband spatial channel model for MIMO systems at 5 GHz in indoor and outdoor environments, *Proceedings of Vehicular Technology Conference, 2006 Spring*, pp. 2916–2921, Melbourne, Victoria, Australia, 2006.
25. M. Steinbauer, A. F. Molisch, and E. Bonek, The double-directional radio channel, *IEEE Antennas and Propagation Magazine*, 43 (4), 51–63, August 2001.
26. C. Schneider, A. Hong, G. Sommerkorn, M. Milojević, and R. S. Thomä, Path loss and wideband channel model parameters for WINNER link and system level evaluation, *Proceedings of Third International Symposium on Wireless Communication Systems, 2006*, Valencia, Spain, September 6–8, 2006.
27. A. Hong, M. Narandžić, C. Schneider, and R. S. Thomä, Estimation of the correlation properties of large scale parameters from measurement data, *Proceedings of IEEE 18th International Symposium on Personal, Indoor, Mobile Communications, PIMRC 2007*, Athens, Greece, December 2007.
28. C. Schneider, M. Narandžić, M. Käske, G. Sommerkorn, and R. S. Thomä, Large scale parameter for the WINNER II channel model at 2.53 GHz in urban macro cell, *Proceedings of IEEE 71st Vehicular Technology Conference, VTC 2010-Spring*, Taipei, Taiwan, May 16–19, 2010.
29. A. Böttcher, C. Schneider, M. Narandžić, P. Vary, and R. S. Thomä, Power and delay domain parameters of channel measurements at 2.53 GHz in an urban macro cell scenario, *Proceedings of the Fourth European Conference on Antennas and Propagation, EuCAP 2010*, Barcelona, Spain, April 12–16, 2010.
30. M. Narandžić, P. Kyösti, J. Meinilä, L. Hentila, M. Alatossava, R. Rautiainen, Y. L. K. de Jonk, C. Schneider, and R. S. Thomä, Advances in "Winner" wideband MIMO system-level channel modelling, *Proceedings of the Second European Conference on Antennas and Propagation*, pp. 1–7, Edinburgh, U.K., 2007.
31. M. Narandžić, M. Käske, C. Schneider, M. Milojević, M. Landmann, G. Sommerkorn, and R. S. Thomä, 3G-Antenna array model for IST-WINNER channel simulations, *Proceedings of IEEE Vehicular Technology Conference (VTC'07-Spring)*, pp. 319–323, Dublin, Ireland, April 2007.
32. L. Correia, *Mobile Broadband Multimedia Networks. Techniques, Models and Tools for 4G*, Academic Press, Oxford, U.K., 2006.
33. M. Narandžić, C. Schneider, and R. S. Thomä, WINNER wideband MIMO system-level channel model—Comparison with other reference models, *Proceedings of 54th Internationales Wissenschaftliches Kolloqioum*, Ilmenau, Germany, 2009.
34. M. Narandzic, C. Schneider, R. Thoma, T. Jamsa, P. Kyosti, and X. Zhao, Comparison of SCM, SCME, and WINNER channel models, *Proceedings of the IEEE 65th Vehicular Technology Conference, VTC 2007-Spring*, Dublin, Ireland, April 22–25, 2007.
35. P. Kyösti, MATLAB SW documentation of WIM2 model, IST-4-027756, Wireless initiative new radio—WINNER II, September 23, 2008.
36. WINNER+ Evaluation Group, http://projects.celtic-initiative.org/winner+/WINNER+%20Evaluation%20Group.html (accessed March 2014).

37. C-Code for report ITU-R M.2135 channel model implementation, http://www.itu.int/oth/R0A06000024/en (accessed March 2014).

38. J. Ellenbeck, O. Mushtaq, F. Sheikh, and Q. S. Shahrukh, Calibrating an efficient C++ implementation of the ITU-R M.2135 channel model for use in system-level simulations, *International Workshop on Propagation and Channel Modeling for Next-Generation Wireless Networks*, Lyon, France, March 2011.

39. P. Popovski, V. Braun, H.-P. Mayer et al., ICT-317669-METIS, Deliverable D1.1, Scenarios, requirements and KPIs for 5G mobile and wireless system, April 2013.

40. P. Agyapong, V. Braun, M. Fallgren et al., Simulation guidelines, ICT METIS, Deliverable D6.1, October 30, 2013, https://www.metis2020.com/wp-content/uploads/deliverables/METIS_D6.1_v1.pdf.

41. The METIS 2020 Project, https://www.metis2020.com/documents/simulations/ (accessed March 2014).

14 Enhancing Simulation Environments with TRAFIL

Christos Bouras, Savvas Charalambides, Michalis Drakoulelis,
Georgios Kioumourtzis, and Kostas Stamos

CONTENTS

ABSTRACT

This chapter presents TRAFIL, a comprehensive tool for enhancing execution of simulations. It provides an overview of the tool, its architecture, and its functionalities. It explains how TRAFIL enhances the entire simulation procedure including graphical setup of simulation scenarios, automated execution of simulations, flexible handling and storage of simulation trace files, and presentation of plots based on processing of simulation results. It presents the concept of metafiles that provides TRAFIL with the flexibility to handle heterogeneous simulation environments. The chapter also compares TRAFIL performance with other similar tools and finds that it offers significantly improved performance. It, therefore, concludes that TRAFIL offers both a rich set of simulation enhancement functionalities and top performance.

14.1 INTRODUCTION

This chapter presents TRAFIL (TRAce *FIL*e [1]), a comprehensive tool for enhancing execution of simulations. It wraps the entire simulation procedure from scenario setup to results analysis, including graphical setup of simulation scenarios, automated execution of simulations, flexible handling and storage of simulation trace files, and presentation of plots based on processing of simulation results. TRAFIL is based on the NS-2 (network simulator) but is built on an extensible architecture that allows any NS-2 additional plug-in to be configured and supported and can even be extended to process trace files from other simulators.

This chapter focuses on TRAFIL architecture and features and intends to serve both as an introduction and presentation of the tool and also as a general discussion of processing techniques for simulation results, by detailing the reasoning for the choices made in the tool architecture. It also introduces the design considerations that enable the tool to be extensible, which may be useful for building flexible and usable tools around existing environments.

TRAFIL aims to make the execution of a great number of network simulations quicker and the extraction of results from a large amount of data more flexible and productive. It offers the possibility to design, create, execute, and review NS-2 simulation scenarios and also offers postsimulation trace analysis functionalities. It is therefore a complete wrapper around the NS-2 simulator, allowing the user to perform all steps from presimulation design to actual simulation execution in an automated way and fast and convenient postsimulation analysis of potentially large amount of data.

In order to accomplish the postsimulation tasks, TRAFIL presents a novel way of interpreting, parsing, reading, and eventually using NS-2 trace files. It introduces the notion of *metafiles* and *submetafiles* throughout the procedures of trace file recognition and parsing, making the overall analysis operation substantially efficient and faster than alternative approaches. Metafiles and submetafiles are used to encode NS-2 trace file structures enabling a more abstract approach to the trace file processing operation. Furthermore, TRAFIL facilitates the overall trace file analysis task by offering the opportunity to store each trace file as well as every quality of service (QoS) measurement produced for each trace file. Following the trace file recognition and processing operations, the information contained in a trace file is presented through a graphical user interface (GUI) offered by TRAFIL along with a variety of data, metrics, and statistics related to simulation results. Finally, the tool offers the opportunity to execute custom structured query language (SQL) queries to the local database and to completely automate the simulation procedure by enabling the user to execute NS-2 scripts as well as perform a simulation of a video transmission using the Evalvid-RA framework.

The rest of this chapter is structured as follows: Section 14.2 presents related work for other similar network simulation enhancement tools. Section 14.3 gives an overview of TRAFIL and Section 14.4 focuses on the metafiles concept. Section 14.5 presents in detail the graphical simulation setup capabilities of the tool, Section 14.6 its trace file analysis functionalities, Section 14.7 its simulation execution wrapper, and Section 14.8 the way that TRAFIL can present simulation plots and results, as well as its performance compared to other similar tools. Section 14.9 summarizes our conclusions and Section 14.10 closes the chapter with our future work plans.

14.2 RELATED WORK

Similar work [2] and tools have been developed that produce statistics of a simulated network's behavior. Some of these projects ([3,4]) have integrated NS-2 unlike TRAFIL that uses NS-2 trace files to produce the requested statistics. Also there are tools like jTrana, Trace Graph, and NS-2 trace analyzer that offer the opportunity to analyze NS-2 trace files by producing numerous statistics, measurements, and charts.

Trace Graph ([5,6]) is an NS-2 trace file analysis tool. This tool provides many options for analysis, including a variety of charts and statistical reports. It is implemented in MATLAB® 6.0 and can be compiled to run without MATLAB. This tool also gives the user the ability to extract from a given NS-2 trace file useful statistics through a GUI. The kind of statistics that can be obtained includes node statistics, network statistics, and QoS metrics. It also produces 2D and 3D graphs for measurements like cumulative sums, throughput, throughput versus delay, jitter, packet ID's, and other common statistics. Finally, this tool supports the following NS-2 trace file formats: old wireless, new wireless, and wired.

jTrana [7] is a Java-based NS-2 wireless trace analyzer. It can be used to analyze the NS-2 wireless simulation trace files through a GUI. Features of jTrana include production of overall network information and plotting of numerous charts regarding that information. jTrana supports both wired and wireless trace files and uses a MySQL database to store the trace file that is subject to analysis at a given time.

NS-2 trace analyzer ([8,9]) is a command line tool written in C/C++ and is designed for use in all OS platforms and Cygwin. As the previous analysis tools, NS-2 trace analyzer can be used to obtain common network statistics using the trace file from a simulation. This tool does not offer the opportunity to create charts regarding the statistics that the user retrieves about a simulation.

The aforementioned tools provide useful information regarding only a specific simulation scenario, and therefore, all the metrics and results refer to only one trace file. Furthermore, these tools do not provide the user with the opportunity to store each trace file he or she has analyzed locally, for instance, in a database, so that he or she can reuse it without having to reopen it. In order to extract information regarding another simulation, the user has to load another trace file. This is a rather slow task and it can be acceptable for small-sized trace files, but when it comes to simulations that produce large trace files, this process can be considerably time consuming. Also, in order to alter the contents of a trace file and compare the results with a previous analysis, a user has to reload the trace file every time. Adding to this is the fact that the results of the analysis that a trace file is subjected are only saved to text files. It is up to the user to keep them organized and safe so that he or she can be able to reuse them.

In terms of performance when it comes to loading a trace file, the earlier mentioned tools behave well for small-sized trace files, although when it comes to serious simulations that produce trace files in the orders of MB, the performance deteriorates significantly.

Finally, there is J-Sim ([10,11]), a Java-based open-source, component-based simulation environment based on the idea of the autonomous component architecture (ACA). As a result, components are one of the basic entities of J-Sim and they can be individually designed, implemented, and tested. Also the way components interact and act in regard to the data transfers is specified at system design time. J-Sim as is the case with TRAFIL is platform independent due to the programming language it is implemented in. In addition, J-Sim can be used in conjunction with scripting languages like Tcl or Python. The scripting languages are used to combine and hold together in one sense the different Java classes as it is done with NS-2 C++ classes and OTcl. In order for a simulation to be created, the basic package is the drcl.inet that contains the base classes defined in the abstract network model, as well as a set of utility functions and classes that facilitate creation of simulation scenarios.

Furthermore, except the Tcl/Java scripting that a user can incorporate to create and orchestrate a simulation, J-Sim can be used in combination with gEditor. gEditor is a GUI that serves as a front for J-Sim enabling the user to create the simulation plane without using Tcl/Java; gEditor takes up the responsibility to interpret the parameters and objects that have been requested and create the corresponding objects. It passes the appropriate components to J-Sim via its console and runs the whole simulation on behalf of the user. This is a very useful feature that gives the ability to rapidly and conveniently create and execute a simulation.

As it will be shown in the following sections, TRAFIL aims to offer a similar feature but is targeted though toward NS-2. TRAFIL enables the user to input a script that describes the simulation plane through a GUI. The description and parameters are used to produce trace files that are eventually subjected to analysis akin to the procedures of the aforementioned postsimulation analysis tools. In a few words, the final objective is to succeed in offering a tool that can be used to perform a complete simulation and its analysis in a flexible, effortless, and robust manner.

14.3 TRAFIL OVERVIEW

TRAFIL is a tool that can be used to automate the complete network simulation procedure. Namely, it offers the ability to graphically design and execute a network simulation as well as analyze its results.

TRAFIL supports network simulations by utilizing NS-2, and it comes with out-of-the-box support for designing NS-2 simulation scenarios, execution of these scenarios by invoking NS-2, and analysis of the produced trace files.

One of TRAFIL's most important features is that it disengages the user of the effort to learn NS-2's specifics in order to design a simulation. Although it offers the ability to execute an NS-2 OTcl simulation script (OTcl is the scripting language used for writing NS-2 simulations scripts), TRAFIL enables users to design the simulation plane using a GUI. Via this GUI, users can select all network components that make up a simulation and place them inside the simulation plane.

Moreover, in terms of postsimulation analysis, TRAFIL is not strictly bound to NS-2 trace files. Although TRAFIL supports every different trace file produced by NS-2, it can also be extended to support a variety of trace files. This is accomplished by creating custom *metafiles* (presented in detail in the next chapter) that describe a new trace file's structure. Metafiles were introduced in the postsimulation analysis domain by TRAFIL in order to offer a more abstract approach to processing simulations' results. They describe a trace file's structure and are used by TRAFIL to mine specific information required for producing charts and measurements.

Finally, TRAFIL executes the simulation scenario by invoking NS-2. However, TRAFIL also offers the ability to simulate video transmission scenarios by utilizing the Evalvid-RA framework [12,13]. Evalvid-RA is an added module to NS-2 that is based on the generation of a trace file (regarding a video file) and can support its rate-adaptive multimedia transfer. In order to use Evalvid-RA along with NS-2, specific preprocessing and postprocessing steps are required before and after the NS-2 simulation, respectively, that involve the usage of certain tools for video processing. TRAFIL automates the video transmission simulation procedure by incorporating Evalvid-RA's utility tools. Therefore, the user is able to define all the parameters as he or she would have done when executing a normal Evalvid-RA simulation procedure and specify the simulation script he or she wants to test. TRAFIL carries out the simulation procedure returning its results in a user-friendly manner.

This section gave a brief introduction to TRAFIL's main operations. The next section presents TRAFIL's actual architecture and discusses each component's role in performing TRAFIL's main operations.

14.3.1 TRAFIL Architecture

TRAFIL's architecture follows a three-tier model as can be seen in Figure 14.1, and the three basic tiers are presentation layer, business layer, and database access layer.

The presentation layer introduces the user to TRAFIL's functionality and is depicted in Figure 14.2. Via the presentation layer, the user can invoke any utility offered by TRAFIL given that there is a trace file loaded. A trace file can be loaded either by opening a new trace file from the file system

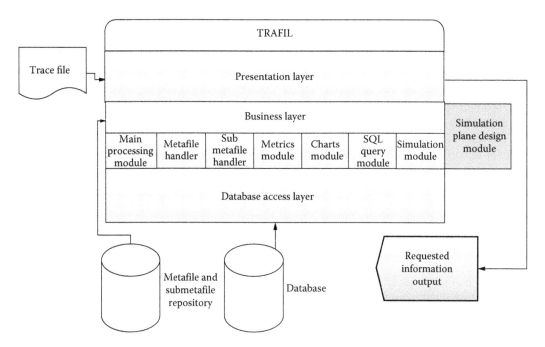

FIGURE 14.1 TRAFIL architecture.

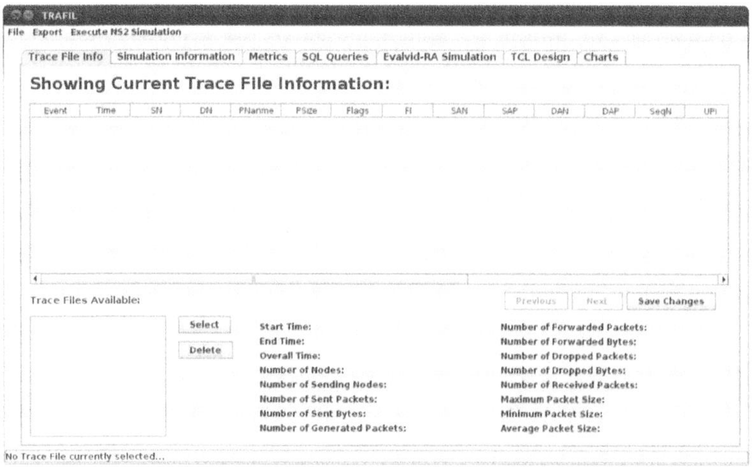

FIGURE 14.2 TRAFIL presentation layer.

or by loading a preexisting one from TRAFIL's database. As shown in Figure 14.2, no trace file is currently selected and thus no data are shown.

Every user request is carried out in the business layer and the results are returned and depicted at the presentation layer. As shown in Figure 14.2, TRAFIL can produce metrics and charts (at the appropriate tabs) based on the trace file information as well as general simulation information that summarizes the events that took place in the specific simulation scenario. The most important part of the general simulation information is also depicted in TRAFIL's Trace File Info tab along with the selected trace file's first 50,000 lines. Moreover, via the SQL Queries tab, users can issue queries directly to TRAFIL's database if they require more specific measurements that are not currently supported by TRAFIL. Finally, TRAFIL supports the novel feature of graphically designing simulation scenarios via the Tcl tab and their execution via the Execute NS-2 Simulation menu item. Finally, another unique feature is the ability to execute video transmission simulations using Evalvid-RA in conjunction with NS-2 via the Evalvid-RA Simulation tab.

The aforementioned TRAFIL features are handled by distinct business layer modules as shown in Figure 14.1. The core module is the Main Processing module and is responsible for the identification of the selected trace file's type. Namely, it determines whether it is a normal, old wireless, new wireless, or a user-specified trace file. Moreover, it is responsible for parsing, processing, and storing the trace file's data to TRAFIL's database if it is a new trace file or loading its data from the database if it is a preexisting one. The metafile and submetafile modules, as their name suggests, are responsible for providing an interface to other modules that need to access metafile data for their operations, that is, during the trace file parsing and identification procedures as we mentioned earlier. The Metrics module is responsible for calculating the following QoS measurements: packet delivery rate, throughput, minimum end-to-end delay, maximum end-to-end delay, average end-to-end delay, delay jitter, average delay jitter, minimum delay jitter, maximum delay jitter, and packet loss ratio. The Charts module plots specific charts that refer to either specific simulation nodes or the communication between two nodes. The SQL Query module is responsible for validating and executing the user-issued SQL queries. Finally, the Simulation and Simulation Plane Design modules support TRAFIL's two novel features. The latter is responsible for the operations required for producing a valid NS-2 simulation script based on the graphical design of the simulation plane. The former is responsible for handling either the execution of a simple NS-2 script that is specified by the user or a video transmission simulation. In the video simulation case, it is responsible for coordinating the execution of a series of steps that involve using external programs that process the input video based on user-specified parameters.

The database access layer interacts solely with the business layer and is responsible for storing data to the local database or returning the data requested by the business layer.

Finally, as Figure 14.1 depicts, TRAFIL uses a MySQL database to store each trace file's data. The database holds each trace file's data in a distinct table and its design is optimized for faster data transfer and retrieval. As shown in Ref. [14], TRAFIL manages to outperform similar tools in terms of trace file processing and storage speeds. Moreover, TRAFIL has a metafile and submetafile repository that contains the actual metafiles and their submetafiles. Each metafile's submetafiles are stored in a unique folder named after the metafile they belong to. This way the loading of the submetafiles for a specific metafile becomes simpler. Thus, in the case that a user wants to introduce his or her own metafile, this convention should be followed in order for TRAFIL to support it. Namely, the metafile should be added to the repository and its submetafiles (if any) to a unique folder named after the metafile.

14.4 METAFILES

Metafiles have been used in various applications to describe the structure or the content of another file. The most useful aspect of metafiles is the fact that they render the applications that use them more generic or abstract as they become independent of the content or the format of the input. It is evident that the use of metafiles is indeed popular and causes applications to be more robust and adaptable, but all trace file analysis tools until now have not made any use of them. On the contrary, they encode the structure of a trace file internally. TRAFIL on the other hand makes use of metafiles during the trace file parsing, processing, analysis, and storage procedures.

In order to identify and analyze a trace file, there must be a way to know its structure and to expect in some degree the input. That is why TRAFIL uses metafiles. Metafiles encode the structure of each different trace format produced by NS-2; they contain information about the number of fields, number of columns, the names of each column, and what data type each column is. In other words, they contain all the necessary information to describe the data of a trace file.

Using each metafile, TRAFIL is no longer dependent on the structure or the data types of an input. If there is a metafile that is constructed correctly so that it can describe a specific trace file format, the tool will be able to process it. TRAFIL is not even dependent on the number of metafiles;

```
NumberOfFields 14
NumberOfColumns 12
UniqueCounter 1
-name event -type char(1) -index 0 -unique +
-name time -type double -index 1
-name SourceNode -type int -index 2
-name DestinationNode -type int -index 3
-name PacketName -type varchar(20) -index 4
-name PacketSize -type int -index 5
-name Flags -type varchar(7) -index 6
-name DestinationAddress -type int -index 9 -delimiter .
-name FlowID -type int -index 7
-name SourceAddress -type int -index 8 -delimiter .
-name SourcePort -type int -index 8
-name DestinationPort -type int -index 9
-name SequenceNumber -type int -index 10
-name UniquePacketID -type int -index 11
-TimeRelated -column time
-NodeRelated -column SourceNode -column DestinationNode
-PacketSize -column PacketSize
-SendingNodes -column SourceNode -column SourceAddress
-GeneratedPackets -column SourceNode -column SourceAddress -column UniquePacketID
-ReceivedPackets -column DestinationNode -column DestinationAddress -column UniquePacketID
-ForwardedPackets -column SourceNode -column SourceAddress -column DestinationAddress -column UniquePacketID
-SentPackets -column SourceNode -column SourceAddress -column UniquePacketID
-DroppedPackets -column UniquePacketID
```

FIGURE 14.3 Metafile structure.

if there is not a metafile present that can be matched to an input, TRAFIL will acknowledge it and report the issue so that the user will construct the appropriate metafile.

The structure of a metafile is depicted in Figure 14.3. As Figure 14.3 shows, there are a number of different fields contained in a specific metafile. The metafile in Figure 14.3 is the one that encodes the structure of a normal trace file and it is utilized when TRAFIL needs to process a trace file of that format. The first three fields are mandatory for every metafile and are extremely important in the trace file recognition phase. The *NumberOfColumns* element is used to demonstrate the actual number of elements that are present in each line read from a trace file of normal trace format and are separated by white spaces.

The *NumberOfFields* parameter refers to the number of fields that must be extracted from the data that are read from each line based on the NS-2 manual. In order to extract these data fields, some modifications must take place on some of the elements contained in each line. These modifications are described by the metafile using special flags as will be explained. Furthermore, the *NumberOfFields* is used to ensure that the metafile's structure is correct. The value defined by the *NumberOfFields* parameter must be the same as the number of lines in the metafile that start with the *-name* flag since these lines are the ones who describe the trace file's structure. These lines contain information about the actual elements of a trace file of a specific format as described by the NS-2 manual, and if their number is not equal to the *NumberOfFields*, the structure of the metafile is considered corrupt and the metafile cannot be used.

The last element of the first three is the *UniqueCounter* flag; this flag is used to define the number of unique characters that must be matched in each line read from the trace file during the trace file recognition process. If a trace file's lines contain the number of unique characters that a metafile defines, then the trace file is matched with it and TRAFIL can start the actual trace file processing. The *UniqueCounter* though only defines the number of unique characters that must be matched; the actual unique characters are marked by the *-unique* flag as it shown in Figure 14.3.

Following the first three parameters are the fields that describe the columns of the trace file. For a normal trace file, these lines are shown in the preceding figure. There are 14 different columns and the *-name* flag is used to define the name of the column, the *-type* flag is used to define the data type of this column, and the *-index* flag is used to show its index inside the actual trace file. These three fields are mandatory for every line of the metafile that describes a trace file's column because these fields are also used by TRAFIL to create the table that will contain the trace file's data in the database. For this reason, the data types that will be defined using the *-type* flag must conform to MySQL's supported data type syntax.

Besides these three flags, there are also some other flags that serve specific purposes like declaring a unique sequence of characters that must be present in this column. For this purpose, the *-unique* flag is used; this flag defines a character sequence that must be matched for all elements of a column for which it is set. Also there must be as many *-unique* flags as are defined using the *UniqueCounter* parameter. This way if all the unique character sequences are found, they can be verified using the *UniqueCounter* parameter and the match can be established.

Another utility flag is the *-delimiter* flag; it is employed to signal a sequence of characters that will be used to divide a complex element into two other components. This is the case that was mentioned earlier in which the number of columns is different than the number of fields that must be extracted from the line. In these cases, some elements are connected by a character sequence that is signaled by the *-delimiter* flag, and that way, TRAFIL can separate them.

Finally, there are two other flags: the *-startsWith* and *-endsWith*. These two as their name suggests are used to define some character sequences that elements of any column for which they are set either start or end with. These character sequences are not useful and must be removed in order to retrieve the useful information; using these flags, TRAFIL can remove these characters.

The remaining lines are the *metric fields*. These fields are used to denote the standard metrics the tool produces for each trace file. Each flag states the metric itself and is followed by the columns that will be used to produce that specific metric. The column names are recognized by the *-column* flag that precedes them. In each metafile, all these metric flags must always be present; if they are not, the metafile is considered to be corrupt and the metric production phase cannot be completed. It is obvious that for different kinds of trace files, the number and type of columns used to extract each metric might be different.

14.4.1   SUBMETAFILES

As mentioned in the previous subsection in order to process each trace file, TRAFIL introduces the idea of using metafiles to describe the format of the input. It is often thought necessary for a metafile to have some other utility files that can be used in situations where a metafile alone is not adequate. In the trace file processing procedure, these situations occur when a trace file may include an arbitrary number of different header fields in each line and therefore have a lot of alternative forms.

The structure of a trace file depends greatly on the simulation scenario. Even in its own content, a trace file may contain lines that are different with each other in the number of elements they contain. This is usually the case when the scenario involves traffic with different packet types traveling along the simulated network and using different routing protocols.

That is why TRAFIL uses a number of submetafiles to complement the use of metafiles. Each trace file can log a number of different header fields, so a very straightforward way to handle all the different patterns is also to enable each metafile to have a number of different submetafiles.

Actually for every one of the three different trace file formats that NS-2 produces, TRAFIL has a different metafile, and for each metafile, there are a number of submetafiles that is the same as the number of different header fields a trace file of a specific format can contain.

The structure of the submetafiles that are used along with the metafile that encodes the structure of a normal trace file is shown here.

Figure 14.4 depicts the structure of the submetafile used to represent a satellite packet's header information that is logged in a normal trace file. Figure 14.5 shows the structure of a submetafile that represents a TCP packet's header information. The structure follows exactly the same conventions

```
NumberOfFields 4
NumberOfColumns 4
UniqueCounter 4
-name SourceLatitude -type double -index 1 -unique .
-name SourceLongitude -type double -index 2 -unique .
-name DestinationLatitude -type double -index 3 -unique .
-name DestinationLongtitude -type double -index 4 -unique .
```

FIGURE 14.4   Satellite submetafile.

```
NumberOfFields 4
NumberOfColumns 4
UniqueCounter 1
-name AckNumber -type int -index 1
-name FlagsTCP -type varchar(7) -index 2 -unique 0x
-name HeaderLength -type int -index 3
-name SocketAddressLength -type int -index 4
```

FIGURE 14.5   TCP submetafile.

as were described earlier for a metafile. The same flags are used as in a metafile and the same first three mandatory fields must always be present as in a metafile. The submetafiles shown earlier are only used after the trace file is matched with the normal metafile. They are not used in the trace file recognition process; they are used after this process to enable the correct transfer of the trace file to the local database. They are also used when a user wishes to load a trace file from the database that was matched with the normal metafile.

14.5   GRAPHICAL SIMULATION SCENARIO SETUP

TRAFIL has taken successive steps toward automating the NS-2 simulation experience. An advanced step of that process was simplifying and automating most of the scenario creation process.

All simulations in NS-2 are presented in an OTcl script that describes the topology and the simulation parameters. This has to be prepared beforehand and requires knowledge of OTcl language (or even C++) from the user, while also being familiar to typical and advanced NS-2 objects and their parameters. Therefore, we introduced a new functionality to TRAFIL that enables the user to design the simulation using a graphical interface.

This interface is designed in such a way that it is easy to learn and allows for quick scenario setup. Through the GUI, users can pick network components and place them inside the simulation topology panel. These components can be typical network objects such as wired or wireless nodes and links that are formed between them. Each of them can be selected from a palette of components next to the simulation topology panel, and after its placement, it can be customized via a pop-up menu that contains any available parameters for that object. All these network components are identical to the objects supported by NS-2; therefore, after the user has finished designing the topology, it is translated to an OTcl script that can be executed by NS-2.

This new module is not separate from the rest of TRAFIL functions. Rather, it is integrated in a way that it synergizes with the rest functionalities. After describing the topology, users can generate the OTcl script and, if NS-2 is available in the current environment, proceed to the simulation execution and postsimulation analysis using TRAFIL tools only. This gives the whole process better transparency, since it omits the underlying procedures, jumping from scenario design to execution, and ultimately to results analysis. Given TRAFIL's performance in fast trace file analysis, this integration also allows for quick changes in the scenario, which in a sense makes it easier for users to follow a trial-and-error method until they reach the desired scenario results.

14.5.1   Simulation Design Plane

The simulation design feature enables the user to describe a network in a way that is closer to designing rather than programming. Figure 14.6 describes its architecture that is organized in four layers.

The first (and most important) layer is the design layer. A key part of this is the simulation topology panel or simulation design plane, a design environment similar in terms of layout to most modern design programs. It consists of the component palette, the design panel, and several pop-up menus.

The palette includes all of the input options for the design panel, such as node types, links, lists of connected objects, and a few buttons that are responsible for other scenario parameters. The rest of the simulation design plane is filled with the design panel, where the network topology schematic representation will be displayed. Depending on the network object type selected from the palette, an

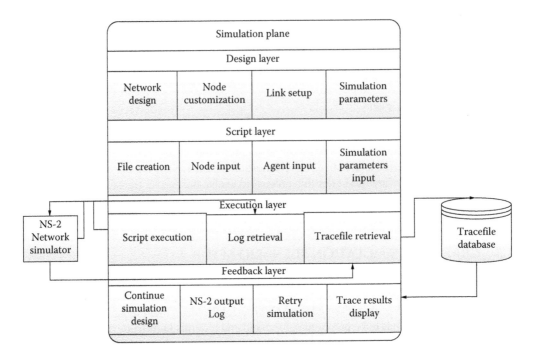

FIGURE 14.6 Simulation plane architecture.

according shape will be painted there, resembling a network component. That shape also provides access to the menu related to its individual component.

All menus appear in separate windows and allow for node customization, connecting existing wired nodes using links or specifying parameters necessary for the scenario that are not part of the visual design. These parameters include simulation scheduling and output file names. Users can keep the menu windows open while still accessing the design panel to edit or add more features to the network.

The next layer is the script layer. It transforms the network topology as shown, and its underlying data, and creates a simulation script file appropriate for NS-2, using the OTcl language's rules. The script file generation follows a specific format native to TRAFIL, which allows for a network layout reconstruction in the design panel, for later use. This allows users to save and load again later their work or manipulate scripts that were previously created by TRAFIL without having to rearrange components in the topology layout. Nonetheless, the specific format does not mean a specific file type too. The output files are still normal Tcl file types.

The next layer is the execution layer. This is an optional layer; depending on the user's operational system as well as whether NS-2 is installed, the user can immediately test the new scenario. The script generated in the previous layer is forwarded to NS-2, using the required syntax to perform the simulation. After the simulation is complete, the execution layer retrieves the output log as well as the files produced by NS-2 and sends them to the next layer. The resulting trace file is automatically parsed and saved in TRAFIL's database, producing metrics along the way.

The last layer is the feedback layer. Here TRAFIL displays a menu with the option to display the data or the output log produced by the simulation. Moreover, there are options to rerun the simulation or return to the design layer and continue where the user left off.

14.5.2 NODE CREATION AND CONFIGURATION

One of the most important components of a network simulation is a node. A node can represent a variety of entities, but its importance lies in the fact that it is the means of introducing data traffic in a network.

NS-2 supports both wired and wireless nodes, each with its own unique parameters. Therefore, TRAFIL enables users to create any of these node types using the GUI. The node is selected from a palette and can be either wired or wireless. The selected node can be placed and dragged anyway inside the design panel.

TRAFIL's new module gives users the opportunity to create a whole network pattern of nodes and assign them any available options, using a pop-up menu designed for easy node parameter configuration. In order to simplify the process, we avoided manual entry of values as much as possible, using drop-down menus and preset values. This way the user can maintain control of the topology without having to deal with OTcl node, or even NS-2 documentation, to find out what his or her options are. All NS-2 available node parameters can be edited and correspond to the ones that are set using OTcl. The difference lies in the fact that the user is not obligated to have any previous knowledge of the actual commands to set them. In order to edit the parameters of a specific node, the user can simply double-click or right-click a node in the design panel, revealing the parameter edit pop-up window.

14.5.3 AGENTS: TRAFFIC GENERATORS/APPLICATIONS

In the previous subsection, we explained how to create and configure network nodes. Although nodes and their proper setup are a major part of a simulation topology, they are not enough to power up a scenario, since they need a way to send packets to each other. For this purpose, each node has to be associated with an agent, and possibly with a traffic source as well.

Figure 14.7 illustrates how an agent and a traffic source are connected. NS-2 offers a variety of agents (sending or receiving) as well as traffic sources. Traffic sources are divided in two major categories: traffic generators (like the one used in Figure 14.7) and already simulated applications. Traffic sources can be attached to almost any type of agent. This is not the case though for simulated applications, which send their packets through a TCP transport agent only.

TRAFIL uses each node's configuration menu to assign agents and traffic sources to that node. Currently it provides support to all current NS-2 traffic sources and the most common agents, which are TCP, User Datagram Protocol (UDP), and Null. The support integration is designed in such a way so that when a user selects a type of agent for a protocol, TRAFIL shows the appropriate applications. Similarly, selecting an application shows you all of its specific NS-2 parameters.

14.5.4 LINKS

However, creating and configuring nodes alone is not always enough to define a topology. Some simulation scenarios use wireless nodes, while others define wired networks. In the latter case, links have to be defined between nodes. This can be done using topology panel, and all link details are

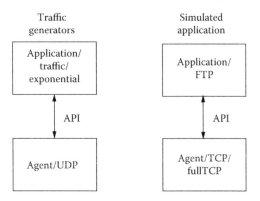

FIGURE 14.7 Connection between agent and traffic source in NS-2.

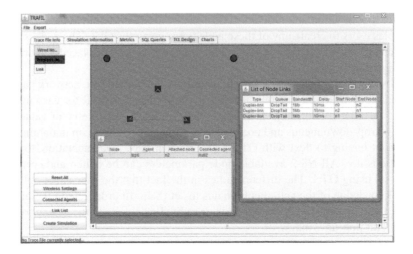

FIGURE 14.8   TRAFIL simulation design.

shown in a table window that displays existing links. Specific link parameters, such as bandwidth, delay, and queue type, are configured using this window. There is a variety of link types available in NS-2 that are supported by TRAFIL.

In order to establish a link between two wired-type nodes, a user has to select a new link from the palette and then click on the two nodes to be linked. The new link is shown as a straight line between the nodes, and its parameters can then be edited in the window described earlier.

14.5.5   SIMULATION DESIGN EXAMPLE

Figure 14.8 depicts the GUI of TRAFIL's design module. In the example shown, three wired nodes (squares) are set up and linked together. There is also a wireless node present (circle). Both links are listed in the open link window, showing their editable parameters.

14.6   TRACE FILE ANALYSIS

This section describes one of TRAFIL's core procedures and functionalities. Before any analysis on the results of a simulation, a trace file is firstly classified in terms of type and properly processed in order for its data to be retrieved and stored. Following these operations, users can then retrieve measurements and plot charts based on the trace file data.

14.6.1   IDENTIFICATION, PROCESSING, AND STORAGE

In order for a meaningful analysis of a trace file's data to take place, the trace file in question has to be classified. A trace file can be either one of NS-2's supported formats (normal, old wireless, or new wireless) or a user-specified trace file format. The identification of a trace file is essentially a procedure of matching it with the appropriate metafile. In other words, TRAFIL identifies the metafile whose encoding describes the trace file's structure. In the case that no such metafile exists, then either the trace file structure is wrong or the metafile repository does not contain the appropriate metafile and the user should create one.

Having identified a metafile that properly describes the trace file in question, TRAFIL proceeds to parse, process, and finally store its data to the local database. Parsing and processing requires the use of all submetafiles that belong to the correct metafile since a trace file contains a variety of lines depending on the packet type or the events.

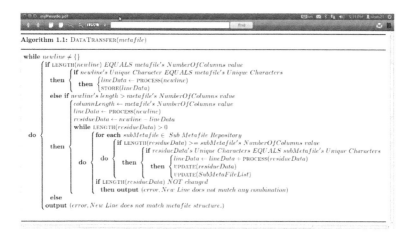

FIGURE 14.9 Trace file processing algorithm.

Using the metafile and submetafile data, the trace file is processed line by line. Figure 14.9 presents the pseudo code for the trace file processing operation, which is one of the most important of TRAFIL's procedures since without properly and efficiently retrieving the data, no further analysis can be made.

As shown in Figure 14.9, each line's number of elements is the first condition checked. That number is first checked if it matches that of the fields of the metafile, and if that condition holds, then the metafile is used to process the line in question. In addition to the number of elements criterion, an additional check is made to establish that indeed the line is described by the metafile using some *unique characters* that are defined in the metafile as it is also shown in Figure 14.9.

If the number of elements in a line is greater than the number specified by the metafile, the submetafiles are used. TRAFIL attempts to identify a combination of submetafiles that can be used to process the additional fields. Thus, TRAFIL matches the additional fields with submetafiles until no fields are left that have not been processed. If there is a subset of additional fields that cannot be matched to any submetafile, then the line is flagged as erroneous.

The identification and processing phase terminates when all the trace file lines have been processed. The next step is the data transfer to TRAFIL's local database and their storage. A new table is created and is named after the trace file. Its structure is based on the metafile that was used along with all its submetafiles.

14.6.2 STATISTICS CALCULATION

Having correctly processed and stored a trace file or loaded a preexisting one, a variety of information can then be retrieved. TRAFIL automatically calculates and presents the trace file's general simulation information without any user involvement. This information refers to the simulation as a whole and also to each specific simulation node. However, the user should specify the node for which he or she wants to view the general simulation information. More specific QoS metrics are calculated when the user requests them since such metrics refer to the communication between two specific nodes. The remainder of this subsection describes the exact general simulation information that can be retrieved using TRAFIL.

14.6.2.1 General Simulation Information

The general simulation information considers the simulation as a whole and can be viewed via TRAFIL's Simulation Information tab. It includes information such as the simulation's start time, end time, and overall duration time. General simulation information about the simulation's nodes includes the number of nodes and number of sending nodes. Moreover, the general simulation information includes communication information in terms of packets and bytes, more precisely, information such as the number of sent packets, number of received packets, number of dropped packets, number of generated packets, and number of forwarded packets. For each one of the aforementioned values, there is a corresponding one calculated in bytes.

In the case of trace files that refer to wireless scenario simulations, TRAFIL presents the following additional information: the number of generated packets and bytes as well as number of received packets and bytes for each of the AGT (application layer), RTR (routing layer), and MAC (medium access layer) trace levels that NS-2 supports. Thus, users who are interested only in the packets and bytes regarding a specific communication layer can specify that layer and retrieve the information they want.

14.6.2.2   Node-Specific Information

TRAFIL can also calculate similar information to the general simulation information, but for specific nodes again via the Simulation Information tab. Users can specify the node of interest and TRAFIL will calculate the general information that is specific to that node. This kind of information includes statistics such as the number of sent, received, dropped, forwarded, and generated packets. Similar to the general simulation information, the same information is also shown in terms of bytes for that node. Furthermore, for wireless scenario simulations, TRAFIL provides additional metrics for the AGT, RTR, and MAC trace levels as it does for the general simulation information mentioned in Section 14.6.2.1.

14.6.2.3   QoS Parameters

In addition to general simulation information, TRAFIL enables users to calculate more specific metrics via the Metrics tab. These are metrics that refer to the communication between simulation nodes. Thus, users specify the node pair of interest and TRAFIL provides the metrics that were mentioned in Section 14.3.1 (TRAFIL architecture). We specify here that the calculation of the delay jitter–related measurements is based on the RFC 3550 [15] for RTP packets. Furthermore, if the trace file refers to a wired simulation, the user must select the layer for which the calculation will be conducted. The options in this case are link layer or physical layer. If the simulation refers to a wireless scenario, then the user must select between the three trace levels: AGT, RTR, and MAC. After all the appropriate parameters have been set, the measurements can then be calculated. If there is no communication between the specified nodes at the particular layer, then all the metrics are set to zero.

14.6.3   Plotting Charts

Chart plotting takes place via TRAFIL's Charts tab. TRAFIL supports the following charts: packet delivery rate in packet/s, throughput in byte/s, delay fitter, and packet loss ratio. These charts can be drawn either for a node pair or for a single node. In addition, users should also specify the sampling rate in seconds and the trace level. For wired simulations, the appropriate levels are link layer and physical layer, and for wireless scenarios, the appropriate trace levels are AGT, MAC, and AGT. If there is no communication in the specified layer, then an empty chart is displayed.

14.6.4   User-Initiated SQL Queries

TRAFIL's database stores each trace file in its own table. Each table's structure is based on the metafile that was used to process it. Thus, a table's columns are as many as the metafile's fields and all its corresponding submetafiles' fields. Having each trace file stored in the database allows for faster loading and on-demand processing. Although the general simulation information and QoS metrics produced by TRAFIL cover the most common statistics, there is the possibility that a user will be interested in a measurement that currently is not offered by TRAFIL. Therefore, TRAFIL allows SQL queries directly to the database to alleviate this issue.

14.7 SIMULATION EXECUTION

Although TRAFIL was originally created as a postsimulation front-end framework, many presimulation features have since been added. For all these features to work, a function that communicates with NS-2 was created, allowing the user to execute any OTcl script using TRAFIL's GUI. The results of such simulations are automatically imported in the local trace file database, allowing direct and seamless access to the results. Of course, it is required that the operational system natively supports NS-2 (i.e., UNIX systems).

There are several circumstances where TRAFIL uses this function, which are listed and explained in the following.

14.7.1 SIMULATION DESIGN PLANE SCENARIOS

Following the process described in Section 14.5 (Graphical Simulation Scenario Setup), the user can use the generated script file immediately for simulation. TRAFIL's generated scripts use a special extra notation in the form of comments, so that a user can save and later load the work he or she has done. They also have a standard structure, which consists of four parts: script parameters (such as file names), node list, link list (and their parameters), agent information, and the simulation schedule.

If the Tcl script file creation is successful, TRAFIL saves it in a folder dedicated for that purpose. Then, the user is given the option to run it right away through TRAFIL's innate NS-2 script execution function. The results will be directly imported in TRAFIL's database. However, if the script creation and execution was not successful, for example, any parameters were missing or if the network layout was incomplete, the output of NS-2 will be shown in a report window, informing the user about the errors.

14.7.2 SIMULATION EXECUTION

After a script has been created, TRAFIL sends it to NS-2 and retrieves the outcome, displaying it to the user accordingly. There is no direct interaction of the user with NS-2, since a successful simulation with proper outcome will automatically be parsed by TRAFIL and stored in its local trace file database.

Of course, it is possible that the user can input his or her own script file for simulation, bypassing the simulation design stage. This can be done using the top menu bar, where the user specifies the script file to be simulated and starts the rest of the procedure right away, in a similar way to design plane scenarios.

14.7.3 VIDEO SIMULATION SCENARIOS

One of the most popular NS-2 add-ons is Evalvid-RA, which enables video stream simulation across a network. Being an add-on, it also uses similar OTcl script files (adding a few extra parameters, such as video input file), making the simulation result and metrics extraction fairly similar as that of any other NS-2 simulation.

To that end, TRAFIL has also a special module that specializes in Evalvid-RA simulations and their specific parameters. Due to the complexity that an Evalvid-RA script file might have, the user has to input his or her own script file that describes the simulation topology and schedule. However, TRAFIL significantly simplifies the process of simulating such a scenario, by taking over all pre- and postsimulation stages, such as media file conversion and NS-2 simulation parameters.

Figure 14.10 shows the Evalvid-RA module of TRAFIL and the parameters described earlier. As shown, the simulation procedure is split in two halves: the presimulation part (which includes the simulation itself) and the postsimulation. In the first part, the user sets the parameters for FFmpeg

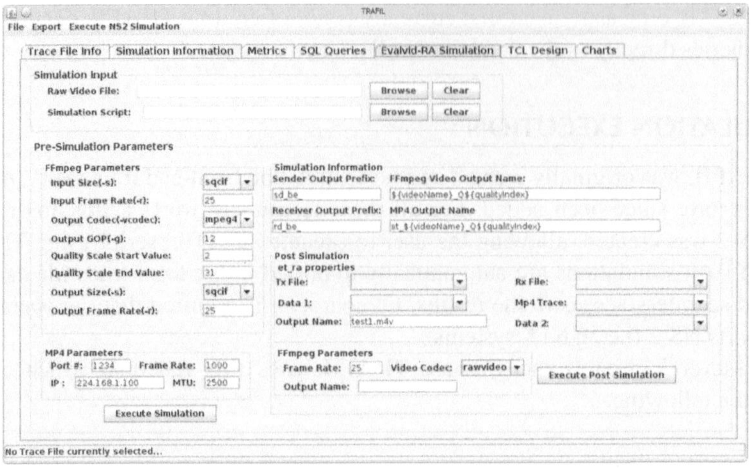

FIGURE 14.10   TRAFIL Evalvid-RA module.

and MP4, as well as the video file and script path files. The input video file, which can be in any video format (*.yuv, *.mov, *.mp4, etc.), is converted to MPEG-4 (*.m4v format) using FFmpeg and then gets prepared for simulation by the Evalvid MP4 tool. The preparation produces many different possible frame traces, which will be used by NS-2.

The second part of the procedure takes over the reconstruction of the video file and the reencoding to a common video format file. The reconstruction is done using *et_ra*, a modified version of the original Evalvid *et.exe* tool, which reads the produced packet Tx and Rx trace files as well as some .dat files that contain information that assists in assembling the resulting MPEG-4 file (in .m4v format again). Afterwards, FFmpeg is used to decode it back to its original format (presumably *.yuv). Then, peak signal-to-noise ratio (PSNR) is used to compare the decoded YUV file [16]. Finally, a report window shows the user the results of the simulation, as well as the produced files, which are typically stored in a default folder.

14.8   PLOTS AND RESULTS

In this section, we present a usage example of TRAFIL in which we open a new trace file and show the information that the tool can extract. We present the calculation of the general simulation information, general node information, QoS measurements between nodes, and charts that can be extracted using TRAFIL. The scenario that was used to create the sample trace file involved four wireless nodes of which three were stationary (nodes 0, 1, 2) and one mobile (node 3). The communication was between nodes 0–3 and 1–3 and the length of the simulation was 400 s. During this simulation, node 3 moved every 50 s either closer or away from nodes 0 and 1, and finally, at 300 s, it moved as far as it could reach from nodes 0 and 1 inside the topology grid. This simulation is clearly a small one and it is only presented in order to introduce TRAFIL and portray its capabilities.

The resulting trace file was given as an input to TRAFIL, and the procedures described in Section 14.6 took place in order to identify the input's format. When the trace file has been analyzed and processed, its contents are visible via the tool from a table as depicted in Figure 14.11. This is another feature that is unique in TRAFIL, and its purpose is to enable users to have all the information they could possibly need centralized and ready to use. The table's columns are created based on the metafile and submetafiles that were used to identify and process the trace file. In addition, the general simulation information, consisting of simple statistics regarding the scenario, is also

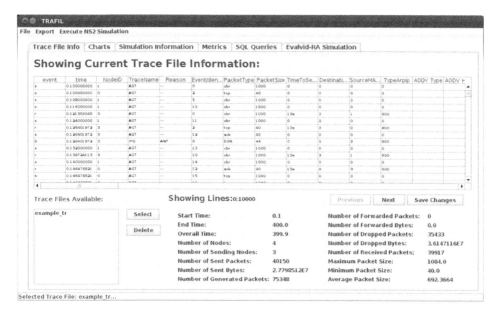

FIGURE 14.11　TRAFIL main view.

presented. The only nontrivial information is the number of generated packets and its difference to the number of sent packets. We consider as number of generated packets every packet that was produced by a node and as sent packets the number of generated packets in every node minus the number of packets that were dropped in the same node they were created. The general information can also be viewed along with some extra fields regarding packets and bytes in each trace level of wireless scenarios in another part of TRAFIL named simulation information. The measurements in that area of the tool are created automatically after the trace file has been successfully transferred to the database. When a trace file that was the result of a wireless scenario is given as an input for analysis, for every trace level, it contains information the corresponding extra fields are filled with the appropriate measurements. Thus, we calculate the same information for each trace level, and that is the reason why in the simulation information part of TRAFIL are three fields for the number of generated packets as well as for the number of generated bytes.

14.8.1　QoS Simulation Parameters Extraction

Once a trace file has been added to TRAFIL or loaded from the database, it can be used to extract various QoS parameters such as the throughput, end-to-end delay, jitter, and packet loss ratio. These parameters are some of the most commonly calculated after a simulation and at the same time the most useful and informative in order to evaluate the performance of a network simulation. The results for our experiment are shown in Figure 14.12.

　　Packet delivery rate is calculated by dividing the number of packets that were successfully sent and received between the selected nodes at the time of arrival of the first and last packets between the nodes. To calculate throughput, we divide the number of bits that were successfully sent and received between the selected nodes at the time of arrival of the first and last bits between the nodes. Each value is calculated in the space defined by the sampling rate.

　　The end-to-end delay for each packet between the two nodes is calculated by finding the difference in the arrival time and transmission time of every packet that was exchanged from the sender node to the receiver node.

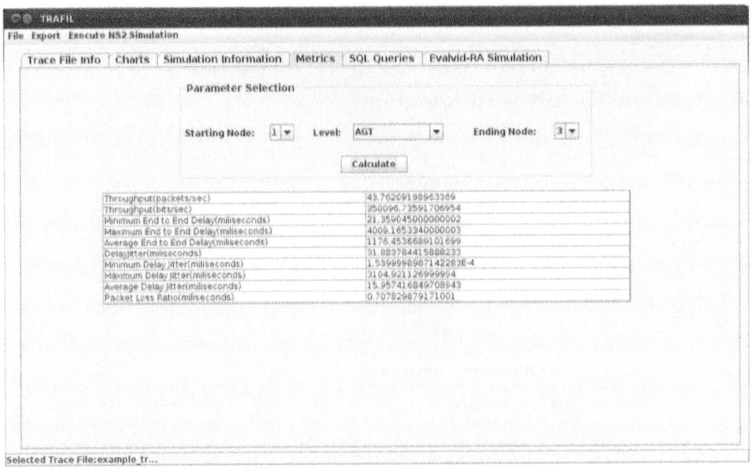

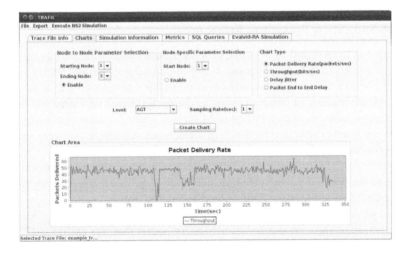

FIGURE 14.12 QoS parameter extraction.

The jitter-related metrics as we have mentioned earlier are calculated based on the RFC 3550. Finally, the packet loss ratio is calculated by finding all the packets that were sent and received between the two nodes, dividing their difference by the packets sent, and multiplying by 100.

14.8.2 Chart Plotting

A useful utility offered by TRAFIL is to plot various charts based on the information contained in a trace file as shown in Figures 14.13 and 14.14. The charts can refer either to a pair of nodes (Figure 14.13) or to a specific node (Figure 14.14). There are four types of charts that can be extracted using TRAFIL, which are the packet delivery rate, throughput, delay jitter, and packet loss ratio. Furthermore, the calculation in each case can be made in two distinct sampling rates: 1 and 5 s. The sampling rate defines the time interval in which we calculate the value of the selected chart. Finally, there is also the opportunity to define the communication layer for which the information will be collected. Namely, for wired scenarios that yield normal trace files, the

FIGURE 14.13 Chart plotting.

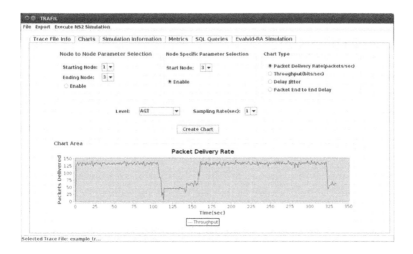

FIGURE 14.14 Specific node chart.

communication levels are link and network layers, and for wireless scenarios, the corresponding levels are MAC, RTR, and AGT layers. Thus, the user can define various parameters regarding the chart and obtain a more accurate result.

14.8.3 USER SQL QUERY EXECUTION

TRAFIL has been designed in such a manner that the user will be able to conduct his or her post-simulation analysis and retrieve results with the least possible work. Nevertheless, there is no way to predict and implement all the different functionalities that a user might require during the analysis of a trace file. Therefore, TRAFIL offers the ability to execute SQL queries directly to the database in order to retrieve information from trace files that is currently not offered by the tool. The only queries that are supported are select queries, and the reason is to protect the database from user errors that might lead to corrupting the system. An example of executing a query to retrieve all received packets from a trace file is shown in Figure 14.15.

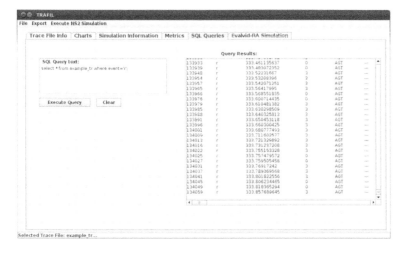

FIGURE 14.15 SQL query execution.

14.8.4   NS-2 VIA TRAFIL

14.8.4.1   Executing OTcl Scripts

Every NS-2 simulation scenario is described and constructed using the OTcl scripting language. A user has the ability to create custom wired or wireless scenarios with an arbitrary number of nodes, protocols, application clients, and traffic generators. Besides the commands that are used to create the simulation plane, the user can create a number of objects that can be used to control and monitor the actual simulation like monitor objects and random generators. Finally, after creating the OTcl script in order for it to be executed, the user must invoke NS-2 giving the script as a parameter. After the simulation has concluded in the majority of cases, the most important data reside in the created trace files. Thus, a user must find a way to process these trace files either using scripting languages like AWK and Perl or using postsimulation analysis tools like TRAFIL, jTrana, or Tracegraph.

Although these are the steps for executing a certain simulation scenario, TRAFIL enables the user to execute an OTcl script through the tool. TRAFIL will execute the specified simulation scenario, locate the resulting trace file, and start the trace file analysis procedure described in Section 14.6.

14.8.4.2   Simulating Video Transmission Using Evalvid-RA

Simulating video transmission is one of the most common uses of NS-2 and it is usually implemented using the Evalvid-RA framework. Therefore, TRAFIL has automated this procedure as shown in Figure 14.16. The only requirement is that NS-2 is installed on the system and Evalvid-RA has been incorporated correctly. The simulation procedure is broken into two specific steps: the presimulation and postsimulation phases. In the presimulation phase, the raw video file as well as the simulation scenario must be specified. In addition, the presimulation phase includes the raw video's processing using FFmpeg and its transmission using the MP4 tool. Thus, their parameters must be specified; TRAFIL has already set some default parameters, which are the ones defined by Evalvid's own examples that accompany its source code. Finally, in order to conduct the postsimulation phase, the user must specify the names of the files he or she uses in his or her Tcl script to read the video traces produced by MP4 and the names of the output receiver and sender files of his or her script. The file names must be specified because these files are crucial in the successful execution of the whole simulation procedure. The files are given as parameters at FFmpeg and et_ra, and if they are not specified in advance, there is no way for TRAFIL to complete the simulation without problems.

FIGURE 14.16   Video simulation.

When all the required parameters have been set, the simulation can be executed. If the simulation has been successful, then the postsimulation parameters become available, namely, the sender and receiver files, the .dat files, as well as the MP4 video traces. Thus, the user is able to select various combinations of these files and retrieve QoS statistics for his or her video transmission. The first phase of the video simulation, the presimulation phase, is shown earlier in Figure 14.16. In this phase, all the necessary parameters for the video encoding using FFmpeg must be defined.

Then using the MP4 tool, trace files for the video transmissions are created, and finally, the execution of the simulation takes place by invoking NS-2. The results of a sample video simulation are shown in Figure 14.17. The results include the actual console output of all the involved tools (FFmpeg, MP4, NS-2) as well as all the created files of the simulation. These files are available in a specific folder in TRAFIL's file hierarchy and its path is also given. In addition, the opportunity is given to save the files to any folder in the operating system or to delete them from TRAFIL without accessing the folder in which they are stored. After the presimulation phase has concluded, all the parameters in order for the postsimulation phase to be executed become available as shown in Figure 14.18. All the created files are now available to be selected, and they include the Tx (sender) and Rx (receiver) files, the data files (Data1, Data2), and MP4's output trace files. These files are all input parameters for the et_ra tool and the final parameter is the output video name.

In addition to the et_ra inputs, the user must specify FFmpeg's parameters in order to reconstruct the video. Again some default parameters are available by TRAFIL.

Figure 14.19 illustrates the results of the postsimulation phase including the PSNR calculation of the video transmission example simulation. All the resulting files including the console output are available to the user. The resulting files are included in the same directory as the previously created file during the postsimulation phase. The user is thus able to reuse them in order to conduct further process and extraction of additional QoS parameters.

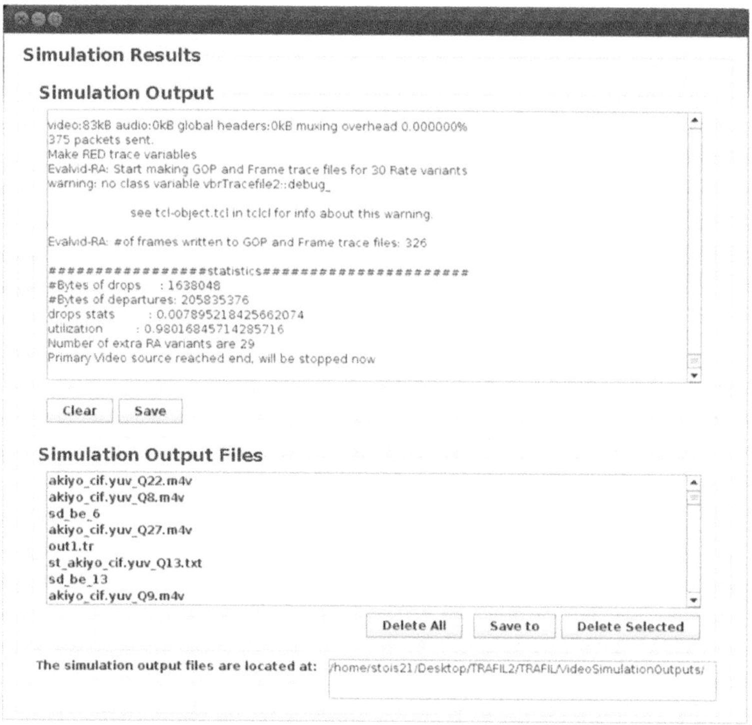

FIGURE 14.17 Presimulation phase results.

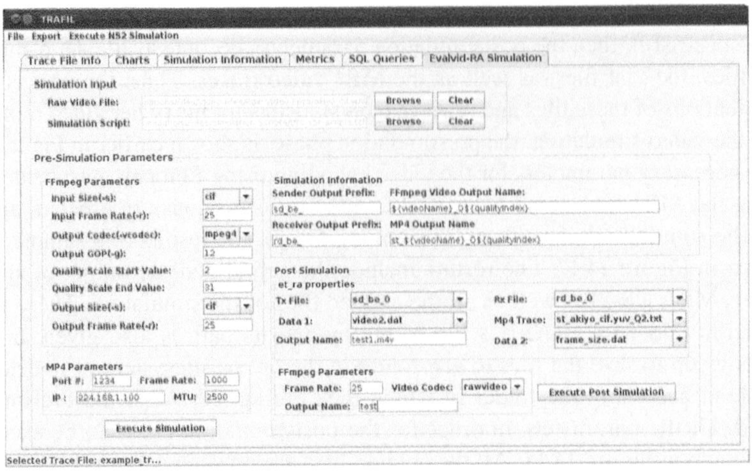

FIGURE 14.18   Postsimulation phase.

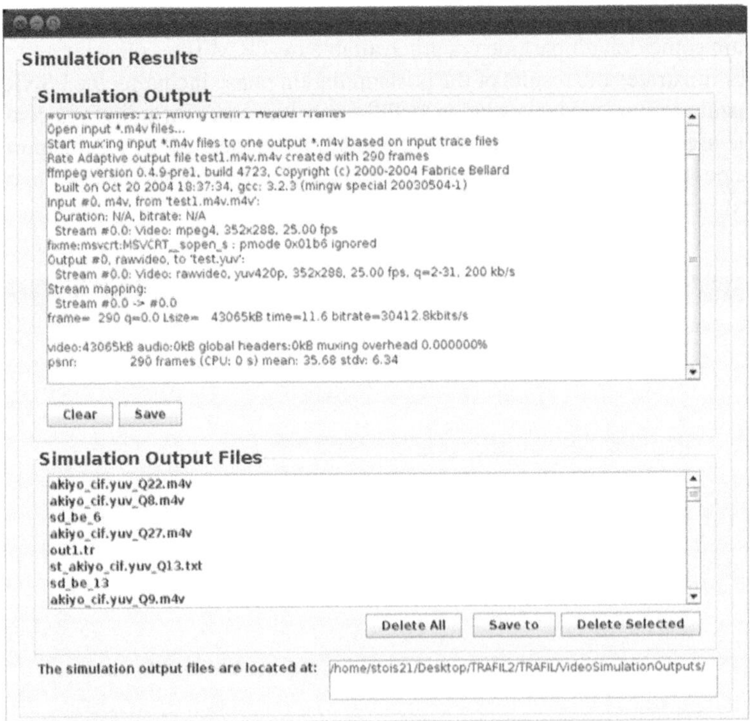

FIGURE 14.19   Postsimulation results.

14.8.5   TRAFIL RESULTS

In this section, we present timing results about the core procedures described in Section 14.6, which include identifying a trace file's type, parsing, processing, and finally transferring its content to the local database. We have argued that TRAFIL, by using metafiles and submetafiles, not only accomplishes to cope with all different kinds of trace files, but it does so exceptionally quick. Therefore, we present in Table 14.1 results regarding the aforementioned procedures for a variety of trace file sizes.

TABLE 14.1
Processing Time Results

Trace File Size (MB)	Execution Time (ms)
0.005	13
0.126	42
0.402	217
1	378
2	998
3	1,569
7	3,558
11	5,186
16	4,887
58	25,351
118	51,087

Based on the preceding results, it is apparent that TRAFIL manages even for very large trace files to keep their processing time extremely low. For trace files that have size up to 10 MB, the execution time is very small and it does not increase significantly, although for trace file sizes that exceed 10 MB, the processing time starts to grow following a near linear increase. Another observation that can be deducted from Table 14.1 is that the trace file with size 16 MB has a smaller processing time than the 11 MB one and the reason is the trace file format. The trace file with size 11 MB was produced using a wireless scenario and the 16 MB trace file resulted from a wired scenario. In order to examine why the trace file format affects the processing time, we measured TRAFIL's execution time for trace files of the same size both for wireless and wired scenarios, and the results are shown in Table 14.2.

From the these results, we can draw the conclusion that certainly trace files that are produced from wired scenarios take less time to be processed by TRAFIL rather than ones that result from a wireless scenario simulation. The reason is that during the trace file processing and parsing phases, the number of submetafiles that are tested against each trace file line is smaller for normal trace files (produced by wired scenarios) compared to the number that is tested for wireless trace file formats. The metafile used to process normal trace files has two utility submetafiles, and the metafiles used to process old wireless or new wireless trace files have eight and nine submetafiles, respectively. Nevertheless, although there is a difference in the processing time, it is not a large one and the most important conclusion is that for both cases, TRAFIL manages to keep the processing time extremely low.

TABLE 14.2
Trace File Format Execution Comparison

Trace File Size (MB)	Wireless Scenario Trace File Processing Time (ms)	Wired Scenario Trace File Processing Time (ms)
2	1,967	1,456
6	2,499	2,215
13	4,973	3,635
24	8,699	5,613
35	13,095	8,406
47	17,816	11,767
62	20,906	16,048

TABLE 14.3
Trace File Analysis Tool Comparison

Trace File Size (MB)	TRAFIL Processing Time (ms)	Tracegraph Processing Time (ms)	jTrana Processing Time (ms)
2	1,967	27,000	29,700
6	2,499	75,400	68,000
13	4,973	197,600	135,200
24	8,699	495,100	177,700
35	13,095	949,300	272,000
47	17,816	1,090,700	362,700
62	20,906	2,097,100	486,800

Finally, we present a comparison between TRAFIL and other popular tools that are used for trace file analysis: Tracegraph 2.02 and jTrana 1.0. These tools have been described in Section 14.2; they are among the most known for processing simulation trace files, but one of their main drawbacks is the amount of time they need to open a trace file [17]. As we have demonstrated earlier, TRAFIL behaves exceptionally well at this task; therefore, we present in Table 14.3 a detailed comparison between these tools and TRAFIL.

The trace files used to create the measurements referred to wireless simulations and were the same for all three tools. In order to retrieve the timings both for Tracegraph 2.02 and jTrana 1.0, we manually measured the time it took for each tool to open a trace file. Furthermore, to obtain more accurate results, we made each measurement multiple times and calculated the mean value. As it can be easily deducted from Table 14.3, TRAFIL yields timings that are considerably smaller than the other two tools (ranging from 15 times smaller to 100 times smaller). This fact becomes more obvious if we consider that TRAFIL's largest processing time is for the trace file with 62 MB size and even then that processing time is smaller than the time it takes the other two tools to process the trace file with 2 MB size.

14.9 CONCLUSION

In this chapter, we presented a new tool named TRAFIL that aims first to assist and support the procedure of analyzing simulation trace files and second to automate the execution of NS-2 simulations along with the execution and analysis of video simulations. To accomplish these goals, TRAFIL introduces the novel idea of using metafiles and submetafiles that render the tool and the process of identifying trace file types more abstract and robust. The task of creating a metafile and adding it to TRAFIL's metafile and submetafile repository is trivial; therefore, TRAFIL is independent of trace file format and can be used with a variety of trace file formats. Moreover, one of the main objectives of TRAFIL was to speed up the identification and processing phases of opening a trace file. Similar tools have not been very effective during this task and needed a fair amount of time to open a trace file. As we thoroughly presented in Section 14.8, the use of metafiles and submetafiles enabled TRAFIL to carry out this task in significantly reduced time compared to other popular tools (up to 100 times faster). Another unique characteristic of TRAFIL in regard to the trace file analysis domain was the ability to store each trace file in a local database, alleviating this way the wearing task of having to reopen each trace file from the disk. Furthermore, TRAFIL gave the opportunity to retrieve a variety of simulation statistics, information, QoS measurements, and charts. Every single piece of information that was produced could also be exported from the tool including the actual trace files in their parsed form either in text or Excel file.

Apart from the analysis of trace files, TRAFIL gave the opportunity to execute OTcl simulation scripts in order to have the results passed to TRAFIL immediately and kept organized without needing any further involvement. This feature although useful serves more as a utility functionality in the grater procedure of executing video transmission simulations. NS-2 and Evalvid-RA have been used extensively for video simulations; therefore, with the development of TRAFIL, we aspired to automate the procedure starting from the video preprocessing until the production of QoS measurements and evaluation of the simulation.

14.10 FUTURE WORK

Our future work will include extensions in the simulation design module that will enable a more complete support of NS-2 functionalities and add-ons. We also plan to investigate the possibility of generalizing the framework to operate around other simulators such as NS-3, in order to be able to provide a generic simulation facilitation framework.

REFERENCES

1. Research Unit 6, CTI, TRAFIL website, 2014, http://ru6.cti.gr/ru6/research_tools.php#TRAFIL (accessed July 14, 2014).
2. J. F. Borin and N. L. S. da Fonseca, Simulator for WiMAX networks, *Simulation Modelling Practice and Theory*, 16(7), 817–833, August 2008. ISSN 1569-190X, 10.1016/j.simpat.2008.05.002.
3. S. Jansen, Network simulation cradle, 2008, http://www.wand.net.nz/~stj2/nsc/software.html (accessed July 14, 2014).
4. C. Cicconetti, E. Mingozzi, and G. Stea, NS2Measure, 2009, http://cng1.iet.unipi.it/wiki/index.php/Ns2measure (accessed July 14, 2014).
5. J. Malek and K. Nowak, Trace graph-data presentation system for network simulator ns. In *Proceedings of the Information Systems—Concepts, Tools and Applications (ISAT 2003)*, Poland, September 2003.
6. J. Malek, The Trace File analysis tool Trace Graph, 2002, http://www.angelfire.com/al4/esorkor/ (accessed January 17, 2013).
7. H. Qian, and W. Fang, jTrana: A Java-based NS2 wireless trace analyzer, 2008, http://sites.google.com/site/ns2trana/ (accessed July 14, 2014).
8. A. U. Salleh, Z. Ishak, N. M. Din, and M. Z. Jamaludin, Trace analyzer for NS-2. In *Proceedings of the 4th Student Conference on Research and Development (SCOReD 2006)*, Shah Alam, Selangor, Malaysia, June 27–28, 2006, pp. 29–32.
9. B. Bartosz and P. Machan, The Trace File analysis tool Trace Analyzer, 2013, http://trace-analyzer.sourceforge.net/ (accessed July 14, 2014).
10. A. Sobeih, W.-P. Chen, J. C. Hou, L.-C. Kung, N. Li, H. Lim, H.-Y. Tyan, and H. Zhang, J-Sim: A simulation environment for wireless sensor networks. In *Proceedings of the 38th Annual Symposium on Simulation*, April 4–6, 2005, pp. 175–187. Washington, DC: IEEE Computer Society.
11. J. Hou, J-Sim, 2013, http://sites.google.com/site/jsimofficial/ (accessed July 14, 2014).
12. A. Lie and J. Klaue, Evalvid-RA: Trace driven simulation of rate adaptive MPEG-4 VBR video, *Multimedia Systems*, 14(1), 33–50, 2008.
13. A. Lie and J. Klaue, Evalvid-RA website, 2011, http://www.item.ntnu.no/~arnelie/Evalvid-RA.htm (accessed July 14, 2014).
14. C. Bouras, S. Charalambides, G. Kioumourtzis, and K. Stamos, TRAFIL: A tool for enhancing simulation TRAce FILes processing. In *International Conference on Data Communication Networking DCNET 2012*, Rome, Italy, July 24–27, 2012, pp. 61–64.
15. H. Schulzrinne, S. Casner, R. Frederick, and V. Jacobson, RTP: A transport protocol for real-time applications, RFC 3550, July 2003.
16. A. Lie, Trace driven simulation of rate adaptive MPEG-4 video, 2005, http://www.item.ntnu.no/~arnelie/evalvid_test/Presentation_Dec05.pdf (accessed July 14, 2014).
17. R. Ben-El-Kezadri, F. Kamoun, and G. Pujolle, XAV: A fast and flexible tracing framework for network simulation. In *Proceedings of the 11th International Symposium on Modeling, Analysis and Simulation of Wireless and Mobile Systems*, Vancouver, British Columbia, Canada, October 27–31, 2008, pp. 47–53.

15 Comparison of Non-Probing-Based Routing Metrics for Static Multihop Underwater Acoustic Networks

Saiful Azad, Mashiour Rahman,
Khandaker Tabin Hasan, and Dip Nandi

CONTENTS

ABSTRACT

Selecting a suitable routing metric is crucial in attaining high-throughput in any wireless network architecture. A wrong choice may lead to selecting lossy links. In this chapter, we investigate the performance of various non-probing-based routing metrics over underwater acoustic networks (UANs). Non-probing-based routing metrics are chosen since most of the nodes in UANs operate on battery power and these routing metrics use less energy. Although, there are a couple of different routing metrics proposed for UANs, a few investigations have been performed to understand the behavior of various routing metrics over underwater network architecture. In this chapter, we take an initiative to investigate and analyze the performance of various non-probing-based routing metrics by designing a simple underwater simulator using an underwater channel model. The findings of the investigation could be applied in selecting a suitable routing metric as well as in designing a new high-throughput routing metric for UANs.

15.1   INTRODUCTION

A routing metric is a value that is assigned to a path and utilized by a routing algorithm to extract a suitable path or a subset of paths from a set of discovered paths [1]. It can be classified as active-probing-based routing metric or non-probing-based routing metric. Routing metrics that exchange probing packets to select a suitable path or a subset of paths are called active-probing-based routing metrics, for example, etx [2], rtt [3], and ett [4]. Conversely, in case of non-probing-based routing metrics, no probing packet is exchanged among the communicating nodes, for example, minimum hop count, minimum delay, and average signal-to-noise ratio (SNR) [4].

A routing metric assists a routing protocol to achieve one or more goals, like minimize hop count, maximize throughput, minimize delay, maximize end-to-end packet delivery, minimize energy consumption, and distribute traffic load equally. An unwisely chosen routing metric may select an inefficient link and thus may fail to attain desired goal(s). Therefore, it is crucial to embed a suitable routing metric within a routing protocol. Again, the suitability of a routing metric may be influenced by the network architecture. Therefore, a routing metric that is preferred in one network architecture may not be desirable at all in another architecture. For instance, an underwater acoustic network (UAN) prefers those routing metrics that require lower energy to deliver a packet since most of the nodes operate on battery power, whereas a wired network does not have this constraint. Therefore, UANs generally prefer non-probing-based routing metrics over active-probing-based routing metrics since the highest amount of energy is depleted in transmission and UANs conserve energy by avoiding unnecessary transmission.

Among all the non-probing-based routing metrics, the minimum hop count is embedded in most of the routing protocols of any network architecture. Analogously, routing protocols proposed in Refs. [5–7] for UANs also employ minimum hop count metric. However, for a static network, when a hop count is minimized, the distances between adjacent nodes are maximized, which is likely to maximize the loss ratio by minimizing the signal strength [2,8]. This phenomenon is more prominent in UAN architecture since the acoustic signal can travel several kilometers. Currently, there are many modems available in operations that support transmissions over several kilometers. For instance, the LinkQuest UWM2000 and UWM3000 modems support 1.5 and 3 km, respectively [9], AquaComm supports ranges of 3 km [10], and the EvoLogics S2C R 48/78 modem supports 3.2 km [11]. Moreover, the higher bit error rate of the underwater channels [12,13] suggests an alternative routing metric for UANs.

However, there are a few non-probing-based routing metrics proposed for UANs. In Ref. [14], a non-probing-based routing metric is proposed where a path is chosen based on low propagation delay. Analogous to minimum hop count metric, it also may fail to attain high throughput. Because when the delay is minimized, implicitly, the distance between adjacent nodes is maximized and thus may select a lossy link. Average SNR is another non-probing-based metric that is employed by the routing protocols proposed in Refs. [15,16]. However, a dominant SNR value within a link may influence average SNR calculation and may fail to select a desirable path. These routing metrics are described in Section 15.2 in details.

Although there are a couple of different routing metrics proposed for UANs, a few investigations have been performed to understand the behavior of various routing metrics over underwater network architecture. In this chapter, we take an initiative to investigate and analyze the performance of various non-probing-based routing metrics by designing a simple underwater simulator using an underwater channel model [17]. The findings of the investigation could be applied in selecting a suitable routing metric as well as in designing a new high-throughput routing metric for UANs.

The rest of the chapter is organized as follows. Routing metrics considered in the comparison are described in Section 15.2. In Section 15.3, the underwater channel model that is employed in the performance evaluation is presented. For performance investigation, we design and develop a simple scenario-dependent underwater simulator using C++ that is described in Section 15.4. The results that are acquired from the investigation are portrayed in Section 15.5. Our discussions end with concluding remarks in Section 15.6.

15.2 ROUTING METRICS

For better understanding of the metrics, let us consider an underwater sensor network as an undirected graph $G = (V, E)$, where V represents a set of underwater sensor nodes and E represents their edges. A path in G can be represented as a sequence of edges $P = \{e_{(1,2)}, e_{(2,3)}, ..., e_{((n-1),n)}\}$, where $e(i, j) \in E$; $i = 1, 2, 3, ..., n - 1$; and $j = 2, 3, 4, ..., n$. Any $e(i, j) \in E$ has an associated weight, $\omega(e_{(i,j)})$, which may vary according to the selection of the routing metric. By substituting every edge with their respective associative weight, we can find $P = \{\omega(e_{(1,2)}), \omega(e_{(2,3)}), ..., \omega(e_{((n-1),n)})\}$. This realization is utilized in the following sections to elaborate all the non-probing-based routing metrics that are considered in the performance investigation.

15.2.1 MINIMUM HOP COUNT

This is the routing metric that is preferred by most of the routing protocols designed for any network architecture [1,2]. However, the implicit assumption of this metric is that every link is an error-free link, which is almost always the case for the wired networks, but not for the UANs. The factors like multipath fading, interference, and environmental factors [12,13,18] that are affecting underwater channels are inexistent in wired channels. Nevertheless, due to its simplicity in both understanding and implementation, several routing protocols design for UANs still utilize this routing metric as mentioned in Section 15.1.

The objective of this metric is to minimize the hop count between a sender–receiver pair. Therefore, a path, P', is chosen if its hop count is minimum among all the discovered paths, that is, $P' = \forall_P \min(|Pi|)$. No additional measurements are required for calculating hop count. One major disadvantage of this metric is that it may select lossy links, thus lowering the throughput performance of the network.

15.2.2 END-TO-END DELAY

It is another simple non-probing-based metric that minimizes the end-to-end delivery time.* In terrestrial wireless network, generally, those paths that ensure the minimum hop count also ensure minimum end-to-end delay since propagation delay is negligible. However, this assumption is not equitable for UANs because of the long propagation delay that is several magnitudes higher than that of terrestrial wireless networks. Consequently, a path chosen by employing minimum hop count metric may not ensure the minimum delay between a source–destination pair.

Recalling the graph, which is described at the start of this segment, the delay between two adjacent nodes is considered as $\omega(e_{(i,j)})$ in this case. A path, P', is selected from a set of Π, if it experiences the lowest delivery time to the destination, that is, $P' = \forall_P \min(\Sigma\omega(e_{(i,j)}))$ where $\omega(e_{(i,j)}) \in P_k$ and $P_k \in \Pi$.

15.2.3 AVERAGE SNR

It is yet another non-probing-based routing metric that does not exchange any probing packet to discover a high-throughput path. However, it utilizes the knowledge of signal strength, that is, SNR, which it could acquire by observing every packet it receives from various adjacent nodes. The SNR of an edge, $e_{(i,j)}$, is considered as the associative weight, $\omega(e_{(i,j)})$, of that edge. Since the SNR is a local parameter of the physical layer, a cross-layer message exchange is essential between the routing layer and the physical layer to acquire the SNR of a received signal.

---

* End-to-end delivery time is the duration between the generation of a packet and the delivery of that packet at the destination.

Recalling the graph example, a path, P', is selected if it has the highest average SNR in Π, that is, $P' = \left(\sum \omega(e(i,j))/|P_k| \right)$ where $\omega(e_{(i,j)}) \in P_k$ and $P_k \in \Pi$. Although this metric considers the signal strength in discovering a suitable path, it may still select a lossy link. A sufficiently large SNR value in a path may influence the calculation of the average and thus may get selected.

15.3 UNDERWATER CHANNEL MODEL

In the underwater channel model, both the attenuation and the ambient noise need to be accounted to compute the SNR at certain receiving end. Attenuation experienced by a signal is calculated based on the spreading loss [17] and on the absorption coefficient, which further can be approximated through Thorp's formula [19]. Consequently, attenuation can be found as [21]

$$10 \log A(d,f) = \xi \cdot 10 \log d + d \cdot \log a(f) \tag{15.1}$$

where the first term in Equation 15.1 is the spreading loss that describes the geometry of the propagation and $\xi = 1.75$ provides a practical value for a typical underwater channel [17]. The second one is the absorption loss. At a given frequency, f in kHz, Thorp's absorption coefficient, $a(f)$, can be given as follows [20]:

$$10 \log a(f) = \begin{cases} 0.11 \dfrac{f^2}{1+f^2} + 44 \dfrac{f^2}{4100+f^2} + 2.75 \dfrac{f^2}{10^4} + 0.003 \\[3mm] 0.11 \dfrac{f^2}{1+f^2} + 0.11 f^2 + 0.002 \end{cases} \tag{15.2}$$

where $a(f)$ is in dB/km. The realization of the absorption coefficient for various f is shown in Figure 15.1. It increases rapidly with frequency, thus limiting the maximum usable frequency for an acoustic link of a given distance.

Four major factors, such as turbulence, shipping, wind, and thermal, mainly influence the overall power spectral density (psd) of the ambient noise, $N(f)$, of the underwater channel, which can be expressed as [21]

$$N(f) = N_t(f) + N_s(f) + N_w(f) + N_{th}(f) \tag{15.3}$$

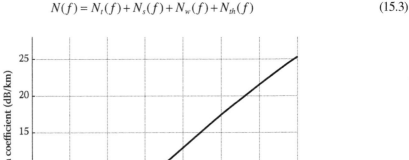

FIGURE 15.1 Absorption coefficient, $a(f)$, obtained for various f.

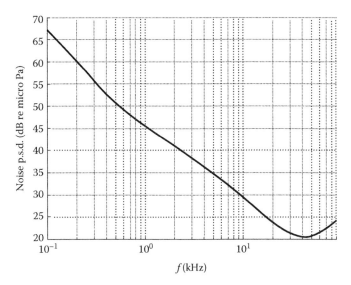

FIGURE 15.2 The psd of ambient noise obtained for various f.

where

$N_t(f)$ is turbulence noise

$N_s(f)$ is the noise that is generated from the shipping activity

$N_w(f)$ is the noise that is generated because of surface motion caused by wind-driven waves

$N_{th}(f)$ is the thermal noise

The interested reader can find details of various noises in Ref. [21]. In Figure 15.2, the relationship between the psd of the ambient noise and f is illustrated.

Combining both Equations 15.1 and 15.3, the overall decays of a signal experienced before arriving at the receiving end can be computed. Therefore, the SNR of an acoustic signal [22], which is transmitted from a distance d with a frequency f, can be found as

$$\mathrm{SNR}(d, f) = \frac{P_T}{A(d, f)N(f)} \tag{15.4}$$

where P_T is the transmission power. From Equation 15.4, it can be observed that the SNR is inversely proportional to the attenuation-noise factor for a given value of P_T, which is also illustrated in Figure 15.3.

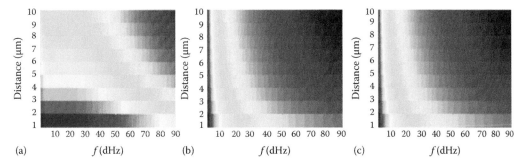

FIGURE 15.3 SNR profiles obtained for various f and d for three different PT values: (a) PT = 70 dB re µ Pa, (b) PT = 80 dB re µ Pa, and (c) PT = 90 dB re µ Pa.

15.4   SIMPLE SCENARIO-DEPENDENT UNDERWATER SIMULATOR

The discussions in this section are separated into two subsections: scenario description and design and implementation of the simulator. Subsection 15.4.1 illustrates the scenario and its parameters that are considered in the performance investigation, whereas Subsections 15.4.2 and 15.4.3 demonstrate the design and the implementation of the simulator.

15.4.1   SCENARIO DESCRIPTION

For investigation, we take into account a network of 12 bottom-mounted nodes, which are deployed within a 3D area of 8000 m × 4000 m × 80 m. The conferred area is uniformly partitioned into 12 cells so that a single node could be placed randomly within a single cell. A sink is also installed at a random location along one of the short sections, illustrated in Figure 15.4. All the nodes involved in the communication send their respective packets to the sink and hence, it is obligated to find a suitable path from every node to the sink. A routing protocol can employ an efficient routing metric to extract a suitable path from a set of available paths that exist between a node and the sink.

Every node in the network communicates at a bit rate of 4500 bps in 25 kHz band like the WHOI Micro-Modem [23]. Transmission range of the modem is 2.5 km and the payload of every data packet is 1000 bits. In physical layer, the binary phase-shift keying (BPSK) modulation technique is utilized with independent channel error. A collision avoidance MAC protocol, like the one in Ref. [24], is embedded in the medium access control layer for collision-free packet exchange. An application generates packets for the sink in a fixed interval. Whenever a packet reaches the sink, it unearths the packet and fetches necessary information before erasing. All the required statistics for the performance comparison are also collected in the sink. Every simulation is run for a whole day and all the results presented in this chapter are obtained by averaging over 100 realizations of the network topology for every routing metric.

15.4.2   DESIGN OF THE SIMULATOR

If we analyze the aforementioned scenario, it can be observed that a preventive measure has been taken to avoid collisions in the network. Consequently, the primary factor that influences the packet drops in the network is the signal strength that decays with the distance, which can be realized from the SNR in Equation 15.4. The acquired SNR can be further applied to calculate the probability of packet errors over that link. Let p be the probability of success for a packet containing L bits. Assuming additive

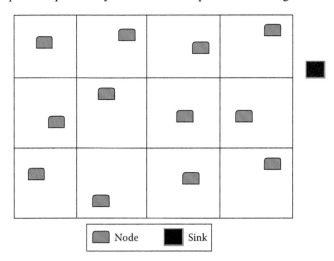

FIGURE 15.4   Underwater network with 12 nodes placed randomly inside a grid.

noise channel, like the one described in Section 15.3, and considering BPSK modulation technique in the physical layer with independent channel error on the L transmitted symbols, we can have

$$p = \left(1 - \frac{1}{2}\text{erfc}\sqrt{\text{SNR}}\right)^L \tag{15.5}$$

Using Equation 15.4, we get

$$p = \left(1 - \frac{1}{2}\text{erfc}\sqrt{\frac{P_T}{A(d,f)N(f)}}\right)^L \tag{15.6}$$

The probability of packet loss in a single link can be obtained from Equation 15.6. However, a path is generally comprised of one or more links. Consequently, the total number of packet drops experienced through a path could be calculated by aggregating probable packet drops of each link in that path. Note that, since most of the modems available to date utilize a fixed f and a fixed P_T and transmit a fixed length payload packet L throughout a communication session, the only remaining parameter that affects p in Equation 15.6 is d.

From the aforementioned discussions, it is understandable that the performance investigation of various metrics could be likely without creating the conventional protocol stack and sending any data packet. It can be attained by calculating the distances of the adjacent nodes, which are present in the path and by knowing the payload of the packet. Therefore, the simulator, which is developed for the performance comparison in this chapter, does not include any protocol stack and no dummy data packet is exchanged. However, in such circumstance, one of the major challenges in the design of the simulator is to discover the paths between every node and the sink without broadcasting any discovery packet. Further challenge is to extract a suitable path according to the given routing metric from a set of discovered paths.

Generally, whenever a sender is unable to find a path in the routing table, it injects a broadcasting packet in the network. Any intermediate node that receives that packet rebroadcasts it and a similar procedure is followed until it reaches the sink or the hop count reaches zero. After receiving an initial discovery packet, the sink launches a timer. All the packets that arrive correctly are received and deposited until the timer is over. When the timeout occurs, the sink selects a path from a set of discovered paths according to the given metric and notifies it to the sender through a unicast packet.

Since our design target is not to employ any conventional protocol stack, therefore, we apply a slightly modified breadth-first search (BFS) [25] search algorithm with a limit of 7 to discover available paths between a node and a sink. In our design, after beginning the search at a node, it inspects all the neighboring nodes and keeps visiting neighbors until the sink is discovered or it reaches the threshold. All the discovered paths are stored in a container from where a suitable path is extracted employing a routing metric. Details of the implementation of the simulator are given in the following Subsection 15.4.3.

15.4.3 Implementation Details

We have implemented the aforementioned scenario-dependent simulator using an object-oriented programming language, namely, C++ [26]. We start our implementation by creating the nodes and the sink of the network. Every node and the sink in the network must contain the following information: a unique ID to identify it individually, the Cartesian coordinates that hold the position information, and the genre of a device whether it is a node or a sink. According to the arrangement in Figure 15.4, all the nodes are placed sequentially from left to right and bottom to top fashion and the sink is placed at one of the short sections. Inside the respective cell, every node is placed randomly applying randomValue() function. The algorithms of the node and the sink modules are given in Algorithms 15.1 and 15.2.

Algorithm 15.1: *CreateNode (NULL)*
global maxX = 8000,maxY = 4000,maxZ = 80
comment: Area of the network considered
global numOfNodes = 12, nodeArray[numOfNodes + 1], id = 0
local numRow = 3, numCol = 4, cellInX, cellInY
local nodeId, nodeMode, xAxis, yAxis, zAxis

cellInX = maxX/numCol
cellInY = maxY/numRow

for i ← 0 to numOfNodes
do
　　nodeClass node
　　node.nodeId ←++id
　　node.xAxis ← ((i%numCol) × cellInX + randomV alue(0, cellInX))
if i < 4
then {node.yAxis ← randomValue(0, cellInY)
else if i > = 4 and i < 8
then {node.yAxis ← (cellInY + randomValue(0, cellInY))
else
then {node.yAxis ← (2 × cellInY + randomValue(0, cellInY))
node.zAxis ← randomValue(0, cellInZ)
node.nodeMode ← SENSOR
nodeArray[i] ← node

Algorithm 15.2: *CreateSink (NULL)*
global maxY = 4000, numOfNodes = 12, nodeArray[numOfNodes + 1]
local nodeId, nodeMode, xAxis, yAxis, zAxis

nodeClass sink
sink.nodeId ← numOfNodes
sink.xAxis ← randomValue(0, 100)
sink.yAxis ← randomValue(0,maxY)
sink.zAxis ← randomValue(0, 5)
sink.nodeMode ← SINK
nodeArray[numOfNodes] ← sink

Since the BFS algorithm demands neighbors' information for path discovery, it is essential to discover the neighbors of every node before starting path discovery procedure. Neighbors of a node could be discovered by calculating the Euclidean distance between the given node and the other nodes and by comparing the distance with the transmission range. Every node that is within the transmission range of the modem (e.g., 2500 m in this performance investigation) could be considered as the neighbor of that node. All the nodes and their respective neighbors are stored in a container for future use. Details of the algorithm are given in Algorithm 15.3.

Algorithm 15.3: *NeighborDiscovery (node)*
global numOfNodes = 12, neighbor[nodeId][neighborId], txRange = 2500

for j ←0 to (numOfNodes + 1)
do
if node ! = j

then
if *Distance(i, j) < txRange*
then *{neighbor[i][j] ← true*
else
then *{neighbor[i][j] ← false*
else
then *{neighbor[i][j] ← false*

After discovering the neighbors of all the nodes and the sink, a modified BFS is applied to discover all the paths between any node and the sink. The searching of the BFS starts from the source node and ends when the threshold limit of searching is reached, which is analogous to the hop count field in a packet. The BFS start searching by enqueuing the source node and then checks every neighbor of the source node. Alike receiving a broadcasting packet by every neighbor, a separate queue is created for every neighbor except those that are already visited. Every queue first copies the ID of the nodes that are already visited by the originator and afterward enqueue the ID of the node that it is currently visiting. For instance, let us assume that the unique ID of a sender *A* is 1, which has two neighbors *B* and *C* with unique IDs 2 and 3, respectively. In a queue, the ID of *A*, that is, 1, is stored and starts looking for the destination. Since *S* has two neighbors, two queues are created, and in every queue, the ID 1 is copied. Afterward, the ID of the currently visited node is stored in the queue. If anyone among the current neighbors is the destination, a path is discovered and that queue is not extended anymore, whereas the other queues keep extending both in size and in the count through considering the neighbors of other levels until they reach the sink or the threshold limit is reached. When the search ends, all the queues that do not contain the destination information are erased. From the search, it is likely to acquire multiple queues that may hold the destination information. Every queue here represents a single path and thus multiple paths could be found at the end of the searching. Algorithm 15.4 demonstrates the details of the modified BFS algorithm that is applied in the simulator.

Algorithm 15.4: *PathDiscovery (node)*
global *typedef vector < int > PathVec*
comment: *In PathVec container, all the nodeIds in a path are stored*
global *typedef vector < vector < nodeId >> MulPathVec*
comment: *MulPathVec container holds multiple paths of a source-destination pair*
global *map < int,MulPathVec > allPath*
comment: *It holds multiple paths of all the nodes involve in the communication*
local *pathSearchLimit = 7, size*

PathVec tempVec, tVec, nVec, adjNodes
MulPathVec tempMulPath,mulPath,mPath
tempVec.push_back(node)
mulPath.push_back(tempVec)

for *i ← 0* ***to*** *pathSearchLimit*
do
tempMulPath ← mulPath
Clear mulPath vector
for *MulPath:: iterator it ← tempMulPath.begin()* ***to*** *tempMulPath.end()*
do
    *tVec ← *it*
adjNodes ← NeighborDiscovery(tVec.back())
size ← adjNodes.size()
for *j ← 0* ***to*** *size*

do
if adjNodes[j] = = sinkId
then
   nVec ← tVec
   nVec.push_back(adjNodes[j])
map < int,MulPath >:: iterator it1 ← allPath.find(node)
if it1 = = allPath.end()
then
     mPath.push_back(nVec)
allPath.insert(node,mPath)
else
then
     (it1– > second).push_back(nVec)
else
then
nVec ← tVec
if adjNodes[j] is not already present in the paths
then
        nVec.push_back(adjNodes[j])
        mulPath.push_back(nVec)
Clear tempMulPath container

From the set of discovered paths between a node and the sink, a suitable path is extracted according to the given routing metric. Different routing metrics have separate objectives to achieve and hence selected paths could be diversified. Following set of Algorithms 15.5 through 15.7 demonstrate the techniques adopted by different metrics to select suitable paths. The first routing metric that is illustrated in the succeeding text is the *hop count*. This is considered as one of the simplest non-probing-based routing metrics. When a list of paths is given to extract a suitable path, it selects that path that has the lowest hop count among all of them.

Algorithm 15.5: *SelectPath (hop count)*
global *typedef vector < int > PathVec*
comment: *In PathVec container, all the nodeIds in a path are stored*
global *typedef vector < vector < nodeId >> MulPathVec*
comment: *MulPathVec container holds multiple paths of a source-destination pair*
global *map < int,MulPathV ec > allPath*
comment: *It holds multiple paths of all the nodes involve in the communication*
local *node*

MulPathVectempPath
PathVecselPath, tempV ec

for *map < int,MulPath >:: iterator it ← allPath.begin()* **to** *allPath.end()*
do
   node ← it → first
   tempPath ← it → second
selPath ← tempPath.front()

for *MulPath:: iterator it1 ← tempPath.begin()* **to** *tempPath.end()*
do
      *tempVec ← *it1*
if selPath.size() > tempV ec.size()

then
Clear selPath container
        selPath ← tempVec

Again, a similar discovered set of paths that are considered for the *hop count* metric are also considered in the selection process of the delay metric. Since the objective of this metric is to minimize the delay, a packet transmitted through the selected path must experience the lowest delay among all. In delay calculation of a path, the total propagation delay and total transmission delay are considered, that is, $d = \tau + \varsigma$, where $\tau = \sum_{i=1}^{m} \tau_i$ is the propagation delay and $\varsigma = \sum_{i=1}^{m} \varsigma_i$ is the transmission delay. Algorithm details are given in Algorithm 15.6.

Algorithm 15.6: *SelectPath (delay)*
global *typedef vector < int > PathVec*
comment: *In PathVec container, all the nodeIds in a path are stored*
global *typedef vector < vector < nodeId >> MulPathVec*
comment: *MulPathVec container holds multiple paths of a source-destination pair*
global *map < int,MulPathV ec > allPath*
comment: *It holds multiple paths of all the nodes involve in the communication*
local *node, delay, minDelay*

MulPathVectempPath
PathVecselPath, tempV ec

for *map < int,MulPath >:: iterator it ← allPath.begin() **to** allPath.end()*
do
    node ← it → first
    tempPath ← it → second
minDelay ← HUGE_VALUE

for *MulPath:: iterator it1 ← tempPath.begin() **to** tempPath.end()*
do
        *tempVec ← *it1*
        delay ← Calculate the Delay of the path in tempVec
if *minDelay > delay*
then
minDelay ← delay
Clear selPath container
        selPath tempVec

Selecting a path applying the average SNR is more challenging than the other two routing metrics considered in this chapter. Detailed descriptions of the SNR calculation are given in Subsection 15.4.2. From Equation 15.4, it can be observed that all the numerators and denominators for calculating the SNR of two adjacent nodes are either given in the scenario or they could be calculated from the information given. For finding the average SNR of a path, the SNR of every link of that path are aggregated and then divided with the number of links exist in that path as discussed in the following Algorithm 15.7.

Algorithm 15.7: *SelectPath (average SNR)*
global *typedef vector < int > PathVec*
comment: *In PathVec container, all the nodeIds in a path are stored*

global typedef vector < vector < nodeId >> MulPathVec
comment: MulPathVec container holds multiple paths of a source-destination pair
global map < int,MulPathV ec > allPath
comment: It holds multiple paths of all the nodes involve in the communication
local node,, avgSnr1, avgSnr2, MulPathVec tempPath

PathVecselPath, tempV ec

for map < int,MulPath >:: iterator it ← allPath.begin() to allPath.end()
do
   node ← it → first
   tempPath ← it → second
avgSnr1 ← 0
for MulPath:: iterator it1 ← tempPath.begin() to tempPath.end()
do
     *tempVec *it1*
avgSnr2 ← Calculate average SNR of the path in tempV ec
if avgSnr1 < avgSnr2
then
avgSnr1 ← avgSnr2
Clear selPath container
selPath ← tempVec

15.5   RESULTS AND DISCUSSIONS

Since the network that is considered in the performance investigation is a static network, the path discovery procedure is initiated only once at the beginning of the simulation. Two performance metrics are considered to evaluate the performance of the routing metrics: packet drop ratio and normalized throughput. The packet drop ratio is the ratio between the number of packets dropped and the number of packets generated, whereas normalized throughput is defined as the number of packets correctly delivered in the network per packet transmission time.

Figures 15.5 and 15.6, respectively, illustrate the normalized throughput and the packet drop ratio of three routing metrics. From the graphs, it can be observed that throughput is increased for all the routing metrics with increasing power. The objective of the minimum hop count metric is to minimize the number of hops a packet must visit to reach to the destination. It generally selects those paths that experience long distance between the pairs of adjacent nodes. As it is mentioned earlier, the longer the distance, the higher the probability of dropping a packet because of erroneous bit(s) reception at the receiving end. Consequently, the drop ratio of the minimum hop count is higher and therefore, the normalized throughput is lower when transmission power is low. When transmission power is 75 dB re μ Pa, the packet drop ratio is about 75%, which declines to zero at 90 dB re μ Pa.

On the other hand, the objective of end-to-end delay metric is to minimize the delay between a source–destination pair. Analogous to the minimum hop count metric, the end-to-end delay metric usually chooses a lossy link by maximizing the distances of the adjacent nodes. From Figure 15.6, it can be noticed that the packet drop ratio of the end-to-end delay metric is higher when transmission power is lower and decreases with increasing power. Among the three compared metrics, end-to-end delay performs slightly better than other metrics in terms of normalized throughput as well as packet drop ratio when signal transmission power is 80 dB re μ Pa.

In case of the average SNR metric, throughput is the lowest and packet drop ratio is the highest among all the metrics. A large SNR value in a link may influence the calculation of the average, whereas other SNRs are still poor. In that circumstance, a lossy link may be selected by the average SNR metric instead of a suitable link.

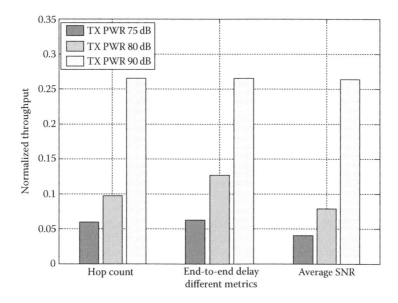

FIGURE 15.5 Normalized throughput of three considered metrics for various transmission power.

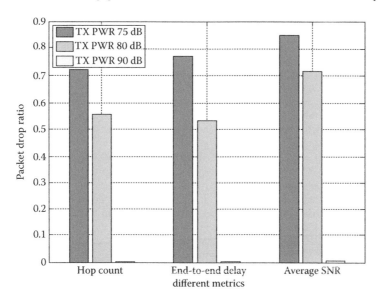

FIGURE 15.6 Packet drop ratio of three metrics for various transmission power.

15.6 CONCLUSIONS

In UANs, active-probing-based routing metrics are not preferred because of their higher packet overheads. Conversely, non-probing-based routing metrics are preferred in UANs, which help to conserve battery power by avoiding unnecessary transmission. In this chapter, we investigate the performance of three non-probing-based routing metrics, namely, hop count, end-to-end delay, and average SNR. For investigation, we design and develop a scenario-dependent simulator using an object-oriented programming language. From the acquired results, it could be observed that most of the non-probing-based routing metrics are unable to select a high-throughput path between a source–destination pair especially in lower transmission power. Among the three compared metrics, end-to-end delay performs slightly better than other metrics in terms of normalized throughput as well as packet drop ratio.

REFERENCES

1. G. Parissidis, M. Karaliopoulos, R. Baumann, and T. Spyropoulos, Routing metrics for wireless mesh networks, in *Guide to Wireless Mesh Networks*, Eds. S. Misra, S.C. Misra, and I. Woungang, Springer, London, U.K., 2008, pp. 199–230.
2. D.S.J.D. Couto, D. Aguayo, J. Bicket, and R. Morris, A high-throughput path metric for multi-hop wireless routing, in *Proceedings of the Ninth Annual International Conference on Mobile Computing and Networking (MobiCom'03)*, San Diego, CA, September 14–19, 2003, pp. 134–146.
3. A. Adya, P. Bahl, J. Padhye, A. Wolman, and L. Zhou, A multi-radio unification protocol for IEEE 802.11 wireless networks, in *Proceedings of First Annual International Conference on Broadband Networks*, San Jose, CA, October 25–29, 2004, pp. 344–354.
4. R. Draves, J. Padhye, and B. Zill, Routing in multi-radio, multi-hop wireless mesh networks, in *Proceedings of the 10th Annual International Conference on Mobile Computing and Networking (MobiCom'03)*, New York, September 26–October 1, 2004, pp. 114–128.
5. P. Xie, J.H. Cui, and L. Lao, VBF: Vector-based forwarding protocol for underwater sensor networks, in *Proceedings of IFIP Networking*, Waterloo, Ontario, Canada, May 2–6, 2005, pp. 1216–1221.
6. N. Nicolaout, J.H. Cui, and K. Maggiorini, Improving the robustness of location-based routing for underwater sensor networks, in *Proceedings of IEEE/OES Oceans*, Aberdeen, Scotland, June 18–21, 2007, pp. 1–6.
7. M. Goetz, S. Azad, P. Casari, I. Nissen, and M. Zorzi, Jamming-resistant multi-path routing for reliable intruder detection in underwater networks, in *Proceedings of WUWNet*, Seattle, WA, December 1–2, 2011, pp. 1–5.
8. R. Draves, J. Padhye, and B. Zill, Comparison of routing metrics for static multi-hop wireless networks, in *Proceedings of the 2004 Conference on Applications, Technologies, Architectures, and Protocols for Computer Communications (SIGCOMM'04)*, Portland, OR, August 30–September 3, 2004, pp. 133–144.
9. LinkQuest Inc., Underwater acoustic modem models. Available online: http://www.link-quest.com/html/models1.html (accessed on May 26, 2013).
10. AquaComm. Underwater wireless modem. Available online: http://www.dspcomm.com/products_aquacomm.html (accessed on May 26, 2013).
11. S2C R. 48/78 Underwater acoustic modem. Available online: http://www.evologics.de/en/products/acoustics/s2cr_48_78.html (accessed on May 26, 2013).
12. M. Chitre, S. Shahabudeen, and M. Stojanovic, Underwater acoustic communications and networking: Recent advances and future challenges, *Marine Technology Society Journal*, 42(1), 2008, 103–116.
13. I.F. Akyildiz, D. Pompili, and T. Melodia, Underwater acoustic sensor networks: Research challenges, *Ad Hoc Networks (Elsevier)*, 3(3), 2005, 257–279.
14. Q. Wu, X. Zhou, J. Wang, Z. Yin, and L. Jiang, A low-cost delay-constrained routing and wavelength assignment algorithm in WDM networks with sparse wavelength conversions, in *Symposium on Photonics and Optoelectronics*, Wuhan, China, August 14–16, 2009, pp. 1–4.
15. E.H. Cherkaoui, S. Azad, P. Casari, L. Toni, N. Agoulmine, and M. Zorzi, Packet error recovery via multipath routing and Reed-Solomon codes in underwater networks, in *Proceedings of MTS/IEEE Oceans*, Hampton Roads, VA, October 14–19, 2012, pp. 1–6.
16. G. Toso, R. Masiero, P. Casari, O. Kebkal, M. Komar, and M. Zorzi, Field experiments for dynamic source routing: S2C EvoLogics modems run the SUN protocol using the DESERT Underwater libraries, in *Proceedings of MTS/IEEE Oceans*, Hampton Roads, VA, October 14–19, 2012, pp. 1–10.
17. R. Urick, *Principles of Underwater Sound*, McGraw-Hill, New York, 1983.
18. W.H. Thorp, Analytical description of the low-frequency attenuation coefficient, *Journal of the Acoustical Society of America*, 42, 1967, 270–271.
19. S. Azad, P. Casari, C. Petrioli, R. Petroccia, and M. Zorzi, On the impact of the environment on MAC and routing in shallow water scenarios, in *Proceedings of IEEE/OES OCEANS*, Santander, Spain, June 6–9, 2011, pp. 1–8.
20. R. Coates, *Underwater Acoustic Systems*, Wiley, New York, 1989.
21. L.M. Brekhovskikh and Yu.P. Lysanov, *Fundamentals of Ocean Acoustics*, Springer, New York, 1982.
22. M. Stojanovic, On the relationship between capacity and distance in an underwater acoustic communication channel, in *Proceedings of ACM WUWNET*, Los Angeles, CA, September 25, 2006, pp. 41–47.
23. Micro-Modem related documentation. Available online: http://www.acomms.whoi.edu/umodem/documentaion.html (accessed on August 25, 2013).
24. N. Chirdchoo, W.S. Soh, and K.C. Chua, Aloha-based MAC protocols with collision avoidance for underwater acoustic networks, in *Proceedings of IEEE INFOCOM 2007*, Anchorage, AK, May 6–12, 2007, pp. 2271–2275.
25. G.T. Heineman, G. Pollice, and S. Selkow, *Algorithms in a Nutshell*, O'Reilly Media, Inc., Sebastopol, CA, 2008.
26. H. Schildt, *C: The Complete Reference*, 4th ed., McGraw-Hill, Berkeley, CA.

Section IV

16 Performance Evaluation of Flooding Algorithms for Wireless Sensor Networks Based on EffiSen

The Custom-Made Simulator

Bartosz Musznicki and Piotr Zwierzykowski

CONTENTS

ABSTRACT

Oftentimes, a computer simulation is the only way to test and verify new ideas long before hardware and software implementations are designed and developed. Moreover, many flaws can be eliminated at the initial stage, which is undoubtedly less expensive and less time-consuming than preparing a fix to an existing sensor platform. Therefore, this chapter begins with a presentation of the principles of wireless sensor networks and of the nature and tasks of routing in sensor networks. Available research methods are also given attention. However, the main emphasis is put on discussing the project rationale, objectives, and functionalities that are related to the development of a dedicated custom-made simulator called EffiSen. A guided description of application structure and operation is given to present the reader with a concept of implementing a wireless sensor network simulator. A methodology for a comparison of energy efficiency of routing algorithms for sensor networks is introduced. Furthermore, performance evaluation of flooding algorithms based on experiments conducted using EffiSen is presented.

16.1 INTRODUCTION

The main aim of this chapter is to present the reader with the concept, the rationale, and the results of creating a custom-made wireless sensor network (WSN) simulator called EffiSen designed to fulfill requirements of a specific study. The general task of the software is the simulation and statistical analysis of the influence of the minimal allowed distance between nodes on the efficiency of the sensor network (the time the network is considered operational) in relation to energy demands of flooding routing algorithms that are being compared. The two remaining evaluated parameters are the time to live (TTL) of an event notification and the capacity of ingress message buffer of every sensor in the simulated network. The most characteristic feature of the simulator is the ability to perform a series of automated uncorrelated pseudorandom simulations aimed at the collection and the analysis of the data obtained in the process. The simulator has a modular structure and consists of a random number generator, topology generation module, node definition module, as well as simulation, visualization, and statistical results modules. Graphical user interface implemented in C# programming language provides all the controls required to conduct and tune the experiments.

To give a clear and in-depth view of the project the focus is not on the presentation of the code of the application but the requirements, idea, implementation, and the simulation results are in the center of attention. Therefore, the text is chiefly directed to WSN researchers and designers, especially those who consider creating a simulator on their own. However, selected sections will be of interest to general readers who are eager to get familiar with the basics of sensor networks and with the types of research methods that are being used. The next section covers the principles of WSNs discussing the basic components of a WSN and the role each component plays. This is followed by an overview of a typical sensor network structure and of the nature of routing in WSNs.

Attention is given to research methods used in studies on routing in WSNs. Subsequently, the flooding algorithms selected to be implemented and simulated are presented in Section 16.3. The custom-made simulator EffiSen is introduced in Section 16.4. The project rationale and objectives are discussed, followed by a detailed description of the simulator structure and operation. Section 16.5 presents how the simulator has been used to perform studies on energy efficiency of flooding routing algorithms. Hence, to better understand the approach the methodology and simulation parameters are presented. Finally, the simulation results are the basis for the detailed discussion of the factors that influence the performance of supported flooding algorithms. The chapter is summed up with conclusions.

16.2 PRINCIPLES OF WSNS

WSNs are the next step from ad hoc networks toward more specific applications. The overall basics of the network are quite similar to those introduced by their predecessor. However, in WSNs, the devices are usually smaller, more mobile, and directed at observing or measuring certain phenomena. Hence, most of sensor networks are designed for applications related to various kinds of monitoring. The first and most basic applications of WSNs were related to measuring simple physical quantities such as temperature, pressure, and light intensity. Thus, one could think that WSN was just a bit more sophisticated and distributed home weather station developed by some enthusiasts. It must be noted that WSN pioneers needed to have had a vision for future applications and enjoy their work. Currently, it can be said that the usage of sensor networks already is, or will be in the near future, limited only by developers' imagination. The idea of future smart homes controlled by a flexible integrated system, or the so-called Internet of things [1], requires a novel approach to communication. Environmental monitoring and emergency systems pose many challenges as well. WSNs composed of small sensing devices will undoubtedly constitute an important part of the solution.

16.2.1 COMPONENTS OF A SENSOR NETWORK

The basic components of a WSN are a sensor node and a gateway. These elements are presented in the following sections.

16.2.1.1 Sensor Node

The term *node* has been well established in the network theory nomenclature as a definition of the device that network paths pass through. Therefore, a *sensor node* is a popular name for a wireless electronic device designed, in general, for monitoring certain physical phenomena. The device, for simplicity usually called a *sensor* or a *node*, uses a built-in radio module to transmit the measurements and to relay the data received from adjacent appliances. Especially in North America, authors often refer to a sensor node as a *mote*, which relates to the very nature of WSNs.

A WSN node, as a small energy-constrained wireless router, has to be energy efficient. Hence, developers put much effort into optimization of every possible hardware and software aspect of a sensor. Each sensor node, as discussed in Ref. [2], consists of five main components:

- *Controller*: a microcontroller or a microprocessor that controls the operation of a WSN node. It runs programs, processes the data, and performs message routing. A controller should be energy aware, so usually various sleep and energy-saving modes are implemented.
- *Communication device*: a module enabling wireless communication with other nodes to gather topology information, send measurement data, relay messages, etc. Most of the transceivers operate on radio waves, but there are concepts of optical and ultrasound transmission as well. The vast majority of sensors use omnidirectional radio antennas, although nothing stands in the way of using directional antenna if required.

- *Sensors and actuators*: at least one device capable of observing (sensor) or controlling (actuator) some parameters of the environment the node operates in. Typically, physical quantities like temperature, humidity, and pressure are measured. More advanced solutions allow the acceleration, carbon dioxide concentration, and noise level to be monitored or even perform distributed video and audio surveillance. In other words, almost everything can be monitored by a properly equipped WSN node. The actuator interface of a node is usually limited to a type of a simple switch or a relay controlling external device, that is, a motor, a pump, or a fan.
- *Memory*: a storage for the program code and the data. Aside from random access memory (RAM), typically, electrically erasable programmable read-only memory (EEPROM) or flash memory is used.
- *Power supply*: the part of a sensor node that has virtually no logic in it. Nonetheless, a node will not operate without any kind of power source. Often AA and AAA batteries are utilized, but due to node miniaturization, button batteries and capacitors are also used. As an alternative or support, capacitors and power harvesting elements such as solar panels and inertial power supplies are introduced to sensor platforms. Some batteries may be also charged wirelessly.

16.2.1.2 Gateway

A gateway, also referred to as a sink node or a base station, is the element that connects WSN to an external network or directly to user device. The first type of gateways is a dedicated appliance, designed specifically for its tasks. It is equipped with various wired and wireless communication interfaces providing remote access to the network. The device may also act as an integrating unit, connecting different sensor networks and allowing them to constitute a larger structure. The base station can affect control over subordinate sensor nodes adjusting the network topology. Due to higher computational and energy resources, as compared to sensors, the gateway may be able to perform advanced data processing, aggregation, and compression.

A base station of the second type is a sensor node responsible for the tasks of a dedicated gateway. It may be used as a cheaper solution, but the node has to provide at least one long-range communication channel, so that the network could be reached from a distance. A node should be chosen when the gateway is required to perform sensing. The main limitation of a sensor-based gateway is, in most cases, an energy-constrained power source.

The third type of gateways includes devices based on *WSN adapters*. As opposed to stand-alone base stations, their main function is to allow personal computers, automation controllers, etc., to operate as data sinks for a WSN. Therefore, data collection and network management software platforms may be used directly at the edge of WSN. The instant control of industrial processes can be based on measurements performed by wireless sensors. Many of commercially available sensor nodes provide USB ports or interface modules that can be used to turn an ordinary computer into a sensor network controller. The devices are usually powered through the connection cable, so developers do not need to provide any additional power sources.

16.2.2 STRUCTURE OF A SENSOR NETWORK

The number of sensors, in a typical WSN model, varies from a few devices to thousands of nodes. Usually, the network has one gateway, but, in larger or more distributed structures, it is reasonable to use several base stations. WSN may be homogenous as well as heterogeneous, depending on the tasks the network is designed for. Following nodes location and logical structure, WSN may operate in any network topology known from other types of computer networks. This includes point-to-point communication, star topologies with the gateway as a central node, and tree structures with base stations at higher levels of hierarchy. Advanced solutions will involve hierarchical hybrid

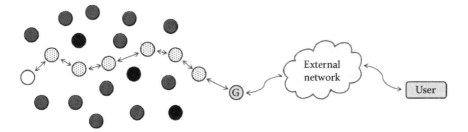

FIGURE 16.1 General structure of a WSN.

networks benefiting from advantages of each basic topology. The most efficient and reliable flat topology, but also the most challenging to design, would be a mesh composed of nodes deployed optimally for particular use.

In general, because of limited radio range, node devices have to constitute a multihop network with each node relaying messages from and to its neighbors. In this way, messages may go around obstacles and traverse long distances. As presented in Figure 16.1, the message to get from the source (the white node) to the gateway (the orange *G* node) has to be passed on by each relay node in the path (the yellow black-dotted nodes). Obviously, the transmission may be performed in the opposite direction, from the base station to the white node. Except for sensors taking part in message exchange, there may also be, and usually are, nodes that do not transmit any messages. Depending on the implementation, they may stay active, waiting for messages and events or switch to a power saving sleep mode. Those nodes are depicted as red circles. The last group of nodes, the black circles, includes sensors that have run out of energy or got damaged. The representation also applies to sensors that have been intentionally disabled. There are topology control mechanisms that support WSN self-healing and decide whether a node should perform its tasks or rather stay excluded from the network structure [3]. Therefore, disabled nodes save energy waiting for the instruction to join active topology and take over the tasks of depleted nodes when needed.

The deployment of WSN nodes is random in many cases. For example, sensors can be thrown out of an airplane over the operation area, which will undoubtedly scatter them. More static applications, like traffic surveillance or building constraints monitoring, will involve nodes placed intentionally by hand in precise locations. There is also a range of projects that involve mobility, that is, focusing on wildlife or livestock monitoring with devices carried by animals. The gateway can be also mobile, acting as a probe moving throughout the area connecting to sensors within the radio range.

A single node will, most likely, have no initial information about the network topology, and therefore, a WSN has to be self-organized. Moreover, sensors need to adapt to often unpredictable changes taking place in the environment. There are two ways a network can operate in such conditions. A node may have no knowledge about neighboring sensors and may depend solely on implemented routing mechanisms in fulfilling its tasks. On the other hand, nodes can exchange information that will make it possible to create, at least local, images of the topology. The process may be supported by global positioning system (GPS) receivers and radio beacon stations that help to estimate node locations.

16.2.3 NATURE AND TASKS OF ROUTING IN WSNs

The general idea of routing in sensor networks is exactly the same as in any other kind of multihop networks. Therefore, it concerns a process of directing a message, a packet, or simply a data unit from the source to the destination with the assistance of intermediate nodes. However, there are many differences and challenges that need to be faced. The great possible number of nodes in WSNs, high scalability, energy constraints, and omnidirectional radio communication force researchers to introduce different forwarding techniques than those used in wired or even ad hoc networks. Following Ref. [4], another aspect and factor influencing routing can be pointed out. Unfortunate node deployment may seriously affect routing performance shortening network lifetime. If the power demands related to a routing method are too high, the network will lose faster its accuracy and become fragmented. Sensor devices may be designed for different tasks and node heterogeneity will have to be considered. Environmental conditions change dynamically and some nodes may run out of energy or get damaged. The routing mechanism needs to be designed to overcome failures so that the network could perform its main tasks. Issues like network coverage maximization, node mobility, and quality of service should be also addressed.

16.2.4 RESEARCH METHODS USED IN STUDIES ON ROUTING IN WSNs

Different approaches have been developed in the area of WSNs research focused on routing analysis and optimization. In most cases, they include studies in a real sensor network, involve the usage of a laboratory testbed, or are based on computer simulations. Each way has its advantages that make it the most suitable one for investigating and solving particular tasks. These approaches may be used separately, as well as in a combined manner, to enable researchers and developers to study the problems thoroughly. Each study method is discussed in the next subsections.

16.2.4.1 Studies Performed in a Real Sensor Network

The direct approach to the research on routing in WSNs is based on working with sensor devices. While taking a decision to study WSNs in such a way, developers need to have the highest project budget, as compared to other research methodologies. There are a few manners of coping with the task. On the one hand, ready-to-go sensor modules, gateways, or complete sensor sets are readily available on the market. So the researcher has to decide which products to buy, pay for the devices, and wait for the package. Usually, the manufacturers offer operating systems and management software for free or for an additional charge. Plainly, those opportunities have some important advantages. Users get already tested devices prepared to operate in the physical environment. The efforts needed for designing hardware and implementing one's own control software can be limited to an absolute minimum, but at the expense of growing costs. On the other hand, one can choose open-source operating systems, together with visualization and management environments like Octopus [5] that cooperates with sensors running TinyOS [6]. Wireless nodes can be designed by the research team, yet it would seriously complicate the project aimed, after all, at proposing and evaluating routing algorithms.

When the hardware platform has been prepared and the first version of routing software is implemented, nodes are deployed in the environment. Afterward field tests begin. When malfunctions occur, researchers start the process of debugging. The solutions for problems have to be found, so that the updated version of firmware files could be loaded to the sensors. The procedure is repeated and if the modifications prove to be successful, making the network exactly how required, the processes of optimization and profiling may be started.

Real-time studies conducted in the outer environment entail the risk of network being damaged before final conclusions could be drawn. Furthermore, routing algorithms can be analyzed only one by one, which increases the amount of time needed to compare proposed variants or different algorithms. Nonetheless, the tests performed in a target environment usually will be the most credible. To obtain reliable results at least tens or hundreds of sensor nodes should be used. Depending on the research area, the deployment of sensing devices may cause a lot of problems itself. An example of such a case is the rainforest monitoring system consisting of 175 nodes deployed in the Springbrook National Park in Australia [7]. In similar projects, it needs to be assumed that devices can stop responding, get destroyed, or stolen, so replacements should be kept in stock. Likewise, spare batteries have to be prepared in case sensors do not use solar panels, radio-frequency energy-harvesting modules, or inertial power supplies, etc. If researchers cannot afford to prepare the required number of wireless nodes because of production costs remaining high, other study method should be chosen.

16.2.4.2   Usage of a Laboratory Testbed

Laboratory testbeds consist of sensor nodes deployed in a controlled environment in a single site, that is, a room or a building [8,9], and they can be composed of a number of federated sites [10,11] even of both real and virtual (simulated) sites [10]. Each network node may be connected to an external power supply, so there would be no need for replacing or charging the batteries. Aside from wireless communication, often every single device can be reached by some type of wired serial connection or through a programming interface. Some of the projects locate nodes above the ceiling in different rooms. Other sensor network laboratories create arrays of nodes mounted on top of special tables, stands, or racks. System interfaces facilitate the general maintenance and direct programming of sensors over a wired connection.

The usage of a test network brings noticeable benefits. Developers have an easy access to all of the devices in case of modifications or repairs. Environmental conditions can be closely controlled, avoiding any interruptions in the process of designing applications. If a suitable location for the testbed is available, the total outlays related to the project should be lower than that of a network deployed in a real environment. Moreover, the obtained results will be still very useful for observation and evaluation of introduced routing mechanisms.

16.2.4.3   Computer Simulation of a Sensor Network

Computer simulation techniques allow researchers to model any aspects of WSNs. Obviously, the analysis is possible only if appropriate simulators are implemented. One of the most significant advantages of WSNs simulation is its time efficiency. New ideas can be tested and verified long before the hardware implementation. With the use of various debugging and profiling tools, many flaws will be eliminated at the initial stage, which, in the case of working with real sensors, would be more time consuming. A complete designing task may be divided into smaller sections, each studied independently. Therefore, researchers can focus their attention solely on the aspects they are interested in.

To begin studies on routing in WSNs, we do not have to consider the basic functions of electronics or the matters of radio communication, which in the case of other research methods would be indispensable. Moreover, the simulation will gain on scalability if the modelled process concerns strictly a given level of abstraction. Since some studies may require a more general research, not only directed at routing algorithms, a cross level simulator may be used.

A precise comparison of various solutions for a given problem requires the repeatability of simulation conditions. The parameters of natural environment continually change, the temperature and pressure vary, and even a slight difference can affect the results. Thus, only virtual environment can fulfil the requirement of providing the exact set of stimuli when required.

A simulation platform is the optimal solution for limited expenses. As the present authors reviewed in Ref. [12], there are at least 36 free WSN simulators available free of charge to be used in the studies, apart from a few commercial products. Therefore, a free open-source program can be chosen and modified, or if none of them fits the requirements, a dedicated software may be designed. In addition, there already are simulators working in a hybrid mode, successfully cooperating with real sensor devices. Therefore, a handful of hardware sensors can communicate with hundreds of simulated nodes providing them with the data gathered in a real environment. Additionally, considering the fact that simulators operate well on personal computers, the simulation is available to anyone—not only professional researchers but also students, enthusiasts, etc.

16.3   FLOODING ALGORITHMS SELECTED TO BE IMPLEMENTED AND SIMULATED

Sensor devices usually utilize omnidirectional antennas, which is the reason why each local message transmission is broadcast to every node in the radio range. The simplest possible routing algorithm, network flooding, is based directly on this principle. When a node observes an event or receives a measurement query, a message is broadcast to every neighbor. Nodes that receive the packet relay it further, and so on. Eventually, after a few cycles of transmission, the whole network will be flooded with multiplied messages and a sink or a node interested in the data will receive the notification. This simple scheme requires no addressing and no message processing. Nonetheless, there is no way of controlling the number of messages in the network. To avoid endless circulation and replication of flooded packets, modifications like message expiration time and traversed nodes verification should be introduced to the method. The flooding algorithms chosen for the research are introduced in the next sections. An example of the event notification routing has been provided as well.

16.3.1   Choice Motivation

In the development of WSNs, much weight is attached to expanding their overall operating time. The ongoing research for smaller power sources, yet offering higher capacities, together with more sophisticated electronics provides means for increasing the network lifetime. Thus, thanks to the growing computing power of microcontrollers, complicated routing methods may be implemented without overloading the network. On the one hand, novel routing algorithms and protocols allow the information to be directed to the destination making use of fewer relay nodes. On the other hand, the number of addresses and metrics stored in routing tables increases, as well as the set of information exchanged in the process of determining the path.

    However, in some applications, the simplicity of implemented routing algorithm is crucial. Networks consisting of simple devices operating on limited resources entail the need of using basic routing strategies. Hence, flooding the network with event notifications appears to be the most direct procedure. Virtually no routing information needs to be carried to propagate the payload to the sink. Each sensor simply relays any received data to all of its neighbors. The major drawback of such an approach is the tremendous energy demand. Fortunately, the efficiency of flooding can be noticeably improved. Therefore, the range of analyzed routing algorithms has been limited solely to flooding techniques and modifications to the algorithm have been introduced. The decision follows the belief that applying changes supported by complex analyses will result in a significant performance improvement.

    The researchers from the Virtual Extension company have drawn similar conclusions and developed a technology called Diversity Path Mesh [13]. The concept is based on using synchronized flooding in mesh WSNs. As stated on the company's website, the system translates into network design simplification, deployment time reduction, and the limitation of required processing resources. Another example that introduces energy-greedy quasi-flooding approach, yet not intended for typical WSNs but designed primarily for power-line communication (PLC) or ultra-high frequency (UHF) transmission techniques, was presented in Ref. [14].

16.3.2   Descriptions of Chosen Flooding Algorithms

The routing algorithms chosen for the simulation studies are as follows:

- Flooding
  The basic and extremely inefficient flooding has been extended to support the TTL parameter, defining the maximum number of relay nodes on a message path. The initial TTL value is decremented by one after each hop. Value equal to one means that the message got to the last node and should not be relayed any further. In this simple way, the number of redundant transmissions is limited, which increases the efficiency of a network. For convenience, this algorithm will be referred to as flooding.

- Flooding with the source node check

  The algorithm is based on flooding, with the difference that each message contains the identification number of an originating node. When a node receives a message, it compares the source node identifier address (ID) with its own ID. If both addresses are equal, a loop occurs so the message is dropped.

- Flooding with the source and last node check

  This flooding enhancement involves not only the source node check but also the comparison of a node that previously relayed the message. It is a slight but very significant modification introducing local loop avoidance. In flooding, a notification will circulate between two nodes as long as the TTL value allows it. Here, such a situation will never take place, which noticeably improves algorithm performance.

Please follow the example of routing single event notification, with TTL set to three, shown in Figures 16.2 through 16.6. In the first step, node number 8 observes an event and sends the notification to all of the nodes within its radio range. Then, sensors 2, 9, 13, 15, and 20 relay a copy of the message to their neighbors. Node number 8 drops received messages, being the originator of the notification. Please keep in mind that the TTL of a cloned message is decreased by one with each transmission.

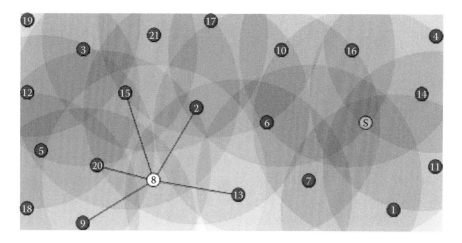

FIGURE 16.2   Flooding with the source and last node check—first step.

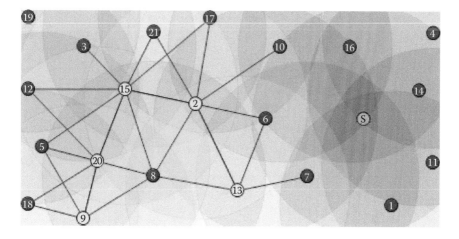

FIGURE 16.3   Flooding with the source and last node check—second step.

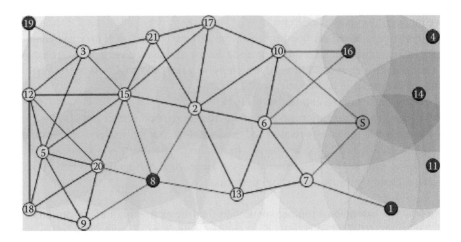

FIGURE 16.4   Flooding with the source and last node check—third step.

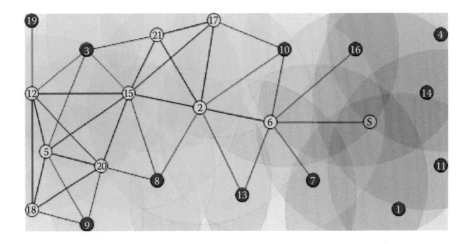

FIGURE 16.5   Flooding with the source and last node check—fourth step.

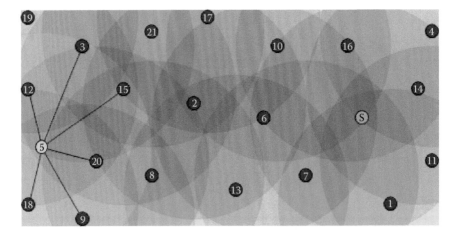

FIGURE 16.6   Flooding with the source and last node check—fifth step.

In the third step, see Figure 16.4, notifications keep on flooding the network, reaching the S sink node. At this stage, the gateway gets notifications about exactly the same event from three different devices replicated. It is an obvious advantage should high notification reliability be required. Nodes 1, 16, and 19 only receive notifications and are not allowed to relay them due to exceeded TTL. This is why, in step four, sensor nodes 4, 11, and 14 do not get any messages. In Figure 16.5, local loop prevention and message expiration can be observed. The sink receives one more event message from node number 6. In the fifth step, the sensor with an ID equal to 5 performs the last transmission and event notification originated by node 8 disappears from the network.

16.4 EFFISEN: THE SIMULATOR OF FLOODING ALGORITHMS FOR WSNs

The following sections are aimed to present the rationale behind the development of the custom-made simulator EffiSen, software components, parameters, operation, and simulation visualization.

16.4.1 PROJECT RATIONALE AND OBJECTIVES

A significant number of research articles the authors are familiar with present conclusions based on the results of a single simulation or measurement. Unfortunately, this is one of the most unacceptable mistakes a researcher can make. As a result, the reader gets a very limited overview of the studied problem and the presented charts usually do not provide any useful information. Such an approach should be publicly criticized at every opportunity to make authors be aware of the problem and to improve the quality of the results of their hard work. Each and every researcher should consider whether the graphs he or she presents bring any added value to a project or only fill the empty space with a few colorful curves.

In the field of WSNs, the low quality of simulation results may be partially caused by the fact that available simulators neither support a series of hundreds of automated pseudorandom simulations nor provide the tools for collecting and analyzing the data generated in each run [12]. Hence, to be able to obtain reliable and objective results, a simulator of routing algorithms for WSNs called EffiSen has been implemented. The program is strongly based on a pseudorandom number generator, which is the starting point for everything that takes place during the simulation experiment. Each experiment comprises a number of uncorrelated simulations allowing the researcher to calculate the average results and corresponding confidence intervals. The program is mainly focused on simulating the influence of the distance between network nodes on the efficiency of routing algorithms. EffiSen also allows the performance of implemented notification delivery algorithms to be compared. What is more, the parameters of the network may be manually tuned, and thus, the program helps to find the settings that will extend the network life to the maximum.

16.4.2 SIMULATOR STRUCTURE

The simulator consists of different modules responsible for particular tasks. The components of EffiSen have been described in the following sections.

16.4.2.1 Random Number Generator

A multiplicative random number generator is one of the most basic elements of the simulator. Pseudorandom number generation is possible for integers ranging from 0 to 2147483647. During simulations, the range is narrowed down to generate numbers defined by the minimal and the maximal values. The generator is provided with a text file containing 2000 seeds obtained by saving every millionth output of a uniform distribution generator. During every simulation, the next unused seeds are loaded from the file and assigned to each instance of the random number generator involved in the simulation. Therefore, the independence of generated pseudorandom sequences can be provided.

16.4.2.2 Topology Generation Module

The module responsible for generating random network topologies is an important part of the simulator. Various sets of mesh-like WSN topologies it provides are used in simulation experiments. The topology generator consists of two logical parts—the node placer and the link builder. The node placer operates on a rectangular area capable of holding $m \times n$ sensor nodes. Each node is located in a coordinate slot provided by the uniform distribution generator. The module is able to keep minimal node spacing and place the maximum number of nodes for particular network parameters.

In terms of radio communication and topology construction, bidirectional links between nodes are available. Those are constructed by the link builder verifying if two nodes can maintain wireless communication. The verification is based on checking whether the propagation range of a node includes the radio interfaces of other sensors. It has been assumed that the antenna is located in the center of each node.

16.4.2.3 Node Definition Module

A separate module is used for performing the tasks of a sensor node. The module also defines the three implemented notification flooding algorithms. Hardware parameters as well as the current state are stored there. A node can be in one of the four states:

- *Idle*: a node is alive and does not send any message.
- *Event observed*: a node observes an event and transmits a message to its neighbors.
- *Active*: a node relays a message.
- *Depleted*: a node run out of energy and is excluded from routing.

16.4.2.4 Simulation Module

The simulation module is responsible for everything related to the network simulation and its progress. The module generates new events, propagates the messages between network nodes, and collects the data for statistical analyses. In general, it is the engine that makes conducting experiments possible. A basic time unit in every simulation is called a cycle. In a single cycle, each node can observe an event, send the notification, or relay a message waiting in the ingress buffer. Also, processing of newest messages received in previous cycle may be performed.

16.4.2.5 Visualization Module

The visualization module provides both a graphical representation of network topologies and a real-time animation of routing processes taking place in the simulated WSN. The user is able to evaluate the structure of generated networks and adjust the parameters before the simulation begins. Later, during the experiment, the correctness of routing algorithm implementation can be verified by slowing down or pausing the simulation and analyzing it step by step. The animation of routing in a basic WSN topology may be used as a valuable teaching aid presenting the principles of WSN operations. If a complex experiment is going to be conducted, the visualization can be disabled to improve the performance of the simulator.

16.4.2.6 Statistics and Results Module

The module collects data and performs a statistical analysis of the simulation experiment. Values stored during the simulation are used for calculating the averages, standard deviations, and confidence intervals. These values are later presented in the form of diagrams that compare simulated parameters. The 95% confidence intervals are calculated with the use of Student's t-distribution quantiles stored in a text file. If the number of values exceeds 50, the quantiles are approximated by the normal distribution and 1.97 is used as the quantile value.

16.4.3   SOFTWARE IMPLEMENTATION

EffiSen has been implemented in the C# programming language on the Microsoft.NET Framework. The environment has been chosen not only because the authors have been familiar with it but also because C# provides various basic classes that simplify the development. Simulation diagrams are generated using ZedGraph [15], a free library offering a vast range of supported charts.

The simulator uses two threads—one for control panel maintenance and one for the experiment and the visualization. This is why the user interface does not freeze when complicated and long-lasting simulations are performed. The user is able to control the simulation and adjust the speed of the process when the visualization is enabled.

16.4.4   GRAPHICAL USER INTERFACE

The user interface of EffiSen enables the configuration of various simulation parameters. These parameters, network visualization, and result graphs are discussed in the following sections.

16.4.4.1   Control Panel

In the control panel, presented in Figure 16.7, the user defines the settings and starts or aborts the experiment. The sections grouping the parameters are as follows:

1. Area dimensions
   a.   Width—the width of the simulation area
   b.   Height—the height of the simulation area
2. Network visualization
   a.   Enable—checked to enable the visualization
   b.   Show the grid—checked to display the grid
   c.   Range visibility—the visibility of the node radio range (0% to 100%)
   d.   Node size—the diameter of a node, used as a scaling factor for the visualization

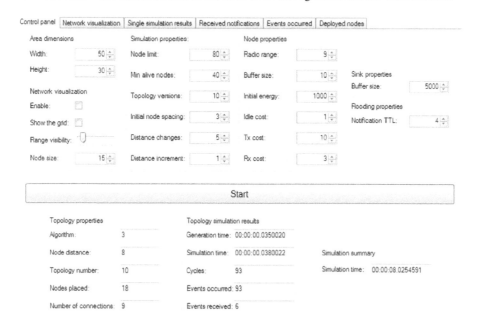

FIGURE 16.7   EffiSen—control panel.

3. Simulation properties
   a. Node limit—the maximal number of nodes to be placed in the simulation area
   b. Min. alive nodes—the minimal number of alive nodes defining how long a topology simulation will proceed
   c. Topology versions—defines how many topologies will be examined for each simulated node distance
   d. Initial node spacing—the initial minimal distance between nodes
   e. Distance changes—the number of node distances examined during the experiment
   f. Distance increment—defines a single increase of the minimal node distance
4. Node properties
   a. Radio range—the radio range of a node
   b. Buffer size—the size of the node ingress message (packet) buffer
   c. Initial energy—the energy stored in the battery
   d. Idle cost—the cost of a simulation cycle when a node is not transmitting nor receiving a message
   e. Tx cost—the cost of the message transmission
   f. Rx cost—the cost of the message reception
5. Sink properties
   a. Buffer size—the size of the sink ingress message (packet) buffer
6. Flooding properties
   a. Notification TTL—the TTL value for an event notification

When experiment parameters are set, the simulation can be started by clicking the *Start* button. To cancel the experiment, the same button, but with the text changed to *Abort*, has to be clicked.

The following information concerning the experiment is displayed at the bottom of the window:

1. Topology properties
   a. Algorithm—the number of simulated routing algorithm
   b. Node distance—the minimal node distance for current topology
   c. Topology number—the number of simulated topology version
   d. Nodes placed—the number of nodes placed in the simulation area
   e. Number of connections—the number of internode radio connections
2. Topology simulation results
   a. Generation time—the time that took the simulator to generate the topology
   b. Simulation time—the time the topology has been simulated
   c. Cycles—the number of network operation cycles
   d. Events occurred—the number of events that occurred during the simulation of the network
   e. Events received—the number of event notifications received by the sink during the simulation of the topology
3. Simulation summary
   a. Simulation time—the complete time of an experiment (displayed when the simulation experiment is completed)

16.4.4.2 Network Visualization

The tab *Network Visualization*, presented in Figure 16.8, contains a graphical representation of the network. The user is able to control the simulation using Start/Pause/Resume and Stop buttons. When the visualization is enabled, the simulation speed may be adjusted using horizontal trackbar. To save current frame, the user has to right-click on the visualization area and choose *Save*. A PNG file, called network.png, will be stored inside EffiSen's directory.

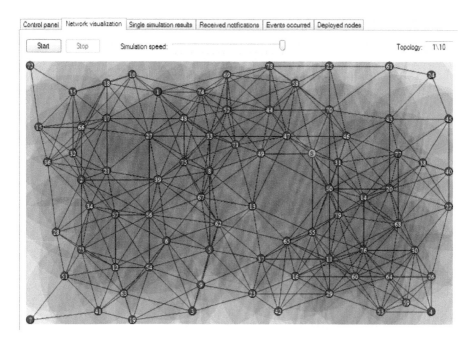

FIGURE 16.8 EffiSen—network visualization.

The gateway, depicted as the orange circle with the letter *S* inside it, does not observe events and is only the sink for event notifications. Each sensor is also displayed as a circle containing the node's ID. One of the four overlaying semitransparent colors is used to draw the sensor:

- *Red*—a node is in the idle state
- *Yellow*—a node transmits a message
- *White*—a node observed an event and transmits a message to its neighbors
- *Black*—the node is depleted and excluded from routing

The radio range of a node is presented as a semitransparent circle surrounding the node. The level of transparency may be controlled by the user. There are two colors used to display the radio range of a node:

- *Blue*—a node is in the idle state
- *Yellow*—a node transmits a message

After a topology is generated, each connection available in the network is visualized. In the topology preview, the links between nodes are displayed as the lines connecting centers of adjacent nodes. During the simulation, only the connections used for transmission are being drawn. If desired, a grid depicting the node slots can be displayed.

16.4.4.3 Single Simulation Results

When a simulation experiment is completed, a bar graph is generated (see Figure 16.9). Each bar corresponds to the number of event notifications received by the sink during the simulation of a particular network topology version. The user may analyze the partial results of the simulation of minimal node distance equal to the initial node spacing value. The results are presented using three colors, respectively, to the simulated routing algorithm:

- *Red*—Flooding
- *Green*—Flooding with the source check
- *Yellow*—Flooding with the source and last hop check

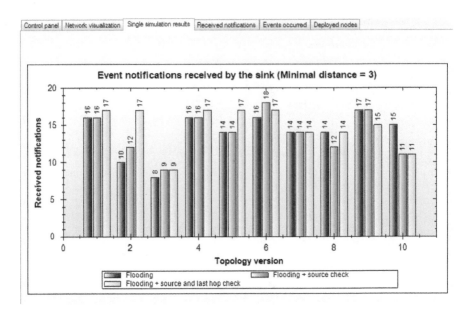

FIGURE 16.9   EffiSen—single simulation results.

16.4.4.4   Complete Experiment Result Diagrams

When the experiment is completed, the *Received Notifications* tab displays a graph including mean values of event notifications received by the sink for the simulated range of minimal node distances. To allow the user to evaluate the reliability of obtained results, the confidence intervals corresponding to all of the values are presented. The three curves depicting the changes of the efficiency of routing algorithms are colored according to the convention introduced in Section 16.6.4.3. The tab called *Events Occurred* shows a diagram of the average number of events occurring during the simulation. The last graph gathering the data on the average number of deployed nodes can be found in the tab *Deployed Nodes*. Each chart can be zoomed in, saved, or printed using the right-click context menu. Examples of the graphs can be seen in Section 16.7.4.

16.5   ANALYSIS OF ENERGY EFFICIENCY OF FLOODING ROUTING ALGORITHMS FOR WSNs

Beginning the research on routing in WSNs, the authors decided to determine which aspects have influence on the efficiency of the message delivery process. The observations suggested that the implemented routing algorithm would be responsible for the number of transmissions taking place in the network, and therefore it would define the energy consumption. As it turned out in further studies, not only the concept of a particular algorithm should be taken into consideration. The performance of a routing algorithm may be also highly related to the topology properties and the node parameters.

16.5.1   Aims and Guidelines for the Experiments

It has been decided to compare the efficiency of three flooding algorithms introduced in Section 16.4 and examine the relation between their performance and properties of the networks they operate in. Most of the emphasis is placed on verifying the influence of the minimal distance between sensor nodes. Various values of TTL and message buffer capacities have been also tested. The sink counts only the unique event notifications it receives, and on this ground, efficiency graphs are generated.

16.5.2 EXPERIMENT METHODOLOGY

To be able to compare various routing algorithms, an experiment methodology needs to be introduced. The proposed procedure focuses on verifying how many event occurrence notifications will reach the destination until the sensor network runs out of energy. The sink has access to an unlimited energy resource, as opposed to network nodes that operate on batteries of a given capacity. The remaining energy is decreased according to the cost of the operation types the node performs in a particular cycle.

Reliable results can be obtained only when a significant number of simulations are conducted. This is the key to the minimization of confidence intervals and the only way of preparing an accurate overview of the simulated algorithms. All of routing algorithms have to be simulated in exactly the same set of network topologies. Uncorrelated network structures are prepared by the random topology generator. The events observed by the sensors occur pseudorandomly throughout the topology. Every simulation of a single topology is performed as long as the defined percentage of network nodes has enough energy to operate. If the limit of depleted nodes is reached, the simulation of next topology begins. When all topologies are examined, the experiment comes to an end and the statistical results are calculated.

16.5.3 SIMULATION PARAMETERS

Most of the simulation parameters discussed in this section are common for every conducted experiment. If any value has been changed, a comment explaining the reason has been included. The parameters influence both the simulation details and the network topologies. A sample generated mesh-like topology deployed in the area of 100 by 50 units that is composed of 59 nodes and 152 connections with minimum node spacing of 8 units and node radio range equal 15 units is shown in Figure 16.10.

- Area dimensions
  The basic simulation area is a rectangle of 100 by 50 units; therefore, it offers 5000 slots for sensor nodes. The shape and dimensions have been chosen to support the deployment of networks corresponding to those deployed in real environments (fields, buildings, etc.).
- Number of nodes
  Each network topology can be composed of, at the maximum, 200 nodes. This value comes from the observation that operating real networks and testbed sites usually comprise up to about 200 devices [7–11].
- Minimum active nodes
  The minimum number of alive nodes the network is still considered operational is set to 40% of sensors making up the network. When 60% of nodes are depleted, the simulation of a topology is finished. This value has been chosen due to the observation that, for higher values, the WSN is too fragmented and simulation results may be blurred.

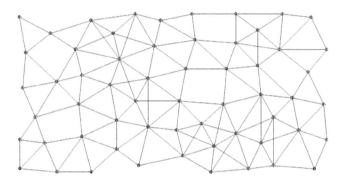

FIGURE 16.10   Mesh-like topology with 59 nodes and 152 bidirectional internode connections.

- Topology versions
  The parameter defines the number of random topologies belonging to a set of networks simulated for each node spacing. Experiments conducted on 300 topologies allow to obtain valuable results and useful widths of the 95% confidence intervals.
- Node spacing
  Simulated node spacing has been set to vary from 0 to 15, with the increment of one unit, to cover every value that is smaller than, or equal to, the node radio range.
- Node radio range
  The radio range of a node has been set to 15 units to provide wireless coverage suitable for the simulation area.
- Message buffer size
  The default buffer capacity has been set to 10 packets, but in a number of experiments, different values were assigned.
- Initial energy
  The initial energy of a node (capacity of the battery) has been set to one hundred thousand units. It allows the network to relay a few thousands of event notification, which results in graphs presenting properties that can be easily scaled and applied for any power source.
- Operation costs
  The amount of energy required by a node during each simulation cycle depends on the types of performed operations. The basic cost of a cycle is set to one energy unit. The reception of a packet costs 3 units and the transmission consumes 10 units. The values have been chosen to follow the average energy demands of commercial hardware sensor platforms.
- Notification TTL
  The basic TTL has been set to five hops, but influence of other values has been also studied.

16.5.4 FACTORS INFLUENCING THE PERFORMANCE OF FLOODING ALGORITHMS

There are a number of factors that can have the influence on the efficiency of flooding algorithms for WSNs. The most crucial parameters have been discussed in the next sections.

16.5.4.1 Minimal Distance between Nodes

Each flooding routing algorithm has been simulated for minimal node distance values ranging from zero units to the number of units equal to the value of the node radio range. The scope ensures that all applicable implementations of deployment have been represented. The higher the spacing, the fewer wireless sensors can be deployed in the area and the topology starts to be mesh-like. The distances equal or greater than the propagation range result in the lack of any connectivity and the nodes do not constitute a network.

Figure 16.11 shows that for short internode distances, the number of deployed nodes remains at the maximum level, and when the free space goes up, the sensor count rapidly decreases, yet slowing down with each distance value.

The influence of the node deployment on the efficiency of flooding algorithms is shown in Figure 16.12. The number of event notifications delivered to the sink is closely connected with the distance between nodes; therefore, it is related to the total number of sensors in the network as well. For spaces between 0 and 4 units, the results differ slightly and hold a constant level, just like the number of sensor nodes in Figure 16.11. When free space in the topology increases, the growth begins and reaches the maximum for the distance of 11 units. The variance of possible network structures rises due to the increased number of degrees of freedom because there are fewer nodes to be scattered in the same area. The confidence intervals are getting wider, although, thanks to the hundreds of simulated topologies, the results are still very reliable. From that point, the sink receives fewer notifications, reaching zero messages when the nodes are disconnected.

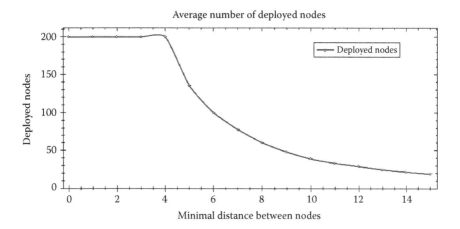

FIGURE 16.11 Number of nodes deployed in the 100 × 50 simulation area.

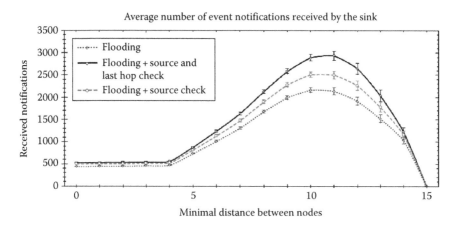

FIGURE 16.12 Received event notifications. Area, 100 × 50; TTL, 5; buffer size, 10.

The experiment carried out in the area of 50 by 30 units with the radio range equal to 8 units resulted in the characteristics corresponding to the results depicted in Figure 16.12.

Figure 16.13 presents the results of the experiment carried out in the area of 50 by 30 units with the radio range equal to 8 units. As it is visible, the characteristics correspond to the results depicted in Figure 16.12.

The performed experiments clearly indicate that to achieve the maximum efficiency of flooding-based routing, the nodes should be deployed with the distances of about two-thirds of the node radio range. The results of the simulations allow us to conclude that the presented pattern can be applied to a sensor network of any size. Such properties of flooding come from the fact that when many nodes are near one another, every transmitted event notification is multiplied many times and sent to all of the neighbors. This causes the nodes to be busy relaying messages that, most likely, already have reached the destination and only occupy the resources that could be used to route genuine messages. If possible, a right decision would be to design the network to form a kind of mesh. A regular topology will simplify the processes of choosing the distances or tuning the radios. When the spaces between sensors get increased, the wireless links are reduced and, as a result, fewer redundant messages will be generated. The efficiency of flooding rises dramatically and a limited number of devices may be used to provide full connectivity in the area. Besides, investors save the money that would otherwise be spent on unnecessary hardware, crippling the routing performance if deployed.

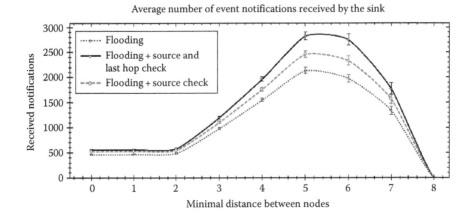

FIGURE 16.13 Received event notifications. Area, 50 × 30; TTL, 5; buffer size, 10.

Network developers need to remember that radio ranges and minimal distances should be selected considering the properties of the deployment area. Sensors should not be scattered too much, because distances close to the propagation range will probably result in sections of the network being disconnected from the sink. Suggested recommendations apply when there is no difference whether the monitored area would be surveyed with hundreds or only tens of sensors. This especially concerns applications such as desert humidity measurement or seismic activity detection. However, if the density of WSN is the most crucial aspect, instead of decreasing the number of nodes, the transmission power of radio modules should be lowered. Undesired message storms will be avoided improving the overall efficiency.

16.5.4.2 Time to Live of an Event Notification

The experiments have been performed in the main simulation area of this study (100 by 50 units), with the buffers capable of storing 10 packets. Six distinct TTL values have been simulated: 1, 2, 3, 5, 10, and 20. We believe this scope of experiments provides sufficient data required to conclude on the general relation between the TTL of an event notification and the performance of routing.

When the TTL is set to one, only the direct communication of originating node and the sink will allow an event notification to reach the destination. The WSN is forced to operate as a structure often called a single-hop network. The same result may be achieved if the network is very dense or when increased radio ranges allow each node to send messages directly to the sink. In this case, virtually no routing needs to be used, since only the gateway will be interested in the data.

The efficiency of each of the three flooding methods is equal, because in single-hop networks, the algorithm-specific rules simply do not apply. The general network lifetime may be significantly prolonged due to the fact that each distinct message is transmitted only once. The chart in Figure 16.14 shows the number of messages received by the gateway when TTL has been set to one. The curve begins at about 1000 received notifications and its gentle slope ends at 600 messages. The efficiency decreases with the growing distances, because it is less likely that a node will be located inside the radio range of the sink. Thus, many sensors will not be able to inform about any events they observe.

The relation between routing performance, minimal node spacing, and TTL values is visible in Figure 16.15 that presents the results for TTL equal two. However, it intensifies when TTL allows the message to visit three network devices.

The graph depicted in Figure 16.16 shows that the discussed relation gets stronger, significantly enhancing the efficiency for larger minimal distances. For dense networks, though, the performance goes down. At this level of TTL, basic flooding falls behind a bit.

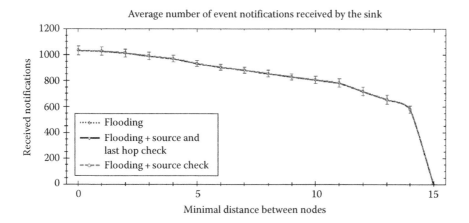

FIGURE 16.14 Received event notifications. Area, 100 × 50; TTL, 1; buffer size, 10.

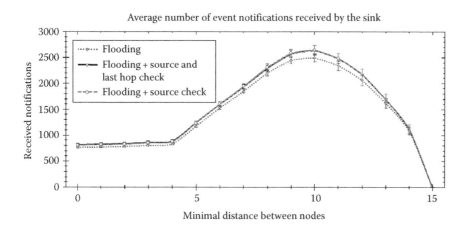

FIGURE 16.15 Received event notifications. Area, 100 × 50; TTL, 2; buffer size, 10.

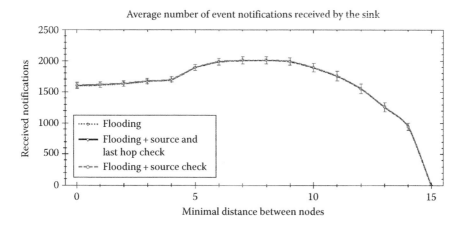

FIGURE 16.16 Received event notifications. Area, 100 × 50; TTL, 3; buffer size, 10.

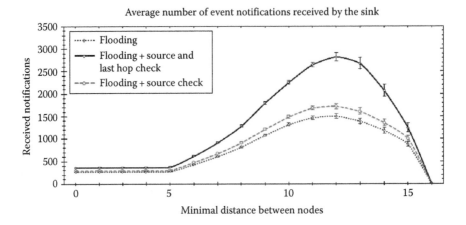

FIGURE 16.17 Received event notifications. Area, 100×50; TTL, 10; buffer size, 10.

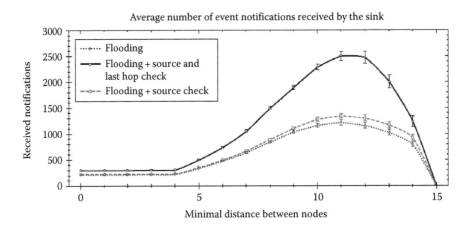

FIGURE 16.18 Received event notifications. Area, 100×50; TTL, 20; buffer size, 10.

To find the results of the simulation of TTL equal to five, see Figure 16.12. The differences in efficiency of the algorithms are visible. Finally, for TTLs equal to 10 (Figure 16.17) and 20 (Figure 16.18), the number of received events falls down if the last hop check is not applied.

16.5.4.3 Capacity of Ingress Message Buffer

The size of ingress packet buffer can have greater effect on the performance of notification flooding than it may seem. Usually, the memory embedded in microcontrollers allows designers to implement packet buffers that will be able to store at least 1000 basic messages. Therefore, a decision has been made that the simulations concerning the buffer should begin at the length of 1000 messages. The results of the experiment are presented in Figure 16.19.

However, if we shorten the capacity by 10 times, to 100 messages, see Figure 16.19, and then compare the performance, we can notice an improvement. It is not spectacular, but every increase in the efficiency should be considered a success. Decreasing the buffer size of one more order of magnitude, see Figure 16.12, brings only a minimal change (Figure 16.20).

The highest gain of efficiency, depicted in Figure 16.21, can be achieved when buffer capacity is set to one. This means that only a single incoming packet may be stored in the memory during each

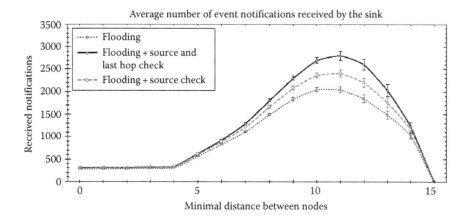

FIGURE 16.19   Received event notifications. Area, 100 × 50; TTL, 5; buffer size, 1000.

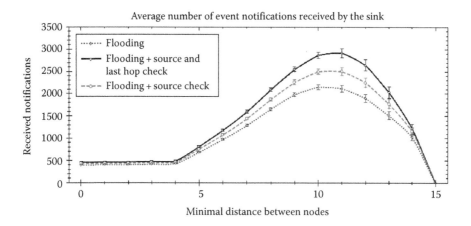

FIGURE 16.20   Received event notifications. Area, 100 × 50; TTL, 5; buffer size, 100.

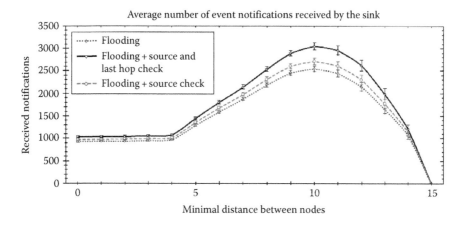

FIGURE 16.21   Received event notifications. Area, 100 × 50; TTL, 5; buffer size, 1.

cycle. Therefore, the number of duplicated notifications spreading throughout the network is limited. It is more likely that a unique notification will be received or relayed, as opposed to networks having longer input queues.

16.5.5 COMPARISON OF ENERGY EFFICIENCY OF FLOODING ALGORITHMS

The diagram in Figure 16.22 shows the average number of events occurring in the network during the simulation of each internode space. Because a new event takes place in every simulation cycle, this graph also presents the lifetime of the network. The time the network is able to operate is directly connected with the density and the number of deployed sensors. The fewer the nodes, the longer the network is alive because the number of redundant transmissions is limited and the energy consumption decreases. As it can be predicted, the lifetime is the longest when all nodes are disconnected, but then it is not a WSN anymore. For dense topologies, no matter which flooding algorithm would be chosen, there will be no difference in the performance of the network. The results start to differ when the distances between the nodes are getting closer to the radio range limit.

The efficiency of flooding may be defined as the ratio of received notifications to occurred events. If we compare Figures 16.20 and 16.22, we will see that for short distances, only about 30% of notifications have reached the sink. For optimal node spacing, the ratio goes to 45% when the source and last hop checks are performed. The efficiency grows by half and, what needs to be noticed, network lifetime is four times longer than for adjoining nodes. The TTL and buffer size values may have an influence on the shape of obtained characteristics, but the general performance properties remain valid.

To increase the number of collected unique notifications, the usage of more than one sink node should be considered. Unfortunately, it might result in the need for further data processing and synchronization, complicating the network. Another solution may be equipping the sink with a radio module having a larger range. More nodes will be connected to the gateway, so the probability of message reaching its destination will increase.

The experiments prove that, in most cases, it is worthwhile to implement flooding with the source and last hop check. The efficiency is always equal or higher than the performance of algorithms checking only the source or performing no verification. What is more, the increase is achieved owing to a simple processing introduced to notification flooding. Yet in particularly rare applications, where every additional executed command will be perceived as wasting of resources, the simplest possible choice would be basic flooding. Although, in such a situation, at least TTL handling needs to be implemented in flooding, because without this limitation, the number of messages replicated in WSN will be enormous. The energy will be quickly drained and the buffers filled up to the maximum—what certainly would paralyze the network.

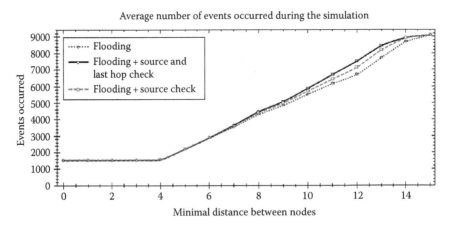

FIGURE 16.22 Events occurring during the simulation. Area, 100 × 50; TTL, 5; buffer size, 100.

16.6 FUTURE RESEARCH DIRECTIONS

EffiSen, the introduced simulator, is at the early stage of the development as compared to advanced network simulation platforms. There is a range of features and optimizations that may be incorporated into EffiSen that could broaden its scope of application, for example, more routing algorithms, additional topology generation approaches, and the support of parallelized hardware acceleration by the usage of stream processors. Thus, EffiSen could be probably publicly distributed among the WSN society as a free open-source program. The presented software can be used also as a teaching aid visualizing the principles of notification dissemination in the WSNs provided that corresponding course materials are prepared. The introduced experiment methodology can be applied in further researches involving more complex routing methods designed not only for WSNs but also for other types of energy-constrained wireless networks.

16.7 CONCLUSIONS

The objectives, structure, and the usage of the custom-made simulator have been described in this chapter. EffiSen fulfills its requirements well and allows to perform flooding algorithms comparisons. The proposed experiment methodology is based on simulating a sensor network until it reaches a certain level of depleted nodes. The same set of stimuli is reconstructed for each simulated algorithm. The data collected during the simulation of hundreds of uncorrelated network topologies allow statistical analysis of experiment results to be performed. Hence, the average network lifetimes related to flooding algorithms can be directly compared. The experiments have proved the implemented analysis method to be useful and noteworthy. The simulation results allow to state that the flooding algorithm performing source and last hop node check is the most cost-effective choice. The work also concerns the analysis of factors influencing the performance of selected flooding algorithms. It has been proved that the node spacing can significantly affect the efficiency of an algorithm. The minimal distances between sensors equal to about two-thirds of the radio range radius have yielded the maximum gain of event notification delivery efficiency. The TTL parameter and the length of node ingress queue may have some influence on routing performance as well.

REFERENCES

1. A. Bassi and G. Horn. Internet of things in 2020: A roadmap for the future. Information Society and Media Directorate General of the European Commission (DG INFSO) and European Technology Platform on Smart Systems Integration (EPoSS), September 5, 2008.
2. H. Karl and A. Willig. *Protocols and Architectures for Wireless Sensor Networks*. John Wiley & Sons Ltd, Hoboken, NJ, 2005.
3. P. Santi. *Topology Control in Wireless Ad Hoc and Sensor Networks*. John Wiley & Sons Ltd, Hoboken, NJ, 2005.
4. J.N. Al-Karaki and A. E. Kamal. Routing techniques in wireless sensor networks: A survey, *IEEE Wireless Communications*, 11(6):6–28, December 20, 2004.
5. Octopus: A dashboard for sensor networks visual control, University College Dublin, School of Computer Science and Informatics, http://www.csi.ucd.ie/content/octopus-dashboard-sensor-networks-visual-control (accessed July 2014).
6. TinyOS operating system, http://www.tinyos.net (accessed July 2014).
7. T. Wark, W. Hu, P. Corke, J. Hodge, A. Keto, B. Mackey, G. Foley, P. Sikka, and M. Brünig. Springbrook: Challenges in developing a long-term, rainforest wireless sensor network, in *Proceedings of ISSNIP 2008, 4th International Conference on Intelligent Sensors, Sensor Networks and Information Processing*, Sydney, Australia, December 15–18, 2008, pp. 599–604.
8. M. Doddavenkatappa, M.C. Chan, and A.L. Ananda. Indriya: A low-cost, 3D wireless sensor network testbed, in *Proceedings of TridentCom 2011, 7th International ICST Conference*, Shanghai, China, April 17–19, 2011, pp. 302–316.
9. E. Ertin, A. Arora, R. Ramnath, M. Nesterenko, V. Naik, S. Bapat, V. Kulathumani, M. Sridharan, H. Zhang, and H. Cao. Kansei: A testbed for sensing at scale, in *Proceedings of the 4th Symposium on Information Processing in Sensor Networks (IPSN/SPOTS Track)*, Nashville, TN, April 2006, pp. 399–406.

10. G. Coulson, B. Porter, I. Chatzigiannakis, C. Koninis, S. Fischer et al. Flexible experimentation in wireless sensor networks, *Communications of the ACM*, 55(1), 2012, 82–90.
11. C. Burin des Roziers, G. Chelius, T. Ducrocq, E. Fleury, A. Fraboulet et al. Using SensLAB as a first class scientific tool for large scale wireless sensor network experiments, in *Proceedings of 10th International IFIP TC 6 Networking Conference*, Valencia, Spain, May 9–13, 2011, pp. 147–159.
12. B. Musznicki and P. Zwierzykowski. Survey of simulators for wireless sensor networks, *International Journal of Grid and Distributed Computing*, 5(3), September 2012, 23–50.
13. Virtual extension's diversity path mesh innovative wireless network technology, Virtual Extension Ltd., http://www.virtual-extension.com/technology.htm (accessed July 2014).
14. P. Kiedrowski, B. Dubalski, T. Marciniak, T. Riaz, and J. Gutierrez. Energy greedy protocol suite for smart grid communication systems based on short range devices, in *Image Processing and Communications Challenges 3*, Advances in Intelligent and Soft Computing, Ed. R.S. Choraś, Vol. 102, Springer, Berlin, Germany, pp. 493–502.
15. ZedGraph, http://sourceforge.net/projects/zedgraph/ (accessed July 2014).

17 Wireless Sensor Network Simulations Using Castalia and a Data-Centric Storage Case Study

Khandakar Ahmed and Mark A. Gregory

CONTENTS

ABSTRACT

Castalia has been built using the OMNeT++ platform and can be used to simulate wireless sensor networks, body area networks, and other networks of low-power embedded devices. Castalia's scalable module structure provides a stable and flexible framework for simulation that incorporates new and innovative distributed algorithms and protocols. This chapter starts out by explaining the structure of Castalia. Then, the procedure to create a new module within Castalia is presented to illustrate the investigation of new algorithms or protocols. Finally, a specific framework incorporating data centric storage is used as a case study to show a step-by-step implementation through the realization of two modules in the application and routing layers. A distributed metric–based data centric similarity storage scheme is implemented in the node application layer. The technique was inspired by the disk-based data centric storage scheme and uses a disc track and sector analogy to map data locations and hence distance. The distributed metric–based data centric similarity storage scheme incorporates the sector based distance routing protocol, which is implemented in the node routing layer.

17.1 INTRODUCTION

Castalia [1] provides a reliable and flexible framework for first-order validation of an algorithm, protocol, or framework. The modularization and configurable nature of Castalia makes it suitable for adaptation and expansion. Sections 17.2 and 17.3 illustrate Castalia's architecture and the procedure that can be followed to create a new module. The Castalia Manual [2] has been used as a guide for the work presented in this chapter so that the example provided can be replicated and used when developing new algorithms and protocols.

17.1.1 CASTALIA NODE COMPOSITE MODULES AND CONNECTIONS

Castalia is built on top of OMNeT++. It follows the modular structure of OMNeT++ [3] and modules communicate with each other by passing messages. Each node, as shown in Figure 17.1b, provides an interaction between the physical processes and wireless channels.

As shown in Figure 17.1a, nodes are composed of five modules including the Sensor Manager, Application, Communication, Resource Manager, and Mobility Manager. The Communication module is a composite module consisting of three submodules—Routing, medium access control (MAC), and Radio. Resource Manager is solely responsible for battery, CPU state, and memory of each node, while Mobility Manager keeps a node mobile by updating its location periodically. Similar to the TCP/IP model, each node has four basic layers including application, routing, MAC, and radio. Sensor Manager provides an interface to sensing devices. A node might have multiple sensing devices that could be affected by either a single or multiple distinct physical processes.

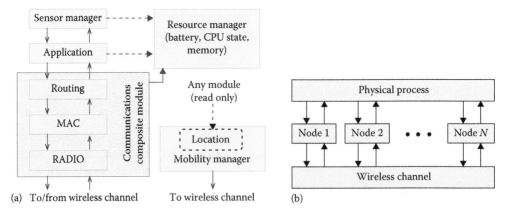

(a) The node composite module. (b) Communication among modules.

FIGURE 17.1 (a) The node composite module. (b) Communication among modules.

However, Sensor Manager establishes a one-to-one correspondence between sensing device type and a physical process module when there are complex interactions.

Information flows between the physical processes and the wireless channels occur using module interfaces and messages. This process of information flow should be familiar to developers of state machines for telecommunication and network devices. Information flows and message passing to the Resource and Mobility Managers occur from modules utilizing a separate structure from the sensor to radio information flow process.

In Castalia, nodes are not directly connected to each other and communicate through a wireless channel module (Figure 17.1b). The wireless channel module permits the specification of a range of wireless channels within Castalia, and this provides an approach for the simulation of real-world wireless communications between low-power sensor devices. The Castalia design permits communication between more than one wireless channel types, though it is more common that simulations would use wireless channel modules of one type.

17.1.2 Castalia Directory and File Structure

Figure 17.2 shows the Castalia Version 3.2 directory and file structure. *Castalia-3.2* is the root directory containing three subdirectories *bin*, *simulations*, and *src*. *Castalia-3.2/bin* contains all the scripts provided by Castalia to run simulations, parse and plot simulation results, and carry out other management tasks. *Castalia* script is used to run a simulation, while *CastaliaResults* and *CastaliaPlot* are used to parse the simulation output file in order to process, summarize, and plot the result. All of the simulations are stored under the directory *Castalia-3.2/Simulations*. Each simulation directory contains a file called *omnetpp.ini*, simulation result, and trace file. *Castalia-3.2/src* contains the module source files and has four subdirectories including *helpStructures, node, physicalProcess*, and *wirelessChannel*. *Castalia-3.2/src/node* again contains five subdirectories including *application, communication, mobilityManager, resourceManager*, and *sensorManager*. All the application layer modules are defined under the *Castalia-3.2/src/node/application* directory. *Castalia-3.2/src/node/communication* consists of three communication subdirectories, that is, *mac, radio*, and *routing*. The subdirectories *Castalia-3.2/src/node/communication/mac* and *Castalia-3.2/src/node/communication/routing* contain corresponding MAC and routing protocols. *Castalia-3.2/src/physicalProcess* consists of two subdirectories defining two physical processes.

17.1.3 List of Modules

Figure 17.3 shows the Castalia Version 3.2 module structure [1]. Referring to Figures 17.2 and 17.3, there are six Application modules: *bridgeTest, connectivityMap, simpleAggregation, throughputTest, valuePropagation*, and *valueReporting*. The six modules share a common interface, that is, *iApplication.ned*. Similar to the Application module, the MAC, Routing, and MobilityManager modules also share a corresponding interface. The Castalia Version 3.2 modules are shown at the bottom of Figure 17.3 within curly brackets.

17.2 CREATING MODULES

Modules that implement new distributed algorithms or protocols can be added to Castalia utilizing Castalia's flexible framework that includes an easy-to-follow approach for module development and placement within the code tree. The Castalia directory structure is hierarchical, for example:

```
Application: src/node/application/
Routing: src/node/communication/routing/
MAC: src/node/communication/mac/
MobilityManager: src/node/mobilityManager/
```

FIGURE 17.2 Castalia directory and file structure.

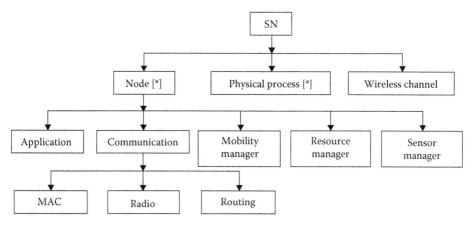

Application —{BridgeTest, ConnectivityMap, SimpleAggregation, ThroughTest, ValuePropagation, ValueReporting}

MAC—{BaselineBANMAC, BypassMAC, Mac802154, TMAC, TunableMAC}

Routing —{BypassRouting, MultipathRingsRouting}

MobilityManager —{LineMobilityManager, Customizable PhysicalProcess}

FIGURE 17.3 Module list.

Modules are placed in subdirectories and consist of three files with extensions *.ned*, *.h* and *.cc*. The three files are given the new module's name. The *.ned* file is written in OMNeT++'s NED language that defines input/output gates and parameters; the *.h* file declares all variables and a new C++ class, which will implement the new module; and the *.cc* file includes the module code and defines (or redefines) methods based on virtual class methods.

Assume a new module to be created is called *NewProtocol*. Using the Castalia naming scheme, the module directory will be *newProtocol* (note the naming convention; the first letter is lowercase). Hence, the name of the three files within *newProtocol* will be *NewProtocol.ned*, *NewProtocol.h*, and *NewProtocol.cc*.

- NewProtocol.ned

The basic structure of *NewProtocol.ned* is shown in Figure 17.4. The first line includes a package definition, for example, assuming *newProtocol* is an Application module:

```
package node.application.newProtocol;
```

It is to be noted that each module's parent directory contains a *.ned* file, referred to as an interface file, which has a name starting with *i*. To include *newProtocol* into Castalia's structure, a linkage to the appropriate interface module is made using the *like* keyword. For example, to recognize *newProtocol* as an Application module, the definition would be as follows:

```
simple NewProtocol like node.application.iApplication
```

Lists of parameters that will be passed to the module at runtime from the simulation configuration are declared in *NewProtocol.ned*. An example of *NewProtocol.ned* is provided in Figure 17.4.

The first five parameters are also defined in the *iApplication.ned* interface file, and hence they must be included within *NewProtocol.ned*. Two new parameters are also defined in *NewProtocol. ned*. It is to be noted that a module loads *parameter* values from the *omnetpp.ini* configuration file;

```
package node.application.newProtocol;
simple NewProtocol like node.application.iApplication {
        parameters:
                string applicationID = default ("newProtocol");
                bool collectTraceInfo = default (false);
                int priority = default (1);
                int packetHeaderOverhead = default (8);
                int constantDataPayload = default (8);

                int newParameter1;
                string newParameter2 = default("default value");

        gates:
                output toCommunicationModule;
                output toSensorDeviceManager;
                input fromCommnicationModule;
                input fromSensorDeviceManager;
                input fromResourceManager;
}
```

FIGURE 17.4 NewProtocol.ned.

however, if they are not assigned there, then it loads the default value assigned in the *.ned* file. For example, *newParameter* must be assigned a value from either the *.ini* or the *.ned* file, and hence this a good practice to provide default values for all parameters in the *.ned* file.

Gates are fully defined within the interface and *.ned* files, and it is normal for the *gates* section to be entirely copied from interface file and placed in the *.ned* file.

- NewProtocol.h

The module *.h* file includes a C++ class declaration, and the class name should match the module name, *.cc* and *.ned* filename. The class inherits the corresponding virtual class:

```
Application: class NewCastaliaModule: public VirtualApplication{
Routing: class NewCastaliaModule: public VirtualRouting {
MAC: class NewCastaliaModule: public VirtualMac {
MobilityManager: class NewCastaliaModule: public
VirtualMobilityManager{
```

It should also be noted that according to the convention, all of the variables are declared private and methods declared protected. An example *NewProtocol.h* is shown in Figure 17.5.

- NewProtocol.cc

In the *NewProtocol.cc* file, *NewProtocol.h* is included and then the *Define_Module* macro is used to register *NewProtocol* as an OMNeT++ module.

```
Define_Module(NewProtocol);
```

Then either inherited methods (at least one) are redefined or new methods are defined. At this stage, the basic task of creating a new module is complete and the code can be compiled:

```
Castalia $ make clean
Castalia $./makemake
Castalia $ make
```

```
#ifndef _NewProtocol_H_
#define _NewProtocol_H_

#include "VirtualApplication.h"

class NewProtocol:public VirtualApplication {
 private:
      int variable1;
      double variable2;
      ----------------
      ----------------
 protected:
      void startup();
      void finishSpecific();
      void fromNetworkLayer(ApplicationPacket *, const char *,
double, double);
      -------------------------
      void timerFiredCallback(int);
      void handleSensorReading(SensorReadingMessage *);
};

#endif                              // _NewProtocol_H_
```

FIGURE 17.5   NewProtocol.h.

The first two commands ensure that new modules will be added to a code build. If code changes are made within an existing module, then only using the *make* command will ensure the fastest code build occurs.

The functionality of virtual class methods is briefly discussed in Sections 17.3.1 through 17.3.4.

17.2.1   APPLICATION

Methods provided by VirtualApplication classes (Figure 17.6) can be categorized into two main kinds: callback and predefined methods.

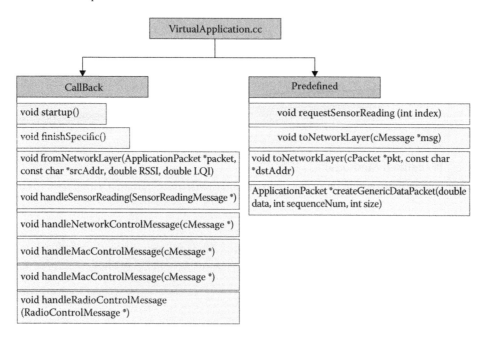

FIGURE 17.6   Methods provided by virtual application class.

17.2.1.1 Callback Methods

Callback methods are called when certain events occur. For example, *fromNetworkLayer* methods are called when a packet is received from the communication stack.

- `void startup()`

This method is called when the module is first used and includes app-specific parameter initialization and assignment to class variables.

- `void finishSpecific()`

Output collection and memory release are carried out by this method that is called at the end of a simulation.

- `void fromNetworkLayer(ApplicationPacket *packet, const char *srcAddr, double RSSI, double LQI)`

fromNetworkLayer is called when a packet is received from the network layer. It receives a packet, received signal strength indicator (RSSI), link quality indicator (LQI), and the network source address as arguments. RSSI and LQI are measured by the Radio module and a network address is a string.

- `void handleSensorReading(SensorReadingMessage *)`

When a sensor reading is returned from the sensing device, *handleSensorReading* is called to handle the sensor values.

- `void handleNetworkControlMessage(cMessage *)`

handleNetworkControlMessage is called when the module receives a control message from the network layer.

- `void handleMacControlMessage(cMessage *)`

handleMacControlMessage is called after receiving a control message from the MAC layer.

- `void handleRadioControlMessage(RadioControlMessage *)`

handleRadioControlMessage provides access to control messages from various communication layers.

17.2.1.2 Predefined Methods

Predefined methods are called by Application modules to perform common actions.

- `void requestSensorReading(int index)`

The sensor reading result is returned after calling the *handleSensorReading* callback method. The result is provided by calling the *requestSensorReading* predefined method.

- `void toNetworkLayer(cMessage *msg)`

toNetworkLayer is used to send a control packet from the application layer to lower layers of the communication layer stack. Unlike for a data packet, the packet transmitted is used to send commands, for example, setting parameters in the network or MAC layers and changing the RX mode or TX power of the radio.

- `void toNetworkLayer(cPacket *pkt, const char *dstAddr)`

toNetworkLayer can also be used to send a data packet to the network layer. The second argument refers to the destination address, which is a literal string. Hence, the address can be either macros like SINK_NETWORK_ADDRESS and BROADCAST_NETWORK_ADDRESS or simply a node's ID. BROADCAST_NETWORK_ADDRESS is used to send packets to all 1-hop neighbors, while SINK_NETWORK_ADDRESS is used to send a packet to the sink node. An extensive use of this method will be shown in Section 17.3.4.

- `ApplicationPacket *createGenericDataPacket(double data, int sequen-ceNum, int size)`

createGenericDataPacket is used to create an application data packet. It takes three arguments— a double-type number as data, a sequence number, and an optional size. The *constantDataPayload* parameter of the application determines the default packet size.

As mentioned earlier, *iApplication.ned* defines five parameters, which are parsed and used by *VirtualApplication*. The five parameters are as follows:

- `string applicationID`

applicationID is used to filter packets from different applications. In Castalia, multiple applications could be run in a deployed network, and in this case, *applicationID* allows developer to filter packets based on *ID*. However, if the value of this parameter is left as null, that is, an empty string, then no filtering is done and *fromNetworkLayer()* is called for all packets received from the routing layer.

- `bool collectTraceInfo`

collectTraceInfo is used to enable or disable trace collection.

- `int priority`

priority is defined to indicate the implied quality of service priority.

- `int packetHeaderOverhead`

packetHeaderOverhead is used to add overhead, measured in bytes, to all application packets before going to the routing layer. The overhead is added in the *VirtualApplication* code and applies to both the constant payload case and the case where the size is given by the user for each packet.

- `int constantDataPayload`

If the size of a packet is not defined or zero, then *VirtualApplication* sets the packet size to *constantDatapayload*.

17.2.2 ROUTING

The base of the Castalia Routing module is defined by *VirtualRouting*. Like *VirtualApplication*, methods provided by *VirtualRouting* classes (Figure 17.7) can also be categorized into two main kinds: callback and predefined methods.

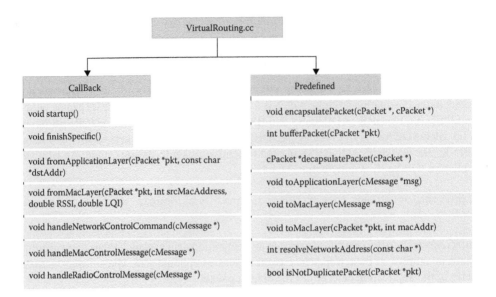

FIGURE 17.7 VirtualRouting class methods.

17.2.2.1 Callback

- `void startup()`

The *startup* method is called at initialization and loads router-specific parameters for assignment to class variables.

- `void finishSpecific()`

Output collection and memory release are carried out by this method, which is called at the end of a simulation.

- `void fromApplicationLayer(cPacket *pkt, const char *dstAddr)`

This mandatory method is called when a Routing module receives a packet from the application layer. This packet is intended to be forwarded to MAC layer of the communication stack. The arguments of this method include application-level packet and destination address, which is a literal string.

- `void fromMacLayer(cPacket *pkt, int srcMacAddress, double RSSI, double LQI)`

fromMacLayer is a mandatory method that is called when the Routing module receives a data packet coming from the MAC layer. It is an incoming packet and is supposed to be either passed to the application layer or forwarded via MAC layer or discarded. Its arguments include the packet itself, source MAC address, RSSI, and LQI of the received packet.

- `void handleNetworkControlCommand(cMessage *)`

handleNetworkControlCommand is called when a *NETWORK_CONTROL_COMMAND* is received and is used to change operational parameters of a routing protocol. This type of message could come from both the application and MAC layers. It is to be noted that unlike *NETWORK_LAYER_PACKET*, this method is defined to exchange messages with other communication stack layers of the same node to control the behavior of the Routing module.

- `void handleMacControlMessage(cMessage *)`

This method reacts when it receives a control message coming from the MAC layer. By default, Castalia passes any MAC control message to the application layer, but it is possible to redefine the method to handle them in the network layer.

- `void handleRadioControlMessage(cMessage *)`

handleRadioControlMessage reacts to a control message coming from the Radio. For example, the Radio might send a carrier sense interrupt (which the MAC has forwarded). By default, Castalia passes any MAC control message to the application layer, but it is possible to redefine the method to handle them in the network layer.

17.2.2.2 Predefined Method

Predefined methods are called by protocol modules to perform common actions.

- `void encapsulatePacket(cPacket *, cPacket *)`

encapsulatePacket encapsulates an outgoing application packet into a routing packet. It also sets the different routing layer fields to the appropriate values. This method is usually called by the *fromApplicationLayer* callback method.

- `int bufferPacket(cPacket *pkt)`

bufferPacket is used to store an encapsulated packet in the buffer. This method is usually called by the *fromApplicationLayer* callback method. If the buffer becomes full, it automatically creates and sends a control message to the application.

- `cPacket *decapsulatePacket(cPacket *)`

decapsulatePacket decapsulates a routing packet to extract an application packet and sets appropriate fields in that packet (RSSI, LQI, source). This method is usually called by the *fromMacLayer* callback method.

- `void toApplicationLayer(cMessage *msg)`

This method sends control messages and packets to the application layer.

- `void toMacLayer(cMessage *msg)`

This method sends control messages to the MAC layer.

- `void toMacLayer(cPacket *pkt, int macAddr)`

This method passes a data packet to the MAC layer. The argument of the method includes the destination MAC address and the data packet. The MAC destination is essentially the next hop. The routing protocol will decide on the next hop based on the information it has. The next hop address will be in a string format (routing address). The MAC address used is an integer and *resolveNetworkAddress* is used for the string to int conversion.

- `int resolveNetworkAddress(const char *)`

This converts network layer address to MAC layer address.

- `bool isNotDuplicatePacket(cPacket *pkt)`

isNotDuplicatePacket performs a valuable test by checking whether the packet has been received or seen before (extracting the source address and sequence number). It can be used to avoid sending duplicate packets to the application layer.

17.2.3   MAC

The *VirtualMac* class defines the basic methods used to operate MAC modules within the Castalia framework. Similar to *VirtualApplication* and *VirtualRouting*, there are two kinds of methods defined by *VirtualMac*: callback and predefined (Figure 17.8).

17.2.3.1   Callback
  • `void startup()`

The *startup* method is called at initialization and loads MAC-specific parameters to be assigned to class variables.

  • `void finishSpecific()`

Output collection and memory release are carried out by this method, which is called at the end of a simulation.

  • `void fromNetworkLayer(cPacket *pkt, int dstAddr)`

fromNetworkLayer is called when the node MAC layer receives an outgoing packet from the network layer. The method arguments include the packet and destination address as an integer.

  • `void fromRadioLayer(cPacket *pkt, double RSSI, double LQI)`

fromRadioLayer is a mandatory method that is called when the MAC module receives a data packet coming from physical or radio layer. It is an incoming packet and is supposed to be either passed to the network layer or forwarded via the radio layer or discarded. Its arguments include the packet, RSSI, and LQI of the received packet.

  • `int handleControlCommand(cMessage * msg)`

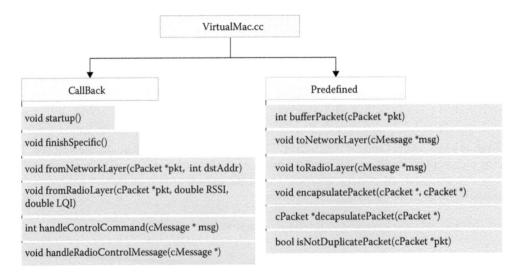

FIGURE 17.8   VirtualMac class methods.

handleControlCommand is called when *MAC_CONTROL_COMMAND* is received. These messages are defined to alter MAC protocol operational parameters and are usually issued from the application or routing layers. However, the messages need to be customized or defined by the developer based on the requirements of the MAC protocol to achieve the desired control outcome.

- `void handleRadioControlMessage(cMessage *)`

handleRadioControlMessage is triggered when the MAC module receives a control message from the Radio, for example, a carrier sense interrupt sent from the Radio. By default, Castalia passes any radio control message to the routing layer, but it is possible to redefine the method to handle radio control messages in the MAC layer.

17.2.3.2 Predefined
Predefined methods can be called from protocols to perform common actions:

- `int bufferPacket(cPacket *pkt)`

bufferPacket is used to store an encapsulated packet in the buffer. It is usually called by the *fromRadioLayer* callback method.

- `void toNetworkLayer(cMessage *msg)`

toNetworkLayer sends packets or messages received from the radio layer to the routing layer. It is usually called from the *fromRadioLayer* callback method.

- `void toRadioLayer(cMessage *msg)`

This method sends packets or messages received from the network layer to radio.

- `void encapsulatePacket(cPacket *, cPacket *)`

This method encapsulates a routing packet in a MAC packet.

- `cPacket *decapsulatePacket(cPacket *)`

This method decapsulates a MAC packet to extract a routing packet.

- `bool isNotDuplicatePacket(cPacket * pkt)`

This method performs a valuable test by checking whether this packet has been received or seen before (extracting the source address and sequence number). It can be used to avoid sending duplicate packets to the network layer. *TunableMAC* provides an example of how to use this method.

17.2.4 MOBILITY MANAGER

Similar to Application, Routing, and MAC, the MobilityManager module is based on the *VirtualMobilityManager* class. However, in contrast to Communication modules, MobilityManager has no extra layer of abstraction with methods like *startup()*, *fromXYlayer()*, and *timeFiredCallback()*, *initialize()* or *finishSpecific()* methods needing to be redefined for specific initialization or finalization. Furthermore, MobilityManager has not defined OMNeT++'s *handleMessage*, and hence it is user defined. A close inspection of the *LineMobilityManager()* module provides a better understanding.

17.2.4.1 Predefined

- `void notifyWirelessChannel()`

notifyWirelessChannel notifies the wireless channel about the node's location. Usually a periodic notification of the node's position is sent to the wireless channel by calling this method. The periodic event could be created by scheduling a self-message in *handleMessage()*.

- `void setLocation(double x, double y, double z)`
- `void setLocation(NodeLocation_type)`

These methods are used to change the location of a node, send automatic notifications to the wireless channel, and may be altered to provide enhanced location information.

17.3 DATA-CENTRIC STORAGE CASE STUDY

Distributed data-centric storage (DCS) [4,5] is a promising and efficient alternative to external storage (ES) [6–8] and local storage (LS) [9–13]. DCS guarantees longevity under the specified energy constraint compared to LS and ES. In DCS, data are hashed to generate a key range for similar data that are stored at the node closest to the originating geographic location. As a result, queries are forwarded directly to the specific node or zone avoiding flooding [9]. In a forthcoming publication [14], a distributed metric-based data-centric similarity storage scheme (DMDCS) is presented. The technique is inspired by disk-based data-centric storage (DBDCS) and uses a disk track and sector analogy to map data location and hence distance. Two exemplar similarity queries, K-nearest neighbor (KNN) and Range, were chosen to test the hypothesis and showed that the technique performs well in terms of bandwidth utilization and latency. DMDCS uses an optimally synchronized routing algorithm, referred to as sector-based distance (SBD) [15,16], which is designed for low-power DCS implementation. SBD can achieve high power efficiency while reducing updates and query traffic, end-to-end delay, and collisions. Time division multiple access (TDMA) schedules are derived using a simple grid coloring algorithm (GCA), which guarantees reliability and optimum end-to-end latency. In subsequent subsections, DBDCS, DMDCS, and SBD will be discussed briefly for readers' convenience; further details are available in the references.

17.3.1 DISK-BASED DATA-CENTRIC STORAGE

In the DBDCS architecture, as shown in Figure 17.9a, the rectangular wireless sensor network (WSN) field is mapped to an $m \times n$ matrix, where m is the number of tracks and n is the number of sectors for each track, generating $S(m \times n)$ sectors. Each sector consists of a sector head (*SH*) and member nodes (*MNs*). The intrasector communication between *SH* and *MN* is one hop, while the intersector communication among *SHs* is multihop. Figure 17.9b shows the intersector communication, while Figure 17.9c shows the intrasector communication. In Figure 17.9b, the sensor nodes inside each sector are not shown explicitly; instead, an aggregated link (see Figure 17.9c) is shown to represent the total traffic from *MNs* to *SH*.

17.3.2 DISTRIBUTED METRIC-BASED DATA-CENTRIC SIMILARITY STORAGE SCHEME

A sensed event E can be defined by an l-dimensional tuple $(A_1, A_2, A_3, \ldots, A_l)$, where $A_g, \forall_g \in [1,l]$ denotes the gth attribute and D_{A_g} is the domain of attribute A_g. Attributes are ordered according to their significance in defining an event, that is, A_1 is the most significant, while A_l is the least. Each *MN* of a sector transmits the sensed event as an l-tuple $\langle v_{i1}, v_{i2}, \ldots, v_{il} \rangle_k$, where $1 \le i \le M_k$, M_k is the total number of *MNs* in kth sector and v_{ij} denotes the value of the jth attribute received from ith

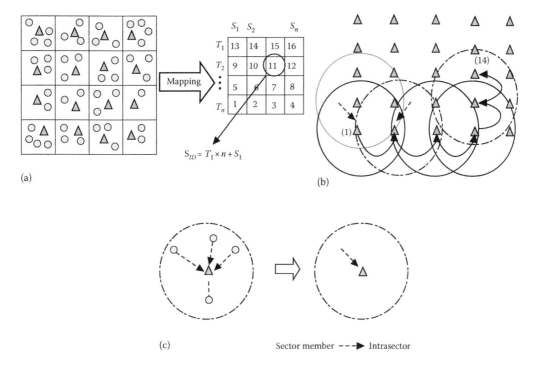

(a)

(b)

(c)   Sector member ---▶ Intrasector

FIGURE 17.9   (a) DBDCS mapping. (b) Intersector communication. (c) Intrasector *MN* to *SH* communication.

MN of *k*th sector. The corresponding *SH*, after collecting tuples from all the *MNs*, aggregates and normalizes them at the end of each epoch before finding the target *SH* mapping.

Hence, after aggregation at epoch *t*,

$$E_k(\mathrm{Agg}(t)) = \int_{\mathrm{Agg}_{i=1}}^{M_k} \langle v_{i1}, v_{i2}, \ldots, v_{il} \rangle(t) = \langle \psi_1, \psi_2, \ldots, \psi_l \rangle(t) \tag{17.1}$$

Here,

$$\psi_j = \left\{ \max_{i=1}^{M_k} v_{ij}, \min_{i=1}^{M_k} v_{ij}, \mathrm{avg}_{i=1}^{M_k} v_{ij} \right\}$$

$$\text{Here, } j \in [1, l], k \in [1, S], 0 \le |\psi_j(\mathrm{avg}/\max/\min)| \le 1 \tag{17.2}$$

Weights are assigned to different attributes according to their significance in the event characterization. Hence, the weight matrix can be defined as

$$w = \{x^i\}_{i=1}^l \tag{17.3}$$

In this equation, the value of *x* is chosen based on the application or size of the deployed network. For the experiments conducted in this chapter, the value of *x* was set as 0.25. Hence, given *l* attributes in an attribute list associated with weight w_j ($1 \le j \le l$) in a WSN application, the source SH_k generates the hash value by

$$h = \sum_{j=1}^{l} \left(\mathrm{avg}_{i=1}^{M_k} v_{ij} \times w_j \right) \tag{17.4}$$

The domain of the derived hash key of an l-dimensional event, denoted by H_D, can be defined by

$$\alpha\left(\alpha_{min} = \sum_{j=1}^{l}\left(min_{i=1}^{M_k} v_{ij} \times w_j\right), \quad \alpha_{max} = \sum_{j=1}^{l}\left(max_{i=1}^{M_k} v_{ij} \times w_j\right)\right) \tag{17.5}$$

Pivot points can be generated dividing H_D into different segments as follows:

$$P_i = \left[\alpha_{min} : \frac{(\alpha_{max} - \alpha_{min})}{S} : \alpha_{max}\right] \tag{17.6}$$

Hence, after each epoch, SH_k forwards the aggregated event $E_k = \langle[\psi_1, \psi_2, \ldots, \psi_l], [t, h]\rangle$, where t denotes epoch number, to the destination SH denoted by SH_i where $P_i \leq h \leq P_{i+1}$ and P_i, P_{i+1} are the lower and upper limits of the ith subinterval, respectively.

17.3.3 SECTOR-BASED DISTANCE

In order to carry out the data store and query operations, DMDCS uses the SBD routing protocol proposed in Ref. [16]. The operation of SBD is divided into rounds, each composed of a learning phase followed by a relaying phase.

17.3.3.1 Learning Phase

The learning phase is again divided into three stages including (1) intersector SH-TDMA slot assignment stage using GCA, (2) *MN-SH* association stage, and (3) intrasector TDMA slot assignment stage to *MNs* managed by the *SH*.

At the first stage of learning phase, each *SH* finds the nonoverlapping slot of operation for corresponding sector using GCA. All sectors of any grid size could be assigned with conflict-free TDMA slot by reusing only four time slots ($C_0 \sim C_4$, see Figure 17.10).

In the *MN-SH* association stage, each *SH* broadcasts a beacon frame. Beacon frames received by *MNs* from more than one *SH* are stored along with the RSSI. *MNs* then select an *SH* from which it receives a beacon frame with highest RSSI and send associating request to it. The *SH* nodes create an *MN* table listing all the *MNs* from which they receive association request.

In the third stage of learning phase, *SH* nodes broadcast a packet containing $C_k (0 \leq k \leq 3)$, Δt (length of C_k), and an array γ, where $\gamma = \{m_1, m_2, m_3, \ldots, m_i\}$ and $|\gamma| = M_k$. In γ, m_i and i denote the *MN* id and index of this *MN* in the array, respectively. The *MNs* then calculate the intrasector slot of transmission based on their position in the array γ by $t_i = \{(i \times \ell) + (C_{k-1} \times \Delta t) | \gamma[i] = M_{S-ID}, \gamma[i] \neq \gamma[j]\}$, where ℓ and M_{S-ID} are the length of the intrasector TDMA time slot and the node's self-network address, that is, node's self-ID, respectively. The number of *MNs* in a sector varies due to the dynamic nature of the *MN-SH* association procedure. Hence, the length of intrasector TDMA time slot of a sector can be defined by

$$\ell = \frac{\Delta t}{(\text{size of}(\gamma) + 1)}$$

17.3.3.2 Relaying Phase

In the relaying phase, all *MNs* report their buffered or aggregated sensed data to their associated *SH* during their allocated intrasector TDMA transmission slot. An *SH*, after each epoch, that is,

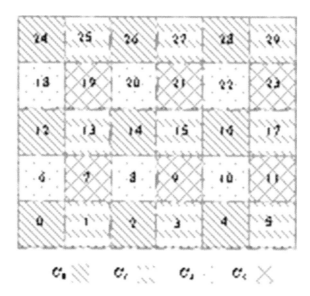

FIGURE 17.10 Conflict-free slot allocation using GCA.

after collecting data from all *MNs*, forwards the mapped event data (according to Section 17.3.2) in a multihop fashion to the corresponding sector for storage. In this intersector communication, *SHs* continue forwarding their packets to their immediate neighbor *SH*, which lies on the same row in the virtual grid (Figure 17.9a) until the packet reaches the *SH* that is on the same column as the destination sector. The packet is then being forwarded vertically up or down until it reaches the destination (Figure 17.9b). The same process of routing is followed for query request and response.

17.3.4 DMDCS MODULE

Figure 17.11 shows the activity diagram of the DMDCS Application module. One single module is created for the *SH* and *MN* application layer; however, depending on the type, different corresponding code segments are executed. A node starts through the initialization of its primary knowledge by loading parameter values from the *.ini* file. All *SH* nodes then run *Pivot Selection* and *Segment Pivot Selection* algorithms according to (17.5) and (17.6) by triggering corresponding events at the 0th and 10th second. At the same time, *MNs* sample different parameters sensed through the associated sensor devices periodically at 0.1 second. Each time the *Request Sample* event is triggered, it aggregates the current sample with the previous sample and generates max, min, and average of each attribute and stores the value in the *aggregated sample* parameter. The *Send Packet to SH* event is triggered periodically at 0.2 second that sends the last *aggregated sample* to their associated *SH*. An *SH* aggregates all the packets received from *MNs* and sends the generated packet to the target mapped sector according to Section 17.3.2. The *Generate Query* event is triggered in a few randomly selected nodes at the 60th second so that random samples are made and sent to the target mapped *SH* for retrieval.

The correct location for the Application module is

```
src/node/application/
```

According to OMNeT's naming convention, the name of the module is *dMDCS*. The *DMDCS. ned* file is created to define the basic structure of a module by defining input/output gates and

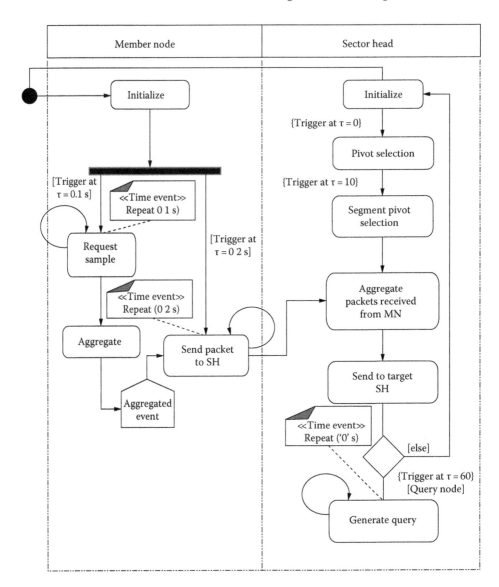

FIGURE 17.11 DMDCS Application module activity diagram.

parameters. In the *NED* file, default values for the parameters can also be defined. The structure of the *DMDCS.ned* file is given for developers' or readers' convenience.

The definition of the module package is always included in the first line. For DMDCS to be recognized as a Castalia Application module, it has been declared using the *like* keyword in the next line, referencing the *iApplication* interface module, which is a *.ned* interface file and belongs to the parent directory of the module (highlighted in Figure 17.12).

```
package node.communication.mac.newCastaliaModule;
simple DMDCS like node.application.iApplication {
```

A list of parameters are declared and initialized with default values that will be passed to the module at runtime from the simulation configuration. Interface files including the *iApplication. ned* file include a set of parameters that are mandatory for all Application modules. In *DMDCS. ned* (Figure 17.12), the first five parameters are mandatory and also declared in *iApplication.ned*.

```
//**************************************************************
//* Copyright: Khandakar Ahmed, RMIT University, 2012-2013    *
//* Author(s): Khandakar Entenam Unayes Ahmed                 *
//* This file is distributed under the terms in the attached  *
//* LICENSE file.                                             *
//* If you do not find this file, copies can be found by      *
//* writing to:                                               *
//*     Khandakar Entenam Unayes Ahmed, SECE, RMIT University *
//*     e-mail: khandakar.ahmed@rmit.edu.au                   *
//*     Attention:  License Inquiry.                          *
//*                                                           *
//**************************************************************/
package node.application.dMDCS;

// The sensor node module. Connects to the wireless channel in
// order to communicate with other nodes. Connects to physical
// processes so it can sample.

simple DMDCS like node.application.iApplication{

  parameters:
      string applicationID = default ("DMDCS");
      bool collectTraceInfo = default (false);
      int priority = default (1);
      int packetHeaderOverhead = default (8);
      int constantDataPayload = default (8);

      double packetSpacing = default (100);    // in ms
      double reportingTime = default (200); string
      applicationID = default ("DMDCS");
      bool collectTraceInfo = default (false);
      ------------------------------------------------
  gates:
      output toCommunicationModule;
      output toSensorDeviceManager;
      input fromCommunicationModule;
      input fromSensorDeviceManager;
      input fromResourceManager;
}
```

FIGURE 17.12 DMDCS.ned.

The remainders of the parameters are specific to the *DMDCS* module and are not shown here. Modules use parameter values from the configuration input file if defined therein, that is, *omnetpp. ini*; otherwise, the default value provided in the definition is used. Gates are the methods to communicate with other modules and should match the definition included in the interface *iApplication.ned* file.

To write the module code, two files *DMDCS.h* and *DMDCS.cc* are created in the directory *src/ node/application/dMDCS*. Castalia provides help through virtual classes or base classes, and hence *DMDCS* needs to be represented by a C++ object, which is inherited from appropriate base classes. Where possible, it is recommended that the methods and structure defined by the virtual classes are used or overloaded. In the *DMDCS.h* file (Figure 17.13), a new C++ class is declared that will implement our module. The class name matches the name of the module and *.ned* file.

```
class DMDCS : public VirtualApplication{
```

In the *DMDCS.cc* file (Figure 17.14), the *DMDCS.h* file is included and then *Define_ Module(DMDCS)* is called to register the new directory as an OMNeT++ module. Then methods are defined (or redefined) based on virtual class methods. The methods included in *DMDCS.cc* are also declared in *DMDCS.h*.

As mentioned earlier, the *startup()* callback method is used to initialize variables and parse parameters from the *.ned* and *omnetpp.ini* files. However, since the *DMDCS* module includes

```
//*****************************************************************
//* Copyright: Khandakar Ahmed, RMIT University, 2012-2013        *
//* Author(s): Khandakar Entenam Unayes Ahmed                     *
//* This file is distributed under the terms in the attached      *
//* LICENSE file.                                                 *
//* Copies can also be found by writing to:                       *
//*     Khandakar Entenam Unayes Ahmed, SECE, RMIT University     *
//*     e-mail: khandakar.ahmed@rmit.edu.au                       *
//*     Attention:  License Inquiry.                              *
//*****************************************************************/
#ifndef _DMDCS_H_
#define _DMDCS_H_

#include "VirtualApplication.h"
#include "DMDCS_m.h"

using namespace std;
struct neighborRecord {
        int id;
        int timesRx;
        int receivedPackets;
};
class DMDCS: public VirtualApplication {
 private:
        int priority;
        int packetHeaderOverhead;
        int constantDataPayload;
        -----------------------
protected:
        void startup();
        void fromNetworkLayer(ApplicationPacket *, const char *,
double, double);
        void timerFiredCallback(int);
        void handleSensorReading(SensorReadingMessage *);
        void requestSensorReading(const char *);
        void updateNeighborTable(int nodeID, DCSApplicationPacket
*appPacket);
        void updateStorageTable(int nodeID, DCSApplicationPacket
*appPacket);
        bool updateSectorDataCollectionTable(int nodeID, int
theSN);
        bool updateQueryRequestTable(int nodeID, int theSN);
        bool updateQueryResponseTable(int nodeID, int theSN);
        void transmitBufferedPacket();
        void pushInteger(vector<int>&, int);
        void parseStringParams(void);
        void finishSpecific();
};
#endif                          // _DMDCS_APPLICATIONMODULE_H_
```

FIGURE 17.13 DMDCS.h.

```
//*****************************************************************
//* Copyright: Ahmed Khandakar, RMIT University, 2012-2013        *
//* Author(s): Khandakar Entenam Unayes Ahmed                     *
//* This file is distributed under the terms in the attached      *
//* LICENSE file.                                                 *
//* If you do not find this file, copies can be found by          *
//* writing to:                                                   *
//*     Khandakar Entenam Unayes Ahmed, SECE, RMIT University     *
//*     e-mail: khandakar.ahmed@rmit.edu.au                       *
//*     Attention:  License Inquiry.                              *
//*****************************************************************

#include "DMDCS.h"
#define DESTINATION_ADDRESS destAddress.c_str()

Define_Module(DMDCS);

void DCSApplication::startup()
{
    -----------------------
    -----------------------
}
```

FIGURE 17.14 DMDCS.cc.

a significant number of parameters, a separate method *parseStringParams()* is called from *startup()* to parse parameters during runtime.

```
void DCSApplication::startup()
{
        neighborTable.clear();
        parseStringParams();
        -----------------------------
}

void DCSApplication::parseStringParams(void){
        const char *parameterStr, *token, *test;
        -----------------------------
}
```

The *timerFiredCallback()* method reacts to timers expiring and handles all eight timers. The *REQUEST_SAMPLE* action prompts *MNs* to return a sample. The action is first set from the *startup()* method at 0 s (i.e., start immediately) as the timer value. However, it is then repeatedly set with the timer value as *sampleInterval* second (in this case study, the value of *sampleInterval* was set to 0.1 second in the *.ned* file) each time at its last trigger. It is to be noted that once a timer is set, it becomes active. When a timer expires, it creates an event, which triggers the *timerFiredCallback()* method with the index of the expired timer as the argument. *SEND_PACKET_TO_SH* is also triggered periodically at 10 s. This timer is also first set in the *startup()* method with *reportingTime* second (in this case study, the value of *reportingTime* was 0.2 s). *PIVOT_SELECTION*, *SEGMENT_PIVOT_SELECTION*, and *QUERY_ GENERATION* are set once from the *startup()* method at 0th, 10th, and 60th second of each round.

```
void DCSApplication::timerFiredCallback(int timerIndex)
{
    switch (timerIndex) {
        case SEND_PACKET_TO_SH:{
            -----------------------------
            setTimer(SEND_PACKET_TO_SH,10);
            break;
        }
        case REQUEST_SAMPLE:{
            requestSensorReading("Temperature");
            -----------------------------
            setTimer(REQUEST_SAMPLE, sampleInterval);
            break;
        }
        case PIVOT_SELECTION:{
            -------------------------
            break;
        }
        case SEGMENT_PIVOT_SELECTION:
        {
            -------------------------
            break;
        }
        case QUERY_GENERATION:
        {
            -------------------------
            break;
        }
    }
}
```

fromNetworkLayer() is a callback method that is called when the Application module receives a packet from the network layer. There are five types of application packets that could come to the *DMDCS* module. *DCS_LOCAL_DATA_REPORT* comes from *MNs* to *SHs*. *SHs* buffer and aggregate the packets until its transmission slot occurs when it transmits to target data storage sector. *DCS_REMOTE_DATA_UPDATE* comes from a remote sector, and if an *SH* receives this packet, then it forwards the packet to an *MN* of a particular segment to store the packet within its memory. An *SH*, after receiving a *DCS_REMOTE_RANGE_QUERY* packet, sends a request to the *MN* of the corresponding segment to retrieve the answer to the query. *DCS_REMOTE_RANGE_QUERY_RESPONSE_LOCAL* is the response that a target *SH* receives from *MN* nodes of a particular segment in response to a query. *DCS_REMOTE_RANGE_QUERY_RESPONSE_REMOTE* is the final response that a query node receives in response to its query from the target *SH*.

```
void DCSApplication::fromNetworkLayer(ApplicationPacket * rcvPacket,
const char *source, double rssi, double lqi)
{
        bool flag = false;
        DCSApplicationPacket *appPacket = check_and_cast<DCSApplicationPac
ket*>(rcvPacket);
        switch(appPacket->getDCSApplicationPacketKind()){
                case DCS_LOCAL_DATA_REPORT:{
                        --------------------------
                        break;
                }
                case DCS_REMOTE_DATA_UPDATE:{
                        --------------------------
                        break;
                }
                case DCS_REMOTE_RANGE_QUERY:{
                        --------------------------
                        break;
                }
                case DCS_REMOTE_RANGE_QUERY_RESPONSE_LOCAL:{
                        --------------------------
                        break;
                }
                case DCS_REMOTE_RANGE_QUERY_RESPONSE_REMOTE:{
                        --------------------------
                        break;
                }
        }
}
```

requestSensorReading() is a predefined method provided by Castalia. It takes the sensor type as an argument. This method is called from the *REQUEST_SAMPLE* action block in *timerFiredCallback()*.

```
void DCSApplication::requestSensorReading(const char * type){
        SensorReadingMessage *reqMsg =  new SensorReadingMessage("App to
Sensor Mgr: sample request", SENSOR_READING_MESSAGE);
        ---------------------------------------------------------
}
```

This callback method is called after a sensor reading is returned from the sensing device via *requestSensorReading()*. For the case of the *DMDCS* module, *index*, *sensType*, and *sensValue* for the sensor type are stored in a data structure.

```
void DCSApplication::handleSensorReading(SensorReadingMessage
*sensReading)
{
        int Index = sensReading->getSensorIndex();
        ------------------------------------
}
```

updateSectorDataCollectionTable() is a new method defined specifically for the *DMDCS* module. This method is called from the *DCS_LOCAL_DATA_REPORT* switch-case block of the *fromNetworkLayer()* method, that is, an *SH* calls this method to store local data packets received from *MNs*.

```
bool DCSApplication::updateSectorDataCollectionTable(int nodeID, int
serialNum){
        int i = 0, pos = -1;
        --------------------------
}
```

updateNeighborTable() is defined in the *DMDCS* module to store packets received from a remote *SH*. This method is called from the *DCS_REMOTE_DATA_UPDATE* switch-case block of the *fromNetworkLayer()* method. It is to be noted that this method is executed only by *SH*.

```
void DCSApplication::updateNeighborTable(int nodeID, DCSApplicationPacket
*appPacket)
{
        int serialNum = appPacket->getSequenceNumber();
        ---------------------------------------------
}
```

updateStorageTable() is defined in the *DMDCS* module to store remote packets in *MNs* received from their associated *SH*. This method is called from the *DCS_REMOTE_DATA_UPDATE* switch-case block of the *fromNetworkLayer()* method. It is to be noted that this method is executed only by *MNs*.

```
void DCSApplication::updateStorageTable(int nodeID, DCSApplicationPacket
*appPacket)
{
        int serialNum = appPacket->getSequenceNumber();
        ---------------------------------------------
}
```

updateQueryRequestTable() is defined in the *DMDCS* module to track the number of query request received by an *SH*. This method is called from the *DCS_REMOTE_RANGE_QUERY* switch-case block of the *fromNetworkLayer()* method and is executed by both *SH* and *MN*.

```
bool DCSApplication::updateQueryRequestTable(int nodeID, int serialNum){
        int i = 0, pos = -1;
        ---------------------
}
```

updateQueryResponseTable() is defined in the *DMDCS* module to track the number of query responses received by a query node against the number of query requests generated. This method is called from the *DCS_REMOTE_RANGE_QUERY_RESPONSE_REMOTE* switch-case block of the *fromNetworkLayer()* method. This method is executed by *SH* query nodes.

```
bool DCSApplication::updateQueryResponseTable(int nodeID, int serialNum){
      int i = 0, pos = -1;
      ----------------------
}
```

bufferPacket() is a predefined method to buffer *DCS_LOCAL_DATA_REPORT*, *DCS_REMOTE_DATA_UPDATE*, and *DCS_REMOTE_RANGE_QUERY* types of application layer packets. Hence, this method is called from corresponding switch-case blocks of *fromNetworkLayer()* method.

```
int DCSApplication::bufferPacket(cPacket *rcvFrame)
{
      trace() << "Buffering Packet";
      --------------------------------
}
```

transmitBufferedPacket() is a predefined method to transmit *DCS_LOCAL_DATA_REPORT*, *DCS_REMOTE_DATA_UPDATE*, and *DCS_REMOTE_RANGE_QUERY* types of application layer packets from a buffer during the corresponding node transmission slots. Hence, this method is called from the corresponding switch-case blocks of the *fromNetworkLayer()* method.

```
void DCSApplication::transmitBufferedPacket(){
      while(!TXBuffer.empty()){---------------------------
}
```

finishSpecific() is a callback method, which is used to collect output and release memory. Debug output messages to the application trace file occur within this method.

```
void DCSApplication::finishSpecific()
{
      //Collection of Remote Packets
      declareOutput("Remote Packets received");
      -----------------------------
}
```

17.3.5   *SBD* ROUTING MODULE

Figure 17.15 shows the activity diagram of the *SBD* Routing module. One module is used for the *SH* and *MN* node routing layer; however, depending on the node type, different code segments within *SBD* are used. A node initializes by loading parameter values from the *.ini* file. All nodes at this stage also set the following two values:

$$roundCounter = 0 \tag{17.7}$$

$$\text{Current round time (CRT)} = roundLength \times roundCounter \tag{17.8}$$

Here, *roundCounter* refers to the current round number and is incremented after each round. *CRT* refers to the start of the CRT and is used as a reference when timers are set and triggered by calling the *timerFiredCallback* method with the index of the expired timer as the argument. *roundLength* (Section 17.3.3) refers to the length of each round that is calculated in the following way:

The *SH* triggers *Sector Slot Selection Round* at *CRT*, that is, at the beginning of each round. Then, the *SH* broadcasts a beacon packet at (*CRT*+2) seconds. An *MN* builds *SHCandidate* array after

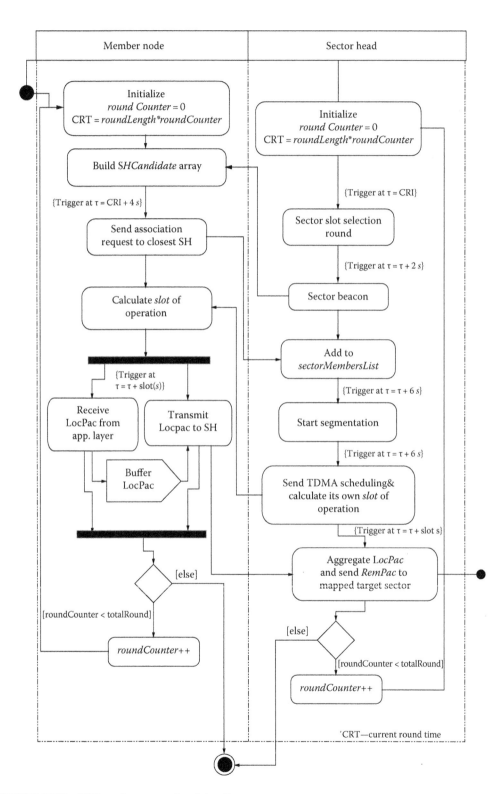

FIGURE 17.15 SBD routing protocol activity diagram.

receiving beacon packets from the nearby *SHs*. *MNs* then send an association request at (*CRT*+4) seconds to the candidate *SH* from which they received a beacon packet with the highest RSSI. *SHs* then build *sectorMembersList* with the ID of *MNs* from which they receive the association request. After this, the *SHs* trigger *segmentation* at (*CRT*+8) seconds. At (*CRT*+14) seconds, *SHs* send TDMA scheduling to their *MNs* and calculate transmission slots. *MNs*, after receiving a TDMA scheduling packet, calculate their *slot* of operation that refers to the time in seconds when an *MN* will transmit a sensed event to the *SH* and *SHs* transmit their aggregated packets of data received from the *MNs* to the target mapped destination (Section 17.3.2). It is to be noted that *MNs* continue to buffer sensed events received from the application layer until its transmission slot occurs.

The correct location for the Application module is

```
src/node/communication/routing/
```

According to OMNeT++'s naming convention, the name of the module is *sBDRouting*. The *SBDRouting.ned* file is created to define the basic structure of a module by defining input and output gates and parameters. In the NED file, default values for all parameters are defined. The structure of the *SBDRouting.ned* file is provided in Figure 17.16.

The definition of the module package is included in the first line. For *SBDRouting* to be recognized as a Castalia Routing module, it has been declared using the *like* keyword in the next line, referencing the *iRouting* interface module, which is a *.ned* interface file and belongs to the parent directory of the module (highlighted in Figure 17.16).

```
package node.communication.routing.sBDRouting;
simple SBDRouting like node.communication.routing.iRouting {
```

```
//*****************************************************************
//* Copyright: Khandakar Ahmed, RMIT University, 2012-2013      *
//* Author(s): Khandakar Entenam Unayes Ahmed                   *
//* This file is distributed under the terms in the attached    *
//* LICENSE file.                                               *
//* If you do not find this file, copies can be found by        *
//* writing to:                                                 *
//*     Khandakar Entenam Unayes Ahmed, SECE, RMIT University   *
//*     e-mail: khandakar.ahmed@rmit.edu.au                     *
//*     Attention:  License Inquiry.                            *
//*                                                             *
//*****************************************************************
/

package node.communication.routing.sBDRouting;

simple SBDRouting like node.communication.routing.iRouting {
 parameters:
     bool collectTraceInfo = default (false);
     int maxNetFrameSize = default (0); // bytes, 0 ->no limit
     int netBufferSize = default (32);  // number of messages
     int netDataFrameOverhead = default (14);  // bytes

     int segmentsPerSector = default (3);// Segments Number
     int nodesPerSegment = default(3);    // nodes per segment
     int packetsPerMemberNode = default (0); // 0 no limit
     double sectorSlotLength = default(4000); // ms
     int totalRound = default(1);
     double MemberNodeSelectionRoundTime = default(20);
     ------------------------------------------------------
 gates:
     output toCommunicationModule;
     output toMacModule;
     input fromCommunicationModule;
     input fromMacModule;
     input fromCommModuleResourceMgr;
}
```

FIGURE 17.16 SBDRouting.ned.

```
//*************************************************************
//* Copyright: Khandakar Ahmed, RMIT University, 2012-2013    *
//* Author(s): Khandakar Entenam Unayes Ahmed                 *
//* This file is distributed under the terms in the attached  *
//* LICENSE file.                                             *
//* If you do not find this file, copies can be found by      *
//* writing to:                                               *
//*     Khandakar Entenam Unayes Ahmed, SECE, RMIT University *
//*     e-mail: khandakar.ahmed@rmit.edu.au                   *
//*     Attention:  License Inquiry.                          *
//*************************************************************
#ifndef _SBDROUTING_H_
#define _SBDROUTING_H_

#include "VirtualRouting.h"
#include "SBDRouting_m.h"

using namespace std;

class SBDRouting: public VirtualRouting {
  private:
        bool isSectorHead;
        int sectorHeadID;
        -------------------
  protected:
        void startup();
        void fromApplicationLayer(cPacket *, const char *);
        void fromMacLayer(cPacket *, int, double, double);
        void timerFiredCallback(int);
        bool updateRoutingTable(int,int,int,double,double);
        --------------------------
  public:
        --------------------------
        int getSegmentsPerSector(void);
        bool cmpRSSI(nodeInfo a, nodeInfo b);
        bool cmpTALLY(storageInMemberNode a, storageInMemberNode
b);
#endif                             //SBDROUTING
```

FIGURE 17.17   SBDRouting.h.

A list of parameters are then declared and initialized with default values that will be passed to the module at runtime from the simulation configuration file. Like each interface, *iRouting.ned* file already includes a set of parameters that are mandatory for all Routing modules. In *SBDRouting. ned* (Figure 17.16), the first four parameters are mandatory and also declared in *iRouting.ned*. The remainders of the parameters are specific to the *SBDRouting* module. A module uses the value of the parameter from the configuration input file, that is, *omnetpp.ini*; otherwise, the default value assigned here will be used. Gates are the methods to communicate with other modules and should match the definition provided in the interface *iRouting.ned*.

To write the module code, two files *SBDRouting.h* and *SBDRouting.cc* are created in *src/node/ communication/routing/sBDRouting*. In the *SBDRouting.h* file (Figure 17.17), a new C++ class is declared that will implement the *SBDRouting* module. The name of the class matches the name of the module and the .*ned* file.

```
class SBDRouting: public VirtualRouting{
```

In the *SBDRouting.cc* (Figure 17.18) file, *SBDRouting.h* is included and then *Define_ Module(SBDRouting)* is called to register the new folder as an OMNeT++ module. Then methods are defined (or redefined) based on virtual class methods. The list of methods included in *SBDRouting.cc* is also declared in *SBDRouting.h* (Figure 17.17).

Like *DMDCS* and all other modules, *SBDRouting* also has a *startup()* callback method, which is used to initialize variables and parse parameters from the *SBDRouting.ned* and *omnetpp.ini* files.

```
//*********************************************************************
//* Copyright: Khandakar Ahmed, RMIT University, 2012-2013    *
//* Author(s): Khandakar Entenam Unayes Ahmed                 *
//* This file is distributed under the terms in the attached  *
//* LICENSE file.                                             *
//* If you do not find this file, copies can be found by      *
//* writing to:                                               *
//*    Khandakar Entenam Unayes Ahmed, SECE, RMIT University  *
//*    e-mail: khandakar.ahmed@rmit.edu.au                    *
//*    Attention:  License Inquiry.                           *
//*********************************************************************

#include "SBDRouting.h"
#include "algorithm"

#define NEXT_HOP_ADDRESS nextHopAddress.c_str()
#define SECTOR_HEAD_ADDRESS sectorHeadAddress.c_str()
#define MEMBER_NODE_ADDRESS memberNodeAddress.c_str()
#define SEGMENT_ADDRESS segmentAddress.c_str()
#define SELF_SEGMENT_ADDRESS selfSegmentID.c_str();

Define_Module(SBDRouting);

void SBDRouting::startup()
{
    -----------------------
    -----------------------

}
```

FIGURE 17.18 SBDRouting.cc.

However, since the *SBDRouting* module includes a significant number of parameters, a separate method is called from *startup()* to parse parameters during runtime.

```
void SBDRouting::startup()
{
    neighborTable.clear();
    -----------------------------
}
```

parseStringParams() is defined to parse parameters during runtime and is called from *startup()*.

```
void SBDRouting::parseStringParams(void){
    cStringTokenizer gridTokenizer(getParentModule()-
>getParentModule()->getParentModule()->par("deployment"), ";");
    -------------------------------------
}
```

The *timerFiredCallback()* method of the *SBDRouting* module reacts to timers expiring and handles eight timers. The *START_ROUND* action is set for all nodes at the start of each round. For the first round, *START_ROUND* is called from *startup()*. At this switch-case block, *SHs* set the *SECTOR_SLOT_SELECTION_ROUND* timer at *currentRoundStartTime* second of each round, while all other nodes set *MEMBER_NODE_SELECTION_ROUND* at the *MemberNodeSelectionRoundTime* second of each round. However, *START_ROUND* is set with the timer value as *roundLength* second (in this case study, *roundLength = no. of nonoverlapping slot * sectorSlotLength + learningPhaseLength + queryTimeLength + 5.0*; 5 s is added as roundSpacing to keep a safe space between two rounds) every time the event triggers. The *SECTOR_SLOT_SELECTION_ROUND* switch case utilizes GCA to assign a conflict-free TDMA time slot to each sector (reuses only four slots). In the *MEMBER_NODE_SELECTION_ROUND* switch case, *SHs* schedule two actions *SEND_BEACON* and *SEGMENTATION_ROUND* at *sendBeaconTime* and *segmentationTime*, respectively, and *MNs* schedule *JOIN_SECTOR* at *joinSectorTime*. In the *SEND_BEACON* switch case, all *SHs* broadcast a beacon packet. In the *JOIN_SECTOR* switch-case block, *MNs* send a joining request to *SH*, from which they receive the beacon packet with the highest RSSI. In the

SEGMENTATION_ROUND switch-case block, *SH* applies segmentation to divide associated *MNs* into different segments and schedules *TDMA_SCHEDULING* at *TDMASchedulingTime* of each round. At the *TDMA_SCHEDULING* switch-case block, each *SH* broadcasts its *Intersector Slot Number*, *Member Array*, and *Segment Number* for each *MN* and also determines its own transmission slot. At *TRANSMISSION_SLOT*, each node transmits their buffered packets.

```
void SBDRouting::timerFiredCallback(int index){
    double timer = uniform(0,1);
    switch(index){
        case START_ROUND:{
            ------------------------
            if(isSectorHead){
                setTimer(SECTOR_SLOT_SELECTION_ROUND, currentRoundStartTime)
            }
            setTimer(MEMBER_NODE_SELECTION_ROUND, MemberNodeSelectionRoundTime);
            setTimer(START_ROUND, roundLength);
            break;
        }
        case SECTOR_SLOT_SELECTION_ROUND:{
            ------------------------
            break;
        case MEMBER_NODE_SELECTION_ROUND:{
            if(isSectorHead){
                setTimer(SEND_BEACON,sendBeaconTime);
                setTimer(SEGMENTATION_ROUND, segmentationTime);
            }
            else{
            setTimer(JOIN_SECTOR,joinSectorTime);
            }
            break;
        }
        case SEND_BEACON:{
            ------------------------
            break;
        }
        case JOIN_SECTOR:{
            ------------------------
            break;
        }
        case SEGMENTATION_ROUND:{
            ------------------------
            break;
        }
        case TDMA_SCHEDULING:{
            ------------------------
            setTimer(TDMA_SCHEDULING,TDMASchedulingTime);
            break;
        }
        case TRANSMISSION_SLOT:{
            transmitBufferedPacket();
            break;
        }
    }
}
```

In the *fromApplicationLayer()* method, *MNs* receive sensed events and range query responses from the application layer, set packet types, and encapsulate application layer packets into network layer packets, then finally push the packets into *TXBuffer*. The *SH* receives and processes three types of packets: Update Received from Remote Sector for Storage, Remote Update to Forward, and Query Packets from the Application Layer. For the first packet type, *SH* forwards them to the corresponding *MNs* to store, the second packet type is relayed to the adjacent or target *SH*, and the third packet type is sent to the corresponding segment. After processing the first two packet types, the *SH* will push them to the MAC layer, while the third packet type is buffered in *TXBuffer*.

```
void SBDRouting::fromApplicationLayer(cPacket * pkt, const char
*destination)
{
        string dst(destination);
        if(!isSectorHead){
                SBDRoutingPacket *netPacket = new SBDRoutingPacket("Local
Data Update Packet", NETWORK_LAYER_PACKET);
                if(atoi(destination)<0){
                        netPacket->setSBDRoutingPacketKind(SBD_DATA_PACKET_
LOCAL_UPDATE);
                }
                else{
                        netPacket->setSBDRoutingPacketKind(SBD_RANGE_LOCAL_QUERY);
                }
                netPacket->setSource(SELF_NETWORK_ADDRESS);
                netPacket->setDestination(SECTOR_HEAD_ADDRESS);
                encapsulatePacket(netPacket,pkt);
                if((int)TXBuffer.size() >= netBufferSize){
                        TXBuffer.pop();
                }
                else
                        bufferPacket(netPacket);
        }
        else{
                if(atoi(destination) == atoi(SELF_NETWORK_ADDRESS)){
                        toApplicationLayer(pkt);
                }
                else if(atoi(destination)<0){
                        SBDRoutingPacket *netPacket = new
SBDRoutingPacket("Remote Data Update Packet", NETWORK_LAYER_PACKET);
                        stringstream convert;
                        switch(-atoi(destination)/addPadDecider){
                                case REMOTE_UPDATE:{
                                        ------------------
                                        break;
                                }
                                case RANGE_QUERY:{
                                        ------------------
                                        break;
                                }
                        }
                }
                else{
                        ----------------------------
                        bufferPacket(netPacket);
                }
        }
}
```

fromMacLayer() is a callback method that is called once the *SBDRouting* module receives a packet from the MAC layer. There are seven types of network layer packets that could come to the *SBDRouting* module from the MAC layer. *SHs* push all *SBD_DATA_PACKET_ LOCAL_UPDATE* received from their associated *MNs* to the application layer. *SBD_RANGE_ LOCAL_QUERY* is the query response from *MNs* to *SHs*. If the packet is destined for this node, then it is pushed to an upper layer, that is, the application layer. When an *SH* receives *SBD_DATA_PACKET_REMOTE_UPDATE* from the MAC layer, there could be two different actions. In the first case, the packet is received by a target or relay *SH* from a remote *SH*, while in the second case, the packet is received by *MNs* from their corresponding *SH* to store. In the first case, if the *SH* is a target node, then it sends the packet to an upper communication stack; otherwise, it forwards the packet to the next hop. In the second case, *MNs* send the packet to the application layer to be stored. *SBD_RANGE_QUERY_PACKET* is the range query request, which is pushed to the application layer, received by *MNs* from their target *SH* for query result retrieval. *MNs*, after receiving *SBD_BEACON_PACKET* from *SHs*, store the packetsm in an array along with the source of the beacon frame and RSSI as a possible parent candidate (*SH*). *SH* adds a member to its member list after receiving an *SBD_JOIN_PACKET* joining request. *MNs* receive *TDMA_SCHEDULING_PACKET* from *SH* and schedule their intrasector conflict-free slot.

```
void SBDRouting::fromMacLayer(cPacket * pkt, int srcMacAddress, double
rssi, double lqi)
{
      SBDRoutingPacket *netPacket = dynamic_cast
<SBDRoutingPacket*>(pkt);
      string destination(netPacket->getDestination());
      if (!netPacket) {
          return;
      }
      switch(netPacket->getSBDRoutingPacketKind()){
          case SBD_DATA_PACKET_LOCAL_UPDATE:{
              ------------------------
              break;
          }
          case SBD_RANGE_LOCAL_QUERY:{
              ------------------------
              break;
          }
          case SBD_DATA_PACKET_REMOTE_UPDATE:{
              ------------------------
              break;
          }
          case SBD_RANGE_QUERY_PACKET:{
              ------------------------
              break;
          }
          case SBD_BEACON_PACKET:{
              ------------------------
              break;
          }
          case SBD_JOIN_PACKET:{
              ------------------------
              break;
          }
```

```
        case SBD_TDMA_SCHEDULE_PACKET:{
             ------------------------
             break;
        }
    }
}
```

The *searchNextHop()* method is defined to find the target *SH* or relay *SH* as the next hop address for a remote packet.

```
int SBDRouting::searchNextHop(int destination){
     int destCol, destRow, currCol, currRow, nextHopCol,  nextHopRow;
     . . . . . . . .
     . . . . . . . .
}
```

transmitBufferedPacket() is a predefined method used to transmit *SBD_DATA_PACKET_LOCAL_ UPDATE*, *SBD_RANGE_LOCAL_QUERY*, and *SBD_DATA_PACKET_REMOTE_UPDATE* types of network layer packets from the buffer during corresponding node transmission slots. Hence, this method is called from corresponding switch-case blocks of the *timerFiredCallback()* method.

```
void SBDRouting::transmitBufferedPacket(){
     int numberOfPacketsSent = 0;
     . . . . . . . .
     . . . . . . . .
}
```

The *destin4MemberNode()* method finds the candidate *MN* where the event is to be stored.

```
int SBDRouting::destin4MemberNode(int destinationSegmentPivotIndex){
     int index=0;
     . . . . . . . .
     . . . . . . . .
}
```

The *sortSegmentMemberNode()* method sorts the *MNs* inside a segment in ascending order based on the number of packets that are stored within.

```
int SBDRouting::sortSegmentMemberNode(int index){
     int k,j,temp_tally, temp_ID;
     double temp_rssi;
     . . . . . . . .
     . . . . . . . .
}
```

17.3.6 RUNNING A SIMULATION

To run the simulation, a *DMDCSSimulation* directory has been created in the following location: `Castalia/Simulations/`

Hence, *Castalia/Simulations/DMDCSSimulation* should include a simulation configuration file, that is, *omnetpp.ini*, which defines the simulation scenario. Figures 17.19 and 17.20 show *omnetpp.ini*.

The simulation is run using the following command:

```
~/Castalia-3.2/Simulations/DMDCSSimulation$ Castalia -c SMAC
Running configuration 1/1
```

```
[General]

# ============================================================
# Always include the main Castalia.ini file
# ============================================================
include ../Parameters/Castalia.ini

sim-time-limit = 300s
SN.field_x = 60                                 # meters
SN.field_y = 60                                 # meters

# Specifying number of nodes and their deployment
SN.deployment = "[0..15]->4x4;[16..79]->8x8"
SN.numNodes = 80

# Removing variability from wireless channel
SN.wirelessChannel.bidirectionalSigma = 0
SN.wirelessChannel.sigma = 0
#SN.node[*].Communication.Radio.mode = "IDEAL"
#SN.node[*].Communication.Radio.collisionModel = 0

# Select a Radio and a default Tx power
SN.node[*].Communication.Radio.RadioParametersFile =
"../Parameters/Radio/CC2420.txt"
SN.node[0..15].Communication.Radio.TxOutputPower = "0dBm"
SN.node[16..79].Communication.Radio.TxOutputPower = "-5dBm"
#SN.node[80..84].Communication.Radio.TxOutputPower = "0dBm"

# Configuring Physical Process of Sensing Data
SN.numPhysicalProcesses = 5
SN.physicalProcessModuleName = "CustomizablePhysicalProcess"

SN.physicalProcess[0].printDebugInfo = true
SN.physicalProcess[0].description = "Environment"
SN.physicalProcess[0].inputType = 0
SN.physicalProcess[0].directNodeValueAssignment = "(250) 17:260
18:280 22:375 28:290 31:375 42:260 53:282"

SN.physicalProcess[1].printDebugInfo = true
SN.physicalProcess[1].description = "Environment"
SN.physicalProcess[1].inputType = 0       #0 = direct node value
assignment 1 = Scenario Based
SN.physicalProcess[1].directNodeValueAssignment = "(20) 17:25
18:23 22:21 28:24 31:23 42:21 53:24"

SN.physicalProcess[2].printDebugInfo = true
SN.physicalProcess[2].description = "Environment"
SN.physicalProcess[2].inputType = 0       #0 = direct node value
assignment 1 = Scenario Based
SN.physicalProcess[2].directNodeValueAssignment = "(3.5) 17:4
18:4.5 22:3.5 28:5 31:5 42:5.5 53:5"

SN.physicalProcess[3].printDebugInfo = true
SN.physicalProcess[3].description = "Environment"
SN.physicalProcess[3].inputType = 0
SN.physicalProcess[3].directNodeValueAssignment = "(250) 17:260
18:280 22:375 28:290 31:375 42:260 53:282"
```

FIGURE 17.19 omnetpp.ini.

An output file (*131105-103946.txt*) and a trace file (*Castalia-Trace.txt*) were created.

```
~/Castalia-3.2/Simulations/DMDCSSimulation$ ls

131105-103946.txt    omnetpp.ini   Castalia-Trace.txt
```

In order to process and summarize results, Castalia provides the script *CastaliaResults*.

```
~/Castalia-3.2/Simulations/DMDCSSimulation$ CastaliaResults
```

```
SN.physicalProcess[4].printDebugInfo = true
SN.physicalProcess[4].description = "Environment"
SN.physicalProcess[4].inputType = 0
SN.physicalProcess[4].directNodeValueAssignment = "(250) 17:260
18:280 22:375 28:290 31:375 42:260 53:282"

SN.node[*].SensorManager.numSensingDevices = 5
SN.node[*].SensorManager.pwrConsumptionPerDevice = "0.02 0.02
0.02 0.02 0.02"
SN.node[*].SensorManager.sensorTypes = "Light Temperature Smoke
Smoke2 Smoke3"
SN.node[*].SensorManager.corrPhyProcess = "0 1 2 3 4"
SN.node[*].SensorManager.maxSampleRates = "1 1 1 1 1"
SN.node[*].SensorManager.devicesBias = "20 5 0.5 20 20"
SN.node[*].SensorManager.devicesDrift = ""
SN.node[*].SensorManager.devicesNoise = "5 0.5 0.2 5 5"
SN.node[*].SensorManager.devicesHysterisis = ""
SN.node[*].SensorManager.devicesSensitivity = "0 0 0 0 0"
SN.node[*].SensorManager.devicesResolution = "0.001 0.001 0.001
0.001 0.001"
SN.node[*].SensorManager.devicesSaturation = "1000 1000 1000
1000 1000"

# Using SectorBasedDistanceRouting and setting parameters
SN.node[*].Communication.RoutingProtocolName = "SBDRouting"
SN.node[*].Communication.Routing.sectorSlotLength = 5000
SN.node[*].Communication.Routing.totalRound = 5
SN.node[16..79].Communication.Routing.packetsPerMemberNode = 2

# Adding Mobility
SN.node[*].MobilityManager.collectTraceInfo = true

# Using DCSApplication application module
SN.node[*].ApplicationName = "DCSApplication"
SN.node[0..15].Application.attributeWeight = "0.34 0.33 0.33
0.33 0.33"
SN.node[*].Application.attributeRange = "100:250:400:20
0:20:30:5 0:3.5:7:0.5 100:250:400:20 100:250:400:20" #
Min:Avg:Max:Theta
SN.node[*].Application.weightValue = 0.75
#SN.node[16..79].Application.sampleInterval = 1000
SN.node[0..15].Application.isSectorHead = true
SN.node[0..15].Application.packetsPerNode = 80
SN.node[16..79].Application.packetsPerNode = 80

[Config SMAC]
SN.node[*].Communication.MACProtocolName = "TMAC"
SN.node[*].Communication.MAC.listenTimeout = 61
SN.node[*].Communication.MAC.disableTAextension = true
SN.node[*].Communication.MAC.conservativeTA = false
SN.node[*].Communication.MAC.collisionResolution = 0
[Config varyTxPower]
SN.node[*].Communication.Radio.TxOutputPower =
${TXpower="0dBm","-1dBm","-3dBm","-5dBm")
[Config varySigma]
SN.wirelessChannel.sigma = ${Sigma=0,1,3,5)
```

FIGURE 17.20   omnetpp.ini (continued).

```
Castalia output files in current directory:
+---------------------+----------------+---------------------+
|                     | Configuration | Date                 |
+---------------------+----------------+---------------------+
| 131105-103946.txt  | SMAC (30)     | 2013-11-05 10:39    |
+---------------------+----------------+---------------------+
```

Running the *CastaliaResults* script without any argument gives a list of valid Castalia output files (also called result files) with information about the configuration that created them and the date they were created. The number in the parentheses (30) denotes the repetitions (with different seeds) for each of the executed configurations.

```
$ CastaliaResults -i 131105-103946.txt
```

```
+---------------------+------------------------------------------+-----------+
|             Module  |                                   Output  | Dimensions|
+---------------------+------------------------------------------+-----------+
|         Application |              Local Packets Received      | 16x64     |
|                     | Local Packets Received (aggregated)      | 16x1      |
|                     |       Query Response Packets Received    | 16x16     |
|                     |       Remote Packets Storage Received    | 64x16     |
|                     |              Remote Packets Received     | 16x16     |
|                     | Remote Packets Received (aggregated)     | 16x1      |
|   Communication.MAC |                Sent packets breakdown    | 80x1(5)   |
| Communication.Radio |                    RX pkt breakdown      | 80x1(6)   |
|                     |                         TXed pkts        | 80x1      |
|      ResourceManager|                  Consumed Energy         | 80x1      |
+---------------------+------------------------------------------+-----------+
```

CastaliaResults parses the file and identifies the outputs recorded by each of the different modules. In this case study, the *DMDCS* Application module produces six kinds of output: (1) *Local Packets Received* provides statistics of the sensed packets sent from *MNs* to *SH*; (2) *Local Packets Received (aggregated)* provides the total number of local packets received by each of the *SH*; (3) *Query Response Packets Received* provides the number of query requests received by a target *SH*; (4) *Remote Packets Storage Received* provides statistics of remote packets received by each of the *MN* for storage; (5) *Remote Packets Received* provides outputs relating to the remote packets received by each of the *SH* with the source to destination mapping; and (6) *Remote Packets Received (aggregated)* provides the total number of remote packets received by *SH*. The rest of the output is provided by Castalia default settings. It is to be noted that sensor-MAC (SMAC) [17] is used as the preferred MAC layer protocol.

17.4   CONCLUSION

Castalia provides a stable, flexible, and easy-to-use framework for simulating WSN, body area network (BAN), and other networks of low-power embedded devices. A description of Castalia modules and the development of a new module have been provided as a guide for algorithm and protocol development and testing. Castalia is used to implement a module that includes storage and routing for an implementation of DCS in a WSN. This chapter provides a successful approach that can be used for the simulation of new techniques and demonstrates the reliability and accuracy of the results achieved.

REFERENCES

1. A. Boulis. Castalia: A simulator for wireless sensor networks and body area networks: Version 3.2 [Online]. Available: http://castalia.npc.nicta.com.au/ [Accessed: February 16, 2014].
2. A. Boulis. (2011). Castalia: A simulator for wireless sensor networks and body area networks: Version 3.2 User's Manual. Available: http://castalia.npc.nicta.com.au/pdfs/Castalia%20-%20User%20Manual. pdf [Accessed: February 16, 2014].
3. OMNeT++Community. (2001). OMNET++. Available: http://www.omnetpp.org/ [Accessed: February 16, 2014].
4. Á. C. Rumín, M. U. Pascual, R. R. Ortega, and D. L. López. (2010). Data centric storage technologies: Analysis and enhancement, *Sensors*, 10, 3023–3056.
5. K. Ahmed and M. A. Gregory. (2012). Techniques and challenges of data centric storage scheme in wireless sensor network, *Journal of Sensor and Actuator Networks*, 1, 59–85.

6. G. J. Pottie and W. J. Kaiser. (2000). Wireless integrated network sensors, *Communications of the ACM*, 43, 51–58.
7. S. Saroiu, P. K. Gummadi, and S. D. Gribble. A measurement study of peer-to-peer file sharing systems, presented at the *Proceedings of the Multimedia Computing and Networking (MMCN)*, San Jose, January 2002.
8. Y. Yao, X. Tang, and E.-P. Lim. In-network processing of nearest neighbor queries for wireless sensor networks, presented at the *Proceedings of the 11th International Conference on Database Systems for Advanced Applications*, Singapore, April 12–25, 2006.
9. C. Intanagonwiwat, R. Govindan, and D. Estrin. Directed diffusion: A scalable and robust communication paradigm for sensor networks, presented at the *Proceedings of the 6th Annual International Conference on Mobile Computing and Networking*, Boston, MA, August 6–11, 2000.
10. W. Zhang, G. Cao, and T. L. Porta. (2007). Data dissemination with ring-based index for wireless sensor networks, *IEEE Transactions on Mobile Computing*, 6, 832–847.
11. S. R. Madden, M. J. Franklin, J. M. Hellerstein, and W. Hong. (2005). TinyDB: An acquisitional query processing system for sensor networks, *ACM Transactions on Database Systems (TODS)*, 30, 122–173, 2005.
12. F. Ye, H. Luo, J. Cheng, S. Lu, and L. Zhang. A two-tier data dissemination model for large-scale wireless sensor networks, presented at the *Proceedings of the 8th Annual International Conference on Mobile Computing and Networking*, Atlanta, GA, September 23–28, 2002.
13. F. Ye, G. Zhong, S. Lu, and L. Zhang. (2005). Gradient broadcast: A robust data delivery protocol for large scale sensor networks, *Wireless Networks*, 11, 285–298.
14. K. Ahmed and M. A. Gregory. Distributed data centric similarity storage scheme in wireless sensor network, in *The 11th Annual IEEE Consumer Communications & Networking Conference*, Las Vegas, NV, January 10–13, 2014.
15. K. Ahmed and M. A. Gregory. Wireless sensor network data centric storage routing using castalia, in *Australasian Telecommunication Networks and Applications Conference*, Brisbane, Australia, November 7–9, 2012.
16. K. Ahmed and M. A. Gregory. Optimized TDMA based distance routing for data centric storage, in *Networked Embedded Systems for Every Application (NESEA), 2012 IEEE 3rd International Conference on*, December 13–14, 2012, pp. 1–7.
17. Y. Wei, J. Heidemann, and D. Estrin. An energy-efficient MAC protocol for wireless sensor networks, in INFOCOM 2002. *Twenty-First Annual Joint Conference of the IEEE Computer and Communications Societies. Proceedings IEEE*, 2002, pp. 1567–1576, vol. 3.

Recent Developments in Simulation Tools for WSNs

An Analytical Study

Anuj Kumar Dwivedi and Om Prakash Vyas

CONTENTS

ABSTRACT

Wireless sensor network (WSN) is a type of wireless ad hoc network formed by the tiny wireless sensor nodes deployed in a region of interest (RoI) to gather some physical environmental phenomena or parameter. Currently, WSN is gaining a lot of attention from the researchers worldwide due to associated challenges and new application dimensions. Rapid advances in the application domain and criticalness of results obtained from the RoI, desired profound study and experiment before deployment of real setup. Analytical methods, use of simulators/emulators, and test bed implementations are the three major available options in research domain. Uses of simulator/emulators are the best choice among them from various aspects. Through this research contribution, the objective is to present an analytical study on the recent developments in simulation tools specifically designed or modified for gathering feasible and nearby realistic experimental results before actual deployment. Starting from the basics of simulation/simulators, the objective is to explore on evaluation criteria, necessity of simulation/simulators, type of simulation, classification/categorization of simulation, phases in simulation study, study and advances on existing simulators/simulation environments with a glimpse on emulators/emulation environments, and finally the future research directions in the area of simulation/emulation related to WSN.

18.1   INTRODUCTION

WSN is a discrete-event system [1]. Research activities in the area of WSNs need expositive performance statistics about systems, scenario, protocols, applications, and gathered data. Various experimental tools are in existence for fulfilling these requirements; some are in practical use though others are in literature. Three main popular techniques are used for analyzing the performance of wired and wireless networks:

- *Analytical methods*: Analytical methods may provide quick insight but fail to give near about realistic results for the reasons like limited energy, memory, and processing power and sheer number and unattended and harsh deployment environments of sensor nodes [2].
- *Simulation/emulation*: These provide a good approximation to verify different schemes and applications developed for WSNs in less time and at low cost. To have realistic results through simulations, correct modeling and the selection of simulation tool plays a vital role.
- *Test bed implementation and physical measurement*: These are the most accurate ways to study WSNs but require huge effort, time, and money.

Currently, due to high cost of large number of sensor nodes, most researches in WSNs area are performed by using experimental tools at various research centers, institutes, and universities before implementing the real one. Also, the statistics gathered from experimental tools can be realistic and convenient. The aforementioned three experimental tools provide the better alternative for studying the behavior of WSNs before and after implementing the physical setup. Evidently, some limitations are associated with experimental tools, but these cannot demean uses of it.

A simulator is a type of software that imitates selected parts of the behavior of the real world and is normally used as a tool for research and development. Depending on the intended usage of the simulator, various parts of the real-world systems are modeled and imitated. The parts that are modeled can also be of varying abstraction level. Initially, simulators specifically designed for WSN imitate the wireless media as well as the constraints nodes in the network but at present, sensor network simulators have a detailed model of the wireless media including effects of obstacles between nodes, while other simulators have a more abstract model.

Simulators are an effective experimental tool at present and commonly used for rapid prototyping. In addition, simulators are also used for the evaluation of new network protocols and algorithms as well as enable repeatability because they are independent of the physical world and its impact on the objects. In addition, simulations enable nonintrusive debugging at the desired level of detail.

When beginning to work with WSNs, it is an important aspect to select a simulation environment that will be up to requirements and will allow the researcher to conduct experiments in a given area. Appropriately, researchers worldwide face a necessity of finding and getting oneself familiar with various simulators, often designed for specific applications. The selection of best simulator for a specific research demand is a critical task.

18.2 PRIOR RESEARCH WORK

Authors of a research contribution [3] identify key properties of WSN evaluation environments and studies. As per the authors, scalability, flexibility, accuracy, repeatability, visibility, cross-environment validity, and reusability are the clearly idealistic key features but all together are very hard to achieve in practice. Authors also present a high-level comparative table based on different WSN evaluation environments. As per the authors, simulation can offer great visibility, repeatability, flexibility, and unlimited scalability.

Authors of a research contribution [4] compare some simulation tools using an identical simulation scenario that can be easily implemented in the different simulation environments. Results show differences while each simulator is of the same order of magnitude, despite using diverse models.

Simulating WSNs requires more specific properties to reflect the real operational environment, to effectively simulate WSNs; requirements from simulators [5] are discussed in a research contribution. These requirements are categorized as general and specific.

General requirements are those that are expected from each simulator and are essential to simulate ordinary WSNs. These requirements are again categorized as nonfunctional and functional requirements. Nonfunctional requirements are open source, platform independence, and visualization support (everyone is familiar with these terms), whereas functional requirements are identified as hardware simulation coverage, battery and power models, propagation modeling, protocols modeling, physical environment modeling, and emulation.

Specific requirements are those that are required when a new protocol for WSNs needs to be tested and evaluated:

- Simulator has to be open source with some guidelines provided to contribute a new feature because various modules will be implemented and then linked to the existing system, in case of a new protocol.
- Simulators have battery and power models that could reflect the energy status of a sensor at any time.
- Simulators have to be able to support other propagation models except RF.
- Simulators have to be able to provide interconnection with the internet, which is necessarily required when remote monitoring or the use of sensory data is required.
- Simulators have to be able to nicely understand protocol behaviors and able to nicely visualized and animated results.
- Simulators should allow researchers to verify new ideas and compare the proposed solutions in a virtual environment helping to avoid unnecessary, time-consuming, or expensive hardware implementations.

From a broad variety of simulation tools, the selection of a single simulation tool for experiments is mainly based on the experience of the researcher rather than on rational arguments [4]. The objective of this chapter will speed up the decision process and facilitate researchers to focus their attention on the simulation tool/framework/environment that fulfills their specific requirements.

The authors of a research contribution [5] compare some popular simulators based on these properties: open source, battery model, power model, visualization model, emulation property, and propagation model.

Key features and limitations of some popular WSN simulators are also presented in a research contribution [6]. According to the authors of a research paper [6], some most popular WSN-specific simulators are compared to these aspects: resources, energy efficiency, decentralized collaboration, simulation scenarios, fault tolerance, global behavior, etc. The authors also technically explore selected simulators under subtitles: summary, environment, simulation language, key features, and limitations.

The term co-simulation is used in a research paper [7], which means to extend the desired features of a general-purpose simulator by inducing some software patches or by the use of desired binaries in libraries.

A good comparison table is available in a master thesis [8], which primarily highlights energy consumption model, supported MAC protocols, and supported wireless channel models.

In a research paper [9], the authors present the findings of a conducted survey. Here, the authors compare different simulators on the following parameters: level of details, timing, soft license, popularity, simulation platform, WSN platform, GUI support, available models and protocols (for wireless channel, PHY, MAC, network, transport, sensing), and energy consumption model.

Comparison of case studies and simulation results are also discussed in a research contribution [3]. Comparison of simulators based on the level of abstraction and scalability is described in a research contribution [6].

Modeling techniques for discrete-event-driven-systems (including WSNs) are described in a research contribution [1]. In the same research contribution, the authors also focused on parallel simulation. As per the authors, parallel simulations (for instance, GloMoSim, SENSE, QualNet, SSFNet) should perform and scale better than sequential simulators, which means it can be a better solution to study/simulate large-scale WSNs.

The design of a WSN-specific simulator is discussed in a research contribution [6]. As per the authors, a WSN simulator consists of various modules like the following:

- *Event*—It is an abstract base class that provides basic functionality for all events.
- *Medium*—It models the wireless medium.
- *Environment*—It has properties that relate to the physical phenomenon modeled.
- *Node*—It serves as a container for all of the components, both hardware and software, in a node.
- *Transceiver*—It models the hardware transceiver on each sensor node.
- *Physical protocol*—It provides services for changing the state of the transceiver, carrier sensing, sending and receiving packets, received energy detection on received packets, and changing channels on physical layers that support multiple channels.
- *MAC protocol*—It provides services for changing the state of the MAC layer (i.e., low-power mode), setting and getting parameters for protocol, as well as sending and receiving packets.
- *Routing protocol*—It provides services for routing messages over multiple hops between nodes that cannot communicate directly.
- *Application layer*—It interfaces with the lower layers and is used to implement a WSN application.

As per the authors of a research paper [10], every simulation is based on the simplified model of a system. Good model allows to investigate system features and possibly predicts its behavior under various conditions. The authors present a general network model, a WSN network model, and a structure of WSN node model with the help of clear figures.

18.3 SIMULATION EVALUATION CRITERIA

In a research paper [9], the authors define a set of criteria to evaluate and compare the simulators, which is as follows:

S.No.	Evaluation Criteria		Highlights/Description
1.	Level of details	Generic simulator	Focuses on high-level aspects of WSN like networking, sensing, and data processing excluding low-level detailing like OS and hardware architecture of sensor nodes. These are useful for evaluation of high-level protocols and algorithms.
		Code level simulator	It uses the same code in simulation as in real sensor node. Application code and OS code (device drivers have to be altered, because there are no real hardware devices) are compiled for the machine that is running the simulator, while hardware architecture of sensor nodes is not taken into account. These simulators can be used to find bugs that are not related to timing or hardware architecture.
		Firmware level simulator	Uses hardware emulation to execute deployable application and OS code compiled for the target platform. Most types of bugs can be found and timing-sensitive software can be tested by the use of this kind of simulators.
2.	Timing	Discrete-event simulators	Events that affect state of system are chronologically ordered into event queue, and event scheduler executes them one by one.
		Continuous simulators	Concerned with modeling a set of equations that represents the system over time.
3.	Software license		Availability of simulators as proprietary/free software/under open-source licenses.
4.	Popularity		Number of hits on popular search engines.

Because of the rapid and widespread scope of WSN applications, the authors of a research contribution [10] state that it is very difficult to give a fixed list of attributes of the ideal WSN simulator but these are the most important features in the context of WSN: reusability and availability, performance and scalability, operating system portability, scripting language semantic, and graphic, debug, and trace support.

Authors of another research paper [6] also focused on same capabilities mentioned by the authors of research paper [10].

18.4 NECESSITY OF SIMULATION

The emergence of WSNs brought many new emerging issues to the network designers. Conventionally, the three popular techniques for analyzing the performance of wired as well as wireless networks are (1) analytical methods, (2) computer simulation, and (3) test bed/physical measurement. Simulation is one of the feasible approaches for the quantitative analysis of WSNs.

The large-scale deployment and due to inaccessibility of deployed WSNs mandate that the code installed in sensor nodes be hardly tested prior to real deployment. This kind of testing is primarily done using software simulators. Simulators are an effective experimental tool commonly used for rapid prototyping and also used for the evaluation of new network protocols and algorithms and enable repeatability because they are independent of the physical world and its impact on the objects. Moreover, simulations enable nonintrusive debugging at the desired level of detail.

Simulation can be a best choice before building a new system or it can be used when an existing system required alternation, to prevent under- or overutilization of resources, to eliminate unforeseen bottlenecks, to reduce the chances of failure to meet specifications, and to optimize system performance.

Highlights of simulation tools are as follows:

- Software that imitates selected parts of the behavior of the real world and is normally used as a tool for research and development.
- Depending on the intended usage of the simulator, different parts of the real-world system are modeled and imitated.
- The challenge of developing, deploying, and debugging applications on the realistic environment will be unmet with simulations.
- Many of the current simulators are unable to model many essential characteristics of the real world. Simulations are based on common simplified assumptions and these do not produce accurate results.
- Results obtained from simulation are only as good as clear models and they are still only estimated or projected outcomes.
- Especially for WSNs, simulation models do not capture the radio and sensor irregularity.
- As the simulation scales, the performance and usability of the simulation decreases, limiting the ability to simulate large WSN deployments.
- Simulation is one of the feasible approaches to the quantitative analysis of sensor networks.

Simulation can be useful to find answers like the best design for a new WSN, the associated resource requirements, the evaluation of WSN performance under specific load conditions and new routing algorithms, the evaluation of protocol's optimization performance, and the evaluation of link failure.

As per a research paper [6], from the WSN system's point of view, two important features of simulators are extremely valuable, these are

- Reproducible experimentation
- Dynamic environment modeling

Simulation is a powerful technique that can be used at several stages of system development and thus can be a better choice, but some limitations are also associated with simulation as well. The limitations of simulation are discussed in research contributions [11,12]. As per the authors of another research contribution [3], simulation suffers greatly from the poor accuracy of implemented simulation models and protocols and has very low reusability in general (especially general use network simulators) and the results gathered on one simulator are hardly comparable to the results from others. Also, as per the authors of the same research paper [3], the major disadvantage of WSN simulations is their low credibility due to oversimplified simulation models. As per the authors [3], the full and safe simulation credibility can be achieved only by implementing deterministic, parameter-free simulation models.

18.5   TYPE OF SIMULATION

Simulators run as either in an event-triggered mode or in asynchronous mode or in synchronous mode, where events happen in parallel in fixed time slots [13]:

- *Synchronous simulation*: This type of simulation is based on rounds. At the start of each round, global time is incremented by one unit by the simulators. And then simulators move the nodes as per their mobility models and update the connections based on connectivity model. Finally, the simulation framework iterates over the set of nodes and performs the aforementioned steps for each node.
- *Asynchronous simulation*: The asynchronous simulation is purely event based. The simulator holds a list of message events and timer events, which is sorted by the time when these events should happen (arrival of message, execution of timer handler). The simulator repeatedly picks the most recent event and executes it.

In general, the asynchronous simulation mode runs much faster than the synchronous mode. The main reason lies in the fact that the synchronous simulation mode loops over all nodes and performs for each node the set of fixed steps even if most of the nodes may not do anything at all, whereas in asynchronous mode, only message and timer events are processed and no unnecessary cycles are wasted.

18.6   CLASSIFICATION/CATEGORIZATION OF SIMULATORS

Simulation platforms/simulators vary in many aspects [14]; thus, according to their features and main applications, simulators have been divided into categories. In a research contribution, WSN simulators are categorized [15] as follows:

- *Generic network simulators*: Generic network simulators simulate systems with a focus on networking aspects. The user of the simulator typically writes the simulation application in a high-level language different from the one used for the real sensor network. Since the focus of the simulation is on networking, the simulator typically provides detailed simulation of the radio medium but less detailed simulation of the nodes. These types of simulators are useful for evaluation of high-level protocols and algorithms.
- *Code level simulators*: Code level simulators use the same code in simulation as in real sensor network nodes. The code is compiled for the machine that is running the simulator, typically a PC workstation that is magnitudes faster than the sensor node. Typically, code level simulators are operating systems specific, since they need to replace driver code for the sensors and radio chips available on the node with driver code that instead have hooks into the simulator. These types of simulators can be used to find bugs that are not related to timing or hardware architecture.

- *Firmware level simulators*: These simulators are based on emulation of the sensor nodes, and the software that runs in the simulator is the actual firmware that can be deployed in the real sensor network. This approach gives the highest level of detail in the simulation and enables accurate execution statistics. This type of simulation provides emulation of microprocessor, radio chip, and other peripherals and simulation of radio medium. Due to the high level of detail provided by firmware level simulators, they are usually slower than code level or generic network simulators. With the help of these types of simulators, most types of bugs can be found and timing-sensitive software can be tested.

In a research contribution [9], two other types of simulator categorization are available:

- *Discrete-event simulators*: Discrete-event simulation is the most common approach used to simulate WSNs. By its very nature, discrete-event simulation models the activity of network nodes and the software components on those nodes as discrete events in time.

  In discrete-event simulators, events that affect the state of the system are chronologically ordered into event queue, and event scheduler executes them one by one. Discrete-event simulators, by their very nature, mask race conditions in the code since simulated interrupts never interrupt running code; an additional limitation of most such simulators is that all simulated nodes execute the same application code, at variance with common practice in actual deployments.

  Due to this event-driven approach, an interrupt that can cause incorrect behavior in a physical node will not cause the same error in simulation. This situation limits discrete-event simulation as a reliable tool to provide strong guarantees about the suitability of software for deployment in the field.
- *Continuous simulators*: Continuous simulators are concerned with modeling a set of equations that represent system over time.

In another research contribution [16], simulators have been classified into the following three major categories based on complexity:

- *Algorithm level simulators*: Some simulators focus on the logic, data structure, and presentation of the algorithms. For example, AlgoSenSim analyzes specific algorithms in WSNs, for example, localization, distributed routing, and flooding. Shawn is targeted to simulate the effect caused by a phenomenon, improve scalability, and support free choice of the implementation model. Sinalgo offers a message passing view of the network, which captures well the view of actual network devices.
- *Packet level simulators*: Some simulators implement the data link and physical layers in a typical OSI network stack. The most popular and widely used network simulator ns-2 is not originally targeted to WSNs but IP networks. SensorSim is an extension to ns-2 that provides battery models, radio propagation models, and sensor channel models. J-Sim adopts loosely coupled, component-based programming model, and it supports real-time process-driven simulation. GloMoSim simulator is designed on the PARSEC's concept of parallel discrete-event simulation capability.
- *Instruction-level simulators*: Some simulators model the CPU execution at the level of instructions or even cycles. They are often regarded as emulators. They compute the power of a particular sensor's hardware platform in WSNs. TOSSIM simulates the TinyOS network stack at the bit level. ATEMU is an emulator that can run nodes with distinct applications at the same time.

Authors of a research paper [14] categorize simulators in different types according to their features and main applications. A summarized list is as follows:

S.No.	Classification of Simulators	Description	Simulators [or Simulation Environments]
1.	Emulators and code level simulators	Focused on to emulate the sensor hardware or process the provided program code in a manner it would be executed on a real device.	ATEMU, Avrora, EmSim, Freemote Emulator, MSPsim, TOSSIM, VMNet, WSim
2.	Topology control simulator	Focused on topology construction and topology maintenance.	Atarraya
3.	Environment and wireless medium simulators	Focused on the simulation of the physical environment of a node and the wireless medium used for the transmission.	Prowler, WSN localization simulator, WSNet
4.	Network and application level simulators	Focused on the simulation of the network as a structure transporting and processing gathered data.	AlgoSenSim, NetTopo, SENSE, Sensor Security Simulator (S3), Shawn, SIDnet-SWANS, Sinalgo, TRMSim-WSN, WSN Simulator, WSNSimPy
5.	Cross-level simulators	Simulators are capable of simulating nodes and networks at various levels of abstraction.	COOJA, J-Sim, SENS, WSNSim
6.	ns-2-based simulators		Mannasim, NRL Sensorsim, RTNS
7.	OMNET++-based simulators		Castalia, MiXiM, NesCT, PAWiS, SENSIM
8.	Ptolemy-II-based simulator		Viptos, VisualSense

18.7 PHASES IN A SIMULATION STUDY

Authors of a chapter [1] focused on the basic steps that should be considered during a simulation study. As per the authors of this chapter, the simulation study life cycle consists these eight phases (Figure 18.1).

The simulation study life cycle does not have to be strictly sequential, it can be iterative, some of the steps can be skipped, and sometimes transitions in opposite directions can also appear.

The authors of a research paper [91] also focused on the steps involved in simulation study with the help of a flow chart. As per the authors [91], a total of 12 steps are involved in simulation study; these are as follows (Figure 18.2).

18.8 SIMULATORS (OR SIMULATION ENVIRONMENTS) FOR WSNs: STUDY AND RECENT ADVANCES

Several simulators exist that are either adjusted or developed specifically for WSNs. Researchers developed new ones; other existing tools and frameworks have departed or evolved causing some of the information to be out of date.

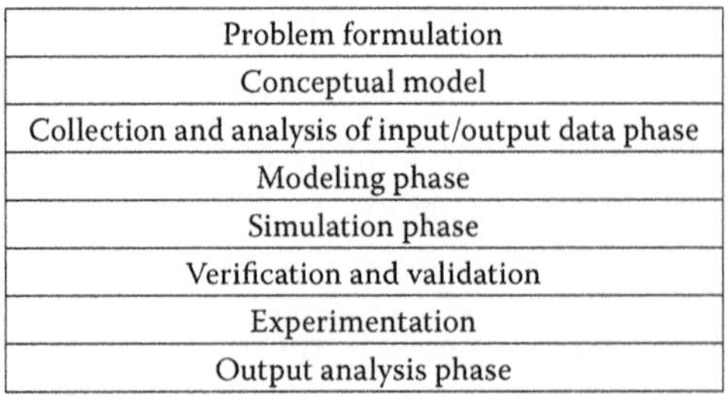

Problem formulation
Conceptual model
Collection and analysis of input/output data phase
Modeling phase
Simulation phase
Verification and validation
Experimentation
Output analysis phase

FIGURE 18.1   Simulation study life cycle for WSNs.

Step 1	Problem formation
Step 2	Setting of objective and overall project plan
Step 3	Model conceptualization
Step 4	Data collection
Step 5	Model translation
Step 6	Verification
Step 7	Validation
Step 8	Experimental design
Step 9	Production runs and analysis
Step 10	More runs (if required)
Step 11	Documentation and reporting
Step 12	Implementation

FIGURE 18.2 Steps involved in simulation study for WSNs.

The authors of a research paper [91] focused on the design of WSN-specific simulators. As per the authors of this research paper [91], a WSN-specific simulator consists of various independent modules glued with each other obtaining design goals and the modules are event, medium, environment, node, transceiver, and finally physical/MAC/routing/application layer protocols.

Reusability and availability, performance and scalability, support for scripting languages having rich semantics to define experiments and process results, and graphical, debug, and trace supports should be the capabilities for the WSN-specific simulators (Table 18.1) [91]. Choosing the right simulator from a big pool of currently available simulators for a given application is very important. To make this choice/selection easier for the researchers, an exploratory study has been conducted on several simulators and it has been found significant and interesting Table 18.2.

18.9 EMULATORS (OR EMULATION ENVIRONMENTS) FOR WSNs: CURRENT STATUS

An emulator or an emulation environment is a special type of simulator that aims to enable realistic and sensible performance evaluation for WSN applications. A networked embedded system includes a WSN application that involves associated sensor node hardware and drivers, operating systems, as well as networking protocols. As a result, the performance of the WSN-specific application depends on the aforementioned factors in addition to its implementation. Emulators or emulation environments are the good preference, in which WSN applications can be directly run for testing, performance evaluation, and debugging. Furthermore, studies on the lower layers (e.g., hardware and associated drivers, OS, and networking support) and cross-layer techniques can also be done in this environment by plugging the target modules into the emulator. For example, emulators can compute the power/energy level of a particular sensor's hardware platform in WSNs [92,93] (Table 18.3).

18.10 FUTURE RESEARCH DIRECTIONS

Significance of simulators/simulation environments for understanding real-world WSN-specific scenarios is an ongoing challenge. It is very crucial to have a single specific tool for analyzing and evaluating WSN's behavior accurately. As per the WSN-specific requirements, design and implementation of a single simulator that leverages with these capabilities (fast, precise, scalable, flexible, GUI-based, low-level detailing capability; proven realistic implementation on energy monitoring and profiling; debugging features; modular, open source, platform independent, etc.) is a challenge and researchers worldwide are working on it. It has been also observed that simulator designers need more attention and focus on implementation of application, transport, and physical layers in WSN-specific simulators. Implementation of cross-layer approaches in WSN-specific simulators is also a challenge.

TABLE 18.1

Comparative Study of WSN-Specific Simulators/Simulation Frameworks

S.No.	Simulator	Properties or Features
1.	Network Simulator (ns) [17,18]	• In original ns-2, a single aspect related to WSN was not present; that is how a sensor will react once it detects its target physical condition in RoI (region of interest). • It is used to evaluate WSNs but the accuracy of results is questionable with lower versions. • The MAC protocols, packet formats, and energy models are very different in this simulator.
2.	Mannasim (ns-2 extension for WSNs) [19]	• Mannasim upgrades ns-2 by introducing new modules for design, development, and analysis of different WSN applications. • Mannasim has inbuilt script generator tool (SGT) for TCL script creator.
3.	TOSSIM [20,21]	• This simulator is useful for both testing the algorithms and its implementations. • It does not simulate the physical phenomena. • It uses hardware abstraction to model the device components.
4.	TOSSF [22]	• It is very similar to and inspired by TOSSIM. • It has one limitation, that is, it no longer simulates the devices as accurately as TOSSIM.
5.	PowerTOSSIM z [23]	• It is a power modeling extension to TOSSIM. • It accurately models the power consumed by tiny operating systems applications.
6.	ATEMU [24]	• It was the first instruction-level sensor node simulator. • It is the most accurate simulator for a particular hardware platform. • The main limitation of ATEMU is that it is dependent on the Mica2 mote hardware architecture.
7.	COOJA [25]	• It is a ContikiOS simulator that allows for cross-level simulation. • It is a novel type of WSN simulator that enables simultaneous simulation at different levels. • In COOJA, one simulation can contain nodes from several different abstraction levels, that is, the network level, the operating system level, and the machine code level.
8.	GloMoSim (global mobile information systems simulation) [26]	• It also suffers the packet formats, energy models, and MAC protocols problems like in ns. • GloMoSim is also used to evaluate the WSNs but the accuracy of results is questionable here.
9.	QualNet [27]	• It is the commercial version of GloMoSim. • It has upgraded features such as providing a comprehensive environment for designing protocols, creating and animating experiments, and analyzing the results of the experiments.
10.	SENSE [28]	• It does not support sensors, physical phenomena, and the environmental effects. • It is totally supported by the MAC protocol. • The radio propagation makes SENSE less than ideal for accurate evaluation of WSNs research.
11.	VisualSense [29]	• It is a good framework. • It does not provide any protocols above the wireless medium nor sensor or physical phenomena. • It provides on only sound.
12.	AlgoSenSim [30]	• It is a framework that is used to simulate distributed algorithms. • It is algorithm oriented but not protocol stack oriented. • It focuses on network-specific algorithms like localization, distributed routing, and flooding. • It uses XML configuration file easily and modularly. • It is an efficiency-oriented simulator, but optimizations are hidden to the user. • The main purpose is to facilitate the implementation and quality analysis of new algorithms.

(Continued)

TABLE 18.1 (*Continued*)
Comparative Study of WSN-Specific Simulators/Simulation Frameworks

S.No.	Simulator	Properties or Features
13.	Georgia Tech Network Simulator (GTNetS) [31]	• GTNetS is a full-featured network simulation environment. • In computer networks, it allows the researchers to study the behavior of moderate- to large-scale networks, under a variety of conditions. • The design philosophy of GTNetS is to create a simulation environment that is much like how the actual networks are structured.
14.	OMNet++ [32,33]	• It is an extensible, modular, component-based C++ simulation library and framework. • It has Eclipse-based IDE and a graphical discrete-event simulator.
15.	Castalia [34]	• For WSNs, it is an extension of OMNet++. • It can be used by researchers and developers who want to test their distributed algorithms and protocols. • Castalia can also be used to evaluate different platform characteristics for specific applications. • It is highly parametric and can simulate a wide range of platforms.
16.	J-Sim (formerly JavaSim) [35]	• It provides a truly platform neutral, component-based, compositional simulation environment. • J-Sim provides support for sensors as well as for physical phenomena. • Energy modeling, with the exception of radio energy consumption, is also appropriate for sensor networks. • Accuracy of simulations still suffers because the only MAC protocol provided for wireless networks is IEEE 802.11.
17.	JiST/SWANS [36]	• JiST is a high-performance discrete-event simulation engine that runs over a standard Java virtual machine. • It is a prototype of a new general-purpose approach to building discrete-event simulators, called virtual-machine-based simulation. • SWANS is a scalable wireless network simulator built atop the JiST platform. • Its capabilities are similar to ns-2 and GloMoSim but are able to simulate much larger networks.
18.	JiST/SWANS++ [37]	• It is an extended version of JiST/SWANS. • It provides more realistic and meaningful simulation results.
19.	Avrora [38]	• It is a cycle-accurate, highly scalable, fast, and accurate instruction-level sensor network simulator. • It is used to scale networks of up to 10,000 nodes. • Handles 25 nodes in a real time. • Avrora's ability is to measure detailed time critical phenomena and can shed new light on design issues for large-scale sensor networks.
20.	Sidh [39]	• Sidh scales the simulate networks with thousands of nodes faster than real time on a typical desktop computer. • It is component based and easily reconfigurable to adapt to different levels of simulation detail and accuracy, communication media, sensors and actuators, environmental conditions, protocols, and applications.
21.	Prowler [40]	• Prowler is a probabilistic WSN simulator. • Prowler is written in MATLAB® and also running under MATLAB. • Providing an easy way of application prototyping with nice visualization capabilities.
22.	(J) Prowler [41]	• It is a discrete-event simulator similar to Prowler but written in Java. • The simulator supports pluggable radio models and MAC protocols and multiple application modules. • Currently, two radio models are implemented (Gaussian and Rayleigh) and one MAC protocol (Mica2 with no acknowledgment). • It could be modified to simulate more general systems. • It does not provide support for sensors or physical phenomena.

(Continued)

TABLE 18.1 (*Continued*)

Comparative Study of WSN-Specific Simulators/Simulation Frameworks

S.No.	Simulator	Properties or Features
23.	LecsSim [42]	• LecsSim is a simulator for large wireless networks and provides an easy way to simulate distributed algorithms in wireless networks. • It includes propagation models, modules for common node functionality, and documentation.
24.	OPNET [43]	• OPNET is slightly different from ns and GlomoSim. • It supports the use of modeling different sensor-specific hardware, for example, physical link transceivers and antennas. • It can also be used to define custom packet formats. • OPNET suffers from the same object-oriented scalability problems as ns.
25.	SENS [44]	• SENS is a component-based simulator with four main components: application, network, physical, and environment. • The former three components make up the sensor node. • SENS is less customizable than any other simulator, providing no opportunity to change the MAC protocol, along with other low-level network protocols.
26.	EmStar/Em* [45,46]	• EmStar/Em* is a software environment for developing and deploying complex WSN applications. • It is of 32-bit embedded microserver platforms and integrated with networks of motes. • EmStar consists of libraries that implement message passing IPC primitives; tools that support simulation, emulation, and visualization of live systems, both real and simulated; and services that support networking, sensing, and time synchronization.
27.	EmTOS [45,46]	• EmTOS is an extension to EmStar. • It can be used either for deployment or simulation of WSNs. • It enables a complete NesC/TOS app to run unmodified under EmStar.
28.	SenQ [47]	• SenQ is an accurate and scalable evaluation framework for sensor networks. • It integrates sensor network operating systems with a very high fidelity simulation of wireless networks such that sensor network applications and protocols can be executed, without modifications, in a repeatable manner under a diverse set of scalable environments. • SenQ extends beyond the existing suite of simulators and emulators in four key aspects: it supports emulation of WSN applications and protocols in an efficient and flexible manner; it provides an efficient set of models of diverse sensing phenomena; it provides accurate models of both battery power and clock drift effect that have been shown to have a significant impact on sensor network studies; and finally, it provides an efficient kernel that allows it to run experiments that provide substantial scalability in both the spatial and temporal contexts.
29.	H-MAS (Heterogeneous mobile ad hoc sensor-network simulation environment) [48]	• It provides a convenient platform on which to evaluate a variety of MAS (mobile ad hoc sensor nets) configurations at the physical, medium access, network, and application layers and to extract meaningful design rules from the experimental data. • It also provides an intuitive visualization that can give insight to the design engineer and casual observer alike.
30.	SensorSim [49]	• It is a simulation framework that inherits the core features of traditional event-driven network simulators. • It builds up new features that include the ability to model power usage in sensor nodes, hybrid simulation that allows the interaction of real and simulated nodes, new communication protocols, and real-time user interaction with graphical data display.
31.	Shawn [50]	• It is a discrete-event simulator for sensor networks. • Due to high customizability, it is extremely fast. • It can be tuned to any accuracy that is required by the simulation or application.

(Continued)

TABLE 18.1 (*Continued*)
Comparative Study of WSN-Specific Simulators/Simulation Frameworks

S.No.	Simulator	Properties or Features
32.	NetTopo [16]	• It is a research-oriented sensor network simulator. • NetTopo has the functions of general sensor networks but specially reflects the research results of the following: Streaming Data Gathering and Topology Prediction in WSNs within Expected Lifetime, Reward Oriented Packet Filtering Algorithm for Heterogeneous Sensor Networks, VIP Bridge: Integrating Several Sensor Networks into One Virtual Sensor Network, Transmitting Streaming Data in Wireless Sensor Networks with Holes, and many more.
33.	Atarraya [51]	• This simulator is specifically focused on the evaluation of topology control protocols in WSNs.
34.	SSFNet (Scalable Simulation Framework) [52]	• It is a command-line-based simulator. • Accordingly, the realization of specific application scenarios and the user interaction is difficult. • SSFNet focuses on static application scenarios. • An important feature of SSFNet is the possibility to parallelize the simulation. • This speedup enables the analysis of large-scale network behavior. • Both toolkits are limited to a single communication interface per node.
35.	WiseNet [53]	• It is a software simulator that can be very useful to carefully plan and select the right type of motes and sensors in a cost-effective manner. • WiSeNet simulates random distribution of sensors. • Through repeated experimentation, it is possible to arrive at an optimal spatial configuration of the sensors that is most effective for a given application. • WiSeNet also allows the wireless range of a sensor to be varied and study the effects on the application.
36.	SimGate [54]	• It is a full-system simulator for the Intel Stargate, intermediate-level, resource-constrained, sensor network device.
37.	SimSync [55]	• It is a time synchronization simulator for WSNs. • It models the distribution of packet delay and the frequency of crystal oscillator as Gaussian.
38.	SNetSim [56]	• It is an event-driven simulation software running on Windows-based operating systems.
39.	SensorMaker [57]	• It supports scalable and fine-grained instrumentation of the entire sensor networks.
40.	TRMSim-WSN [58]	• It is a Java-based simulator. • Its main aim is to test trust and reputation models for WSNs.
41.	PAWiS [59]	• A simulation framework that provides functionality to simulate the network nodes with their internal structure as well as the network between the nodes. • One main feature is the contemporaneous simulation of the power consumption of every single node. • The framework is based on the discrete-event simulator OMNeT++. • The user-defined model (expressed with C++ classes) is compiled to an executable simulator.
42.	OLIMPO [60]	• It is a discrete-event simulator for WSN. • It is designed to be easily reconfigured by the user. • It is providing a way to design, develop, and test communication protocols.
43.	DiSenS (distributed sensor network simulation) [61]	• It is a complete scalable and extensible distributed simulation system for sensor networks. • It provides a cycle-accurate device emulator that is extendable by various fidelity enhancing models (radio, power, etc.) for tunable simulation accuracy. • A key distinguishing feature of DiSenS is that it is implemented for distributed memory parallel cluster systems.

(Continued)

TABLE 18.1 (*Continued*)
Comparative Study of WSN-Specific Simulators/Simulation Frameworks

S.No.	Simulator	Properties or Features
44.	WISDOM [62]	• A simulator is written in Java and uses recursive porous agent simulation toolkit (Repast) O as the simulation engine to perform discrete-event-driven simulations. • WISDOM can be used to simulate and verify middleware services for routing, sensing activity scheduling, group formation and management, target detection and tracking, and collaborative classification and fusion in a WSN. • The versatility and performance of WISDOM for middleware service protocol development and evaluation have proven to be valuable.
45.	Sinalgo [13]	• It is a simulation framework for testing and validating network algorithms for WSN. • It is unlike most other network simulators, which spend most time simulating the different layers of the network stack. • Sinalgo focuses on the verification of network algorithms and abstracts from the underlying layers. • It offers a message passing view of the network, which captures well the view of actual network devices.
46.	SENSORIA [63]	• It is a fully fledged simulator for WSNs that has considerable differences to all other existing simulators. • Sensoria is very powerful in simulating a range of small- to large-scale WSNs based on a simple and complete graphical user interface (GUI). • Sensoria's GUI allows users to design various simulation scenarios and display the simulation results graphically with many formats. • Sensoria is a component-based simulator and it can be easily reconfigured to adapt to different levels of simulation details and accuracy.
47.	Capricorn [64]	• It is a large-scale discrete-event WSN simulator developed at Wayne State University.
48.	SIDnet-SWANS [65]	• It is a simulation-based environment that enables run time interactions with the network. • It is used for the purpose of observing the behavior of algorithm protocols in the presence of various conditions such as phenomena fluctuations or a sudden loss of service both at an individual node and a collection of nodes.
49.	Stargate simulator (starsim) [66]	• It is a full-system simulator for Stargate, the XScale-based gateway device for WSN. • It also boots original Linux image from crossbow. • It also features an XScale pipeline simulator to provide cycle estimation.
50.	SNSim [67]	• It is a prototype software tool, designed to support and balance the lifetime of a WSN and the quality of data (QoD) that is sampled and processed. • It includes elements of power consumption characteristics and is built to mimic real performances of Mica motes. • This graphical interface tool is created toward investigating various aspects of development, as well as building applications/simulations for such networks. • Both SNSim and its event-driven simulation engine are written in Java, which offers an enhanced portability and efficiency of development time.
51.	SNIPER-WSNSim [68]	• It is a less known simulator that is specifically designed for WSNs. • It takes benefits from the richness of the .NET framework 3.5 and from the portability of the C# language. • It is a graphical-interface-based simulator that deals with a particular sector of WSN development such as sensor nodes distribution, routing protocols, and clustering.
52.	SNAP [69]	• It is defined as an integrated hardware simulation and deployment platform. • It is a microprocessor that can be used in two ways: as the core of a deployed sensor or as a part of an array of processors that performs parallel simulation. • Again, *real* code for sensors can be simulated. • By combining arrays of SNAPs (called network on a chip), it is claimed to be able to simulate networks on the order of 100,000 nodes.

(Continued)

TABLE 18.1 (*Continued*)
Comparative Study of WSN-Specific Simulators/Simulation Frameworks

S.No.	Simulator	Properties or Features
53.	SimPy [70]	• It is a bare simulation written in Python. • In SimPy, the basic simulation entities are processes. • They are executed in parallel and may exchange Python objects among each other. • Most processes include an infinite loop in which the main actions of the process are performed. • Besides abstractions for processes and the related exchange of objects, SimPy provides instructions for the synchronization of simulation processes and commands for the monitoring of simulation data. • Unlike the other simulators, there is no public available network models that exist for it.
54.	Mule [71]	• It is a hybrid simulator. • It combines the ease of debugging multiple simulated motes on a host PC with high fidelity of message transmission and sensor data acquisition of physical motes.
55.	CaVi [72]	• It provides a uniform interface to state of the art simulation methods and formal verification methods for WSN. • Due to the probabilistic behavior of WSNs systems, however, the simulation covers only a small fraction of all possible behaviors. • Formal model checking techniques, based on Markov decision processes, use less detailed and more abstract models and compute exact probabilities and expected values for the entire behavior, where simulation can only give averages. • It allows for creating a single model for simulation, Monte Carlo simulation, and model checking.
56.	Ptolemy [73]	• It is a discrete-event simulator and a design tool for concurrent, real-time, embedded systems. • It could be used to simulate WSNs. • In fact, VisualSense, which is a framework built on top of Ptolemy, is intended to assist researchers in the design, visualization, and simulation of WSNs. • In VisualSense, sensor nodes could be defined using either the discrete event blocks or the continuous time and real-time blocks available on Ptolemy. • In addition, the sensor nodes could also be written in Java to meet specific needs.
57.	Maple [74]	• It is a simulator that allows researchers and WSN developers to focus on particular aspects such as sensor nodes distribution and WSN lifetime estimation. • Moreover, the parallel processing capability presents an important feature for further distributed simulations. • Furthermore, the intuitive and convivial user interface makes the simulator accessible for the average users while being flexible and scalable for improvements for advanced users.
58.	WISENES (wireless sensor network simulator) [75]	• It simulates high-level WSN protocol and application designs and provides accurate information about their performance in a real environment. • The WISENES framework implements models for transmission medium (for modeling wireless communications), sensing channel (for physical phenomena), and nodes (for physical node platforms). • The designer selects the protocols from the library or implements new ones in SDL and integrates them to the WISENES framework. • The framework components and node protocols communicate using SDL signals. • A node model can be dynamically instantiated separately for each simulated node. Thus, virtually, any number of nodes can be simulated simultaneously.

(Continued)

TABLE 18.1 (*Continued*)

Comparative Study of WSN-Specific Simulators/Simulation Frameworks

S.No.	Simulator	Properties or Features
59.	WSNet—Worldsens and WSim [76]	• WSNet is an event-driven, large-scale WSN simulator. • WSNet uses models for applications, protocols, and radio medium communication with a parameterized accuracy. • WSim can be connected to WSNet, in place of the application and protocol models used during the high-level simulation to achieve a full distributed application simulation. • WSNet and WSNet+WSim allow a continuous refinement from high-level estimations down to low-level real-time validation.
60.	LSU SensorSimulator [77]	• It is a framework for simulating WSNs. • It is a customizable and extendible simulator, which allows testing and analyzing software for WSNs. • The users can subclass the framework classes and customize the behavior of various network layers. • This subclassing gives a way to the developers and an opportunity to analyze and investigate phenomenological, networking, robustness, and scaling issues, to explore arbitrary algorithms for distributed sensors, independent of hardware constraint.
61.	WSNGE [78]	• It is a flexible and extensible environment. • It provides a highly scalable simulator with unique characteristics such as focusing on user friendliness, providing every function in both scriptable and visual way, and allowing the researcher to define simulations and view results in an easy to use graphical environment. • WSNGE does not distinguish between different scenario types, allowing multiple different protocols to run at the same time. • It enables rich online interaction with running simulations, allowing parameters, topologies, or the whole scenario to be altered at any point in time.
62.	TikTak [79]	• It is a scalable simulator for WSNs including hardware/software interaction. • It specifically allows the design exploration and the complete microprocessor instruction-level debug of network formation, data congestion, and nodes interaction, all-in-one simulation environment. • An innovative feature is the co-emulation of selected nodes at clock-cycle-accurate hardware processing level, allowing code debug and exact execution latency evaluation, together with other nodes at abstract protocol level, meeting a designer's needs of simulation speed, scalability, and reliability. • The simulator is centered on the Zigbee protocol and can be retargeted for different node microarchitectures.
63.	Mote simulator (motesim) [66]	• It is a full-system, cycle-accurate simulator for Mica2 and Micaz motes. • It runs with original TinyOS binaries.
64.	Boris [80]	• An extensible Java simulator for WSNs.
65.	SmartSim [81]	• Provides a useful all-in-one solution. • It has new, unique features such as detailed graphical presentation of topology, ability to proceed through network events either forward or backward and a comprehensive power usage report generation. • It eases the algorithm implementation in TinyOS-based WSN and provides troubleshooting tools and linear and nonlinear energy usage of the network.
66.	WSNsim [82]	• A new discrete-event-based network simulator, specifically developed for performance analysis of WSNs deployed on farm animals. • Useful for evaluation of novel protocol ideas customized to the needs of cattle herd monitoring.

(Continued)

TABLE 18.1 (*Continued*)
Comparative Study of WSN-Specific Simulators/Simulation Frameworks

S.No.	Simulator	Properties or Features
67.	EnergySim [83]	• It is a novel, fast, extensible WSN-specific MAC protocol simulator for evaluating energy efficiency. • This simulator makes implementation of any protocol simple by meeting the requirements of varying users especially those concerned with energy consumption in WSNs.
68.	MOB-TOSSIM [84]	• An extension framework for TOSSIM simulator to support mobility in WSNs and WSANs. • It implements three components: (1) a set of probabilistic and controlled mobility models, (2) an extended and modified version of PowerTOSSIM that includes the energy consumption due to mobility and radio transmission, and (3) a radio component based on a realistic propagation model.
69.	AEON [85]	• A novel tool built on top of Avrora's energy model and used to evaluate the energy consumption and to accurately predict the lifetime of sensor networks. • It makes use of the cycle-accurate execution of sensor node applications for precise energy measurements. • It uses accurate measurements of node current draw and the execution of real code to enable accurate prediction of the actual power consumption of sensor nodes. • AEON's prediction of power consumption is based on the execution of real code, capturing all low-level events of the application in a cycle-accurate way.
70.	Sensor security simulator (S3) [86]	• A research simulator for selected security problems in large-scale WSNs. • Developed for Microsoft Windows 2K/XP/Vista/7 operating systems and written in C++. • Support for probabilistic key predistribution.
71.	Wireless sensor network localization simulator [87]	• It is an easy, extendable, scalable discrete-event simulator that supports multithreading and is based on modular design. • It has been validated using different types of localization algorithms: range-free, range-based, and hybrid localization for WSNs. • It supports a lot of mobility models such as random waypoint, modified random waypoint, random direction, modified boundless, Manhattan, Freeway, and RPGM. • It supports two-ray ground and shadowing propagation models. • Supports Monte Carlo localization (MCL) scheme, improved Monte Carlo localization (IMCL) scheme, and Kalman filter localization (KFL) scheme besides mobility prediction localization (MPL) scheme and its secure version using elliptic curve cryptography (ECC). • Supports wormhole, Sybil, spoofing, and replay attacks and their defense algorithms. • Supports two types of sensor models: MICA2 and TelosB.
72.	Xen WSN simulator [88]	• Heterogeneous sensor networks can be emulated. • A framework supporting different radio models has been provided to enable realistic radio transmissions for use in simulations. • Enabling the emulation of large-scale sensor fields. • Three WSN-specific OSs have been ported: TinyOS, Contiki, and InceOS.
73.	UWSim [89]	• It is a simulator used for underwater sensor networks (UWSN) and simulates the acoustic network. • Its focus on handling low bandwidth, low frequency, high transmission power, limited memory, effect of salinity, and temperature with depth. • It is based on a network-component-based approach. • Currently, support for limited number of functionalities. • A disadvantage is that it cannot be used for simulating any other sensor network except UWSN.
74.	Network in a box (NAB) [90]	• Simulator for large-scale sensor networks.

TABLE 18.2
Categorization of Various WSN-Specific Simulators

S.N.	Simulators	Generic Network Simulator	Code Level Simulator	Firmware Level Simulator	Algorithm Level Simulator	Packet Level Simulator	Instruction-Level Simulator
1.	Network Simulator (ns)	✓				✓	
2.	Mannasim (ns-2 extension for WSNs)	✓					
3.	TOSSIM		✓		✓		✓
4.	TOSSF		✓				
5.	PowerTOSSIM z		✓				
6.	ATEMU			✓			✓
7.	COOJA	✓	✓	✓			
8.	GloMoSim (global mobile information systems simulation)	✓				✓	
9.	QualNet	✓			✓	✓	
10.	SENSE					✓	
11.	VisualSense					✓	
12.	AlgoSenSim				✓		
13.	Georgia Tech Network Simulator (GTNetS)	✓					
14.	OMNet++	✓					
15.	Castalia	✓			✓		
16.	J-Sim (formerly JavaSim)					✓	
17.	JiST/SWANS	✓		✓		✓	
18.	JiST/SWANS++	✓		✓		✓	
19.	Avrora						✓
20.	Sidh		✓				
21.	Prowler			✓			
22.	(J) Prowler			✓		✓	
23.	LecsSim	✓			✓		
24.	OPNET			✓		✓	
25.	SENS	✓	✓			✓	
26.	EmStar/Em*	✓		✓			✓
27.	EmTOS	✓			✓		
28.	SenQ	✓	✓	✓		✓	
29.	SIDnet-SWANS				✓		
30.	SensorSim					✓	
31.	Shawn		✓				
32.	SSFNet (Scalable Simulation Framework)		✓			✓	
33.	Atarraya						
34.	NetTopo				✓		
35.	WiseNet		✓				
36.	SimGate		✓				
37.	SimSync					✓	
38.	SNetSim		✓				
39.	SensorMaker		✓				
40.	TRMSim-WSN			✓			
41.	PAWiS	✓					

(Continued)

TABLE 18.2 (*Continued*)
Categorization of Various WSN-Specific Simulators

S.N.	Simulators	Generic Network Simulator	Code Level Simulator	Firmware Level Simulator	Algorithm Level Simulator	Packet Level Simulator	Instruction-Level Simulator
42.	OLIMPO					✓	
43.	DiSenS (distributed sensor network simulation)	✓					
44.	WISDOM			✓		✓	
45.	Sinalgo				✓	✓	
46.	SENSORIA		✓				
47.	Capricorn	✓					
48.	H-MAS (heterogeneous mobile ad hoc sensor-network simulation environment)	✓				✓	
49.	Stargate simulator (starsim)		✓				
50.	Mote simulator (motesim)		✓				
51.	SNSim			✓			
52.	SNIPER-WSNSim			✓		✓	
53.	SNAP			✓			
54.	SimPy			✓			
55.	Mule	✓					
56.	CaVi				✓		
57.	Ptolemy			✓			
58.	Maple				✓		
59.	WISENES (wireless sensor network simulator)					✓	
60.	WSNet-Worldsens and WSim					✓	
61.	LSU SensorSimulator		✓				
62.	WSNGE	✓					
63.	TikTak			✓			

TABLE 18.3
List of WSN-Specific Emulators

1.	VMNET	2.	Freemote	3.	UbiSec&Sens
4.	ATEMU	5.	EMPro	6.	Emuli
7.	Emstar	8.	NetTopo	9.	MSPsim
10.	TOSSIM	11.	OCTAVEX	12.	MEADOWS
13.	AvroraZ/Avrora	14.	SENSE	15.	WiEmu

18.11 CONCLUSION

Now, simulation becomes a de facto standard tool for the evaluation of WSNs. Based on the study related to simulation, it is found that most of the currently available simulation efforts focus on protocols and algorithm-level issues, sometimes intentionally or sometimes by mistake; simulation tools are powerless on issues related to hardware resource consumption and radio channel utilization. Emulators are a type of simulator that can be the answer of these hardware-associated issues. But emulators are quite laborious since extensive prior profiling is required. Numerous simulation tools/simulators are available either by extending widely used tools or developing a new one to aid researchers to understand the behavior and performance of WSNs. This chapter presents content, based on an analytical and exploratory study about various simulators and simulation platforms with a focus on associated main features. This chapter can assist/help to a WSN researcher to explore/find a best simulator at its initial stage of research.

REFERENCES

1. S. Mahlknecht, S. A. Madani, and J. Kazm, Wireless sensor networks: Modeling and simulation, in *Discrete Event Simulations*, A. Goti (ed.), InTech, Croatia, 2010.
2. G. Chen, J. Branch, M. Pflug, L. Zhu, and B. Szymanski2, Sense: A wireless sensor network simulator, in *Advances in Pervasive Computing and Networking*, B. K. Szymanski and B. Yener (eds.), Springer, New York, 2006, pp. 249–267.
3. K. Garg, A. Förster, D. Puccinelli, and S. Giordano, Towards realistic and credible wireless sensor network evaluation, *Proceedings of ADHOCNETS 2011, Third International ICST Ad Hoc Networks Conference*, Paris, France, September 21–23, 2011.
4. A. Timm-Giel, K. Murray, M. Becker, C. Lynch, C. Görg, and D. Pesch, Comparative simulations of WSN, *Proceedings of the ICT-MobileSummit*, Stockholm, Sweden, 2008.
5. I. Khemapech, A. Miller, and I. Duncan, Simulating wireless sensor networks, Technical Reports, School of Computer Science, University of St Andrews, 2005.
6. H. Sundani, H. Li, V. K. Devabhaktuni, M. Alam, and P. Bhattacharya, Wireless sensor network simulators: A survey and comparisons, *International Journal of Computer Networks*, 2, 249–256, February 2011.
7. K.-E. Arzén, M. Ohlin, A. Cervin, P. Alriksson, and D. Henriksson, Holistic simulation of mobile robot and sensor network applications using TrueTime, *Proceedings of the European Control Conference 2007*, Kos, Greece, July 2–5, 2007.
8. M. Stehlík, Comparison of simulators for wireless sensor networks, Master Thesis, Faculty of Informatics, Masaryk University, Brno, Czech Republic, 2011.
9. M. Jevtic, N. Zogovic, and G. Dimic, Evaluation of wireless sensor network simulators, *17th Telecommunications forum TELFOR*, Belgrade, Serbia, 2009.
10. P. Moravek, D. Komosny, and M. Simek, Specifics of WSN simulations, *ElektroRevue*, 2 (3), 15–21, September 2011.
11. A. K. Dwivedi and O. P. Vyas, An exploratory study of experimental tools for wireless sensor networks, *Wireless Sensor Network*, 3 (7), 215–240, July 2011. doi:10.4236/wsn.2011.37025.
12. D. Sakamuri, NetEye: A wireless sensor network testbed, Thesis for Master of Science Degree, Wayne State University, Detroit, MI, 2008.
13. Distributed Computing Group's. Sinalgo—Simulator for network algorithms. http://disco.ethz.ch/projects/sinalgo/tutorial/tuti.html (accessed November 5, 2013).
14. B. Musznicki and P. Zwierzykowski, Survey of simulators for wireless sensor networks, *International Journal of Grid and Distributed Computing*, 5 (3), 23–50, September 2012.
15. J. Eriksson, Detailed simulation of heterogeneous wireless sensor networks, Dissertation for Licentiate of Philosophy in Computer Science, Uppsala University, Uppsala, Sweden, April 2009.
16. L. Shu, M. Hauswirth, Y. Zhang, S. Mao, N. Xiong, and J. Chen, NetTopo: A framework of simulation and visualization for wireless sensor networks, *Proceedings of the ACM/Springer Mobile Networks and Applications*, Vilamoura, Portugal, 2009.
17. Network Simulator. http://www.isi.edu/nsnam/ns/index.html (accessed November 5, 2013).
18. I. Downard, Simulating sensor networks in ns-2, Naval Research Laboratory Formal Report 5522, Washington, DC, April 2004.

19. Mannasim Simulator. http://www.mannasim.dcc.ufmg.br (accessed November 5, 2013).
20. P. Levis, N. Lee, M. Welsh, and D. Culler, TOSSIM: Accurate and scalable simulation of entire TinyOS applications, *Proceedings of the First ACM Conference on Embedded Networked Sensor Systems*, Los Angeles, CA, 2003.
21. UC Berkeley TOSSIM. www.cs.berkeley.edu/~pal/research/tossim.html (accessed November 5, 2013).
22. L. F. Perrone and D. M. Nicol, A scalable simulator for Tinyos applications, *Proceedings of the 2002 Winter Simulation Conference*, San Diego, CA, 2002.
23. E. Perla, A. Ó. Catháin, R. S. Carbajo, M. Huggard, and C. M. Goldrick, PowerTOSSIM z: Realistic energy modelling for wireless sensor network environments, *Proceedings of the Third ACM Workshop on Performance Monitoring and Measurement of Heterogeneous Wireless and Wired Networks*, Vancouver, British Columbia, Canada, 2008.
24. J. Polley, D. Blazakis, J. McGee, D. Rusk, J. S. Baras, and M. Karir, ATEMU: A fine-grained sensor network simulator, *Proceedings of the First Annual IEEE Communications Society Conference on Sensor and Ad Hoc Communications and Networks*, Santa Clara, CA, 2004, pp. 145–152.
25. F. Osterlind, A. Dunkels, J. Eriksson, N. Finne, and T. Voigt, Cross-level sensor network simulation with COOJA, *Proceedings of the 31st IEEE Conference on Local Computer Networks*, Tampa, FL, November 2006, pp. 641–648.
26. X. Zeng, R. Bagrodia, and M. Gerla, GloMoSim: A library for parallel simulation of large-scale wireless networks, *Proceedings of the 12th Workshop on Parallel and Distributed Simulation (PADS)*, Banff, Alberta, Canada, 1998.
27. QualNet Simulator by Scalable Network Technologies. http://www.scalable-networks.com/products/qualnet/ (accessed November 5, 2013).
28. G. Chen, J. Branch, M. J. Pflug, L. Zhu, and B. Szymanski, SENSE: A sensor network simulator, in B. K. Szymanski and B. Yener, eds., *Advances in Pervasive Computing and Networking*, Springer, New York, 2004, pp. 249–267.
29. P. Baldwin, S. Kohli, E. A. Lee, X. Liu, and Y. Zhao, VisualSense: Visual modeling for wireless and sensor network systems, Technical Memorandum UCB/ERL M05/25, University of California, Oakland, CA, July 2005. http://ptolemy.eecs.berkeley.edu/papers/05/visualsense/.
30. AlgoSenSim. http://tcs.unige.ch/doku.php/algosensim (accessed November 5, 2013).
31. Georgia Tech Network Simulator (GTNetS). http://www.ece.gatech.edu/research/labs/MANIACS/GTNetS/ (accessed November 8, 2013).
32. OMNeT++ Simulation System. http://www.omnetpp.org/ (accessed November 8, 2013).
33. A. Varga, The OMNeT++ discrete event simulation system, *Proceedings of the European Simulation Multiconference (ESM'2001)*, Prague, Czech Republic, June 2001.
34. D. Pediaditakis, S. H. Mohajerani, and A. Boulis, Poster abstract: Castalia: The difference of accurate simulation in WSN, *Proceedings of the Fourth European conference on Wireless Sensor Networks (EWSN)*, Delft, the Netherlands January 2007. http://castalia.npc.nicta.com.au/index.php.
35. A. Sobeih, M. Viswanathan, D. Marinov, and J. C. Hou, J-Sim: An integrated environment for simulation and model checking of network protocols, *Proceedings of the IEEE International Parallel and Distributed Processing Symposium*, Long Beach, CA, 2007, pp. 1–6. http://www.j-sim.zcu.cz/.
36. JiST/SWANS. http://jist.ece.cornell.edu/ (accessed November 8, 2013).
37. JiST/SWANS++. http://www.aqualab.cs.northwestern.edu/projects/swans++/ (accessed November 8, 2013).
38. B. L. Titzer, D. K. Lee, and J. Palsberg, Avrora: Scalable sensor network simulation with precise timing, *Proceedings of the Fourth International Symposium on Information Processing in Sensor Networks*, Los Angeles, CA, April 2005.
39. T. W. Carley, Sidh: A wireless sensor network simulator, ISR Technical Report, Department of Electrical and Computer Engineering, University of Maryland, College Park, MD, 2004. http://drum.lib.umd.edu/bitstream/1903/6565/1/TR_2005-88.pdf (accessed November 10, 2013).
40. Prowler, Institute for Software Integrated Systems, Vanderbilt University, Nashville, TN, 2002. http://www.isisvanderbilt.edu/Projects/nest/prowler (accessed November 10, 2013).
41. (J) Prowler, Institute for Software Integrated Systems, Vanderbilt University, Nashville, TN, 2011. http://w3.isis.vanderbilt.edu/Projects/nest/jprowler/index.html (accessed November 10, 2013).
42. A. Cerpa and D. Braginsky. LecsSim-Wireless Network Simulator. 2002. http://sourceforge.net/projects/lecssim (accessed November 10, 2013).
43. OPNET Modeller. OPNET Technologies Inc., Bethesda, MD. http://www.opnet.com (accessed November 10, 2013).

44. S. Sundresh, W. Kim, and G. Agha, SENS: A sensor, environment and network simulator, *Proceedings of the 37th Annual Symposium on Simulation*, Arlington, VA, 2004, pp. 221.

45. L. Girod, N. Ramanathan, J. Elson, T. Stathopoulos, M. Lukac, and D. Estrin, Emstar: A software environment for developing and deploying heterogeneous sensor-actuator networks, *ACM Transactions on Sensor Networks*, 3 (3), 13, August 2007.

46. L. Girod, T. Stathopoulos, N. Ramanathan, J. Elson, D. Estrin, E. Osterweil, and T. Schoellhammer, A system for simulation, emulation and development of heterogeneous sensor networks, *Proceedings of the Second ACM Conference on Embedded Networked Sensor Systems (SenSys'04)*, Baltimore, MD, November 2004.

47. M. Varshney, D. Xu, M. Srivastava, and R. Bagrodia, SenQ: A scalable simulation and emulation environment for sensor networks, *Proceedings of the Sixth International Conference on Information Processing in Sensor Networks*, Cambridge, MA, April 2007.

48. B. C. Mochocki and G. R. Madey, H-MAS: A heterogeneous, mobile, ad-hoc sensor-network simulation environment, *Proceedings of the Seventh Annual SWARM Users/Researchers Conference*, Notre Dame, IN, April 2003.

49. S. Park, A. Savvides, and M. B. Srivastava, SensorSim: A simulation framework for sensor networks, *Proceedings of the Third ACM International Workshop on Modeling, Analysis and Simulation of Wireless and Mobile Systems*, Boston, MA, August 2000, pp. 104–111.

50. S. P. Fekete, A. Kroller, S. Fischer, D. Pfisterer, and C. Braunschweig, Shawn: The fast, highly customizable sensor network simulator, *Proceedings of the Fourth International Conference on Networked Sensing Systems*, Brainschweig, Germany, 2007.

51. P. M. Wightman and M. A. Labrador, Atarraya: A simulation tool to teach and research topology control algorithms for wireless sensor networks, *Proceedings of the SIMUTools*, Rome, Italy, 2009.

52. SSFNet. http://www.ssfnet.org/ (accessed December 1, 2013).

53. WiSeNet: A Software Simulator for Wireless Sensor Network Applications. https://groups.google.com/d/forum/wisenet-simulator (accessed December 1, 2013).

54. Y. Wen, S. Gurun, N. Chohan, R. Wolski, and C. Krintz, SimGate: Full-system, cycle-close simulation of the stargate sensor network intermediate node, *Proceedings of the International Conference on Architectures, Modeling and Simulation*, Samos, Greece, July 2006.

55. X. Chao-Nong, Z. Lei, X. Yong-Jun, and L. Xiao-Wei, Simsync: A time synchronization simulator for sensor networks, *Acta Automatica Sinica*, 32 (6), 1008–1014, November 2006.

56. SNetSim. Naval Science and Engineering Institute, Istanbul, Turkey. http://www.dho.edu.tr/enstitunet/snetsim/index.htm (accessed December 1, 2013).

57. S. Yi, H. Min, Y. Cho, and J. Hong, SensorMaker: A wireless sensor network simulator for scalable and fine-grained instrumentation, in *Computational Science and Its Applications*, O. Gervasi, B. Murgante, A. Laganà, D. Taniar, and Y. Mun (eds.), Springer, Berlin, Germany, 2008.

58. F. Gómez and G. Martínez, TRMSim-WSN: Trust and reputation models simulator for wireless sensor networks, *Proceedings of the IEEE International Conference on Communications*, Dresden Germany, June 2009.

59. D. Weber, J. Glaser, and S. Mahlknecht, Discrete event simulation framework for power aware wireless sensor networks, *Proceedings of the Fifth IEEE International Conference on Industrial Informatics*, Vienna, Austria, July 2007, Vol. 1, pp. 335–340.

60. J. Barbancho, F. J. Molina, C. León, J. Ropero, and A. Barbancho, OLIMPO: An ad-hoc wireless sensor network simulator for optimal SCADA-applications, *Proceedings of the Communication Systems and Networks*, Marbella, Spain, 2004.

61. Y. Wen, R. Wolski, and G. Moore, DiSenS: Scalable distributed sensor network simulation, UCSB Computer Science Technical Report, University of California, Oakland, CA, 2005.

62. H. B. Lim, B. Wang, C. Fu, Phull, and A. D. Ma, WISDOM: Simulation framework for middleware services in wireless sensor networks, *Proceeding of the Fifth IEEE Consumer Communications and Networking Conference*, Las Vegas, NV, 2008.

63. J. N. Al-Karaki and G. A. Al-Mashaqbeh, SENSORIA: A new simulation platform for wireless sensor networks, *Proceedings of the International Conference on Sensor Technologies and Applications*, Valencia, Spain, 2007, pp. 424–429.

64. K. Sha, J. Du, and W. Shi, Capricorn: A large scale wireless sensor network simulator, Technical Report MIST-TR-2004- 001, Wayne State University, Detroit, MI, January 2004.

65. O. C. Ghica, G. Trajcevski, P. Scheuermann, Z. Bischof, and N. Valtchanov, SIDnet-SWANS: A simulator and integrated development platform for sensor networks applications, *Proceedings of the Sixth ACM Conference on Embedded Network Sensor Systems*, Raleigh, NC, 2008.

66. Stargate Simulator (starsim) and Mote Simulator (motesim). http://www.cs.ucsb.edu/~wenye/ (accessed December 1, 2013).

67. L. Chuan-Wen, G. Yu, L. Fang-Fang, X. Jia, and Y. Ge, SNSim: Study and implementation of a wireless sensor network simulator, *Journal of Chinese Computer Systems*, 31 (6), 1025–1029, 2010.

68. S. N. Sinha, Z. Chaczko, and R. Klempous, SNIPER: A wireless sensor network simulator, *Proceedings of the EUROCAST*, Canary Islands, Spain, 2009, pp. 913–920.

69. C. Kelly, V. Ekanayake, and R. Manohar, SNAP: A sensor network asynchronous processor, *Proceedings of the Ninth ACM International Symposium on Asynchronous Circuits and Systems*, Washington, DC, 2003.

70. K. Mueller. SimPy Documentation. http://simpy.sourceforge.net/discuss.htm (accessed December 1, 2013).

71. D. Watson and M. Nesterenko, Mule: Hybrid simulator for testing and debugging wireless sensor networks, *Proceedings of the Second International Workshop on Sensor and Actor Network Protocols and Applications*, Boston, MA, 2004, pp. 67–71.

72. A. Boulis, A. Fehnker, M. Fruth, and A. McIver, CaVi—Simulation and model checking for wireless sensor networks, *Proceedings of the Quantitative Evaluation of Systems (QEST)*, St Malo, France, 2008.

73. P. Baldwin, S. Kohli, E. A. Lee, X. Liu, and Y. Zhao, Modeling of sensor nets in Ptolemy II, *Proceedings of the Information Processing in Sensor Networks (IPSN)*, Berkeley, CA, 2004, pp. 359–368.

74. Maple Simulator. http://www.maplesoft.com/ (accessed December 1, 2013).

75. WISENES Simulator. http://www.tkt.cs.tut.fi/research/daci/daci_wsn_wisenes.html (accessed December 1, 2013).

76. A. Fraboulet, G. Chelius, and E. Fleury, Worldsens: Development and prototyping tools for application specific wireless sensors networks, *Proceedings of the Sixth International Conference on Information Processing in Sensor Networks*, Cambridge, MA, April 2007.

77. A. Suri, Simulation study for wireless sensor networks and load sharing routing protocol to increase network life and connectivity, Master's Thesis, Department of Computer Science, Louisiana State University, Baton Rouge, LA, December 2005.

78. M. Karagiannis, I. Chatzigiannakis, and J. Rolim, WSNGE: A platform for simulating complex wireless sensor networks supporting rich network visualization and online interactivity, *Proceedings of the Spring Simulation Multiconference (SpringSim'09)*, San Diego, CA, 2009.

79. F. Menichelli and M. Olivieri, TikTak: A scalable simulator of wireless sensor networks including hardware/software interaction, *Wireless Sensor Network*, 2 (11), 815–822, November 2010.

80. P. Downey, Boris: An extensible java simulator for wireless sensor networks, January 2004.

81. M. Safaei and A. S. H. Ismail, SmartSim: Graphical sensor network simulation based on TinyOS and Tossim, *2012 Third International Conference on Intelligent Systems Modelling and Simulation (ISMS)*, Kingston, Ontario, Canada, February, 2012, pp. 611–615. doi:http://doi.ieeecomputersociety.org/10.1109/ISMS.2012.19.

82. S. Sarkar, L. Stankovic, and I. Andonovic, Protocol design for farm animal monitoring using simulation, *Ad-hoc, Mobile, and Wireless Networks*, Springer-Verlag, Berlin, Germany, 2012, Vol. 7363/2012, pp. 126–138. doi:10.1007/978-3-642-31638-8_10.

83. A. Nasir and B.-H. Soong, EnergySim—A novel, fast, extensible wireless sensor network MAC protocol simulator for evaluating energy efficiency, *Proceeding of Eighth International Conference on Information, Communications and Signal Processing (ICICS)*, Singapore, December 2011, pp. 1–5. doi:10.1109/ICICS.2011.6173526.

84. A. Derhab, F. Ounini, and B. Remli, MOB-TOSSIM: An extension framework for TOSSIM simulator to support mobility in wireless sensor and actuator networks, *IEEE Eighth International Conference on Distributed Computing in Sensor Systems (DCOSS)*, Hangzhou, China, May 2012, pp. 300–305. doi:10.1109/DCOSS.2012.34.

85. O. Landsiedel and K. Wehrle, AEON: Accurate prediction of power consumption in sensor networks, *Proceedings of the 2nd IEEE workshop on Embedded Networked Sensors*, April 30–May 1, 2005, pp. 37–44.

86. P. Svenda, The link key security in wireless sensor networks, PhD Thesis, Faculty of Informatics—Masaryk University, Brno, Czech Republic, September 2008.

87. A. Naguib, Wireless sensor network localization simulator v1.1, Systems and Computers Engineering Department, Al-Azhar University, Cairo, Egypt, 2011.

88. P. Harvey and J. Sventek, Wireless sensor network simulation with Xen, *Proceedings of the 46th Annual Simulation Symposium (ANSS 13)*, 2013, Article No. 4.

89. S. Dhurandher, S. Misra, M. Obaidat, and S. Khairwal, UWSim: A simulator for underwater sensor networks, *Simulation*, 84 (7), 327–338, 2008.

90. NAB (Network in A Box). http://nab.epfl.ch/ (accessed December 1, 2013), 2004.
91. S. Mishra, S. Mishra, and A. Kayal, and S. R. Chudi, Simulation in wireless sensor networks, *International Journal of Electronics Communication and Computer Technology (IJECCT)*, 2 (4), 176–182, July 2012.
92. H. Wu, Q. Luo, P. Zheng, and L. M. Ni, VMNet: Realistic emulation of wireless sensor networks, *IEEE Transaction on Parallel Distributed System*, 18 (2), 277–288, February 2007.
93. VMNET: A WSN Emulator, 2004. http://www.cse.ust.hk/vmnet/.

19 Implementation and Performance Analysis of EM-Based Underwater Sensor Networks in QualNet

Jun Li, Mathieu Déziel, Mylène Toulgoat, and Simon Perras

CONTENTS

ABSTRACT

Wireless communications and networking based on electromagnetic (EM) wave propagation is considered as an alternative to acoustic communications in seawater. Compared to acoustic waves, EM communications for underwater sensing applications are less susceptible to multipath distortion and environmental noise. Key factors affecting the performance of EM-based underwater networks include EM channel characteristics and protocols in the upper layers (e.g., the media access control [MAC] layer). In this chapter, EM-based underwater surveillance networks, with relatively low data generation rates, were developed, and a simulation analysis was carried out in QualNet. Since there was no existing QualNet implementation of the physical layer for underwater EM communications, it was implemented in the simulation platform. The EM-based physical model included an EM channel communication model, an environmental noise model for seawater, and a bit-error rate (BER) computation method. Based on the physical layer implementation of EM radios, simulation analyses were carried out to determine the performance of MAC protocols for an EM-based underwater surveillance network. The MAC schemes selected for this simulation study included ALOHA, multiple access with collision avoidance (MACA), carrier sense multiple access without acknowledgement (CSMAWithoutACK), and carrier sense multiple access with acknowledgement (CSMAWithACK). Simulation results showed that CSMAWithoutACK and CSMAWithACK outperforms ALOHA and MACA in terms of the packet average delay, the packet delivery ratio, and the scheme overhead. In addition, CSMAWithoutACK achieved lower packet delay and scheme overhead values compared to CSMAWithACK. Thus, CSMAWithoutACK is the most appropriate MAC scheme for EM-based underwater surveillance networks.

19.1 INTRODUCTION

Autonomous sensors in the form of networked sensor arrays will play an important role in improved surveillance of coastal waters due to their self-organizing capability. Acoustic waves have been traditionally used for underwater communications [6,12], but they are prone to multipath distortion and environmental noise, which limits their use in shallow waters where most of the underwater sensor networks are deployed. Recently, electromagnetic (EM) waves have been considered to replace acoustic communications in underwater sensor networks since EM waves are less susceptible to multipath distortion and environmental noise.

Research on EM-based underwater communications showed that EM channel properties in seawater are very different from these in the air. For instance, compared with terrestrial wireless communications, underwater EM channels experience slower propagation speed, lower transmission rates, and more severe signal attenuation. Studies showed that EM channels in seawater can achieve improved throughput performance and higher data rates than acoustic channels [4,7,17]. Propagation properties of EM waves in seawater were investigated at carrier frequencies ranging from kilohertz to gigahertz [2,3,14,15,19]. Simulation analysis on underwater multihop sensor networks using EM channels was carried out in Ref. [16].

The performance of an EM-based underwater sensor network can be significantly affected by the channel properties, the physical layer (PHY), and the media access control (MAC) layer protocols [8]. The PHY and propagation model are used to characterize the properties of EM channels in seawater and to quantify the variations of the received signal power, which in turn determines the signal-to-noise ratio (SNR). The receiver's SNR allows to compute the bit error rate (BER) for a given signal modulation scheme and determine the limit of reliable information rate of the underwater EM channel. The MAC protocols regulate the access of multiple nodes to a shared medium (e.g., an EM channel) during the process of packet transmission. In the past two decades, a wide range of MAC schemes have been proposed for terrestrial networks. However, an existing MAC scheme may not perform the same in seawater as in air, given that EM signal propagation in seawater is very different. For instance, since the transmission delay of a packet in EM-based underwater networks is in seconds, which is thousands of times that of a terrestrial wireless network, a MAC protocol with request-to-send (RTS)/ clear-to-send (CTS) handshake mechanism, which is suitable for a terrestrial network, may not be appropriate for an EM-based underwater sensor network. In addition, in contrast to studies that evaluated the performance of MAC protocols for acoustic underwater networks (e.g., [11,18] and references wherein), little work has been reported in the literature on simulation implementation and analysis of EM-based underwater sensor networks.

This chapter presents simulation development, implementation, and performance evaluation of an EM-based underwater sensor network. The simulator platform chosen for this work is QualNet, a discrete-event simulator platform widely used in the area of wireless communications and networking [13]. The development and implementation for this study includes an EM channel communication model, an environmental noise model, a BER computation, and a surveillance traffic model. Figure 19.1 shows a general view of the bottom layers (i.e., MAC layer, PHY, and channel model)

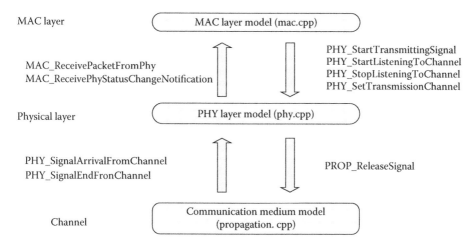

FIGURE 19.1 MAC/PHY/channel model stack.

of the QualNet protocol stack along with respective application programming interfaces (APIs) for each layer. The communication medium model quantizes the characteristics (e.g., attenuation, delay, fading, and shadowing of signals) of the physical medium. The PHY model conducts signal transmission and reception and reflects the effects of all aspects of the transceiver including modulation, medium and antenna parameters, noise, and interference. The MAC model regulates the access of sensor nodes (SNs) to the medium (i.e., an EM channel) for packet transmission.

Based on the simulation implementations, the MAC layer performance is evaluated for EM-based underwater sensor networks. The simulation results and performance are compared for four MAC protocols: ALOHA, multiple access with collision avoidance (MACA), carrier sense multiple access (CSMA) without acknowledgment (CSMAWithoutACK), and CSMA with acknowledgment (CSMAWithACK). The results show that CSMAWithoutACK and CSMAWithACK outperform ALOHA and MACA in terms of the packet average delay, the packet delivery ratio, and the scheme overhead. In addition, CSMAWithoutACK achieves smaller packet delay and better scheme overhead results compared to CSMAWithACK for EM-based underwater surveillance networks with low generation rates of sensed data. Thus, CSMAWithoutACK is the most appropriate MAC scheme for EM-based underwater surveillance networks. Nevertheless, contributions for this work were twofold: the development and implementation of the EM channel model and the PHY for a realistic EM-based underwater sensor network and the performance evaluation and comparisons of various MAC protocols, in order to determine the most appropriate MAC protocol.

The rest of this chapter is organized as follows: development and implementation of the PHY for networking EM-based underwater sensors is detailed in Section 19.2; configurations of an EM-based underwater sensor network are elaborated in Section 19.3; simulation results on the MAC protocols are presented and discussed in Section 19.4; and concluding remarks are given in Section 19.5.

19.2 IMPLEMENTATION OF EM CHANNEL MODEL AND PHYSICAL LAYER

In this section, the implementation of channel model and the PHY model in QualNet is discussed. It includes the creation of a model library, an EM channel communication model, an environmental noise model, BER computation, and an EM-based physical model. Based on the implementation the EM radio PHY models, simulation studies on EM-based underwater sensor networks are carried out in this work.

19.2.1 CREATING UNDERWATER NETWORK MODEL LIBRARY

To implement the EM radio PHY models in QualNet, a model library is created, which contains source code files to be developed for EM-based underwater sensor networks. The model library is named "UWEMMAC" located in the "qualnet/addons" directory, where "qualnet/" represents the directory where all QualNet files are located. The procedure of creating "UWEMMAC" is described as follows:

1. Create a directory "UWEMMAC", and in the directory, create a subdirectory "src" for source code files to be programmed.
2. In the directory "UWEMMAC", create two files, "Makefile-windows" and "Makefile-common", which are used when the Microsoft Visual C++ compiler links and compiles the source files. The file "Makefile-windows" contains the following lines.

```
# Define Underwater EM Networks library models.
include../addons/UWEMMAC/Makefile-common
ADDON_OPTIONS = $(ADDON_OPTIONS) $(UWEMMAC_OPTIONS)
ADDON_SRCS = $(ADDON_SRCS) $(UWEMMAC_SRCS)
ADDON_INCLUDES = $(ADDON_INCLUDES) $(UWEMMAC_INCLUDES)
The file "Makefile-common" contains the following lines:
UWEMMAC_OPTIONS = -DADDON_UWEM
UWEMMAC_DIR =../addons/UWEMMAC/src
# common sources
UWEMMAC_SRCS = \
$(UWEMMAC_DIR)/app_uwtm.cpp \
$(UWEMMAC_DIR)/phy_UWEM.cpp \
$(UWEMMAC_DIR)/prop_UWEM.cpp

UWEMMAC_INCLUDES = \
-I$(UWEMMAC_DIR)
```

3. Add the following lines in the file "qualnet/main/Makefile-addons-windows" to include the model library "UWEMMAC" when compiling source code files.

```
# INSERT ADDONS HERE
include../addons/UWEMMAC/Makefile-windows
```

19.2.2 EM Radio Communication Model

An EM communication model mimics how EM signals propagate between nodes by taking into account propagation delays and pass losses. Since there are negligible effects of signal fading and shadowing on underwater EM signals, only a path loss model "UWEM" is implemented in the EM communication model. A path loss model refers to signal attenuation due to effects such as absorption, reflection, and refraction on signals in transit from a transmitter to a receiver. Implementation of the "UWEM" model is based on the following result [5]:

$$\text{pathloss (dB)} = 10 \log_{10}\left(\left(\frac{4\pi d}{\lambda\kappa}\right)^2 \exp\left(\frac{2d}{\delta}\right)\right) = 20 \log_{10}\left(\frac{4\pi d}{\kappa\lambda}\right) + \frac{8.686d}{\delta}, \qquad (19.1)$$

where
λ is the wavelength of EM signals
d is the distance
δ is the skin depth of seawater
κ is the compensation factor of the signal wavelength

In addition, $\lambda = v/f$, where v and f are the speed (in meters per second) and the frequency (in Hz), respectively, of EM signals. For calculating δ, the following approximate expression is used:

$$\delta = \sqrt{\frac{1}{\pi\sigma\mu f}}, \tag{19.2}$$

where σ and μ are the conductivity and the permeability, respectively, of seawater. In this work, $\sigma = 4$, which roughly corresponds to the Atlantic Ocean, and $\mu = 4\pi \times 10^{-7}$, which is the same as the permeability in the air. The propagation speed v of EM signals is given by

$$v = \sqrt{\frac{4\pi f}{\mu\sigma}}. \tag{19.3}$$

The wavelength compensation factor κ is given by

$$\kappa = \sqrt{\frac{1}{4\delta^2 d^4} + \frac{1}{2\delta^3 d^3} + \frac{1}{2\delta^4 d^2} + \frac{1}{\delta^5 d}}. \tag{19.4}$$

The header and source code files implementing the underwater EM channel path loss model "UWEM" are "prop_UWEM.h" and "prop_UWEM.cpp", respectively. These are located in the directory "qualnet/addons/UWEMMAC/src/". In "prop_UWEM.h", it is assumed that two functions are declared: UWEMInitialize as an initialization function for the pass loss model and PathlossUWEM being a path loss calculation function for the pass loss model. In the following are the procedures for implementation of the "UWEM" model in QualNet:

1. Add the following statement in the source file "qualnet/libraries/wireless/src/propagation. cpp" in order for generic functions to call corresponding functions of the underwater EM channel path loss model.

```
#ifdef ADDON_UWEM
#include ``prop_UWEM.h''
#endif
```

2. Add the "UWEM" model to the enumeration "PathlossModel" defined in "qualnet/include/ propagation.h" to include the underwater EM path loss model in the list of path loss models of QualNet.

```
enum PathlossModel {
  FREE_SPACE = 0,
  TWO_RAY,
  ...,
  UWEM,    //underwater EM path loss model
  FLAT_BINNING
};
```

3. Add the following code in the initialization function "PROP_GlobalInit" in the code file "qualnet/libraries/wireless/src/propagation.cpp" in order for the function to call "UWEMInitialize" of the "UWEM" model for initialization.

```
void PROP_GlobalInit(PartitionData *partitionData, NodeInput *nodeInput) {
  ...
  for (i = 0; i < numChannels; i++) {
    ...
```

```
        //Set pathlossModel
        IO_ReadStringInstance(ANY_NODEID, ANY_ADDRESS, nodeInput,
          ``PROPAGATION-PATHLOSS-MODEL", channelIndex, TRUE, &wasFound, buf);
        if (wasFound) {
            if (strcmp(buf, "FREE-SPACE") = = 0) {
                propProfile->pathlossModel = FREE_SPACE;
            }
            ...
            else if (strcmp(buf,``UWEM'') = =0) {
                propProfile->pathlossModel = UWEM;
#ifdef ADDON_UWEM
                UWEMInitialize(&(propChannel[channelIndex]),
                channelIndex,nodeInput);
#else
                ERROR_ReportMissingLibrary(buf, ``UWEM'');
#endif   //ADDON_UWEM
            }//if
            ...
        }//if
        ...
    }//for
    ...
}
```

4. Add the following code block in the function "PROP_CalculatePathloss" in the source file "qualnet/libraries/wireless/src/propagation.cpp" in order for the function to call "PathlossUWEM" for path loss calculation.

```
void PROP_CalculatePathloss(Node* node, NodeId txNodeId,...){
    ...
    switch (propProfile->pathlossModel) {
        ...
        case ITM: {
            ...
            return;
        }
        case UWEM: {
#ifdef ADDON_UWEM
            *pathloss_dB = PathlossUWEM(pathProfile->distance, frequency,
            wavelength);
#endif//ADDON_UWEM
            return;
        }//UWEM
        ...
    }//switch
    ...
}
```

5. Modify the description file, "channel_properties.prt" in the directory "qualnet/gui/settings/ protocol models/", as follows to integrate the new path loss model into QualNet Architect.

```
<?xml version = "1.0" encoding = "ISO-8859-1"?>
<category name = "NODE CONFIGURATION">
  <subcategory name = "Channel Properties">
    <variable name = "Number of Channels" key =...>
        ...
```

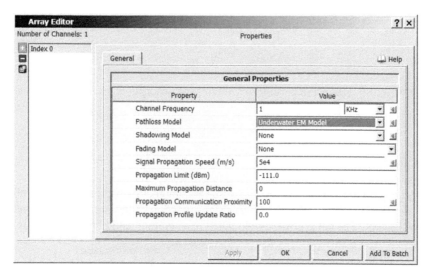

FIGURE 19.2 Configuration GUI for underwater EM channels.

```
<variable name = "Pathloss Model" key =...>
  <option value = "TWO-RAY" name =.../>
  <option value = "FREE-SPACE" name =.../>
  <option value = "UWEM" name = "Underwater EM Model" addon =
  "wireless"/>
  ...
```

After recompiling all source files, the QualNet configuration graphical user interface (GUI) for the underwater EM channel properties is shown in Figure 19.2.

19.2.3 Environmental Noise

Due to limited studies in the literature on the noise in the seawater, the atmospheric noise model [9] is used for modeling the noise power in the seawater. That is, the noise power is given by

$$P_N(\text{mW}) = 1000(F_s + 1)k_B T_0 B_W, \tag{19.5}$$

where
F_s is the noise factor at the carrier frequency in the seawater
$k_B = 1.38 \times 10^{-23}$ (J/K) is the Boltzmann constant
$T_0 = 290$ (K) is the reference temperature
B_W is the signal bandwidth in hertz

The noise factor, F_s, is the product of the noise factor in the air F_a, the refraction coefficient at the sea surface A_{ref}, and the absorption coefficient in the seawater A_{abs}. That is,

$$F_s = F_a A_{ref} A_{abs}. \tag{19.6}$$

From Ref. [5], for a given frequency f in Hertz,

$$10 \log_{10}(F_a) = \begin{cases} 260 - 28 \log_{10}(f), & \text{if } f \in (0, 100]; \\ 313 - 54 \log_{10}(f), & \text{if } f \in (100, 3{,}000]; \\ 125, & \text{if } f \in (3{,}000, 5{,}000]; \\ 305 - 116 \log_{10}(f), & \text{if } f \in (5{,}000, 10{,}000]; \end{cases} \tag{19.7}$$

The refraction coefficient is given by

$$A_{ref} = 4\sqrt{\frac{\pi\varepsilon_0 f}{\sigma}}, \tag{19.8}$$

where $\varepsilon_0 = 8.85 \times 10^{12}$ F/m is the permittivity of air. The absorption coefficient is given by

$$A_{abs} = \exp\left(-\frac{2D}{\delta}\right), \tag{19.9}$$

where D is the depth (in meters) below the sea surface for which the underwater communication nodes are located.

The implementation of the underwater noise power computation is provided in the following text and can be found in the "PHY_CreateAPhyForMac" function of the source file "qualnet/libraries/wireless/src/phy.cpp". The noise power is computed in a communication node when it is initialized.

```
void PHY_CreateAPhyForMac(Node *node,...., int* phyNumber) {
  char buf[10*MAX_STRING_LENGTH];
  ...
  //Get the noise factor
  IO_ReadDouble(node, node->nodeId, interfaceIndex, nodeInput, "PHY-
NOISE-FACTOR", &wasFound, &noiseFactor);
  if (wasFound = = FALSE) {
     noiseFactor = PHY_DEFAULT_NOISE_FACTOR;
  }
  else {
     assert(wasFound = = TRUE);
  }
//Noise factor is recalculated if PHY_UWEM is used
#ifdef ADDON_UWEM
  if (phyModel = = PHY_UWEM) {
     PropProfile* propProfile;
     double depthBelowSeasurface;
     double frequency_hz;
     double extenalFactor_db;
     double extenalFactor;
     double absorptionFactor;
     double refrectionFactor;
     //get the depth below the sea surface
     IO_ReadDouble(node, node->nodeId, interfaceIndex, nodeInput,
       "PHY-DEPTH-BELOW-SEASURFACE", &wasFound, &depthBelowSeasurface);
     if (wasFound = = FALSE) {
       depthBelowSeasurface = PHY_UWEM_DEFAULT_DEPTH_BELOW_SEASURFACE;
     }
     else {
       assert(wasFound = = TRUE);
     }
     //get the frequency in Hz
     propProfile = node->partitionData->propChannel[0].profile;
     frequency_hz = propProfile->frequency;
     assert(frequency_hz > 0);
     assert(frequency_hz < = 10000.0);

     if (frequency_hz < = 100.0) {
       extenalFactor_db = 260.0 - 28.0 * log10(frequency_hz);
     }
```

```
        else if (frequency_hz>100.0 && frequency_hz< = 3000.0) {
          extenalFactor_db = 313.0 - 54.0 * log10(frequency_hz);
        }
        else if (frequency_hz>3000.0 && frequency_hz< = 5000.0) {
          extenalFactor_db = 125.0;
        }
        else if (frequency_hz>5000.0 && frequency_hz< = 10000.0) {
          extenalFactor_db = (-1)*305.0+116.0*log10(frequency_hz);
        }
        extenalFactor = pow(10.0, (extenalFactor_db/10.));

        //absorption (A.12) in report
        double sigma = 4.0;
        double muAir = 4.0 * PI * pow(10.0, -7);
        double muSea = muAir;
        double delta = 1.0/sqrt(PI * muSea * sigma * frequency_hz);
        double expon = (-1) * 2.0 * depthBelowSeasurface/delta;
        absorptionFactor = exp(expon);

        //refrection (A.19) in report
        double epsilonAir = 8.854 * pow(10.0, -12);
        refrectionFactor = 4.0*sqrt(PI*epsilonAir*frequency_hz/sigma);

        //overall
        noiseFactor = extenalFactor*absorptionFactor*refrectionFactor;
        noiseFactor = noiseFactor+1;
    }
#endif//ADDON_UWEM
    //
    //Calculate thermal noise
    //
    noise_mW_hz = BOLTZMANN_CONSTANT*temperature*noiseFactor*1000.0;
    thisPhy->noise_mW_hz = noise_mW_hz;
    thisPhy->noiseFactor = noiseFactor;
    //
    //Set PHY-RX-MODEL
    ...
}
```

19.2.4   BER-BASED RECEPTION

A packet reception model specifies how packets are received by the PHY model of a receiver. The BER-based reception model is used in this work. Once a node is in range, the BER-based reception model calculates the signal to interference plus noise ratio (SINR) and determines an associated BER by looking up a predefined BER table. The obtained BER is used in the function "PhyAbstractCheckRxPacketError()" to determine if a packet is in error.

19.2.5   EM RADIO PHYSICAL LAYER MODEL

The source files implementing the EM radio PHY are "phy_UWEM.h" and "phy_UWEM.cpp" located in the directory "qualnet/addons/UWEMMAC/". Two data structures (i.e., "PhyUWEMStats" and "PhyDataUWEM") and the prototypes for interface functions implemented in "phy_UWEM.cpp" are defined in "phy_UWEM.h". "PhyDataUWEM" is the data structure of the underwater PHY model, and "PhyUWEMStats" is a collection of the statistics for the underwater physical model. The following is the procedure to embed the EM radio PHY model into QualNet.

1. Add the following statement in the file "qualnet/libraries/wireless/src/phy.cpp" in order for functions in "phy.cpp" to call the initialization, event handler, and finalization functions of the underwater PHY model.

```
#ifdef ADDON_UWEM
#include ``phy_UWEM.h''
#include ``prop_UWEM.h''
#endif//ADDON_UWEM
```

2. Add the EM radio PHY model name (i.e., "PHY_UWEM") to the enumeration "PhyModel" defined in "qualnet/include/phy.h" as follows to include the implemented underwater PHY model in the list of PHY models.

```
enum PhyModel {
  PHY802_11a,
  ...
  PHY_NONE,
  PHY_UWEM
};
```

3. Modify the function "AddNodeToSubnet" in "qualnet/main/mac.cpp" as follows so that the PHY initialization function "PHY_CreateAPhyForMac" is called and "phyModel" is set to "PHY_UWEM" when "PHY_UWEM" is specified as the PHY model.

```
static void//inline//AddNodeToSubnet(Node *node,..., short subnetIndex)
{
  ...
  if (strncmp(macProtocolName, "FCSC-", 5) = = 0) {
    ...
  }
  else {
    ...
    if (strncmp(phyModelName, "FCSC-", 5) = = 0) {
      ...
    }
    else if (strcmp(phyModelName, "PHY802.11a") = = 0) {
      ...
    }
    else if (strcmp(phyModelName, "PHY_UWEM") = = 0) {
      int phyNum(-1);
      PHY_CreateAPhyForMac(node, nodeInput, interfaceIndex,
      &address, PHY_UWEM, &phyNum);
      node->macData[interfaceIndex]->setPhyNum(phyNum, 0);
      phyModel = PHY_UWEM;
    }
    ...
  }
}
```

4. Modify the function "PHY_CreateAPhyForMac" in "qualnet/libraries/wireless/src/phy.cpp" as follows so that when the function is called, the channel and PHY parameters specified in the configuration file are read and stored, the used reception model ("BER_BASED"

in this work) and receiver parameters (e.g., BER tables for "BER_BASED" reception model) are set, and the initialization function (i.e., "PhyUWEMInit") is executed.

```
void PHY_CreateAPhyForMac(Node *node,...., int* phyNumber)
{
  char buf[10*MAX_STRING_LENGTH];
  ...
  thisPhy->phyModel = phyModel;
  assert(phyModel = = PHY802_11a ||
         ...
         phyModel = = PHY_LTE ||
         phyModel = = PHY_UWEM);
  ...
  //Set PHY-RX-MODEL
  //No change on this part due to chosen BER-BASED model.
  ...
  //PHY model initialization
  switch(thisPhy->phyModel) {
#ifdef WIRELESS_LIB
  case PHY802_11b:
  case PHY802_11a: {
        ...
  }
#endif//WIRELESS_LIB
#ifdef ADDON_UWEM
  case PHY_UWEM: {
        PhyUWEMInit(node, phyIndex, nodeInput);
        break;
  }
#endif//ADDON_UWEM
  ...
}
```

5. Modify the **PHY** event dispatch function "PHY_ProcessEvent" in source file "qualnet/libraries/wireless/src/phy.cpp" as follows to trigger a timer event indicating the end of transmission of a packet in a node (i.e., "MSG_PHY_TransmissionEnd") (handled by the function "PhyUWEMTransmissionEnd" in "qualnet/addons/UWEMMAC/phy_UWEM.h").

```
void PHY_ProcessEvent(Node *node, Message *msg) {
  int phyIndex = MESSAGE_GetInstanceId(msg);
  ...
  switch(node->phyData[phyIndex]->phyModel) {
#ifdef WIRELESS_LIB
    case PHY802_11b:
    ...
#endif//WIRELESS_LIB
#ifdef ADDON_UWEM
    case PHY_UWEM: {
      switch (msg->eventType) {
        case MSG_PHY_TransmissionEnd: {
          PhyUWEMTransmissionEnd(node, phyIndex);
          MESSAGE_Free(node, msg);
          break;
        }
```

```
              case MSG_PHY_CollisionWindowEnd: {
                PhyUWEMCollisionWindowEnd(node, phyIndex);
                MESSAGE_Free(node, msg);
                break;
              }
              default: abort();
            }
          break;
        }
  #endif//ADDON_UWEM
        ...
      }//END switch
  }
```

6. Modify the generic functions in "qualnet/libraries/wireless/src/phy.cpp" so that appropriate functions defined in "qualnet/addons/UWEMMAC/phy_UWEM.h" are called when "PHY_UWEM" is chosen as the PHY model running at the interface.

7. Modify the function "PHY_Finalize" in the source file "qualnet/libraries/wireless/src/phy.cpp" as follows in order for the finalization function "PhyUWEMFinalize" to be called at the end of simulation.

```
void PHY_Finalize(Node *node) {
  int phyNum;
  ...
  for (phyNum = 0; (phyNum < node->numberPhys); phyNum++)
  {
      ...
      switch(node->phyData[phyNum]->phyModel) {
#ifdef WIRELESS_LIB
         case PHY802_11b:
             ...
#endif//WIRELESS_LIB
#ifdef ADDON_UWEM
         case PHY_UWEM: {
            PhyUWEMFinalize(node, phyNum);
            break;
         }
#endif//ADDON_UWEM
      ...
      }
  }
  ...
}
```

8. Modify the PHY description file, "phy_layer.prt" in the directory "qualnet/gui/settings/protocol_models/", to integrate the EM radio PHY model into the QualNet GUI.

```
<?xml version = "1.0" encoding = "ISO-8859-1"?>
<category name = "NODE CONFIGURATION">
  <subcategory name = "Physical Layer" class =...>
    ...
    <variable name = "Radio Type" key = "PHY-MODEL" type =...>
      <option value = "None" name = "None"/>
      <option value = "PHY802.11a" name =...>
```

```
...
</option>
<option value = "PHY_UWEM" name = "UWEM" addon = "wireless">
  <variable key = "PHY-UWEM-DATA-RATE" type = "Fixed
  multiplier"
  name = "Data Rate" default = "128 bps" help = "bandwidth in bps"
  unit = "bps" maxunit = "Gbps" minunit = "bps"/>

  <variable name = "Frequency Bandwidth" key = "PHY-UWEM-
  BANDWIDTH"
  type = "Fixed multiplier" default = "128 Hz" help = "channel
  frequency bandwidth in Hz" unit = "Hz"/>

  <variable key = "PHY-DEPTH-BELOW-SEASURFACE" type = "Fixed"
  name = "Depth Below Seasurface (m)" default = "50" help =
  "The depth of
  seawater where UWEM nodes are deployed (in meters)"/>

  <variable key = "PHY-UWEM-TX-POWER" type = "Fixed" name =
  "Transmission
  Power (dBm)" default = "15.0" help = "transmission power in
  dBm"/>

  <variable key = "PHY-UWEM-RX-SENSITIVITY" type = "Fixed"
  name = "Reception
  Sensitivity (dBm)" default = "-91.0" help = "sensitivity in
  dBm"/>

  <variable key = "PHY-UWEM-RX-THRESHOLD" type = "Fixed" name
  = "Reception
  Threshold (dBm)" default = "-81.0" help = "Minimum power for
  received packet(in dBm)"/>

  <variable key = "PHY-RX-MODEL" type = "Selection" name =
  "Packet
  Reception Model" default = "SNR-THRESHOLD-BASED">

    <option value = "SNR-THRESHOLD-BASED" name = "SNR-based
    Reception Model"
    help = "If the Signal to Noise Ratio (SNR) is more than
    the SNR Threshold (in dB), it
    receives the signal without error. Otherwise the packet is
    dropped."><!- visibilityrequires =
    "[MAC-PROTOCOL] ! = 'MAC-CES-WNW-MDL'">-  >
    <variable key = "PHY-RX-SNR-THRESHOLD" type = "Fixed"
        name = "SNR Threshold" default = "10.0"/>
    </option>

    <option value = "BER-BASED" name = "BER-based Reception
    Model" help = "If specified, the Simulator will look up
    Bit Error Rate (BER) in the SNR - BER table
    specified by BER-TABLE-FILE." >
    <!- visibilityrequires = "[MAC-PROTOCOL] ! =
    'MAC-CES-WNW-MDL'">-  >
        <variable name = "Number of BER Tables" key =
        "DUMMY-NUM-PHY-RX-BER-TABLE-ABSTRACT" default = "1"
        type = "Array">
          <variable key = "PHY-RX-BER-TABLE-FILE" type =
          "File" name = "BER Table" default = "[Required]"
```

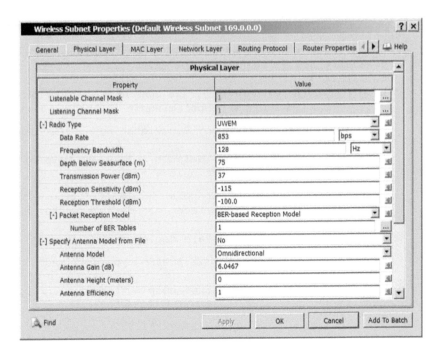

FIGURE 19.3   Configuration GUI for the PHY model.

```
         filetype = "ber" />
       </variable>
   </option>
 </variable>
</option>
<option value = "PHY802.11b" name =...>
...
```

As discussed in Ref. [5], the antenna device in each communication node consists of two cylinders of radius of a meters and height of h meters separated with a distance of l meters. The antenna gain is given as follows:

$$G = \frac{\ell^2 h}{2(\ln(\ell - a) - \ln a)}.$$ (19.10)

For $l = 5$ m, $h = 2.0$ m, and $a = 0.01$ m, the antenna gain $G = 4.0241$ or 6.0467 dB.

The QualNet configuration GUI for the PHY properties of underwater EM sensor networks is shown in Figure 19.3.

19.3   CONFIGURATIONS OF EM-BASED UNDERWATER NETWORKS

The simulation settings are elaborated in this section. They include an underwater sensor network model, an underwater surveillance traffic model at the application layer, MAC layer configuration, and PHY configuration, for simulation analysis of the MAC layer performance.

19.3.1   NETWORK MODEL

A fixed number of communication units are placed on the seabed of a choke point at a depth of 75 m to form an underwater surveillance network. For the purpose of this work, a communication unit can be an SN, which integrates the communication unit with a sensing unit (generating surveillance information,

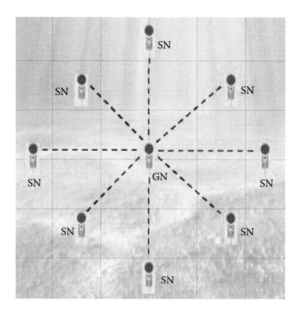

FIGURE 19.4 Network model.

such as location, speed, size, and depth of a target), or a gateway node (GN), which is capable of forwarding surveillance data to a surface station through another communication media. This work considers a single-hop network, star architecture, as shown in Figure 19.4. In the network, eight SNs are within the transmission range of a unique GN, and all surveillance data packets generated by the SNs are destined to the GN. Each SN generates 24-byte data packets according to a Poisson process with rate ν packets per minute. The packet generation processes in two different SNs are assumed to be independent. After adding an 8-byte MAC header and an 8-byte PHY header, each data packet is assumed to have a fixed packet size of 40 bytes when it is ready for transmission via an EM channel, while each MAC control packet (i.e., RTS, CTS, or acknowledgment [ACK]) is assumed to have a fixed packet size of 24 bytes.

19.3.2 Underwater Surveillance Traffic Model

The source files of the underwater surveillance traffic model include "app_uwtm.h" and "app_uwtm.cpp" located in the directory "qualnet/addons/UWEMMAC/". The protocol configuration format is given as:

```
UWTM <src> <dest> <items_to_send> <item_size> <mean_interval> <start_
time> <end_time>
```

where <src> is the client node's identifier or IP address, <dest> is the server node's identifier or IP address, <items_to_send> is the number of items to send, <item_size> is the size of each item, <mean_interval> is the mean time between transmissions of successive items, <start_time> is the transmission start time, and <end_time> is the transmission end time. For example, the configuration line "UWTM 10 1 0 24 10M 100S 3600M" specifies that node 10 sends 24-byte items to node 1, starting at 100 simulation seconds, sending an item after each exponential period of mean 10 min, and ending at 3600 simulation minutes. The following are steps to develop the underwater traffic model in QualNet:

1. Add the model names (i.e., APP_UWTM_CLIENT and APP_UWTM_SERVER) to the enumeration "AppType" defined in "qualnet/include/application.h" to include the underwater surveillance traffic model in the list of application layer protocols.

```
enum AppType {
  APP_EFTP_SERVER_DATA = 20,
  ...
```

```
APP_LINK_16_CBR_CLIENT,
APP_UWTM_CLIENT,
APP_UWTM_SERVER,
...
APP_PLACEHOLDER
};
```

2. Add the model (i.e., TRACE_UWTM) to the enumeration "TraceProtocolType" defined in "qualnet/include/trace.h" to provide detailed traces of packets as they traverse the protocol stack at each node.

```
enum TraceProtocolType {
  TRACE_UNDEFINED = 0,
  ...
  TRACE_UWTM,
  //Must be last one!!!
  TRACE_ANY_PROTOCOL
};
```

3. Include the underwater surveillance traffic model in the source file "qualnet/main/application.cpp" in order for generic functions to call corresponding functions of the model.

```
#include ``app_uwtm.h''
```

4. Modify the function "APP_InitializeApplications" in the file "qualnet/main/application. cpp" as follows to read the configuration parameters from the input file and to call the initialization functions for the client and server when the underwater surveillance data traffic model is initialized.

```
void APP_InitializeApplications(Node *firstNode, const NodeInput
*nodeInput){
  NodeInput appInput;
  ...
  for (i = 0; i < appInput.numLines; i++){
    ...
    else if (strcmp(appStr, ``CBR'') = = 0){
      ...
    }
    else if (strcmp(appStr, ``UWTM'') = = 0){
      ...
      //split input string to multiple words
      numValues = sscanf(appInput.inputStrings[i],
      ``%*s%s%s%d%d%s%s%s%s%s'', sourceString, destString,
      &itemsToSend, &itemSize, intervalStr, startTimeStr,
      endTimeStr, optionToken1, optionToken2);
      ...
      //get source and destination id and address
      IO_AppParseSourceAndDestStrings(firstNode, appInput.
      inputStrings[i], sourceString,
        &sourceNodeId, &sourceAddr, destString, &destNodeId,
        &destAddr);
      //set source node pointer
      node = MAPPING_GetNodePtrFromHash(nodeHash, sourceNodeId);
      ...
```

```
                    //call UWTM client initialization function
                    AppUwtmClientInit(node, sourceAddr, destAddr, itemsToSend,
                    itemSize, interval, startTime, endTime, tos);
                    ...
                    //Handle Loopback Address
                    if (node = = NULL || !APP_SuccessfullyHandledLoopback(
                        node, appInput.inputStrings[i], destAddr, destNodeId,
                        sourceAddr, sourceNodeId)){
                            node = MAPPING_GetNodePtrFromHash(nodeHash, destNodeId);
                    }
                    if (node ! = NULL){
                        AppUwtmServerInit(node);
                    }
                }//end UWTM
                ...
            }
            ...
    }
```

5. Modify the function "APP_ProcessEvent" in the file "qualnet/main/application.cpp" to
   implement the application layer event dispatcher that informs the appropriate application
   protocol of received events.

```
void APP_ProcessEvent(Node *node, Message *msg){
    ...
    switch(protocolType){
        ...
        case APP_CBR_SERVER: {
            AppLayerCbrServer(node, msg);
            break;
        }
        case APP_UWTM_CLIENT: {
            AppLayerUwtmClient(node, msg);
            break;
        }
        case APP_UWTM_SERVER: {
            AppLayerUwtmServer(node, msg);
            break;
        }
        ...
    }
}
```

6. Modify the function "APP_Finalize" in the file "qualnet/main/application.cpp" to print the
   protocol statistics at the end of simulation.

```
void APP_Finalize(Node *node){
    ...
    AppInfo *appList = NULL;
    AppInfo *nextApp = NULL;
    ...
    for (appList = node->appData.appPtr;appList! = NULL;appList =
    nextApp){
        switch (appList->appType) {
            ...
```

```
        case APP_CBR_SERVER: {
          AppCbrServerFinalize(node, appList);
          break;
          }
        case APP_UWTM_CLIENT: {
          AppUwtmClientFinalize(node, appList);
          break;
          }
        case APP_UWTM_SERVER: {
          AppUwtmServerFinalize(node, appList);
          break;
          }
        ...
        }//end switch
        nextApp = appList->appNext;
      }//end for
      ...
    }
```

7. Create a new component file "uwtm.cmp" containing the following contents and add it in the directory "qualnet/gui/settings/components/".

```
<?xml version = "1.0" encoding = "ISO-8859-1"?>
<root version = "1.0">
<category name = "UWTM Properties" singlehost = "false" loopback =
"enabled" propertytype = "UWTM">
  <variable name = "Source" key = "SOURCE" type = "SelectionDynamic"
  keyvisible = "false" optional = "false"/>

  <variable name = "Destination" key = "DESTINATION" type =
  "Selection Dynamic" keyvisible = "false" optional = "false"/>

  <variable name = "Items to Send" key = "ITEM-TO-SEND" type = "Integer"
  default = "100" min = "0" keyvisible = "false" help = "Number of
  items to send" optional = "false"/>

  <variable key = "ITEM-SIZE" type = "Integer" name = "Item Size
  (bytes)"
  default = "512" min = "24" max = "65023" keyvisible = "false" help =
  "Item size in bytes" optional = "false"/>

  <variable name = "Mean Interval" key = "INTERVAL" type = "Time"
  default = "1S" keyvisible = "false" optional = "false"/>

  <variable name = "Start Time" key = "START-TIME" type = "Time"
  default = "1S" keyvisible = "false" optional = "false"/>

  <variable name = "End Time" key = "END-TIME" type = "Time" default =
  "25S" keyvisible = "false" optional = "false"/>

  <variable name = "Priority" key = "PRIORITY" type = "Selection"
  default = "PRECEDENCE" keyvisible = "false">
     <option value = "PRECEDENCE" name = "Precedence" help = "value
     (0-7) of the Precedence bits in the IP header">
        <variable name = "Precedence Value" key = "PRECEDENCE-BITS"
        type = "Integer" default = "0" min = "0" max = "7" keyvisible =
        "false" optional = "false"/>
```

```
        </option>
    </variable>
  </category>
</root>
```

8. Create an icon image named "uwtm.png" and put the image file in the directory "qualnet/gui/icons/3Dvisualizer/icons".

9. Modify the file "Standard.xml" in "C:/Users/< username >/.qualnetUserDir/qualnet_5_2/Toolsets/" to display the new traffic model button in the standard toolset of QualNet Architect panel.

```
<root>
  ...
  <category name = ``Applications''>
    <subcategory categorytype = "Applications" tooltip =.../>

    <subcategory categorytype = "Applications" tooltip = "Underwater
    Traffic Model"
    icon = "uwtm.png" type = "App" name = "UWTM" propertytype = "UWTM"/>

    <subcategory categorytype = "Applications" tooltip =.../>
    ...
</category>
```

Following the aforementioned steps, the QualNet standard toolset panel including the underwater surveillance traffic model is shown in Figure 19.5.

19.3.3 MAC Layer Configuration

The MAC schemes selected in this study are ALOHA, MACA, CSMAWithoutACK, and CSMAWithACK. These schemes were selected to evaluate the MAC layer performance because

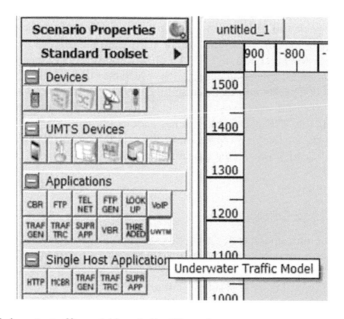

FIGURE 19.5 Underwater traffic model icon in QualNet toolset.

they are fundamental building blocks that have been broadly used to build complete MAC proto-cols in the existing terrestrial wireless networks. For instance, the MAC protocol used in the IEEE 802.11 family of standards is a combination of CSMAWithACK and MACA. The following is an overview of these four MAC schemes to be configured in the MAC layer:

1. ALOHA: In the ALOHA MAC scheme, a terminal transmits a packet and waits for an ACK packet [1]. If the ACK packet is received and another packet is available, the terminal transmits the next packet. On the other hand, if the acknowledgment is not received, the terminal retransmits the original packet after a random time delay. The packet will be retransmitted up to a predefined maximum number of times before it is discarded.

2. MACA: MACA uses control packets to announce data transmissions [10]. After a trans-mission and the transmission deferral time expires, the transmitter sends an RTS control packet to indicate that it has data to send. The receiver sends back a CTS control packet to indicate that the transmission can begin. If the CTS packet is not received, the transmit-ter defers the transmission to a later time. Upon receiving the CTS packet, the transmitter sends the packet and waits for an ACK packet from the receiver. If the ACK does not arrive in a specified amount of time, the packet is retransmitted. Both CTS and RTS packets contain the information about the length of the transmission, so that other transmitters can calculate the end time of the transmission. With the timing information, the nodes can set up a *virtual* carrier sense. A transmitter that is not currently transmitting uses the calcu-lated end time of the transmission (when the virtually sensed carrier ends) to add a random deferral time for its next transmission. The transmitter with the shortest deferral time will send the RTS packet first, restarting the process.

3. CSMAWithoutACK and CSMAWithACK: In the CSMA scheme, a transmitter needs to sense the carrier to see if there are other transmissions taking place. If the channel is clear, the transmitter sends the packet. If the transmitter detects an ongoing transmission on the channel, it yields until the channel becomes free. When the channel becomes free, the transmitter defers its transmission for a random amount of time before transmitting the packet. In the CSMAWithoutACK scheme, a transmitter does not wait for an ACK packet to transmit the next packet. That is, after the transmitter sends out a packet and if it has another packet to transmit, the transmitter immediately starts sensing the carrier for preparation of the next packet transmission disregarding whether or not the previous packet is successfully received. In CSMAWithACK, however, a transmitter always waits for an ACK packet after sending out a packet, which means that if the ACK packet is not received, the same packet may be retransmitted in the next transmission cycle. That is, if the receiver successfully receives the packet, it sends an ACK packet to the transmitter. After receiving the ACK packet and if it has another packet to transmit, the transmitter starts sensing the carrier for preparation of the next packet transmission. If the packet is not acknowledged, however, the transmitter retransmits the same packet starting with carrier sense.

19.3.4 PHYSICAL LAYER CONFIGURATION

Each communication unit operates at a frequency of $f = 1000$ Hz with the channel bandwidth $B_w = 200$ Hz. A quadrature amplitude modulation (QAM) scheme is used to achieve a modem date rate R between 200 and 900 bps. The transmission power P_t is fixed to 5 W (or 37 dBm), and the data reception sensitivity threshold is set to −110 dBm. The one-hop distance d is set to a fixed value 40 m, and the conductivity $\sigma = 4$ S/m, which roughly corresponds to the conductivity of the Atlantic Ocean.

After the received power using (19.1) and the noise power using (19.5) are obtained, the receiver i determines the SINR for signals transmitted from node j according to the following formula:

$$\text{SINR}_i = \frac{P_r^{(j)}(\text{mW})}{P_n(\text{mW}) + \sum_{k \in \Phi_i \setminus j} P_r^{(k)}(\text{mW})}, \tag{19.11}$$

where
  $P_r^{(x)}(\text{mW})$ is the received power in mW at node i of signals transmitted from node x
  Φ_i is the set of the communication units from which the received power at node i is greater than or equal to the data reception sensitivity threshold

Once the SINR is computed, an associated BER is determined by looking up the QAM-based BER table (i.e., "qam64.ber") given in QualNet simulation data set. Finally, the BER is used to determine whether a packet is received successfully or in error.

19.3.5 PERFORMANCE METRICS

The metrics defined in the following are used in this study to evaluate the MAC layer performance:

- Packet average delay is defined as the average duration between the time a surveillance data packet is generated at a SN and the time at which the packet reaches the application layer of the GN.
- Packet delivery ratio is defined as the ratio of the total number of surveillance data packets received at the application layer of the GN to the total number of surveillance data packets generated by the SNs.
- Scheme overhead is defined as the percentage of nonapplication layer data (including both control packets and header information in a data packet) processed (received from or sent to the PHY) at the GN.

The metrics defined earlier are key performance indicators (KPIs) for a MAC scheme. The packet average delay shows how fast a MAC scheme is transferring data to the media. The packet delivery ratio helps evaluate how much data loss occurs when using a given MAC scheme. The scheme overhead helps measure how efficiently a MAC scheme can utilize the media, given a fixed channel bandwidth. These KPIs will shed light on how MAC schemes can be affected by the unique characteristics of EM underwater channels, which consequently will help determine whether an individual MAC scheme should be deployed in a real-world underwater communication system.

19.4   SIMULATION RESULTS

This section presents simulation results with respect to two network parameters: the modem data rate R (bits per second) and the packet generation rate ν (packets per minute).

19.4.1 PACKET AVERAGE DELAY

Figure 19.6 plots the packet average delay results when R varies from 200 to 900 bps and ν is set to 0.1. Two observations are drawn from Figure 19.6. First, the packet average delay decreases with an increase of the data rate for a fixed ν. This is caused by a decreasing packet transmission delay as the

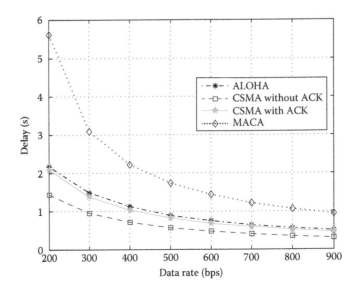

FIGURE 19.6 Packet delay vs. R ($\nu = 0.1$).

data rate increases. The second observation from Figure 19.6 is that among the four MAC schemes, CSMAWithoutACK has the smallest packet delay. CSMAWithACK performs slightly better than ALOHA in terms of the packet average delay.

The MAC scheme with the worst delay performance is MACA, whose packet average delay values are significantly larger than the other three MAC schemes due to additional transmission delay of control packets. This observation is also confirmed in Figure 19.7, which plots the packet average

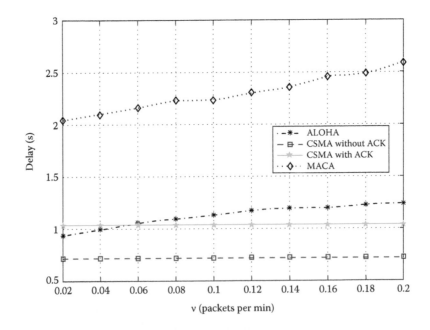

FIGURE 19.7 Packet delay vs. ν ($R = 400$ bps).

delay results when R is set to 400 bps and v varies from 0.02 to 0.2. Therefore, from the point of view of the packet average delay, CSMAWithoutACK performs the best, while MACA performs the worst, with the traffic scenario considered in the network.

19.4.2 Packet Delivery Ratio

The packet delivery ratio results are plotted in Figure 19.8 for R varying from 200 to 900 bps and v equal to 0.1. As shown, for a given packet generation rate, the packet delivery ratio for all four MAC schemes increases with the modem date rate. This is caused by the fact that for a larger data rate, the transmission delay of a packet becomes smaller, rendering a smaller packet collision probability. MACA and CSMAWithoutACK and CSMAWithACK achieve comparable performance in terms of the packet delivery ratio, which is not less than 95% for a data rate between 200 and 900 bps. Even though ALOHA outperforms other MAC schemes when the data rate becomes larger (i.e., $R \geq 600$ bps), it achieves relatively small percentage ($\leq 50\%$) of successful packet delivery when the data rate is smaller than 300 bps.

The simulation results of the packet delivery ratio are provided in Figure 19.9 where R is set to 400 bps and v varies from 0.02 to 0.2. As observed, the packet delivery ratio for all four MAC schemes decreases by increasing v. This is expected, since the probability of transmission collisions becomes larger as the intensity of surveillance data traffic increases. This results in more packets being dropped at the MAC layer when the maximum number of retransmissions is reached. The results for ALOHA substantially decrease (from above 95% to below 55%) with v, but the packet delivery ratio values for other schemes are only slightly reduced (less than 5% from the highest to the lowest value) with increasing v. For each v, CSMAWithoutACK, CSMAWithACK, and MACA achieve a comparable packet delivery ratio result being at least 95%. Since the packet delivery ratio for ALOHA is enormously affected by different values of R and v, it is not a proper option for the MAC protocol used in EM-based underwater networks. Note in Figure 19.9 that packet generation rates that could result in a more congested network are not considered in this study, because the targeted application operates within the simulated values of v.

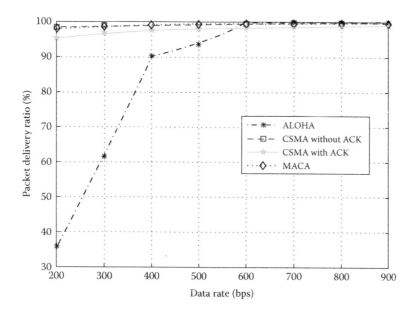

FIGURE 19.8 Packet delivery ratio vs. R ($v = 0.1$).

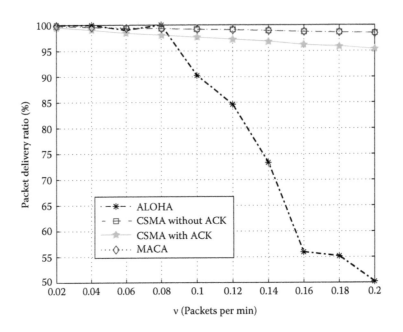

FIGURE 19.9   Packet delivery ratio vs. v ($R = 400$ bps).

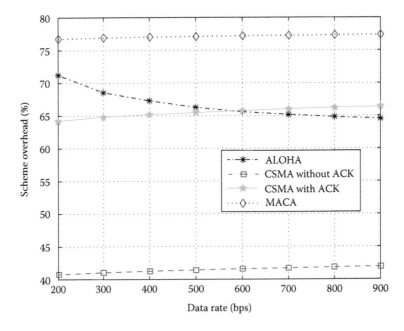

FIGURE 19.10   Scheme overhead vs. R ($v = 0.1$).

19.4.3   SCHEME OVERHEAD

Similarly, the scheme overhead results are shown in Figures 19.10 and 19.11 with respect to R and v, respectively. From the plots, it is observed that the scheme overhead has only small changes with varying values of either the modem data rate or the packet generation rate. The overhead results for both CSMAWithoutACK and CSMAWithACK slowly increase with R but slowly decrease with v. In addition, compared to other MAC schemes, CSMAWithoutACK achieves significantly reduced overhead results.

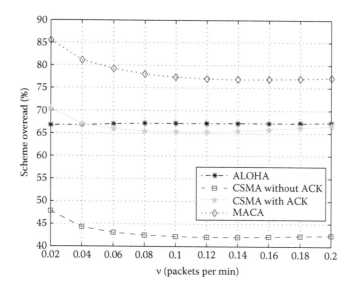

FIGURE 19.11 Scheme overhead vs. ν ($R = 400$ bps).

19.5 CONCLUDING REMARKS

The development and implementation of EM-based radio models at the PHY and an underwater surveillance traffic model at the application layer using QualNet was provided in this chapter. The performance of four MAC protocols in an EM-based underwater sensor network was evaluated. Simulation results showed that CSMAWithoutACK and CSMAWithACK outperform ALOHA and MACA in terms of the packet average delay, the packet delivery ratio, and the scheme overhead. CSMAWithoutACK achieves lower packet delay and scheme overhead values than CSMAWithACK. Thus, CSMAWithoutACK is the most appropriate MAC scheme for EM-based underwater surveillance networks with a relatively low data generation rate. It is noted that given a short transmission range of an EM-based communication unit in seawater, a multihop network is apparently more realistic than a single-hop network model used in this work. Future simulation studies on multihop EM-based underwater networks will be conducted to evaluate the network performance in a multihop network environment.

ACKNOWLEDGMENTS

This work reported herein was supported by Defence Research and Development Canada (DRDC).

REFERENCES

1. N. Abramson, The ALOHA system: Another alternative for computer communications, *Proceedings of Fall Joint Computer Conference, AFIPS Conference*, New York, November 17–19, 1970, pp. 281–285.
2. A.I. Al-Shamma'a, A. Shaw, and S. Saman, Propagation of electromagnetic waves at MHz frequencies through seawater, *IEEE Transactions on Antennas and Propagation*, 52(11), 2843–2849, November 2004.
3. U.M. Cella, R. Johnstone, and N. Shuley, Electromagnetic wave wireless communication in shallow water coastal environment: Theoretical analysis and experimental results, *Proceedings of the 4th ACM International Workshop on Underwater networks (WUWNet'09)*, New York, November 3, 2009, pp. 91–98.
4. X. Che, I. Wells, G. Dickers, P. Kear, and X. Gong, Reevaluation of RF electromagnetic communication in underwater sensor networks, *IEEE Communications Magazine*, 48(12), 143–151, December 2010.
5. P. Djukic and M. Toulgoat, Achieving high throughput and long range with electromagnetic radio in seawater, CRC Report, No. CRC-RP-2012-003, September 2012.

6. M. Erol-Kantarci, H.T. Mouftah, and S. Oktug, Localization techniques for underwater acoustic sensor networks, *IEEE Communications Magazine*, 48(12), 152–158, December 2010.

7. M.R. Frater, M.J. Ryan, and R.M. Dunbar, Electromagnetic communications within swarms of autonomous underwater vehicles, *Proceedings of the 1st ACM International Workshop on Underwater Networks (WUWNet'06)*, New York, September 2006, pp. 64–70.

8. J. Hao, B. Zhang, and H. Mouftah, Routing protocols for duty cycled wireless sensor networks: A survey, *IEEE Communications Magazine*, 50(12), 116–123, December 2012.

9. ITU-T, Recommendation P.372-10 radio noise, International Telecommunication Union Std., Rev. 10/2009, 2009, http://www.itu.int/dms_pubrec/itu-r/rec/p/R-REC-P.372-10-200910-S!!PDF-E.pdf.

10. P. Karn, MACA: A new channel access method for packet radio, *Proceedings of the 9th ARRL Computer Networking Conference*, London, Canada, September 1990, pp. 134–140.

11. B. Peleato and M. Stojanovic, Distance aware collision avoidance protocol for ad-hoc underwater acoustic sensor networks, *IEEE Communications Letters*, 11(12), 1025–1027, December 2007.

12. D. Pompili, T. Melodia, and I.F. Akyildiz, Distributed routing algorithms for underwater acoustic sensor networks, *IEEE Transactions on Wireless Communications*, 9(9), 2934–2944, September 2010.

13. Scalable Network Technologies, QualNet, http://www.scalable-networks.com/content/ products/qualnet (accessed on January 10, 2014).

14. D. Shin, D. Hwang, and D. Kim, DFR: An efficient directional flooding-based routing protocol in underwater sensor networks, *Wireless Communications and Mobile Computing*, 12(17), 1517–1527, December 2012.

15. C. Uribe and W. Grote, Radio communication model for underwater WSN, in *Proceedings of the 3rd International Conference on New Technologies, Mobility and Security (NTMS 2009)*, Cairo, Egypt, December 2009, pp. 1–5.

16. I. Wells, X. Che, P. Kear, G. Dickers, X. Gong, and M. Rhodes, Node pattern simulation of an undersea sensor network using RF electromagnetic communications, in *Proceedings of International Conference on Ultra Modern Telecommunications and Workshops (ICUMT '09)*, St.-Petersburg, Russia, October 12–14, 2009, pp. 1–4.

17. Z. Wu, J. Xu, and B. Li, A high-speed digital underwater communication solution using electric current method, in *Proceedings of the 2nd International Conference on Computer and Communication (ICFCC 2010)*, Vol. 2, Wuhan, China, May 2010, pp. 14–16.

18. J. Zheng, N. Ansari, C. Li, and B. Zhang, Underwater sensor networks: Architectures and protocols, *Wireless Communications and Mobile Computing*, 8(8), 973–975, October 2008.

19. A. Zoksimovski, C. Rappaport, D. Sexton, and M. Stojanovic, Underwater electromagnetic communications using conduction - Channel characterization, in *Proceedings of the 7th ACM International Conference on Underwater Networks and Systems (WUWNet'12)*, Los Angeles, CA, November 2012.

Section V

20 Simulation of Communications and Networking in Vehicular Ad Hoc Networks

Jorge I. Blanco and Houbing Song

CONTENTS

ABSTRACT

Vehicular ad hoc networks (VANETs), also known as connected vehicle, or the Internet of vehicles (IOV), have been identified as a key technology for improving traffic safety, mobility, and environmental protection and providing Internet access on the move to ensure wireless ubiquitous connectivity. By leveraging vehicle-to-vehicle (V2V) communications and vehicle-to-infrastructure (V2I) communications based on short- and medium-range communication like dedicated short-range communications (DSRC) or Wi-Fi as well as on long-range cellular systems, VANET has the potential to enable a wide range of applications, including safety applications, mobility applications, environmental applications, and other applications and systems involving communication to and between vehicles. Before large-scale deployment of VANET technologies, these technologies and applications must be tested in a real-world environment. Field tests are the best way to serve the needs of testing activities. However, high cost is usually required for establishing test beds and performing field tests, particularly if a large number of vehicles are involved. A low-cost alternative to field tests is simulation, which could evaluate the performance of most VANET applications under close to real-world operating conditions. In practice, field tests and simulation complement each other to complete a verification, validation, and evaluation (VV&E) process in which simulation is conducted for verification firstly and then field tests are conducted for validation. Typically a complete VANET simulation environment consists of two major components: a traffic simulator that simulates vehicle mobility in VANET and a network simulator that simulates message dissemination in VANET. This chapter is focused on the latter, that is, simulation of communications and networking in VANET. It presents the opportunities and challenges of VANET simulation and analyzes the strengths and weaknesses of various network simulators.

20.1 INTRODUCTION

We are in the era of pervasive communications and networking. One important application of wireless communications and networking could be found in the field of vehicular transportation. The application of advanced wireless technologies will bring connectivity to transportation and enable transformative change. Wireless vehicular communications and networking has been identified as a key technology for improving traffic safety, mobility, and environmental protection and providing Internet access on the move to ensure wireless ubiquitous connectivity. By leveraging vehicle-to-vehicle (V2V) communications and vehicle-to-infrastructure (V2I) communications based on short- and medium-range communication like dedicated short-range communications (DSRC) or Wi-Fi as well as on long-range cellular systems, vehicular networking has the potential to enable a wide range of applications, including safety applications, mobility applications, environmental applications, and other applications and systems involving communication to and between vehicles.

In 2004, the *First ACM International Workshop on Vehicular Ad Hoc Networks* took place in Philadelphia, for which the term *VANET* was coined by Hartenstein and Ken Laberteaux [1]. The term VANET as an acronym for vehicular ad hoc networks was originally adopted to reflect the ad hoc nature of these highly dynamic networks. However, because the term *ad hoc network* was associated widely with unicast routing-related research, the acronym VANET has been redefined to deemphasize ad hoc networking. The state-of-the-art VANET consists of three types of connectivity: V2V, vehicle to roadside (or vehicle to infrastructure, V2I), and vehicle to Internet (V2Internet) [2]. VANET is also known as connected vehicle, or the Internet of vehicles (IOV).

VANET is based on the use of DSRC technology, which is a two-way short- to medium-range wireless communications capability that permits very high data transmission critical in communications-based active safety applications. In Report and Order FCC-03-324, the Federal Communications Commission (FCC) allocated 75 MHz of spectrum in the 5.9 GHz band for use by intelligent transportation systems (ITS) vehicle safety and mobility applications [3]. DSRC is preferred over Wi-Fi because the proliferation of Wi-Fi handheld and hands-free devices that occupy the 2.4 and 5 GHz bands, along with the projected increase in Wi-Fi hot spots and wireless mesh extensions, could cause intolerable and uncontrollable levels of interference that could hamper the reliability and effectiveness of active safety applications [3].

Before large-scale deployment of VANET technologies, these technologies and applications must be tested in a real-world environment. Field tests are the best way to serve the needs of testing activities. However, high cost is usually required for establishing test beds and performing field tests, particularly if a large number of vehicles are involved. A low-cost alternative to field tests is simulation, which could evaluate the performance of most VANET applications under close to real-world operating conditions. In practice, typically two phases are needed before the deployment of a VANET application [4]. In the first phase, simulation is used to evaluate the performance of the VANET application and validate the assumptions. In the second phase, a test bed is used to conduct field tests to double check the simulation results from the first phase. In this way, with the two-phase test procedure, field tests and simulation complement each other to complete a verification, validation, and evaluation (VV&E) process in which simulation is conducted for verification firstly and then field tests are conducted for validation. Typically a complete VANET simulation environment consists of two major components: a traffic simulator that simulates vehicle mobility in VANET and a network simulator that simulates message dissemination in VANET.

The organization of this chapter is as follows: Section 20.2 presents the state of the art and practice of VANET. Section 20.3 gives an overview of communications and networking in VANET. Next, the simulation of communications and networking in VANET is presented in Section 20.4. Section 20.5 concludes this chapter with a discussion of open issues and an outlook.

20.2   STATE OF THE ART AND PRACTICE OF VANET

According to the U.S. Department of Transportation (DOT), with transportation connectivity brought by VANET, a transformed transportation system that features a connected transportation environment among vehicles of all types, the infrastructure, and portable devices that will serve the public good by leveraging technology to maximize safety, mobility, and environmental performance could be envisioned [5]:

- A system in which highway crashes and their tragic consequences are rare because vehicles of all types can sense and communicate the events and hazards happening around them.
- A fully connected, information-rich environment within which travelers, transit riders, freight managers, system operators, and other users are aware of all aspects of the system's performance.
- Travelers who have comprehensive and accurate information on travel options—transit travel times, schedules, cost, and real-time locations; driving travel times, routes, and travel costs; parking costs, availability, and ability to reserve a space; and the environmental footprint of each trip.
- System operators who have full knowledge of the status of every transportation asset.
- Vehicles of all types that can communicate with traffic signals to eliminate unnecessary stops and help people drive in a more fuel-efficient manner.
- Vehicles that can communicate the status of onboard systems and provide information that can be used by travelers and system operators to mitigate the vehicle's impact on the environment or make more informed choices about travel modes.

The applications of VANET can be classified into three categories: safety applications, mobility applications, and environmental applications.

Safety applications of VANET are designed to increase situational awareness and reduce or eliminate crashes through V2V and V2I data transmission that supports driver advisories, driver warnings, and vehicle and/or infrastructure controls. These technologies may potentially address up to 82% of crash scenarios with unimpaired drivers, preventing tens of thousands of automobile crashes every year [6]. The safety applications of VANET based on V2V communications include [7]

- Emergency brake light warning
- Forward collision warning
- Intersection movement assist
- Blind spot and lane change warning
- Do not pass warning
- Control loss warning

The safety applications of VANET based on V2I communications include [8]

- Intersection safety
- Run-off-road
- Speed management
- Commercial/transit vehicle enforcement and operations for safety

Mobility applications of VANET provide a connected, data-rich travel environment. One application is real-time data capture and management in which the VANET captures real-time data from equipment located onboard vehicles (automobiles, trucks, and buses) and within the infrastructure [9]. The other application is dynamic mobility applications in which the data are transmitted wirelessly and are used by transportation managers in a wide range of dynamic, multimodal applications to manage the transportation system for optimum performance [10].

Environmental applications of VANET both generate and capture environmentally relevant real-time transportation data and use this data to create actionable information to support and facilitate "green" transportation choices. They also assist system users and operators with "green" transportation alternatives or options, thus reducing the environmental impacts of each trip. For instance, informed travelers may decide to avoid congested routes, take alternate routes and public transit, or reschedule their trip—all of which can make their trip more fuel efficient and eco-friendly. Data generated from connected vehicle systems can also provide operators with detailed, real-time information on vehicle location, speed, and other operating conditions. This information can be used to improve system operation. Onboard equipment (OBE, or onboard unit, OBU) may also advise vehicle owners on how to optimize the vehicle's operation and maintenance for maximum fuel efficiency [6]. One application is Applications for the Environment: Real-Time Information Synthesis (AERIS), which contributes to mitigating some of the negative environmental impacts of surface transportation [11]. The other application is road weather applications for connected vehicles, which is the next generation of applications and services that assess, forecast, and address the impacts that weather has on roads, vehicles, and travelers [12].

These applications of VANET are summarized in Table 20.1.

As of September 2013, there are 400 VANET projects being conducted by academia, government, and industry all over the world [13], as summarized in Table 20.2.

In the United States, VANET is more commonly known as connected vehicle. In early January 2013, the National Science Foundation (NSF) and the Federal Highway Administration (FHWA) decided to coordinate on cyber-physical systems (CPSs) for highway transportation, and connected vehicle has been integrated into the framework of CPSs, which are engineered systems that are built from and depend upon the synergy of computational and physical components [14]. On January 23 and 24, 2014, the NSF sponsored 2014 national workshop on transportation CPS, which gathered leaders from industry, research laboratories, academic institutions, and government agencies to assess the state of transportation CPS and continue discussion on future directions for research and development (R&D) [15]. On February 4, 2014, the U.S. DOT National Highway Traffic Safety

TABLE 20.1
VANET Applications

Category	Application
Safety	V2V communications for safety
	V2I communications for safety
Mobility	Real-time data capture and management
	Dynamic mobility applications
Environmental	Applications for the AERIS
	Road weather applications for connected vehicles

Source: U.S. Department of Transportation (US DOT), Connected vehicle research, http://www.its.dot.gov/connected_vehicle/connected_vehicle. htm [last accessed on January 1, 2014].

TABLE 20.2
Geographical Summary of VANET Projects

Continent	Country	Projects	Total by Continent
Asia	China	15	85
	India	1	
	Israel	6	
	Japan	44	
	Singapore	1	
	South Korea	17	
	Turkey	1	
Europe	Austria	2	159
	Belgium	10	
	Finland	2	
	France	14	
	Germany	43	
	Greece	2	
	Italy	12	
	The Netherlands	21	
	Norway	2	
	Portugal	1	
	Romania	1	
	Spain	6	
	Sweden	15	
	United Kingdom	9	
	Europe-wide	19	
North America	Canada	5	149
	United States	144	
Oceania	Australia	6	7
	New Zealand	1	
Total		400	400

Source: Michigan Department of Transportation & Center for Automotive Research, International Survey of Best Practices in Connected and Automated Vehicle Technologies: 2013 Update, September 2013.

Administration (NHTSA) announced that it will begin taking steps to enable V2V communication technology for light vehicles [16].

To test VANET applications in a real-world environment, the U.S. DOT has established seven test beds in Michigan, Virginia, Florida, California, New York, Arizona, and Tennessee [17]. These VANET test beds provide the V2V and V2I communication system that others can utilize to test and demonstrate traveler services through applications that interface within this framework. For example, the VANET test bed includes a number of features that support in-vehicle signage (the display of messages to drivers), including [17]

- OBE that stores messages that should be displayed when a vehicle enters a geographic area and tracks the vehicle's position to display messages at appropriate locations
- Roadside equipment (RSE) that broadcasts vehicle messaging data to vehicles and OBE that receives the data and adds new messages to the list of messages that should be displayed
- Back office servers that receive requests to post in-vehicle messages from other applications and transmit those messages to the appropriate RSE

There are many ways that VANET applications could use these in-vehicle signage features to provide traveler services. For example, a traveler information application could use these features to provide congestion information to drivers. If an incident occurs, this type of application could transmit information about that incident to the VANET test bed back office servers that would then push that information to appropriate RSE and, from there, to OBE-equipped vehicles. The vehicles could then display information about the incident if the vehicle was in the vicinity of the incident [17]. Other support features provided by the VANET test bed include [17]

- Probe data services
- Signal phase and timing (SPaT) services
- V2I communication services
- V2V communication services
- Tolling transaction services
- OBE application hosting
- RSE application hosting

In addition, VANET test beds have been established in Belgium, Italy, and other countries and areas [13].

20.3 COMMUNICATIONS AND NETWORKING IN VANET

As a special case of networks, VANET follows the ISO Open Systems Interconnection (OSI) reference model [15]: physical layer, media access control (MAC) layer, network layer, transport layer, and application layer. The protocols of the physical layer and MAC layer are based on IEEE Standard 802.11p, which specifies the physical and MAC layers for wireless communications in a vehicular environment. The protocols of higher layers are based on the IEEE 1609 Family of Standards for wireless access in vehicular environments (WAVE), which defines an architecture and a complementary, standardized set of services and interfaces that collectively enable secure V2V and V2I wireless communications [18]. These standards also define how applications that utilize WAVE will function in the WAVE environment. These standards and how they function are illustrated and described in the Draft Guide for WAVE Architecture (IEEE 1609.0) [19], based on the security protocols defined in IEEE Standard 1609.2-2013, the networking service protocols defined in IEEE Standard 1609.3-2010, extensions to the physical and MAC defined in IEEE Standard 802.11-2012 to support the multichannel WAVE standards in IEEE Standard 1609.4-2010, and the use of identifier allocations in IEEE Standard 1609.12-2012. The relationship between VANET communications and networking protocol stack and OSI layers is shown in Table 20.3 [20]. The resource manager and security services do not fit easily within the layered structure of the OSI model (Table 20.4).

There are different types of wireless communication systems that can be used for information exchange in VANET. In general, VANET can make use of all kinds of wireless systems: broadcast systems, cellular systems, local wireless systems like wireless local area network (WLAN) or infrared, and near field communication (NFC) systems [4]. To establish the relationship between VANET

TABLE 20.3
IEEE Standards, VANET Protocols, and OSI Layers

IEEE Standard	VANET Protocols	Purpose of VANET Protocols	OSI Layer
1609.4	Multichannel operations	Provides enhancements to the IEEE 802.11 MAC to support multichannel WAVE operations.	2
1609.3	Networking services	Defines network and transport layer services, including addressing and routing, in support of secure WAVE data exchange. It also defines WAVE short messages, providing an efficient WAVE-specific alternative to IPv6 that can be directly supported by applications. Further, this standard defines the management information base (MIB) for the WAVE protocol stack.	2, 3, and 4
1609.2	Security services for applications and management messages	Defines secure message formats and processing. This standard also defines the circumstances for using secure message exchanges and how those messages should be processed upon receipt.	NA
1609.1	Resource manager	Describes an application that allows the interaction of on-board units (OBUs) with limited computing resources and complex processes running outside the OBUs in order to give the impression that the processes are running in the OBUs.	NA
802.11p	PHY and MAC	Specifies the PHY and MAC functions required of an IEEE 802.11 device to work in the rapidly varying vehicular environment.	1 and 2

TABLE 20.4
VANET Protocols

OSI Model	Data Plane		Management Plane	
	Resource Manager (IEEE 1609.1)			Security Services
Layer 4	UDP/TCP (IEEE 1609.3)	WSMP (IEEE 1609.3)	WME (IEEE 1609.3)	(IEEE 1609.2)
Layer 3	IPv6 (IEEE 1609.3)			
Layer 2	LLC (IEEE 1609.3)			
	Multichannel Operation (IEEE 1609.4)		MLME Extension (IEEE 1609.4)	
	WAVE MAC (802.11p)		MLME (802.11p)	
Layer 1	WAVE PHY (802.11p)		PLME (802.11p)	

Source: U.S. Department of Transportation (US DOT), The connected vehicle test bed, http://www.its.dot.gov/testbed.htm [last accessed on January 1, 2014].

applications and VANET communications and networking technologies, VANET applications can be classified into [4]

- Driving-related applications, which aim at improving traffic on the road and enhancing both the mobility and safety of moving vehicles by providing support and assistance to the driver or by improving the cooperation of the traffic participants
- Vehicle-related applications, which is related to the improved operation, management, and simplified configuration of the actual vehicle
- Passenger-related applications, which are focused on the comfort, convenience, and entertainment of the passenger

In addition to V2V and V2I, VANET communications and networking technologies can include [4] the following:

- Vehicle-to-mobile (V2M) typically operates either via a local pairing of Bluetooth or NFC, for example, or via an in-vehicle WLAN hot spot. It is worth mentioning that mobile devices comprise common consumer electronics device but also RFID-based solutions.
- Vehicle-to-central (V2Central) infrastructure cannot be clustered with other communication forms.
- V2Internet and vehicle-to-private network (V2Private) are typically realized on IP-based communication protocols, resulting in the same requirements for the communication technologies.

The relationship between VANET applications and VANET communications and networking technologies is summarized in Table 20.5.

A central challenge of VANET is that no communication coordinator can be assumed. Although some VANET applications likely will involve infrastructure (e.g., traffic signal violation warning, toll collection), several applications will be expected to function reliably using decentralized communications [21]. Therefore, neither centralized client/server systems nor infrastructure-based peer-to-peer communication can be used directly for VANET communications and networking [4].

20.4   SIMULATION OF COMMUNICATIONS AND NETWORKING IN VANET

VANET research aims to develop communication systems between vehicles for the dissemination of data or information in a timely and efficient way to provide safety and comfort to its occupants and drivers. Many network vehicular scenarios require a wide variety of communications protocols particularly in the data link and network layers.

While it is important to test and evaluate protocol implementations in a real world, the use of simulation tools is the first step to be taken in research to develop and implement networking protocols of VANET. Vehicular networking simulation is fundamentally different from the simulation of mobile ad hoc networks (MANETs) due to its particular characteristics such as restricted topologies, signal fading, high mobility, nodes with different speeds, traffic congestion, and different behaviors of drivers. For this reason and the promising future VANET, various simulation tools have been developed, including network simulators, mobility generators, and hybrid simulators.

Figure 20.1 shows a description of popular VANET simulation tools. Nowadays these tools are still being used and improved continuously. The ultimate purpose of this section is to present a proper selection of simulators for VANET, which will help researchers in the field of VANET in the selection of simulation software tools.

TABLE 20.5
Relationship between VANET Applications and VANET Communications and Networking Technologies

Applications		V2V V2I	V2M	V2Central	V2Internet V2Private
Driving related	Navigating	Cellular WLAN Auto-WLAN	Cellular WLAN Auto-WLAN NFC	Broadcast Cellular	Cellular WLAN
	Maneuvering	(Cellular) Auto-WLAN	NFC	Broadcast Cellular	Cellular WLAN
	Stabilizing	(Cellular) Auto-WLAN	NFC	Broadcast Cellular	Cellular WLAN
Vehicle related	Fleet based		WLAN NFC		Cellular WLAN
	Vehicle centered	Cellular WLAN Auto-WLAN	WLAN NFC		Cellular WLAN
Passenger related	Information		Cellular WLAN NFC	Broadcast Cellular	Cellular WLAN
	Entertainment	Cellular (WLAN)	Cellular WLAN NFC	Broadcast Cellular	Cellular WLAN
	Communication	Cellular (WLAN)	Cellular WLAN NFC		Cellular WLAN
	Comfort and convenience	Cellular (WLAN)	Cellular WLAN NFC		Cellular WLAN

Source: Kosch, T. et al., *Automotive Inter-networking*, Wiley, Hoboken, NJ, 2012.

Mobility generators	Network simulators	Hybrid simulators
SUMO	OPNET	GrooveNet
MOVE	OMNET	TraNS
STRAW	NS 2–3	MobiREAL
FreeSim	QualNet	NCTUns
VANETMobiSim	SWANS	Mobitools
CityMob	OMNEST	
NetStream		
BonnMotion		

FIGURE 20.1 Popular simulation tools for VANET.

20.4.1 NETWORK SIMULATORS

A network simulator must implement widely accepted communication standards, such as the different IEEE 802.11 specifications. IEEE 802.11, also called WLAN, specifies both the physical layer and the MAC layer of the ISO OSI reference model. By and large, a network simulator is a software tool that predicts the behavior of a network. The basic purpose of network simulation is to monitor the behavior of the network. The behavior of the network is calculated either by network entities using mathematical formulas or by capturing and playing back observations from a production network [22]. Attributes can also be modified in a controlled manner to assess how the network would behave under different conditions, typically with the support of most popular protocols and metric networks used at different layers since all layers permit the evaluation of parameters like data traffic transmission, packet level of source or destinations, reception, background load, routes, links, control channels, throughput, end-to-end delay, delivery of packets, and packet delay variation (PDV), among many other metrics and parameters. Most existing network simulators are developed for MANETs, and hence extensions to VANET are required, such as using the vehicular mobility generators before they can be used to simulate VANET. This section will present some of the most promising network tools to simulate VANET scenarios together with their principal characteristics to analyze communication between vehicles and vehicles to infrastructure.

20.4.1.1 Optimized Network Engineering Tools

Optimized Network Engineering Tools (OPNET) [23] is a simulation tool with many nice features including packet format to define and test protocols, nodes, and process models for the behavior of a network component [24]. It includes project window to design network topologies and their respective links and contains a window to capture or display the simulation results. OPNET has a development environment in which scripts based on the programming language C++ are made to modify, build, or model different components of a network like links, nodes, and antennas and to implement technologies such as Transmission Control Protocol (TCP), Internet Protocol version 6 (IPv6), MPLS, and VoIP. OPNET has several predefined modules for MANET including some protocols but has no dedicated module to VANET. For evaluating VANET network protocols, it is necessary to integrate mobility models because OPNET mobility has only random mobility and trajectory mobility. Trajectory mobility only allows to set a path node by node, which limits the development of complex and realistic scenarios by limiting the interaction between them. Some recent research works have been developed using OPNET for VANET network simulation [25–27]. OPNET is a commercial tool with a very powerful network analyzer that offers solutions in the performance management and analysis for R&D.

20.4.1.2 OMNeT++

OMNeT++ [28] is a network simulator based on discrete events and is modular, object oriented, and developed for different operating systems such as Linux, Mac OS/X, and Windows. Its main area of application is in the simulation of wire and wireless communication networks. OMNeT++ provides an architecture of modules programmed in C++ and alternative programming languages like Java and C#. Such modules are reusable because OMNeT++ supports the interaction with the user through a graphical interface, an eclipse-based IDE, and a graphical runtime environment, due to its modular architecture, and the simulation kernel can be embedded within different applications. OMNeT++ is gaining more and more popularity in the scientific community for the following capabilities:

- Modeling protocols
- Modeling wireless and wired communication networks
- Validation of hardware architecture
- Performance evaluation of software
- Modeling queuing networks

- Modeling multiprocessor and distributed systems
- Modeling discrete event systems where entities need to communicate through the exchange of messages or packages

Vehicular networks simulator (VNS) is one of the most recent frameworks released by OMNeT. VNS is a simulation framework that integrates the mobility component with network component in a transparent and efficient way, reducing the overhead of communication and synchronization between different simulators. VNS provides bidirectional interaction between a microscopic mobility model and a network simulator. OMNeT++ is a widely used platform in the global scientific community because it is free for academic purpose and nonprofit use. Some research works have been conducted to evaluate vehicular networks using OMNeT++ [29,30].

20.4.1.3 ns-2 and ns-3

ns-2 [31] and its successor ns-3 [32] are discrete event simulators. They are open source and licensed under GNU GPLv2. ns-2 is being partially maintained but not being considered for journal publications. It still has a big user base. ns-3 is still under constant development and offers some interesting characteristics for developers.

ns-2 network simulations are composed of C++ code, which is used to model the behavior of the simulation nodes, and oTcl scripts that control the simulation and specify further aspects, for instance, the network topology. This design choice was originally made to avoid unnecessary recompilations if changes are made to the simulation setup. Back in 1996 when the first version of ns-2 was released, this was a reasonable intent, as the frequent time-consuming recompilation of C++ programs slowed down the research cycle. However, from today's perspective, the design of ns-2 trades off simulation performance for the saving of recompilations, which is questionable if one is interested in conducting scalable network simulations.

A team led by Tom Henderson at the University of Washington received funding from the U.S. NSF to build a replacement for ns-2, called ns-3. In the process of developing ns-3, it was decided to completely abandon backward compatibility with ns-2.

ns-3 [33] is one of the network simulators widely used in academic and research community. The ns-3 simulation core supports research on both IP and non-IP-based networks. However, the large majority of its users focus on wireless/IP simulations, which involve models for Wi-Fi, WiMAX, or LTE for layers one and two and a variety of static or dynamic routing protocols for MANET IP-based applications.

Like its predecessor, ns-3 relies on C++ for the implementation of the simulation models. However, ns-3 no longer uses oTcl scripts. ns-3 is slated to support the integration of real implementation code by providing standard application programming interface (API), such as Berkeley sockets or POSIX threads, which are transparently mapped to the simulation.

Font et al. [34] provide some meaningful remarks about current differences between these two tools from developer's point of view. Leaving performance and resources consumption aside, technical issues might help to choose one or another, depending on simulation and project management requirements.

20.4.1.4 QualNet

QualNet [35] is a commercial version of Global Mobile Information System Simulator (GloMoSim), and it is designed to be extensible. QualNet permits the use of different protocols in a variety of standard or user-configured network components and applications running on the network. In QualNet, a specific network topology is referred to as a scenario. Each scenario allows to specify all the network components and conditions under which the network will operate. This includes terrain details; channel propagation effects including path loss, fading, and shadowing; wired and wireless subnets; and network devices such as switches, hubs, and routers. The protocol stack for wireless networks is divided into a set of layers, each with its own API. The modular implementation allows consistent comparison of multiple protocols in each layer in the ISO OSI reference model.

QualNet can model thousands of nodes by taking advantage of the latest hardware and parallel computing techniques. It can permit to model large networks with high fidelity by running multicore, on cluster, and multiprocessor systems. QualNet can support real-time speed to enable software-in-the-loop, network emulation, and human-in-the-loop modeling. Faster speed enables model developers and network designers to run multiple analyses by varying model, network, and traffic parameters in a short time.

The most significant QualNet components are as follows [36]:

- *QualNet analyzer.* A statistical graphing tool that displays hundreds of metrics collected during simulation of a network scenario. This module permits to see predesigned reports or customize graphs with their own statistics. Multiexperiment reports are also available. All statistics are exportable to spreadsheets in comma-separated values (CSV) format.
- *QualNet packet tracer.* Provides a visual representation of packet trace files generated during the simulation of a network scenario. Trace files are text files in XML format that contain information about packets as they move up and down through the protocol stack.
- *QualNet file editor.* A text editing tool that displays the contents of the selected file in text format and allows the user to edit files.
- *QualNet command line interface.* Enables a user to run QualNet from a DOS prompt (in Windows) or from a command window (in Linux or Mac OS X). When QualNet is run from the command line, input to QualNet is in the form of text files that can be created and modified using any text editor.
- *QualNet external interfaces.* QualNet can interact with a number of external tools in real time. The QualNet STK interface, which is a part of the Developer Model Library, provides a way to interface QualNet with the Satellite Tool Kit (STK) developed by Analytical Graphics, Inc. (AGI) and function in a client–server environment.

In QualNet, when simulations are running, users can watch packets at various layers flowing through the network and view dynamic graphs of critical performance metrics. Real-time statistics are also an option, where it is possible to view dynamic graphs while a network scenario simulation is running.

20.4.1.5   Scalable Wireless Ad Hoc Network Simulator

Scalable Wireless Ad Hoc Network Simulator (SWANS) is a wireless network simulator built on Java in Simulation Time (JiST). SWANS is organized as an independent software component that can be complemented to sensor or wireless networks, and it has capabilities that are similar to ns-2 and GloMoSim. SWANS [37] is also able to simulate larger network topologies by making efficient use of memory. The latest version nowadays available was released in 2007. There is no version for Windows environment. SWANS is a repository for the development of extensions for the JiST/SWANS wireless network simulation test bed. It currently includes a mobility model for VANET simulations and a visualization tool. Its programming language is in Java and has high scalability. SWANS is an open source that includes graphical user interface (GUI) and supports parallel processing but does not support IEEE 802.11p. In [38], the authors propose enhancements to AODV protocol by minimizing its control messages overhead. The simulation was done using JiST/SWANS simulator. Another interesting development made with SWANS [39] is a module called ASH (application-aware SWANS with highway mobility), which makes a significant contribution by allowing for the needed two-way communication between the mobility model and the network model because most VANET simulators do not allow for feedback between the vehicle mobility model and the network simulator. This adds to the scalable SWANS simulator allowing for realistic VANET simulations of important safety and traffic information applications.

20.4.1.6　OMNEST

OMNEST corresponds to the commercial version of OMNeT++. It is used principally by R&D researchers and engineers to probe different scenarios and design technologies like wireless/wired protocols, architectural designs, ad hoc networks, and queuing-based network.

OMNEST [40] and OMNeT++ are almost alike. Simulation models written for OMNEST are guaranteed to compile and run with OMNeT++, and vice versa. Differences are limited basically to licensing, packaging, and some subtle features. OMNEST is an object-oriented discrete event network simulator. Its framework has a generic architecture, which allows it to be applied to various problem domains where complex behavior needs to be simulated with high performance. Components regularly do not interact with each other directly on the C++ level, merely via means delivered by OMNEST, such as messages. In terms of flexibility, OMNEST simulates and models in an extended and flexible way: unusual scenarios such as interfacing with other simulators and external systems, parallel simulation, emulation, and combinations can be computed; OMNEST also offers the opportunity to study and debug source code when required.

OMNEST bases its operation on component models, which are composed of nested modules that primarily communicate by exchanging messages. Models are built from reusable components and can be combined to form more complex structures. The depth of module nesting is not limited. Modules communicate primarily by message passing, via connections, or by direct sending. Module behavior can be programmed in C++, using the simulation infrastructure that OMNEST provides.

Table 20.6 compares the major features of various VANET network simulators.

20.4.2　Integration of Traffic Simulator with Network Simulator

Analyzing VANET is a costly and time-consuming job because the deployment of various VANET applications differs in their features, design, working, and testing. Research community around the world has developed and studied many protocols and applications with various simulation software

TABLE 20.6
Principal Features in VANET Network Simulators

Key Features	OPNET	QualNet	OMNEST	ns-2	ns-3	OMNeT++	SWANS
Open-source code	Not	Not	Not	Yes	Yes	Yes	Yes
Programming language	C++	C++	C++	oTcl C++	Python C++	C++	Java
Runs natively on the principal OS	Microsoft Windows/ Linux	Microsoft Windows/ Linux	Microsoft Windows/ Linux/ Mac OS X	Linux	Linux/ Mac OS X	Microsoft Windows/ Linux/ Mac OS X	Linux
Parallel processing	Yes	Yes	Yes	Not	Yes	Yes	Yes
IEEE 802.11p support	Not	Not	Not	Not	Yes	Not	Not
Graphical user interface (GUI)	Yes	Yes	Yes	Yes	Yes	Yes	Yes
Continuous enhancements	Yes	Yes	Yes	Not	Yes	Yes	Not
Papers related with VANET in IEEE/ ACM database	5%	9%	1.7%	48%	14%	10%	9.6%
Scalability	High	High	High	Reduced	High	High	High

tools to model different conditions. VANET simulators help to perform real-world activities under varying conditions while reducing high cost and time. To develop correct simulations as realistic as possible in VANET, mobility models need to be included before any real implementations. The accuracy of the simulation results depends on the models used in the simulation and the realism of the parameters selected.

Mobility models like random way point model (RWP) or random walk model (RWM) [41] are used in MANET. These mobility models used in MANET simulation assume a topography without obstacles and nodes to be able to move freely in the whole stimulation area. These parameters cannot be used in VANET because vehicles in VANET move along defined roads with different speeds without random movement, and communications between VANET nodes do not rely on dedicated infrastructure as MANET nodes do. These are some reasons why mobility generators or traffic simulators are necessary to use because they produce better traffic traces that are close to real world and traffic simulators generate traces containing node locations and timing details that can be used as input to a network simulator.

To integrate two components, that is, network and mobility simulators, it is necessary to simulate VANET. These two simulators provide independent functionalities. On one hand, network simulators permit topology building; on the other hand, mobility simulators produce the traces that contain the file with the coordinates specifying the movements of vehicles, which can be loaded to the network simulator. In this case, the network simulator imports the trace files generated by mobility simulator, but there is no interaction in real time between these two simulators. Nowadays there exist several simulators that integrate network simulation with traffic simulation. These integrated simulators do not use trace files. In this kind of simulation, it is possible that the mobility and network models can communicate, providing feedback from the network simulator to adjust some parameters in real time like node movements in traffic congestion notification and collision avoidance systems. The majority of VANET research has been conducted in separate simulators.

20.4.3 MOBILITY MODELS

One of the most significant components for VANET simulation is a mobility model that produces right decisions from simulation experiments and thus will carry through to real deployments; in other words, the mobility model is the pattern that describes the movements of mobile nodes during a simulation time within the simulated area. A significant component of VANET simulation is the movement pattern of vehicles, more commonly called mobility model. The mobility model manages a series of rules that express the form of node movement in ad hoc networks. These mobility models determine the location of nodes in the topology at any given instant, which strongly affects metrics like network connectivity, PDV, and throughput. By using this information, network simulators create random or predefined topologies based on nodes position and perform some tasks between the nodes. Mobility model needs to take into account some constraints like streets, traffic lights, roads, buildings, pedestrians, vehicles, car movements, and driver behaviors and many more restrictions that a VANET researcher has to consider before a simulation.

There are primarily two types of mobility models that are based on the node movements; they are random node movement and real-world mobility models. The first one presents mainly the early approaches of modeling mobility. They were based on randomly moving nodes in any direction at any speed. A good classification and description can be found in [42]. Although their focus is ad hoc networks, these models clearly do not describe real car movement on highways or roads. The real-world mobility models are based on the data recorded from simulation traces or real-world interpretations by utilizing means such as Global Positioning System (GPS). These opinions include information from long periods of time involving many vehicles [43]. Furthermore, publicly available street map data can be used in modeling realistic mobility, like Google Map or Google Earth, OpenStreetMap (OSM), and the Topologically Integrated

Geographic Encoding and Referencing (TIGER) database. Vehicular traffic can be modeled on these maps to create a mobility model that can be applied in a specific geographical area or modified to suit similar regions.

More recently, there have been other mobility models, such as bidirectional coupled simulators and artificial traces of car movements. It is evident that the intrinsic complexity of such traffic models is significantly increasing. More complicated solutions based on bidirectional coupled simulators and artificial traces of car movements have been developed.

Other authors classify mobility models into microscopic, macroscopic, and mesoscopic, which are described in the following.

20.4.3.1 Mesoscopic Flow Model

This model combines elements of microscopic and macroscopic traffic flow models. Mesoscopic model may take different forms such as the modeling of the headway distribution or the size or density of a cluster of vehicles and can describe the interactions between vehicles at an individual level. A specific model may describe the velocity distribution at a specific time and space or the vehicular arrival rate. Mesoscopic flow also models the mobility at the flow level where a number of cars are characterized by certain averaging properties like average speed or time headway at a specific time and space and at the same time controls the behavior of a vehicle as a function of this information, with the characteristic that the flows are distinguishable. Initially the mesoscopic model did not attract as much attention as the macroscopic or microscopic did but recently has shown an efficient use between the modeling of large quantities of vehicles and individual vehicles as a practical advantage.

20.4.3.2 Microscopic Flow Model

The major characteristic of this VANET traffic flow is to model vehicular traffic avoiding accidents by controlling each individual vehicle to maintain a safe interdistance between cars, a safe time headway, or both. Really microscopic models simulate traffic by following each vehicle from moment to moment or executing events in time sequence. In addition, microscopic models are in charge of modeling the location, velocity, and acceleration of each vehicle that participates in the simulated scenario, and they are doubtless the most popular class of driver models because microscopic flow models, generally, represent time, position, speed, and even acceleration as continuous functions, but most have been extended to run in discrete time.

20.4.3.3 Macroscopic Flow Model

The macroscopic flow model governs vehicular traffic, like the road topology, defining speed limits, constraining cars movement, characterization and number of lanes, overtaking and safety rules for each street, or the traffic sign description establishing the intersection crossing rules. In general terms, a macroscopic flow model simulates the behavior of vehicular traffic flow such as traffic density, rather than that of each individual vehicle.

It is evident that microscopic simulation that models the behavior of single vehicles and the interactions between them is the most appropriate mobility model for simulating VANET. Microscopic simulation offers the level of details that macroscopic and mesoscopic models cannot offer without neglecting the flexibility model mesoscopic traffic flow offers.

Table 20.7 shows the most significant advantages and disadvantages related to mobility models for VANET.

20.4.4 Mobility Simulators

The traffic modeling is an area of knowledge fairly investigated by the civil engineering for the correct modeling of vehicular traffic in the design phase and the construction phase of new roads and intersections. To increase the level of reality in the simulation of VANET, the use of vehicular

TABLE 20.7

Advantages and Disadvantages of Mobility Models

	Random Movement	Real-World Traces	Artificial Mobility Traces	Bidirectional Coupled Simulators
Advantages	Easy to configure and friendly	Reusable traces, most realistic node movement	Realistic node movement, free parameterization, reusable traces	Realistic node movement, free parameterization feedback on driver behavior
Drawbacks	Do not offer greater precision	Costly and time consuming and no permit-free parameterization	Do not offer feedback on driver behavior	No reusable traces
Major support simulators	Majority of simulators have this option	OPNET, ns-2, Shawn, OMNEST, JiST/ SWANS, OMNeT++, GloMoSim, QualNet	OPNET, ns-2, ns-3, Shawn, JiST/SWANS, OMNeT++, GloMoSim, QualNet, OMNEST	In development ns-3, Shawn, JiST/ SWANS, OMNeT++, OMNEST

mobility generators through traffic models is necessary. Mobility simulators are a necessary tool to test the impact of vehicular movements even for really complex scenarios like roads and highways. Among the most relevant options offered by traffic simulators are to allow varying the number of lanes and the shape of the roads, including traffic lights and intersections. In terms of vehicular mobility, they usually implement a car-following model, as well as a multilane changing model to simulate overtaking among vehicles. At the moment, several simulation software environments exist and they are capable of generating trace files reflecting vehicles' trajectory or movements.

Simulation environments currently exist, which are capable to generate trace files that reflect the movement of vehicles. These trace files must be exported to the network simulation programs. Some of the most important mobility simulators will be briefly described as follows.

20.4.4.1 Simulation of Urban Mobility

It is a microscopic and open-source traffic generator capable to handle vehicle environments up to 10,000 streets. Among its main features are handling movements, collision-free vehicle, and different types of cars. Each vehicle has its own path and is simulated individually. One of the major drawbacks is the fact that the generated traces cannot be directly utilized by any available network simulator; for this reason, the creation of an extension or script that converts data files generated by Simulation of Urban Mobility (SUMO) to be understood and interpreted by the corresponding network simulator is needed. SUMO [44], developed and maintained by the Institute of Transportation Systems at the German Aerospace Center, runs on all major well-known operating systems, and its basic programming language is C++.

20.4.4.2 Mobility Model Generator for Vehicular Networks

Mobility model generator for vehicular networks (MOVE) can generate realistic mobility models for vehicular network simulation. MOVE works on SUMO. MOVE [45] basically consists of two components: map editor and editor of vehicle movement. The map editor is used for topology that routes can be created manually or automatically or by importing maps from a database like TIGER, OSM, or Google Earth. In turn, the vehicle editor is used to generate the movements of vehicles; the output of previous editors is a trace file that contains the information of the moving vehicles used by the network simulator like ns-2 or QualNet [46]. All settings of automobile movement are performed in a static way. This model does not consider the characteristics of microscopic mobility, which is related to model the behavior of each vehicle. MOVE provides a set of graphical interfaces that allow users to quickly generate simulation scenarios without the need to build scripts or many internal details.

20.4.4.3 Street Random Waypoint

Street random waypoint (STRAW) is a traffic modeler that basically consists of two kinds of implementations, route management and execution, one of which is a modified version of the model *random waypoint* that requires source and destination information. The model *random waypoint* determines a vehicle's path at each intersection, for example, the vehicle can make a turn at an intersection with a certain probability that can be independently assigned to each node. STRAW [47] uses information origin–destination and interarrival time to lead mobility in the network. Further, the pair of information origin–destination is selected by each car, and the paths are initially calculated by a metric or minimum cost as fastest time or shortest distance, so the model can be configured to recalculate the route of the vehicle if the cost of a route along or near its precalculated route changes significantly, enabling each node to react to changes in the traffic information.

20.4.4.4 FreeSim (STRAW)

FreeSim [48] is licensed under the GNU (General Public License), and its code is freely available. FreeSim is a macroscopic or microscopic traffic simulator that can be customized. FreeSim allows multiple systems to be easily represented by highways and loaded into the simulator as a graphical data structure with a weight at the edges of the roads determined by the current speed of the mobile nodes. The traffic and graph algorithms can be created and executed by the entire network or vehicle individually, and the traffic data used by the simulator can be generated by the user or collected in real time or provided by any entity of transportation management. Nodes in FreeSim have the ability to communicate with the monitoring highway traffic system that makes it an ideal traffic modeler for the simulation of ITS.

20.4.4.5 VanetMobiSim

VanetMobiSim [49] is an extension of CanuMobiSim that corresponds to a framework for the modeling of vehicular mobility. It is based on Java and can generate traces of movement in different formats. Meanwhile, VanetMobiSim, as an extension, focuses on vehicular mobility at the macroscopic and microscopic levels in the context macroscopic level. VanetMobiSim has the ability to import maps from American Census Bureau TIGER and provides implementations of various random mobility models as well as models of physics and vehicle dynamics. It supports multiline restriction in differentiated speeds and traffic signals and intersections avenues, at the microscopic level, and implements new mobility models for communications between vehicles and vehicle infrastructure. According to these models, the nodes can regulate its speed depending on the closeness between vehicles and act according to the traffic signals in the presence of an intersection. One feature that distinguishes VanetMobiSim is the concept of the shortest path algorithm *randomized Dijkstra* for handling multiple available routes with the same metric. *Speed-based shortest path* allows to set a threshold between segments of high-speed roads and distance from the target with the purpose to generate longer but faster route.

20.4.4.6 CityMob

CityMob attempts to resolve one of the most critical issues in simulation studies of VANET related to the mobility models in order to get reliable results. CityMob permits to create urban mobility scenarios, including the possibility of modeling vehicular accidents. The modeling was designed to integrate with the network simulator ns-2. In CityMob, three types of mobility models are implemented: (1) simple model, (2) Manhattan model, and (3) downtown model. Specifically, downtown model shows that it is a fairly realistic model for the simulation of vehicular traffic accidents.

It also shows that a moderate number of vehicles is required for flooding algorithm to be effective. Martinez et al. [50] show the performance of an analysis to obtain more detailed results by comparing the behavior of the system with each mobility model.

20.4.4.7 NetStream

NetStream (*net*work *s*imulator for *tr*affic *e*fficiency and *m*obility) has been developed to simulate traffic over wide areas to predict the effects of ITS in order to reduce traffic congestion, reduce pollution, and eventually measure the traffic density. In order to calculate the flow of traffic in large areas with high-speed vehicular movement, NetStream [51] uses a density method block that computes the movement of the nodes by means of an approximation of the traffic flow and estimates the rate of guided vehicles. NetStream uses a traffic flow model that is able to calculate the movement of each vehicle, as required in ITS. Newest version of NetStream applies predicting traffic conditions in real time as guidelines in dynamic routing.

Overall, NetStream was developed to evaluate the efficiency of ITS, on one hand, and provide traffic information and related measures in traffic restrictions, on the other hand.

20.4.4.8 BonnMotion (Mobility Scenario Generation and Analysis Tool)

BonnMotion is a Java software that creates and analyzes mobility scenarios and is most commonly used as a tool for the investigation of MANET characteristics. Scenarios can be exported to several network simulators, such as ns-2, ns-3, and GloMoSim/QualNet. BonnMotion [52] is being jointly developed by the Communication Systems group at the University of Bonn, Germany; the Toilers group at the Colorado School of Mines, Golden, CO; and the Distributed Systems group at the University of Osnabrück, Germany.

There are two possibilities to feed input parameters into the BonnMotion scenario generation. The first one is to enter the parameters on the command line, and the second one is to have a file containing the parameters. These two methods can also be combined. In this case, the command line parameters override those given in the input file. The scenario generator writes all parameters used to create a certain scenario to a file. In this way, settings are saved and particular scenario parameters can be varied without the need to reenter all other parameters. BonnMotion supports (protocol independent) two classes of metrics: pure movement metrics and link-based metrics. As pure movement metrics, velocity, relative mobility, and dwell time are supported. As link-based metrics, BonnMotion supports link duration, time to link break, node degree, partitions, and k-connectivity. For these metrics, only symmetric (bidirectional) links are considered. The statistics can be calculated over the simulation time, parts of it, as well as averaged.

The most important mobility models for VANET have been described. To summarize this work, the trend is definitely to go toward an environment with capability to create realistic mobility models with high degree of realism in the modeling of vehicular mobility.

Network and traffic simulators have been gaining much attention because they are the tools used by the researchers and other actors such as the automotive industry to find or get how these VANETs behave in different scenarios.

20.4.5 HYBRID SIMULATORS

The network and mobility models have frequently been decoupled in two separated simulation tools. In fact, high-quality simulators exist in each of these areas like those explained earlier. The simulation of VANET applications not only requires simulating aspects related to routing protocols or wireless communication between the vehicles but also requires simulating the mobility of the vehicles. A simple method to achieve this goal is to integrate a mobility model in a network simulator, but without allowing the network messages to feedback to the mobility model. This kind of simulation is generally called one way; in other words, they just take a movement trace file as input, without allowing the interaction between the network and the mobility simulators.

Another important aspect in a simulation model for an intervehicle communication system is the drivers' response to VANET applications. The reaction of drivers in different situations could affect the performance of VANET applications. There are many reasons why VANET researchers are interested in developing new simulators in which mobility and communications are coupled. With this purpose, there is a clear need for an integrated mobility and network simulator in order to evaluate effectively the performance of VANET systems. Considering the fact that hybrid simulators need higher computational resources, the distributed computing could be an alternative. The state-of-the-art hybrid or integrated simulators are presented in the following.

20.4.5.1   Traffic and Network Simulation Environment

Traffic and Network Simulation Environment (TraNS) links two open-source simulators, ns-2 and SUMO, which make it a simulator that allows to use real mobility models, in addition to its ability to influence or configure traffic behavior based on intervehicle communications. The purpose of TraNS is to avoid simulation results to differ significantly from those obtained in real environments. TraNS [53] has two modes of operation: (1) network-centric mode that is used to realistically evaluate the mobility of nodes and communication protocols for VANET, which do not influence real-time mobility of nodes such as when exchanging or distributing content like music or travel user information, and (2) application-centric mode that is used to evaluate applications for VANET, which influence the mobility of the node in real time and thus in simulation time. One example of such applications is the prevention of collisions or unexpected braking. The development of TraNS is suspended. Hence, TraNS does not support the latest version of both SUMO and ns-2. Currently, TraNS cannot provide any support. But it is possible to generate mobility traces for ns-2 using TraNSLite.

20.4.5.2   GrooveNet

GrooveNet is an integrated simulator that allows communication between a simulated vehicle and a real vehicle to model intervehicular communication with a topology map. Through its modular architecture, GrooveNet allows to model different parameters like trajectory, mobility, broadcast message delivery, and test safety applications, vehicle interactions, and new security models. GrooveNet [54] is a modular event-based simulator with well-defined model interfaces that make adding models easy. It supports multiple vehicles, trips, and mobility models over a variety of network links and physical layer models, including simple car-following, traffic lights, lane changing, and simulated GPS models. GrooveNet supports three types of simulated nodes: vehicles that are capable of multihopping data over one or more DSRC channels, fixed infrastructure nodes, and mobile gateways capable of V2V and V2I communications. Also GrooveNet supports multiple network interfaces for real V2V and V2I communications. GrooveNet is able to connect to the vehicle's onboard computer and read OBD-II diagnostic codes. Events such as sudden deceleration, braking, air bag deployment, and signals from the antilock braking system can trigger alert or warning messages. GrooveNet is built on C++ and runs on Linux environments.

20.4.5.3   NCTUns

NCTUns is a software for the modeling of microscopic traffic and network simulation that works as emulator with the ability to simulate various protocols used in both wire and wireless networks in IP networks. Regarding network devices and protocols, NCTUns supports for wireless LAN networks, wireless mesh networks, GPRS networks, quality of service (QoS) DiffServ networks, RTP/RTCP/SIP VoIP protocols, and several other wireless standards such as IEEE 802.11(e). QoS protocols have been added to NCTUns [55]. The most important feature is the fact that it can be used as an emulator, that is, an external host in the real world can exchange packets as required for configuring a TCP connection with nodes like routers, hosts, or mobile stations. This feature is very useful to test VANET under various conditions in the simulated network. Another important aspect is that NCTUns supports emulations distributed on multiple computers in a large network, for situations in which the emulated network has many nodes with real-world applications that need to connect to the network. NCTUns can partition an emulated network into smaller parts so that they can be simulated on different computer machines automatically and transparently. In addition to the preceding characteristics, NCTUns also uses the TCP protocol and Linux to achieve greater flexibility in the simulation results.

20.4.5.4 MobiREAL

MobiREAL is a network simulator for ubiquitous environments through mobile devices that can simulate real movements of people and vehicles with the ability to change their behavior depending on the context of the application with which a more detailed evaluation of performance can be achieved for VANET. MobiREAL makes modeling nodes through C++ and adopts a probabilistic rule-based model to describe the behavior of the nodes. MobiREAL [56] provides a suite of useful tools with a visualization tool called an animator. The animator can visually animate the movement of nodes, network topology, and packet propagation, and so on and can also show statistical information like node density and the packet error rate observed in each subregion. Therefore, it can easily be applied to create trace-based mobility scenarios supported by many other network simulators. This feature allows users of the other simulators to receive the benefit of realistic mobility scenarios. MobiREAL is basically a MANET simulator and provides a methodology to model and simulate more realistic movement of the nodes. Although it is focused on the applications of MANET, it can be applied to VANET.

20.4.5.5 MobiTools

MobiTools is an integrated VANET tool chain [57]. MobiTools is composed of MobiView, a QualNet mobility visualizer based on NASA World Wind; MobiMap, a web front end that graphically, via Google Maps API, supports the use of TIGER census data; MobiDense, a vehicular traffic simulator that builds on known flow intensities on streets and turns probabilities at intersections to create mobility traces; and MobiRay, a propagation simulator tool capable of predicting the effects of reflections on buildings and on the ground. The main idea behind this tool is to provide the research community with an easy-to-use, accurate mobility engine. The creation of MobiTools should benefit all those who design new protocols and applications for vehicular networks. MobiTools attempts to capture the effects that impact the performance of VANET, but this clearly leaves space to a number of improvements that may also be added. Accuracy can always be improved in describing a physical system such as VANET. Future work can span in a number of different directions. A great amount of accuracy, for example, has been reached in modeling car behavior on streets. Another possible direction of improvement is in the estimation of signal propagation effects. Other possible directions of work include exploring network census data, such as estimating the impact of a cellular or Wi-Fi infrastructure that would impact the performance of VANET.

At first, VANET was simulated by using network simulators that received as input just a mobility file with the positions and directions of the vehicles along the simulation. However, nowadays the trend is to integrate a vehicle mobility simulator into a network traffic simulator with bidirectional communication. The major challenge is how to address the interaction between them. Thus, new integrated simulators have been developed in order to meet these new requirements.

This section did not include various aspects related to radio interferences caused by both static and dynamic obstacles. This is a key requirement that is crucial to accurate design and analysis of next-generation VANET.

20.5 CONCLUSIONS

VANET is an important application of communications and networking. VANET leverages the potentially transformative capabilities of wireless communications and networking to make surface transportation safer, smarter, and greener. Various VANET applications have been developed to improve traffic safety, mobility, and environmental protection. Before large-scale deployment of VANET technologies, these technologies and applications must be tested in a real-world environment. Field tests are the best way to serve the needs of testing activities. However, high cost is usually required for establishing test beds and performing field tests, particularly if a large number of vehicles are involved. A low-cost alternative to field tests is simulation, which could evaluate the performance of most VANET applications under close to real-world operating conditions. In practice, field tests and simulation complement each other to complete a VV&E process in which simulation is conducted for verification firstly and then field tests are conducted for validation. The simulation of VANET consists of a communication network simulator and a traffic simulator. In this chapter, firstly the state of the art and practice of VANET is given; then the communications and networking of VANET is presented. The main body of this chapter is focused on the review of various communication network simulators and traffic simulators for VANET. For each simulator, its mechanism, strength, and weakness are reviewed. It is expected that simulation remains a powerful tool to evaluate the performance of a VANET application and validate the assumptions.

REFERENCES

1. H. Hartenstein and K. Laberteaux, *VANET: Vehicular Applications and Inter-Networking Technologies*, Wiley, Chichester, U.K., 2010.
2. ACM SIGMOBILE, The Tenth ACM International Workshop on VehiculAr Inter-NETworking, Systems, and Applications, http://www.uwicore.umh.es/vanet2013/ [last accessed on January 1, 2014].
3. U.S. Department of Transportation (US DOT), Overview of dedicated short range communications (DSRC) technology, http://www.its.dot.gov/DSRC/index.htm [last accessed on January 1, 2014].
4. T. Kosch, C. Schroth, M. Strassberger, and M. Bechler, *Automotive Inter-Networking*, Wiley, Hoboken, NJ, 2012.
5. U.S. Department of Transportation (US DOT), ITS Strategic Research Plan, 2010–2014, Executive Summary, http://www.its.dot.gov/strategic_plan2010_2014/index.htm [last accessed on January 1, 2014].
6. U.S. Department of Transportation (US DOT), Connected vehicle research, http://www.its.dot.gov/connected_vehicle/connected_vehicle.htm [last accessed on January 1, 2014].
7. U.S. Department of Transportation (US DOT), Connected vehicle applications: Vehicle-to-Vehicle (V2V) communications for safety, http://www.its.dot.gov/research/v2v.htm [last accessed on January 1, 2014].
8. U.S. Department of Transportation (US DOT), Connected vehicle applications: Vehicle-to-Infrastructure (V2I) communications for safety, http://www.its.dot.gov/research/v2i.htm [last accessed on January 1, 2014].
9. U.S. Department of Transportation (US DOT), Connected vehicle applications: Real-time data capture and management, http://www.its.dot.gov/data_capture/data_capture.htm [last accessed on January 1, 2014].
10. U.S. Department of Transportation (US DOT), Connected vehicle applications: Dynamic mobility applications, http://www.its.dot.gov/dma/index.htm [last accessed on January 1, 2014].
11. U.S. Department of Transportation (US DOT), Connected vehicle applications: Applications for the Environment: Real-Time Information Synthesis (AERIS), http://www.its.dot.gov/aeris/index.htm [last accessed on January 1, 2014].
12. U.S. Department of Transportation (US DOT), Connected vehicle applications: Road weather connected vehicle applications, http://www.its.dot.gov/connected_vehicle/road_weather.htm [last accessed on January 1, 2014].
13. Michigan Department of Transportation & Center for Automotive Research, International Survey of Best Practices in Connected and Automated Vehicle Technologies: 2013 Update, Ann Arbor, MI, September 2013.
14. National Science Foundation, Dear Colleague Letter: NSF-FHWA Coordination on Cyber Physical Systems for Highway Transportation, NSF 13-034, Arlington, VA, January 2, 2013.
15. National Science Foundation, *2014 National Workshop on Transportation Cyber-Physical Systems*, Arlington, VA, January 23–24, 2014.
16. National Highway Traffic Safety Administration (NHTSA), U.S. Department of Transportation Announces Decision to Move Forward with Vehicle-to-Vehicle Communication Technology for Light Vehicles, NHTSA 05-14, Washington, DC, February 3, 2014.

17. U.S. Department of Transportation (US DOT), The connected vehicle test bed, Washington, DC, http://www.its.dot.gov/testbed.htm [last accessed on January 1, 2014].

18. U.S. Department of Transportation (US DOT), Fact Sheet about IEEE 1609—Family of Standards for Wireless Access in Vehicular Environments (WAVE), http://www.standards.its.dot.gov/Factsheets/Factsheet/80 [last accessed on January 1, 2014].

19. IEEE Draft Guide for Wireless Access in Vehicular Environments (WAVE)—Architecture, IEEE P1609.0/D6.0, pp. 1–96, June 13, 2013.

20. R. Uzcategui and G. Acosta-Marum, WAVE: A tutorial, *IEEE Communications Magazine*, 47 (5), 126–133, May 2009.

21. H. Hartenstein and K.P. Laberteaux, A tutorial survey on vehicular ad hoc networks, *IEEE Communications Magazine*, 46 (6), 164–171, June 2008.

22. A. Mahmood et al., Key features and optimum performance of network simulators: A brief study, *International Journal of Computer Trends and Technology (IJCTT)*, 4 (9), 3207–3213, September 2013.

23. Riverbed Technology, Network simulation, http://www.opnet.com. [last accessed on January 1, 2014].

24. N.S. Nafi and J.Y. Khan, A VANET based intelligent road traffic signaling system, *Telecommunication Networks and Applications Conference (ATNAC)*, Brisbane, Queensland, Australia, pp. 1–6, November 7–9, 2012.

25. X. Fang, K.K. Chai, Y. Alfadhl, and Y. Sun, Evaluation of ad-hoc routing protocols in vehicular ad-hoc network using OPNET, *2011 11th International Conference on ITS Telecommunications (ITST)*, St. Petersburg, Russia, pp. 39–44, August 23–25, 2011.

26. F. Kaisser et al., A framework to simulate VANET scenarios with SUMO. Opnetwork, 2011.

27. C. Sharma and B. Tyagi, Performance evaluation of TCP variants under different node speeds using OPNET simulator, *2013 IEEE Third International Advance Computing Conference (IACC)*, Ghaziabad, India, pp. 302–307, February 22–23, 2013.

28. OMNeT++ Community, Network simulator, http://www.omnetpp.org/ [last accessed on January 1, 2014].

29. J. Zuo, Y. Wang, Y. Liu, and Y. Zhang, Performance evaluation of routing protocol in VANET with vehicle-node density, *2010 Sixth International Conference on Wireless Communications Networking and Mobile Computing (WiCOM)*, Chengdu, China, pp. 1–4, September 23–25, 2010.

30. M. Baguena, S.M. Tornell, A. Torres, C.T. Calafate, J.-C. Cano, and P. Manzoni, VACaMobil: VANET car mobility manager for OMNeT++, *2013 IEEE International Conference on Communications Workshops (ICC)*, Budapest, Hungary, pp. 1057–1061, June 9–13, 2013.

31. ns-2 wiki, Network simulator, http://nsnam.isi.edu/nsnam/index.php/User_Information [last accessed on January 1, 2014].

32. ns-3 wiki, Network simulator, http://www.nsnam.org/overview/media-kit/ [last accessed on January 1, 2014].

33. T.R. Henderson et al., Network simulations with the ns-3 simulator, SIGCOMM demonstration, 2008.

34. J.L. Font et al., Architecture, design and source code comparison of ns-2 and ns-3 network simulators, *Proceedings of the 2010 Spring Simulation Multiconference*, Orlando, FL, 2010.

35. Scalable Network Technologies, Inc., Network simulator, http://web.scalable-networks.com/content/qualnet, 2013 [last Accessed on January 1, 2014].

36. Scalable Network Technologies, Inc., QualNet 5.1 User's Guide, http://rainbow.sunmoon.ac.kr/qualnet/menuals/Documents/QualNet-5.1-UsersGuide.pdf [last accessed on January 1, 2014].

37. Swans++ wiki, http://aqualab.cs.northwestern.edu/legacy/swans++/ [last accessed on January 1, 2014].

38. M. Ayash, M. Mikki, and K. Yim, Improved AODV routing protocol to cope with high overhead in high mobility MANETs, *2012 Sixth International Conference on Innovative Mobile and Internet Services in Ubiquitous Computing (IMIS)*, Palermo, Italy, pp. 244–251, July 4–6, 2012.

39. K. Ibrahim, and M.C. Weigle, ASH: Application-aware SWANS with highway mobility, *IEEE INFOCOM Workshops 2008*, Phoenix, AZ, pp. 1–6, April 13–18, 2008.

40. Simulcraft, Inc., Network simulator, http://www.omnest.com/index.php [last accessed on January 1, 2014].

41. F. Bai and A. Helmy, A survey of mobility models. Wireless Ad Hoc Networks, University of Southern California, Los Angeles, CA, p. 206, 2004.

42. R.R. Roy, *Handbook of Mobile Ad Hoc Networks for Mobility Models*, Springer, New York, 2011.

43. C. Sommer and F. Dressler, Progressing toward realistic mobility models in VANET simulations, *Communications Magazine, IEEE*, 46 (11), 132–137, 2008.

44. J.E. Daniel Krajzewicz, M. Behrisch, and L. Bieker, Recent development and applications of SUMO—Simulation of Urban MObility. *International Journal on Advances in Systems and Measurements*, 5 (3&4), 128–138, 2012.

45. Laboratory for Experimental Network and System (LENS) at National Cheng Kung University, Rapid Generation of Realistic Simulation for VANET, http://lens.csie.ncku.edu.tw/Joomla_version/index.php/research-projects/past/18-rapid-vanet [last accessed on January 1, 2014].

46. F.K. Karnadi, Z. H. Mo, and K.-C. Lan, Rapid generation of realistic mobility models for VANET, *IEEE Wireless Communications and Networking Conference (WCNC)*, Sheraton, Hong Kong, pp. 2506–2511, March 11–15, 2007.

47. D. R. Choffnes and F. E. Bustamante, An integrated mobility and traffic model for vehicular wireless networks,. *Proceedings of the Second ACM International Workshop on Vehicular Ad Hoc Networks (VANET'05)*, New York, pp. 69–78, 2005.

48. J. Miller, and E. Horowitz, FreeSim—A free real-time freeway traffic simulator, *IEEE Intelligent Transportation Systems Conference (ITSC)*, Seattle, WA, pp. 18–23, September 30–October 3, 2007.

49. J. Harri and M. Fiore, VanetMobiSim—Vehicular Ad hoc Network mobility extension to the CanuMobiSim framework, Institut Eurécom Department of Mobile Communication, Sophia Antipolis, France, p. 6904, 2006.

50. F.J. Martinez, J.-C. Cano, C.T. Calafate, and P. Manzoni, CityMob: A mobility model pattern generator for VANETs, *IEEE International Conference on Communications (ICC) Workshops*, Beijing, China, pp. 370–374, May 19–23, 2008.

51. H. Mori, H. Kitaoka, and E. Teramoto, Traffic simulation for predicting traffic situations at Expo 2005, *R&D Review of Toyota CRDL*, 41 (4), 45–51, 2006.

52. N. Aschenbruck, R. Ernst, and P. Martini, Evaluation of wireless multi-hop networks in tactical scenarios using BonnMotion, *2010 European Wireless Conference (EW)*, Lucca, Italy, pp. 810–816, April 12–15, 2010.

53. M. Piorkowski, M. Raya, A. Lezama Lugo, P. Papadimitratos, M. Grossglauser, and J.-P. Hubaux, TraNS: Realistic joint traffic and network simulator for VANETs, *SIGMOBILE Mobile Computing and Communications Review*, 12 (1), 31–33, January 2008.

54. R. Mangharam, D. Weller, R. Rajkumar, P. Mudalige, and F. Bai, GrooveNet: A hybrid simulator for vehicle-to-vehicle networks, *Third Annual International Conference on Mobile and Ubiquitous Systems—Workshops*, San Jose, CA, pp. 1–8, July 17–21, 2006.

55. S.Y. Wang and Y.B. Lin, NCTUns network simulation and emulation for wireless resource management, *Wireless Communications and Mobile Computing*, 5 (8), 899–916, 2005.

56. T. Umedu, H. Urabe, J. Tsukamoto, K. Sato, and T. Higashinoz, A MANET protocol for information gathering from disaster victims, *Fourth Annual IEEE International Conference on Pervasive Computing and Communications Workshops 2006 (PerCom Workshops 2006)*, Pisa, Italy, pp. 5–446, March 13–17, 2006.

57. G. Marfia, P. Lutterotti, and G. Pau, MobiTools: An integrated toolchain for mobile Ad Hoc networks, University of California, Los Angeles, CA, Technical Report, 2007.

21 Performance Evaluation of Realistic Vehicular Networks
A MAC Layer Perspective

*Ali Balador, Carlos T. Calafate,
Juan-Carlos Cano, and Pietro Manzoni*

CONTENTS

ABSTRACT

Vehicular ad hoc networks (VANETs) have attractive potential in order to reduce traffic jams and avoid transportation disasters. They are also able to provide infotainment services like web browsing, e-mail, or using social networks on the road. Achieving a well-designed medium access control (MAC) protocol is a challenging issue to improve communications efficiency due to the dynamic nature of VANETs. The CSMA-based MAC protocol adopted by the IEEE 802.11p standard was selected as the best choice for the current generation of VANETs considering its availability, maturity, and cost. Despite these benefits, the common problem in all IEEE 802.11-based protocols is scalability, exhibiting performance degradation in highly variable network scenarios. This chapter provides an overview of 802.11/802.11p, as well as alternatives presented by the research community to address the detected limitations in vehicular scenarios. In addition, performance evaluation results highlight the need for contention window (CW) adjustment schemes in order to avoid performance degradation, especially in dense and highly mobile scenarios. Realistic vehicular environment clearly affects the obtained results, so we tried to choose appropriate and realistic network and traffic simulators. In this chapter, we also explain why the simulator choice is a relevant issue, and how to properly configure simulation settings for vehicular environments. Analysis and simulation results using OMNeT++ in vehicular scenarios, including highway and urban scenarios, show that alternatives to 802.11p able to perform CW adjustments are able to improve the overall performance, even for high network density scenarios.

21.1   INTRODUCTION

With advances in wireless communication technologies, VANETs have emerged as an attractive research area for both academia and industry. Vehicular networks are formed by vehicles equipped with wireless devices, called on-board units, which allow communicating with other vehicles in infrastructure-less wireless networks (vehicle-to-vehicle, V2V) or with roadside units (vehicle-to-infrastructure, V2I). Vehicular environments integrating intervehicle communications can be assumed as a special form of mobile ad hoc networks (MANETs) in which node's mobility is road-constrained [1].

The main differences between VANETs and MANETs have to do with rapid topology changes and diversity of network scenarios. Also, it must be taken into account that, in contrast to MANETs, vehicle's density is highly dynamic. For example, in highway scenarios, vehicles have mostly one straightforward direction, but high relative speeds up to 100 km/h, compared to the cars driving in the opposite direction that can lead to a challenging situation. On the other hand, in urban scenarios, high-density networks appear during rush hours. Also, network disconnection can occur in the late night hours or idle daytime hours. Therefore, designing protocols able to take all of these characteristics into account is still a challenging open issue.

MAC protocols play an important role since critical communications rely on them. Unfortunately, research results [2] highlight that the topic of MAC support in VANETs has received less attention than in other research fields (we further mention some of the existing works). Also, most of the relatively few research works are dedicated to V2I communications; therefore, MAC support for V2V communication needs more attention. MAC layer design challenges in VANET environments can be summarized as follows [3]: (1) Achieving an effective channel access coordination in the presence of changing vehicle locations and variable channel characteristics, (2) supporting scalability in the presence of various traffic densities, and (3) supporting a diverse set of application requirements.

According to the definition in [4], a system is said to be scalable if it can handle the addition of users and resources without suffering a noticeable loss of performance or increase in administrative complexity. The number of vehicles is neither restricted nor predictable in vehicular environments, so the MAC protocol must be scalable in order to fulfill end-to-end delay and packet delivery ratio (PDR) requirements.

Applications for VANETs can be categorized into three groups: safety, convenience (traffic efficiency), and commercial applications (infotainment) [5]; each of these classes of applications has its own quality of service (QoS) requirements. Safety applications represent the main target of intervehicle communications. Their goal is to increase each vehicle's awareness about its neighborhood. For this purpose, they use small messages that are broadcasted to a close neighborhood and are limited in time.

The assumed objectives for traffic efficiency applications can be considered as reducing the time each vehicle spends on the road in order to reduce fuel consumption and air pollution. Contrary to safety applications, these applications do not require a high penetration ratio, and infrastructure requirements can be met by existing 3G/4G networks. Finally, the third group of applications consists of infotainment applications. Although they do not have any effect on road traffic, they provide convenience and comfort to drivers and/or passengers. These applications, similar to traffic efficiency applications, are infrastructure-based. The delay is not considered as critical as in safety applications, but the transferred data volume is much larger than in safety applications.

IEEE 802.11p [6], which is an amendment for wireless access in vehicular environments (WAVE), basically proposes physical layer changes, while the MAC layer has remained mostly the same as in IEEE 802.11.

A lot of research has been done by the research community in order to improve the performance of IEEE 802.11p concerning safety applications (e.g., [7–10]). The results of [5] confirm that some traffic efficiency applications, and also the majority of commercial applications, rely on unicast communications. This classification clarifies that although safety applications have attracted special interest and require more scrutiny, other types of applications need to be attended as well. Also, as previous studies have shown [11–13], VANETs introduce complex dynamics that have not emerged in other types of MANETs.

Since the IEEE 802.11 MAC protocol was basically designed for static wireless networks, it achieves poor performance in VANET environments. A well-known problem in IEEE 802.11-based MAC protocol is scalability, which becomes more challenging in VANETs in the presence of high and variable network densities, as a result of higher mobility. Several works have been proposed and carried out for MANET environments [14–21] to either solve or reduce this problem. In particular, those solutions that estimate the channel density by keeping a history of channel transmission states, and then adapting the CW size based on the density estimations, can be quite effective.

The rest of this chapter is organized as follows. Section 21.2 presents the basic principles of IEEE 802.11/802.11p standards, challenges, and issues in VANET simulation, and reviews developed tools and simulators for vehicular environments. Section 21.3 reviews some alternatives for these standards in MAC layer. Section 21.4 describes the simulation requirements including scenarios, the measurements, and the metrics used for performance evaluation. Simulation results are provided in Section 21.5, including results in both highway and urban scenarios. Finally, Section 21.6 concludes the chapter.

21.2 LITERATURE REVIEW AND RELATED WORK

21.2.1 VANET SIMULATION CHALLENGES

Although VANETs are a kind of MANETs, its characteristics make MANET simulation tools hardly applicable for VANET environments. While using common wireless network simulators (such as NS-2 [22]) in conjunction with widely used mobility models (such as random way point [23]) gives acceptable results, modifications to simulators are required to introduce new characteristics. In the following, we summarize the important challenges for VANET simulation: First of all, *vehicular mobility* must be modeled in VANET simulation frameworks. The widely used models for vehicular simulation are microscopic models that represent traffic loads by individually considering each vehicle behavior, based on the state of the neighboring cars, and also the characteristics of the driver.

The first step to model vehicular environments is adapting a road infrastructure, which can be manually specified in the simulator or by using real-world maps. The next step is identifying the map topology (such as stop signs or traffic lights), which has an important impact on local vehicle density [24]. Two commonly used examples of microscopic models are car-following and cellular automaton.

In the *car-following* model, accelerating and decelerating of each vehicle depends on the car in front, which is called leader. When the distance from the leader decreases below a certain threshold, the follower starts decelerating; on the contrary, when the distance increases, the follower can increase its velocity until it reaches its own maximum speed. In the *cellular automaton* model, the road is divided into different sections. Each such section can be occupied at any given moment by one single car. Each car moves from one section to another and decelerates when another car has already occupied the next section. Although the cellular automaton model is not as accurate as the car-following model, it can be used in simulations that do not require a high simulation detail to decrease the simulation time.

Simulating vehicle's properties like car length, acceleration limit, and maximum speed allows us to present a vehicular environment as accurate as possible. In addition to the vehicle's properties, driver characteristics also play an important role in the mobility pattern of a car. Notice that each driver has different perception-response time, braking time, sign visibility and legibility, hazard recognition, and identification. The intelligent driver model (IDM) [25] is a model that improved upon the car-following model to take into account some of these parameters. The lane-changing policy is another issue that can be considered in a microscopic model, and it depends on the driver attributes.

Vehicle generation is another issue that must be taken into account. It can be modeled randomly or by identifying the entry and destination points. In randomly generated vehicles, they are randomly placed on the map; however, this method is not realistic, so it is necessary to wait for a warm-up period before producing reliable results. The other method, which is normally based on macroscopic statistical data, assigns entry and destination points for each vehicle.

Another challenge for VANET simulation is modeling the effects of various obstacles on radio wave propagation. These obstacles vary from one scenario to another scenario: in an urban scenario, they consist of buildings and trees surrounding the road, and in highway scenarios, these are tunnels or other cars.

Usually the signal power is calculated at the receiver side after applying a series of additions and subtractions to the signal strength at the sender side. Then the decision is made by determining whether the received signal power is above or below a certain threshold. Several models have been introduced for calculating radio wave propagation. In the easiest one, which is called the free space model, the signal power at the receiver side is only calculated based on the distance between communicating entities. The two-ray ground model enhanced the free space model with a multipath propagation, which causes self-interference. Also, the Nakagami model uses Nakagami distributions to model effects of scattered signals that reach a receiver through multiple paths.

In the majority of network simulation scenarios, memory and CPU consumption of network simulators grow linearly when increasing the number of nodes, even when some nodes do not contribute to any communication. In the case of V2V network, where each car can be a source of traffic and its destination may be any other car, resource consumption grows with the square of the number of cars. Therefore, when the simulation environment assumes as a city with more than 10,000 nodes, its simulation with current software becomes a challenging issue.

21.2.2 VANET MODELS, TOOLS, AND PLATFORMS

21.2.2.1 Network Simulation Tools

Currently, we can find several network simulators that can be used for VANET simulations. In the following, we will discuss about these simulators and attempt to determine which one can be more adequate for vehicular environment scenarios.

21.2.2.1.1 OMNeT++

OMNeT++ [26] is a C++ based, open-source, and discrete event network simulator with strong graphical user interface (GUI) support. It can be used for modeling any system that can be described as a queuing system. Since OMNeT++ is a simulation framework, it does not provide models for wireless network simulation. Therefore, several frameworks have been developed on OMNeT++ for simulating MANETs. Among all of these frameworks, INETMANET [27] and MiXiM [28], which include propagation models specially designed for vehicular environments, are the most appropriate ones for VANET simulation [29].

21.2.2.1.2 Network Simulator 2/3 (NS-2/NS-3)

NS-2 [22] is a discrete event simulator developed by the VINT project research group at the University of California at Berkeley, with the support of the US Defense Advanced Research Projects Agency. It is assumed as the standard simulator for wired networks due to the large number of models that have been developed for this simulator. It was extended by the Monarch research group in Carnegie Mellon University in order to be able to model node mobility, and more realistic PHY and MAC layers (IEEE 802.11 DCF) for wireless networks. Several mobility and radio propagation models exist in NS-2 that can be used, but none is designed for VANET simulation. Some new models have been produced for VANET simulation in NS-2, but still there are some disadvantages of using NS-2, including its high complexity that makes the implementation of vehicular mobility models difficult inside the framework; also, its memory and CPU consumption do not allow simulating large scenarios.

NS-3 [30] is an optimized version of NS-2, which introduces less complexity by removing the C++/TCL interactions used in NS-2, thereby allowing it to handle large-scale scenarios. However, it still cannot use all the models that have been modeled in NS-2, which is a negative point for NS-3.

21.2.2.1.3 Global Mobile Information System Simulator and QualNET

Global Mobile Information System Simulator (GloMoSim) is an open-source and scalable simulator for wireless and wired environments developed at the University of California in Los Angeles. The layered design of this simulator, with different APIs between them, eases the integration of different layers by different people. The GloMoSim project stopped in 2000, and a commercial version, QualNET, has been maintained since then. The QualNET framework [31] is able to simulate large-scale scenarios with several thousand nodes. Also, a new VANET mobility model, called CORNER [32], has been recently added, which makes it as a good choice for VANET simulation, but its commercial orientation limits its wide usage.

21.2.2.1.4 Java in Simulation Time /Scalable Wireless Ad Hoc Network Simulator [33]

Java in Simulation Time (JiST) is a general-purpose discrete-event simulator written in Java, developed at the Cornell University, and Scalable Wireless Ad Hoc Network Simulator (SWANS) has been specially designed for MANETs. The SWANS is able to simulate large scenarios of more than 10,000 nodes. The interesting point of this project is that the real applications written in Java can be directly tested with this simulator. The disadvantages are that the project is no longer maintained and it does not include any mobility and radio propagation model for VANET simulation.

21.2.2.2 Traffic Simulation Tools

There are several traffic simulators that are able to generate vehicle's movements. However, in this section, we just consider simulators that are still updated and supported in different operating systems. For example, CORridor SIMulation (CORSIM) [34] and Verkehr In Stadten SIMulationsmodell (VISSIM) [35] are just supported in a Microsoft Windows platform, which is a negative point for network researchers that usually prefer the Linux OS. Also, MObility model generator for VEhicular networks (MOVE) [36], which was built on top of Simulation of Urban MObility (SUMO), is no longer maintained.

21.2.2.2.1 Simulation of Urban Mobility

SUMO [37] is an open-source, microscopic traffic simulator specially designed to handle large-scale vehicular mobility. It supports different vehicle types, multilane streets with lane changing, junction-based right-of-way rules, hierarchy of junction types, a powerful GUI, and dynamic routing. SUMO can also import different network formats.

21.2.2.2.2 VanetMobiSim [38]

This traffic simulator is an extension to the CANU Mobility Simulation Environment (CanuMobiSim) [39], and it was specially designed for modeling vehicular mobility. It has the ability to import maps and produce mobility traces with different formats, supporting different simulation and emulation tools for mobile networks. Vehicular mobility patterns can be based on random trips or an origin–destination approach. It supports an enhanced version of IDM including both Intersection Management (IDM/IM) and Lane Changing (IDM/LC), and also an overtaking model (MOBIL) [40].

21.2.2.3 Interlinking Tools

In this section, we describe different modules used to make an interaction between a network and a traffic simulator. As mentioned before, one of the important challenges in VANET simulation is that the movement of each node is not individual, and it is affected by its surroundings. Therefore, the traffic simulator must be able to change the initial trace file during the simulation based on feedbacks received from the network simulator. In the following, we summarize some modules that are produced to link some commonly used simulators like NS-2, OMNeT++, and SUMO.

21.2.2.3.1 Vehicles in Network Simulation

Vehicles in Network Simulation (Veins) [41] is assumed as the best package for VANET simulation, especially at the MAC level. It produces a TCP connection in order to provide an interaction between SUMO and the INET framework from OMNeT++. In Veins, the network simulator is able to influence the vehicles' movements produced by traffic simulator by exchanging some commands.

21.2.2.3.2 Traffic and Network Simulation Environment and Integrated Wireless and Traffic Platform for Real-Time Road Traffic Management Solutions

Traffic and Network Simulation Environment (TraNS) [42] was the first tool using a link between a network and a traffic simulator, but it is not a high-performance framework, it does not support new versions of SUMO, and it is no longer maintained (since 2008). The Integrated Wireless and Traffic Platform for Real-Time Road Traffic Management Solutions (iTetris) [43] is seen as the successor of TraNS, so that it couples NS-3 and SUMO through a central control block named iTetris Control System (iCS).

21.2.2.3.3 V2X Simulation Runtime Infrastructure [44]

Unlike the aforementioned interlinking tools, V2X Simulation Runtime Infrastructure (VsimRTI) is a generic framework that can be used by different simulators such as VISSIM, SUMO, and JiST/SWANS. Also, this project can work as an emulator for directly testing real applications in V2X environments.

21.2.2.4 VANET Simulation Tools

As we mentioned before, VANET simulations require modeling the drivers' behavior in detail, as well as radio channels. Also, integrated frameworks for simulating both mobility and networking are clearly needed for effectively evaluating the performance of vehicular systems. In this section, we describe the simulators that have been specially designed for vehicular communications.

21.2.2.4.1 GrooveNet [45]

This simulator was developed at the Carnegie Mellon University, and it is a hybrid simulator that enables communication between real and simulated vehicles. It has a powerful GUI and the ability to import maps. Although it has several communication protocols implemented, its complexity and lack of documentation make its development a rather difficult task.

21.2.2.4.2   National Chiao Tung University Network Simulation

National Chiao Tung University Network Simulation [46] is also a hybrid simulator like GrooveNet, and it was the first simulator to implement the complete IEEE WAVE architecture. The advantages of this simulator are that it can directly run real applications because it uses the real Linux TCP/IP protocol stack, and it supports parallel simulation on multicore machines. The disadvantages are that it is only compatible with Fedora 9 Linux distribution, and it has been commercial since 2011.

21.2.2.4.3   Integrated Wireless Intersection Simulator [47]

Integrated wireless intersection simulator is a tool that was not widely adopted because it is specially designed for intersection management. It provides detailed information for propagation models by considering urban elements. Also, it models several VANET MAC layer protocols. As a conclusion, it can be used as a good simulator, especially for MAC protocols in highway scenarios.

21.2.3   Overview of IEEE 802.11 and 802.11p

Although initially proposed for wireless LAN environments, IEEE 802.11 has expanded in order to fulfill the communication requirements of different environments. IEEE 802.11-based MAC protocols rely on the carrier sense multiple access with collision avoidance (CSMA/CA) mechanism, so that each node listens to the channel before transmission in order to prevent collisions. Since they cannot guarantee a collision-free MAC, two main mechanisms are proposed to handle medium access collisions among nodes while also avoiding the hidden node problem. First, the Request-to-Send (RTS)/Clear-to-Send (CTS) mechanism is proposed in order to reserve the medium by sending small packets before transmitting large data packets, basically to avoid the hidden node problem. However, due to the high overhead involved when using this mechanism, it is inactive by default. Second, each node must select a random time (called backoff) before sending each packet to avoid packet collisions. Also, IEEE 802.11 [48] proposes three different Inter Frame Space (IFS) time intervals, such as SIFS, DIFS, and PIFS, in order to prioritize access to the wireless channel.

When a new data packet is waiting in the buffer for transmission, the node is allowed to transmit the packet if it finds the channel free after waiting for a time equal to DIFS. Otherwise, if the channel is found busy after this period, it chooses a backoff value and must wait before attempting to transmit again. The next time the channel is found idle, the node must wait for a time equal to DIFS, plus a backoff time before transmission. The Binary Exponential Backoff (BEB) algorithm uniformly selects the backoff time from the interval (0, CW). The IEEE 802.11 standard initializes the CW to the predefined value, CW_{min}, and doubles it upon transmission failures up to the predefined value, CW_{max}. Also, The CW is reset to the CW_{min} by a successful transmission. The node senses the channel and, if it finds the channel idle, decreases the backoff timer by one; otherwise it pauses the timer. The backoff timer is resumed when the channel again remains idle for a period longer than DIFS.

The IEEE 802.11 uses positive acknowledgments (ACK) to confirm the correctness of the current transmission to the transmitter. Such mechanism is mandatory since wireless interfaces cannot transmit and listen to the channel simultaneously. Notice that its own signal is too strong, masking any other incoming signals, and so collision detection becomes very difficult. Second, collision detection at the transmitter side does not allow inferring about collisions at the receiver side. Therefore, if the transmitter cannot receive a correct ACK packet, it considers that a collision has occurred, and the current data packet must be retransmitted (up to a predefined number of times). After receiving a correct data packet, the receiver waits for a SIFS time and sends an ACK to the transmitter. Figure 21.1 shows the routines for IEEE 802.11 MAC protocol.

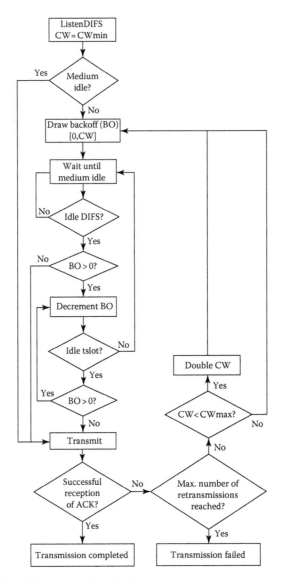

FIGURE 21.1   IEEE 802.11 MAC protocol mechanisms.

As previously mentioned, IEEE 802.11 introduces specifications for wireless networks, but it cannot provide peak efficiency in vehicular environments. The main standards and specifications for vehicular environments are short-range communications (DSRC), IEEE 802.11p, and IEEE 1609. DSRC includes specifications and standards for communications between vehicles in close neighborhood. Both the United States and Europe have allocated spectrum in the 5.9 GHz range for DSRC. IEEE developed IEEE 802.11p [6], which is an amendment to the original IEEE 802.11 standard for WAVE to better support vehicular communications. IEEE 802.11p defines specifications for link layer and physical layer, also IEEE 1609.4 specifies higher layers that will operate on top of IEEE 802.11p. IEEE 802.11p is an IEEE 802.11-based MAC protocol offering a priority scheme in a similar way to IEEE 802.11e EDCA, while IEEE 1609.4 manages channel coordination and supports MAC service data unit delivery. IEEE 802.11p proposed a multichannel operation so that physical layer consists of seven 10 MHz channels where one of them is control channel (CCH), used for safety communications, and the remaining are called service channels (SCHs), and they are used for nonsafety applications, shown in Figure 21.2.

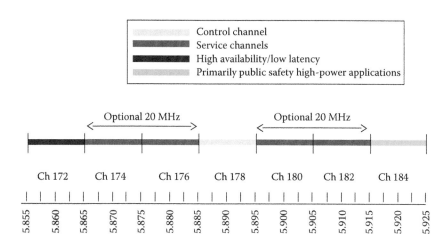

FIGURE 21.2 DSRC frequency plan.

TABLE 21.1
Different Application Categories in IEEE 802.11p

AC	CW$_{min}$	CW$_{max}$	AIFSN
Video traffic (AC3)	3	7	2
Voice traffic (AC2)	3	7	3
Best effort (AC1)	7	225	6
Background (AC0)	15	1023	9

An orthogonal frequency division multiplexing modulation scheme is used to multiplex data, similarly to the IEEE 802.11a standard. However, the bandwidth that is used in each channel by IEEE 802.11p is half of the bandwidth used in the IEEE 802.11a. The MAC layer follows the same approach as used by IEEE 802.11e EDCA in order to provide QoS support. The EDCA mechanism defines four different access categories (ACs) for each channel compared to just one in IEEE 802.11. Different ACs provide different access priorities, and based on that, they are assigned different contention parameters. For example, AC3, which has the highest priority when accessing medium, has the lowest arbitrary IFS (AIFS) and CW values, whereas AC0, which has the lowest priority, has the highest values. Table 21.1 shows the default parameter settings used in IEEE 802.11p for different traffic types. There are six SCHs and one CCH, and each of them has four different ACs. Also, there are two contention procedures including internal contentions between different ACs in each channel, and the contention between channels to access the medium.

21.3 ALTERNATIVES TO IEEE 802.11P FOR VEHICULAR ENVIRONMENTS

The current IEEE 802.11p standard, which applies to communication in wireless vehicular environments, does not propose an explicit mechanism in order to dynamically adapt the CW size based on the network density. Research results focusing on IEEE 802.11p performance show a lack of scalability, meaning that the CW size, which is calculated by the current static approach, is not optimal and causes more collisions as the network density increases.

To solve this problem in IEEE 802.11-based MAC protocols, some schemes [14–17] have been proposed based on network density estimation where the CW size is selected based on that network density estimation. The AOB mechanism [14] measures the network contention level by using two simple values: the slot utilization, and the average size of transmitted frames, based on the information provided by the carrier sensing mechanism. In [15], a three-level estimator is used for computing the optimal CW size, and in [16], a method was proposed to estimate the number of active nodes by means of a Kalman filter to optimize the CW size. Also, in [17], a novel access method was proposed called idle sense to adjust the CW size based on the network load. The CW is derived after measuring the number of idle slots between two transmission attempts and comparing it with the optimal number, which is equal to 5.68 for IEEE 802.11b. Considering the high overhead of these methods to estimate the network density, these approaches are not highly acceptable in ad hoc networks where power and memory resources are restricted.

In order to avoid the high overhead of density estimation, other schemes [18–21] introduce static mechanisms to optimize CW size selection instead of harshly changing the CW size by resetting it to the minimum value (or doubling it) upon a successful or unsuccessful transmission, respectively. For example, the MILD scheme [18] increases the CW size by multiplying it by 1.5 and decreasing the CW size by 1 unit. In [19], when the retry counter reaches the limit, the CW is not rest to the minimum CW. After a successful transmission, CW is set to the value *max[CW/2, CWmin+1]*, and upon a transmission failure, CW is set to the value *min[2CW, CWmax+1]*. EIED, which is proposed in [20], modifies the performance of BEB and MILD under low traffic conditions. The EIED scheme is based on an exponential distribution for both increasing and decreasing the CW size. In [21], the new scheme is proposed, which, following a successful transmission, reduces the CW to a value near the old one based on some factor that is evaluated during simulation. While these schemes work better than the BEB algorithm for high-density networks, the basic problem remains since the static behavior persists, and so it cannot adapt to different network densities.

To fill the gap between the two aforementioned groups of methods, a low-overhead and dynamic solution called HBCWC is proposed in [49]. This algorithm does not need complex computation to estimate the network density. Instead, HBCWC obtains the network density information by observing the channel status, and it dynamically adapts the CW size based on the channel condition history. It shows a significant improvement in the presence of high network densities, while not requiring significant changes to the original IEEE 802.11 MAC.

To the best of our knowledge, very few studies address this subject. A fuzzy logic-based enhanced 802.11p mechanism is proposed in [50], which adapts the CW size based on a nonlinear control law, which relies on channel observation. Also, in [51], DBM-ACW was proposed in order to improve the performance of IEEE 802.11p by adapting the CW size based on network density for unicast communications. It proposes a new scheme that allows selecting the CW size based on the network traffic density. In this scheme, the channel conditions are observed based on the packet transmission status, and the result is stored into a channel state (CS) vector. If the transmitter receives an ACK frame from the receiver, a value of 1 is inserted into the CS vector. Otherwise, if a collided/faulty frame is received, or if the transmitter waiting timer expires before receiving the acknowledgment, a value of 0 is inserted into the CS vector. The CS vector is updated by shifting before setting the CS_0 value.

In DBM-ACW, upon each packet loss, timer expiration, or collision, the CW size is multiplied by 2 in order to obtain the highest PDR, except for the case in which the CS array contains two consecutive ones before the new state; that particular situation means that we observed two successful transmissions before detecting an unsuccessful transmission. In that case, the CW is multiplied by parameter A. Also, the CW size is set to the minimum CW, CW_{min}, upon each acknowledgment reception, except for the case in which the CS array contains two consecutive zeroes before the new state, which means that two unsuccessful transmissions are observed before detecting a successful transmission. In that case, the CW is multiplied by parameter B. The value of parameters A and B was set to 1.7 and 0.8, respectively.

Up to now, we have introduced background information about vehicular networks and some proposed MAC layer protocols for vehicular environments. In the following sections, we first define the simulation settings and the generated traffic loads in our scenarios. Then, we evaluate the performance of the aforementioned MAC protocols in the proposed vehicular scenarios.

21.4   SIMULATION REQUIREMENTS

This section introduces briefly the OMNeT++ network simulator, stating how it must be configured in order to be used for vehicular network simulation. Then, traffic simulation methods and the SUMO traffic simulator are presented. Finally, we illustrate how to connect these two simulators to achieve a realistic vehicular network simulation.

21.4.1   NETWORK SIMULATION (OMNeT++)

Each model in OMNeT++ is composed of reusable components termed modules. OMNeT++ allows users to use two types of components: simple and compound modules. The functionality of each simple module is defined by C++ code, and it uses a simulation library. These simple modules can be combined unlimitedly like LOGO blocks to make compound modules. Simple modules can be connected with links, which connect two gates of two different modules to each other. Gates are input and output interfaces of modules. Connections are created within a single level of module hierarchy. Connections spanning different hierarchy levels are not permitted, as they would prevent model reuse. Modules communicate with each other through messages, which carry arbitrary data. Messages travel through a chain of connections, starting and arriving in simple modules. Messages can contain some information from other modules or be sent to the same module (self-messages) in order to implement a timer.

Some of the features of OMNeT++ are as follows:

- Possibility of designing modular simulation models, which can be combined and reused flexibly.
- Composing models with any granular hierarchy.
- Availability of extensive simulation libraries that include support for input/output, statistics, data collection, random number generation, and data structures.
- C++-based simulation kernel, which allows using it in larger applications.
- Graphical-based simulation configuration using NED and omnetpp.ini without requiring the use of scripts.

The main problems of this simulator in order to be a suitable simulator for VANETs are lack of models for wireless networks and the lack of integrated mobility manager. Therefore, a mobility manager and a framework in order to support models for wireless communications must be installed to provide these key functionalities when attempting to achieve a realistic VANET simulation environment. We chose this simulator coupled with the INETMANET framework and SUMO in order to provide a realistic vehicular scenario. The INETMANET framework provides detailed models for simulating wireless networks in OMNeT++ such as wireless channels, connectivity, mobility, and MAC layer protocols. Also, SUMO is used to generate real vehicular traffic in road networks.

In summary, for each simulation using OMNeT++, the following parts must be described:

1. Simple modules (provided by some packages like INETMANET, MiXiM, etc.)
2. Topology (using NED files)
3. Simulation configuration (using omnetpp.ini)

21.4.1.1 INETMANET

INETMANET is an open-source package that provides network simulation models for OMNeT++. Although it focuses on the high level of the protocol architecture for wired and wireless communications, it also includes MAC layer protocols similar to IEEE 802.11p. Another option is to use this package in conjunction with the MiXiM package. MIXIM implemented detailed wireless NICs, so using these two packages together introduces detailed implementation at all levels of the protocol architecture.

21.4.1.2 Scenario Definition Using NED Files

We use the INETMANET framework in order to be able to use different network protocols, but in order to connect and make a node structure in OMNeT++, NED files and omnetpp.ini are used. We generated two NED files, one for building a node (vehicle) and the other one for network scenario description. Each NED file includes four parts: parameters, gates, submodules, and connections (Figure 21.3). The parameters and gates sections are optional, and they can be left out if there is no gate or parameter. In the submodule section, we defined simple modules, and the connection section describes how these simple modules must be connected in order to build an architecture.

21.4.1.3 omnetpp.ini File

Each simple module defines some parameters to customize its behavior. These parameters can be assigned in either the NED files or the configuration file, omnetpp.ini. In Figure 21.4, you can find that a reduced version of the omnetpp.ini file is used for simulation. When the program is executed in OMNeT++, it first reads all the NED files in order to build the network topology, and then it uses omnetpp.ini for extracting the settings to control the simulation execution. In omnetpp.ini example, the simulation is repeated 10 times, and each time it takes 300 seconds and adapts different seeds. The output results are written into output files including in both vectorial and scalar formats. Notice that, at the application layer, the destination address is randomly chosen from the list of current nodes using "random_name(host)" or "moduleListByPath("**.host[*]")."

```
module Car
{
    parameters:
    gates:
        input radioIn @directIn;
    submodules:
        notificationBoard: NotificationBoard
        ac_wlan: HostAutoConfigurator
        interfaceTable: InterfaceTable
        app: UDPBasicBurstNotification
        mobility: TraCIMobility
        routingTable: RoutingTable
        udp: UDP
        networkLayer: NetworkLayer
        nic: Ieee80211Nicext
        manetrouting: AODV
    connections allowunconnected:
        udp.appOut++ --> app.udpIn;
        udp.appIn++ <-- app.udpOut;
        udp.ipOut --> networkLayer.udpIn;
        udp.ipIn <-- networkLayer.udpOut;
        nic.upperLayerOut --> networkLayer.ifIn[0];
        nic.upperLayerIn <-- networkLayer.ifOut[0];
        networkLayer.manetOut --> manetrouting.from_ip if routingProtocol != "";
        networkLayer.manetIn <-- manetrouting.to_ip if routingProtocol != "";
        radioIn --> nic.radioIn;
}
```

FIGURE 21.3 NED file.

```
[General]
repeat = 10
seed-set = ${repetition}
...
sim-time-limit = 300s
#########################################
# TraCIScenarioManagerLaunchd
*.manager.updateInterval = 0.1s
*.manager.host = "localhost"
*.manager.port = 9999
...
*.manager.launchConfig = xmldoc("erlangen.launchd.xml") #default
#########################################
# udp apps (on)
**.host[*].app.destAddresses = moduleListByPath("**.host[*]")
**.app.localPort = 1234
**.app.destPort = 1234
**.app.messageLength = 512B # Bytes
**.app.sendInterval = 0.5s + uniform(-0.001s,0.001s,0)
...
**.app.startTime = simTime()+1s
**.app.p = ${0.4, 0.5, 0.6, 0.7, 0.8, 0.9, 1.0}
###############################
# nic settings
**.wlan*.bitrate = 18Mbps
**.wlan*.opMode = "p"
...
**.wlan*.mac.retryLimit = 7
**.wlan*.mac.cwMinData = 7
**.wlan*.mac.cwMaxData = 1023
...
**.propagationModel = "NakagamiModel"
**.radio.nak_m = 0.7
###############################
# parameters : AODVUU
**.routingProtocol = "AODVUU"
**.log_to_file = false
**.hello_jittering = true
**.optimized_hellos = true
...
###############################
```

FIGURE 21.4   omnetpp.ini file.

21.4.2 TRAFFIC SIMULATION

Traffic flow simulations can be divided into three groups depending on the level of details. In macroscopic traffic models, the flow is assumed as the basic entity. The next group includes microscopic traffic models in which more details are implemented, so that vehicles are the smallest entities. This type of traffic models is common in vehicular network simulations. The last group includes submicroscopic traffic model in which the inner parts of each vehicle are also implemented, including engine, gearbox, and so on. However, since these types of traffic models require too high computational overhead, they are not considered as a suitable option for large-scale network simulations, like VANETs.

21.4.2.1 SUMO

SUMO uses a microscopic traffic flow model that allows defining three types of elements: vehicle types, trips, and routes. A vehicle type specifies physical properties of a typical vehicle in the simulator. A trip defines the departure time and the destination edge, while route expands trip by defining all the edges through which a vehicle will pass. In general, we provide different files in SUMO in order to define the simulation map, the obstacles, and the routes used. To provide a simulation map, we extract the map from openstreetmap.org [52] in osm version. To import the extracted map into SUMO, it must be converted into an xml file by the NETCONVERTER [53] application. Also, routes are generated by the DUAROUTER [54] application using the shortest path algorithm. DUAROUTER requires the xml file already generated by the NETCONVERTER application, and it takes into account several data such as street length, speed limit, lane count, and street type for the shortest path computation. To make a more realistic simulation, buildings, city furniture, trees, or other obstacles can also be added through the POLYCONVERT [55] application in SUMO.

In conclusion, to prepare mobility traces for network simulation using SUMO, the following steps must be followed:

1. Extract the map from openstreetmap.org.
2. Create a road network file (using NETCONVERTER).
3. Generate random trips (using randomTrips.py).
4. Convert the trips into routes and traffic flows (using DUAROUTER).
5. Generate an obstacles file (using POLYCONVERT).
6. Create a configuration file.

After following these steps, the configuration file, which includes the road network file, the routes file, the simulation time steps, as well as begin-and-end simulation times, is used in order to introduce mobility in network simulation experiments.

21.4.3 TRAFFIC CONTROL INTERFACE

As described earlier, OMNeT++ does not have any integrated mobility manager components, so we propose using Veins, which is an open-source intervehicular communication simulation framework that integrates both OMNeT++ and SUMO. In order to allow SUMO to control traffic mobility during the simulation, it connects both elements through the Traffic Control Interface (TraCI) interface [56]. TraCI opens a TCP connection at port 9999 between both simulators, so that the simulators can interact using a series of commands (e.g., speed, position) in small time steps.

The vehicle and mobility generation are handled by SUMO. However, when a vehicle reaches its destination in SUMO, it must leave the simulation, and so we cannot ensure a constant number of vehicles throughout the entire simulation time. Therefore, the VACaMobil [57] tool is used to handle this issue, inserting new vehicles in the simulation when other vehicles leave it, thereby maintaining the same number of vehicles throughout the simulation time. In our experiments, the number of nodes varies from 50 to 200.

21.5 CASE STUDY

This section presents OMNeT++ simulation results illustrating the performance of the MAC protocols under analysis according to the simulation setup explained in the previous sections. In order to simulate these MAC protocols in vehicular environments, we used both urban and highway scenarios, which are described in this section, and also the simulation parameters defined in Section 21.5.2.

21.5.1 SIMULATION SCENARIOS

21.5.1.1 Urban Scenario

In contrast to highway scenarios, where vehicles always drive in a same direction, and where obstacles are mostly nonexistent, urban scenarios not only offer more flexibility in terms of mobility but also introduce more obstacles like buildings, vehicles, and urban furniture. This issue leads to lower transmission ranges for urban scenarios. As a representative urban vehicular environment, we will focus on an area of 1500×1500 m^2 that is extracted by using digital maps, freely available in OpenStreetMap, from the downtown area of Valencia (Spain) and integrating real obstacles. Figure 21.5 shows two map views: the OpenStreetMap view and the SUMO view.

21.5.1.2 Highway Scenario

The highway scenario models a 4 km highway with three lanes, where the lane width is 3 m. As assumed in the previous section, the destination is randomly chosen, so it can be a car ahead or behind of the transmitter. In order to model a realistic scenario, we assumed that vehicles can be

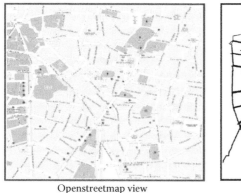
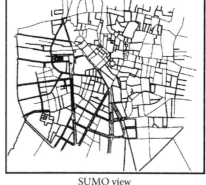

Openstreetmap view	SUMO view

FIGURE 21.5   Map layout for urban scenario (Valencia, Spain).

TABLE 21.2
Vehicle Types

Vehicle Type	Accel (ms-2)	Decel (ms-2)	Driver Imperfection	Max Speed (ms-1)	Probability (%)
Fast	4	6	0.2	36	10
Normal	2	4	0.3	28	80
Slow	2	4	0.5	20	10

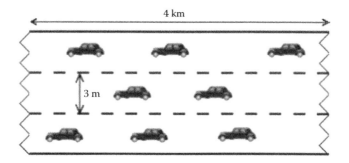

FIGURE 21.6   Map layout for highway scenario.

selected based on a normal distribution from three different categories that are summarized in Table 21.2. Driver imperfection is selected in the range from zero to one. Also, vehicles are injected in the highway according to a Poisson process with a mean interval time of 2 s, where the best lane is assigned to each vehicle. In contrast to the urban scenario, the transmission rate is 2 packets per second in this scenario. Figure 21.6 depicts the highway scenario.

21.5.2   SIMULATION PARAMETERS AND METRICS

The general simulation parameters are as follows: each vehicle generates constant bit rate traffic using UDP datagrams (we chose UDP and not TCP for the same reasons stated in [12]). The 512-byte datagrams were transmitted at a rate of 4 packets per second.

Considering the routing protocol, we assessed different routing protocols (i.e., AODV, OLSR, DYMO, DSR), and despite the different overall performance level obtained by these protocols, they have the same impact on different MAC protocols evaluated in this chapter. Thus, the results

TABLE 21.3

Simulation Parameters

Simulation Parameter	Value
Traffic type	CBR
CBR packet size	512 byte
CBR data rate	4 packet/s
Transport protocol	UDP
Routing protocol	AODV
MAC protocol	802.11p, HBCWC, DBM-ACW
Max. and min. of CW	7, 1023
Max. number of retransmissions	7
Max. queue size	14
RTS/CTS threshold	2346 byte
Slot time	13 μs
Max. transmission range	250 m
Propagation model	Nakagami
Nakagami-m	0.7
Simulation time	300 s
Number of repetitions	10

presented in this chapter are obtained using the AODV routing protocol; notice that, since several researchers use this protocol as well, it makes comparison against other proposals easier. We used the Nakagami radio propagation model, commonly used by the VANET community, in order to achieve a more realistic signal propagation model for vehicular environments [58]. Parameter m for this propagation model is set to 0.7. Table 21.3 summarizes the simulation parameters.

The metrics used to evaluate the performance of the MAC solutions under study were similar to those used in previous studies. They are summarized as follows: (1) PDR, which represents the ratio of the total number of packets received by the final destination and the packets originated by the source; (2) average end-to-end delay, which represents the average time required for a packet to travel from source to destination; (3) standard deviation of end-to-end delay, which shows how much variation exists from the average end-to-end delay; and (4) average MAC collisions, which shows the average number of collisions experienced per source.

21.5.3 Urban

In this first experiment, we evaluate DBM-ACW, HBCWC, and IEEE 802.11p in an urban scenario using V2V communications. Since we are interested in high-congestion environments, we define a large number of connections, so that each vehicle, immediately after joining the network, starts a new connection and maintains it active until the destination leaves the network. When this occurs, a new destination will be chosen by the transmitter. This experiment represents a stressing situation for a MAC protocol due to the large number of simultaneous connections. Thus, the experimental setup is adequate to assess how these protocols are able to overcome a high number of collisions to obtain a suitable throughput.

Figure 21.7 shows the PDR when varying the number of source nodes. This figure shows poor performance of IEEE 802.11p in vehicular environments, which achieves the lowest performance compared to the other solutions analyzed: DBM-ACW and HBCWC. As mentioned in the previous chapter, IEEE 802.11/802.11p do not introduce any solution for the dynamic adaptation of CW, and also the CW is reset to CW_{min} upon each successful transmission, which causes a critical problem. As a result, the PDR for IEEE 802.11p shows faster decreases than alternative protocols when increasing the number of nodes. DBM-ACW, which is specially designed for vehicular

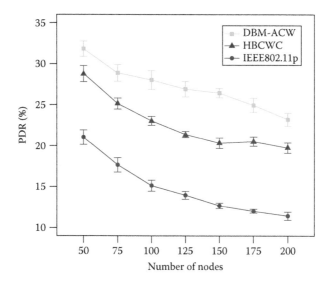

FIGURE 21.7 PDR for the urban scenario.

environments, outperforms both IEEE 802.11p and HBCWC. We can observe that the improvement ratio for DBM-ACW in high-density networks (more than 100 nodes) is higher than for low-density networks. This stems from the fact that DBM-ACW avoids resetting the CW to the minimum value and mostly maintains the average CW size at high values in the presence of frequent collisions. Therefore, it is able to decrease the number of dropped packets, but under low densities, when the collision frequency is low, the degree of efficiency is not comparable to high-density situations. Also, HBCWC shows a good performance because it also estimates the network density to choose the optimal CW size. However, the results show that resetting the CW size upon a successful transmission, like 802.11p and HBCWC, has a significant cost in terms of PDR.

The average number of MAC collisions, shown in Figure 21.8, offers a hint on how to achieve improvements in terms of PDR. As can be observed in this figure, DBM-ACW shows that the

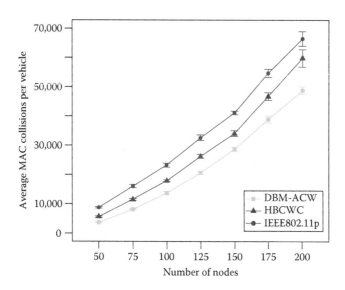

FIGURE 21.8 Average number of collisions for the urban scenario.

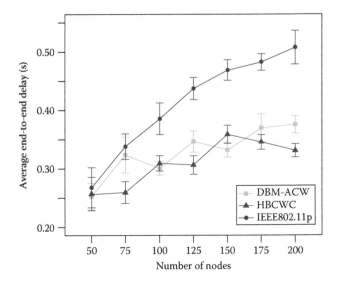

FIGURE 21.9 Average end-to-end delay for the urban scenario.

optimal CW was chosen so that it decreases the probability of picking the same backoff value, and consequently the number of collisions.

One of the key differences between IEEE 802.11p and DBM-ACW is that 802.11p resets the CW size to the minimum value when the retransmission limit is reached, without taking into account that this event is possibly associated to channel collisions; thus, it assigns a minimum CW size for the next packet. Considering this behavior, one can expect that the new packet will experience fewer success changes in order to be sent and will also need more retransmissions on average. Therefore, by avoiding to reset the CW size in this situation, DBM-ACW achieves lower end-to-end delay than IEEE 802.11p, as shown in Figure 21.9.

As we expected, the end-to-end delay increases for IEEE 802.11p under high densities, but the alternative solutions does not follow the same trend as 802.11p. They show a higher bound for delay so that, as the network density increases, the average end-to-end delay values remain low, fluctuating at values close to 0.35 s, as shown in Figure 21.9. Therefore, in this way, they improve the scalability of IEEE 802.11p, which represents an important challenge in VANETs.

Moreover, Figure 21.9 evidences the difference between DBM-ACW and HBCWC, which is further clarified in Table 21.4. In particular, DBM-ACW avoids resetting the CW size for the case in which it observes two consecutive successful transmissions before detecting an unsuccessful transmission. As a result, it is able to achieve improvements in terms of end-to-end delay, as well as improved standard deviation of delays for DBM-ACW compared to HBCWC. In HBCWC, a few packets are sent with very low delay, and this decreases the total end-to-end delay, while in DBM-ACW, most packets experience a similar delay.

TABLE 21.4
Standard Deviation of Delays for the Urban Scenario

	50	75	100	125	150	175	200
DBM-ACW	0.42	0.45	0.46	0.47	0.49	0.52	0.53
HBCWC	0.44	0.44	0.48	0.50	0.55	0.54	0.53

21.5.4 HIGHWAY

In the assumed highway scenario, the number of collisions is lower than in the urban scenario (due to the lower transmission rate), depicted in Figure 21.10. This figure shows that IEEE 802.11p has the highest number of collisions when compared to alternative solutions. Among them, DBM-ACW introduces more improvement ratios under high densities than HBCWC. Consequently, DBM-ACW cannot achieve a high improvement ratio compared to HBCWC in terms of PDR under lower densities, as depicted in Figure 21.11. However, for high densities, DBM-ACW achieves a higher improvement ratio compared to HBCWC when simultaneously considering both the PDR and the number of collisions. While DBM-ACW shows a lower improvement ratio

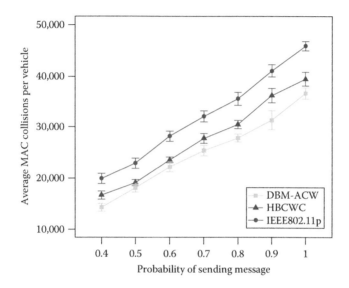

FIGURE 21.10 Average number of collisions for the highway scenario.

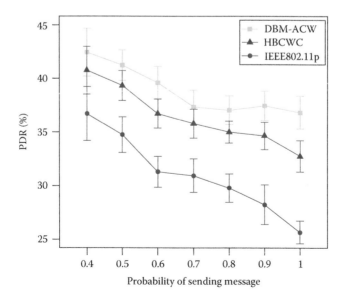

FIGURE 21.11 PDR for the highway scenario.

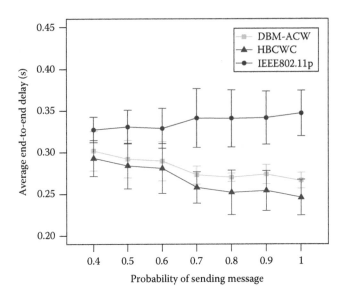

FIGURE 21.12   Average end-to-end delay for the highway scenario.

when compared with the urban scenario, it still shows better results than HBCWC and IEEE 802.11p considering the PDR and the number of collisions.

In terms of end-to-end delay, there is a big gap between IEEE 802.11p and the other alternatives, as shown in Figure 21.12. Also, while IEEE 802.11p has an increasing trend, DBM-ACW and HBCWC show a decreasing trend, meaning that it does not increase the delay when the number of connections increases, as desired.

Figure 21.12 shows that HBCWC achieves slightly better delay when compared with DBM-ACW. However, DBM-ACW achieves a better trade-off between PDR and end-to-end delay, meaning that the total PDR improvement ratio is higher than the delay degradation ratio when compared with HBCWC.

21.6   CONCLUSIONS

The CSMA-based MAC technique, and in particular the IEEE 802.11p protocol, is the method of choice for the current generation of VANETs. Although this scheme has its merits, it suffers from scalability problems in dynamic environments with a high node density, like vehicular environments, which affect its performance. In this chapter, we demonstrate the impact of the CW adjustment on the overall performance of unicast communications in VANETs. For these purposes, we evaluate the performance of three different MAC protocols: DBM-ACW, HBCWC, and IEEE 802.11p in two different vehicular scenarios including urban and highway. We clearly detailed the procedures to follow in order to properly compare these protocols through simulation, and also we selected the best network and vehicular traffic simulators to make a realistic vehicular environment. Extensive simulation results prove that IEEE 802.11p is not able to offer optimal performance in different congestion states, and for both urban and highway scenarios, we find that alternatives able to dynamically adapt the CW size according to channel conditions are a better choice, allowing to improve the PDR and the end-to-end delay for different vehicular environments.

ACKNOWLEDGMENTS

This work was partially supported by the *Ministerio de Ciencia e Innovación*, Spain, under Grant TIN2011-27543-C03-01.

REFERENCES

1. H. Hartenstein and K.P. Laberteaux, A tutorial survey on vehicular ad hoc networks, *Communications Magazine, IEEE*, 46 (6), 164–171, June 2008.
2. M.J. Booysen, S. Zeadally, and G.-J. van Rooyen, Survey of media access control protocols for vehicular ad hoc networks, *Communications, IET*, 5 (11), 1619–1631, July 22, 2011.
3. J. Kenney, Standards and regulations, in H. Hartenstein and K.P. Laberteaux (eds.), *VANET: Vehicular Applications and Inter-Networking Technologies*, Wiley, Chichester, U.K., Chapter 10, pp. 365–428, 2010.
4. B. Clifford Neuman, Scale in distributed systems, in T.L. Casavant and M. Singhal (eds.), *Readings in Distributed Computing Systems*, IEEE Computer Society Press, Los Alamitos, CA, 1994.
5. F. Bai, T. Elbatt, G. Hollan, H. Krishnan, and V. Sadekar, Towards characterizing and classifying communication-based automotive applications from a wireless networking perspective, *First IEEE Workshop on Automotive Networking and Applications (AutoNet)*, San Francisco, CA, 2006.
6. 802.11p-2010, *IEEE Standard for Information Technology—Telecommunications and Information Exchange between Systems—Local and Metropolitan Area Networks Specific Requirements—Part 11: Wireless LAN Medium Access Control (MAC) and Physical Layer (PHY) Specifications Amendment 6: Wireless Access in Vehicular Environments*, Institute of Electrical and Electronics Engineers, New York, pp. 1–51, July 2010.
7. S. Asadallahi and H.H. Refai, Modified R-ALOHA: Broadcast MAC protocol for vehicular ad hoc networks, *Wireless Communications and Mobile Computing Conference (IWCMC), 2012 Eighth International*, Limassol, Cyprus, pp. 734–738, August 27–31, 2012.
8. H. Omar, W. Zhuang, and L. Li, VeMAC: A TDMA-based MAC protocol for reliable broadcast in VANETs, *Mobile Computing, IEEE Transactions on*, 12 (9), 1724–1736, 2013. doi:10.1109/TMC.2012.142.
9. S.-S. Wang, H.-C. Chen, and J.-K. Chang, A distributed adaptive MAC protocol for efficient broadcasting in vehicular ad hoc networks, *Wireless Communications and Networking Conference (WCNC), 2012 IEEE*, Paris, France, pp. 1555–1560, April 1–4, 2012.
10. R. Stanica, E. Chaput, and A.-L. Beylot, Enhancements of IEEE 802.11p protocol for access control on a VANET control channel, *Communications (ICC), 2011 IEEE International Conference on*, Kyoto, Japan, pp. 1–5, June 5–9, 2011.
11. M. Wellens, B. Westphal, and P. Mahonen, Performance evaluation of IEEE 802.11-based WLANs in vehicular scenarios, *Vehicular Technology Conference, 2007. VTC2007-Spring. IEEE 65th*, Dublin, Ireland, pp. 1167–1171, April 22–25, 2007.
12. D.N. Cottingham, I.J. Wassell, and R.K. Harle, Performance of IEEE 802.11a in vehicular contexts, *Vehicular Technology Conference, 2007. VTC2007-Spring. IEEE 65th*, Dublin, Ireland, pp. 854–858, April 22–25, 2007.
13. J. Jansons and A. Barancevs, Using wireless networking for vehicular environment: IEEE 802.11a standard performance, *Digital Information Processing and Communications (ICDIPC), 2012 Second International Conference on*, Klaipeda, Lithuania, pp. 5–9, July 10–12, 2012.
14. L. Bononi, M. Conti, and E. Gregori, Runtime optimization of IEEE 802.11 wireless LANs performance, *Parallel and Distributed Systems, IEEE Transactions on*, 15 (1), 66–80, January 2004.
15. F. Cali, M. Conti, and E. Gregori, Dynamic tuning of the IEEE 802.11 protocol to achieve a theoretical throughput limit, *Networking, IEEE/ACM Transactions on*, 8 (6), 785–799, December 2000.
16. G. Bianchi and I. Tinnirello, Kalman filter estimation of the number of competing terminals in an IEEE 802.11 network, *INFOCOM 2003. 22nd Annual Joint Conference of the IEEE Computer and Communications. IEEE Societies*, San Francisco, CA, vol. 2, pp. 844–852, March 30–April 3, 2003.
17. M. Heusse, F. Rousseau, R. Guillier, and A. Duda, Idle sense: An optimal access method for high throughput and fairness in rate diverse wireless LANs, *ACM SIGCOMM*, Philadelphia, PA, 2005.
18. V. Bharghavan, A. Demers, S. Shenker, and L. Zhang, MACAW: A media access protocol for wireless LANs, *Proceedings of ACM SIGCOMM 1994*, London, U.K., pp. 212–225, 1994.
19. H. Wu, S. Cheng, Y. Peng, K. Long, and J. Ma, IEEE 802.11 distributed coordination function (DCF): Analysis and enhancement, *Communications, 2002. ICC 2002. IEEE International Conference on*, New York, vol. 1, pp. 605–609, 2002.
20. N.-O. Song, B.-J. Kwak, J. Song, and M.E. Miller, Enhancement of IEEE 802.11 distributed coordination function with exponential increase exponential decrease backoff algorithm, *Vehicular Technology Conference, 2003. VTC 2003-Spring. The 57th IEEE Semiannual*, Orlando, FL, vol. 4, pp. 2775–2778, April 22–25, 2003.

21. Q. Ni, I. Aad, C. Barakat, and T. Turletti, Modeling and analysis of slow CW decrease IEEE 802.11 WLAN, *Personal, Indoor and Mobile Radio Communications, 2003. PIMRC 2003. 14th IEEE Proceedings on*, Beijing, China, vol. 2, pp. 1717–1721, September 7–10, 2003.

22. The Network Simulator—ns-2, http://www.isi.edu/nsnam/ns/ (accessed in November 2013).

23. T. Camp, J. Boleng, and V. Davies, A survey of mobility models for ad hoc network research, *Wireless Communications & Mobile Computing: Special Issue on Mobile Ad Hoc Networking: Research, Trends and Applications*, 2 (5), 483–502, 2002.

24. A. Mahajan, N. Potnis, K. Gopalan, and A. Wang, Evaluation of mobility models for vehicular ad-hoc network simulations, Technical Report N. 051220, Florida State University, Tallahassee, FL, pp. 1–13, 2005.

25. M. Treiber, A. Hennecke, and D. Helbing, Congested traffic states in empirical observations and microscopic simulations, *Physical Review E*, 62 (2), 1805–1824, 2000.

26. OMNeT++ home page, http://www.omnetpp.org (accessed in November 2012).

27. INETMANET home page, http://inet.omnetpp.org/ (accessed in November 2012).

28. MiXiM home page, http://mixim.sourceforge.net/ (accessed in November 2012).

29. C. Sommer, D. Eckhoff, R. German, and F. Dressler, A computationally inexpensive empirical model of IEEE 802.11p radio shadowing in urban environments, Technical Report CS-2010-06, University of Erlangen, Department of Computer Science, Erlangen, Germany, September 2010.

30. The Network Simulator—ns-3, http://www.nsnam.org/ (accessed in November 2013).

31. Scalable Networks Technologies—QualNET Simulator, http://www.scalable-networks.com/products/qualnet/ (accessed in November 2013).

32. E. Giordano, R. Frank, G. Pau, and M. Gerla, CORNER: A step towards realistic simulations for VANET, *Proceedings of the Seventh ACM International Workshop on Vehicular Internetworking*, Chicago, IL, pp. 41–50, September 2010.

33. Java in Simulation Time—Scalable wireless ad hoc network simulator, http://jist.ece.cornell.edu/ (accessed in November 2013).

34. Corridor Simulation (CORSIM)—Microscopic traffic simulation model, http://mctrans.ce.ufl.edu/featured/tsis/Version5/corsim.htm (accessed in November 2013).

35. Verkehr In Stadten SIMulationsmodell (VISSIM), http://www.vissim.de (accessed in November 2013).

36. F. Karnadi, Z. Mo, and K. Lan, Rapid generation of realistic mobility models for VANET, *Proceedings of the IEEE Wireless Communications and Networking Conference*, Kowloon, Hong Kong, pp. 2506–2511, March 2007.

37. M. Behrisch, L. Bieker, J. Erdmann, and D. Krajzewicz, SUMO—Simulation of urban mobility: An overview. *SIMUL 2011, The Third International Conference on Advances in System Simulation*, Barcelona, Spain, 2011.

38. M. Fiore, J. Haerri, F. Filali, and C. Bonnet, Vehicular mobility simulation for VANETs, *Proceedings of the 40th Annual Simulation Symposium*, Norfolk, VA, pp. 301–309, March 2007.

39. CANU mobility simulation environment, http://canu.informatik.uni-stuttgart.de/mobisim/ (accessed in November 2013).

40. M. Treiber and A. Kesting, Modeling lane-changing decisions with MOBIL, *Proceedings of the Traffic and Granularity Flow Conference*, Paris, France, pp. 211–221, June 2007.

41. Veins home page, http://veins.car2x.org/ (accessed in November 2012).

42. M. Piorkowski, M. Raya, A. Lezama Lugo, P. Papadimitratos, M. Grossglauser, and J.-P. Hubaux, TraNS—Realistic joint traffic and network simulator for VANETs, *ACM SIGMOBILE Mobile Computing and Communications Review*, 12 (1), pp. 31–33, January 2008.

43. iTetris Project, http://www.ict-itetris.eu/ (accessed in November 2013).

44. N. Naumann, B. Schuenemann, and I. Radusch, VSimRTI—Simulation runtime infrastructure for V2X communication scenarios, *Proceedings of the 16th World Congress and Exhibition on Intelligent Transport Systems and Services*, Stockholm, Sweden, pp. 1–10, September 2009.

45. R. Mangharam, D. Weller, R. Rajkumar, P. Mudalige, and F. Bai, GrooveNet: A hybrid simulator for vehicle-to-vehicle networks, *Proceedings of the Third Annual International Conference on Mobile and Ubiquitous Systems—Workshops*, San Jose, CA, pp. 1–8, July 2006.

46. NCTUns 6.0, http://nsl.csie.nctu.edu.tw/nctuns.html (accessed in November 2013).

47. B. Jarupan, Y. Balcioglu, E. Ekici, F. Ozguner, and U. Ozguner, An integrated wireless intersection simulator for collision warning systems in vehicular networks, *Proceedings of the IEEE International Conference on Vehicular Electronics and Safety*, Columbus, OH, pp. 340–345, September 2008.

48. ANSI/IEEE Std 802.11, 1999 Edition (R2003), *IEEE Standard for Information Technology—Telecommunications and Information Exchange between Systems—Local and Metropolitan Area Networks- Specific Requirements—Part 11: Wireless LAN Medium Access Control (MAC) and Physical Layer (PHY) Specifications*, Institute of Electrical and Electronics Engineers, New York, pp. i–513, 2003.

49. A. Balador, A. Movaghar, and S. Jabbehdari, History based contention window control in IEEE 802.11 MAC protocol in error prone channel, *Journal of Computer Science*, 6 (2), 205–209, 2010.

50. C. Chrysostomou, C. Djouvas, and L. Lambrinos, Applying adaptive QoS-aware medium access control in priority-based vehicular ad hoc networks, *Computers and Communications (ISCC), 2011 IEEE Symposium on*, Corfu, Greece, pp. 741–747, June 28–July 1, 2011.

51. A. Balador, C.T. Calafate, J. Cano, and P. Manzoni, Reducing channel contention in vehicular environments through an adaptive contention window solution, *Wireless Days (WD), 2013 IFIP*, Valencia, Spain, November 15–17, 2013.

52. OpenStreetMap, http://www.openstreetmap.org/ (accessed in November 2012).

53. SUMO NETCONVERTER application, http://sourceforge.net/apps/mediawiki/sumo/index.php?title=NETCONVERT (accessed in November 2012).

54. SUMO DUAROUTER application, http://sourceforge.net/apps/mediawiki/sumo/index.php?title=DUAROUTER (accessed in November 2012).

55. SUMO POLYCONVERT application, http://sourceforge.net/apps/mediawiki/sumo/index.php?title=POLYCONVERT (accessed in November 2012).

56. TraCI Modules, http://veins.car2x.org/documentation/modules/traci/ (accessed in November 2012).

57. M. Baguena, S. Tornell, A. Torres, C.T. Calafate, J.C. Cano, and P. Manzoni, VACaMobil: VANET car mobility manager for OMNeT++, *IEEE International Conference on Communications 2013 Third IEEE International Workshop on Smart Communication Protocols and Algorithms (SCPA 2013)*, Budapest, Hungary, June 2013.

58. M. Baguena, C.T. Calafate, J. Cano, and P. Manzoni, Towards realistic vehicular network simulation models, *Wireless Days (WD), 2012 IFIP*, Dublin, Ireland, pp. 1–3, November 21–23, 2012.

22 Modeling and Simulation of Vehicular Networks

Network Simulators, Traffic Simulators, and Their Interworking

Kaveh Shafiee, Jinwoo (Brian) Lee,
Victor C.M. Leung, and Garland Chow

CONTENTS

ABSTRACT

Even though the actual testing of protocols in field experiments provides the most realistic results, it is often impractical due to cost, access, and scalability constraints, particularly in the case of larger networks such as vehicular networks. Hence, simulation has been the tool of choice among protocol and system designers. To simulate a networking scenario in a vehicular setting, both the vehicle movements and the wireless communications between them should be modeled and simulated. The movements of vehicles and the communications between them are modeled via appropriate *road traffic simulators* and *network simulators*, respectively. For a given simulation scenario, the handshaking between the employed traffic and network simulators occurs by means of a *conversion tool* or an *integration tool*. The conversion tool translates the outputs of the traffic simulator into trace files readable by the network simulator, whereas the integration tool supports live interactions between the traffic and network simulators. With a conversion tool, the trace files are generated before the network simulation begins, whereas the integration tool makes the simultaneous interworking and data exchange of the traffic and network simulators possible. Clearly, for scenarios such as safety or traffic where the movements of vehicles are affected by the data received, the use of an integration tool is inevitable. In this chapter, we introduce a comprehensive set of traffic and network simulators as well as conversion and integration tools. We believe that this chapter helps the researchers new to the field of vehicular networking to select the simulation package that best suits their needs and provides them with helpful insights as they run their first simulations.

22.1 INTRODUCTION

As automotive technology progresses into the twenty-first century, the need for wireless communication between vehicles becomes apparent. Wireless data dissemination and data delivery involving vehicles in the form of vehicular networks allow a wide variety of services to provide safety and nonsafety applications (also referred to as comfort or infotainment applications). Safety applications include collision warning/avoidance, dynamic speed limits, traffic congestion mitigation, and cooperative driving. Providing drivers with real-time safety and traffic information via wireless communications saves lives, avoids time and energy wastes, and reduces pollution. Nonsafety applications include Internet access, multimedia streaming, file sharing, and highway toll services to mention a few.

Communication networks formed by moving vehicles are characterized by the intrinsic characteristics of vehicular environments, such as dynamic topologies, high-speed mobility, highly variable channel conditions, and limited mobility due to road constraints. As a result, readily available protocols for nonvehicular wireless networks, such as those for general mobile ad hoc networks (MANETs), may not meet the varying requirements of vehicular applications or may be inefficient in vehicular network settings. Hence, new communication protocols need to be developed with respect to specific needs in vehicular networks and eventually deployed.

Newly designed communication protocols, network architectures, and potential services for vehicular networks should be well investigated and evaluated before being deployed. Two of the most common assessment alternatives are real-world test beds and computer simulations. Even though the actual testing of protocols in field experiments provides the most realistic results, it is often impractical due to cost, access, and scalability constraints, particularly in the case of large networks. Field testing is potentially hazardous when the protocols are supporting safety applications such as collision warning and avoidance. Therefore, simulation has been the tool of choice among protocol and system designers. Even in the case of smaller networks, simulation is widely used as the first step to assess newly developed protocols.

The architectures of vehicular networks are chosen primarily with considerations for the applications and scenarios that are of concern and the characteristics of vehicular traffic on the highway, urban, or rural street found in the considered scenarios. In highways or urban areas, a hybrid architecture is appropriate because vehicle-to-vehicle (V2V) communications may be used collaboratively in conjunction with vehicle-to-roadside (V2R) communications in order to provide ubiquitous connectivity and better performance. In rural areas where roadside infrastructure is sparse, the network is expected to rely in most situations on V2V communications for economical reasons.

In safety applications such as collision prevention or road condition notification, the delay requirements are usually very stringent. Hence, V2V data dissemination may be the only possible way to provide timely safety warnings to avoid the excessive delay of transferring data to and from the infrastructure. To balance road traffic, numerous traffic information systems based on both hybrid and infrastructure-less architectures have been proposed in the literature. In traffic systems with a hybrid architecture, the main goal is to minimize the traveling times of individuals in the network, whereas in infrastructure-less architectures, the objective is to balance the vehicular traffic in different neighborhoods. Various nonsafety applications require different data communication mechanisms and architectures that depend on the application, ranging from data dissemination in sale advertisements to data routing in Internet access, content delivery, etc. Any of these scenarios calls for a different approach in building vehicular network simulation-based test beds.

For simulating a wireless communication scenario in a vehicular environment, both the mobility of vehicles and the wireless communications between them should be modeled using appropriate simulation tools or platforms. For simulating the wireless networking protocol, a *network simulator* should be used, whereas the mobility of vehicles is usually simulated using a road *traffic simulator.* The outputs of traffic simulators, which include all vehicle positions at every time step of the simulation runtime, need to be converted to trace files with a format readable by network simulators. For this purpose, various conversion tools have been developed. Note that in this case, the generation of the trace files takes place before the network simulation begins. However, for some vehicular applications such as safety or traffic applications, the movements of vehicles are affected by the received packets. So for these applications, both traffic and network simulators are expected to be running simultaneously and exchanging data during the simulation runtime.

Previous survey papers on vehicular network simulations either investigate the integration of only one single traffic simulator and one single network simulator for different scenarios and applications [1–3] or only provide an overview of several traffic simulators and (or) network simulators for vehicular networks without studying or even considering the necessity for live interactions between the network and traffic simulators [4,5]. This motivated us to write this chapter in which we introduce a comprehensive set of both traffic and network simulators as well as possible conversion tools and integration alternatives. We believe that this chapter helps the researchers new to the field select appropriate vehicular network simulation platforms and provide them with helpful insights as they run their first vehicular simulations. The chapter is organized as follows.

A survey of the most common network simulators is given in Section 22.2 including the descriptions of the network simulators, followed by their unique features regarding the simulation of vehicular scenarios and their performance comparison. Available network simulators tend to oversimplify the propagation model of the physical channel, which might lead to unrealistic results when modeling wireless networking protocols in vehicular environments. However, the study of more precise channel propagation models for vehicular scenarios requires much further work and is out of the scope of our book chapter. In Section 22.3, we will introduce a number of popular traffic simulators. Also, we will explore the process and principles of generating vehicular movement scenarios in traffic simulators with an emphasis on how to generate road topologies, traffic demands, and desirable outputs. In Section 22.4, a number of conversion tools will be introduced as well as various integrated platforms providing real-time interactions between traffic and network simulators. In Section 22.5, we explain how the simulation setup is generated for a typical vehicular mobility and networking scenario. A brief summary is given in Section 22.6.

22.2 NETWORK SIMULATORS

The number of existing wireless network simulators is increasing every day, making it more difficult for researchers to choose the most appropriate one that best meets their needs. Some of the most
common network simulators in this area of research are Network Simulator 2 (NS-2) [6], Qualnet
[7], OPNET [8], OMNeT++ [9], GloMoSim [10], and JiST/SWANS [11]. Among these network simulators, OPNET and Qualnet are commercial products whose core programs are not open source.
In this section, we briefly explain the most popular open-source network simulators that are used by
researchers for performance evaluation of vehicular networks. NS-2 has been one of the most commonly utilized network simulators over the last decade, whereas simulators such as OMNeT++ and
JiST/SWANS are gaining more popularity as they develop more advanced user-friendly features
and claim to be more efficient in terms of their execution performances. It is worth mentioning that
all these simulators are discrete event based, which means that they use an event scheduler to keep
track of all the events stored in an event queue and their simulation times.

22.2.1 NETWORK SIMULATOR VERSION 2

NS-2 was originally aimed for the simulation of Transmission Control Protocol (TCP), Internet Protocol
(IP), and various routing and multicast protocols over wired and wireless networks. The major engine
of NS-2 is written in the C++ programming language, which defines the operations of different network component objects and the event scheduler in the network. Some examples of network objects
are nodes, links, agents, and protocols such as routing, transport, and application protocols. Aside from
C++, NS-2 also uses the Tcl programming language in the form of object-oriented Tcl (OTcl) scripts as
the user interface to define the network topology as a combination of objects and to initiate the network
scheduler. The use of OTcl scripts makes the topology and scheduling changes in the network convenient and time efficient as it removes the need for unnecessary recompilations of the C++ core as a result
of any slight changes. Due to the popularity of NS-2, many efforts have been made to adapt its modules to vehicular networking scenarios and to mitigate the detailed programming needed to enable the
required level of realism for vehicular networking. Some of these efforts are described in Section 22.4.

22.2.2 OMNeT++

The building blocks in OMNeT++ simulations are called *simple modules*. As in NS-2, these modules,
which form the lowest level of simulator hierarchy, are written in C++. A number of simple modules can
be integrated by the user to form a *compound module*. Subsequently, multiple simple and/or compound
modules can be linked to form a *model* such as a protocol. To define the structure of a model describing
the message exchanges and communications between the corresponding modules, the user takes advantage of a network description language called NED. NED plays the similar interface role as the Tcl language does in NS-2. It should be noted that contrary to NS-2, which is known as a network simulator,
in its website, OMNeT++ is introduced as a general-purpose simulation engine that takes advantage
of independently developed frameworks, also called packages, to support the simulation of wireless
networks. Two popular frameworks that provide a comprehensive set of wireless protocols are INET
and INETMANET frameworks. INET is designed for both wired and wireless networks and contains
models for protocols such as Internet protocol (IP), transmission control protocol (TCP), user datagram protocol (UDP), Ethernet, multiprotocol label switching (MPLS), and 802.11. INETMANET
is derived from INET, but specifically aimed for the implementation of routing protocols in MANETs.

22.2.3 JiST/SWANS

Java in Simulation Time (JiST) is a general-purpose simulation engine running atop a standard
Java virtual machine. A simulation scenario in JiST is composed of a number of entities, each of

them running an independent Java code in parallel. The simulation engine controls the interactions and synchronizations between different entities any time two or more entities need to execute some commands in their codes simultaneously. In order to simulate wireless scenarios, a Scalable Wireless Ad hoc Network Simulator (SWANS) is built over the JiST simulation engine, which supports most of the protocols readily available for ad hoc networks and is also compatible with NS-2 source codes. One of the reasons JiST/SWANS is gaining more acceptance among researchers for the simulation of vehicular networking scenarios is the fact that the street random waypoint (STRAW) [12] mobility model is specifically designed for SWANS. STRAW is an adaptation of the well-known random waypoint mobility model [13] to vehicular environments. It was observed that STRAW yields more accurate simulation results compared to the ordinary random waypoint model when applied to US city maps [12]. Some important features of STRAW include constraining vehicle movements to street layouts defined by street maps, considering the effects of traffic congestion on vehicle mobility, supporting lane changing on multilane streets, and replicating simple traffic control mechanisms. It should be noted that new mobility models are claimed to be included in more recent versions of STRAW [14]. The random waypoint mobility model is described in more detail in Section 22.3.1.

22.2.4 Performance Comparison

Much research has been conducted to compare the effectiveness of different network simulators [15–18]. The most popular evaluation metrics are simulation runtime and memory consumption. As mentioned before, all the network simulators we have introduced in this section are open source and therefore no commercial license is required. The evaluation procedure is that a reference simulation scenario is defined and is implemented in all of the network simulators being evaluated. The results show that JiST has a superior runtime performance that is attributed to the fact that different entities run their codes separately but in parallel. NS-2 has the slowest simulation runtime caused by the interworking between C++ and OTcl codes in its architectural design. JiST also appears to be more efficient in terms of memory consumption compared to NS-2 and OMNeT++ in which the required memory usage grows linearly as the size of the network increases. It is worth mentioning that in terms of memory consumption, OMNeT++ performs slightly better than NS-2. Despite its performance shortcomings, NS-2 is the most commonly used network simulator. A review on 151 research papers involving wireless ad hoc network simulations in 2005 [19] showed that NS-2 had the highest usage, that is, 43.8%, among others. This is because NS-2 has been established for a long time. Since it is open source, a comprehensive set of models and protocols has been developed and contributed by researchers over this long period of time. As a result, besides the popular network protocols present in most other network simulators, there also exist a large number of less frequently used models and protocols contributed to NS-2. Most of these contributions are not included in the basic NS-2 installation packages, but can be found in the *Contributed Code* webpage [20] or in one of the numerous NS-2 forums or newsgroups. Thus, NS-2 is virtually considered as the standard for network simulation. A brief comparison of the network simulators under study is given in Table 22.1.

TABLE 22.1
Comparison of Network Simulators

	JiST	NS-2	OMNeT++
Simulation runtime	Low	High	Medium
Memory consumption	Low	High	High
Library support for existing protocols and models	Low	High	Medium

22.3   ROAD TRAFFIC SIMULATORS

Road traffic simulators also play a key role in the simulation of vehicular networks. In general, traffic simulators can be categorized as macroscopic and microscopic, depending on the level of details describing the traffic flows [21]. Macroscopic models represent vehicles as an aggregated traffic flow for simplicity. On the other hand, microscopic models have the ability to describe each vehicle as an individual object with unique characteristics and so are able to account for the complex interactions among vehicles and between drivers and transportation infrastructures.

Road traffic simulators require a complete set of inputs for reliable modeling of vehicular traffic. One of the key inputs is a road network or *road topology* that is represented as a set of intersections or *nodes* linked to each other via a set of streets or *edges*. Road topology data must include the information of node locations and link lengths as well as the geometric and operational details of each link, for example, number of lanes or allowable directions, curvature radius, and lane configuration at intersections. Also, various traffic control and management measures at intersections could optionally accompany the road topology data including no control, signs, markings, roundabouts, and traffic signals depending on the traffic characteristics and surrounding environments.

Another required input is the *traffic demand* that generates flows of vehicles on the road topology moving from their origins to destinations and assigns the vehicle *mobility models* specifying individual drivers' behaviors such as cars' speeds, accelerations, interactions with other nearby cars, behaviors at intersections, and overtaking. Also, before starting a simulation, user-specified input parameters such as the simulation period, the simulation start and end times, and the simulation step size should be determined. In the subsequent subsections, we will give more details on the provisioning of any of these inputs in a number of traffic simulators.

22.3.1   GENERIC MOBILITY SIMULATION FRAMEWORK

The generic mobility simulation framework (GMSF) [22–24] developed in ETH Zurich is a microscopic traffic simulator that can employ a number of mobility models to generate realistic mobility trace files. The road topologies originally used in the framework are derived from Swiss geographic information system (GIS) landscape model [25], which is not an open-source database. However, the road topologies could also be imported as networks generated by OpenStreetMap (OSM) [26], which is a free road topology database.

The employable mobility models suggested by the developers are GIS-based mobility model [24], multiagent microscopic traffic simulator (MMTS) model [27], Manhattan model [28], and random waypoint model [13]. In the GIS-based mobility model, vehicles define steady-state random trips constrained to the road topology derived from GIS maps [25] with the possibility of keeping a safe distance from the vehicles in front of them. When moving along the trips, vehicles use the intelligent driver model (IDM) [29] to adjust their speeds and perform their car-following mechanism. Also, a simple traffic light model could be used to regulate vehicle traffic at intersections.

MMTS is a microscopic vehicular mobility model also developed at ETH Zurich. It realistically models the behaviors of all individuals living in the region of interest collected by a census. It generates mobility traces according to the trips individuals take during their daily schedules taking their means of transportation into account for any of the three possible rural, urban, or city scenarios. In the Manhattan model, a grid layout is used as the road topology. Vehicles are randomly accelerating and decelerating within a permitted range of speed unless their distance to the vehicle ahead of them falls below a threshold. In this case, the maximum speed in the allowable range of speed will be set to the speed of the front vehicle. The random waypoint model was originally designed for MANETs and may not suit the purpose of realistic vehicular mobility modeling. In this model, a node moves along a straight line from its origins to an intermediate waypoint at a fixed speed in an open field with no restriction, which is randomly selected between a lower and an upper speed limit. At a waypoint, the node pauses for a time randomly selected between zero and a maximum value and resumes its straight-line motion at a new and randomly chosen direction to the next waypoint by repeating this process. Therefore, this model cannot be used for realistic modeling of vehicle movements, which are constrained to street layouts and controlled by traffic regulations.

Many formats for the outputs of GMSF are possible. GMSF can generate trace files immediately usable by NS-2 and Qualnet network simulators. Also, the traces can be stored in a generic simulator-independent Extensible Markup Language (XML) file format that can be easily translated to any format according to users' needs. Some realistic vehicular mobility trace files for NS-2, which are obtained using MMTS mobility model, are available in [27].

22.3.2 VANETMOBISIM

VanetMobiSim [30–32] is a flexible vehicular traffic simulator coded in Java for modeling vehicular mobility and generating realistic vehicular movement traces for use in different network simulators such as NS-2, GloMoSim, and Qualnet.

The road topology to be used in the modeling can either be defined by users or can be imported from roadmap databases. The user-defined topology is a graph in which intersections and streets connecting them are mapped onto vertices and edges of the graph. The imported maps could be according to that geographical data file (GDF) [33] standard, which are not freely available, or else derived from maps obtained from the U.S. Census Bureau TIGER database [34], which are more widely available as the database is open source.

After the topology is determined, road structure characteristics should be specified by introducing the roads with multiple lanes allowing vehicles to flow on any of the lanes in different directions and the definition of speed limits and other traffic signs on streets. Traffic flows at intersections could be regulated by *stop signs* or *traffic lights.*

For modeling the traffic demand, that is, vehicle movement patterns, two modules have been provided in VanetMobiSim. The first module called *trip generating module* is responsible for defining a set of start and stop points on the road topology. The other module called *path computation module* is responsible for selecting the best path between the start and stop points denoted by a sequence of edges on the topology graph. The path computation logic VanetMobiSim employs is to select the path with the minimum sum of edge costs according to Dijkstra's algorithm. The edge costs could be determined by either the lengths, traffic congestion levels, or speed limits of the edges.

To achieve more realistic results, VanetMobiSim has been designed to model a number of individual vehicle behaviors including smooth variations in vehicle speeds, car queues, following traffic regulations at intersections, traffic congestion, and overtaking. This feature requires that vehicle movements are defined as functions of nearby vehicles in their multilane multiflow surroundings. For this purpose, two mobility models have been defined in VanetMobiSim. One of them is referred to as *Intelligent Driver Model with Intersection Management* (IDM-IM), which adapts behaviors of vehicles to the traffic signs at intersections, that is, stop signs or traffic lights. The other one is called *Intelligent Driver Model with Lane Changes* (IDM-LC), which controls vehicle behaviors upon overtaking and changing lanes. It is worth mentioning that VanetMobiSim also has an incident generator feature whereby one may generate incidents by specifying the class of each incident, its location, and its duration.

22.3.3 VISSIM

In the transportation community, a number of commercial microscopic traffic simulation packages have been widely used as an analysis tool for the design and assessment of a variety of transportation systems. AIMSUM, PARAMICS, and VISSIM (*Verkehr In Städten—SIMulationsmodell,* German for *traffic in cities—simulation model*) are full-featured simulation models providing user-friendly modeling interfaces with the ability to model large-scale transportation networks to account for area-wide impacts of localized activities or systems operation, in a sufficiently detailed level to analyze the interactions among vehicles, that is, lane changing or overtaking, and between vehicles and transportation systems. These microscopic simulation software programs also offer the ability to obtain detailed state variable information on each vehicle on time scales with better than second-by-second accuracy [35]. They can simulate surface street networks, freeways, interchanges, weaving and on-off ramp sections, and stop or traffic-actuated controlled intersections.

For the last decade, significant advancements have been made in microscopic traffic simulation, so each software has similar strengths and capabilities nowadays [36]. AIMSUN, PARAMICS, and VISSIM provide a strong customized application development tool through application programming interfaces (APIs) that enable users to develop and test advanced transportation systems and applications including, but not limited to, intelligent transportation systems and wireless communication–based vehicular networks in general.

This section describes some key capabilities of current microscopic simulation software with VISSIM as an example of this class of tools. VISSIM is the traffic simulator developed by PTV AG in Germany [37,38]. VISSIM is based on a time step and behavior-based microsimulation model. Detailed state variable information on each vehicle can be obtained on a time scale as low as 1/10 s. The basic concept of vehicle movement modeling is based on the psychophysical driver behavior model developed by Wiedemann [39]. Speed and spacing decision of each individual vehicle is determined by stochastic distributions, which have been calibrated through multiple field measurements in Germany. The traffic simulator of VISSIM allows drivers to react to not only preceding vehicles but also neighboring vehicles on multiple lane roadways as well as traffic signals in a higher alertness when approaching signalized intersections.

The traffic simulator of VISSIM is a multimodal simulator that allows users to define a full range of vehicle types including passenger cars, buses, trucks, and heavy and light rail vehicles as well as pedestrians and cyclists. Traffic demands can be assigned to the network using the static or dynamic method. The static method requires users to define the traffic demand, trip start and end points, and the (fixed) travel route for the traffic. The dynamic routing method allows drivers to adaptively switch their traveling route among user-definable paths when specific events occur such as incidents or congestions. Traffic demands must be defined using origin and destination matrices that are dependent on time and vehicle class when using the dynamic traffic assignment (DTA) feature.

For the modeling of traffic control measures such as traffic signals, roundabouts, signs, or any combination, VISSIM offers a selection of several alternatives depending on the operation complexity. Simple traffic signal operations (i.e., fixed-time signal control) or sign-controlled intersections can be modeled using the graphical intersection modeling interface of VISSIM. For more complex control measures such as traffic-actuated control, VISSIM provides the vehicle actuated programming (VAP) interface, a C-like traffic control macrolanguage. VAP is supplemented with an error checking and debugging features.

One very important feature of VISSIM that sets it apart from other simulation models is its component object model (COM) programming interface. The COM interface allows users to develop and implement their own applications on the VISSIM network using computer programming language such as C++, Visual Basic, or Python. The COM interface provides user-developed applications with an access to the network topology, signal control, path flows, and vehicle behaviors enabling VISSIM to model complex control logic and sophisticated transportation systems and components [40].

22.3.4 SIMULATION OF URBAN MOBILITY

Simulation of Urban Mobility (SUMO) [41,42] is an open-source microscopic continuous-space and discrete-time vehicular traffic simulator developed by the German Aerospace Centre [43]. Some important features of SUMO are the support for different vehicle types, multilane streets with lane changing, right-of-way rules at intersections, support of a graphical user interface (GUI), dynamic vehicular routing, and support of a large variety of network formats as inputs.

The first step in defining the simulation scenario is building the network. In SUMO, the road topology can be obtained by either generating an XML file manually describing the network, importing from other software packages, or using automatic topology generation functions in SUMO. To generate the road topology by hand, two files are needed: one file for defining the nodes including their IDs and positions and another file for describing the edges between the nodes, including their start and end nodes, number of lanes, maximum speed, length, *functionality*, and some optional parameters such as *priority*. Edges in terms of functionality could be considered as *plain*, *source*, or *sink*. Vehicles are emitted to the network on source edges, whereas they are removed from the network on sink edges. Also, the assigned priorities are used in computing way-giving rules at intersections. Another possibility is to assign *additional weights* to edges to make them more or less attractive for vehicles to choose.

Some of the popular formats to import road topologies to SUMO are as follows. SUMO supports importing of networks from VISUM [44], VISSIM, TIGER, and OSM as well as importing ArcView [45] networks in the form of ArcView databases. Recent TIGER files are stored in the form of ArcView shapefiles, which are directly convertible into network descriptions readable by SUMO. XML files generated by OSM can be directly imported into SUMO. Alternatively, they can be imported to eWorld for editing and enrichment [46]. eWorld is a framework that can visualize and manipulate OSM mapping data before passing it to SUMO or VanetMobiSim. Some possible modifications in mapping data allowed by eWorld include the editing of names, number of lanes, speed ranges, and (or) priorities, adding traffic lights and editing their logics, and generating traffic events.

In addition to the manual generation of topology descriptions by hand and importing them, SUMO provides the possibility of automatic topology generation including three network types: *grid networks*, *spider networks*, and *random networks*. In grid networks, the number of intersections in both horizontal and vertical directions should be determined as well as the relative distances between adjacent intersections. In spider networks, the number of axes, circles, and the distance between circles are to be specified.

After the road topology has been described, traffic demand is generated. Several approaches for generating traffic demand are possible in SUMO. Some of these approaches are the definition of *trips*, *flows*, *turning probabilities*, or *routes*. Both *deterministic* and *random* routes can be defined or can be imported from other simulation tools. The parameters used in defining a *trip* for a typical vehicle are the origin or the starting edge, the destination or the ending edge, and the departure time that determines when the vehicle is emitted to the network. If the trip describes periodic vehicle emissions to the network, other parameters such as period and (or) number of vehicles to be emitted are also required.

Other than using the definition of a trip for periodic vehicle emissions, one may define a *flow* to describe several vehicles using the same trip. Again, in the definition of a flow, the starting and ending edges should be determined. However, instead of the departure time, begin and end times and the number of vehicles to be emitted are determined. The time interval between begin and end times is divided by the number of vehicles to obtain departure times that are uniformly spread within the interval. Another possibility in defining flows is to leave out the ending street in their descriptions and include *turning probabilities* at junctions. In this case, vehicles will leave the network any time they arrive at a sink edge.

As the next step in the process of defining the simulation scenario, which is optional, SUMO can employ the *dynamic user assignment* approach in [47] to avoid congestion by changing vehicle paths adaptively. Upon the completion of all these steps, the topology description and movements of vehicles are available. In the final step, the corresponding files are fed to SUMO along with the simulation begin and end times to perform the simulation. SUMO can use various formats to generate its output files. One possible output format is the raw vehicle state dump, which contains the positions and speeds of all the vehicles in the network at all simulated time steps. This type of output can be imported to NS-2 and is usable by NS-2 after necessary conversions.

22.3.5 Summary

A large number of mobility parameters and features must be enumerated and assigned to different traffic simulators. It is thus evident that selecting an appropriate traffic simulator depends on many factors. Some important factors are the required level of details; the applications and services

TABLE 22.2
Comparison of Traffic Simulators

	GMSF	VanetMobiSim	VISSIM	SUMO
Road topologies imported from	OSM	GDF, TIGER	N/A	VISUM, VISSIM, TIGER, OSM, ArcView
Mobility models supported	IDM, MMTS, Manhattan, random waypoint	IDM-IM, IDM-LC	Wiedemann psychophysical driver behavior model, internal traffic control via VAP interface	Internal car-following model
Outputs usable in	NS-2, Qualnet	NS-2, Qualnet, GloMoSim	SUMO	NS-2

to be offered, which may or may not be affected by different features or mobility models supported by the traffic simulator; and very importantly, the need to interlink the selected traffic simulator to the selected network simulator. Among the traffic simulators reviewed, SUMO is the most popular one, and as described in Section 22.4, SUMO is the most commonly utilized road traffic simulator for integration into and interworking with other simulators. The strengths of SUMO can be summarized as follows: It provides the flexibility of working over various operating systems. Also, its simulation execution speed is much faster than other traffic simulators, which can be attributed to its car-following model. The car-following model employed may not include as many details as the models used in some other traffic simulators do. Another reason of SUMO's popularity is its open-source license and its active open-source community, which improves it and keeps its features updated. A brief comparison of the traffic simulators under study is given in Table 22.2.

22.4 INTERLINKING TRAFFIC AND NETWORK SIMULATORS

The vehicle positions and traffic states that have been obtained via traffic simulators are required to be converted to trace files usable by wireless network simulators. In this regard, many *conversion tools* have been developed. Some popular examples are TraceExporter [48], mobility model generation for vehicular networks (MOVE) [49], and TraNSLite [50], which will be described in Section 22.4.1.

Conversion tools only take into account the effects of vehicle movements on wireless communications. In other words, static vehicle traffic traces are obtained from traffic simulators and imported to network simulators before the network simulation process begins. However, for some vehicular applications, the data received by the vehicles during the simulation runtime also influence drivers' behaviors and, as a result, vehicle movements. Therefore, for a more realistic investigation of these applications, a coupled interlink between the traffic and network simulators, which supports bidirectional exchange of information, is required. Some examples of these applications are the case in which a vehicle receives the updated traffic conditions and recalculates its path to take the path with the minimum travel time or in the case of a collision or an emergency in which the vehicle should stop or change its movement direction to avoid the collision. In this regard, various *integrated platforms* providing real-time interactions between traffic and network simulators will be introduced in Section 22.4.2.

22.4.1 CONVERSION TOOLS

22.4.1.1 TraceExporter

TraceExporter coded in Java converts raw vehicle position dumps obtained as output files in SUMO to trace files readable by NS-2. Since NS-2 only accepts positive x and y coordinates for the positions of nodes, the conversion should shift all node positions with negative coordinates accordingly.

As a result of the conversion, three output files are obtained all in form of OTcl script files. Note that in NS-2, all information about the simulation scenario, for example, the topology and events, should be written in OTcl programming language in the form of OTcl script files. One of the files is a *config* file describing the configuration of the simulation scenario. Another file is an *activity* file specifying the start and stop times of the vehicle movements. The third file is a *mobility* file describing the actual movements of vehicles during the simulation time. The traffic simulation scenario parameters should be read from the OTcl files for use in the main OTcl file controlling all the events in the simulation scenario by using the *source* command.

22.4.1.2 Mobility Model Generation for Vehicular Networks

MOVE is another open-source conversion tool built atop SUMO that generates mobility trace files employable by both NS-2 and Qualnet. One of its features is a set of GUIs that helps generate mobility scenarios more quickly. Also, by using these GUIs, users can avoid having to write scripts from scratch to describe the traffic simulation scenarios, and they do not need to learn about the details of the underlying traffic simulator (SUMO).

22.4.1.3 TraNSLite

TraNSLite is another tool for fast generation of realistic mobility trace files for NS-2 from SUMO. It also supports map cropping and speed rescaling if the maps are imported to SUMO from TIGER. TraNSLite can be conveniently run in any computer supporting the Java runtime environment.

22.4.2 INTEGRATED PLATFORMS

Two types of integrated platforms are distinguishable in vehicular network simulations. In the first type, readily available traffic and network simulators are combined with the help of an *interface tool* interlinking them in a bidirectional way. In this regard, the most commonly used interface tool, that is, traffic control interface (TraCI) [51,52], and its architecture are briefly studied in Section 22.4.2.1, followed by the introduction of a number of integrated platforms of the first type. In the second type of integrated platforms, the traffic and network simulators are exclusively designed to interwork with each other. A couple of these types of platforms are studied in Section 22.4.2.2.

22.4.2.1 Integrated Platforms Utilizing Interface Tool

22.4.2.1.1 Traffic Control Interface

TraCI is an open-source interface that couples a road traffic simulator and a network simulator to enable real-time interactions. One of its important features is that a large variety of traffic and network simulators and different combinations of them can be interlinked via TraCI. The main idea of TraCI is to control and access the traffic simulator via the network simulator by means of an *application* running over the network simulator. For this purpose, a *TraCI-server* component and a *TraCI-client* component have been developed atop the traffic and network simulators, respectively. The two components use a *TCP connection* to exchange data.

TraCI works as follows. At every simulation step, the network simulator sends a simulation command to the traffic simulator including vehicle positions and the simulation times that are used for synchronization purposes. Upon the reception of this simulation command, the traffic simulator performs the next simulation step and sends back the resulting vehicle positions to the network simulator. Based on these newly received positions, the network simulator executes the vehicular networking protocol and any decisions based on the protocol are made, and the respective commands are sent to the traffic simulator in the next step.

So far, TraCI-client implementations for NS-2, OMNeT++, and SWANS++ have been provided. More information regarding implementation details and latest versions of TraCI can be found in the TraCI wiki [51].

22.4.2.1.2 Traffic and Network Simulation

The integrated platform that couples SUMO and NS-2 with the help of TraCI interface is referred to as the traffic and network simulation (TraNS) [50,53]. The developers of TraNS claim that TraNS could be used to couple any traffic simulator with any network simulator [53]. The implementation of TraNS for integrating SUMO and JiST/SWANS is investigated in [54].

TraNS provides the possibility to model traffic and safety events, such as traffic congestion and collisions. The events could be generated either at a specific location or via a specific vehicle. Also, a number of mobility models have been designed in TraNS to simulate individual vehicle behaviors in critical situations including *road danger warning* and *dynamic rerouting* to provide safety and traffic efficiency. However, the development of TraNS is currently suspended [49].

22.4.2.1.3 Vehicle in Network Simulation

Another integrated platform that integrates SUMO as the traffic simulator and OMNeT++ as the network simulator through TraCI interface is the Vehicle in network simulation (Veins) [55]. OMNeT++ has various frameworks, each of them developed for a specific purpose and provides protocols relevant for that purpose. TraCI is fully supported in both INET and INETMANET frameworks. Instructions on how to set up OMNeT++, INET framework, and the respective TraCI-client implementation are given in Sommer [55].

22.4.2.1.4 Multiple Simulation Interlinking Environment for C2CC in VANETs

Generally speaking, each simulator focuses on a specific set of applications and protocols of vehicular networks. Some researchers believe that the possibility of integrating various simulation environments keeps all their advantages while some disadvantages are removed. In this regard, a platform was developed in [56] to integrate NS-2 as the network simulator, VISSIM as the traffic simulator, MATLAB®/Simulink® [57] for application development, and Click [58] for realistic routing protocol simulations. Among all the integrated simulators, NS-2 controls the platform in a centralized manner by communicating with the other simulators. Hence, NS-2 can directly influence the mobility of vehicles and drivers' behaviors.

22.4.2.2 Exclusively Designed Integrated Platforms

22.4.2.2.1 NCTUns

Even though NCTUns [59,60] was originally developed as a network simulator, its more recent version is an integrated platform capable of supporting the interactions between the network and traffic simulators. Some vehicle movement parameters in the traffic simulator are initial and maximum speeds and initial and maximum accelerations. The moving paths and speeds of vehicles at each point in time are determined based on the traffic conditions in their surroundings at that point. Also, an intelligent driver behavior module is included in NCTUns, which provides features such as car following, lane changing, overtaking, turning, and obeying traffic lights. Different agents in the platform communicate by means of the TCP/UDP/IP protocol stack. The fact that road traffic is highly integrated in the platform makes the use of external traffic simulators in the platform difficult if not impossible, which in turn makes it a less favorable option for those researchers who wish to have more freedom in selecting an appropriate traffic simulator.

22.5 SIMULATION SETUP FOR A TYPICAL VEHICULAR SCENARIO

In this section, we generate the simulation setup for a typical vehicular mobility and networking scenario with two wireless vehicular nodes moving in an urban area using a TCP connection. In our scenario, we use NS-2, SUMO, and MOVE as the network simulator, traffic simulator, and conversion tool, respectively. Note that in the present scenario, we only use the protocols that are currently supported in NS-2. For more information regarding how new protocols can be defined in

NS-2, please see the *ns Manual* in [6]. We start with developing the wireless networking part of the scenario in NS-2 for which we start with creating an OTcl file to define the network topology and to initiate the network scheduler. Next, we elaborate on how node mobility traces are generated in SUMO. The generated mobility traces can then be converted to traces readable by NS-2 in MOVE.

First, we create an object of type "Simulator" and assign it the handle "ns" as follows:

```
set ns [new Simulator]
```

Also, we create a trace file "out.tr" in which traces are written and stored. The handle "tf" is assigned to the file:

```
set tf [open out.tr w]
```

By using the following line, the simulator object "ns" will trace all the events over the simulation time and write them in the trace file "out.tr":

```
$ns trace-all $tf
```

Now we continue with defining the networking topology. In NS-2, in order for the nodes to have mobility and wireless networking functionalities, various *network components* should be configured. The most common network components to be configured are ad hoc routing protocol, link layer (LL), address resolution protocol (ARP) model connected to LL, an interface priority queue (IfQ), and the media access control (MAC) layer, channel, and radio propagation model. More details about these components and other configurable components can be found in the ns Manual [6]. In NS-2, before creating wireless mobile nodes, their network components should be configured as follows:

```
$ns node-config     -addressingType flat\#or hierarchical
                    -adhocRouting DSR\#or DSDV or TORA or AODV
                    -llType LL \
                    -macType Mac/802_11 \
                    -ifqType DropTail\#or Queue or PriQueue
                    -ifqLen 50 \
                    -antType Antenna/OmniAntenna \
                    -propType Propagation/TwoRayGround \
                    -phyType Phy/WirelessPhy \
                    -topoInstance $topo \
                    -channelType Channel/WirelessChannel \
                    -wiredRouting ON #or OFF \
                    -mobileIP ON #or OFF \
                    -energyModel <EnergyModel type> \
                    -initialEnergy <specified in Joules> \
                    -rxPower <specified in W> \
                    -txPower <specified in W> \
                    -agentTrace ON #or OFF \
                    -routerTrace ON# or OFF \
                    -macTrace ON #or OFF \
                    -movementTrace ON #or OFF \
```

Note that in the second line, dynamic source routing (DSR), destination-sequenced distance vector routing (DSDV), temporally ordered routing algorithm (TORA), and ad hoc on demand distance vector routing (AODV) are classic routing protocols for wireless ad hoc networks that are frequently used as the benchmark protocol for the purpose of performance evaluations. The last four lines enable us to trace the simulation events at different levels of the network stack, that is, router, MAC, or *agent* level. The definition of agent will be given later in this section.

After configuring the network components, we create two objects of type "node" with these configurations and assign them handles "n0" and "n1" as follows:

```
set n0 [$ns node]
$n0 random-motion 0
set n1 [$ns node]
$n1 random-motion 0
```

Since we intend to import the mobility traces from SUMO, random motion is disabled for both of the nodes. As the next step, the node mobility file generated by SUMO and converted to the right format by MOVE is loaded into the simulation scenario in NS-2 using the following command:

```
Source <movement-scenario-files>
```

More details on how this file is generated are given later in this section.

In NS-2, when a TCP connection is established between two nodes, they would exchange packets as long as they are in the transmission range of each other, which means that the packets are dropped as soon as the nodes go out of the transmission range. In the following, the TCP connection is set up between "n0" and "n1":

```
set tcp [new Agent/TCP]
$ns attach-agent $n0 $tcp
set sink [new Agent/TCPSink]
$ns attach-agent $n1 $sink
$ns connect $tcp $sink
```

Note that in NS-2, the data are always sent from one object of type "Agent" to another "Agent" object. Hence, in the aforementioned, we create one object of type "Agent/TCP" with handle "tcp" and another object of type "Agent/TCPSink" with handle "sink" and connect them to each other.

Now, a traffic source should be generated and should be attached to the "tcp" agent defined earlier. In the following, we create an object of type "Application/FTP" with handle "ftp" over the TCP connection:

```
set ftp [new Application/FTP]
$ftp attach-agent $tcp
```

Now, we need to determine when the "ftp" traffic source should start and stop generating traffic by using these commands:

```
$ns at 1.0 "$ftp start"
$ns at 100.0 "$ftp stop"
```

In order to finish the simulation and close the trace file "out.tr", we tell the simulator object "ns" to execute "finish" procedure at time 120.0 s:

```
$ns at 120.0 "finish"
```

where "finish" procedure is defined as

```
proc finish {} {
        global ns tf
        $ns flush-trace
        close $tf
}
```

Eventually, we start the simulation using the following command:

```
$ns run
```

The only part of the simulation scenario that is still left to be generated is the node mobility file. In the rest of the section, we turn our attention to generating the road topology and traffic demand in SUMO. To manually generate the road topology in an XML file, the intersections, so-called nodes in SUMO, are defined in a .nod.xml file, whereas the streets linking the intersections, so-called edges, are defined in a .edg.xml file. To give an example, several nodes are defined in sample file "example.nod.xml" as follows:

```
<nodes>
        <node id = "1" x = "-500.0" y = "0.0"/>
        <node id = "2" x = "+500.0" y = "0.0"/>
        <node id = "3" x = "+501.0" y = "0.0"/>
</nodes>
```

Also, in file "example.edg.xml", several edges are defined to connect the nodes defined in "example.nod.xml" as follows:

```
<edges>
        <edge fromnode = "1" id = "1to2" tonode = "2"/>
        <edge fromnode = "2" id = "out" tonode = "3"/>
</edges>
```

Now, function "netconvert" should be called to generate the road topology from .nod.xml and .edg.xml files. For the set of nodes and edges defined in "example.nod.xml" and "example.edg.xml", the network topology can be generated in file "example.net.xml" as follows:

```
netconvert --xml-node-files=example.nod.xml --xml-edge-files=example.edg.xml --output-file=example.net.xml
```

The other alternative for generating the road topology is to import it from another software such as VISUM, VISSIM, TIGER, or OSM. In the present scenario, we import the file from OSM. OSM generates .osm.xml files. We can use "netconvert" to import file "example.osm.xml" to SUMO as follows:

```
Netconvert --osm-files=exmple.osm.xml --output-file=example.net.xml
```

The last step is to define the traffic demand. As mentioned before, there are various possibilities for generating traffic demands. One option is to define the vehicles' trips. The common syntax for defining a trip contains the ID, departure time, and origin and destination of the trip. An example is as follows:

```
<trips>
        <tripdef id = "0" depart = "25" from = "edge0" to = "edge1"/>
</trips>
```

Another possibility is to define a traffic flow. The definition of the flow should contain ID, origin, destination, begin and end times, and the number of vehicles that should be generated between the begin and end times. An example is

```
<flows>
        <flow id = "0" from = "edge0" to = "edge1" begin = "0" end = "360"
no = "100"/>
</flows>
```

The other common way to generate traffic demand is by defining turning probabilities, an example of which is as follows:

```
<turn-defs>
    <fromedge id = "myEdge0">
        <toedge id = "myEdge1" probability = "0.2"/>
        <toedge id = "myEdge2" probability = "0.7"/>
        <toedge id = "myEdge3" probability = "0.1"/>
    </fromedge>
    ... any other edges ...
</turn-defs>
```

22.6   SUMMARY

In the study of vehicular communications, due to high costs of real-world experiments and scalability problems, simulation tools have gained a high degree of acceptance in the research community for evaluating new protocols and mechanisms in vehicular networks. The simulation of vehicular scenarios is comprised of the simulation of wireless communications in a network simulator and simulating the movements of vehicles in a traffic simulator. In Section 22.2, we have introduced some of the most popular network simulators with the focus on open-source discrete event-driven simulations. A comparison of the introduced simulators indicates that JiST is both faster and more efficient with respect to memory consumption. However, NS-2 has been the most popular network as it was developed much earlier than other network simulators and also due to the wide range of available models that have been developed for it to cover a wide range of protocols for various types of communication networks. In Section 22.3, a number of popular road traffic simulators have been introduced, and their unique features and the types of scenarios and applications they are intended for have been presented. Among the traffic simulators reviewed, SUMO is most commonly used in relevant studies as it can be easily adjusted to work in various operating systems and due to its faster execution speeds. A survey on available approaches for one-way or bidirectional connection of traffic simulators to network simulators has been given in Section 22.4. A number of conversion tools have been introduced for importing outputs of traffic simulators to network simulators, and integrated platforms have been reviewed for applications that require real-time interactions between traffic and network simulators.

EXERCISES

22.1 Road traffic simulators fall into two categories including microscopic and macroscopic approaches by the level of detail in modeling traffic flows. Describe those two approaches and main differences.

22.2 Road traffic simulators allow static and DTA methods. Describe and compare these two assignment approaches.

22.3 Determine whether a conversion tool or an integrated platform should be used for any of the following simulation scenarios:

    a. The report of a slippery segment on a highway should be notified to every vehicle in a two-kilometer neighborhood.

    b. A driver in a city environment requires Internet access to one of the Wi-Fi hotspots on the roadside to download music.

    c. A driver in a city environment inquires the traffic center about the traffic conditions in its neighborhood in order to avoid congestions.

22.4 For simulating any of the scenarios described in the previous question, specify which set of network simulator, traffic simulator, and (or) integrated platform among those discussed in the book chapter you would choose. Give appropriate reasons to justify your selections.

22.5 Using a deliberate set of the simulation software introduced in this chapter, implement the following scenario:

100 vehicles are moving on a grid street layout network of 2 km × 2 km with average street lengths of 400 m. Upon arrival on each street, the arriving vehicle selects a fixed speed and drives at the same speed until it leaves that street. The speed to be selected for every arriving vehicle has a uniform distribution with minimum and maximum of 15 and 25 m/s, respectively. At intersections, the turning probabilities are the same for all the outgoing streets. Assume that four Wi-Fi access points are available at the four corners of the grid street layout to which vehicles are sending their packet traffic. The simulation time is 200 s, the transmission ranges of vehicles are 200 m, the radio propagation model is two ray ground, and the traffic model is constant bitrate (CBR) over 20 random vehicles at the rate of 4 packets/s with packet size of 1 kB. The data rate is 1 Mbps and the nodes use 802.11 MAC with maximum contention window of 32.

Assuming that the maximum tolerable one-way delay of a generated packet is 10 s, compare the average packet delivery ratio and average packet delivery delay of the packets using DSR and AODV as the ad hoc routing protocols.

ACKNOWLEDGMENTS

This work was supported in part by grants from AUTO21 under the Canadian Network of Centers of Excellence Program and from the NSERC DIVA Strategic Research Network.

REFERENCES

1. R. Fujimoto, H. Wu, R. Guensler, and M. Hunter, Evaluating vehicular networks: Analysis, simulation, and field experiments, in *Modeling and Simulation Tools for Emerging Telecommunication Networks*, A. Nejat Ince and E. Topuz, eds. Springer, New York, 2006.
2. R. Fernandes, P.M. d'Orey, and M. Ferreira, DIVERT for realistic simulation of heterogeneous vehicular networks, in *Mobile Ad hoc and Sensor Systems (MASS), 2010 IEEE Seventh International Conference on*, San Francisco, CA, pp. 721–726, November 8–12, 2010.
3. S. Fontanelli, Vehicular networks: Traffic simulations and communication protocols, Bachelor Degree and Master Degree thesis, 2008.
4. M. Boban and T.T.V. Vinhoza, Modeling and simulation of vehicular networks: Towards realistic and efficient models, in *Mobile Ad-Hoc Networks: Applications*, X. Wang, ed. InTech, 2011.
5. J. Harri, F. Filali, and C. Bonnet, Mobility models for vehicular ad hoc networks: A survey and taxonomy, *Communications Surveys & Tutorials, IEEE*, 11 (4), 19–41, 2009.
6. The Network Simulator—ns-2, http://www.isi.edu/nsnam/ns/ (last accessed February 10, 2014).
7. Scalable Network Technologies, http://www.qualnet.com (last accessed February 10, 2014).
8. OPNET, Application and Network Performance, http://www.opnet.com/ (last accessed February 10, 2014).
9. OMNeT++ Network Simulation Framework, http://www.omnetpp.org/ (last accessed February 10, 2014).
10. GloMoSim, Global Mobile Information System Simulation Library, http://pcl.cs.ucla.edu/projects/glomosim/ (last accessed February 10, 2014).
11. JiST/SWANS, Java in Simulation Time/Scalable Wireless Ad hoc Network Simulator, http://jist.ece.cornell.edu/ (last accessed February 10, 2014).
12. STRAW—STreet RAndom Waypoint, vehicular mobility model for network simulations, http://www.aqualab.cs.northwestern.edu/projects/144-straw-street-random-waypoint-vehicular-mobility-model-for-network-simulations-e-g-car-networks (last accessed February 10, 2014).
13. C. Bettstetter, H. Hartenstein, and X. Pérez-Costa, Stochastic properties of the random waypoint mobility model, *ACM/Kluwer Wireless Networks: Special Issue on Modeling and Analysis of Mobile Networks*, 10 (5), 555–567, 2004.

14. SWANS++, Extensions to the Scalable Wireless Ad-hoc Network Simulator, http://www.aqualab. cs.northwestern.edu/projects/143-swans-extensions-to-the-scalable-wireless-ad-hoc-network-simulator (last accessed February 10, 2014).
15. K.M. Reineck, Evaluation and comparison of network simulation tools, Master Thesis, August 2008.
16. E. Weingartner, H.V. Lehn, and K. Wehrle, A performance comparison of recent network simulators, in *Communications, 2009. ICC'09. IEEE International Conference on*, Dresden, Germany, pp. 1–5, June 14–18, 2009.
17. B. Schilling, Qualitative comparison of network simulation tools, Modeling and Simulation of Computer Systems seminar, Institute of Parallel and Distributed Systems (IPVS), University of Stuttgart, Stuttgart, Germany, 2005.
18. D.M. Nicol, Comparison of network simulators revisited, http://www.ssfnet.org/Exchange/gallery/ dumbbell/dumbbell-performance-May02.pdf (last accessed February 10, 2014).
19. S. Kurkowski, T. Camp, and M. Colagrosso, MANET simulation studies: The incredible, *SIGMOBILE Mobile Computing and Communications Review*, 9 (4), 50–61, 2005.
20. NS-2 Contributed Codes, http://nsnam.isi.edu/nsnam/index.php/Contributed_Code (last accessed February 10, 2014).
21. M. Pursula, Simulation of traffic systems—An overview, *Journal of Geographic Information and Decision Analysis*, 3 (1), 1–8, 1999.
22. P. Sommer, Design and analysis of realistic mobility model for wireless mesh networks, Master Thesis, ETH Zurich, Zürich, Switzerland, 2007.
23. R. Baumann, F. Legendre, and P. Sommer, Generic mobility simulation framework (GMSF), in *Proceeding of the First ACM SIGMOBILE Workshop on Mobility Models (MobilityModels'08)*, Hong Kong, China. ACM, New York, pp. 49–56, 2008.
24. Generic Mobility Simulation Framework (GMSF), http://gmsf.sourceforge.net/ (last accessed February 10, 2014).
25. VECTOR 25—Landscape model of Switzerland, The Swiss Federal Office of Topography (swisstopo), http://www.swisstopo.admin.ch/internet/swisstopo/en/home/products/landscape/vector25.html (last accessed February 10, 2014).
26. Open Street Map, http://www.openstreetmap.org/index.html (last accessed February 10, 2014).
27. Realistic Vehicular Traces, Laboratory for Software Technology, ETH Zurich, Zürich, Switzerland, http://www.lst.inf.ethz.ch/research/ad-hoc/car-traces/ (last accessed February 10, 2014).
28. F. Bai, N. Sadagopan, and A. Helmy, IMPORTANT: A framework to systematically analyze the Impact of Mobility on Performance of RouTing protocols for Adhoc NeTworks, in *INFOCOM 2003. 22nd Annual Joint Conference of the IEEE Computer and Communications. IEEE Societies*, San Francisco, CA, vol. 2, pp. 825–835, March 30–April 3, 2003.
29. M. Treiber, A. Hennecke, and D. Helbing, Congested traffic states in empirical observations and microscopic simulations, *Physical Review E*, 62, 1805–1824, 2000.
30. VanetMobiSim, http://vanet.eurecom.fr/ (last accessed February 10, 2014).
31. J. Härri, M. Fiore, F. Fethi, and C. Bonnet, VanetMobiSim: Generating realistic mobility patterns for VANETs, in *Proceedings of the Third International Workshop on Vehicular Ad hoc Networks (VANET'06)*, Los Angeles, CA. ACM, New York, pp. 96–97, 2006.
32. M. Fiore, J. Härri, F. Fethi, and C. Bonnet, Vehicular mobility simulation for VANETs, in *Simulation Symposium, 2007. ANSS'07. 40th Annual*, Norfolk, VA, pp. 301–309, March 26–28, 2007.
33. Ertico, Intelligent Transport Systems and Services for Europe, http://www.ertico.com (last accessed February 10, 2014).
34. U.S. Census Bureau TIGER system database, http://www.census.gov/geo/www/tiger (last accessed February 10, 2014).
35. D. Gettman and L. Head, Surrogate safety measures from traffic simulation models, Final report prepared for the Federal Highway Administration, U.S. Department of Transportation, Washington, DC, 2003.
36. S.L. Jones, A.J. Sullivan, N. Cheekoti, M.D. Anderson, and D. Malave, Traffic simulation software comparison study, Final report prepared for University Transportation Centre for Alabama, The University of Alabama, Tuscaloosa, AL, UTCA Report 02217, 2004.
37. VISSIM, Multi-Modal Traffic Flow Modeling, http://vision-traffic.ptvgroup.com/en-us/products/ ptv-vissim/ (last accessed February 10, 2014).
38. PTV Vision, VISSIM 5.30–05 User Manual, 2011, http://www.et.byu.edu/~msaito/CE662MS/Labs/ VISSIM_530_e.pdf (last accessed February 10, 2014).
39. R. Wiedemann, Simulation des Straßenverkehrsflusses, Schriftenreihe des Instituts für Verkehrswesen der Universität Karlsruhe, Karlsruhe, Germany, Heft 8, 1974.

40. VISSIM 5.10 COM Interface Manual, PTV Germany, 2009.
41. SUMO, Simulation of Urban MObility, http://sumo-sim.org/wiki/Main_Page (last accessed February 10, 2014).
42. D. Krajzewicz, J. Erdmann, M. Behrisch, and L. Bieker, Recent development and applications of SUMO—Simulation of Urban MObility, *International Journal on Advances in Systems and Measurements*, 5 (3&4), 128–138, December 2012.
43. Deutsches Zentrum fur Luft und Raumfahrt, http://www.dlr.de/ (last accessed February 10, 2014).
44. VISUM, http://vision-traffic.ptvgroup.com/en-us/products/ptv-visum/ (last accessed February 10, 2014).
45. Arc View Network Import, http://sourceforge.net/apps/mediawiki/sumo/index.php?title=Networks/Import/ArcView (last accessed February 10, 2014).
46. eWorld, A Traffic Simulation Connector, http://eworld.sourceforge.net/ (last accessed February 10, 2014).
47. C. Gawron, Simulation-based traffic assignment, Inaugural Dissertation, 1998.
48. TraceExporter, http://sourceforge.net/apps/mediawiki/sumo/index.php?title=Tools/TraceExporter (last accessed February 10, 2014).
49. F.K. Karnadi, Z.H. Mo, and K.C. Lan, Rapid generation of realistic mobility models for VANET, in *Wireless Communications and Networking Conference, 2007. WCNC 2007. IEEE*, Hong Kong, China, pp. 2506–2511, March 11–15, 2007.
50. TraNS, Realistic Simulator for VANETs, http://trans.epfl.ch/ (last accessed February 10, 2014).
51. Wiki for the Traffic Control Interface, http://sourceforge.net/apps/mediawiki/sumo/?title=TraCI (last accessed February 10, 2014).
52. A. Wegener, M. Piorkowski, M. Raya, H. Hellbrück, S. Fischer, and J.P. Hubaux, TraCI: An interface for coupling road traffic and network simulators, in *Proceedings of the 11th Communications and Networking Simulation Symposium (CNS'08)*, Ottawa, Ontario, Canada. ACM, New York, pp. 155–163, 2008.
53. M. Piorkowski, M. Raya, A.L. Lugo, P. Papadimitratos, M. Grossglauser, and J.P. Hubaux, TraNS: Realistic joint traffic and network simulator for VANETs, in *SIGMOBILE Mobile Computing and Communications Review*, 12 (1), 31–33, January 2008.
54. D. Choffnes and F. Bustamante, An integrated mobility and traffic model for vehicular wireless networks, in *Proceedings of the 2nd ACM International Workshop on Vehicular Ad Hoc Networks (VANET'05)*, Cologne, Germany. ACM, New York, pp. 69–78, 2005.
55. C. Sommer, Veins—TraCI modules for SUMO, OMNeT++, and JiST/SWANS, Vehicles in Network Simulation, http://veins.car2x.org/ (last accessed February 10, 2014).
56. Multiple Simulator Interlinking Environment for C2CC in VANETs, http://www.cn.uni-duesseldorf.de/projects/MSIE (last accessed February 10, 2014).
57. Simulink—Simulation and Model-Based Design, http://www.mathworks.com/products/simulink/ (last accessed February 10, 2014).
58. Click, The Click Modular Router Project, http://www.read.cs.ucla.edu/click/ (last accessed February 10, 2014).
59. NCTUns 6.0 Network Simulator and Emulator, http://nsl.csie.nctu.edu.tw/nctuns.html (last accessed February 10, 2014).
60. S.Y. Wang, C.L. Chou, Y.H. Chiu, Y.S. Tseng, M.S. Hsu, Y.W. Cheng, W.L. Liu, and T.W. Ho, NCTUns 4.0: An integrated simulation platform for vehicular traffic, communication, and network researches, in *Vehicular Technology Conference, 2007. VTC-2007 Fall. 2007 IEEE 66th*, Dublin, Ireland, pp. 2081–2085, September 30–October 3, 2007.

Index

Printed and bound by CPI Group (UK) Ltd, Croydon, CR0 4YY

18/10/2024

01776210-0012